Two-dimensional Single-Variable Cubic Nonlinear Systems, Vol. I

Albert C. J. Luo

Two-dimensional Single-Variable Cubic Nonlinear Systems, Vol. I

A Self-univariate Cubic Vector Field

Springer

Albert C. J. Luo
Department of Mechanical and Mechatronics Engineering
Southern Illinois University Edwardsville
Edwardsville, IL, USA

ISBN 978-3-031-48471-1 ISBN 978-3-031-48472-8 (eBook)
https://doi.org/10.1007/978-3-031-48472-8

This Springer imprint is published by the registered company Springer Nature Switzerland AG
The registered company address is: Gewerbestrasse 11, 6330 Cham, Switzerland

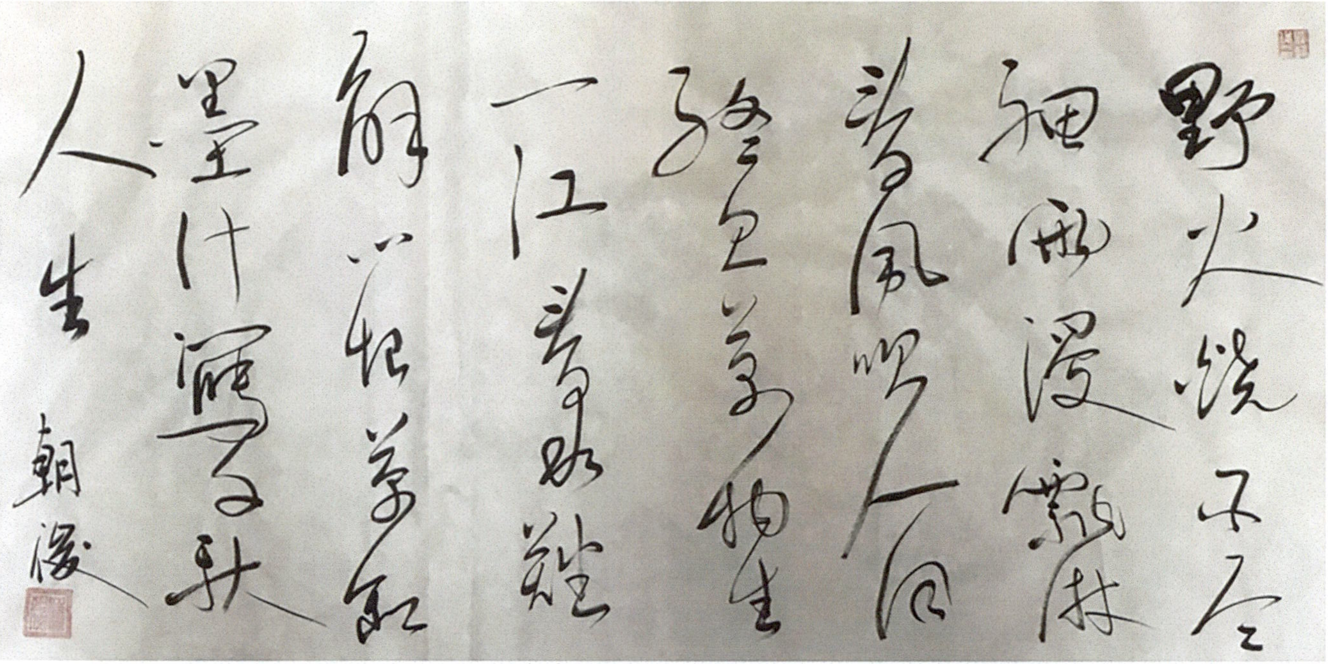

野火烧不尽，细雨漫飘淋。春风吹人泪，终见万物生。一江春水，难解心恨。万缸墨汁，写千秋人生。 — 朝俊于壬寅

Preface

This book is about single-variable cubic nonlinear systems based on a self-univariate cubic vector field as the first part. The second part is about single-variable cubic nonlinear systems based on a non-self-univariate cubic vector field, which will be presented in the second volume. The dynamical behaviors of one-dimensional flows in such single-variable cubic systems are discussed. The first- and third-order sink and source flows with upper-saddle and lower-saddle flows exist. Based on the single-variable systems in the two-dimensional systems, the first- and third-order up-parabola and down-parabola flows with inflection flows exist. The appearing and switching bifurcations for the first- and third-order sink and source flows and the upper-saddle and lower-saddle flows with the first- and third-order up-parabola and down-parabola flows and the inflection flows are discussed. From one-dimensional flows, the global dynamics of cubic nonlinear systems based on a self-univariate cubic field can be better understood.

In Chap. 1, cubic dynamical systems with constant and self-univariate cubic vector fields are discussed. The corresponding stability and bifurcation of the one-dimensional flow are discussed. The first and third-order sink and source flows in such a cubic nonlinear system with single-variable functions are presented. The upper-saddle and lower-saddle flows are the appearing and switching bifurcations of the sink and source flows. The third-order source flow bifurcations are the appearing bifurcations from a source flow to source, sink, and source flows. The third-order sink flow bifurcations are the appearing bifurcations from a sink flow to sink, source, and sink flows. The third-order source flows are also the switching bifurcations for source, sink, and source flows. The third-order sink flows are the switching bifurcations for sink, source, and sink flows.

In Chap. 2, nonlinear dynamical systems possessing linear variable-crossing and self-univariate cubic vector fields are discussed. The corresponding stability and bifurcations of equilibriums are presented. In addition to the first- and third-order sink and source flows with upper-saddle and lower-saddle flows, the parabola flows based on the linear crossing-variable vector fields are discussed. The inflection-sink (source) infinite-equilibriums are for the switching bifurcations of parabola flows and sink (source) flows. The down-up saddles are the appearing bifurcation from an up-parabola flow to the source, down-parabola, and sink flows. The up-down saddles are the appearing bifurcation from a down-parabola flow to source, up-parabola, and sink flows. The third-order-increasing and decreasing asymptotic *source* infinite-equilibriums are the switching bifurcations of the up-parabola and down-parabola with the third-order *source* flow. The third-order-increasing and decreasing asymptotic *sink* infinite-equilibriums are the switching bifurcations of the up-parabola and down-parabola with the third-order *sink* flow.

In Chap. 3, two vector fields in dynamical systems to be crossing-quadratic and self-cubic vector fields are discussed. In addition to the first- and third-order sink and source flows with upper-saddle and lower-saddle flows, the parabola and inflection flows based on the quadratic crossing-variable vector fields are determined. The infinite-equilibriums include parabola-sink and source infinite-equilibriums, the (2: 2)-inflection upper-saddle infinite-equilibrium, and third-order-up and down-logarithmic *source and sink* infinite-equilibriums, which are for the switching bifurcations of sink and source flows with up-parabola and down-parabola flows. The down-up saddles are the appearing bifurcation from an up-parabola flow to source, down-parabola, and sink flows. The up-down saddles are the appearing bifurcation from a down-parabola flow to source, up-parabola, and sink flows.

In Chap. 4, cubic dynamical systems with two cubic univariate vector fields with a single variable are studied. In addition to the first- and third-order sink and source flows with upper-saddle and lower-saddle flows, the first- and third-order parabola and inflection flows based on the cubic crossing-variable vector fields are discussed. The (1:3)-*up-parabola* sink (source) infinite-equilibriums, the down-up and up-down saddles, the (2:2)-inflection upper-saddle infinite-equilibriums, the third-order-increasing and decreasing asymptotic *source and sink* infinite-equilibriums, the third-order-up and down-logarithmic *source and sink* infinite-equilibriums are discussed in the first three chapters. The (3:3)-inflection-source and sink infinite-equilibrium are for the appearing and switching bifurcations of source, sink, source flows and the sink, source, and sink flows with up-parabola, down-parabola, and up-parabola flows and the down-parabola, up-parabola, and down-parabola flows. The (3:3)-inflection-source and sink infinite-equilibrium are also for the switching bifurcations of source and sink flows and the upper-saddle and lower-saddle flows with up-parabola and down-parabola flows, increasing and decreasing-inflection flows.

Finally, the author hopes the materials presented herein can provide a better understanding of nonlinear dynamics in science and engineering.

Edwardsville, IL, USA — Albert C. J. Luo

Contents

Chapter 1
Constant and Self-Cubic Vector Fields

In this chapter, dynamical systems with constant and self-univariate cubic vector fields are considered. The corresponding stability and bifurcation of one-dimensional flows are discussed. The first-order and third-order sink and source flows in such a cubic nonlinear system with single-variable functions are presented from the first integral manifolds. The upper-saddle and lower-saddle flows are the appearing and switching bifurcations of the sink and source flows. The third-order source flow bifurcations are the appearing bifurcations from a source flow to source, sink, and source flows. The third-order sink flow bifurcations are the appearing bifurcations from a sink flow to sink, source, and sink flows. The third-order source flows are the switching bifurcations for source, sink, and source flows. The third-order sink flows are also the switching bifurcations for sink, source, and sink flows.

1.1 Constant and Self-Cubic Systems

The self-univariate cubic dynamical systems with a constant vector field will be discussed, and the sink and source flows in such a cubic dynamical system will be presented. The third-order sink and source flows switching bifurcations will be discussed herein.

Theorem 1.1 *Consider a self-univariate cubic system with a constant vector field as*

$$\begin{aligned}\dot{x}_{j_2} &= A_{j_2 j_1 3},\\ \dot{x}_{j_1} &= A_{j_1 j_1 3} + A_{j_1 j_1 2} x_{j_1} + A_{j_1 j_1 1} x_{j_1}^2 + A_{j_1 j_1 0} x_{j_1}^3\\ &= a_{j_1 j_1 0}(x_{j_1} - b_{j_1 j_1 1})(x_{j_1}^2 + B_{j_1 j_1 1} x_{j_1} + C_{j_1 j_1 1})\end{aligned} \tag{1.1}$$

A. C. J. Luo, *Two-dimensional Single-Variable Cubic Nonlinear Systems, Vol. I*,
https://doi.org/10.1007/978-3-031-48472-8_1

with

$$\Delta_{j_1j_1} = (B_{j_1j_1 1})^2 - 4C_{j_1j_1 1} \text{ for } j_1, j_2 \in \{1, 2\}, j_1 \neq j_2. \tag{1.2}$$

(i) For $A_{j_2j_1 3} = 0$,

$$\begin{aligned} \dot{x}_{j_2} &= A_{j_2j_1 3} = 0, \\ \dot{x}_{j_1} &= a_{j_1j_1 0}(x_{j_1} - b_{j_1j_1 1})(x_{j_1}^2 + B_{j_1j_1 1}x_{j_1} + C_{j_1j_1 1}) \end{aligned} \tag{1.3}$$

The two-dimensional dynamical system becomes a one-dimensional cubic nonlinear system in the x_{j_1}*-direction.*
(i_1) For $\Delta_{j_1j_1} < 0$, *the dynamical system becomes*

$$\begin{aligned} \dot{x}_{j_2} &= A_{j_2j_1 3} = 0, \\ \dot{x}_{j_1} &= a_{j_1j_1 0}(x_{j_1} - a_{j_1j_1 1})\left[(x_{j_1} - a_{j_1j_1 2})^2 + b_{j_1j_1}\right] \end{aligned} \tag{1.4}$$

where

$$a_{j_1j_1 1} = b_{j_1j_1 1}, a_{j_1j_1 2} = -\frac{1}{2}B_{j_1j_1 1}, a_{j_1j_1 0} = A_{j_1j_1 0}, b_{j_1j_1} = \frac{1}{4}(-\Delta_{j_1j_1}). \tag{1.5}$$

The corresponding solution is

$$\begin{aligned} x_{j_2} &= x_{j_2 0}; \\ t - t_0 &= \frac{1}{a_{j_1j_1 0}} \frac{1}{(a_{j_1j_1 1} - a_{j_1j_1 2})^2 + b_{j_1j_1}} \left[\ln \frac{|x_{j_1} - a_{j_1j_1 1}|}{|x_{j_1 0} - a_{j_1j_1 1}|} \right. \\ &\quad - \frac{1}{2} \ln \frac{(x_{j_1} - a_{j_1j_1 2})^2 + b_{j_1j_1}}{(x_{j_1 0} - a_{j_1j_1 2})^2 + b_{j_1j_1}} + \frac{(a_{j_1j_1 2} - a_{j_1j_1 1})}{\sqrt{b_{j_1j_1}}} \\ &\quad \left. \times \left(\arctan \frac{x_{j_1} - a_{j_1j_1 2}}{\sqrt{b_{j_1j_1}}} - \arctan \frac{x_{j_1 0} - a_{j_1j_1 2}}{\sqrt{b_{j_1j_1}}}\right) \right]. \end{aligned} \tag{1.6}$$

The one-dimensional flow has the following properties.

- *For* $a_{j_1j_1 0} > 0$,

$$(a_{j_1j_1 1}, x_{j_2 0}) = \underbrace{(\text{SO}, \text{IC})}_{\text{source}} \tag{1.7}$$

The equilibrium $x_{j_1}^* = a_{j_1j_1 1}$ *at the initial point* $x_{j_2} = x_{j_2 0}$ *is called an* (SO,IC)*-source.*

- *For* $a_{j_1j_10} < 0$,

$$(a_{j_1j_11}, x_{j_20}) = \underbrace{(\mathrm{SI}, \mathrm{IC})}_{\text{sink}} \tag{1.8}$$

The equilibrium $x_{j_1}^* = a_{j_1j_11}$ *at the initial point* $x_{j_2} = x_{j_20}$ *is called an* (SI,IC)-*sink.*

(i$_2$*) For* $\Delta_{j_1j_10} = 0$,

$$\begin{aligned} \dot{x}_{j_2} &= A_{j_2j_13} = 0, \\ \dot{x}_{j_1} &= a_{j_1j_10}(x_{j_1} - a_{j_1j_1s_1})(x_{j_1} - a_{j_1j_1s_2})^2 \end{aligned} \tag{1.9}$$

where

$$\begin{aligned} &a_{j_1j_1s_1}, a_{j_1j_1s_2} \in \{a_{j_1j_11}, a_{j_1j_12}\} \\ &a_{j_1j_11} = b_{j_1j_11}, a_{j_1j_12} = -\frac{1}{2}B_{j_1j_11}, a_{j_1j_10} = A_{j_1j_10}. \end{aligned} \tag{1.10}$$

For $a_{j_1j_1s_1} \neq a_{j_1j_1s_2}$, *the corresponding solution is*

$$\begin{aligned} x_{j_2} &= x_{j_20}; \\ t - t_0 &= \frac{1}{a_{j_1j_10}(a_{j_1j_1s_2} - a_{j_1j_1s_1})^2}\left\{ \ln\frac{|x_{j_1} - a_{j_1j_1s_1}|}{|x_{j_10} - a_{j_1j_1s_1}|} \right. \\ &\quad - (a_{j_1j_1s_2} - a_{j_1j_1s_1})\left[\frac{1}{(x_{j_1} - a_{j_1j_1s_2})} - \frac{1}{(x_{j_10} - a_{j_1j_1s_2})}\right] \\ &\quad \left. - \ln\frac{|x_{j_1} - a_{j_1j_1s_2}|}{|x_{j_10} - a_{j_1j_1s_2}|}\right\}. \end{aligned} \tag{1.11}$$

(i$_{2a}$*) The one-dimensional flow at* $x_{j_1}^* = a_{j_1j_1s_1} \neq a_{j_1j_1s_2}$ *has the following properties.*

- *For* $a_{j_1j_10}(a_{j_1j_1s_1} - a_{j_1j_1s_2})^2 > 0$,

$$(a_{j_1j_1s_1}, x_{j_20}) = \underbrace{(\mathrm{SO}, \mathrm{IC})}_{\text{source}}. \tag{1.12}$$

The equilibrium $x_{j_1}^* = a_{j_1j_1s_1}$ *at the initial point* $x_{j_2} = x_{j_20}$ *is called an* (SO,IC)-*source.*

- *For* $a_{j_1j_10}(a_{j_1j_1s_1}-a_{j_1j_1s_2})^2<0$,

$$(a_{j_1j_1s_1},x_{j_20})=\underbrace{(\mathrm{SI},\mathrm{IC})}_{\text{sink}}. \tag{1.13}$$

The equilibrium $x^*_{j_1}=a_{j_1j_1s_1}$ *at the initial point* $x_{j_2}=x_{j_20}$ *is called an* (SI,IC)-*sink.*

(i_{2b}*) The one-dimensional flow at* $x^*_{j_1}=a_{j_1j_1s_2}\neq a_{j_1j_1s_1}$ *has the following properties.*

- *For* $a_{j_1j_10}(a_{j_1j_1s_2}-a_{j_1j_1s_1})>0$,

$$(a_{j_1j_1s_2},x_{j_20})=\underbrace{(\mathrm{US},\mathrm{IC})}_{\text{upper-saddle}}. \tag{1.14}$$

The equilibrium $x^*_{j_1}=a_{j_1j_1s_2}$ *at the initial point* $x_{j_2}=x_{j_20}$ *is called a* (US,IC)-*upper-saddle, which is called an upper-saddle bifurcation.*

- *For* $a_{j_1j_10}(a_{j_1j_1s_2}-a_{j_1j_1s_1})<0$,

$$(a_{j_1j_1s_2},x_{j_20})=\underbrace{(\mathrm{LS},\mathrm{IC})}_{\text{lower-saddle}}. \tag{1.15}$$

The equilibrium $x^*_{j_1}=a_{j_1j_1s_2}$ *at the initial point* $x_{j_2}=x_{j_20}$ *is called an* (LS,IC)-*lower-saddle, which is called a lower-saddle bifurcation.*

(i_{2c}*) For* $a_{j_1j_1s_1}=a_{j_1j_1s_2}=a_{j_1j_11}$, *the corresponding solution is*

$$\begin{aligned} x_{j_2}&=x_{j_20};\\ t-t_0&=-\frac{1}{2a_{j_1j_10}}\left[\frac{1}{(x_{j_1}-a_{j_1j_11})^2}-\frac{1}{(x_{j_10}-a_{j_1j_11})^2}\right]. \end{aligned} \tag{1.16}$$

The one-dimensional flow at $x^*_{j_1}=a_{j_1j_11}$ *has the following properties.*

- *For* $a_{j_1j_10}>0$,

$$(a_{j_1j_11},x_{j_20})=\underbrace{(3^{\text{rd}}\mathrm{SO},\mathrm{IC})}_{\text{third-order source}}. \tag{1.17}$$

The equilibrium $x^*_{j_1}=a_{j_1j_11}$ *at the initial point* $x_{j_2}=x_{j_20}$ *is called a* (3^{rd}SO,IC)-*third-order source, which is called a third-order source bifurcation.*

- *For* $a_{j_1j_10} < 0$,

$$(a_{j_1j_11}, x_{j_20}) = \underbrace{(3^{\text{rd}}\text{SI, IC})}_{\text{third-order sink}}. \tag{1.18}$$

The equilibrium $x^*_{j_1} = a_{j_1j_11}$ *at the initial point* $x_{j_2} = x_{j_20}$ *is called a* (3^{rd}SI,IC)-*third-order sink, which is called a third-order sink bifurcation.*

(i₃) For $\Delta_{j_1j_1} > 0$, *the dynamical system becomes*

$$\begin{aligned} &\dot{x}_{j_2} = A_{j_2j_12} = 0,\\ &\dot{x}_{j_1} = a_{j_1j_10}(x_{j_1} - a_{j_1j_11})(x_{j_1} - a_{j_1j_12})(x_{j_1} - a_{j_1j_13}) \end{aligned} \tag{1.19}$$

where

$$\begin{aligned} &\{a_{j_1j_11}, a_{j_1j_12}, a_{j_1j_13}\} = \text{sort}\{b_{j_1j_11}, b_{j_1j_12}, b_{j_1j_13}\},\\ &a_{j_1j_1l} < a_{j_1j_1(l+1)}, l = 1, 2;\\ &b_{j_1j_12}, b_{j_1j_13} = -\frac{1}{2}(B_{j_1j_11} \pm \sqrt{\Delta_{j_1j_1}}). \end{aligned} \tag{1.20}$$

For $a_{j_1j_1s_1} \neq a_{j_1j_1s_2} \neq a_{j_1j_1s_3}$ $(s_1, s_2, s_3 \in \{1,2,3\}, s_1 \neq s_2 \neq s_3)$, *the corresponding solution is*

$$\begin{aligned} &x_{j_2} = x_{j_20};\\ &t - t_0 = \frac{1}{a_{j_1j_10}} \sum\nolimits_{s_1=1}^{3} \frac{1}{\prod_{s_2=1, s_2 \neq s_1}^{3} (a_{j_1j_1s_1} - a_{j_1j_1s_2})} \ln \frac{|x_{j_1} - a_{j_1j_1s_1}|}{|x_{j_10} - a_{j_1j_1s_1}|}. \end{aligned} \tag{1.21}$$

The one-dimensional flow at $x^*_{j_1} = a_{j_1j_1s_1} \neq a_{j_1j_1s_2} \neq a_{j_1j_1s_3}$ $(s_1, s_2, s_3 \in \{1,2,3\})$ *has the following properties:*

- *For* $a_{j_1j_10}\Pi^3_{s_2=1, s_2 \neq s_1}(a_{j_1j_1s_1} - a_{j_1j_1s_2}) > 0$,

$$(a_{j_1j_1s_1}, x_{j_20}) = \underbrace{(\text{SO, IC})}_{\text{source}}. \tag{1.22}$$

The equilibrium of $x^*_{j_1} = a_{j_1j_1s_1}$ *at* $x_{j_2} = x_{j_20}$ *is an* (SO,IC)-*source.*

- *For* $a_{j_1j_10}\Pi^3_{s_2=1, s_2 \neq s_1}(a_{j_1j_1s_1} - a_{j_1j_1s_2}) < 0$,

$$(a_{j_1j_1s_1}, x_{j_20}) = \underbrace{(\text{SI, IC})}_{\text{sink}}. \tag{1.23}$$

The equilibrium of $x^*_{j_1} = a_{j_1j_1s_1}$ *and* $x_{j_2} = x_{j_20}$ *is an* (SI,IC)-*sink.*

(ii) If

$$A_{j_2j_13} \neq 0 \text{ and } \Delta_{j_1j_1} < 0, \tag{1.24}$$

the two-dimensional dynamical system does not have any equilibriums for the x_{j_1}-direction on the x_i-direction. The standard form is

$$\begin{aligned} \dot{x}_{j_2} &= A_{j_2j_13} \neq 0, \\ \dot{x}_{j_1} &= a_{j_1j_10}(x_{j_1} - a_{j_1j_11})\left[(x_{j_1} - a_{j_1j_12})^2 + b_{j_1j_1}\right] \end{aligned} \tag{1.25}$$

where

$$a_{j_1j_11} = b_{j_1j_11}, a_{j_1j_12} = -\frac{1}{2}B_{j_1j_11}, a_{j_1j_10} = A_{j_1j_10}, b_{j_1j_1} = \frac{1}{4}(-\Delta_{j_1j_1}). \tag{1.26}$$

The solution is

$$\begin{aligned} x_{j_2} &= A_{j_2j_13}(t - t_0) + x_{j_20} \\ t - t_0 &= \frac{1}{a_{j_1j_10}} \frac{1}{(a_{j_1j_11} - a_{j_1j_12})^2 + b_{j_1j_1}} \Bigg[\ln \frac{|x_{j_1} - a_{j_1j_11}|}{|x_{j_10} - a_{j_1j_11}|} \\ &\quad - \frac{1}{2} \ln \frac{(x_{j_1} - a_{j_1j_12})^2 + b_{j_1j_1}}{(x_{j_10} - a_{j_1j_12})^2 + b_{j_1j_1}} \\ &\quad + \frac{(a_{j_1j_12} - a_{j_1j_11})}{\sqrt{b_{j_1j_1}}} \left(\arctan \frac{x_{j_1} - a_{j_1j_12}}{\sqrt{b_{j_1j_1}}} - \arctan \frac{x_{j_10} - a_{j_1j_12}}{\sqrt{b_{j_1j_1}}} \right) \Bigg]. \end{aligned} \tag{1.27}$$

The first integral manifold is

$$\begin{aligned} \frac{x_{j_2} - x_{j_20}}{A_{j_2j_13}} &= \frac{1}{a_{j_1j_10}} \frac{1}{(a_{j_1j_11} - a_{j_1j_12})^2 + b_{j_1j_1}} \Bigg[\ln \frac{|x_{j_1} - a_{j_1j_11}|}{|x_{j_10} - a_{j_1j_11}|} \\ &\quad - \frac{1}{2} \ln \frac{(x_{j_1} - a_{j_1j_12})^2 + b_{j_1j_1}}{(x_{j_10} - a_{j_1j_12})^2 + b_{j_1j_1}} \\ &\quad + \frac{(a_{j_1j_12} - a_{j_1j_11})}{\sqrt{b_{j_1j_1}}} \left(\arctan \frac{x_{j_1} - a_{j_1j_12}}{\sqrt{b_{j_1j_1}}} - \arctan \frac{x_{j_10} - a_{j_1j_12}}{\sqrt{b_{j_1j_1}}} \right) \Bigg]. \end{aligned} \tag{1.28}$$

The flow at $x^*_{j_1} = a_{j_1 j_1 1}$ *has the following properties:*

- *For* $a_{j_1 j_1 0} > 0$ *and* $A_{j_2 j_1 3} > 0$,

$$(a_{j_1 j_1 1}, \dot{x}_{j_2}) = \underbrace{(\mathrm{SO}, \mathrm{pF})}_{\text{source flow } (+)}. \tag{1.29}$$

The flow at $x^*_{j_1} = a_{j_1 j_1 1}$ *is called an* (SO,pF)*-positive source flow.*

- *For* $a_{j_1 j_1 0} < 0$ *and* $A_{j_2 j_1 3} > 0$,

$$(a_{j_1 j_1 1}, \dot{x}_{j_2}) = \underbrace{(\mathrm{SI}, \mathrm{pF})}_{\text{sink flow } (+)}. \tag{1.30}$$

The flow at $x^*_{j_1} = a_{j_1 j_1 1}$ *is called an* (SI,pF)*-positive sink flow.*

- *For* $a_{j_1 j_1 0} > 0$ *and* $A_{j_2 j_1 3} < 0$,

$$(a_{j_1 j_1 1}, \dot{x}_{j_2}) = \underbrace{(\mathrm{SO}, \mathrm{nF})}_{\text{source flow } (-)}. \tag{1.31}$$

The flow at $x^*_{j_1} = a_{j_1 j_1 1}$ *is called an* (SO,nF)*-negative source flow.*

- *For* $a_{j_1 j_1 0} < 0$ and $A_{j_2 j_1 3} < 0$,

$$(a_{j_1 j_1 1}, \dot{x}_{j_2}) = \underbrace{(\mathrm{SI}, \mathrm{nF})}_{\text{sink flow } (-)}. \tag{1.32}$$

The flow at $x^*_{j_1} = a_{j_1 j_1 1}$ *is called an* (SI,nF)*-negative sink flow.*

(iii) If

$$A_{j_2 j_1 3} \neq 0 \text{ and } \Delta_{j_1 j_1} = 0, \tag{1.33}$$

the corresponding dynamical system becomes

$$\begin{aligned} \dot{x}_{j_2} &= A_{j_2 j_1 3}, \\ \dot{x}_{j_1} &= a_{j_1 j_1 0}(x_{j_1} - a_{j_1 j_1 s_1})(x_{j_1} - a_{j_1 j_1 s_2})^2. \end{aligned} \tag{1.34}$$

For $a_{j_1 j_1 s_1} \neq a_{j_1 j_1 s_2}$, *the solution is*

$$\begin{aligned} & x_{j_2} - x_{j_2 0} = A_{j_2 j_1 3}(t - t_0); \\ & t - t_0 = \frac{1}{a_{j_1 j_1 0}(a_{j_1 j_1 s_2} - a_{j_1 j_1 s_1})^2} \left\{ \ln \frac{|x_{j_1} - a_{j_1 j_1 s_1}|}{|x_{j_1 0} - a_{j_1 j_1 s_1}|} - (a_{j_1 j_1 s_2} - a_{j_1 j_1 s_1}) \right. \\ & \qquad \left. \times \left[\frac{1}{(x_{j_1} - a_{j_1 j_1 s_2})} - \frac{1}{(x_{j_1 0} - a_{j_1 j_1 s_2})} \right] - \ln \frac{|x_{j_1} - a_{j_1 j_1 s_2}|}{|x_{j_1 0} - a_{j_1 j_1 s_2}|} \right\}. \end{aligned} \tag{1.35}$$

The first integral manifold is

$$\begin{aligned}\frac{x_{j_2}-x_{j_20}}{A_{j_2j_13}} &= \frac{1}{a_{j_1j_10}(a_{j_1j_1s_2}-a_{j_1j_1s_1})^2}\Big\{\ln\frac{|x_{j_1}-a_{j_1j_1s_1}|}{|x_{j_10}-a_{j_1j_1s_1}|}-(a_{j_1j_1s_2}-a_{j_1j_1s_1})\\ &\times\Big[\frac{1}{(x_{j_1}-a_{j_1j_1s_2})}-\frac{1}{(x_{j_10}-a_{j_1j_1s_2})}\Big]-\ln\frac{|x_{j_1}-a_{j_1j_1s_2}|}{|x_{j_10}-a_{j_1j_1s_2}|}\Big\}.\end{aligned} \tag{1.36}$$

The two-dimensional dynamical system has two equilibriums for the x_{j_1}-direction, i.e.,

$$a_{j_1j_1s_1}=b_{j_1j_11}, x_{j_1}^*=-\frac{B_{j_1j_11}}{2}=a_{j_1j_1s_2}, a_{j_1j_10}=A_{j_1j_10} \text{ for the } x_{j_1}\text{-direction.} \tag{1.37}$$

(iii$_I$) The flow at $x_{j_1}^*=a_{j_1j_1s_1}$ has the following properties:

- *For $a_{j_1j_10}(a_{j_1j_1s_1}-a_{j_1j_1s_2})^2>0$ and $A_{j_2j_13}>0$,*

$$(a_{j_1j_1s_1},\dot{x}_{j_2})=\underbrace{(\text{SO},\text{pF})}_{\text{source flow }(+)}. \tag{1.38}$$

  The flow at $x_{j_1}^*=a_{j_1j_1s_1}$ is called an (SO,pF)*-positive source flow.*

- *For $a_{j_1j_10}(a_{j_1j_1s_1}-a_{j_1j_1s_2})^2<0$ and $A_{j_2j_13}>0$,*

$$(a_{j_1j_1s_1},\dot{x}_{j_2})=\underbrace{(\text{SI},\text{pF})}_{\text{sink flow }(+)}. \tag{1.39}$$

  The flow at $x_{j_1}^*=a_{j_1j_1s_1}$ is called an (SI,pF)*-positive sink flow.*

- *For $a_{j_1j_10}(a_{j_1j_1s_1}-a_{j_1j_1s_2})^2>0$ and $A_{j_2j_13}<0$,*

$$(a_{j_1j_1s_1},\dot{x}_{j_2})=\underbrace{(\text{SO},\text{nF})}_{\text{source flow }(-)}. \tag{1.40}$$

  The flow at $x_{j_1}^*=a_{j_1j_1s_1}$ is called an (SO,nF)*-negative source flow.*

- *For $a_{j_1j_10}(a_{j_1j_1s_1}-a_{j_1j_1s_2})^2<0$ and $A_{j_2j_13}<0$,*

$$(a_{j_1j_1s_1},\dot{x}_{j_2})=\underbrace{(\text{SI},\text{nF})}_{\text{sink flow }(-)}. \tag{1.41}$$

  The flow at $x_{j_1}^*=a_{j_1j_1s_1}$ is called an (SI,nF)*-negative sink flow.*

(iii$_2$) The equilibrium of $x_{j_1}^* = a_{j_1 j_1 s_2}$ for the x_{j_1}-direction has the following properties.

- *For $a_{j_1 j_1 0}(a_{j_1 j_1 s_2} - a_{j_1 j_1 s_1}) > 0$ and $A_{j_2 j_1 3} > 0$,*

$$(a_{j_1 j_1 s_2}, \dot{x}_{j_2}) = \underbrace{(\text{US}, \text{pF})}_{\text{upper-saddle flow } (+)} . \tag{1.42}$$

  The flow at $x_{j_1}^* = a_{j_1 j_1 s_2}$ for the x_{j_1}-direction is called a (US,pF)*-positive upper-saddle flow.*

- *For $a_{j_1 j_1 0}(a_{j_1 j_1 s_2} - a_{j_1 j_1 s_1}) < 0$ and $A_{j_2 j_1 3} > 0$,*

$$(a_{j_1 j_1 s_2}, \dot{x}_{j_2}) = \underbrace{(\text{LS}, \text{pF})}_{\text{lower-saddle flow } (+)} . \tag{1.43}$$

  The flow at $x_{j_1}^* = a_{j_1 j_1 s_2}$ for the x_{j_1}-direction is called an (LS,pF)*-positive lower-saddle flow.*

- *For $a_{j_1 j_1 0}(a_{j_1 j_1 s_2} - a_{j_1 j_1 s_1}) > 0$ and $A_{j_2 j_1 3} < 0$,*

$$(a_{j_1 j_1 s_2}, \dot{x}_{j_2}) = \underbrace{(\text{US}, \text{nF})}_{\text{upper-saddle flow } (-)} . \tag{1.44}$$

  The flow at $x_{j_1}^* = a_{j_1 j_1 s_2}$ for the x_{j_1}-direction is called a (US,nF)*-negative upper-saddle flow.*

- *For $a_{j_1 j_1 0}(a_{j_1 j_1 s_2} - a_{j_1 j_1 s_1}) < 0$ and $A_{j_2 j_1 3} < 0$,*

$$(a_{j_1 j_1 s_2}, \dot{x}_{j_2}) = \underbrace{(\text{LS}, \text{nF})}_{\text{lower-saddle flow } (-)} . \tag{1.45}$$

  The flow at $x_{j_1}^* = a_{j_1 j_1 s_2}$ for the x_{j_1}-direction is called an (LS,nF)*-negative lower-saddle flow.*

– *The corresponding states are the bifurcation states for sink-flow and source-flow for the x_{j_1}-direction at a specific parameter.*

(iii$_{3c}$) For $a_{j_1 j_1 s_1} = a_{j_1 j_1 s_2} = a_{j_1 j_1 1}$, the solution is

$$\begin{aligned} x_{j_2} - x_{j_2 0} &= A_{j_2 j_1 3}(t - t_0); \\ t - t_0 &= -\frac{1}{2a_{j_1 j_1 0}}\left[\frac{1}{(x_{j_1} - a_{j_1 j_1 1})^2} - \frac{1}{(x_{j_1 0} - a_{j_1 j_1 1})^2}\right]. \end{aligned} \tag{1.46}$$

The first integral manifold is

$$\frac{x_{j_2}-x_{j_2 0}}{A_{j_2 j_1 3}} = -\frac{1}{2a_{j_1 j_1 0}}\left[\frac{1}{(x_{j_1}-a_{j_1 j_1 1})^2} - \frac{1}{(x_{j_1 0}-a_{j_1 j_1 1})^2}\right]. \tag{1.47}$$

The two-dimensional flow at $x_{j_1}^* = a_{j_1 j_1 1}$ *has the following properties.*

- *For* $a_{j_1 j_1 0} > 0$ *and* $A_{j_2 j_1 3} > 0$,

$$(a_{j_1 j_1 1}, \dot{x}_{j_2}) = \underbrace{(3^{\text{rd}}\text{SO, pF})}_{\text{third-order source flow } (+)}. \tag{1.48}$$

The flow at $x_{j_1}^* = a_{j_1 j_1 1}$ *is a* $(3^{\text{rd}}\text{SO,pF})$*-third-order source flow.*

- *For* $a_{j_1 j_1 0} < 0$ *and* $A_{j_2 j_1 3} > 0$,

$$(a_{j_1 j_1 1}, \dot{x}_{j_2}) = \underbrace{(3^{\text{rd}}\text{SI, pF})}_{\text{third-order sink flow } (+)}. \tag{1.49}$$

The flow at $x_{j_1}^* = a_{j_1 j_1 1}$ *is a* $(3^{\text{rd}}\text{SI,pF})$*-third-order positive sink flow.*

- *For* $a_{j_1 j_1 0} > 0$ *and* $A_{j_2 j_1 3} < 0$,

$$(a_{j_1 j_1 1}, \dot{x}_{j_2}) = \underbrace{(3^{\text{rd}}\text{SO, nF})}_{\text{third-order source flow } (-)}. \tag{1.50}$$

The flow at $x_{j_1}^* = a_{j_1 j_1 1}$ *is a* $(3^{\text{rd}}\text{SO,nF})$*-third-order negative source flow.*

- *For* $a_{j_1 j_1 0} < 0$ *and* $A_{j_2 j_1 3} < 0$,

$$(a_{j_1 j_1 1}, \dot{x}_{j_2}) = \underbrace{(3^{\text{rd}}\text{SI, nF})}_{\text{third-order sink flow } (-)}. \tag{1.51}$$

The flow at $x_{j_1}^* = a_{j_1 j_1 1}$ *is a* $(3^{\text{rd}}\text{SI,nF})$*-third-order negative sink flow.*

– *The third-order sink and source flows are the switching bifurcations of the upper-saddle and lower-saddle flows with source and sink flows.*

(iv) If

$$A_{j_2 j_1 3} \neq 0 \text{ and } \Delta_{j_1 j_1} > 0, \tag{1.52}$$

then the dynamical system becomes

$$
\begin{aligned}
\dot{x}_{j_2} &= A_{j_2 j_1 3}, \\
\dot{x}_{j_1} &= a_{j_1 j_1 0}(x_{j_1} - a_{j_1 j_1 1})(x_{j_1} - a_{j_1 j_1 2})(x_{j_1} - a_{j_1 j_1 3}).
\end{aligned} \tag{1.53}
$$

For $a_{j_1 j_1 s_1} \neq a_{j_1 j_1 s_2} \neq a_{j_1 j_1 s_3}$ *(*$s_1, s_2, s_3 \in \{1,2,3\}$, $s_1 \neq s_2 \neq s_3$*), the corresponding solution is*

$$
\begin{aligned}
& x_{j_2} - x_{j_2 0} = A_{j_2 j_1 3}(t - t_0); \\
& t - t_0 = \frac{1}{a_{j_1 j_1 0}} \sum_{s_1=1}^{3} \frac{1}{\Pi_{s_2=1, s_2 \neq s_1}^{3} (a_{j_1 j_1 s_1} - a_{j_1 j_1 s_2})} \ln \frac{|x_{j_1} - a_{j_1 j_1 s_1}|}{|x_{j_1 0} - a_{j_1 j_1 s_1}|}.
\end{aligned} \tag{1.54}
$$

The first integral manifold is

$$
\frac{x_{j_2} - x_{j_2 0}}{A_{j_2 j_1 3}} = \frac{1}{a_{j_1 j_1 0}} \sum_{s_1=1}^{3} \frac{1}{\Pi_{s_2=1, s_2 \neq s_1}^{3} (a_{j_1 j_1 s_1} - a_{j_1 j_1 s_2})} \ln \frac{|x_{j_1} - a_{j_1 j_1 s_1}|}{|x_{j_1 0} - a_{j_1 j_1 s_1}|}. \tag{1.55}
$$

The two-dimensional dynamical system has three simple equilibriums for the x_{j_1}*-direction, i.e.,*

$$
x_{j_1}^* = a_{j_1 j_1 1}, a_{j_1 j_1 2}, a_{j_1 j_1 3} \text{ for the } x_{j_1}\text{-direction.} \tag{1.56}
$$

The equilibrium of $x_{j_1}^* = a_{j_1 j_1 s_1}$ *(*$s_1, s_2, s_3 \in \{1,2,3\}$, $s_1 \neq s_2 \neq s_3$*) in the* x_{j_1}*-direction has the following properties.*

- *For* $a_{j_1 j_1 0} \Pi_{s_2=1, s_2 \neq s_1}^{3} (a_{j_1 j_1 s_1} - a_{j_1 j_1 s_2}) > 0$ *and* $A_{j_2 j_1 3} > 0$,

$$
(a_{j_1 j_1 s_1}, \dot{x}_{j_2}) = \underbrace{(\text{SO}, \text{pF})}_{\text{source flow }(+)}. \tag{1.57}
$$

  The flow at $x_{j_1}^* = a_{j_1 j_1 s_1}$ *is an* (SO,pF)*-positive source flow.*

- *For* $a_{j_1 j_1 0} \Pi_{s_2=1, s_2 \neq s_1}^{3} (a_{j_1 j_1 s_1} - a_{j_1 j_1 s_2}) < 0$ *and* $A_{j_2 j_1 3} > 0$,

$$
(a_{j_1 j_1 s_1}, \dot{x}_{j_2}) = \underbrace{(\text{SI}, \text{pF})}_{\text{sink flow }(+)}. \tag{1.58}
$$

  The flow at $x_{j_1}^* = a_{j_1 j_1 s_1}$ *is an* (SI,pF)*-positive sink flow.*

- *For* $a_{j_1 j_1 0} \Pi_{s_2=1, s_2 \neq s_1}^{3} (a_{j_1 j_1 s_1} - a_{j_1 j_1 s_2}) > 0$ *and* $A_{j_2 j_1 3} < 0$,

$$
(a_{j_1 j_1 s_1}, \dot{x}_{j_2}) = \underbrace{(\text{SO}, \text{nF})}_{\text{source flow }(-)}. \tag{1.59}
$$

  The flow at $x_{j_1}^* = a_{j_1 j_1 s_1}$ *is an* (SO,nF)*-negative source flow.*

- *For* $a_{j_1j_10}\Pi_{s_2=1,s_2\neq s_1}^{3}(a_{j_1j_1s_1}-a_{j_1j_1s_2})<0$ *and* $A_{j_2j_13}<0$,

$$(a_{j_1j_1s_1},\dot{x}_{j_2})=\underbrace{(\text{SI, nF})}_{\text{sink flow }(-)}. \tag{1.60}$$

The flow at $x_{j_1}^*=a_{j_1j_1s_1}$ *is an* (SI,nF)*-negative sink flow.*

(vi) Under

$$A_{j_2j_13}\neq 0 \text{ and } \Delta_{j_1j_1}=0 \text{ with } a_{j_1j_1s_1}=a_{j_1j_1s_2}=a_{j_1j_11}, \tag{1.61}$$

the dynamical system is

$$\begin{aligned}\dot{x}_{j_2}&=A_{j_2j_13},\\ \dot{x}_{j_1}&=a_{j_1j_10}(x_{j_1}-a_{j_1j_11})^3\end{aligned} \tag{1.62}$$

with

$$a_{j_1j_1s_1}^{-}\to a_{j_1j_1s_2}^{+} \text{ and } a_{j_1j_1s_2}^{-}\to a_{j_1j_1s_1}^{+}. \tag{1.63}$$

The corresponding equilibriums of $x_{j_1}^*=a_{j_1j_11}$ *for the* x_{j_1}*-direction possess the following properties.*

- *For* $a_{j_1j_10}>0$ *and* $A_{j_2j_13}>0$,

$$\left\{\begin{array}{c}\underbrace{(a_{j_1j_1s_2}^{-},\dot{x}_{j_2})}_{(\text{LS, pF})\text{-flow }(+)}\\ \underbrace{(a_{j_1j_1s_1}^{-},\dot{x}_{j_2})}_{(\text{SO, pF})\text{-flow }(+)}\end{array}\right\}\rightleftharpoons\underbrace{(a_{j_1j_11},\dot{x}_{j_2})}_{\substack{(3^{\text{rd}}\text{SO, pF})\text{-flow}\\ \text{switching bifurcation}}}\rightleftharpoons\left\{\begin{array}{c}\underbrace{(a_{j_1j_1s_1}^{+},\dot{x}_{j_2})}_{(\text{SO, pF})\text{-flow }(+)}\\ \underbrace{(a_{j_1j_1s_2}^{+},\dot{x}_{j_2})}_{(\text{US, pF})\text{-flow }(+)}\end{array}\right\}. \tag{1.64}$$

The (3rdSO,pF)*-positive source flow at* $x_{j_1}^*=a_{j_1j_11}$ *is a* (3rdSO,pF)*-switching bifurcation for positive upper-saddle and lower-saddle flows with a positive source flow.*

- *For* $a_{j_1j_10}<0$ *and* $A_{j_2j_13}>0$,

$$\left\{\begin{array}{c}\underbrace{(a_{j_1j_1s_2}^{-},\dot{x}_{j_2})}_{(\text{LS, pF})\text{-flow }(+)}\\ \underbrace{(a_{j_1j_1s_1}^{-},\dot{x}_{j_2})}_{(\text{SI, pF})\text{-flow }(+)}\end{array}\right\}\rightleftharpoons\underbrace{(a_{j_1j_11},\dot{x}_{j_2})}_{\substack{(3^{\text{rd}}\text{SI, pF})\text{-flow}\\ \text{switching bifurcation}}}\rightleftharpoons\left\{\begin{array}{c}\underbrace{(a_{j_1j_1s_1}^{+},\dot{x}_{j_2})}_{(\text{SI, pF})\text{-flow }(+)}\\ \underbrace{(a_{j_1j_1s_2}^{+},\dot{x}_{j_2})}_{(\text{US, pF})\text{-flow }(+)}\end{array}\right\}. \tag{1.65}$$

The (3^rd^SI,pF)*-positive source flow at* $x_{j_1}^* = a_{j_1j_1 1}$ *is a* (3^rd^SI,pF)*-switching bifurcation for positive upper-saddle and lower-saddle flows with a positive sink flow.*

- *For* $a_{j_1j_1 0} > 0$ *and* $A_{j_2j_1 3} < 0$,

$$\left\{\begin{array}{c}\underbrace{(a_{j_1j_1s_2}^-, \dot{x}_{j_2})}_{(\mathrm{LS,nF})\text{-flow }(-)}\\ \underbrace{(a_{j_1j_1s_1}^-, \dot{x}_{j_2})}_{(\mathrm{SO,nF})\text{-flow }(-)}\end{array}\right\} \rightleftharpoons \underbrace{(a_{j_1j_1 1}, \dot{x}_{j_2})}_{\substack{(3^{\mathrm{rd}}\mathrm{SO,nF})\text{-flow}\\ \text{switching bifurcation}}} \rightleftharpoons \left\{\begin{array}{c}\underbrace{(a_{j_1j_1s_1}^+, \dot{x}_{j_2})}_{(\mathrm{SO,nF})\text{-flow }(-)}\\ \underbrace{(a_{j_1j_1s_2}^+, \dot{x}_{j_2})}_{(\mathrm{US,nF})\text{-flow }(-)}\end{array}\right\}. \tag{1.66}$$

The (3^rd^SO,nF)*-positive source flow at* $x_{j_1}^* = a_{j_1j_1 1}$ *is a* (3^rd^SO,nF)*-switching bifurcation for negative upper-saddle and lower-saddle flows with a negative source flow.*

- *For* $a_{j_1j_1 0} < 0$ *an* $A_{j_2j_1 3} < 0$,

$$\left\{\begin{array}{c}\underbrace{(a_{j_1j_1s_2}^-, \dot{x}_{j_2})}_{(\mathrm{US,nF})\text{-flow }(-)}\\ \underbrace{(a_{j_1j_1s_1}^-, \dot{x}_{j_2})}_{(\mathrm{SI,nF})\text{-flow }(-)}\end{array}\right\} \rightleftharpoons \underbrace{(a_{j_1j_1 1}, \dot{x}_{j_2})}_{\substack{(3^{\mathrm{rd}}\mathrm{SI,nF})\text{-flow}\\ \text{switching bifurcation}}} \rightleftharpoons \left\{\begin{array}{c}\underbrace{(a_{j_1j_1s_1}^+, \dot{x}_{j_2})}_{(\mathrm{SI,nF})\text{-flow }(-)}\\ \underbrace{(a_{j_1j_1s_2}^+, \dot{x}_{j_2})}_{(\mathrm{LS,nF})\text{-flow }(-)}\end{array}\right\}. \tag{1.67}$$

The (3^rd^SI,pF)*-positive source flow at* $x_{j_1}^* = a_{j_1j_1 1}$ *is a* (3^rd^SI,pF)*-switching bifurcation for negative upper-saddle and lower-saddle flows with a negative sink flow.*

(vi) Under

$$\begin{aligned}&A_{j_2j_1 3} \neq 0, \text{ and } \Delta_{j_1j_1} = (a_{j_1j_1s_2} - a_{j_1j_1s_3})^2 = 0,\\ &a_{j_1j_1s_2} = a_{j_1j_1s_3} \equiv a_{j_1j_1s_1} = a_{j_1j_1 1},\end{aligned} \tag{1.68}$$

a dynamical system is

$$\begin{aligned}\dot{x}_{j_2} &= A_{j_2j_1 3},\\ \dot{x}_{j_1} &= a_{j_1j_1 0}(x_{j_1} - a_{j_1j_1 1})^3\end{aligned} \tag{1.69}$$

with

$$a_{j_1j_1 1}^- \to a_{j_1j_1 3}^+, a_{j_1j_1 3}^- \to a_{j_1j_1 1}^+, \text{ and } a_{j_1j_1 2}^- \to a_{j_1j_1 2}^+. \tag{1.70}$$

The corresponding equilibriums of $x_{j_1}^* = a_{j_1 j_1 1}$ *for the* x_{j_1}*-direction possess the following properties.*

- *For* $a_{j_1 j_1 0} > 0$ *and* $A_{j_2 j_1 3} > 0$,

$$\left\{\begin{array}{c}\underbrace{(a_{j_1 j_1 3}^-, \dot{x}_{j_2})}_{(\text{SO},\text{pF})\text{-flow }(+)}\\ \underbrace{(a_{j_1 j_1 2}^-, \dot{x}_{j_2})}_{(\text{SI},\text{pF})\text{-flow }(+)}\\ \underbrace{(a_{j_1 j_1 1}^-, \dot{x}_{j_2})}_{(\text{SO},\text{pF})\text{-flow }(+)}\end{array}\right\} \rightleftharpoons \underbrace{(a_{j_1 j_1 1}, \dot{x}_{j_2})}_{\substack{(3^{\text{rd}}\text{SO},\,\text{pF})\text{-flow}\\ \text{switching bifurcation}}} \rightleftharpoons \left\{\begin{array}{c}\underbrace{(a_{j_1 j_1 3}^+, \dot{x}_{j_2})}_{(\text{SO},\text{pF})\text{-flow }(+)}\\ \underbrace{(a_{j_1 j_1 2}^+, \dot{x}_{j_2})}_{(\text{SI},\text{pF})\text{-flow }(+)}\\ \underbrace{(a_{j_1 j_1 1}^+, \dot{x}_{j_2})}_{(\text{SO},\text{pF})\text{-flow }(+)}\end{array}\right\}. \tag{1.71}$$

  The $(3^{\text{rd}}\text{SO,pF})$*-positive source flow at* $x_{j_1}^* = a_{j_1 j_1 1}$ *is called a* $(3^{\text{rd}}\text{SO,pF})$*-switching bifurcation for positive source, sink, and source flows.*

- *For* $a_{j_1 j_1 0} < 0$ *and* $A_{j_2 j_1 3} > 0$,

$$\left\{\begin{array}{c}\underbrace{(a_{j_1 j_1 3}^-, \dot{x}_{j_2})}_{(\text{SI},\text{pF})\text{-flow }(+)}\\ \underbrace{(a_{j_1 j_1 2}^-, \dot{x}_{j_2})}_{(\text{SO},\text{pF})\text{-flow }(+)}\\ \underbrace{(a_{j_1 j_1 1}^-, \dot{x}_{j_2})}_{(\text{SI},\text{pF})\text{-flow }(+)}\end{array}\right\} \rightleftharpoons \underbrace{(a_{j_1 j_1 1}, \dot{x}_{j_2})}_{\substack{(3^{\text{rd}}\text{SI},\,\text{pF})\text{-flow}\\ \text{switching bifurcation}}} \rightleftharpoons \left\{\begin{array}{c}\underbrace{(a_{j_1 j_1 3}^+, \dot{x}_{j_2})}_{(\text{SI},\text{pF})\text{-flow }(+)}\\ \underbrace{(a_{j_1 j_1 2}^+, \dot{x}_{j_2})}_{(\text{SO},\text{pF})\text{-flow }(+)}\\ \underbrace{(a_{j_1 j_1 1}^+, \dot{x}_{j_2})}_{(\text{SI},\text{pF})\text{-flow }(+)}\end{array}\right\}. \tag{1.72}$$

  The $(3^{\text{rd}}\text{SI,pF})$*-positive source flow at* $x_{j_1}^* = a_{j_1 j_1 1}$ *is called a* $(3^{\text{rd}}\text{SI,pF})$*-switching bifurcation for positive sink, source and sink flows.*

- *For* $a_{j_1 j_1 0} > 0$ *and* $A_{j_2 j_1 3} < 0$,

$$\left\{\begin{array}{c}\underbrace{(a_{j_1 j_1 3}^-, \dot{x}_{j_2})}_{(\text{SO},\text{nF})\text{-flow }(-)}\\ \underbrace{(a_{j_1 j_1 2}^-, \dot{x}_{j_2})}_{(\text{SI},\text{nF})\text{-flow }(-)}\\ \underbrace{(a_{j_1 j_1 1}^-, \dot{x}_{j_2})}_{(\text{SO},\text{nF})\text{-flow }(-)}\end{array}\right\} \rightleftharpoons \underbrace{(a_{j_1 j_1 1}, \dot{x}_{j_2})}_{\substack{(3^{\text{rd}}\text{SO},\,\text{nF})\text{-flow}\\ \text{switching bifurcation}}} \rightleftharpoons \left\{\begin{array}{c}\underbrace{(a_{j_1 j_1 3}^+, \dot{x}_{j_2})}_{(\text{SO},\text{nF})\text{-flow }(-)}\\ \underbrace{(a_{j_1 j_1 2}^+, \dot{x}_{j_2})}_{(\text{SI},\text{nF})\text{-flow }(-)}\\ \underbrace{(a_{j_1 j_1 1}^+, \dot{x}_{j_2})}_{(\text{SO},\text{nF})\text{-flow }(-)}\end{array}\right\}. \tag{1.73}$$

  The $(3^{\text{rd}}\text{SO,nF})$*-positive source flow at* $x_{j_1}^* = a_{j_1 j_1 1}$ *is called a* $(3^{\text{rd}}\text{SO,nF})$*-switching bifurcation for negative source, sink, and source flows.*

- *For $a_{j_1j_10} < 0$ an $A_{j_2j_13} < 0$,*

$$\left\{\begin{array}{c}\underbrace{(a^-_{j_1j_13}, \dot{x}_{j_2})}_{(\text{SI, nF})\text{-flow } (-)}\\ \underbrace{(a^-_{j_1j_12}, \dot{x}_{j_2})}_{(\text{SO, nF})\text{-flow } (-)}\\ \underbrace{(a^-_{j_1j_11}, \dot{x}_{j_2})}_{(\text{SI, nF})\text{-flow } (-)}\end{array}\right\} \rightleftharpoons \underbrace{(a_{j_1j_11}, \dot{x}_{j_2})}_{\substack{(3^{\text{rd}}\text{SI, nF})\text{-flow}\\ \text{switching bifurcation}}} \rightleftharpoons \left\{\begin{array}{c}\underbrace{(a^+_{j_1j_13}, \dot{x}_{j_2})}_{(\text{SI, nF})\text{-flow } (-)}\\ \underbrace{(a^+_{j_1j_12}, \dot{x}_{j_2})}_{(\text{SO, nF})\text{-flow } (-)}\\ \underbrace{(a^+_{j_1j_11}, \dot{x}_{j_2})}_{(\text{SI, nF})\text{-flow } (-)}\end{array}\right\}. \tag{1.74}$$

The (3rdSI,pF)*-positive source flow at $x^*_{j_1} = a_{j_1j_11}$ is called a* (3rdSI,pF)*-switching bifurcation for negative sink, source and sink flows.*

(vii) Under

$$A_{j_2j_13} \neq 0 \text{ and } \Delta_{j_1j_1} = 0, a_{j_1j_1s_1} = a_{j_1j_1s_2}, s_1 \neq s_2 \in \{1, 2, 3\} \tag{1.75}$$

the dynamical system is

$$\begin{aligned}\dot{x}_{j_2} &= A_{j_2j_13},\\ \dot{x}_{j_1} &= a_{j_1j_10}(x_{j_1} - a_{j_1j_11})^3.\end{aligned} \tag{1.76}$$

The third-order sink and source flows at $x^*_{j_1} = a_{j_1j_1s_1}$ in the x_{j_1}-direction are the appearing bifurcations of the third-order sink and source flows.

- *For $a_{j_1j_10} > 0$ and $A_{j_2j_13} > 0$,*

$$\underbrace{(\text{SO, pF})}_{\text{source flow } (+)} \rightleftharpoons \underbrace{(3^{\text{rd}}\text{SO, pF})}_{\text{appearing bifurcation}} \rightleftharpoons \left\{\begin{array}{c}\underbrace{(\text{SO, pF})}_{\text{source flow } (+)}\\ \underbrace{(\text{SI, pF})}_{\text{sink flow } (+)}\\ \underbrace{(\text{SO, pF})}_{\text{source flow } (+)}\end{array}\right\}. \tag{1.77}$$

The (3rdSO,pF)*-positive source flow at $x^*_{j_1} = a_{j_1j_1s_1}$ is a* (3rdSO,pF)*-appearing bifurcation from a positive source flow to positive source, sink, and source flows.*

- *For* $a_{j_1j_10} < 0$ *and* $A_{j_2j_13} > 0$,

$$\underbrace{(\mathrm{SI,pF})}_{\text{source flow }(+)} \rightleftharpoons \underbrace{(3^{\mathrm{rd}}\mathrm{SI,pF})}_{\text{appearing bifurcation}} \rightleftharpoons \left\{\begin{array}{c} \underbrace{(\mathrm{SI,pF})}_{\text{sink flow }(+)} \\ \underbrace{(\mathrm{SO,pF})}_{\text{source flow }(+)} \\ \underbrace{(\mathrm{SI,pF})}_{\text{sink flow }(+)} \end{array}\right\}. \tag{1.78}$$

The (3rdSI,pF)*-positive source flow at* $x_{j_1}^* = a_{j_1j_1s_1}$ *is a* (3rdSI,pF)*-appearing bifurcation from a positive sink flow to positive sink, source, sink flows.*

- *For* $a_{j_1j_10} > 0$ *and* $A_{j_2j_13} < 0$,

$$\underbrace{(\mathrm{SO,nF})}_{\text{source flow }(-)} \rightleftharpoons \underbrace{(3^{\mathrm{rd}}\mathrm{SO,nF})}_{\text{appearing bifurcation}} \rightleftharpoons \left\{\begin{array}{c} \underbrace{(\mathrm{SO,nF})}_{\text{source flow }(-)} \\ \underbrace{(\mathrm{SI,nF})}_{\text{sink flow }(-)} \\ \underbrace{(\mathrm{SO,nF})}_{\text{source flow }(-)} \end{array}\right\}. \tag{1.79}$$

The (3rdSO,pF)*-positive source flow at* $x_{j_1}^* = a_{j_1j_1s_1}$ *is a* (3rdSO,pF)*-appearing bifurcation from a negative source flow to negative source, sink, source flows.*

- *For* $a_{j_1j_10} < 0$ *and* $A_{j_2j_13} < 0$,

$$\underbrace{(\mathrm{SI,nF})}_{\text{sink flow }(-)} \rightleftharpoons \underbrace{(3^{\mathrm{rd}}\mathrm{SI,nF})}_{\text{appearing bifurcation}} \rightleftharpoons \left\{\begin{array}{c} \underbrace{(\mathrm{SI,nF})}_{\text{sink flow }(-)} \\ \underbrace{(\mathrm{SO,nF})}_{\text{source flow }(-)} \\ \underbrace{(\mathrm{SI,nF})}_{\text{sink flow }(-)} \end{array}\right\}. \tag{1.80}$$

The (3rdSI,pF)*-positive source flow at* $x_{j_1}^* = a_{j_1j_1s_1}$ *is a* (3rdSI,pF)*-appearing bifurcation from a negative sink flow to negative sink, source, sink flows.*

1.2 Proof of Theorem 1.1

Consider

$$\begin{aligned}\dot{x}_{j_2} &= A_{j_2 j_1 3},\\ \dot{x}_{j_1} &= A_{j_1 j_1 3} + A_{j_1 j_1 2} x_{j_1} + A_{j_1 j_1 1} x_{j_1}^2 + A_{j_1 j_1 0} x_{j_1}^3\\ &= a_{j_1 j_1 0}(x_{j_1} - b_{j_1 j_1 1})(x_{j_1}^2 + B_{j_1 j_1 1} x_{j_1} + C_{j_1 j_1 1})\end{aligned}$$

with

$$\Delta_{j_1 j_1} = (B_{j_1 j_1 1})^2 - 4C_{j_1 j_1 1} \text{ for } j_1, j_2 \in \{1,2\}, j_1 \neq j_2.$$

(i) For $A_{j_2 j_1 3} = 0$,

$$\begin{aligned}\dot{x}_{j_2} &= A_{j_2 j_1 3} = 0,\\ \dot{x}_{j_1} &= a_{j_1 j_1 0}(x_{j_1} - b_{j_1 j_1 1})(x_{j_1}^2 + B_{j_1 j_1 1} x_{j_1} + C_{j_1 j_1 1}).\end{aligned}$$

(i_1) For $\Delta_{j_1 j_1} < 0$, the dynamical system becomes

$$\begin{aligned}\dot{x}_{j_2} &= A_{j_2 j_1 3} = 0,\\ \dot{x}_{j_1} &= a_{j_1 j_1 0}(x_{j_1} - a_{j_1 j_1 1})\left[(x_{j_1} - a_{j_1 j_1 2})^2 + b_{j_1 j_1}\right]\end{aligned}$$

where

$$a_{j_1 j_1 1} = b_{j_1 j_1 1}, a_{j_1 j_1 2} = -\frac{1}{2} B_{j_1 j_1 1}, a_{j_1 j_1 0} = A_{j_1 j_1 0}, b_{j_1 j_1} = \frac{1}{4}(-\Delta_{j_1 j_1}).$$

Because

$$\begin{aligned}\dot{x}_{j_2} &= A_{j_2 j_1 3} = 0,\\ dt &= \frac{dx_{j_1}}{a_{j_1 j_1 0}(x_{j_1} - a_{j_1 j_1 1})\left[(x_{j_1} - a_{j_1 j_1 2})^2 + b_{j_1 j_1}\right]}\\ &= \frac{dx_{j_1}}{a_{j_1 j_1 0}} \frac{1}{(a_{j_1 j_1 1} - a_{j_1 j_1 2})^2 + b_{j_1 j_1}} \Big[\frac{1}{x_{j_1} - a_{j_1 j_1 1}}\\ &\quad - \frac{x_{j_1} - a_{j_1 j_1 2}}{(x_{j_1} - a_{j_1 j_1 2})^2 + b_{j_1 j_1}} + \frac{a_{j_1 j_1 2} - a_{j_1 j_1 1}}{(x_{j_1} - a_{j_1 j_1 2})^2 + b_{j_1 j_1}}\Big],\end{aligned}$$

with an initial condition of $(x_{j_1 0}, x_{j_2 0})$ at $t = t_0$, the simple integration gives the corresponding solution, i.e.,

$$
\begin{aligned}
x_{j_2} &= x_{j_2 0};\\
t - t_0 &= \frac{1}{a_{j_1 j_1 0}} \frac{1}{(a_{j_1 j_1 1} - a_{j_1 j_1 2})^2 + b_{j_1 j_1}} \left[\ln \frac{|x_{j_1} - a_{j_1 j_1 1}|}{|x_{j_1 0} - a_{j_1 j_1 1}|} \right.\\
&\quad - \frac{1}{2} \ln \frac{(x_{j_1} - a_{j_1 j_1 2})^2 + b_{j_1 j_1}}{(x_{j_1 0} - a_{j_1 j_1 2})^2 + b_{j_1 j_1}}\\
&\quad \left. + \frac{(a_{j_1 j_1 2} - a_{j_1 j_1 1})}{\sqrt{b_{j_1 j_1}}} \left(\arctan \frac{x_{j_1} - a_{j_1 j_1 2}}{\sqrt{b_{j_1 j_1}}} - \arctan \frac{x_{j_1 0} - a_{j_1 j_1 2}}{\sqrt{b_{j_1 j_1}}}\right) \right].
\end{aligned}
$$

In phase space, at $x_{j_1}^* = a_{j_1 j_1 1}$ and $x_{j_2} = x_{j_2 0}$, the variational equation is

$$
\begin{aligned}
x_{j_2} &= x_{j_2 0},\\
\Delta \dot{x}_{j_1} &= a_{j_1 j_1 0} \left[(a_{j_1 j_1 1} - a_{j_1 j_1 2})^2 + b_{j_1 j_1} \right] \Delta x_{j_1}
\end{aligned}
$$

Because $(a_{j_1 j_1 1} - a_{j_1 j_1 2})^2 + b_{j_1 j_1} > 0$, the flows at $x_{j_1}^* = a_{j_1 j_1 1}$ for $a_{j_1 j_1 0} > 0$ and $a_{j_1 j_1 0} < 0$ are source and sink flows in the x_{j_1}-direction, respectively.

Thus, the one-dimensional flow has the following properties as in Eqs. (1.7) and (1.8).

- For $a_{j_1 j_1 0} > 0$,

$$
(a_{j_1 j_1 1}, x_{j_2 0}) = \underbrace{(\text{SO}, \text{IC})}_{\text{source}}.
$$

  The equilibrium of $x_{j_1}^* = a_{j_1 j_1 1}$ and $x_{j_2} = x_{j_2 0}$ is called an (SO,IC)-source.

- *For* $a_{j_1 j_1 0} < 0$,

$$
(a_{j_1 j_1 1}, x_{j_2 0}) = \underbrace{(\text{SI}, \text{IC})}_{\text{sink}}.
$$

  The equilibrium of $x_{j_1}^* = a_{j_1 j_1 1}$ and $x_{j_2} = x_{j_2 0}$ is called an (SI,IC)-sink.

(i_2) For $\Delta_{j_1 j_1} = 0$, the dynamical system becomes

$$
\begin{aligned}
\dot{x}_{j_2} &= A_{j_2 j_1 3} = 0,\\
\dot{x}_{j_1} &= a_{j_1 j_1 0}(x_{j_1} - a_{j_1 j_1 s_1})(x_{j_1} - a_{j_1 j_1 s_2})^2
\end{aligned}
$$

where

$$a_{j_1j_1s_1}, a_{j_1j_1s_2} \in \{a_{j_1j_1 1}, a_{j_1j_1 2}\}$$
$$a_{j_1j_1 1} = b_{j_1j_1 1}, a_{j_1j_1 2} = -\frac{1}{2}B_{j_1j_1 1}, a_{j_1j_1 0} = A_{j_1j_1 0}.$$

- *For* $a_{j_1j_1s_1} \neq a_{j_1j_1s_2}$,

$$\begin{aligned}
\dot{x}_{j_2} &= A_{j_2j_1 3} = 0,\\
dt &= \frac{dx_{j_1}}{a_{j_1j_1 0}(x_{j_1} - a_{j_1j_1s_1})(x_{j_1} - a_{j_1j_1s_2})^2}\\
&= \frac{dx_{j_1}}{a_{j_1j_1 0}} \frac{1}{(a_{j_1j_1s_2} - a_{j_1j_1s_1})^2} \Big[\frac{1}{x_{j_1} - a_{j_1j_1s_1}} - \frac{x_{j_1} - a_{j_1j_1s_2}}{(x_{j_1} - a_{j_1j_1s_2})^2} + \frac{a_{j_1j_1s_2} - a_{j_1j_1s_1}}{(x_{j_1} - a_{j_1j_1s_2})^2}\Big].
\end{aligned}$$

With an initial condition of $(x_{j_1 0}, x_{j_2 0})$ at $t = t_0$, the simple integration gives the corresponding solution as

$$\begin{aligned}
x_{j_2} &= x_{j_2 0};\\
t - t_0 &= \frac{1}{a_{j_1j_1 0}(a_{j_1j_1s_2} - a_{j_1j_1s_1})^2}\Big\{\ln \frac{|x_{j_1} - a_{j_1j_1s_1}|}{|x_{j_1 0} - a_{j_1j_1s_1}|}\\
&\quad - (a_{j_1j_1s_2} - a_{j_1j_1s_1})\Big[\frac{1}{(x_{j_1} - a_{j_1j_1s_2})} - \frac{1}{(x_{j_1 0} - a_{j_1j_1s_2})}\Big]\\
&\quad - \ln \frac{|x_{j_1} - a_{j_1j_1s_2}|}{|x_{j_1 0} - a_{j_1j_1s_2}|}\Big\}.
\end{aligned}$$

(i_{2a}) For $x^*_{j_1} = a_{j_1j_1s_1}$ and $x_{j_2} = x_{j_2 0}$, the variational equation is

$$\begin{aligned}
x_{j_2} &= x_{j_2 0},\\
\Delta\dot{x}_{j_1} &= a_{j_1j_1 0}(a_{j_1j_1s_1} - a_{j_1j_1s_2})^2 \Delta x_{j_1}.
\end{aligned}$$

Because $(a_{j_1j_1s_1} - a_{j_1j_1s_2})^2 > 0$, the flows at $x^*_{j_1} = a_{j_1j_1 1}$ for $a_{j_1j_1 0} > 0$ and $a_{j_1j_1 0} < 0$ are source and sink flows in the x_{j_1}-direction, respectively.

Thus, the one-dimensional flow at $x^*_{j_1} = a_{j_1j_1s_1} \neq a_{j_1j_1s_2}$ has the following properties as in Eqs. (1.12) and (1.13).

- For $a_{j_1j_1 0}(a_{j_1j_1s_1} - a_{j_1j_1s_2})^2 > 0$,

$$(a_{j_1j_1s_1}, x_{j_2 0}) = \underbrace{(\text{SO}, \text{IC})}_{\text{source}}.$$

The equilibrium $x^*_{j_1} = a_{j_1j_1s_1}$ at the initial point $x_{j_2} = x_{j_2 0}$ is called an (SO,IC)-source.

- For $a_{j_1j_10}(a_{j_1j_1s_1} - a_{j_1j_1s_2})^2 < 0$,

$$(a_{j_1j_1s_1}, x_{j_20}) = \underbrace{(\text{SI}, \text{IC})}_{\text{sink}}.$$

The equilibrium $x_{j_1}^* = a_{j_1j_1s_1}$ at the initial point $x_{j_2} = x_{j_20}$ is called an (SI,IC)-sink.

(i_{2b}) For at $x_{j_1}^* = a_{j_1j_1s_2}$ and $x_{j_2} = x_{j_20}$, the variational equation is

$$\begin{aligned} x_{j_2} &= x_{j_20}, \\ \Delta\dot{x}_{j_1} &= a_{j_1j_10}(a_{j_1j_1s_2} - a_{j_1j_1s_1})(\Delta x_{j_1})^2. \end{aligned}$$

The flows at $x_{j_1}^* = a_{j_1j_1s_2}$ for $a_{j_1j_10} > 0$ and $a_{j_1j_10} < 0$ are upper-saddle and lower-saddle flows in the x_{j_1}-direction, respectively.

Thus, the one-dimensional flow at $x_{j_1}^* = a_{j_1j_1s_2} \neq a_{j_1j_11}$ has the following properties as in Eqs. (1.14) and (1.15).

- For $a_{j_1j_10}(a_{j_1j_1s_2} - a_{j_1j_1s_1}) > 0$,

$$(a_{j_1j_1s_2}, x_{j_20}) = \underbrace{(\text{US}, \text{IC})}_{\text{upper-saddle}}.$$

The equilibrium $x_{j_1}^* = a_{j_1j_1s_2}$ at the initial point $x_{j_2} = x_{j_20}$ is called a (US,IC)-upper-saddle, which is called an upper-saddle bifurcation.

- For $a_{j_1j_10}(a_{j_1j_1s_2} - a_{j_1j_1s_1}) < 0$,

$$(a_{j_1j_1s_2}, x_{j_20}) = \underbrace{(\text{LS}, \text{IC})}_{\text{lower-saddle}}.$$

The equilibrium $x_{j_1}^* = a_{j_1j_1s_2}$ at the initial point $x_{j_2} = x_{j_20}$ is called an (LS,IC)-lower-saddle, which is called a lower-saddle bifurcation.

(i_{2c}) For $a_{j_1j_1s_1} = a_{j_1j_1s_2} \equiv a_{j_1j_11}$,

$$\begin{aligned} \dot{x}_{j_2} &= A_{j_2j_13} = 0, \\ dt &= \frac{dx_{j_1}}{a_{j_1j_10}(x_{j_1} - a_{j_1j_11})^3}, \end{aligned}$$

with an initial condition of (x_{j_10}, x_{j_20}) at $t = t_0$, the simple integration gives the corresponding solution as

$$\begin{aligned} x_{j_2} &= x_{j_2 0}; \\ t - t_0 &= -\frac{1}{2a_{j_1 j_1 0}}\Big[\frac{1}{(x_{j_1} - a_{j_1 j_1 1})^2} - \frac{1}{(x_{j_1 0} - a_{j_1 j_1 1})^2}\Big]. \end{aligned}$$

For $x_{j_1}^* = a_{j_1 j_1 1}$ and $x_{j_2} = x_{j_2 0}$, the variational equation is

$$\begin{aligned} x_{j_2} &= x_{j_2 0}, \\ \Delta\dot{x}_{j_1} &= a_{j_1 j_1 0}(\Delta x_{j_1})^3. \end{aligned}$$

The flows at $x_{j_1}^* = a_{j_1 j_1 1}$ for $a_{j_1 j_1 0} > 0$ and $a_{j_1 j_1 0} < 0$ are third-order source and sink flows in the x_{j_1}-direction, respectively.

Thus, the one-dimensional flow at $x_{j_1}^* = a_{j_1 j_1 1}$ has the properties as in Eqs. (1.17) and (1.18).

- For $a_{j_1 j_1 0} > 0$,

$$(a_{j_1 j_1 1}, x_{j_2 0}) = \underbrace{(3^{\text{rd}}\text{SO}, \text{IC})}_{\text{third-order source}}.$$

  The equilibrium of $x_{j_1}^* = a_{j_1 j_1 1}$ at the initial point $x_{j_2} = x_{j_2 0}$ is called a (3^{rd}SO, IC)-third-order source, which is called a third-order source bifurcation.
- For $a_{j_1 j_1 0} < 0$,

$$(a_{j_1 j_1 1}, x_{j_2 0}) = \underbrace{(3^{\text{rd}}\text{SI}, \text{IC})}_{\text{third-order sink}}.$$

  The equilibrium of $x_{j_1}^* = a_{j_1 j_1 1}$ at the initial point $x_{j_2} = x_{j_2 0}$ is called a (3^{rd}SI,IC)-third-order sink, which is called a third-order sink bifurcation.

(i_3) For $\Delta_{j_1 j_1} > 0$, the dynamical system becomes

$$\begin{aligned} \dot{x}_{j_2} &= A_{j_2 j_1 2} = 0, \\ \dot{x}_{j_1} &= a_{j_1 j_1 0}(x_{j_1} - a_{j_1 j_1 1})(x_{j_1} - a_{j_1 j_1 2})(x_{j_1} - a_{j_1 j_1 3}) \end{aligned}$$

where

$$\begin{aligned} &\{a_{j_1 j_1 1}, a_{j_1 j_1 2}, a_{j_1 j_1 3}\} = \text{sort}\{b_{j_1 j_1 1}, b_{j_1 j_1 2}, b_{j_1 j_1 3}\}, \\ &a_{j_1 j_1 l} \le a_{j_1 j_1 (l+1)}, l = 1, 2; \\ &b_{j_1 j_1 2}, b_{j_1 j_1 3} = -\frac{1}{2}(B_{j_1 j_1 1} \pm \sqrt{\Delta_{j_1 j_1}}). \end{aligned}$$

For $a_{j_1j_1s_1} \neq a_{j_1j_1s_2} \neq a_{j_1j_1s_3}$ $(s_1, s_2, s_3 \in \{1,2,3\},\ s_1 \neq s_2 \neq s_3)$,

$$\begin{aligned}
\dot{x}_{j_2} &= A_{j_2j_12} = 0,\\
dt &= \frac{dx_{j_1}}{a_{j_1j_10}} \frac{1}{(x_{j_1} - a_{j_1j_11})(x_{j_1} - a_{j_1j_12})(x_{j_1} - a_{j_1j_13})}\\
&= \frac{dx_{j_1}}{a_{j_1j_10}} \Big[\frac{1}{(x_{j_1} - a_{j_1j_11})(a_{j_1j_11} - a_{j_1j_12})(a_{j_1j_11} - a_{j_1j_13})}\\
&\quad - \frac{1}{(a_{j_1j_12} - a_{j_1j_11})(x_{j_1} - a_{j_1j_12})(a_{j_1j_12} - a_{j_1j_13})}\\
&\quad - \frac{1}{(a_{j_1j_13} - a_{j_1j_11})(a_{j_1j_13} - a_{j_1j_12})(x_{j_1} - a_{j_1j_13})}\Big].
\end{aligned}$$

With an initial condition of (x_{j_10}, x_{j_20}) at $t = t_0$, the simple integration gives the corresponding solution, i.e.,

$$\begin{aligned}
x_{j_2} &= x_{j_20};\\
t - t_0 &= \frac{1}{a_{j_1j_10}} \sum\nolimits_{s_1=1}^{3} \frac{1}{\Pi_{s_2=1,s_2\neq s_1}^{3}(a_{j_1j_1s_1} - a_{j_1j_1s_2})} \ln \frac{|x_{j_1} - a_{j_1j_1s_1}|}{|x_{j_10} - a_{j_1j_1s_1}|}.
\end{aligned}$$

For $x_{j_1}^{*} = a_{j_1j_1s_1} \neq a_{j_1j_1s_2} \neq a_{j_1j_1s_3}$ $(s_1, s_2, s_3 \in \{1,2,3\})$ and $x_{j_2} = x_{j_20}$, the variational equation is

$$\begin{aligned}
x_{j_2} &= x_{j_20},\\
\Delta\dot{x}_{j_1} &= a_{j_1j_10}(a_{j_1j_1s_1} - a_{j_1j_1s_2})(a_{j_1j_1s_1} - a_{j_1j_1s_3})\Delta x_{j_1}\\
&= a_{j_1j_10}\Pi_{s_2=1,s_2\neq s_1}^{3}(a_{j_1j_1s_1} - a_{j_1j_1s_2})\Delta x_{j_1}.
\end{aligned}$$

The flows at $x_{j_1}^{*} = a_{j_1j_1s_1}$ are source and sink flows in the x_{j_1}-direction for $a_{j_1j_10}\Pi_{s_2=1,s_2\neq s_1}^{3}(a_{j_1j_1s_1} - a_{j_1j_1s_2}) > 0$ and $a_{j_1j_10}\Pi_{s_2=1,s_2\neq s_1}^{3}(a_{j_1j_1s_1} - a_{j_1j_1s_2}) < 0$, respectively.

Thus, the one-dimensional flow at $x_{j_1}^{*} = a_{j_1j_1s_1} \neq a_{j_1j_1s_2} \neq a_{j_1j_1s_3} s_1, s_2, s_3 \in \{1,2,3\})$ has the following properties as in Eqs. (1.22) and (1.23).

- For $a_{j_1j_10}\Pi_{s_2=1,s_2\neq s_1}^{3}(a_{j_1j_1s_1} - a_{j_1j_1s_2}) > 0$,

$$(a_{j_1j_1s_1}, x_{j_20}) = \underbrace{(\mathrm{SO}, \mathrm{IC})}_{\text{source}}.$$

The equilibrium of $x_{j_1}^{*} = a_{j_1j_1s_1}$ at $x_{j_2} = x_{j_20}$ is an (SO,IC)-source.

- For $a_{j_1j_10}\Pi_{s_2=1,s_2\neq s_1}^{3}(a_{j_1j_1s_1} - a_{j_1j_1s_2}) < 0$,

$$(a_{j_1j_1s_1}, x_{j_20}) = \underbrace{(\mathrm{SI}, \mathrm{IC})}_{\text{sink}}.$$

The equilibrium of $x_{j_1}^{*} = a_{j_1j_1s_1}$ and $x_{j_2} = x_{j_20}$ is an (SI,IC)-sink.

(ii) If

$$A_{j_2 j_1 3} \neq 0 \text{ and } \Delta_{j_1 j_1} < 0,$$

the dynamical system becomes

$$\begin{aligned}\dot{x}_{j_2} &= A_{j_2 j_1 3} \neq 0,\\ \dot{x}_{j_1} &= a_{j_1 j_1 0}(x_{j_1} - a_{j_1 j_1 1})\left[(x_{j_1} - a_{j_1 j_1 2})^2 + b_{j_1 j_1}\right]\end{aligned}$$

where

$$a_{j_1 j_1 1} = b_{j_1 j_1 1}, a_{j_1 j_1 2} = -\frac{1}{2} B_{j_1 j_1 1}, a_{j_1 j_1 0} = A_{j_1 j_1 0}, b_{j_1 j_1} = \frac{1}{4}(-\Delta_{j_1 j_1}).$$

Similar to (i_1), with an initial condition of $(x_{j_1 0}, x_{j_2 0})$ at $t = t_0$, the simple integration gives the corresponding solution, i.e.,

$$\begin{aligned}x_{j_2} &= A_{j_2 j_1 3}(t - t_0) + x_{j_2 0},\\ t - t_0 &= \frac{1}{a_{j_1 j_1 0}} \frac{1}{(a_{j_1 j_1 1} - a_{j_1 j_1 2})^2 + b_{j_1 j_1}} \Big[\ln \frac{|x_{j_1} - a_{j_1 j_1 1}|}{|x_{j_1 0} - a_{j_1 j_1 1}|}\\ &\quad - \frac{1}{2} \ln \frac{(x_{j_1} - a_{j_1 j_1 2})^2 + b_{j_1 j_1}}{(x_{j_1 0} - a_{j_1 j_1 2})^2 + b_{j_1 j_1}}\\ &\quad + \frac{(a_{j_1 j_1 2} - a_{j_1 j_1 1})}{\sqrt{b_{j_1 j_1}}} \Big(\arctan \frac{x_{j_1} - a_{j_1 j_1 2}}{\sqrt{b_{j_1 j_1}}} - \arctan \frac{x_{j_1 0} - a_{j_1 j_1 2}}{\sqrt{b_{j_1 j_1}}}\Big)\Big].\end{aligned}$$

The corresponding first integral manifold is

$$\begin{aligned}\frac{x_{j_2} - x_{j_2 0}}{A_{j_2 j_1 3}} &= \frac{1}{a_{j_1 j_1 0}} \frac{1}{(a_{j_1 j_1 1} - a_{j_1 j_1 2})^2 + b_{j_1 j_1}} \Big[\ln \frac{|x_{j_1} - a_{j_1 j_1 1}|}{|x_{j_1 0} - a_{j_1 j_1 1}|}\\ &\quad - \frac{1}{2} \ln \frac{(x_{j_1} - a_{j_1 j_1 2})^2 + b_{j_1 j_1}}{(x_{j_1 0} - a_{j_1 j_1 2})^2 + b_{j_1 j_1}}\\ &\quad + \frac{(a_{j_1 j_1 2} - a_{j_1 j_1 1})}{\sqrt{b_{j_1 j_1}}} \Big(\arctan \frac{x_{j_1} - a_{j_1 j_1 2}}{\sqrt{b_{j_1 j_1}}} - \arctan \frac{x_{j_1 0} - a_{j_1 j_1 2}}{\sqrt{b_{j_1 j_1}}}\Big)\Big].\end{aligned}$$

For $x_{j_1}^* = a_{j_1 j_1 s_1}$, the variational equation is

$$\begin{aligned}\dot{x}_{j_2} &= A_{j_2 j_1 3} \neq 0,\\ \Delta\dot{x}_{j_1} &= a_{j_1 j_1 0}\left[(a_{j_1 j_1 1} - a_{j_1 j_1 2})^2 + b_{j_1 j_1}\right]\Delta x_{j_1}.\end{aligned}$$

Because $(a_{j_1 j_1 1} - a_{j_1 j_1 2})^2 + b_{j_1 j_1} > 0$, the flows at $x_{j_1}^* = a_{j_1 j_1 1}$ for $a_{j_1 j_1 0} > 0$ and $a_{j_1 j_1 0} < 0$ are source and sink flows in the x_{j_1}-direction, respectively.

Thus, the flow at $x_{j_1}^* = a_{j_1 j_1 1}$ has the following properties as in Eqs. (1.29)–(1.32):

- For $a_{j_1 j_1 0} > 0$ and $A_{j_2 j_1 3} > 0$,

$$(a_{j_1 j_1 1}, \dot{x}_{j_2}) = \underbrace{(\mathrm{SO}, \mathrm{pF})}_{\text{source flow }(+)}.$$

  The flow at $x_{j_1}^* = a_{j_1 j_1 1}$ is called an (SO,pF)-positive source flow.
- For $a_{j_1 j_1 0} < 0$ and $A_{j_2 j_1 3} > 0$,

$$(a_{j_1 j_1 1}, \dot{x}_{j_2}) = \underbrace{(\mathrm{SI}, \mathrm{pF})}_{\text{sink flow }(+)}.$$

  The flow at $x_{j_1}^* = a_{j_1 j_1 1}$ is called an (SI,pF)-positive sink flow.
- For $a_{j_1 j_1 0} > 0$ and $A_{j_2 j_1 3} < 0$,

$$(a_{j_1 j_1 1}, \dot{x}_{j_2}) = \underbrace{(\mathrm{SO}, \mathrm{nF})}_{\text{source flow }(-)}.$$

  The flow at $x_{j_1}^* = a_{j_1 j_1 1}$ is called an (SO,nF)-negative source flow.
- For $a_{j_1 j_1 0} < 0$ and $A_{j_2 j_1 3} < 0$,

$$(a_{j_1 j_1 1}, \dot{x}_{j_2}) = \underbrace{(\mathrm{SI}, \mathrm{nF})}_{\text{sink flow }(-)}.$$

  The flow at $x_{j_1}^* = a_{j_1 j_1 1}$ is called an (SI,nF)-negative sink flow.

(iii) If

$$A_{j_2 j_1 3} \neq 0 \text{ and } \Delta_{j_1 j_1} = 0,$$

the corresponding dynamical system becomes

$$\begin{aligned} \dot{x}_{j_2} &= A_{j_2 j_1 3}, \\ \dot{x}_{j_1} &= a_{j_1 j_1 0}(x_{j_1} - a_{j_1 j_1 s_1})(x_{j_1} - a_{j_1 j_1 s_2})^2. \end{aligned}$$

For $a_{j_1 j_1 s_1} \neq a_{j_1 j_1 s_2}$, similar to case (i$_2$), with an initial condition of $(x_{j_1 0}, x_{j_2 0})$ at $t = t_0$, the simple integration gives the corresponding solution as

$$
\begin{aligned}
x_{j_2}-x_{j_2 0} &= A_{j_2 j_1 3}(t-t_0);\\
t-t_0 &= \frac{1}{a_{j_1 j_1 0}(a_{j_1 j_1 s_2}-a_{j_1 j_1 s_1})^2}\Big[\ln\frac{|x_{j_1}-a_{j_1 j_1 s_1}|}{|x_{j_1 0}-a_{j_1 j_1 s_1}|}\\
&\quad -(a_{j_1 j_1 s_2}-a_{j_1 j_1 s_1})\Big(\frac{1}{x_{j_1}-a_{j_1 j_1 s_2}}-\frac{1}{x_{j_1 0}-a_{j_1 j_1 s_2}}\Big)\\
&\quad -\ln\frac{|x_{j_1}-a_{j_1 j_1 s_2}|}{|x_{j_1 0}-a_{j_1 j_1 s_2}|}\Big].
\end{aligned}
$$

The corresponding first integral manifold is

$$
\begin{aligned}
\frac{x_{j_2}-x_{j_2 0}}{A_{j_2 j_1 3}} &= \frac{1}{a_{j_1 j_1 0}(a_{j_1 j_1 s_2}-a_{j_1 j_1 s_1})^2}\Big[\ln\frac{|x_{j_1}-a_{j_1 j_1 s_1}|}{|x_{j_1 0}-a_{j_1 j_1 s_1}|}\\
&\quad -(a_{j_1 j_1 s_2}-a_{j_1 j_1 s_1})\Big(\frac{1}{x_{j_1}-a_{j_1 j_1 s_2}}-\frac{1}{x_{j_1 0}-a_{j_1 j_1 s_2}}\Big)\\
&\quad -\ln\frac{|x_{j_1}-a_{j_1 j_1 s_2}|}{|x_{j_1 0}-a_{j_1 j_1 s_2}|}\Big].
\end{aligned}
$$

The two-dimensional dynamical system has two equilibriums for the x_{j_1}-direction, i.e.,

$$
a_{j_1 j_1 s_1}=b_{j_1 j_1 1},\ x_{j_1}^*=-\frac{B_{j_1 j_1 1}}{2}=a_{j_1 j_1 s_2},\ a_{j_1 j_1 0}=A_{j_1 j_1 0} \text{ for the } x_{j_1}\text{-direction.}
$$

(iii$_1$) At $x_{j_1}^*=a_{j_1 j_1 s_1}$, the variational equation is

$$
\begin{aligned}
\dot{x}_{j_2} &= A_{j_2 j_1 3},\\
\Delta\dot{x}_{j_1} &= a_{j_1 j_1 0}(a_{j_1 j_1 s_1}-a_{j_1 j_1 s_2})^2\Delta x_{j_1}.
\end{aligned}
$$

Because $(a_{j_1 j_1 s_1}-a_{j_1 j_1 s_2})^2>0$, the flows at $x_{j_1}^*=a_{j_1 j_1 s_1}$ for $a_{j_1 j_1 0}>0$ and $a_{j_1 j_1 0}<0$ are source and sink flow in the x_{j_1}-direction, respectively.

Thus, the flow at $x_{j_1}^*=a_{j_1 j_1 s_1}$ has the following properties as in Eqs. (1.38)–(1.41):

- For $a_{j_1 j_1 0}(a_{j_1 j_1 s_1}-a_{j_1 j_1 s_2})^2>0$ and $A_{j_2 j_1 3}>0$,

$$
(a_{j_1 j_1 s_1},\dot{x}_{j_2})=\underbrace{(\mathrm{SO},\mathrm{pF})}_{\text{source flow }(+)}.
$$

The flow at $x_{j_1}^*=a_{j_1 j_1 s_1}$ is called an (SO,pF)-positive source flow.

- For $a_{j_1j_10}(a_{j_1j_1s_1} - a_{j_1j_1s_2})^2 < 0$ and $A_{j_2j_13} > 0$,

$$(a_{j_1j_1s_1}, \dot{x}_{j_2}) = \underbrace{(\mathrm{SI}, \mathrm{pF})}_{\text{sink flow }(+)}.$$

The flow at $x_{j_1}^* = a_{j_1j_1s_1}$ is called an (SI,pF)-positive sink flow.

- For $a_{j_1j_10}(a_{j_1j_1s_1} - a_{j_1j_1s_2})^2 > 0$ and $A_{j_2j_13} < 0$,

$$(a_{j_1j_1s_1}, \dot{x}_{j_2}) = \underbrace{(\mathrm{SO}, \mathrm{nF})}_{\text{source flow }(-)}.$$

The flow at $x_{j_1}^* = a_{j_1j_1s_1}$ is called an (SO,nF)-negative source flow.

- For $a_{j_1j_10}(a_{j_1j_1s_1} - a_{j_1j_1s_2})^2 < 0$ and $A_{j_2j_13} < 0$,

$$(a_{j_1j_1s_1}, \dot{x}_{j_2}) = \underbrace{(\mathrm{SI}, \mathrm{nF})}_{\text{sink flow }(-)}.$$

The flow at $x_{j_1}^* = a_{j_1j_1s_1}$ is called an (SI,nF)-negative sink flow.

(iii$_2$) At $x_{j_1}^* = a_{j_1j_1s_2}$, the variational equation is

$$\begin{aligned}
\dot{x}_{j_2} &= A_{j_2j_13}, \\
\Delta\dot{x}_{j_1} &= a_{j_1j_10}(a_{j_1j_1s_2} - a_{j_1j_1s_1})(\Delta x_{j_1})^2.
\end{aligned}$$

The flows at $x_{j_1}^* = a_{j_1j_1s_2}$ for $a_{j_1j_10}(a_{j_1j_1s_2} - a_{j_1j_1s_1}) > 0$ and $a_{j_1j_10}(a_{j_1j_1s_2} - a_{j_1j_1s_1}) < 0$ are source and sink flow in the x_{j_1}-direction, respectively.

Thus, the equilibrium of $x_{j_1}^* = a_{j_1j_1s_2}$ for the x_{j_1}-direction has the following properties as in Eqs. (1.42)–(1.45):

- For $a_{j_1j_10}(a_{j_1j_1s_2} - a_{j_1j_1s_1}) > 0$ and $A_{j_2j_13} > 0$,

$$(a_{j_1j_1s_2}, \dot{x}_{j_2}) = \underbrace{(\mathrm{US}, \mathrm{pF})}_{\text{upper-saddle flow }(+)}.$$

The flow at $x_{j_1}^* = a_{j_1j_1s_2}$ for the x_{j_1}-direction is called a (US,pF)-positive upper-saddle flow.

- For $a_{j_1j_10}(a_{j_1j_1s_2} - a_{j_1j_1s_1}) < 0$ and $A_{j_2j_13} > 0$,

$$(a_{j_1j_1s_2}, \dot{x}_{j_2}) = \underbrace{(\mathrm{LS}, \mathrm{pF})}_{\text{lower-saddle flow }(+)}.$$

The flow at $x_{j_1}^* = a_{j_1j_1s_2}$ for the x_{j_1}-direction is called an (LS,pF)-positive lower-saddle flow.

- For $a_{j_1j_10}(a_{j_1j_1s_2}-a_{j_1j_1s_1})>0$ and $A_{j_2j_13}<0$,

$$(a_{j_1j_1s_2},\dot{x}_{j_2})=\underbrace{(\text{US},\text{nF})}_{\text{upper-saddle flow }(-)}.$$

The flow at $x_{j_1}^*=a_{j_1j_1s_2}$ for the x_{j_1}-direction is called a (US,nF)-negative upper-saddle flow.

- For $a_{j_1j_10}(a_{j_1j_1s_2}-a_{j_1j_1s_1})<0$ and $A_{j_2j_13}<0$,

$$(a_{j_1j_1s_2},\dot{x}_{j_2})=\underbrace{(\text{LS},\text{nF})}_{\text{lower-saddle flow }(-)}.$$

The flow at $x_{j_1}^*=a_{j_1j_1s_2}$ for the x_{j_1}-direction is called an (LS,nF)-negative lower-saddle flow.

– The corresponding states are the bifurcation states for sink-flow and source-flows for the x_{j_1}-direction at a specific parameter.

(iii$_{3c}$) For $a_{j_1j_1s_1}=a_{j_1j_1s_2}=a_{j_1j_11}$, the dynamical system becomes

$$\begin{aligned}\dot{x}_{j_2}&=A_{j_2j_13},\\ \dot{x}_{j_1}&=a_{j_1j_10}(x_{j_1}-a_{j_1j_11})^3.\end{aligned}$$

Similar to case (i$_{2c}$), with an initial condition of (x_{j_10},x_{j_20}) at $t=t_0$, the solution is

$$\begin{aligned}&x_{j_2}-x_{j_20}=A_{j_2j_13},(t-t_0);\\ &t-t_0=-\frac{1}{2a_{j_1j_10}}\Big[\frac{1}{(x_{j_1}-a_{j_1j_11})^2}-\frac{1}{(x_{j_10}-a_{j_1j_11})^2}\Big].\end{aligned}$$

The corresponding first integral manifold is

$$\frac{x_{j_2}-x_{j_20}}{A_{j_2j_13}}=-\frac{1}{2a_{j_1j_10}}\Big[\frac{1}{(x_{j_1}-a_{j_1j_11})^2}-\frac{1}{(x_{j_10}-a_{j_1j_11})^2}\Big].$$

At $x_{j_1}^*=a_{j_1j_11}$, the variational equation is

$$\begin{aligned}\dot{x}_{j_2}&=A_{j_2j_13},\\ \Delta\dot{x}_{j_1}&=a_{j_1j_10}(\Delta x_{j_1})^3.\end{aligned}$$

The flows at $x_{j_1}^*=a_{j_1j_11}$ for $a_{j_1j_10}>0$ and $a_{j_1j_10}<0$ are the third-order source and sink flows in the x_{j_1}-direction, respectively.

Thus, the two-dimensional flow at $x_{j_1}^* = a_{j_1 j_1 s_2} = a_{j_1 j_1 s_1} = a_{j_1 j_1 1}$ has the following properties as in Eqs. (1.48)–(1.51):

- For $a_{j_1 j_1 0} > 0$ and $A_{j_2 j_1 3} > 0$

$$(a_{j_1 j_1 1}, \dot{x}_{j_2}) = \underbrace{(3^{\text{rd}}\text{SO, pF})}_{\text{third-order source flow } (+)}.$$

The flow at $x_{j_1}^* = a_{j_1 j_1 1}$ is a (3^{rd}SO,pF)-third order positive source flow.

- For $a_{j_1 j_1 0} < 0$ and $A_{j_2 j_1 3} > 0$

$$(a_{j_1 j_1 1}, \dot{x}_{j_2}) = = \underbrace{(3^{\text{rd}}\text{SI, pF})}_{\text{third-order sink flow } (+)}.$$

The flow at $x_{j_1}^* = a_{j_1 j_1 1}$ is a (3^{rd}SI,pF)-third-order positive sink flow.

- For $a_{j_1 j_1 0} > 0$ and $A_{j_2 j_1 3} < 0$

$$(a_{j_1 j_1 1}, \dot{x}_{j_2}) = \underbrace{(3^{\text{rd}}\text{SO, nF})}_{\text{third-order source flow } (-)}.$$

The flow at $x_{j_1}^* = a_{j_1 j_1 1}$ is a (3^{rd}SO,nF)-third-order negative source flow.

- For $a_{j_1 j_1 0} < 0$ and $A_{j_2 j_1 3} < 0$

$$(a_{j_1 j_1 1}, \dot{x}_{j_2}) = \underbrace{(3^{\text{rd}}\text{SI, nF})}_{\text{third-order sink flow } (-)}.$$

The flow at $x_{j_1}^* = a_{j_1 j_1 1}$ is a (3^{rd}SI,nF)-third-order negative sink flow.

– The third-order sink and source flows are the switching bifurcations of the upper-saddle and lower-saddle flows with source and sink flows.

(iv) If

$$A_{j_2 j_1 3} \neq 0 \text{ and } \Delta_{j_1 j_1} > 0,$$

then the dynamical system becomes

$$\begin{aligned}
\dot{x}_{j_2} &= A_{j_2 j_1 3}, \\
\dot{x}_{j_1} &= a_{j_1 j_1 0}(x_{j_1} - a_{j_1 j_1 1})(x_{j_1} - a_{j_1 j_1 2})(x_{j_1} - a_{j_1 j_1 3}).
\end{aligned}$$

For $a_{j_1 j_1 s_1} \neq a_{j_1 j_1 s_2} \neq a_{j_1 j_1 s_3}$ ($s_1, s_2, s_3 \in \{1, 2, 3\}$, $s_1 \neq s_2 \neq s_3$), similar to case (i_3), with an initial condition of $(x_{j_1 0}, x_{j_2 0})$ at $t = t_0$, the corresponding solution is

$$\begin{aligned} & x_{j_2} - x_{j_2 0} = A_{j_2 j_1 3}(t - t_0); \\ & t - t_0 = \frac{1}{a_{j_1 j_1 0}} \sum\nolimits_{s_1 = 1}^{3} \frac{1}{\Pi_{s_2 = 1, s_2 \neq s_1}^{3} (a_{j_1 j_1 s_1} - a_{j_1 j_1 s_2})} \ln \frac{|x_{j_1} - a_{j_1 j_1 s_1}|}{|x_{j_1 0} - a_{j_1 j_1 s_1}|}. \end{aligned}$$

The corresponding first integral manifold is

$$\frac{x_{j_2} - x_{j_2 0}}{A_{j_2 j_1 3}} = \frac{1}{a_{j_1 j_1 0}} \sum\nolimits_{s_1 = 1}^{3} \frac{1}{\Pi_{s_2 = 1, s_2 \neq s_1}^{3} (a_{j_1 j_1 s_1} - a_{j_1 j_1 s_2})} \ln \frac{|x_{j_1} - a_{j_1 j_1 s_1}|}{|x_{j_1 0} - a_{j_1 j_1 s_1}|}.$$

The two-dimensional dynamical system has three simple equilibriums for the x_{j_1}-direction, i.e.,

$$x_{j_1}^* = a_{j_1 j_1 1}, a_{j_1 j_1 2}, a_{j_1 j_1 3} \text{ for the } x_{j_1}\text{-direction.}$$

At $x_{j_1}^* = a_{j_1 j_1 s_1}$ (s_1, s_2, $s_3 \in \{1, 2, 3\}$, $s_1 \neq s_2 \neq s_3$), the variational equation is

$$\begin{aligned} x_{j_2} &= x_{j_2 0}, \\ \Delta \dot{x}_{j_1} &= a_{j_1 j_1 0}(a_{j_1 j_1 s_1} - a_{j_1 j_1 s_2})(a_{j_1 j_1 s_1} - a_{j_1 j_1 s_3}) \Delta x_{j_1} \\ &= a_{j_1 j_1 0} \Pi_{s_2 = 1, s_2 \neq s_1}^{3} (a_{j_1 j_1 s_1} - a_{j_1 j_1 s_2}) \Delta x_{j_1}. \end{aligned}$$

The flows at $x_{j_1}^* = a_{j_1 j_1 s_1}$ are source and sink flows in the x_{j_1}-direction for $a_{j_1 j_1 0} \Pi_{s_2 = 1, s_2 \neq s_1}^{3} (a_{j_1 j_1 s_1} - a_{j_1 j_1 s_2}) > 0$ and $a_{j_1 j_1 0} \Pi_{s_2 = 1, s_2 \neq s_1}^{3} (a_{j_1 j_1 s_1} - a_{j_1 j_1 s_2}) < 0$, respectively.

Thus, the equilibrium of $x_{j_1}^* = a_{j_1 j_1 s_1}$ (s_1, s_2, $s_3 \in \{1, 2, 3\}$, $s_1 \neq s_2 \neq s_3$) in the x_{j_1}-direction has the following properties as in Eqs. (1.57)–(1.60).

- For $a_{j_1 j_1 0} \Pi_{s_2 = 1, s_2 \neq s_1}^{3} (a_{j_1 j_1 s_1} - a_{j_1 j_1 s_2}) > 0$ and $A_{j_2 j_1 3} > 0$,

$$(a_{j_1 j_1 s_1}, \dot{x}_{j_2}) = \underbrace{(\text{SO}, \text{pF})}_{\text{source flow } (+)}.$$

  The flow at $x_{j_1}^* = a_{j_1 j_1 s_1}$ is an (SO,pF)-positive source flow.

- For $a_{j_1 j_1 0} \Pi_{s_2 = 1, s_2 \neq s_1}^{3} (a_{j_1 j_1 s_1} - a_{j_1 j_1 s_2}) < 0$ and $A_{j_2 j_1 3} > 0$,

$$(a_{j_1 j_1 s_1}, \dot{x}_{j_2}) = \underbrace{(\text{SI}, \text{pF})}_{\text{sink flow } (+)}.$$

  The flow at $x_{j_1}^* = a_{j_1 j_1 s_1}$ is an (SI,pF)-positive sink flow.

- For $a_{j_1j_10}\Pi_{s_2=1,s_2\neq s_1}^{3}(a_{j_1j_1s_1}-a_{j_1j_1s_2})>0$ and $A_{j_2j_13}<0$,

$$(a_{j_1j_1s_1},\dot{x}_{j_2})=\underbrace{(\mathrm{SO},\mathrm{nF})}_{\text{source flow }(-)}.$$

The flow at $x_{j_1}^*=a_{j_1j_1s_1}$ is an (SO,nF)-negative source flow.

- For $a_{j_1j_10}\Pi_{s_2=1,s_2\neq s_1}^{3}(a_{j_1j_1s_1}-a_{j_1j_1s_2})<0$ and $A_{j_2j_13}<0$,

$$(a_{j_1j_1s_1},\dot{x}_{j_2})=\underbrace{(\mathrm{SI},\mathrm{nF})}_{\text{sink flow }(-)}.$$

The flow at $x_{j_1}^*=a_{j_1j_1s_1}$ is an (SI,nF)-negative sink flow.

(v) Under

$$A_{j_2j_13}\neq 0 \text{ and } \Delta_{j_1j_1}=0 \text{ with } a_{j_1j_1s_1}=a_{j_1j_1s_2}=a_{j_1j_11},$$

the dynamical system is

$$\begin{aligned}\dot{x}_{j_2}&=A_{j_2j_13},\\ \dot{x}_{j_1}&=a_{j_1j_10}(x_{j_1}-a_{j_1j_11})^3\end{aligned}$$

with

$$a_{j_1j_1s_1}^{-}\to a_{j_1j_1s_2}^{+} \text{ and } a_{j_1j_1s_2}^{-}\to a_{j_1j_1s_1}^{+}.$$

From case (iii), the corresponding equilibriums of $x_{j_1}^*=a_{j_1j_11}$ for the x_{j_1}-direction possess the following properties as in Eqs. (1.64)–(1.67).

- For $a_{j_1j_10}>0$ and $A_{j_2j_13}>0$,

$$\left\{\begin{array}{c}\underbrace{(a_{j_1j_1s_2}^{-},\dot{x}_{j_2})}_{(\mathrm{LS},\mathrm{pF})\text{-flow }(+)}\\ \underbrace{(a_{j_1j_1s_1}^{-},\dot{x}_{j_2})}_{(\mathrm{SO},\mathrm{pF})\text{-flow }(+)}\end{array}\right\}\rightleftharpoons\underbrace{(a_{j_1j_11},\dot{x}_{j_2})}_{\substack{(3^{\mathrm{rd}}\mathrm{SO},\,\mathrm{pF})\text{-flow}\\ \text{switching bifurcation}}}\rightleftharpoons\left\{\begin{array}{c}\underbrace{(a_{j_1j_1s_1}^{+},\dot{x}_{j_2})}_{(\mathrm{SO},\mathrm{pF})\text{-flow }(+)}\\ \underbrace{(a_{j_1j_1s_2}^{+},\dot{x}_{j_2})}_{(\mathrm{US},\mathrm{pF})\text{-flow }(+)}\end{array}\right\}.$$

The (3^{rd}SO,pF)-positive source flow at $x_{j_1}^*=a_{j_1j_11}$ is a (3^{rd}SO,pF)-switching bifurcation for positive upper-saddle and lower-saddle flows with a positive source flow.

- For $a_{j_1 j_1 0} < 0$ and $A_{j_2 j_1 3} > 0$,

$$\left\{\begin{array}{c} \underbrace{(a_{j_1 j_1 s_2}^{-}, \dot{x}_{j_2})}_{(\mathrm{LS,pF})\text{-flow }(+)} \\ \underbrace{(a_{j_1 j_1 s_1}^{-}, \dot{x}_{j_2})}_{(\mathrm{SI,pF})\text{-flow }(+)} \end{array}\right\} \rightleftharpoons \underbrace{(a_{j_1 j_1 1}, \dot{x}_{j_2})}_{\substack{(3^{\mathrm{rd}}\mathrm{SI,pF})\text{-flow} \\ \text{switching bifurcation}}} \rightleftharpoons \left\{\begin{array}{c} \underbrace{(a_{j_1 j_1 s_1}^{+}, \dot{x}_{j_2})}_{(\mathrm{SI,pF})\text{-flow }(+)} \\ \underbrace{(a_{j_1 j_1 s_2}^{+}, \dot{x}_{j_2})}_{(\mathrm{US,pF})\text{-flow }(+)} \end{array}\right\}.$$

The (3^{rd}SI,pF)-positive source flow at $x_{j_1}^* = a_{j_1 j_1 1}$ is a (3^{rd}SI,pF)-switching bifurcation for positive upper-saddle and lower-saddle flows with a positive sink flow.

- For $a_{j_1 j_1 0} > 0$ and $A_{j_2 j_1 3} < 0$,

$$\left\{\begin{array}{c} \underbrace{(a_{j_1 j_1 s_2}^{-}, \dot{x}_{j_2})}_{(\mathrm{LS,nF})\text{-flow }(-)} \\ \underbrace{(a_{j_1 j_1 s_1}^{-}, \dot{x}_{j_2})}_{(\mathrm{SO,nF})\text{-flow }(-)} \end{array}\right\} \rightleftharpoons \underbrace{(a_{j_1 j_1 1}, \dot{x}_{j_2})}_{\substack{(3^{\mathrm{rd}}\mathrm{SO,nF})\text{-flow} \\ \text{switching bifurcation}}} \rightleftharpoons \left\{\begin{array}{c} \underbrace{(a_{j_1 j_1 s_1}^{+}, \dot{x}_{j_2})}_{(\mathrm{SO,nF})\text{-flow }(-)} \\ \underbrace{(a_{j_1 j_1 s_2}^{+}, \dot{x}_{j_2})}_{(\mathrm{US,nF})\text{-flow }(-)} \end{array}\right\}.$$

The (3^{rd}SO,nF)-positive source flow at $x_{j_1}^* = a_{j_1 j_1 1}$ is a (3^{rd}SO,nF)-switching bifurcation for negative upper-saddle and lower-saddle flows with a negative source flow.

- For $a_{j_1 j_1 0} < 0$ an $A_{j_2 j_1 3} < 0$,

$$\left\{\begin{array}{c} \underbrace{(a_{j_1 j_1 s_2}^{-}, \dot{x}_{j_2})}_{(\mathrm{US,nF})\text{-flow }(-)} \\ \underbrace{(a_{j_1 j_1 s_1}^{-}, \dot{x}_{j_2})}_{(\mathrm{SI,nF})\text{-flow }(-)} \end{array}\right\} \rightleftharpoons \underbrace{(a_{j_1 j_1 1}, \dot{x}_{j_2})}_{\substack{(3^{\mathrm{rd}}\mathrm{SI,nF})\text{-flow} \\ \text{switching bifurcation}}} \rightleftharpoons \left\{\begin{array}{c} \underbrace{(a_{j_1 j_1 s_1}^{+}, \dot{x}_{j_2})}_{(\mathrm{SI,nF})\text{-flow }(-)} \\ \underbrace{(a_{j_1 j_1 s_2}^{+}, \dot{x}_{j_2})}_{(\mathrm{LS,nF})\text{-flow }(-)} \end{array}\right\}.$$

The (3^{rd}SI,pF)-positive source flow at $x_{j_1}^* = a_{j_1 j_1 1}$ is a (3^{rd}SI,pF)-switching bifurcation for negative upper-saddle and lower-saddle flows with a negative sink flow.

(vi) Under

$$\begin{aligned} &A_{j_2 j_1 3} \neq 0, \text{and } \Delta_{j_1 j_1} = (a_{j_1 j_1 s_2} - a_{j_1 j_1 s_3})^2 = 0, \\ &a_{j_1 j_1 s_2} = a_{j_1 j_1 s_3} \equiv a_{j_1 j_1 s_1} = a_{j_1 j_1 1}, \end{aligned}$$

a dynamical system is

$$\begin{aligned}\dot{x}_{j_2} &= A_{j_2 j_1 3},\\ \dot{x}_{j_1} &= a_{j_1 j_1 0}(x_{j_1} - a_{j_1 j_1 1})^3\end{aligned}$$

with

$$a^{-}_{j_1 j_1 1} \to a^{+}_{j_1 j_1 3},\; a^{-}_{j_1 j_1 3} \to a^{+}_{j_1 j_1 1},\text{ and } a^{-}_{j_1 j_1 2} \to a^{+}_{j_1 j_1 2}.$$

From case (iv), the corresponding equilibriums of $x^{*}_{j_1} = a_{j_1 j_1 1} = a_{j_1 j_1 2}$ for the x_{j_1}-direction possess the following properties as in Eqs. (1.71)–(1.74).

- For $a_{j_1 j_1 0} > 0$ and $A_{j_2 j_1 3} > 0$,

$$\left\{\begin{matrix}\underbrace{(a^{-}_{j_1 j_1 3}, \dot{x}_{j_2})}_{(\text{SO, pF})\text{-flow }(+)}\\ \underbrace{(a^{-}_{j_1 j_1 2}, \dot{x}_{j_2})}_{(\text{SI, pF})\text{-flow }(+)}\\ \underbrace{(a^{-}_{j_1 j_1 1}, \dot{x}_{j_2})}_{(\text{SO, pF})\text{-flow }(+)}\end{matrix}\right\} \rightleftharpoons \underbrace{(a_{j_1 j_1 1}, \dot{x}_{j_2})}_{\substack{(3^{\text{rd}}\text{SO, pF})\text{-flow}\\ \text{switching bifurcation}}} \rightleftharpoons \left\{\begin{matrix}\underbrace{(a^{+}_{j_1 j_1 3}, \dot{x}_{j_2})}_{(\text{SO, pF})\text{-flow }(+)}\\ \underbrace{(a^{+}_{j_1 j_1 2}, \dot{x}_{j_2})}_{(\text{SI, pF})\text{-flow }(+)}\\ \underbrace{(a^{+}_{j_1 j_1 1}, \dot{x}_{j_2})}_{(\text{SO, pF})\text{-flow }(+)}\end{matrix}\right\}.$$

  The (3^{rd}SO,pF)-positive source flow at $x^{*}_{j_1} = a_{j_1 j_1 1}$ is called a (3^{rd}SO,pF)-switching bifurcation for positive source, sink, and source flows.
- For $a_{j_1 j_1 0} < 0$ and $A_{j_2 j_1 3} > 0$,

$$\left\{\begin{matrix}\underbrace{(a^{-}_{j_1 j_1 3}, \dot{x}_{j_2})}_{(\text{SI, pF})\text{-flow }(+)}\\ \underbrace{(a^{-}_{j_1 j_1 2}, \dot{x}_{j_2})}_{(\text{SO, pF})\text{-flow }(+)}\\ \underbrace{(a^{-}_{j_1 j_1 1}, \dot{x}_{j_2})}_{(\text{SI, pF})\text{-flow }(+)}\end{matrix}\right\} \rightleftharpoons \underbrace{(a_{j_1 j_1 1}, \dot{x}_{j_2})}_{\substack{(3^{\text{rd}}\text{SI, pF})\text{-flow}\\ \text{switching bifurcation}}} \rightleftharpoons \left\{\begin{matrix}\underbrace{(a^{+}_{j_1 j_1 3}, \dot{x}_{j_2})}_{(\text{SI, pF})\text{-flow }(+)}\\ \underbrace{(a^{+}_{j_1 j_1 2}, \dot{x}_{j_2})}_{(\text{SO, pF})\text{-flow }(+)}\\ \underbrace{(a^{+}_{j_1 j_1 1}, \dot{x}_{j_2})}_{(\text{SI, pF})\text{-flow }(+)}\end{matrix}\right\}.$$

  The (3^{rd}SI,pF)-positive source flow at $x^{*}_{j_1} = a_{j_1 j_1 1}$ is called a (3^{rd}SI,pF)-switching bifurcation for positive sink, source and sink flows.

- For $a_{j_1j_10} > 0$ and $A_{j_2j_13} < 0$,

$$\left\{\begin{array}{c}\underbrace{(a^-_{j_1j_13}, \dot{x}_{j_2})}_{(\text{SO, nF})\text{-flow }(-)}\\ \underbrace{(a^-_{j_1j_12}, \dot{x}_{j_2})}_{(\text{SI, nF})\text{-flow }(-)}\\ \underbrace{(a^-_{j_1j_11}, \dot{x}_{j_2})}_{(\text{SO, nF})\text{-flow }(-)}\end{array}\right\} \rightleftharpoons \underbrace{(a_{j_1j_11}, \dot{x}_{j_2})}_{\substack{(3^{\text{rd}}\text{SO, nF})\text{-flow}\\ \text{switching bifurcation}}} \rightleftharpoons \left\{\begin{array}{c}\underbrace{(a^+_{j_1j_13}, \dot{x}_{j_2})}_{(\text{SO, nF})\text{-flow }(-)}\\ \underbrace{(a^+_{j_1j_12}, \dot{x}_{j_2})}_{(\text{SI, nF})\text{-flow }(-)}\\ \underbrace{(a^+_{j_1j_11}, \dot{x}_{j_2})}_{(\text{SO, nF})\text{-flow }(-)}\end{array}\right\}.$$

  The (3^{rd}SO,nF)-positive source flow at $x^*_{j_1} = a_{j_1j_11}$ is called a (3^{rd}SO,nF)-switching bifurcation for negative source, sink, and source flows.
- For $a_{j_1j_10} < 0$ an $A_{j_2j_13} < 0$,

$$\left\{\begin{array}{c}\underbrace{(a^-_{j_1j_13}, \dot{x}_{j_2})}_{(\text{SI, nF})\text{-flow }(-)}\\ \underbrace{(a^-_{j_1j_12}, \dot{x}_{j_2})}_{(\text{SO, nF})\text{-flow }(-)}\\ \underbrace{(a^-_{j_1j_11}, \dot{x}_{j_2})}_{(\text{SI, nF})\text{-flow }(-)}\end{array}\right\} \rightleftharpoons \underbrace{(a_{j_1j_11}, \dot{x}_{j_2})}_{\substack{(3^{\text{rd}}\text{SI, nF})\text{-flow}\\ \text{switching bifurcation}}} \rightleftharpoons \left\{\begin{array}{c}\underbrace{(a^+_{j_1j_13}, \dot{x}_{j_2})}_{(\text{SI, nF})\text{-flow }(-)}\\ \underbrace{(a^+_{j_1j_12}, \dot{x}_{j_2})}_{(\text{SO, nF})\text{-flow }(-)}\\ \underbrace{(a^+_{j_1j_11}, \dot{x}_{j_2})}_{(\text{SI, nF})\text{-flow }(-)}\end{array}\right\}.$$

  The (3^{rd}SI,pF)-positive source flow at $x^*_{j_1} = a_{j_1j_11}$ is called a (3^{rd}SI,pF)-switching bifurcation for negative sink, source, and sink flows.

(vii) Under

$$A_{j_2j_13} \neq 0 \text{ and } \Delta_{j_1j_1} = 0, a_{j_1j_1s_1} = a_{j_1j_1s_2}, s_1 \neq s_2 \in \{1, 2, 3\}$$

the dynamical system is

$$\begin{aligned}\dot{x}_{j_2} &= A_{j_2j_13},\\ \dot{x}_{j_1} &= a_{j_1j_10}(x_{j_1} - a_{j_1j_11})^3.\end{aligned}$$

Therefore, the third-order sink and source flows at $x^*_{j_1} = a_{j_1j_11}$ in the x_{j_1}-direction are the appearing bifurcations of the third-order sink and source flows as in Eqs. (1.75)–(1.78).

- For $a_{j_1 j_1 0} > 0$ and $A_{j_2 j_1 3} > 0$,

$$\underbrace{(\text{SO}, \text{pF})}_{\text{source flow }(+)} \rightleftharpoons \underbrace{(3^{\text{rd}}\text{SO}, \text{pF})}_{\text{appearing bifurcation}} \rightleftharpoons \left\{ \begin{array}{c} \underbrace{(\text{SO}, \text{pF})}_{\text{source flow }(+)} \\ \underbrace{(\text{SI}, \text{pF})}_{\text{sink flow }(+)} \\ \underbrace{(\text{SO}, \text{pF})}_{\text{source flow }(+)} \end{array} \right\}.$$

The (3^{rd}SO,pF)-positive source flow at $x^*_{j_1} = a_{j_1 j_1 1}$ is a (3^{rd}SO,pF)-appearing bifurcation from a positive source flow to positive source, sink, and source flows.

- For $a_{j_1 j_1 0} < 0$ and $A_{j_2 j_1 3} > 0$,

$$\underbrace{(\text{SI}, \text{pF})}_{\text{source flow }(+)} \rightleftharpoons \underbrace{(3^{\text{rd}}\text{SI}, \text{pF})}_{\text{appearing bifurcation}} \rightleftharpoons \left\{ \begin{array}{c} \underbrace{(\text{SI}, \text{pF})}_{\text{sink flow }(+)} \\ \underbrace{(\text{SO}, \text{pF})}_{\text{source flow }(+)} \\ \underbrace{(\text{SI}, \text{pF})}_{\text{sink flow }(+)} \end{array} \right\}.$$

The (3^{rd}SI,pF)-positive source flow at $x^*_{j_1} = a_{j_1 j_1 1}$ is a (3^{rd}SI,pF)-appearing bifurcation from a positive sink flow to positive sink, source, sink flows.

- For $a_{j_1 j_1 0} > 0$ and $A_{j_2 j_1 3} < 0$,

$$\underbrace{(\text{SO}, \text{nF})}_{\text{source flow }(-)} \rightleftharpoons \underbrace{(3^{\text{rd}}\text{SO}, \text{nF})}_{\text{appearing bifurcation}} \rightleftharpoons \left\{ \begin{array}{c} \underbrace{(\text{SO}, \text{nF})}_{\text{source flow }(-)} \\ \underbrace{(\text{SI}, \text{nF})}_{\text{sink flow }(-)} \\ \underbrace{(\text{SO}, \text{nF})}_{\text{source flow }(-)} \end{array} \right\}.$$

The (3^{rd}SO,pF)-positive source flow at $x^*_{j_1} = a_{j_1 j_1 1}$ is a (3^{rd}SO,pF)-appearing bifurcation from a negative source flow to negative source, sink, source flows.

- For $a_{j_1j_10} < 0$ and $A_{j_2j_13} < 0$,

$$\underbrace{(\mathrm{SI,nF})}_{\text{sink flow }(-)} \rightleftharpoons \underbrace{(3^{\mathrm{rd}}\mathrm{SI,nF})}_{\text{appearing bifurcation}} \rightleftharpoons \left\{\begin{array}{c} \underbrace{(\mathrm{SI,nF})}_{\text{sink flow }(-)} \\ \underbrace{(\mathrm{SO,nF})}_{\text{source flow }(-)} \\ \underbrace{(\mathrm{SI,nF})}_{\text{sink flow }(-)} \end{array}\right\}.$$

The (3^{rd}SI,pF)-positive source flow at $x_{j_1}^* = a_{j_1j_11}$ is a (3^{rd}SI,pF)-appearing bifurcation from a negative sink flow to negative sink, source, sink flows.

Therefore, from the above local analysis and the first integral manifolds, Theorem 1.1 is proved. ■

1.3 Source and Sink Flows and Bifurcations

In this section, the one-diemsnional singular flows and the corresponding bifurcations for sink and source flows will be presented for a better understanding of singular flows and flow bifurcations.

Consider a self-univariate quadratic dynamical system with a constant vector filed as

$$\begin{aligned} \dot{x}_1 &= a_{110}(x_1 - a_{111})^3, \\ \dot{x}_2 &= A_{213} \neq 0 \end{aligned} \tag{1.81}$$

and the corresponding integral manifold is

$$a_{110}(x_2 - x_{20}) = -\frac{A_{213}}{2}\left[\frac{1}{(x_1 - a_{111})^2} - \frac{1}{(x_{10} - a_{111})^2}\right]. \tag{1.82}$$

For $x_1^* = a_{111}$, $x_1 = 0$ with the third-order singularity and $\dot{x}_2 = A_{212} \neq 0$. Thus, for $x_1^* = a_{111}$, the flow direction is determined by $\dot{x}_2 \neq 0$. Thus,

$$\begin{aligned} (x_1^*, \dot{x}_2) &= \underbrace{(3^{\mathrm{rd}}\mathrm{SO,pF})}_{\text{third-order source flow }(+)}, \text{ for } a_{110} > 0, A_{213} > 0; \\ (x_1^*, \dot{x}_2) &= \underbrace{(3^{\mathrm{rd}}\mathrm{SI,pF})}_{\text{third-order sink flow }(+)}, \text{ for } a_{110} < 0, A_{213} > 0; \\ (x_1^*, \dot{x}_2) &= \underbrace{(3^{\mathrm{rd}}\mathrm{SO,nF})}_{\text{third-order source flow }(-)}, \text{ for } a_{110} > 0, A_{213} < 0; \\ (x_1^*, \dot{x}_2) &= \underbrace{(3^{\mathrm{rd}}\mathrm{SI,nF})}_{\text{third-order sink flow }(-)}, \text{ for } a_{110} < 0, A_{213} < 0. \end{aligned} \tag{1.83}$$

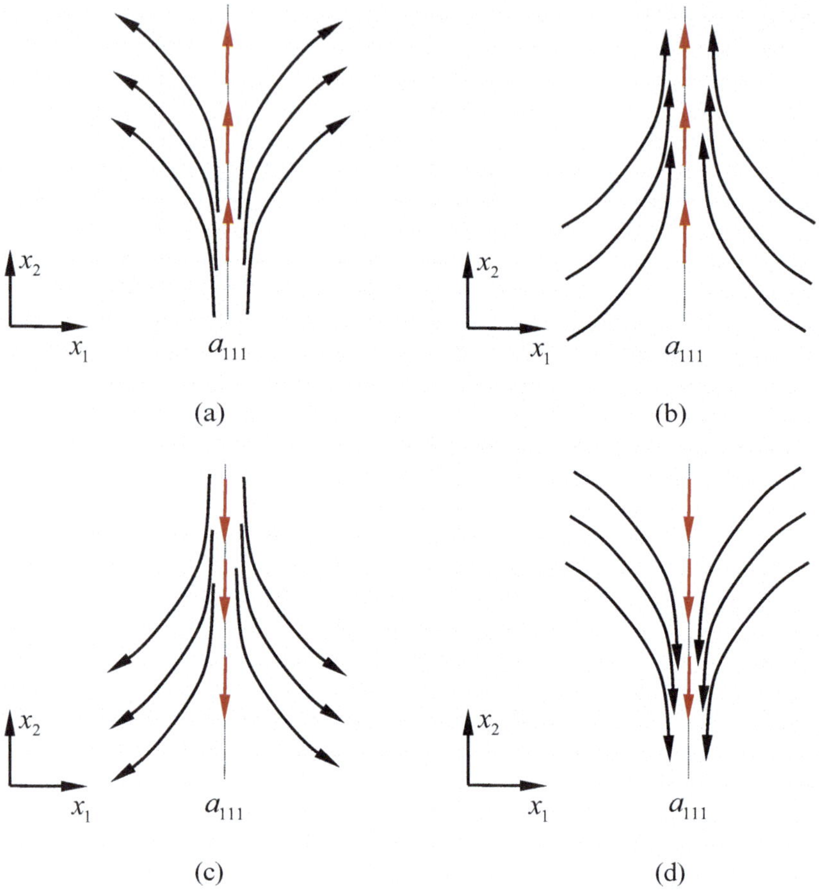

Fig. 1.1 Phase portraits for two-dimensional systems with the triple-repeated equilibrium of $x_1^* = a_{111}$ in the x_1-direction and constant speed in the x_2-direction: (**a**) (3rdSO,pF)-positive source flow ($a_{110} > 0$, $A_{213} > 0$), (**b**) (3rdSI,pF)-positive sink flow ($a_{110} < 0$, $A_{213} > 0$), (**c**) (3rdSO,nF)-negative source flow ($a_{110} > 0, A_{213} < 0$), (**d**) (3rdSI,nF)-negative sink flow ($a_{110} < 0, A_{213} < 0$). *SO* source, *SI* sink. *pF* positive flow, *nF* negative flow

From the first integral manifold, the phase portraits for two-dimensional systems with a triple-repeated equilibrium of $x_1^* = a_{111}$, in the x_1-direction and constant speed in the x_2-direction are presented in Fig. 1.1. In Fig. 1.1a, the phase portrait for $a_{110} > 0$ and $A_{213} > 0$ is presented from the first integral manifold. At $x_1^* = a_{111}$, the flow is toward the x_2-positive direction because of $\dot{x}_2 = A_{213} > 0$. In the x_1 direction, the equilibrium of $x_1^* = a_{111}$ is a third-order source. At the point of $x_1^* = a_{111}$, the flow is a (3rdSO,pF)-positive third-order source flow. Similarly, the phase portraits for the (3rdSI,pF)-positive third-order sink flow ($a_{110} < 0$, $A_{213} > 0$), the (3rdSO,nF)-negative third-order source flow ($a_{110} > 0$, $A_{213} < 0$), and the (3rdSI,nF)-negative third-order sink flow ($a_{110} < 0, A_{213} < 0$) are presented in Fig. 1.1b–d, respectively.

The acronyms "SO" and "SI" are for source and sink, respectively. "pF" and "nF" are for positive and negative flows, respectively. Such third-order source and sink flow states are the appearing bifurcations for the sink and source flows in the two-dimensional dynamical system. In additions, such third-order source and sink flow states can be the switching bifurcations for the sink and source flows in the two-dimensional dynamical system.

Consider a dynamical system with a constant vector field and a cubic univariate vector field in its own direction as

$$\begin{aligned} \dot{x}_1 &= a_{110}(x_1 - a_{111})(x_1 - a_{112})(x_1 - a_{113}), \\ \dot{x}_2 &= A_{213} \neq 0 \end{aligned} \tag{1.84}$$

and the first integral manifold is

$$\begin{aligned} a_{110}(x_2 - x_{20}) = A_{213} \Bigg[& \frac{1}{(a_{111} - a_{112})(a_{111} - a_{113})} \ln \frac{|x_1 - a_{111}|}{|x_{10} - a_{111}|} \\ & + \frac{1}{(a_{112} - a_{111})(a_{112} - a_{113})} \ln \frac{|x_1 - a_{112}|}{|x_{10} - a_{112}|} \\ & + \frac{1}{(a_{113} - a_{111})(a_{113} - a_{112})} \ln \frac{|x_1 - a_{113}|}{|x_{10} - a_{113}|} \Bigg]. \end{aligned} \tag{1.85}$$

Its phase portraits for two-dimensional systems with $x_1^* = a_{111}, a_{112}, a_{113}$ in the x_1-direction and constant speed in the x_2-direction are presented in Fig. 1.2. For $x_1^* = a_{111}, a_{112}, a_{113}$, we have $\dot{x}_1 \neq 0$ and $\dot{x}_2 = A_{213} \neq 0$. Thus, the flow is determined by $\dot{x}_2 \neq 0$. The corresponding sink and source flows are summarized as follows.

$$\begin{Bmatrix} (a_{111}, \dot{x}_2) \\ (a_{112}, \dot{x}_2) \\ (a_{113}, \dot{x}_2) \end{Bmatrix} = \begin{Bmatrix} \underbrace{(\text{SO, pF})}_{\text{source flow }(+)} \\ \underbrace{(\text{SI, pF})}_{\text{sink flow }(+)} \\ \underbrace{(\text{SO, pF})}_{\text{source flow }(+)} \end{Bmatrix}, \text{for } a_{110} > 0, A_{213} > 0; \tag{1.86}$$

$$\begin{Bmatrix} (a_{111}, \dot{x}_2) \\ (a_{112}, \dot{x}_2) \\ (a_{113}, \dot{x}_2) \end{Bmatrix} = \begin{Bmatrix} \underbrace{(\text{SI, pF})}_{\text{sink flow }(+)} \\ \underbrace{(\text{SO, pF})}_{\text{source flow }(+)} \\ \underbrace{(\text{SI, pF})}_{\text{sink flow }(+)} \end{Bmatrix}, \text{for } a_{110} < 0, A_{213} > 0; \tag{1.87}$$

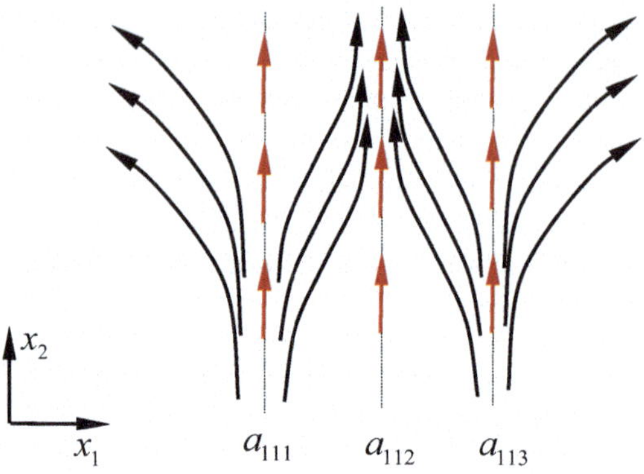

(a)

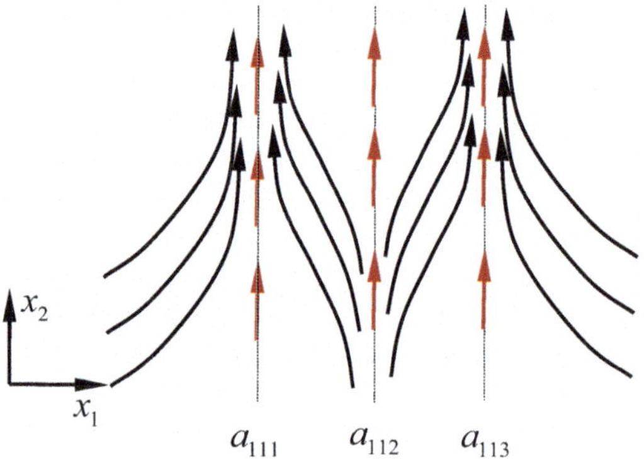

(b)

Fig. 1.2 Phase portraits for two-dimensional systems on the x_1-direction with $x_1^* = a_{111}, a_{112}, a_{113}$: (**a**) one (SI,pF)-positive sink flow and two (SO,pF)-positive source flows ($a_{110} > 0$, $A_{213} > 0$), (**b**) one (SO,pF)-positive source flow and two (SI,pF)-positive sink flows ($a_{110} < 0$, $A_{213} > 0$), (**c**) one (SI,nF)-negative sink flow and two (SO,nF)-positive source flows ($a_{110} > 0$, $A_{213} < 0$), (**d**) one (SO, nF)-negative source flow and two (SI,nF)-negative sink flows ($a_{110} < 0$, $A_{213} < 0$). SI sink, SO source. nF negative flow, pF positive flow

$$\begin{Bmatrix} (a_{111}, \dot{x}_2) \\ (a_{112}, \dot{x}_2) \\ (a_{113}, \dot{x}_2) \end{Bmatrix} = \begin{Bmatrix} \underbrace{(\text{SO, nF})}_{\text{source flow } (-)} \\ \underbrace{(\text{SI, nF})}_{\text{sink flow } (-)} \\ \underbrace{(\text{SO, nF})}_{\text{source flow } (-)} \end{Bmatrix}, \text{for } a_{110} > 0, A_{213} < 0; \tag{1.88}$$

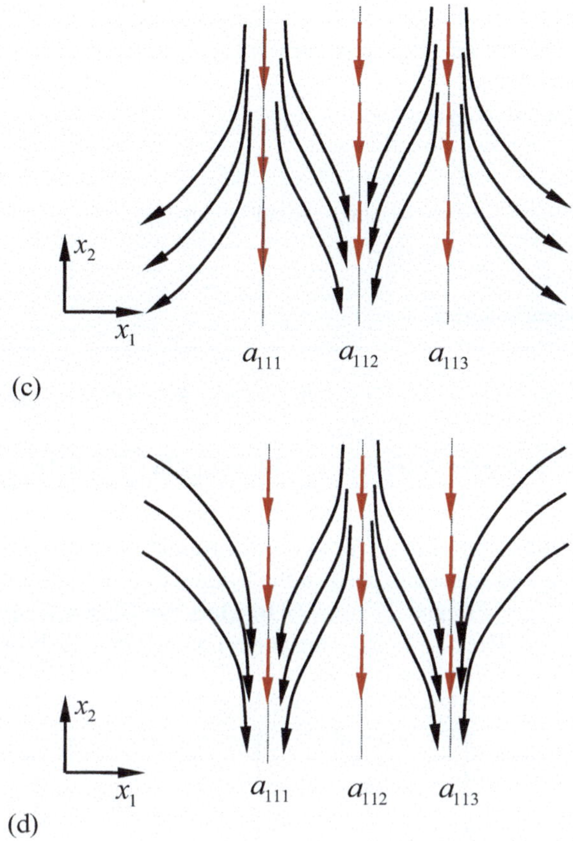

Fig. 1.2 (continued)

$$\begin{Bmatrix} (a_{111}, \dot{x}_2) \\ (a_{112}, \dot{x}_2) \\ (a_{113}, \dot{x}_2) \end{Bmatrix} = \begin{Bmatrix} \underbrace{(\mathrm{SI}, \mathrm{nF})}_{\text{sink flow }(-)} \\ \underbrace{(\mathrm{SO}, \mathrm{nF})}_{\text{source flow }(-)} \\ \underbrace{(\mathrm{SI}, \mathrm{nF})}_{\text{sink flow }(-)} \end{Bmatrix}, \text{for } a_{110} < 0, A_{213} < 0. \tag{1.89}$$

In Fig. 1.2a, the phase portrait of one (SI,pF)-positive sink and two (SO,pF)-positive source flows for $a_{110} > 0$ and $A_{213} > 0$ is presented from the first integral manifold. At $x_1^* = a_{111}, a_{112}, a_{113}$, the flow is toward the x_2-positive direction because of $\dot{x}_2 = A_{213} > 0$. Similarly, the phase portraits for one (SO,pF)-positive source and two (SI,pF)-positive sink flows ($a_{110} < 0$, $A_{213} > 0$), one (SI,nF)-negative sink and two (SO,nF)-negative source flows ($a_{110} > 0$, $A_{213} < 0$), and one (SO,nF)-negative source and two (SI,nF)-negative sink flow ($a_{110} < 0$, $A_{213} < 0$) are presented in Fig. 1.2b–d, respectively.

Consider a dynamical system with a constant vector field and a quadratic univariate vector field in its own direction as

$$\begin{aligned} \dot{x}_1 &= a_{110}(x_1 - a_{111})(x_1 - a_{112})^2, \\ \dot{x}_2 &= A_{213} \neq 0 \end{aligned} \tag{1.90}$$

and the first integral manifold is

$$\begin{aligned} \frac{x_2 - x_{20}}{A_{213}} = & \frac{1}{a_{110}(a_{112} - a_{111})^2} \Big\{ \ln \frac{|x_1 - a_{111}|}{|x_{10} - a_{111}|} - (a_{112} - a_{111}) \\ & \times \Big[\frac{1}{(x_1 - a_{112})} - \frac{1}{(x_{10} - a_{112})}\Big] - \ln \frac{|x_1 - a_{112}|}{|x_{10} - a_{112}|} \Big\}. \end{aligned} \tag{1.91}$$

Its phase portraits ($a_{111} < a_{112}$) for two-dimensional systems with $x_1^* = a_{111}, a_{112}$ in the x_1-direction and constant speed in the x_2-direction are presented in Fig. 1.3. For $x_1^* = a_{111}, a_{112}$, $\dot{x}_1 = 0$ and $\dot{x}_2 = A_{213} \neq 0$. Thus, for $x_1^* = a_{111}, a_{112}$, the flow is determined by $\dot{x}_2 \neq 0$. For $a_{111} < a_{112}$, the corresponding sink and source flows with saddle flows are summarized as follows.

$$\left\{ \begin{array}{c} (a_{111}, \dot{x}_2) \\ (a_{112}, \dot{x}_2) \end{array} \right\} = \left\{ \begin{array}{c} \underbrace{(\text{SO, pF})}_{\text{source flow }(+)} \\ \underbrace{(\text{US, pF})}_{\text{upper-saddle flow }(+)} \end{array} \right\}, \text{ for } a_{110} > 0, A_{213} > 0; \tag{1.92}$$

$$\left\{ \begin{array}{c} (a_{111}, \dot{x}_2) \\ (a_{112}, \dot{x}_2) \end{array} \right\} = \left\{ \begin{array}{c} \underbrace{(\text{SI, pF})}_{\text{sink flow }(+)} \\ \underbrace{(\text{LS, pF})}_{\text{lower-saddle flow }(+)} \end{array} \right\}, \text{ for } a_{110} < 0, A_{213} > 0; \tag{1.93}$$

$$\left\{ \begin{array}{c} (a_{111}, \dot{x}_2) \\ (a_{112}, \dot{x}_2) \end{array} \right\} = \left\{ \begin{array}{c} \underbrace{(\text{SO, nF})}_{\text{source flow }(-)} \\ \underbrace{(\text{US, nF})}_{\text{upper-saddle flow }(-)} \end{array} \right\}, \text{ for } a_{110} > 0, A_{213} < 0; \tag{1.94}$$

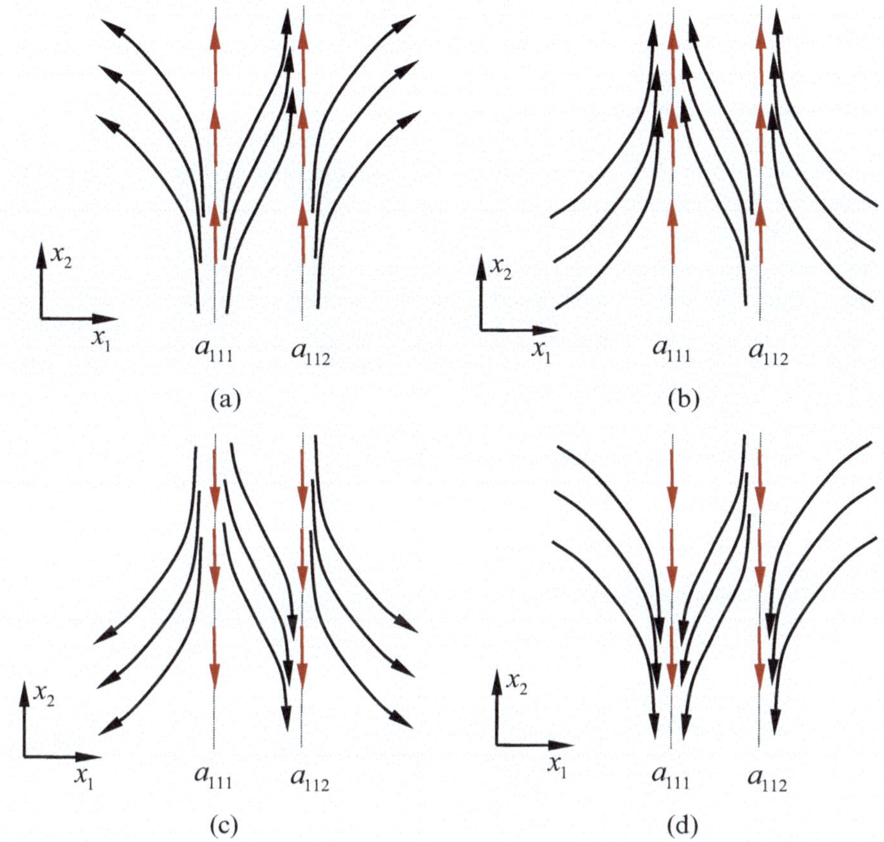

Fig. 1.3 Phase portraits ($a_{111} < a_{112}$) for two-dimensional systems on the x_1-direction with $x_1^* = a_{111}, a_{112}$: (**a**) (SO,pF)-positive source flow and (US,pF)-positive upper-saddle flow ($a_{110} > 0$, $A_{213} > 0$), (**b**) (SI,pF)-positive sink flow and (LS,pF)-positive lower-saddle flow ($a_{110} < 0$, $A_{213} > 0$), (**c**) (SO,nF)-negative source flow and (US,nF)-positive upper-saddle flow ($a_{110} > 0$, $A_{213} < 0$), (**d**) (SI,nF)-negative sink flow and (LS, nF)-negative lower-saddle flow ($a_{110} < 0$, $A_{213} < 0$). SI sink, SO source, nF negative flow, pF positive flow

$$\left\{ \begin{array}{l} (a_{111}, \dot{x}_2) \\ (a_{112}, \dot{x}_2) \end{array} \right\} = \left\{ \begin{array}{c} \underbrace{(\text{SI, nF})}_{\text{sink flow }(-)} \\ \underbrace{(\text{LS, nF})}_{\text{lower-saddle flow }(-)} \end{array} \right\}, \text{for } a_{110} < 0, A_{213} < 0. \tag{1.95}$$

In Fig. 1.3a, the phase portrait of one (SO,pF)-positive source flow and one (US, pF)-positive upper-saddle flows for $a_{110} > 0$ and $A_{213} > 0$ is presented from the first integral manifold. At $x_1^* = a_{111}, a_{112}$, the flow is toward the x_2-positive direction because of $\dot{x}_2 = A_{213} > 0$. Such a (US,pF)-flow appearing bifurcation is for asymmetric appearing bifurcations. Similarly, the phase portraits for (SI,pF)-positive sink flow and (LS,pF)-positive lower-saddle flow ($a_{110} < 0, A_{213} > 0$), (SO,nF)-negative source flow and (US,nF)-negative uppersaddle flow ($a_{110} > 0$, $A_{213} < 0$), and (SI, nF)-negative sink flow and (LS,nF)-negative lower-saddle flow ($a_{110} < 0, A_{213} < 0$) are presented in Fig. 1.3b–d, respectively. The saddle-flow asymmetric appearing bifurcations for sink and source flows can be also for saddle-flow asymmetric switching bifurcations.

Its phase portraits ($a_{111} > a_{112}$) for two-dimensional systems with $x_1^* = a_{111}, a_{112}$ in the x_1-direction and constant speed in the x_2-direction are presented in Fig. 1.4. For $x_1^* = a_{111}, a_{112}$, $\dot{x}_1 = 0$ and $\dot{x}_2 = A_{213} \neq 0$. Thus, for $x_1^* = a_{111}, a_{112}$, the flow is determined by $\dot{x}_2 \neq 0$. For $a_{111} < a_{112}$ the corresponding sink and source flows with saddle flows are summarized as follows.

$$\left\{ \begin{array}{c} (a_{112}, \dot{x}_2) \\ (a_{111}, \dot{x}_2) \end{array} \right\} = \left\{ \begin{array}{c} \underbrace{(\text{LS, pF})}_{\text{lower-saddle flow } (+)} \\ \underbrace{(\text{SO, pF})}_{\text{source flow } (+)} \end{array} \right\}, \text{for } a_{110} > 0, A_{213} > 0; \tag{1.96}$$

$$\left\{ \begin{array}{c} (a_{112}, \dot{x}_2) \\ (a_{111}, \dot{x}_2) \end{array} \right\} = \left\{ \begin{array}{c} \underbrace{(\text{US, pF})}_{\text{upper-saddle flow } (+)} \\ \underbrace{(\text{SI, pF})}_{\text{sink flow } (+)} \end{array} \right\}, \text{for } a_{110} < 0, A_{213} > 0; \tag{1.97}$$

$$\left\{ \begin{array}{c} (a_{112}, \dot{x}_2) \\ (a_{111}, \dot{x}_2) \end{array} \right\} = \left\{ \begin{array}{c} \underbrace{(\text{LS, nF})}_{\text{lower-saddle flow } (-)} \\ \underbrace{(\text{SO, nF})}_{\text{source flow } (-)} \end{array} \right\}, \text{for } a_{110} > 0, A_{213} < 0; \tag{1.98}$$

$$\left\{ \begin{array}{c} (a_{112}, \dot{x}_2) \\ (a_{111}, \dot{x}_2) \end{array} \right\} = \left\{ \begin{array}{c} \underbrace{(\text{US, nF})}_{\text{upper-saddle flow } (-)} \\ \underbrace{(\text{SI, nF})}_{\text{source flow } (-)} \end{array} \right\}, \text{for } a_{110} < 0, A_{213} < 0. \tag{1.99}$$

In Fig. 1.4a, the phase portrait of one (SO,pF)-positive source flow and one (LS, pF)-positive lower-saddle flows for $a_{110} > 0$ and $A_{213} > 0$ is presented from the first integral manifold. At $x_1^* = a_{111}, a_{112}$, the flow is toward the x_2-positive direction because of $\dot{x}_2 = A_{212} > 0$. Such an (LS,pF)-flow appearing bifurcation is for asymmetric appearing bifurcations. Similarly, the phase portraits for (SI,pF)-positive sink

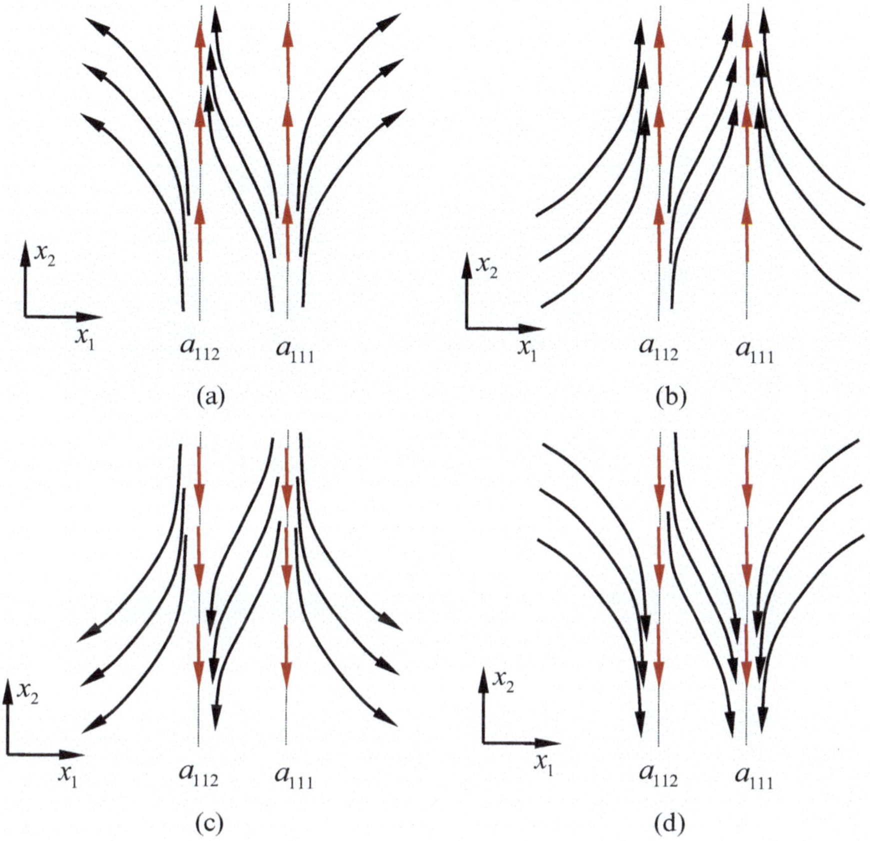

Fig. 1.4 Phase portraits ($a_{111} > a_{112}$) for two-dimensional systems on the x_1-direction with $x_1^* = a_{111}, a_{112}$: (**a**) (SO,pF)-positive source flow and (LS,pF)-positive lower-saddle flow ($a_{110} > 0$, $A_{213} > 0$), (**b**) (SI,pF)-positive sink flow and (US,pF)-positive upper-saddle flow ($a_{110} < 0$, $A_{213} > 0$), (**c**) (SO,nF)-negative source flow and (LS,nF)-positive lower-saddle flow ($a_{110} > 0$, $A_{213} < 0$), (**d**) (SI,nF)-negative sink flow and (US,nF)-negative upper-saddle flow ($a_{110} < 0$, $A_{213} < 0$). SI sink, SO source, US upper-saddle, LS lower-saddle, nF negative flow, pF positive flow

flow and (US,pF)-positive upper-saddle flow ($a_{110} < 0, A_{213} > 0$), (SO,nF)-negative source flow and (LS,F)-negative lower-saddle flow ($a_{110} > 0, A_{213} < 0$), and (SI,nF)-negative sink flow and (US,nF)-negative upper-saddle flow ($a_{110} < 0, A_{213} < 0$) are presented in Fig. 1.4b–d, respectively. The saddle-flow asymmetric appearing bifurcations for sink and source flows can be also for saddle-flow asymmetric switching bifurcations.

Chapter 2
Crossing-Linear and Self-Cubic Vector Fields

In this chapter, consider two vector fields in dynamical systems to be linear variable-crossing and self-univariate cubic vector fields. The corresponding stability and bifurcations of equilibriums are presented. In addition to the first- and third-order sink and source flows with upper-saddle and lower-saddle flows, the parabola flows based on the linear crossing-variable vector fields are determined. The inflection-sink (source) infinite-equilibriums are for the switching bifurcations of parabola flows and sink (source) flows. The down-up saddles are the appearing bifurcation from an up-parabola flow to source, down-parabola, and sink flows. The up-down saddles are the appearing bifurcation from a down-parabola flow to source, up-parabola, and sink flows. The third-order increasing and decreasing asymptotic *source* infinite-equilibriums are the switching bifurcations of the up-parabola and down-parabola with the third-order *source flow*. The third-order increasing and decreasing asymptotic *sink* infinite-equilibriums are the switching bifurcations of the up-parabola and down-parabola with the third-order *sink flow*.

2.1 Crossing-Linear and Self-Cubic Systems

In this section, the dynamics of crossing-linear and self-univariate cubic systems will be presented. The corresponding first integral manifolds will be determined for regular and singular flows and the corresponding switching and appearing bifurcations for the regular flows (i.e., sink, source, and parabola flows) will be discussed.

Theorem 2.1 *Consider a self-univariate cubic system with a linear variable-crossing vector field as*

A. C. J. Luo, *Two-dimensional Single-Variable Cubic Nonlinear Systems, Vol. I*,
https://doi.org/10.1007/978-3-031-48472-8_2

$$\begin{aligned}\dot{x}_{j_2} &= A_{j_2j_1 2} + A_{j_2j_1 1}x_{j_1},\\ \dot{x}_{j_1} &= A_{j_1j_1 3} + A_{j_1j_1 2}x_{j_1} + A_{j_1j_1 1}x_{j_1}^2 + A_{j_1j_1 0}x_{j_1}^3\\ &= a_{j_1j_1 0}(x_{j_1} - b_{j_1j_1 1})(x_{j_1}^2 + B_{j_1j_1 1}x_{j_1} + C_{j_1j_1 1})\end{aligned} \tag{2.1}$$

with

$$\Delta_{j_1j_1} = (B_{j_1j_1 1})^2 - 4C_{j_1j_1 1} \quad \text{for } j_1, j_2 \in \{1,2\}, j_1 \neq j_2. \tag{2.2}$$

(i) If

$$A_{j_2j_1 1} \neq 0, A_{j_2j_1 2} \neq 0 \text{ and } \Delta_{j_1j_1} < 0, \tag{2.3}$$

the corresponding dynamical system is

$$\begin{aligned}\dot{x}_{j_2} &= a_{j_2j_1 0}(x_{j_1} - a_{j_2j_1 1}),\\ \dot{x}_{j_1} &= a_{j_1j_1 0}(x_{j_1} - a_{j_1j_1 1})\left[(x_{j_1} - a_{j_1j_1 2})^2 + b_{j_1j_1}\right]\end{aligned} \tag{2.4}$$

where

$$\begin{aligned}&a_{j_2j_1 1} = -\frac{1}{A_{j_2j_1 1}}A_{j_2j_1 2}, a_{j_2j_1 0} = A_{j_2j_1 1};\\ &a_{j_1j_1 1} = b_{j_1j_1 1}, a_{j_1j_1 2} = \frac{1}{2}A_{j_1j_1 1}, a_{j_1j_1 0} = A_{j_1j_1 0}, b_{j_1j_1} = \frac{1}{4}(-\Delta_{j_1j_1}).\end{aligned} \tag{2.5}$$

(i_1) For $a_{j_1j_1 1} \neq a_{j_2j_1 1}$, the first integral manifold is

$$\begin{aligned}x_{j_2} - x_{j_2 0} = &\frac{a_{j_2j_1 0}}{a_{j_1j_1 0}}\left\{\frac{1}{\sqrt{b_{j_1j_1}}}(\arctan\frac{x_{j_1} - a_{j_1j_1 2}}{\sqrt{b_{j_1j_1}}} - \arctan\frac{x_{j_1 0} - a_{j_1j_1 2}}{\sqrt{b_{j_1j_1}}})\right.\\ &+ \frac{(a_{j_1j_1 1} - a_{j_2j_1 1})}{(a_{j_1j_1 1} - a_{j_1j_1 2})^2 + b_{j_1j_1}}\left[\ln\frac{|x_{j_1} - a_{j_1j_1 1}|}{|x_{j_1 0} - a_{j_1j_1 1}|}\right.\\ &- \frac{1}{2}\ln\frac{|(x_{j_1} - a_{j_1j_1 2})^2 + b_{j_1j_1}|}{|(x_{j_1 0} - a_{j_1j_1 2})^2 + b_{j_1j_1}|}\\ &\left.\left.+ \frac{(a_{j_1j_1 2} - a_{j_1j_1 1})}{\sqrt{b_{j_1j_1}}}(\arctan\frac{x_{j_1} - a_{j_1j_1 2}}{\sqrt{b_{j_1j_1}}} - \arctan\frac{x_{j_1 0} - a_{j_1j_1 2}}{\sqrt{b_{j_1j_1}}})\right]\right\}.\end{aligned} \tag{2.6}$$

(i_{1a}) *The flow at* $x_{j_1}^* = a_{j_1 j_1 1}$ *for the* x_{j_1}*-direction has the following properties.*

- *For* $a_{j_1 j_1 0} > 0$ *and* $a_{j_2 j_1 0}(a_{j_1 j_1 1} - a_{j_2 j_1 1}) > 0$,

$$(a_{j_1 j_1 1}, \dot{x}_{j_2}) = \underbrace{(\mathrm{SO}, \mathrm{pF})}_{\text{source flow } (+)} . \tag{2.7}$$

The flow at $x_{j_1}^* = a_{j_1 j_1 1}$ *is an* (SO,pF)*-positive source flow.*

- *For* $a_{j_1 j_1 0} < 0$ *and* $a_{j_2 j_1 0}(a_{j_1 j_1 1} - a_{j_2 j_1 1}) > 0$,

$$(a_{j_1 j_1 1}, \dot{x}_{j_2}) = \underbrace{(\mathrm{SI}, \mathrm{pF})}_{\text{sink flow } (+)} . \tag{2.8}$$

The flow at $x_{j_1}^* = a_{j_1 j_1 1}$ *is an* (SI,pF)*-positive sink flow.*

- *For* $a_{j_1 j_1 0} > 0$ *and* $a_{j_2 j_1 0}(a_{j_1 j_1 1} - a_{j_2 j_1 1}) < 0$,

$$(a_{j_1 j_1 1}, \dot{x}_{j_2}) = \underbrace{(\mathrm{SO}, \mathrm{nF})}_{\text{source flow } (-)} . \tag{2.9}$$

The flow at $x_{j_1}^* = a_{j_1 j_1 1}$ *is an* (SO,nF)*-negative source flow.*

- *For* $a_{j_1 j_1 0} < 0$ *and* $a_{j_2 j_1 0}(a_{j_1 j_1 1} - a_{j_2 j_1 1}) < 0$,

$$(a_{j_1 j_1 1}, \dot{x}_{j_2}) = \underbrace{(\mathrm{SI}, \mathrm{nF})}_{\text{sink flow } (-)} . \tag{2.10}$$

The flow at $x_{j_1}^* = a_{j_1 j_1 1}$ *is an* (SI,nF)*-negative sink flow.*

(i_{1b}) *The flow at* $x_{j_1}^* = a_{j_2 j_1 1}$ *for the* x_{j_2}*-direction has the following properties.*

- *For* $a_{j_1 j_1 0}(a_{j_2 j_1 1} - a_{j_1 j_1 1}) > 0$ *and* $a_{j_2 j_1 0} > 0$,

$$(\dot{x}_{j_1}, a_{j_2 j_1 1}) = \underbrace{(\mathrm{pF}, \mathrm{UP})}_{\text{up-parabola flow } (+)} . \tag{2.11}$$

The flow at $x_{j_1}^* = a_{j_2 j_1 1}$ *is a* (pF,UP)*-positive up-parabola flow.*

- *For* $a_{j_1 j_1 0}(a_{j_2 j_1 1} - a_{j_1 j_1 1}) < 0$ *and* $a_{j_2 j_1 0} > 0$,

$$(\dot{x}_{j_1}, a_{j_2 j_1 1}) = \underbrace{(\mathrm{nF}, \mathrm{DP})}_{\text{down-parabola flow } (-)} . \tag{2.12}$$

The flow at $x_{j_1}^* = a_{j_2 j_1 1}$ *is an* (nF,DP)*-negative down-parabola flow.*

- *For* $a_{j_1j_10}(a_{j_2j_11}-a_{j_1j_11})>0$ *and* $a_{j_2j_10}<0$,

$$(\dot{x}_{j_1},a_{j_2j_11})=\underbrace{(\mathrm{pF},\mathrm{DP})}_{\text{down-parabola flow }(+)}. \tag{2.13}$$

The flow at $x_{j_1}^*=a_{j_2j_11}$ *is a* (pF,DP)*-positive down-parabola flow.*

- *For* $a_{j_1j_10}(a_{j_2j_11}-a_{j_1j_11})<0$ *and* $a_{j_2j_10}<0$,

$$(\dot{x}_{j_1},a_{j_2j_11})=\underbrace{(\mathrm{nF},\mathrm{UP})}_{\text{up-parabola flow }(-)}. \tag{2.14}$$

The flow at $x_{j_1}^*=a_{j_2j_11}$ *is an* (nF,UP)*-negative up-parabola flow.*

(i_2) For $a_{j_1j_11}=a_{j_2j_11}$, *the first integral manifold is*

$$x_{j_2}-x_{j_20}=\frac{a_{j_2j_10}}{a_{j_1j_10}}\frac{1}{\sqrt{b_{j_1j_1}}}(\arctan\frac{x_{j_1}-a_{j_1j_12}}{\sqrt{b_{j_1j_1}}}-\arctan\frac{x_{j_10}-a_{j_1j_12}}{\sqrt{b_{j_1j_1}}}). \tag{2.15}$$

The equilibrium of $x_{j_1}^*=a_{j_1j_11}=a_{j_2j_11}$ *with* $\overline{x}_{j_2}\in(-\infty,\infty)$ *has the following properties.*

- *For* $a_{j_1j_10}>0$ *and* $a_{j_2j_10}>0$,

$$(a_{j_1j_11},\overline{x}_{j_2})=\underbrace{(\mathrm{SO},\mathrm{II})}_{\text{increasing-inflection source}}. \tag{2.16}$$

The equilibrium of $x_{j_1}^*=a_{j_1j_11}=a_{j_2j_11}$ *is an* (SO,II)*-increasing-inflection source infinite-equilibrium.*

- *For* $a_{j_1j_10}<0$ *and* $a_{j_2j_10}>0$,

$$(a_{j_1j_11},\overline{x}_{j_2})=\underbrace{(\mathrm{SI},\mathrm{DI})}_{\text{decreasing-inflection sink}}. \tag{2.17}$$

The equilibrium of $x_{j_1}^*=a_{j_1j_11}=a_{j_2j_11}$ *is an* (SI,DI)*-decreasing-inflection sink infinite-equilibrium.*

- *For* $a_{j_1j_10}>0$ *and* $a_{j_2j_10}<0$,

$$(a_{j_1j_11},\overline{x}_{j_2})=\underbrace{(\mathrm{SO},\mathrm{DI})}_{\text{decreasing-inflection source}}. \tag{2.18}$$

The equilibrium of $x_{j_1}^*=a_{j_1j_11}=a_{j_2j_11}$ *is an* (SO,DI)*-decreasing-inflection source infinite-equilibrium.*

- *For* $a_{j_1j_10} < 0$ *and* $a_{j_2j_10} < 0$,

$$(a_{j_1j_11}, \bar{x}_{j_2}) = \underbrace{(\text{SI, II})}_{\text{increasing-inflection sink}}. \tag{2.19}$$

The equilibrium of $x_{j_1}^* = a_{j_1j_11} = a_{j_2j_11}$ *is an* (SI,II)*-increasing-inflection sink infinite-equilibrium.*

– *The inflection infinite-equilibriums are the switching bifurcations for parabola flows with source (sink) flows.*

(ii) If

$$A_{j_2j_11} \neq 0, A_{j_2j_12} \neq 0 \text{ and } \Delta_{j_1j_1} = 0, \tag{2.20}$$

the two-dimensional dynamical system becomes

$$\begin{aligned}\dot{x}_{j_2} &= a_{j_2j_10}(x_{j_1} - a_{j_2j_11}),\\ \dot{x}_{j_1} &= a_{j_1j_10}(x_{j_1} - a_{j_1j_1s_1})(x_{j_1} - a_{j_1j_1s_2})^2.\end{aligned} \tag{2.21}$$

(ii$_1$) For $a_{j_2j_11} \neq a_{j_1j_1s_1} \neq a_{j_1j_1s_2}$ $(s_1, s_2 \in \{1, 2\})$*, the first integral manifold is*

$$\begin{aligned}x_{j_2} - x_{j_20} = \frac{a_{j_2j_10}}{a_{j_1j_10}}\Bigg[&-\frac{a_{j_1j_1s_2} - a_{j_2j_11}}{a_{j_1j_1s_2} - a_{j_1j_1s_1}}\left(\frac{1}{x_{j_1} - a_{j_1j_1s_2}} - \frac{1}{x_{j_10} - a_{j_1j_1s_2}}\right)\\ &+\frac{a_{j_1j_1s_1} - a_{j_2j_11}}{(a_{j_1j_1s_1} - a_{j_1j_1s_2})^2}\left(\ln\frac{|x_{j_1} - a_{j_1j_1s_1}|}{|x_{j_10} - a_{j_1j_1s_1}|} - \ln\frac{|x_{j_1} - a_{j_1j_1s_2}|}{|x_{j_10} - a_{j_1j_1s_2}|}\right)\Bigg].\end{aligned} \tag{2.22}$$

(ii$_{1a}$) The equilibrium of $x_{j_1}^* = a_{j_1j_1s_1}$ *possesses the following properties.*

- *For* $a_{j_1j_10}(a_{j_1j_1s_1} - a_{j_1j_1s_2})^2 > 0$ *and* $a_{j_2j_10}(a_{j_1j_1s_1} - a_{j_2j_11}) > 0$,

$$(a_{j_1j_1s_1}, \dot{x}_{j_2}) = \underbrace{(\text{SO, pF})}_{\text{source flow }(+)}. \tag{2.23}$$

The flow at $x_{j_1}^* = a_{j_1j_1s_1}$ *is an* (SO,pF)*-positive source flow.*

- *For* $a_{j_1j_10}(a_{j_1j_1s_1} - a_{j_1j_1s_2})^2 < 0$ *and* $a_{j_2j_10}(a_{j_1j_1s_1} - a_{j_2j_11}) > 0$,

$$(a_{j_1j_1s_1}, \dot{x}_{j_2}) = \underbrace{(\text{SI, pF})}_{\text{sink flow }(+)}. \tag{2.24}$$

The flow at $x_{j_1}^* = a_{j_1j_1s_1}$ *is an* (SI,pF)*-positive sink flow.*

- *For* $a_{j_1j_10}(a_{j_1j_1s_1} - a_{j_1j_1s_2})^2 > 0$ *and* $a_{j_2j_10}(a_{j_1j_1s_1} - a_{j_2j_11}) < 0$,

$$(a_{j_1 j_1 s_1}, \dot{x}_{j_2}) = \underbrace{(\text{SO}, \text{nF})}_{\text{source flow } (-)} . \tag{2.25}$$

The flow at $x^*_{j_1} = a_{j_1 j_1 s_1}$ *is an* (SO,nF)*-negative source flow.*

- *For* $a_{j_1 j_1 0}(a_{j_1 j_1 s_1} - a_{j_1 j_1 s_2})^2 < 0$ *and* $a_{j_2 j_1 0}(a_{j_1 j_1 s_1} - a_{j_2 j_1 1}) < 0$,

$$(a_{j_1 j_1 s_1}, \dot{x}_{j_2}) = \underbrace{(\text{SI}, \text{nF})}_{\text{sink flow } (-)} . \tag{2.26}$$

The flow at $x^*_{j_1} = a_{j_1 j_1 s_1}$ *is an* (SI,nF)*-negative sink flow.*

(ii$_{1b}$) The equilibrium of $x^*_{j_1} = a_{j_1 j_1 s_2}$ *possesses the following properties.*

- *For* $a_{j_1 j_1 0}(a_{j_1 j_1 s_2} - a_{j_1 j_1 s_1}) > 0$ *and* $a_{j_2 j_1 0}(a_{j_1 j_1 s_2} - a_{j_2 j_1 1}) > 0$,

$$(a_{j_1 j_1 s_2}, \dot{x}_{j_2}) = \underbrace{(\text{US}, \text{pF})}_{\text{upper-saddle flow } (+)} . \tag{2.27}$$

The flow at $x^*_{j_1} = a_{j_1 j_1 s_2}$ *is a* (US,pF)*-positive upper-saddle flow.*

- *For* $a_{j_1 j_1 0}(a_{j_1 j_1 s_2} - a_{j_1 j_1 s_1}) < 0$ *and* $a_{j_2 j_1 0}(a_{j_1 j_1 s_2} - a_{j_2 j_1 1}) > 0$,

$$(a_{j_1 j_1 s_2}, \dot{x}_{j_2}) = \underbrace{(\text{LS}, \text{pF})}_{\text{lower-saddle flow } (+)} . \tag{2.28}$$

The flow at $x^*_{j_1} = a_{j_1 j_1 s_2}$ *is an* (LS,pF)*-positive lower-saddle flow.*

- *For* $a_{j_1 j_1 0}(a_{j_1 j_1 s_2} - a_{j_1 j_1 s_1}) > 0$ *and* $a_{j_2 j_1 0}(a_{j_1 j_1 s_2} - a_{j_2 j_1 1}) < 0$,

$$(a_{j_1 j_1 s_2}, \dot{x}_{j_2}) = \underbrace{(\text{US}, \text{nF})}_{\text{upper-saddle flow } (-)} . \tag{2.29}$$

The flow at $x^*_{j_1} = a_{j_1 j_1 s_2}$ *is a* (US,nF)*-negative upper-saddle flow.*

- *For* $a_{j_1 j_1 0}(a_{j_1 j_1 s_2} - a_{j_1 j_1 s_1}) < 0$ *and* $a_{j_2 j_1 0}(a_{j_1 j_1 s_2} - a_{j_2 j_1 1}) < 0$,

$$(a_{j_1 j_1 s_2}, \dot{x}_{j_2}) = \underbrace{(\text{LS}, \text{nF})}_{\text{lower-saddle flow } (-)} . \tag{2.30}$$

The flow at $x^*_{j_1} = a_{j_1 j_1 s_2}$ *is an* (LS,nF)*-negative lower-saddle flow.*

– *The upper (lower)-saddle flows are the appearing and switching bifurcations of the source and sink flows.*

(ii$_{1c}$) The equilibrium of $x^*_{j_1} = a_{j_2 j_1 1}$ *for the* x_{j_2}*-direction possesses the following properties.*

- *For* $a_{j_1j_10}(a_{j_2j_11}-a_{j_1j_1s_1})(a_{j_2j_11}-a_{j_1j_1s_2})^2>0$ *and* $a_{j_2j_10}>0$,

$$(\dot{x}_{j_1},a_{j_2j_11})=\underbrace{(\mathrm{pF},\mathrm{UP})}_{\text{up-parabola flow }(+)}. \tag{2.31}$$

The flow at $x_{j_1}^*=a_{j_2j_11}$ *is a* (pF,UP)*-positive up-parabola flow.*

- *For* $a_{j_1j_10}(a_{j_2j_11}-a_{j_1j_1s_1})(a_{j_2j_11}-a_{j_1j_1s_2})^2<0$ *and* $a_{j_2j_10}>0$,

$$(\dot{x}_{j_1},a_{j_2j_11})=\underbrace{(\mathrm{nF},\mathrm{DP})}_{\text{down-parabola flow }(-)}. \tag{2.32}$$

The flow at $x_{j_1}^*=a_{j_2j_11}$ *is an* (nF,DP)*-negative down-parabola flow.*

- *For* $a_{j_1j_10}(a_{j_2j_11}-a_{j_1j_1s_1})(a_{j_2j_11}-a_{j_1j_1s_2})^2>0$ *and* $a_{j_2j_10}<0$,

$$(\dot{x}_{j_1},a_{j_2j_11})=\underbrace{(\mathrm{pF},\mathrm{DP})}_{\text{down-parabola flow }(+)}. \tag{2.33}$$

The flow at $x_{j_1}^*=a_{j_2j_11}$ *is a* (pF,DP)*-positive down-parabola flow.*

- *For* $a_{j_1j_10}(a_{j_2j_11}-a_{j_1j_1s_1})(a_{j_2j_11}-a_{j_1j_1s_2})^2<0$ *and* $a_{j_2j_10}<0$,

$$(\dot{x}_{j_1},a_{j_2j_11})=\underbrace{(\mathrm{nF},\mathrm{UP})}_{\text{up-parabola flow }(-)}. \tag{2.34}$$

The flow at $x_{j_1}^*=a_{j_2j_11}$ *is an* (nF,UP)*-negative up-parabola flow.*

(ii$_2$*) For* $a_{j_2j_11}=a_{j_1j_1s_1}\neq a_{j_1j_1s_2}(s_1,\,s_2\in\{1,2\})$*, the first integral manifold is*

$$x_{j_2}-x_{j_20}=-\frac{a_{j_2j_10}}{a_{j_1j_10}}\Big[\frac{1}{x_{j_1}-a_{j_1j_1s_2}}-\frac{1}{x_{j_10}-a_{j_1j_1s_2}}\Big]. \tag{2.35}$$

The equilibrium of $x_{j_1}^*=a_{j_1j_1s_1}=a_{j_2j_11}$ *with* $\bar{x}_{j_2}\in(-\infty,\infty)$ *has the following properties.*

- *For* $a_{j_1j_10}(a_{j_1j_1s_1}-a_{j_1j_1s_2})^2>0$ *and* $a_{j_2j_10}>0$,

$$(a_{j_1j_1s_1},\bar{x}_{j_2})=\underbrace{(\mathrm{SO},\mathrm{II})}_{\text{increasing-inflection source}}. \tag{2.36}$$

The equilibrium of $x_{j_1}^*=a_{j_1j_1s_1}=a_{j_2j_11}$ *is an* (SO,II)*-increasing-inflection source infinite-equilibrium.*

- *For* $a_{j_1j_10}(a_{j_1j_1s_1}-a_{j_1j_1s_2})^2<0$ *and* $a_{j_2j_10}>0$,

$$(a_{j_1j_1s_1},\bar{x}_{j_2})=\underbrace{(\mathrm{SI},\mathrm{DI})}_{\text{decreasing-inflection sink}}. \tag{2.37}$$

The equilibrium of $x_{j_1}^*=a_{j_1j_1s_1}=a_{j_2j_11}$ *is an* (SI,DI)-*decreasing-inflection sink infinite-equilibrium.*

- *For* $a_{j_1j_10}(a_{j_1j_1s_1}-a_{j_1j_1s_2})^2>0$ *and* $a_{j_2j_10}<0$,

$$(a_{j_1j_1s_1},\bar{x}_{j_2})=\underbrace{(\mathrm{SO},\mathrm{DI})}_{\text{decreasing-inflection source}}. \tag{2.38}$$

The equilibrium of $x_{j_1}^*=a_{j_1j_1s_1}=a_{j_2j_11}$ *is an* (SO,DI)-*decreasing-inflection source infinite-equilibrium.*

- *For* $a_{j_1j_10}(a_{j_1j_1s_1}-a_{j_1j_1s_2})^2<0$ *and* $a_{j_2j_10}<0$,

$$(a_{j_1j_1s_1},\bar{x}_{j_2})=\underbrace{(\mathrm{SI},\mathrm{II})}_{\text{increasing-inflection sink}}. \tag{2.39}$$

The equilibrium of $x_{j_1}^*=a_{j_1j_1s_1}=a_{j_2j_11}$ *is an* (SI,II)-*increasing-inflection sink infinite-equilibrium.*

– *The inflection source (sink) infinite-equilibriums are the switching bifurcations for parabola flows and source (sink) flows.*

(ii₃) For $a_{j_2j_11}=a_{j_1j_1s_2}\neq a_{j_1j_1s_1}$, *the first integral manifold is*

$$x_{j_2}-x_{j_20}=\frac{a_{j_2j_10}}{a_{j_1j_10}}\frac{1}{a_{j_1j_1s_1}-a_{j_1j_1s_2}}\left(\ln\frac{|x_{j_1}-a_{j_1j_1s_1}|}{|x_{j_10}-a_{j_1j_1s_1}|}-\ln\frac{|x_{j_1}-a_{j_1j_1s_2}|}{|x_{j_10}-a_{j_1j_1s_2}|}\right). \tag{2.40}$$

The infinite-equilibrium of $x_{j_1}^*=a_{j_1j_1s_2}=a_{j_2j_11}$ *with* $\bar{x}_{j_2}\in(-\infty,\infty)$ *has the following properties.*

- *For* $a_{j_1j_10}(a_{j_1j_1s_2}-a_{j_1j_1s_1})>0$ *and* $a_{j_2j_10}>0$,

$$(a_{j_1j_1s_2},\bar{x}_{j_2})=\underbrace{(\mathrm{US},\mathrm{DU})}_{\text{down-up upper-saddle}}. \tag{2.41}$$

The equilibrium of $x_{j_1}^*=a_{j_1j_1s_2}=a_{j_2j_11}$ *is a* (US,DU)-*down-up upper-saddle infinite-equilibrium.*

- *For* $a_{j_1j_10}(a_{j_1j_1s_2}-a_{j_1j_1s_1})<0$ *and* $a_{j_2j_10}>0$,

$$(a_{j_1j_1s_2}, \bar{x}_{j_2}) = \underbrace{(\mathrm{LS}, \mathrm{UD})}_{\text{up-down lower-saddle}}. \tag{2.42}$$

The equilibrium of $x_{j_1}^* = a_{j_1j_1s_2} = a_{j_2j_1 1}$ *is an* (LS,UD)*-up-down lower-saddle infinite-equilibrium.*

- *For* $a_{j_1j_1 0}(a_{j_1j_1s_2} - a_{j_1j_1s_1}) > 0$ *and* $a_{j_2j_1 0} < 0$,

$$(a_{j_1j_1s_2}, \bar{x}_{j_2}) = \underbrace{(\mathrm{US}, \mathrm{UD})}_{\text{up-down upper-saddle}}. \tag{2.43}$$

The equilibrium of $x_{j_1}^* = a_{j_1j_1s_2} = a_{j_2j_1 1}$ *is a* (US,UD)*-up-down upper-saddle infinite-equilibrium.*

- *For* $a_{j_1j_1 0}(a_{j_1j_1s_2} - a_{j_1j_1s_1}) < 0$ *and* $a_{j_2j_1 0} < 0$,

$$(a_{j_1j_1s_2}, \bar{x}_{j_2}) = \underbrace{(\mathrm{LS}, \mathrm{DU})}_{\text{down-up lower-saddle}}. \tag{2.44}$$

The equilibrium of $x_{j_1}^* = a_{j_1j_1s_2} = a_{j_2j_1 1}$ *is an* (LS,DU)*-down-up lower-saddle infinite-equilibrium.*

– *The up-down and down-up saddle infinite-equilibriums are the* switching *bifurcations for parabola flows and upper-saddle (lower-saddle) flows.*
– *The up-down (down-up) saddle infinite-equilibriums are the* appearing *bifurcations from a down (up)-parabola flow to source, up (down)-parabola, and sink flows.*

(ii$_4$) For $a_{j_2j_1 1} \neq a_{j_1j_1s_1} = a_{j_1j_1s_2} \equiv a_{j_1j_1 1}(s_1, s_2 \in \{1,2\})$*, the two-dimensional dynamical system becomes*

$$\begin{aligned} \dot{x}_{j_2} &= a_{j_2j_1 0}(x_{j_1} - a_{j_2j_1 1}), \\ \dot{x}_{j_1} &= a_{j_1j_1 0}(x_{j_1} - a_{j_1j_1 1})^3. \end{aligned} \tag{2.45}$$

The first integral manifold is

$$\begin{aligned} x_{j_2} - x_{j_2 0} = & -\frac{a_{j_2j_1 0}}{a_{j_1j_1 0}}\Big\{\Big(\frac{1}{x_{j_1} - a_{j_1j_1 1}} - \frac{1}{x_{j_1 0} - a_{j_1j_1 1}}\Big) \\ & + \frac{1}{2}(a_{j_1j_1 1} - a_{j_2j_1 1})\Big[\frac{1}{(x_{j_1} - a_{j_1j_1 1})^2} - \frac{1}{(x_{j_1 0} - a_{j_1j_1 1})^2}\Big]\Big\}. \end{aligned} \tag{2.46}$$

(ii$_{4a}$) The equilibrium of $x_{j_1} = a_{j_2j_1 1} \neq a_{j_1j_1 1}$ *in the* x_{j_2}*-direction has the following properties.*

- *For* $a_{j_1j_1 0}(a_{j_2j_1 1} - a_{j_1j_1 1})^3 > 0$ *and* $a_{j_2j_1 0} > 0$,

$$(\dot{x}_{j_1}, a_{j_2j_1 1}) = \underbrace{(\text{pF}, \text{UP})}_{\text{up-parabola flow }(+)}. \tag{2.47}$$

The flow at $x_{j_1} = a_{j_2j_1 1} \neq a_{j_1j_1 1}$ *is a* (pF,UP)*-positive up-parabola flow.*

- *For* $a_{j_1j_1 0}(a_{j_2j_1 1} - a_{j_1j_1 1})^3 < 0$ *and* $a_{j_2j_1 0} > 0$,

$$(\dot{x}_{j_1}, a_{j_2j_1 1}) = \underbrace{(\text{nF}, \text{DP})}_{\text{down-parabola flow }(-)}. \tag{2.48}$$

The flow at $x_{j_1} = a_{j_2j_1 1} \neq a_{j_1j_1 1}$ *is an* (nF,DP)*-negative down-parabola flow.*

- *For* $a_{j_1j_1 0}(a_{j_2j_1 1} - a_{j_1j_1 1})^3 > 0$ *and* $a_{j_2j_1 0} < 0$,

$$(\dot{x}_{j_1}, a_{j_2j_1 1}) = \underbrace{(\text{pF}, \text{DP})}_{\text{down-parabola flow }(+)}. \tag{2.49}$$

The flow at $x_{j_1} = a_{j_2j_1 1} \neq a_{j_1j_1 1}$ *is a* (pF,DP)*-positive down-parabola flow.*

- *For* $a_{j_1j_1 0}(a_{j_2j_1 1} - a_{j_1j_1 1})^3 < 0$ *and* $a_{j_2j_1 0} < 0$,

$$(\dot{x}_{j_1}, a_{j_2j_1 1}) = \underbrace{(\text{nF}, \text{UP})}_{\text{up-parabola flow }(-)}. \tag{2.50}$$

The flow at $x_{j_1} = a_{j_2j_1 1} \neq a_{j_1j_1 1}$ *is an* (nF,UP)*-negative up-parabola flow.*

(ii_{4b}) The equilibrium of $x_{j_1} = a_{j_1j_1 1} \neq a_{j_2j_1 1}$ *in the* x_{j_1}*-direction has the following properties.*

- *For* $a_{j_1j_1 0} > 0$ *and* $a_{j_2j_1 0}(a_{j_1j_1 1} - a_{j_2j_1 1}) > 0$,

$$(a_{j_1j_1 1}, \dot{x}_{j_2}) = \underbrace{(3^{\text{rd}}\text{SO}, \text{pF})}_{\text{third-order source flow }(+)}. \tag{2.51}$$

The flow at $x_{j_1} = a_{j_1j_1 1} \neq a_{j_2j_1 1}$ *is a* (3^{rd}SO,pF)*-positive third-order source flow.*

- *For* $a_{j_1j_1 0} < 0$ *and* $a_{j_2j_1 0}(a_{j_1j_1 1} - a_{j_2j_1 1}) > 0$,

$$(a_{j_1j_1 1}, \dot{x}_{j_2}) = \underbrace{(3^{\text{rd}}\text{SI}, \text{pF})}_{\text{third-order sink flow }(+)}. \tag{2.52}$$

The flow at $x_{j_1} = a_{j_1j_1 1} \neq a_{j_2j_1 1}$ *is a* (3^{rd}SI,pF)*-positive third-order sink flow.*

- *For* $a_{j_1j_1 0} > 0$ *and* $a_{j_2j_1 0}(a_{j_1j_1 1} - a_{j_2j_1 1}) < 0$,

$$(a_{j_1j_1 1}, \dot{x}_{j_2}) = \underbrace{(3^{\rm rd}{\rm SO,\,nF})}_{\text{third-order source flow } (-)}. \tag{2.53}$$

The flow at $x_{j_1} = a_{j_1j_1 1} \neq a_{j_2j_1 1}$ *is a* (3rdSO,nF)*-negative third-order source flow.*

- *For* $a_{j_1j_1 0} < 0$ *and* $a_{j_2j_1 0}(a_{j_1j_1 1} - a_{j_2j_1 1}) < 0$,

$$(a_{j_1j_1 1}, \dot{x}_{j_2}) = \underbrace{(3^{\rm rd}{\rm SI,\,nF})}_{\text{third-order sink flow } (-)}. \tag{2.54}$$

The flow at $x_{j_1} = a_{j_1j_1 1} \neq a_{j_2j_1 1}$ *is a* (3rdSI,nF)*-negative third-order sink flow.*

– *The third-order source and sink flows are the appearing and switching bifurcations of source, sink, and source flows and of sink, source, and sink flows, respectively.*

(ii$_{4c}$) For $a_{j_2j_1 1} = a_{j_1j_1 s_1} = a_{j_1j_1 s_2} \equiv a_{j_1j_1 1}(s_1, s_2 \in \{1, 2\})$*, the first integral manifold is*

$$x_{j_2} - x_{j_2 0} = -\frac{a_{j_2j_1 0}}{a_{j_1j_1 0}}\left[\frac{1}{x_{j_1} - a_{j_1j_1 1}} - \frac{1}{x_{j_1 0} - a_{j_1j_1 1}}\right]. \tag{2.55}$$

The equilibrium of $a_{j_2j_1 1} = a_{j_1j_1 s_1} = a_{j_1j_1 s_2} \equiv a_{j_1j_1 1}$ *with* $\bar{x}_{j_2} \in (-\infty, \infty)$ *has the following properties.*

- *For* $a_{j_1j_1 0} > 0$ *and* $a_{j_2j_1 0} > 0$,

$$(a_{j_1j_1 1}, \bar{x}_{j_2}) = \underbrace{(3^{\rm rd}{\rm SO,\,(3{:}1)IC})}_{\text{third-order increasing-concave source}}. \tag{2.56}$$

The equilibrium of $x^*_{j_1} = a_{j_1j_1 1} = a_{j_2j_1 1}$ *is a* (3rdSO,(3:1)IC)*-third-order increasing-concave source infinite-equilibrium.*

- *For* $a_{j_1j_1 0} < 0$ *and* $a_{j_2j_1 0} > 0$,

$$(a_{j_1j_1 1}, \bar{x}_{j_2}) = \underbrace{(3^{\rm rd}{\rm SI,\,(3{:}1)DC})}_{\text{third-order decreasing-concave sink}}. \tag{2.57}$$

The equilibrium of $x^*_{j_1} = a_{j_1j_1 1} = a_{j_2j_1 1}$ *is a* (3rdSI,(3:1)DC)*-third-order decreasing-concave sink infinite-equilibrium.*

- *For* $a_{j_1j_1 0} > 0$ *and* $a_{j_2j_1 0} < 0$,

$$(a_{j_1j_1 1}, \bar{x}_{j_2}) = \underbrace{(3^{\rm rd}{\rm SO,\,(3{:}1)DC})}_{\text{third-order decreasing-concave source}}. \tag{2.58}$$

The equilibrium of $x^*_{j_1} = a_{j_1j_1 1} = a_{j_2j_1 1}$ *is a* (3rdSO,(3:1)DC)*-third-order decreasing-concave source infinite-equilibrium.*

- *For* $a_{j_1j_10}<0$ *and* $a_{j_2j_10}<0$,

$$(a_{j_1j_11},\bar{x}_{j_2})=\underbrace{(3^{\text{rd}}\text{SI},(3:1)\text{IC})}_{\text{third-order increasing-concave sink}}. \tag{2.59}$$

The equilibrium of $x_{j_1}^*=a_{j_1j_11}=a_{j_2j_11}$ *is a* $(3^{\text{rd}}\text{SI},(3:1)\text{IC})$*-third-order increasing-concave sink infinite-equilibrium.*

- *The third-order increasing and decreasing-concave infinite-equilibriums are the* switching *bifurcations of parabola flows with third-order source (sink) flows.*
- *The third-order increasing and decreasing-concave infinite-equilibriums are the* appearing *bifurcations from parabola and source (sink) flows to source (sink), sink (source), and source (sink) flows with a parabola flow.*

(iii) If

$$A_{j_2j_11}\neq 0, A_{j_2j_12}\neq 0 \text{ and } \Delta_{j_1j_1}>0, \tag{2.60}$$

the two-dimensional dynamical system becomes

$$\begin{aligned}\dot{x}_{j_2}&=a_{j_2j_10}(x_{j_1}-a_{j_2j_11}),\\ \dot{x}_{j_1}&=a_{j_1j_10}(x_{j_1}-a_{j_1j_11})(x_{j_1}-a_{j_1j_12})(x_{j_1}-a_{j_1j_13})\end{aligned} \tag{2.61}$$

where

$$\begin{aligned}&\{a_{j_1j_11},a_{j_1j_12},a_{j_1j_13}\}=\text{sort}\{b_{j_1j_11},b_{j_1j_12},b_{j_1j_13}\}, a_{j_1j_11}<a_{j_1j_12}<a_{j_1j_13};\\ &b_{j_1j_12},b_{j_1j_13}=-\frac{1}{2}(B_{j_1j_11}\pm\sqrt{\Delta_{j_1j_1}}), a_{j_1j_10}=A_{j_1j_10}.\end{aligned} \tag{2.62}$$

(iii₁) For $a_{j_1j_1s_1}\neq a_{j_2j_11}$ $(s_1,\, s_2=1,\,2,\,3$ *and* $s_1\neq s_2)$ *the first integral manifold is*

$$x_{j_2}-x_{j_20}=\frac{a_{j_2j_10}}{a_{j_1j_10}}\sum_{s_1=1}^{3}\frac{a_{j_1j_1s_1}-a_{j_2j_11}}{\Pi_{s_2=1,s_2\neq s_1}^{3}(a_{j_1j_1s_1}-a_{j_1j_1s_2})}\ln\frac{|x_{j_1}-a_{j_1j_1s_1}|}{|x_{j_10}-a_{j_1j_1s_1}|}. \tag{2.63}$$

The equilibriums for the x_{j_1} *and* x_{j_2}*-directions are*

$$\begin{aligned}x_{j_1}^*&=a_{j_1j_11},a_{j_1j_12},a_{j_1j_13} \text{ for the } x_{j_1}\text{-direction},\\ x_{j_1}^*&=a_{j_2j_11} \text{ for the } x_{j_2}\text{-direction}.\end{aligned} \tag{2.64}$$

(iii$_{1a}$) The equilibrium of $x_{j_1}^*=a_{j_1j_1s_1}$ $(s_1,\, s_2\in\{1,2,3\}, s_1\neq s_2)$ *and* $a_{j_1j_1s_1}\neq a_{j_2j_11}$ *for the* x_{j_1}*-direction has the following properties.*

- *For* $a_{j_1j_10}\prod_{s_2=1,s_2\neq s_1}^{3}(a_{j_1j_1s_1}-a_{j_1j_1s_2})>0$ *and* $a_{j_2j_10}(a_{j_1j_1s_1}-a_{j_2j_11})>0$,

$$(a_{j_1j_1s_1}, \dot{x}_{j_2}) = \underbrace{(\text{SO}, \text{pF})}_{\text{source flow } (+)}. \tag{2.65}$$

The flow at $x_{j_1}^* = a_{j_1j_1s_1}$ *is an* (SO,pF)*-positive source flow.*

- *For* $a_{j_1j_1 0}\prod_{s_2=1, s_2\neq s_1}^{3}(a_{j_1j_1s_1} - a_{j_1j_1s_2}) < 0$ *and* $a_{j_2j_1 0}(a_{j_1j_1s_1} - a_{j_2j_1 1}) > 0$,

$$(a_{j_1j_1s_1}, \dot{x}_{j_2}) = \underbrace{(\text{SI}, \text{pF})}_{\text{sink flow } (+)}. \tag{2.66}$$

The flow at $x_{j_1}^* = a_{j_1j_1s_1}$ *is an* (SI,pF)*-positive sink flow.*

- *For* $a_{j_1j_1 0}\prod_{s_2=1, s_2\neq s_1}^{3}(a_{j_1j_1s_1} - a_{j_1j_1s_2}) > 0$ *and* $a_{j_2j_1 0}(a_{j_1j_1s_1} - a_{j_2j_1 1}) < 0$,

$$(a_{j_1j_1s_1}, \dot{x}_{j_2}) = \underbrace{(\text{SO}, \text{nF})}_{\text{source flow } (-)}. \tag{2.67}$$

The flow at $x_{j_1}^* = a_{j_1j_1s_1}$ *is an* (SO,nF)*-negative source flow.*

- *For* $a_{j_1j_1 0}\prod_{s_2=1, s_2\neq s_1}^{3}(a_{j_1j_1s_1} - a_{j_1j_1s_2}) < 0$ *and* $a_{j_2j_1 0}(a_{j_1j_1s_1} - a_{j_2j_1 1}) < 0$,

$$(a_{j_1j_1s_1}, \dot{x}_{j_2}) = \underbrace{(\text{SI}, \text{nF})}_{\text{sink flow } (-)}. \tag{2.68}$$

The flow at $x_{j_1}^* = a_{j_1j_1s_1}$ *is an* (SI,nF)*-negative sink flow.*

(iii$_{1b}$) The equilibrium of $x_{j_1}^* = a_{j_2j_1 1}$ *and* $a_{j_2j_1 1} \neq a_{j_1j_1s_1}$ $(s_1, s_2 \in \{1,2,3\}, s_1 \neq s_2)$ *for the* x_{j_2}*-direction has the following properties:*

- *For* $a_{j_1j_1 0}\prod_{s_1=1}^{3}(a_{j_2j_1 1} - a_{j_1j_1s_1}) > 0$ *and* $a_{j_2j_1 0} > 0$,

$$(\dot{x}_{j_1}, a_{j_2j_1 1}) = \underbrace{(\text{pF}, \text{UP})}_{\text{up-parabola flow } (+)}. \tag{2.69}$$

The flow at $x_{j_1}^* = a_{j_2j_1 1}$ *is a* (pF,UP)*-positive up-parabola flow.*

- *For* $a_{j_1j_1 0}\prod_{s_1=1}^{3}(a_{j_2j_1 1} - a_{j_1j_1s_1}) < 0$ *and* $a_{j_2j_1 0} > 0$,

$$(\dot{x}_{j_1}, a_{j_2j_1 1}) = \underbrace{(\text{nF}, \text{DP})}_{\text{down-parabola flow } (-)}. \tag{2.70}$$

The flow at $x_{j_1}^* = a_{j_2j_1 1}$ *is an* (nF,DP)*-negative down-parabola flow.*

- *For* $a_{j_1j_1 0}\prod_{s_1=1}^{3}(a_{j_2j_1 1} - a_{j_1j_1s_1}) > 0$ *and* $a_{j_2j_1 0} < 0$,

$$(\dot{x}_{j_1}, a_{j_2j_1 1}) = \underbrace{(\text{pF, DP})}_{\text{down-parabola flow }(+)}. \tag{2.71}$$

The flow at $x^*_{j_1} = a_{j_2j_1 1}$ *is a* (pF,DP)*-positive down-parabola flow.*

- *For* $a_{j_1j_1 0}\prod_{s_1=1}^{3}(a_{j_2j_1 1} - a_{j_1j_1 s_1}) < 0$ *and* $a_{j_2j_1 0} < 0$,

$$(\dot{x}_{j_1}, a_{j_2j_1 1}) = \underbrace{(\text{nF, UP})}_{\text{up-parabola flow }(-)}. \tag{2.72}$$

The flow at $x^*_{j_1} = a_{j_2j_1 1}$ *is an* (nF,UP)*-negative up-parabola flow.*

(iii$_2$*) For* $a_{j_2j_1 1} = a_{j_1j_1 s_1} (s_1, s_2, s_3 \in \{1,2,3\}, s_1 \neq s_2 \neq s_3)$*, the first integral manifold is*

$$x_{j_2} - x_{j_2 0} = \frac{a_{j_2j_1 0}}{a_{j_1j_1 0}} \frac{1}{a_{j_1j_1 s_2} - a_{j_1j_1 s_3}} \left(\ln \frac{|x_{j_1} - a_{j_1j_1 s_2}|}{|x_{j_1 0} - a_{j_1j_1 s_2}|} - \ln \frac{|x_{j_1} - a_{j_1j_1 s_3}|}{|x_{j_1 0} - a_{j_1j_1 s_3}|} \right). \tag{2.73}$$

The infinite-equilibrium of $x^*_{j_1} = a_{j_1j_1 s_1} = a_{j_2j_1 1}$ $(s_1, s_2, s_3 \in \{1,2,3\}, s_1 \neq s_2 \neq s_3)$ *with* $\bar{x}_{j_2} \in (-\infty, \infty)$ *possesses the following properties:*

- *For* $a_{j_1j_1 0}\prod_{s_2=1, s_2\neq s_1}^{3}(a_{j_1j_1 s_1} - a_{j_1j_1 s_2}) > 0$ *and* $a_{j_2j_1 0} > 0$,

$$\left\{ \begin{array}{c} \underbrace{(\text{nF, DP})}_{\text{down-parabola flow }(-)} \\ \underbrace{(\text{SO, pF})}_{\text{source flow }(+)} \end{array} \right\} \rightleftharpoons \underbrace{(\text{SO, II})}_{\text{increasing-inflection source}} \rightleftharpoons \left\{ \begin{array}{c} \underbrace{(\text{SO, nF})}_{\text{source flow }(-)} \\ \underbrace{(\text{pF, UP})}_{\text{up-parabola flow }(+)} \end{array} \right\}. \tag{2.74}$$

The equilibrium of $x^*_{j_1} = a_{j_1j_1 s_1} = a_{j_2j_1 1}$ *is an (SO,II)-increasing-inflection source infinite-equilibrium for the switching bifurcation of up- and down-parabola flows with source flows.*

- *For* $a_{j_1j_1 0}\prod_{s_2=1, s_2\neq s_1}^{3}(a_{j_1j_1 s_1} - a_{j_1j_1 s_2}) < 0$ *and* $a_{j_2j_1 0} > 0$,

$$\left\{ \begin{array}{c} \underbrace{(\text{nF, UP})}_{\text{up-parabola flow }(-)} \\ \underbrace{(\text{SO, nF})}_{\text{source flow }(-)} \end{array} \right\} \rightleftharpoons \underbrace{(\text{SI, DI})}_{\text{decreasing-inflection sink}} \rightleftharpoons \left\{ \begin{array}{c} \underbrace{(\text{SO, pF})}_{\text{source flow }(+)} \\ \underbrace{(\text{pF, DP})}_{\text{down-parabola flow }(+)} \end{array} \right\}. \tag{2.75}$$

The equilibrium of $x^*_{j_1} = a_{j_1j_1 s_1} = a_{j_2j_1 1}$ *is an* (SI,DI)*-decreasing-inflection sink infinite-equilibrium.*

- *For* $a_{j_1j_1 0}\prod_{s_2=1, s_2\neq s_1}^{3}(a_{j_1j_1 s_1} - a_{j_1j_1 s_2}) > 0$ *and* $a_{j_2j_1 0} < 0$,

$$\left\{\begin{array}{c}\underbrace{(\text{nF},\text{DP})}_{\text{down-parabola flow }(-)}\\ \underbrace{(\text{SO},\text{pF})}_{\text{source flow }(+)}\end{array}\right\}\rightleftharpoons\underbrace{(\text{SO},\text{DI})}_{\text{decreasing-inflection source}}\rightleftharpoons\left\{\begin{array}{c}\underbrace{(\text{SO},\text{nF})}_{\text{source flow }(-)}\\ \underbrace{(\text{pF},\text{UP})}_{\text{up-parabola flow }(+)}\end{array}\right\} \tag{2.76}$$

The equilibrium of $x_{j_1}^* = a_{j_1 j_1 s_1} = a_{j_2 j_1 1}$ *is an* (SO,DI)*-decreasing-inflection source infinite-equilibrium.*

- *For* $a_{j_1 j_1 0}\prod_{s_2=1, s_2\neq s_1}^{3}(a_{j_1 j_1 s_1} - a_{j_1 j_1 s_2}) < 0$ *and* $a_{j_2 j_1 0} < 0$,

$$\left\{\begin{array}{c}\underbrace{(\text{pF},\text{DP})}_{\text{down-parabola flow }(+)}\\ \underbrace{(\text{SI},\text{nF})}_{\text{sink flow }(-)}\end{array}\right\}\rightleftharpoons\underbrace{(\text{SO},\text{II})}_{\text{increasing-inflection sink}}\rightleftharpoons\left\{\begin{array}{c}\underbrace{(\text{SI},\text{pF})}_{\text{sink flow }(+)}\\ \underbrace{(\text{nF},\text{DP})}_{\text{down-parabola flow }(-)}\end{array}\right\}. \tag{2.77}$$

The equilibrium of $x_{j_1}^* = a_{j_1 j_1 s_1} = a_{j_2 j_1 1}$ *is an* (SO,II)*-increasing-inflection sink infinite-equilibrium.*

(iv) Under

$$A_{j_2 j_1 1} \neq 0, A_{j_2 j_1 2} \neq 0 \text{ and } \Delta_{j_1 j_1} = 0, \tag{2.78}$$

the two-dimensional dynamical system becomes

$$\begin{aligned}\dot{x}_{j_2} &= a_{j_2 j_1 0}(x_{j_1} - a_{j_2 j_1 1}),\\ \dot{x}_{j_1} &= a_{j_1 j_1 0}(x_{j_1} - a_{j_1 j_1 s_1})(x_{j_1} - a_{j_1 j_1 s_2})^2.\end{aligned} \tag{2.79}$$

(iv$_1$) For $a_{j_1 j_1 s_2} \neq a_{j_1 j_1 s_1} \neq a_{j_2 j_1 1}$, *the equilibrium of* $x_{j_1}^* = a_{j_1 j_1 s_2}$ *possesses the appearing bifurcations.*

- *For* $a_{j_1 j_1 0}(a_{j_1 j_1 s_2} - a_{j_1 j_1 s_1}) > 0$ *and* $a_{j_1 j_1 0}(a_{j_1 j_1 s_2} - a_{j_2 j_1 1}) > 0$,

$$(a_{j_1 j_1 s_2}, \dot{x}_{j_2}) = \underbrace{(\text{US},\text{pF})}_{\text{upper-saddle flow }(+)}\rightleftharpoons\left\{\begin{array}{c}\underbrace{(\text{SI},\text{pF})}_{\text{sink flow }(+)}\\ \underbrace{(\text{SO},\text{pF})}_{\text{source flow }(+)}\end{array}\right\} \tag{2.80}$$

The flow at $x_{j_1}^* = a_{j_1 j_1 s_2}$ *is called a* (US,pF)*-flow appearing bifurcation.*

- *For* $a_{j_1 j_1 0}(a_{j_1 j_1 s_2} - a_{j_1 j_1 s_1}) < 0$ *and* $a_{j_1 j_1 0}(a_{j_1 j_1 s_2} - a_{j_2 j_1 1}) > 0$,

$$(a_{j_1j_1s_2}, \dot{x}_{j_2}) = \underbrace{(\text{LS}, \text{pF})}_{\text{lower-saddle flow }(+)} \rightleftharpoons \left\{ \begin{array}{c} \underbrace{(\text{SO}, \text{pF})}_{\text{source flow }(+)} \\ \underbrace{(\text{SI}, \text{pF})}_{\text{sink flow }(+)} \end{array} \right\} \tag{2.81}$$

The flow at $x_{j_1}^* = a_{j_1j_1s_2}$ *is called an* (LS,pF)*-flow appearing bifurcation.*

- *For* $a_{j_1j_10}(a_{j_1j_1s_2} - a_{j_1j_1s_1}) > 0$ *and* $a_{j_1j_10}(a_{j_1j_11} - a_{j_2j_11}) < 0$,

$$(a_{j_1j_1s_2}, \dot{x}_{j_2}) = \underbrace{(\text{US}, \text{nF})}_{\text{upper-saddle flow }(-)} \rightleftharpoons \left\{ \begin{array}{c} \underbrace{(\text{SI}, \text{nF})}_{\text{sink flow }(-)} \\ \underbrace{(\text{SO}, \text{nF})}_{\text{source flow }(-)} \end{array} \right\} \tag{2.82}$$

The flow at $x_{j_1}^* = a_{j_1j_1s_2}$ *is called a* (US,pF)*-flow appearing bifurcation.*

- *For* $a_{j_1j_10}(a_{j_1j_1s_2} - a_{j_1j_1s_1}) < 0$ *and* $a_{j_1j_10}(a_{j_1j_11} - a_{j_2j_11}) < 0$,

$$(a_{j_1j_1s_2}, \dot{x}_{j_2}) = \underbrace{(\text{LS}, \text{nF})}_{\text{lower-saddle flow }(-)} \rightleftharpoons \left\{ \begin{array}{c} \underbrace{(\text{SO}, \text{pF})}_{\text{source flow }(-)} \\ \underbrace{(\text{SI}, \text{pF})}_{\text{sink flow }(-)} \end{array} \right\} \tag{2.83}$$

The flow at $x_{j_1}^* = a_{j_1j_1s_2}$ *is called an* (LS,pF)*-flow appearing bifurcation.*

(iv$_2$*) For* $a_{j_1j_1s_2} = a_{j_2j_11}$, *the two-dimensional dynamical system becomes*

$$\begin{aligned} \dot{x}_{j_2} &= a_{j_2j_10}(x_{j_1} - a_{j_2j_11}), \\ \dot{x}_{j_1} &= a_{j_1j_10}(x_{j_1} - a_{j_1j_1s_1})(x_{j_1} - a_{j_1j_1s_2})^2. \end{aligned} \tag{2.84}$$

(iv$_{2a}$*) The equilibrium of* $x_{j_1}^* = a_{j_1j_1s_2} = a_{j_2j_11}$ *possesses the appearing bifurcations from an up (down)-parabola flow to sink and source flows with a down (up)-parabola flow.*

- *For* $a_{j_1j_10}(a_{j_1j_1s_2} - a_{j_1j_1s_1}) > 0$ *and* $a_{j_2j_10} > 0$,

$$\underbrace{(\text{pF}, \text{UP})}_{\text{up-parabola flow }(+)} \rightleftharpoons \underbrace{(\text{US}, \text{DU})}_{\text{down-up upper-saddle}} \rightleftharpoons \left\{ \begin{array}{c} \underbrace{(\text{SI}, \text{nF})}_{\text{sink flow }(-)} \\ \underbrace{(\text{nF}, \text{DP})}_{\text{down-parabola flow }(-)} \\ \underbrace{(\text{SO}, \text{pF})}_{\text{source flow }(+)} \end{array} \right\}. \tag{2.85}$$

The infinite-equilibrium is called a (US,DU)*-appearing bifurcation.*

- *For* $a_{j_1j_10}(a_{j_1j_1s_2}-a_{j_1j_1s_1})<0$ *and* $a_{j_2j_10}>0$,

$$\underbrace{(\text{nF},\text{DP})}_{\text{up-parabola flow }(-)} \rightleftharpoons \underbrace{(\text{LS},\text{UD})}_{\text{up-down lower-saddle}} \rightleftharpoons \left\{\begin{array}{c}\underbrace{(\text{SO},\text{nF})}_{\text{source flow }(-)}\\ \underbrace{(\text{pF},\text{UP})}_{\text{up-parabola flow }(+)}\\ \underbrace{(\text{SI},\text{pF})}_{\text{sink flow }(+)}\end{array}\right\}. \tag{2.86}$$

The infinite-equilibrium is called an (LS,UD)*-appearing bifurcation.*

- *For* $a_{j_1j_10}(a_{j_1j_1s_2}-a_{j_1j_1s_1})>0$ *and* $a_{j_2j_10}<0$,

$$\underbrace{(\text{pF},\text{DP})}_{\text{down-parabola flow }(+)} \rightleftharpoons \underbrace{(\text{US},\text{UD})}_{\text{up-down upper-saddle}} \rightleftharpoons \left\{\begin{array}{c}\underbrace{(\text{SI},\text{pF})}_{\text{sink flow }(+)}\\ \underbrace{(\text{nF},\text{UP})}_{\text{up-parabola flow }(-)}\\ \underbrace{(\text{SO},\text{nF})}_{\text{source flow }(-)}\end{array}\right\}. \tag{2.87}$$

The infinite-equilibrium is called a (US,UD)*-appearing bifurcation.*

- *For* $a_{j_1j_10}(a_{j_1j_1s_2}-a_{j_1j_1s_1})<0$ *and* $a_{j_2j_10}<0$,

$$\underbrace{(\text{nF},\text{UP})}_{\text{up-parabola flow }(-)} \rightleftharpoons \underbrace{(\text{LS},\text{DU})}_{\text{down-up lower-saddle}} \rightleftharpoons \left\{\begin{array}{c}\underbrace{(\text{SO},\text{pF})}_{\text{source flow }(+)}\\ \underbrace{(\text{pF},\text{DP})}_{\text{down-parabola flow }(+)}\\ \underbrace{(\text{SI},\text{nF})}_{\text{sink flow }(-)}\end{array}\right\}. \tag{2.88}$$

The infinite-equilibrium is called an (LS,DU)*-appearing bifurcation.*

(iv$_{2b}$) The equilibrium of $x^*_{j_1}=a_{j_1j_1s_2}=a_{j_2j_11}$ *possesses the switching bifurcations from up-parabola and down-parabola flows and up (down)-saddle flows:*

- *For* $a_{j_1j_10}(a_{j_1j_1s_2} - a_{j_1j_1s_1}) > 0$ *and* $a_{j_2j_10} > 0$,

$$\left\{\begin{array}{c}\underbrace{(\text{US},\text{nF})}_{\text{upper-saddle flow }(-)}\\ \underbrace{(\text{pF},\text{UP})}_{\text{up-parabola flow }(+)}\end{array}\right\} \rightleftharpoons \underbrace{(\text{US},\text{DU})}_{\text{down-up upper-saddle}} \rightleftharpoons \left\{\begin{array}{c}\underbrace{(\text{pF},\text{UP})}_{\text{up-parabola flow }(+)}\\ \underbrace{(\text{US},\text{pF})}_{\text{upper-saddle flow }(+)}\end{array}\right\}. \tag{2.89}$$

The infinite-equilibrium is called a (US,DU)-*switching bifurcation.*

- *For* $a_{j_1j_10}(a_{j_1j_1s_2} - a_{j_1j_1s_1}) < 0$ *and* $a_{j_2j_10} > 0$,

$$\left\{\begin{array}{c}\underbrace{(\text{LS},\text{nF})}_{\text{lower-saddle flow }(-)}\\ \underbrace{(\text{nF},\text{DP})}_{\text{down-parabola flow }(-)}\end{array}\right\} \rightleftharpoons \underbrace{(\text{LS},\text{UD})}_{\text{up-down lower-saddle}} \rightleftharpoons \left\{\begin{array}{c}\underbrace{(\text{nF},\text{DP})}_{\text{down-parabola flow }(-)}\\ \underbrace{(\text{LS},\text{pF})}_{\text{lower-saddle flow }(+)}\end{array}\right\}. \tag{2.90}$$

The infinite-equilibrium is called an (LS,UD)-*switching bifurcation.*

- *For* $a_{j_1j_10}(a_{j_1j_1s_2} - a_{j_1j_1s_1}) > 0$ *and* $a_{j_2j_10} < 0$,

$$\left\{\begin{array}{c}\underbrace{(\text{US},\text{pF})}_{\text{upper-saddle flow }(+)}\\ \underbrace{(\text{pF},\text{DP})}_{\text{down-parabola flow }(+)}\end{array}\right\} \rightleftharpoons \underbrace{(\text{US},\text{UD})}_{\text{up-down upper-saddle}} \rightleftharpoons \left\{\begin{array}{c}\underbrace{(\text{pF},\text{DP})}_{\text{down-parabola flow }(+)}\\ \underbrace{(\text{US},\text{nF})}_{\text{upper-saddle flow }(-)}\end{array}\right\}. \tag{2.91}$$

The infinite-equilibrium is called a (US,UD)-*switching bifurcation.*

- *For* $a_{j_1j_10}(a_{j_1j_1s_2} - a_{j_1j_1s_1}) < 0$ *and* $a_{j_2j_10} < 0$,

$$\left\{\begin{array}{c}\underbrace{(\text{LS},\text{pF})}_{\text{lower-saddle flow }(+)}\\ \underbrace{(\text{nF},\text{UP})}_{\text{up-parabola flow }(-)}\end{array}\right\} \rightleftharpoons \underbrace{(\text{LS},\text{DU})}_{\text{down-up lower-saddle}} \rightleftharpoons \left\{\begin{array}{c}\underbrace{(\text{nF},\text{UP})}_{\text{up-parabola flow }(-)}\\ \underbrace{(\text{LS},\text{nF})}_{\text{lower-saddle flow }(-)}\end{array}\right\}. \tag{2.92}$$

The infinite-equilibrium is called an (LS,DU)-*switching bifurcation.*

(v) Under

$$A_{j_2j_11} \neq 0, A_{j_2j_12} \neq 0 \text{ and } \Delta_{j_1j_1} = 0 \text{ with } a_{j_1j_1s_1} = a_{j_1j_1s_2} \equiv a_{j_1j_11} \tag{2.93}$$

the two-dimensional dynamical system becomes

$$\begin{aligned}\dot{x}_{j_2} &= a_{j_2 j_1 0}(x_{j_1} - a_{j_2 j_1 1}),\\ \dot{x}_{j_1} &= a_{j_1 j_1 0}(x_{j_1} - a_{j_1 j_1 1})^3.\end{aligned} \tag{2.94}$$

(v_{1a}) The equilibrium of $a_{j_2 j_1 1} \neq a_{j_1 j_1 1}$ has the following appearing bifurcations.

- *For $a_{j_1 j_1 0} > 0$ and $a_{j_2 j_1 0}(a_{j_1 j_1 1} - a_{j_2 j_1 1}) > 0$,*

$$\underbrace{(\text{SO}, \text{pF})}_{\text{source flow }(+)} \rightleftharpoons \underbrace{(3^{\text{rd}}\text{SO}, \text{pF})}_{\text{third-order source flow }(+)} \rightleftharpoons \left\{\begin{array}{c}\underbrace{(\text{SO}, \text{pF})}_{\text{source flow }(+)}\\ \underbrace{(\text{SI}, \text{pF})}_{\text{sink flow }(+)}\\ \underbrace{(\text{SO}, \text{pF})}_{\text{source flow }(+)}\end{array}\right\}. \tag{2.95}$$

The $(3^{\text{rd}}\text{SO,pF})$*-positive third-order source flow of $x^*_{j_1} = a_{j_1 j_1 1} \neq a_{j_2 j_1 1}$ is a* $(3^{\text{rd}}\text{SO,pF})$*-flow appearing bifurcation.*

- *For $a_{j_1 j_1 0} < 0$ and $a_{j_2 j_1 0}(a_{j_1 j_1 1} - a_{j_2 j_1 1}) > 0$,*

$$\underbrace{(\text{SI}, \text{pF})}_{\text{sink flow }(+)} \rightleftharpoons \underbrace{(3^{\text{rd}}\text{SI}, \text{pF})}_{\text{third-order sink flow }(+)} \rightleftharpoons \left\{\begin{array}{c}\underbrace{(\text{SI}, \text{pF})}_{\text{sink flow }(+)}\\ \underbrace{(\text{SO}, \text{pF})}_{\text{source flow }(+)}\\ \underbrace{(\text{SI}, \text{pF})}_{\text{sink flow }(+)}\end{array}\right\}. \tag{2.96}$$

The $(3^{\text{rd}}\text{SI,pF})$*-positive third-order sink flow of $x^*_{j_1} = a_{j_1 j_1 1} \neq a_{j_2 j_1 1}$ is a* $(3^{\text{rd}}\text{SI,}$ $\text{pF})$*-flow appearing bifurcation.*

- *For $a_{j_1 j_1 0} > 0$ and $a_{j_2 j_1 0}(a_{j_1 j_1 1} - a_{j_2 j_1 1}) < 0$,*

$$\underbrace{(\text{SO}, \text{nF})}_{\text{source flow }(-)} \rightleftharpoons \underbrace{(3^{\text{rd}}\text{SO}, \text{nF})}_{\text{third-order source flow }(-)} \rightleftharpoons \left\{\begin{array}{c}\underbrace{(\text{SO}, \text{nF})}_{\text{source flow }(-)}\\ \underbrace{(\text{SI}, \text{nF})}_{\text{sink flow }(-)}\\ \underbrace{(\text{SO}, \text{nF})}_{\text{source flow }(-)}\end{array}\right\}. \tag{2.97}$$

The $(3^{\text{rd}}\text{SO,nF})$*-negative third-order source flow of $x^*_{j_1} = a_{j_1 j_1 1} \neq a_{j_2 j_1 1}$ is a* $(3^{\text{rd}}\text{SO,nF})$*-flow appearing bifurcation.*

- *For* $a_{j_1j_10}<0$ *and* $a_{j_2j_10}(a_{j_1j_11}-a_{j_2j_11})<0$,

$$\underbrace{(\mathrm{SI,nF})}_{\text{sink flow }(-)} \rightleftharpoons \underbrace{(3^{\mathrm{rd}}\mathrm{SI,nF})}_{\text{third-order sink flow }(-)} \rightleftharpoons \left\{\begin{array}{c}\underbrace{(\mathrm{SI,nF})}_{\text{sink flow }(-)}\\ \underbrace{(\mathrm{SO,nF})}_{\text{source flow }(-)}\\ \underbrace{(\mathrm{SI,nF})}_{\text{sink flow }(-)}\end{array}\right\}. \tag{2.98}$$

The $(3^{\mathrm{rd}}$SI,nF)-*negative third-order sink flow of* $x_{j_1}^*=a_{j_1j_11}\neq a_{j_2j_11}$ *is a* $(3^{\mathrm{rd}}$SI,nF)-*flow appearing bifurcation.*

(v_{1b}) *The equilibrium of* $a_{j_2j_11}\neq a_{j_1j_11}$ *has the following appearing bifurcations.*

- *For* $a_{j_1j_10}>0$ *and* $a_{j_2j_10}(a_{j_1j_11}-a_{j_2j_11})>0$,

$$\left\{\begin{array}{c}\underbrace{(\mathrm{SO,pF})}_{\text{source flow }(+)}\\ \underbrace{(\mathrm{SI,pF})}_{\text{sink flow }(+)}\\ \underbrace{(\mathrm{SO,pF})}_{\text{source flow }(+)}\end{array}\right\} \rightleftharpoons \underbrace{(3^{\mathrm{rd}}\mathrm{SO,pF})}_{\text{third-order source flow }(+)} \rightleftharpoons \left\{\begin{array}{c}\underbrace{(\mathrm{SO,pF})}_{\text{source flow }(+)}\\ \underbrace{(\mathrm{SI,pF})}_{\text{sink flow }(+)}\\ \underbrace{(\mathrm{SO,pF})}_{\text{source flow }(+)}\end{array}\right\}. \tag{2.99}$$

The $(3^{\mathrm{rd}}$SO,pF)-*positive third-order source flow of* $x_{j_1}^*=a_{j_1j_11}\neq a_{j_2j_11}$ *is a* $(3^{\mathrm{rd}}$SO,pF)- *flow switching bifurcation.*

- *For* $a_{j_1j_10}<0$ *and* $a_{j_2j_10}(a_{j_1j_11}-a_{j_2j_11})>0$,

$$\left\{\begin{array}{c}\underbrace{(\mathrm{SI,pF})}_{\text{sink flow }(+)}\\ \underbrace{(\mathrm{SO,pF})}_{\text{source flow }(+)}\\ \underbrace{(\mathrm{SI,pF})}_{\text{sink flow }(+)}\end{array}\right\} \rightleftharpoons \underbrace{(3^{\mathrm{rd}}\mathrm{SI,pF})}_{\text{third-order sink flow }(+)} \rightleftharpoons \left\{\begin{array}{c}\underbrace{(\mathrm{SI,pF})}_{\text{sink flow }(+)}\\ \underbrace{(\mathrm{SO,pF})}_{\text{source flow }(+)}\\ \underbrace{(\mathrm{SI,pF})}_{\text{sink flow }(+)}\end{array}\right\}. \tag{2.100}$$

The $(3^{\mathrm{rd}}$SI,pF)-*positive third-order sink flow of* $x_{j_1}^*=a_{j_1j_11}\neq a_{j_2j_11}$ *is a* $(3^{\mathrm{rd}}$SI,pF)-*flow switching bifurcation.*

- *For* $a_{j_1j_10}>0$ *and* $a_{j_2j_10}(a_{j_1j_11}-a_{j_2j_11})<0$,

$$\left\{\begin{array}{c}\underbrace{(\mathrm{SO},\mathrm{nF})}_{\text{source flow }(-)}\\ \underbrace{(\mathrm{SI},\mathrm{nF})}_{\text{sink flow }(-)}\\ \underbrace{(\mathrm{SO},\mathrm{nF})}_{\text{source flow }(-)}\end{array}\right\}\rightleftharpoons\underbrace{(3^{\mathrm{rd}}\mathrm{SO},\mathrm{nF})}_{\text{third-order source flow }(-)}\rightleftharpoons\left\{\begin{array}{c}\underbrace{(\mathrm{SO},\mathrm{nF})}_{\text{source flow }(-)}\\ \underbrace{(\mathrm{SI},\mathrm{nF})}_{\text{sink flow }(-)}\\ \underbrace{(\mathrm{SO},\mathrm{nF})}_{\text{source flow }(-)}\end{array}\right\}. \tag{2.101}$$

The (3^{rd}SO,nF)*-negative third-order source flow of* $x_{j_1}^* = a_{j_1j_1 1} \neq a_{j_2j_1 1}$ *is a* (3^{rd}SO,nF)*-flow switching bifurcation.*

- *For* $a_{j_1j_1 0} < 0$ *and* $a_{j_2j_1 0}(a_{j_1j_1 1} - a_{j_2j_1 1}) < 0$,

$$\left\{\begin{array}{c}\underbrace{(\mathrm{SI},\mathrm{nF})}_{\text{sink flow }(-)}\\ \underbrace{(\mathrm{SO},\mathrm{nF})}_{\text{source flow }(-)}\\ \underbrace{(\mathrm{SI},\mathrm{nF})}_{\text{sink flow }(-)}\end{array}\right\}\rightleftharpoons\underbrace{(3^{\mathrm{rd}}\mathrm{SI},\mathrm{nF})}_{\text{third-order sink flow }(-)}\rightleftharpoons\left\{\begin{array}{c}\underbrace{(\mathrm{SI},\mathrm{nF})}_{\text{sink flow }(-)}\\ \underbrace{(\mathrm{SO},\mathrm{nF})}_{\text{source flow }(-)}\\ \underbrace{(\mathrm{SI},\mathrm{nF})}_{\text{sink flow }(-)}\end{array}\right\}. \tag{2.102}$$

The (3^{rd}SI,nF)*-negative third-order sink flow of* $x_{j_1}^* = a_{j_1j_1 1} \neq a_{j_2j_1 1}$ *is a* (3^{rd}SI,nF)*-flow switching bifurcation.*

(v_{2a}) The equilibrium of $x_{j_1}^* = a_{j_1j_1 1} = a_{j_2j_1 1}$ *with* $\bar{x}_{j_2} \in (-\infty, \infty)$ *has the following appearing bifurcations.*

- *For* $a_{j_1j_1 0} > 0$ *and* $a_{j_2j_1 0} > 0$,

$$\left\{\begin{array}{c}\underbrace{(\mathrm{SO},\mathrm{nF})}_{\text{source flow }(-)}\\ \underbrace{(\mathrm{pF},\mathrm{UP})}_{\text{up-parabola flow }(+)}\end{array}\right\}\rightleftharpoons\underbrace{(3^{\mathrm{rd}}\mathrm{SO},(3:1)\mathrm{IC})}_{\text{third-order increasing-concave source}}\rightleftharpoons\left\{\begin{array}{c}\underbrace{(\mathrm{SO},\mathrm{nF})}_{\text{source flow }(-)}\\ \underbrace{(\mathrm{pF},\mathrm{UP})}_{\text{up-parabola flow }(+)}\\ \underbrace{(\mathrm{SI},\mathrm{pF})}_{\text{sink flow }(+)}\\ \underbrace{(\mathrm{SO},\mathrm{pF})}_{\text{source flow }(+)}\end{array}\right\}. \tag{2.103}$$

and

$$\left\{\begin{array}{c}\underbrace{(\text{nF}, \text{DP})}_{\text{down-parabola flow }(-)}\\ \underbrace{(\text{SO}, \text{pF})}_{\text{source flow }(+)}\end{array}\right\} \rightleftharpoons \underbrace{(3^{\text{rd}}\text{SI}, (3:1)\text{IC})}_{\text{third-order increasing-concave source}} \rightleftharpoons \left\{\begin{array}{c}\underbrace{(\text{SO}, \text{nF})}_{\text{source flow }(-)}\\ \underbrace{(\text{SI}, \text{nF})}_{\text{sink flow }(-)}\\ \underbrace{(\text{nF}, \text{DP})}_{\text{down-parabola flow }(-)}\\ \underbrace{(\text{SO}, \text{pF})}_{\text{source flow }(+)}\end{array}\right\}. \tag{2.104}$$

The $(3^{\text{rd}}\text{SO},(3:1)\text{IC})$*-third-order increasing-concave source infinite-equilibrium of* $x_{j_1}^* = a_{j_1j_1 1} = a_{j_2j_1 1}$ *is a* $(3^{\text{rd}}\text{SO},(3:1)\text{IC})$*-appearing bifurcation.*

- *For* $a_{j_1j_1 0} < 0$ *and* $a_{j_2j_1 0} > 0$,

$$\left\{\begin{array}{c}\underbrace{(\text{SI}, \text{nF})}_{\text{sink flow }(-)}\\ \underbrace{(\text{nF}, \text{DP})}_{\text{down-parabola flow }(-)}\end{array}\right\} \rightleftharpoons \underbrace{(3^{\text{rd}}\text{SI}, (3:1)\text{DC})}_{\text{third-order decreasing-concave sink}} \rightleftharpoons \left\{\begin{array}{c}\underbrace{(\text{SI}, \text{nF})}_{\text{source flow }(-)}\\ \underbrace{(\text{nF}, \text{DP})}_{\text{up-parabola flow }(-)}\\ \underbrace{(\text{SO}, \text{pF})}_{\text{source flow }(+)}\\ \underbrace{(\text{SI}, \text{pF})}_{\text{sink flow }(+)}\end{array}\right\}. \tag{2.105}$$

and

$$\left\{\begin{array}{c}\underbrace{(\text{pF}, \text{UP})}_{\text{up-parabola flow }(+)}\\ \underbrace{(\text{SI}, \text{pF})}_{\text{sink flow }(+)}\end{array}\right\} \rightleftharpoons \underbrace{(3^{\text{rd}}\text{SI}, (3:1)\text{DC})}_{\text{third-order decreasing-concave sink}} \rightleftharpoons \left\{\begin{array}{c}\underbrace{(\text{SI}, \text{nF})}_{\text{source flow }(-)}\\ \underbrace{(\text{SO}, \text{nF})}_{\text{source flow }(-)}\\ \underbrace{(\text{pF}, \text{UP})}_{\text{up-parabola flow }(+)}\\ \underbrace{(\text{SI}, \text{pF})}_{\text{sink flow }(+)}\end{array}\right\}. \tag{2.106}$$

The $(3^{\text{rd}}\text{SI}, (3:1)\text{DC})$*-third-order decreasing-concave sink infinite-equilibrium of* $x_{j_1}^* = a_{j_1j_1 1} = a_{j_2j_1 1}$ *is a* $(3^{\text{rd}}\text{SI}, (3:1)\text{DC})$*-appearing bifurcation.*

- *For $a_{j_1j_1 0} > 0$ and $a_{j_2j_1 0} < 0$,*

$$\left\{\begin{array}{c}\underbrace{(\text{SO},\text{pF})}_{\text{sink flow}(+)}\\ \underbrace{(\text{pF},\text{DP})}_{\text{down-parabola flow}(+)}\end{array}\right\} \rightleftharpoons \underbrace{(3^{\text{rd}}\text{SO},(3:1)\text{DC})}_{\text{third-order decreasing-concave source}} \rightleftharpoons \left\{\begin{array}{c}\underbrace{(\text{SO},\text{pF})}_{\text{source flow}(+)}\\ \underbrace{(\text{pF},\text{DP})}_{\text{down-parabola flow}(+)}\\ \underbrace{(\text{SI},\text{nF})}_{\text{sink flow}(-)}\\ \underbrace{(\text{SO},\text{nF})}_{\text{source flow}(-)}\end{array}\right\}. \tag{2.107}$$

and

$$\left\{\begin{array}{c}\underbrace{(\text{nF},\text{UP})}_{\text{up-parabola flow}(-)}\\ \underbrace{(\text{SO},\text{nF})}_{\text{source flow}(-)}\end{array}\right\} \rightleftharpoons \underbrace{(3^{\text{rd}}\text{SO},(3:1)\text{DC})}_{\text{third-order decreasing-concave source}} \rightleftharpoons \left\{\begin{array}{c}\underbrace{(\text{SO},\text{pF})}_{\text{source flow}(+)}\\ \underbrace{(\text{SI},\text{pF})}_{\text{sink flow}(+)}\\ \underbrace{(\text{nF},\text{UP})}_{\text{up-parabola flow}(-)}\\ \underbrace{(\text{SO},\text{nF})}_{\text{source flow}(-)}\end{array}\right\}. \tag{2.108}$$

The (3^{rd}SO,(3:1)DC)-*third-order decreasing-concave source infinite-equilibrium of $x^*_{j_1} = a_{j_1j_1 1} = a_{j_2j_1 1}$ is a* (3^{rd}SO, (3:1)DC)-*appearing bifurcation.*

- *For $a_{j_1j_1 0} < 0$ and $a_{j_2j_1 0} < 0$,*

$$\left\{\begin{array}{c}\underbrace{(\text{SI},\text{pF})}_{\text{sink flow}(+)}\\ \underbrace{(\text{nF},\text{UP})}_{\text{up-parabola flow}(-)}\end{array}\right\} \rightleftharpoons \underbrace{(3^{\text{rd}}\text{SI},(3:1)\text{IC})}_{\text{third-order increasing-concave sink}} \rightleftharpoons \left\{\begin{array}{c}\underbrace{(\text{SI},\text{pF})}_{\text{sink flow}(+)}\\ \underbrace{(\text{nF},\text{UP})}_{\text{up-parabola flow}(-)}\\ \underbrace{(\text{SO},\text{nF})}_{\text{source flow}(-)}\\ \underbrace{(\text{SI},\text{nF})}_{\text{sink flow}(-)}\end{array}\right\}. \tag{2.109}$$

and

$$\left\{\begin{array}{c}\underbrace{(\mathrm{pF},\mathrm{DP})}_{\text{down-parabola flow }(+)}\\ \underbrace{(\mathrm{SI},\mathrm{nF})}_{\text{sink flow }(-)}\end{array}\right\}\rightleftharpoons\underbrace{(3^{\mathrm{rd}}\mathrm{SI},(3:1)\mathrm{IC})}_{\text{third-order increasing-concave sink}}\rightleftharpoons\left\{\begin{array}{c}\underbrace{(\mathrm{SI},\mathrm{pF})}_{\text{sink flow }(+)}\\ \underbrace{(\mathrm{SO},\mathrm{pF})}_{\text{source flow }(+)}\\ \underbrace{(\mathrm{pF},\mathrm{DP})}_{\text{down-parabola flow }(+)}\\ \underbrace{(\mathrm{SI},\mathrm{nF})}_{\text{sink flow }(-)}\end{array}\right\}. \quad (2.110)$$

The (3rdSI, (3:1)IC)*-third-order decreasing-concave sink infinite-equilibrium of* $x_{j_1}^* = a_{j_1j_1 1} = a_{j_2j_1 1}$ *is a* (3rdSI, (3:1)IC)*-appearing bifurcation.*

(v_{2b}) The equilibrium of $x_{j_1}^* = a_{j_1j_1 1} = a_{j_2j_1 1}$ *with* $\overline{x}_{j_2} \in (-\infty, \infty)$ *has the following switching bifurcations.*

- *For* $a_{j_1j_1 0} > 0$ *and* $a_{j_2j_1 0} > 0$,

$$\left\{\begin{array}{c}\underbrace{(3^{\mathrm{rd}}\mathrm{SO},\mathrm{nF})}_{\text{source flow }(-)}\\ \underbrace{(\mathrm{pF},\mathrm{UP})}_{\text{up-parabola flow }(+)}\end{array}\right\}\rightleftharpoons\underbrace{(3^{\mathrm{rd}}\mathrm{SO},(3:1)\mathrm{IC})}_{\text{third-order increasing-concave source}}\rightleftharpoons\left\{\begin{array}{c}\underbrace{(\mathrm{nF},\mathrm{DP})}_{\text{down-parabola flow }(-)}\\ \underbrace{(3^{\mathrm{rd}}\mathrm{SO},\mathrm{pF})}_{\text{source flow }(+)}\end{array}\right\}. \quad (2.111)$$

The (3rdSO,(3:1)IC)*-third-order increasing-concave source infinite-equilibrium of* $x_{j_1}^* = a_{j_1j_1 1} = a_{j_2j_1 1}$ *is a* (3rdSO,(3:1)IC)*-switching bifurcation.*

- *For* $a_{j_1j_1 0} < 0$ *and* $a_{j_2j_1 0} > 0$,

$$\left\{\begin{array}{c}\underbrace{(3^{\mathrm{rd}}\mathrm{SI},\mathrm{nF})}_{\text{sink flow }(-)}\\ \underbrace{(\mathrm{nF},\mathrm{DP})}_{\text{down-parabola flow }(-)}\end{array}\right\}\rightleftharpoons\underbrace{(3^{\mathrm{rd}}\mathrm{SI},(3:1)\mathrm{DC})}_{\text{third-order decreasing-concave sink}}\rightleftharpoons\left\{\begin{array}{c}\underbrace{(\mathrm{pF},\mathrm{UP})}_{\text{up-parabola flow }(+)}\\ \underbrace{(3^{\mathrm{rd}}\mathrm{SI},\mathrm{pF})}_{\text{sink flow }(+)}\end{array}\right\}. \quad (2.112)$$

The (3rdSI,(3:1)DC)*-third-order decreasing-concave sink infinite-equilibrium of* $x_{j_1}^* = a_{j_1j_1 1} = a_{j_2j_1 1}$ *is a* (3rdSI, (3:1)DC)*-switching bifurcation.*

- *For* $a_{j_1j_1 0} > 0$ *and* $a_{j_2j_1 0} < 0$,

$$\left\{\begin{array}{c}\underbrace{(3^{\mathrm{rd}}\mathrm{SO},\mathrm{nF})}_{\text{sink flow }(-)}\\ \underbrace{(\mathrm{pF},\mathrm{DP})}_{\text{down-parabola flow }(+)}\end{array}\right\}\rightleftharpoons\underbrace{(3^{\mathrm{rd}}\mathrm{SO},(3:1)\mathrm{DC})}_{\text{third-order decreasing-concave source}}\rightleftharpoons\left\{\begin{array}{c}\underbrace{(\mathrm{nF},\mathrm{UP})}_{\text{up-parabola flow }(-)}\\ \underbrace{(3^{\mathrm{rd}}\mathrm{SO},\mathrm{nF})}_{\text{ssource flow }(-)}\end{array}\right\}. \tag{2.113}$$

The $(3^{\mathrm{rd}}\mathrm{SO},(3{:}1)\mathrm{DC})$*-third-order decreasing-concave source infinite-equilibrium of* $x_{j_1}^* = a_{j_1j_1 1} = a_{j_2j_1 1}$ *is a* $(3^{\mathrm{rd}}\mathrm{SO}, (3{:}1)\mathrm{DC})$*-switching bifurcation.*

- *For* $a_{j_1j_1 0} < 0$ *and* $a_{j_2j_1 0} < 0$,

$$\left\{\begin{array}{c}\underbrace{(3^{\mathrm{rd}}\mathrm{SI},\mathrm{pF})}_{\text{sink flow }(+)}\\ \underbrace{(\mathrm{nF},\mathrm{UP})}_{\text{up-parabola flow }(-)}\end{array}\right\}\rightleftharpoons\underbrace{(3^{\mathrm{rd}}\mathrm{SI},(3:1)\mathrm{IC})}_{\text{third-order increasing-concave sink}}\rightleftharpoons\left\{\begin{array}{c}\underbrace{(\mathrm{pF},\mathrm{DP})}_{\text{down-parabola flow }(+)}\\ \underbrace{(3^{\mathrm{rd}}\mathrm{SI},\mathrm{nF})}_{\text{sink flow }(-)}\end{array}\right\}. \tag{2.114}$$

The $(3^{\mathrm{rd}}\mathrm{SI}, (3{:}1)\mathrm{IC})$*-third-order decreasing-concave sink infinite-equilibrium of* $x_{j_1}^* = a_{j_1j_1 1} = a_{j_2j_1 1}$ *is a* $(3^{\mathrm{rd}}\mathrm{SI}, (3{:}1)\mathrm{IC})$*-switching bifurcation.*

2.2 Proof of Theorem 2.1

Consider

$$\begin{aligned}\dot{x}_{j_2} &= A_{j_2j_1 2} + A_{j_2j_1 1}x_{j_1},\\ \dot{x}_{j_1} &= a_{j_1j_1 0}(x_{j_1} - b_{j_1j_1 1})(x_{j_1}^2 + B_{j_1j_1 1}x_{j_1} + C_{j_1j_1 1})\end{aligned}$$

with

$$\Delta_{j_1j_1} = (B_{j_1j_1 1})^2 - 4C_{j_1j_1 1} \quad \text{for } j_1, j_2 \in \{1,2\}, j_1 \neq j_2.$$

(i) If

$$A_{j_2j_1 1} \neq 0, A_{j_2j_1 2} \neq 0 \text{ and } \Delta_{j_1j_1} < 0,$$

the corresponding dynamical system is

$$
\begin{aligned}
\dot{x}_{j_2} &= a_{j_2j_1 0}(x_{j_1} - a_{j_2j_1 1}),\\
\dot{x}_{j_1} &= a_{j_1j_1 0}(x_{j_1} - a_{j_1j_1 1})\left[(x_{j_1} - a_{j_1j_1 2})^2 + b_{j_1j_1}\right]
\end{aligned}
$$

where

$$
\begin{aligned}
&a_{j_2j_1 1} = -\frac{1}{A_{j_2j_1 1}}A_{j_2j_1 2}, a_{j_2j_1 0} = A_{j_2j_1 1};\\
&a_{j_1j_1 1} = b_{j_1j_1 1}, a_{j_1j_1 2} = \frac{1}{2}A_{j_1j_1 1}, a_{j_1j_1 0} = A_{j_1j_1 0}, b_{j_1j_1} = \frac{1}{4}(-\Delta_{j_1j_1}).
\end{aligned}
$$

(i_1) For $a_{j_1j_1 1} \neq a_{j_2j_1 1}$, in phase space,

$$
\begin{aligned}
\frac{dx_{j_2}}{dx_{j_1}} &= \frac{a_{j_2j_1 0}}{a_{j_1j_1 0}}\frac{(x_{j_1} - a_{j_2j_1 1})}{(x_{j_1} - a_{j_1j_1 1})\left[(x_{j_1} - a_{j_1j_1 2})^2 + b_{j_1j_1}\right]}\\
&= \frac{a_{j_2j_1 0}}{a_{j_1j_1 0}}\left\{\frac{1}{(x_{j_1} - a_{j_1j_1 2})^2 + b_{j_1j_1}} + \frac{a_{j_1j_1 1} - a_{j_2j_1 1}}{(a_{j_1j_1 1} - a_{j_1j_1 2})^2 + b_{j_1j_1}}\right.\\
&\quad \left.\times\left[\frac{1}{x_{j_1} - a_{j_1j_1 1}} - \frac{x_{j_1} - a_{j_1j_1 2}}{(x_{j_1} - a_{j_1j_1 2})^2 + b_{j_1j_1}}\frac{a_{j_1j_1 2} - a_{j_1j_1 1}}{(x_{j_1} - a_{j_1j_1 2})^2 + b_{j_1j_1}}\right]\right\}.
\end{aligned}
$$

With an initial condition of $(x_{j_1 0}, x_{j_2 0})$ at $t = t_0$, the first integration gives the first integral manifold as

$$
\begin{aligned}
x_{j_2} - x_{j_2 0} = \frac{a_{j_2j_1 0}}{a_{j_1j_1 0}}&\left\{\frac{1}{\sqrt{b_{j_1j_1}}}\left(\arctan\frac{x_{j_1} - a_{j_1j_1 2}}{\sqrt{b_{j_1j_1}}} - \arctan\frac{x_{j_1 0} - a_{j_1j_1 2}}{\sqrt{b_{j_1j_1}}}\right)\right.\\
&+\frac{(a_{j_1j_1 1} - a_{j_2j_1 1})}{(a_{j_1j_1 1} - a_{j_1j_1 2})^2 + b_{j_1j_1}}\left[\ln\frac{|x_{j_1} - a_{j_1j_1 1}|}{|x_{j_1 0} - a_{j_1j_1 1}|}\right.\\
&-\frac{1}{2}\ln\frac{|(x_{j_1} - a_{j_1j_1 2})^2 + b_{j_1j_1}|}{|(x_{j_1 0} - a_{j_1j_1 2})^2 + b_{j_1j_1}|}\\
&\left.\left.+\frac{(a_{j_1j_1 2} - a_{j_1j_1 1})}{\sqrt{b_{j_1j_1}}}\left(\arctan\frac{x_{j_1} - a_{j_1j_1 2}}{\sqrt{b_{j_1j_1}}} - \arctan\frac{x_{j_1 0} - a_{j_1j_1 2}}{\sqrt{b_{j_1j_1}}}\right)\right]\right\}.
\end{aligned}
$$

(i_{1a}) In vicinity of $x^*_{j_1} = a_{j_1j_1 1}$, the variational equation is

$$
\begin{aligned}
\dot{x}_{j_2} &= a_{j_2j_1 0}(a_{j_1j_1 1} - a_{j_2j_1 1}),\\
\Delta\dot{x}_{j_1} &= a_{j_1j_1 0}\left[(a_{j_1j_1 1} - a_{j_1j_1 2})^2 + b_{j_1j_1}\right]\Delta x_{j_1}.
\end{aligned}
$$

Because of $(a_{j_1j_1 1} - a_{j_1j_1 2})^2 + b_{j_1j_1} > 0$, the flows at $x^*_{j_1} = a_{j_1j_1 1}$ in the x_{j_1}-direction are source and sink flows for $a_{j_1j_1 0} > 0$ and $a_{j_1j_1 0} < 0$, respectively.

However, for $a_{j_2j_10}(a_{j_1j_11}-a_{j_2j_11})>0$ and $a_{j_2j_10}(a_{j_1j_11}-a_{j_2j_11})<0$, we have $\dot{x}_{j_2}>0$ and $\dot{x}_{j_2}<0$, respectively.

Therefore, the flow at $x^*_{j_1}=a_{j_1j_11}$ for the x_{j_1}-direction has the following properties as in Eqs. (2.7)–(2.10).

- For $a_{j_1j_10}>0$ and $a_{j_2j_10}(a_{j_1j_11}-a_{j_2j_11})>0$,

$$(a_{j_1j_11},\dot{x}_{j_2})=\underbrace{(\mathrm{SO},\mathrm{pF})}_{\text{source flow }(+)}.$$

  The flow at $x^*_{j_1}=a_{j_1j_11}$ is an (SO,pF)-positive source flow.

- For $a_{j_1j_10}<0$ and $a_{j_2j_10}(a_{j_1j_11}-a_{j_2j_11})>0$,

$$(a_{j_1j_11},\dot{x}_{j_2})=\underbrace{(\mathrm{SI},\mathrm{pF})}_{\text{sink flow }(+)}.$$

  The flow at $x^*_{j_1}=a_{j_1j_11}$ is an (SI,pF)-positive sink flow.

- For $a_{j_1j_10}>0$ and $a_{j_2j_10}(a_{j_1j_11}-a_{j_2j_11})<0$,

$$(a_{j_1j_11},\dot{x}_{j_2})=\underbrace{(\mathrm{SO},\mathrm{nF})}_{\text{source flow }(-)}.$$

  The flow at $x^*_{j_1}=a_{j_1j_11}$ is an (SO,nF)-negative source flow.

- For $a_{j_1j_10}<0$ and $a_{j_2j_10}(a_{j_1j_11}-a_{j_2j_11})<0$,

$$(a_{j_1j_11},\dot{x}_{j_2})=\underbrace{(\mathrm{SI},\mathrm{nF})}_{\text{sink flow }(-)}.$$

  The flow at $x^*_{j_1}=a_{j_1j_11}$ is an (SI,nF)-negative sink flow.

(i_{1b}) In phase space, at $x^*_{j_1}=a_{j_2j_11}$ and $\overline{x}_{j_1}\neq a_{j_1j_11}$,

$$\frac{dx_{j_2}}{dx_{j_1}}\bigg|_{x^*_{j_1}=a_{j_2j_11}}=\frac{a_{j_2j_10}}{a_{j_1j_10}}\frac{(x_{j_1}-a_{j_2j_11})}{(x_{j_1}-a_{j_1j_11})\left[(x_{j_1}-a_{j_1j_12})^2+b_{j_1j_1}\right]}\bigg|_{x^*_{j_1}=a_{j_2j_11}}=0.$$

If

$$\frac{d^2x_{j_2}}{dx^2_{j_1}}\bigg|_{x^*_{j_1}=a_{j_2j_11}}=\frac{a_{j_2j_10}}{a_{j_1j_10}}\frac{1}{(a_{j_2j_11}-a_{j_1j_11})\left[(a_{j_2j_11}-a_{j_1j_12})^2+b_{j_1j_1}\right]}>0,$$

then there is an up-parabola flow in the x_{j_2}-direction at $x^*_{j_1}=a_{j_2j_11}$ and $\overline{x}_{j_1}\neq a_{j_1j_11}$. If

$$\frac{d^2x_{j_2}}{dx_{j_1}^2}\bigg|_{x_{j_1}^*=a_{j_2j_1 1}} = \frac{a_{j_2j_1 0}}{a_{j_1j_1 0}}\frac{1}{(a_{j_2j_1 1}-a_{j_1j_1 1})\left[(a_{j_2j_1 1}-a_{j_1j_1 2})^2+b_{j_1j_1}\right]} < 0,$$

then there is a down-parabola flow in the x_{j_2}-direction at $x_{j_1}^* = a_{j_2j_1 1}$ and $\overline{x}_{j_1} \neq a_{j_1j_1 1}$.

Because of $(a_{j_2j_1 1}-a_{j_1j_1 2})^2+b_{j_1j_1} > 0$, for $a_{j_1j_1 0}(a_{j_2j_1 1}-a_{j_1j_1 1}) > 0$,

$$\dot{x}_{j_2} = a_{j_1j_1 0}(a_{j_2j_1 1}-a_{j_1j_1 1})\left[(a_{j_2j_1 1}-a_{j_1j_1 2})^2+b_{j_1j_1}\right] > 0,$$

and for $a_{j_1j_1 0}(a_{j_2j_1 1}-a_{j_1j_1 1}) > 0$,

$$\dot{x}_{j_2} = a_{j_1j_1 0}(a_{j_2j_1 1}-a_{j_1j_1 1})\left[(a_{j_2j_1 1}-a_{j_1j_1 2})^2+b_{j_1j_1}\right] < 0.$$

Therefore, the flow at $x_{j_1}^* = a_{j_2j_1 1}$ for the x_{j_2}-direction has the following properties as in Eqs. (2.11)–(2.14).

- For $a_{j_1j_1 0}(a_{j_2j_1 1}-a_{j_1j_1 1}) > 0$ and $a_{j_2j_1 0} > 0$,

$$(\dot{x}_{j_1}, a_{j_2j_1 1}) = \underbrace{(\text{pF}, \text{UP})}_{\text{up-parabola flow }(+)}.$$

The flow at $x_{j_1}^* = a_{j_2j_1 1}$ is a (pF,UP)-positive up-parabola flow.

- For $a_{j_1j_1 0}(a_{j_2j_1 1}-a_{j_1j_1 1}) < 0$ and $a_{j_2j_1 0} > 0$,

$$(\dot{x}_{j_1}, a_{j_2j_1 1}) = \underbrace{(\text{nF}, \text{DP})}_{\text{down-parabola flow }(-)}.$$

The flow at $x_{j_1}^* = a_{j_2j_1 1}$ is an (nF,DP)-negative down-parabola flow.

- For $a_{j_1j_1 0}(a_{j_2j_1 1}-a_{j_1j_1 1}) > 0$ and $a_{j_2j_1 0} < 0$,

$$(\dot{x}_{j_1}, a_{j_2j_1 1}) = \underbrace{(\text{pF}, \text{DP})}_{\text{down-parabola flow }(+)}.$$

The flow at $x_{j_1}^* = a_{j_2j_1 1}$ is a (pF,DP)-positive down-parabola flow.

- For $a_{j_1j_1 0}(a_{j_2j_1 1}-a_{j_1j_1 1}) < 0$ and $a_{j_2j_1 0} < 0$,

$$(\dot{x}_{j_1}, a_{j_2j_1 1}) = \underbrace{(\text{nF}, \text{UP})}_{\text{up-parabola flow }(-)}.$$

The flow at $x_{j_1}^* = a_{j_2j_1 1}$ is an (nF,UP)-negative up-parabola flow.

(i_2) For $a_{j_1j_1 1} = a_{j_2j_1 1}$, at $\overline{x}_{j_1} \neq a_{j_1j_1 1} = a_{j_2j_1 1}$, in phase space,

$$\frac{dx_{j_2}}{dx_{j_1}} = \frac{a_{j_2j_1 0}}{a_{j_1j_1 0}} \frac{1}{(x_{j_1} - a_{j_1j_1 2})^2 + b_{j_1j_1}}$$

With an initial condition of $(x_{j_1 0}, x_{j_2 0})$ at $t = t_0$, the first integral manifold is

$$x_{j_2} - x_{j_2 0} = \frac{a_{j_2j_1 0}}{a_{j_1j_1 0}} \frac{1}{\sqrt{b_{j_1j_1}}} (\arctan \frac{x_{j_1} - a_{j_1j_1 2}}{\sqrt{b_{j_1j_1}}} - \arctan \frac{x_{j_1 0} - a_{j_1j_1 2}}{\sqrt{b_{j_1j_1}}}).$$

At $x_{j_1}^* = a_{j_1j_1 1} = a_{j_2j_1 1}$, if

$$\frac{dx_{j_2}}{dx_{j_1}}\Big|_{x_{j_1}^* = a_{j_1j_1 1} = a_{j_2j_1 1}} = \frac{a_{j_2j_1 0}}{a_{j_1j_1 0}} \frac{1}{(x_{j_1} - a_{j_1j_1 2})^2 + b_{j_1j_1}} > 0,$$

the flow is the first-order increasing-inflection flow, and if

$$\frac{dx_{j_2}}{dx_{j_1}}\Big|_{x_{j_1}^* = a_{j_1j_1 1} = a_{j_2j_1 1}} = \frac{a_{j_2j_1 0}}{a_{j_1j_1 0}} \frac{1}{(x_{j_1} - a_{j_1j_1 2})^2 + b_{j_1j_1}} < 0,$$

the flow is the first-order decreasing-inflection flow.

At $x_{j_1}^* = a_{j_1j_1 1} = a_{j_2j_1 1}$, the variational equations are

$$\begin{aligned} \Delta\dot{x}_{j_2} &= a_{j_2j_1 0}\Delta x_{j_1}, \\ \Delta\dot{x}_{j_1} &= a_{j_1j_1 0}\left[(a_{j_1j_1 1} - a_{j_1j_1 2})^2 + b_{j_1j_1}\right]\Delta x_{j_1}. \end{aligned}$$

The flows for $a_{j_1j_1 0} > 0$ and $a_{j_1j_1 0} < 0$ are source and sink flows, respectively.

Therefore, the equilibrium of $x_{j_1}^* = a_{j_1j_1 1} = a_{j_2j_1 1}$ with $\overline{x}_{j_2} \in (-\infty, \infty)$ has the following properties as in Eqs. (2.16)–(2.19).

- For $a_{j_1j_1 0} > 0$ and $a_{j_2j_1 0} > 0$,

$$(a_{j_1j_1 1}, \overline{x}_{j_2}) = \underbrace{(\text{SO}, \text{II})}_{\text{increasing-inflection source}}.$$

  The equilibrium of $x_{j_1}^* = a_{j_1j_1 1} = a_{j_2j_1 1}$ is an (SO,II)-increasing-inflection source infinite-equilibrium.
- For $a_{j_1j_1 0} < 0$ and $a_{j_2j_1 0} > 0$,

$$(a_{j_1j_1 1}, \overline{x}_{j_2}) = \underbrace{(\text{SI}, \text{DI})}_{\text{decreasing-inflection sink}}.$$

  The equilibrium of $x_{j_1}^* = a_{j_1j_1 1} = a_{j_2j_1 1}$ is an (SI,DI)-decreasing-inflection sink infinite-equilibrium.

- For $a_{j_1j_10} > 0$ and $a_{j_2j_10} < 0$,

$$(a_{j_1j_11}, \bar{x}_{j_2}) = \underbrace{(\text{SO}, \text{DI})}_{\text{decreasing-inflection source}}.$$

The equilibrium of $x_{j_1}^* = a_{j_1j_11} = a_{j_2j_11}$ is an (SO,DI)-decreasing-inflection source infinite-equilibrium.

- For $a_{j_1j_10} < 0$ and $a_{j_2j_10} < 0$,

$$(a_{j_1j_11}, \bar{x}_{j_2}) = \underbrace{(\text{SI}, \text{II})}_{\text{increasing-inflection sink}}.$$

The equilibrium of $x_{j_1}^* = a_{j_1j_11} = a_{j_2j_11}$ is an (SI,II)-increasing-inflection sink infinite-equilibrium.

(ii) If

$$A_{j_2j_11} \neq 0, A_{j_2j_12} \neq 0 \text{ and } \Delta_{j_1j_1} = 0,$$

the two-dimensional dynamical system becomes

$$\begin{aligned}\dot{x}_{j_2} &= a_{j_2j_10}(x_{j_1} - a_{j_2j_11}),\\ \dot{x}_{j_1} &= a_{j_1j_10}(x_{j_1} - a_{j_1j_1s_1})(x_{j_1} - a_{j_1j_1s_2})^2.\end{aligned}$$

(ii$_1$) For $a_{j_2j_11} \neq a_{j_1j_1s_1} \neq a_{j_1j_1s_2}(s_1, s_2 \in \{1, 2\})$,

$$\begin{aligned}\frac{dx_{j_2}}{dx_{j_1}} &= \frac{a_{j_2j_10}(x_{j_1} - a_{j_2j_11})}{a_{j_1j_10}(x_{j_1} - a_{j_1j_1s_1})(x_{j_1} - a_{j_1j_1s_2})^2}\\ &= \frac{a_{j_2j_10}}{a_{j_1j_10}}\Big\{\frac{1}{(x_{j_1} - a_{j_1j_1s_2})^2} + \frac{(a_{j_1j_1s_1} - a_{j_2j_11})}{(a_{j_1j_1s_1} - a_{j_1j_1s_2})^2}\\ &\quad\times\Big[\frac{1}{x_{j_1} - a_{j_1j_1s_1}} + \frac{(a_{j_1j_1s_2} - a_{j_1j_1s_1})}{(x_{j_1} - a_{j_1j_1s_2})^2} - \frac{1}{x_{j_1} - a_{j_1j_1s_2}}\Big]\Big\}.\end{aligned}$$

With an initial condition of (x_{j_10}, x_{j_20}), the first integral manifold is

$$\begin{aligned}x_{j_2} - x_{j_20} = \frac{a_{j_2j_10}}{a_{j_1j_10}}\Big[&-\frac{a_{j_1j_1s_2} - a_{j_2j_11}}{a_{j_1j_1s_2} - a_{j_1j_1s_1}}\Big(\frac{1}{x_{j_1} - a_{j_1j_1s_2}} - \frac{1}{x_{j_10} - a_{j_1j_1s_2}}\Big)\\ &+\frac{(a_{j_1j_1s_1} - a_{j_2j_11})}{(a_{j_1j_1s_2} - a_{j_1j_1s_1})^2}\Big(\ln\frac{|x_{j_1} - a_{j_1j_1s_1}|}{|x_{j_10} - a_{j_1j_1s_1}|} - \ln\frac{|x_{j_1} - a_{j_1j_1s_2}|}{|x_{j_10} - a_{j_1j_1s_2}|}\Big)\Big].\end{aligned}$$

(ii$_{1a}$) At $x_{j_1}^* = a_{j_1 j_1 s_1}$, the variational equation is

$$\begin{aligned}\dot{x}_{j_2} &= a_{j_2 j_1 0}(a_{j_1 j_1 s_1} - a_{j_2 j_1 1}),\\ \Delta\dot{x}_{j_1} &= a_{j_1 j_1 0}(a_{j_1 j_1 s_1} - a_{j_1 j_1 s_2})^2 \Delta x_{j_1}.\end{aligned}$$

The flows at $x_{j_1}^* = a_{j_1 j_1 s_1}$ are source and sink flows for $a_{j_1 j_1 0} > 0$ and $a_{j_1 j_1 0} < 0$ in the x_{j_1}-direction, and the flows are positive and negative in the x_{j_2}-direction for $a_{j_2 j_1 0}(a_{j_1 j_1 s_1} - a_{j_2 j_1 1}) > 0$ and $a_{j_2 j_1 0}(a_{j_1 j_1 s_1} - a_{j_2 j_1 1}) < 0$.

Therefore, the equilibrium of $x_{j_1}^* = a_{j_1 j_1 s_1}$ possesses the following properties as in Eqs. (2.23)–(2.26).

- For $a_{j_1 j_1 0}(a_{j_1 j_1 s_1} - a_{j_1 j_1 s_2})^2 > 0$ and $a_{j_2 j_1 0}(a_{j_1 j_1 s_1} - a_{j_2 j_1 1}) > 0$,

$$(a_{j_1 j_1 s_1}, \dot{x}_{j_2}) = \underbrace{(\text{SO}, \text{pF})}_{\text{source flow } (+)}.$$

The flow at $x_{j_1}^* = a_{j_1 j_1 s_1}$ is an (SO,pF)-positive source flow.

- For $a_{j_1 j_1 0}(a_{j_1 j_1 s_1} - a_{j_1 j_1 s_2})^2 < 0$ and $a_{j_2 j_1 0}(a_{j_1 j_1 s_1} - a_{j_2 j_1 1}) > 0$,

$$(a_{j_1 j_1 s_1}, \dot{x}_{j_2}) = \underbrace{(\text{SI}, \text{pF})}_{\text{sink flow } (+)}.$$

The flow at $x_{j_1}^* = a_{j_1 j_1 s_1}$ is an (SI,pF)-positive sink flow.

- For $a_{j_1 j_1 0}(a_{j_1 j_1 s_1} - a_{j_1 j_1 s_2})^2 > 0$ and $a_{j_2 j_1 0}(a_{j_1 j_1 s_1} - a_{j_2 j_1 1}) < 0$,

$$(a_{j_1 j_1 s_1}, \dot{x}_{j_2}) = \underbrace{(\text{SO}, \text{nF})}_{\text{source flow } (-)}.$$

The flow at $x_{j_1}^* = a_{j_1 j_1 s_1}$ is an (SO,nF)-negative source flow.

- For $a_{j_1 j_1 0}(a_{j_1 j_1 s_1} - a_{j_1 j_1 s_2})^2 < 0$ and $a_{j_2 j_1 0}(a_{j_1 j_1 s_1} - a_{j_2 j_1 1}) < 0$,

$$(a_{j_1 j_1 s_1}, \dot{x}_{j_2}) = \underbrace{(\text{SI}, \text{nF})}_{\text{sink flow } (-)}.$$

The flow at $x_{j_1}^* = a_{j_1 j_1 s_1}$ is an (SI,nF)-negative sink flow.

(ii$_{1b}$) At $x_{j_1}^* = a_{j_1 j_1 s_2}$, the variational equation is

$$\begin{aligned}\dot{x}_{j_2} &= a_{j_2 j_1 0}(a_{j_1 j_1 s_2} - a_{j_2 j_1 1}),\\ \Delta\dot{x}_{j_1} &= a_{j_1 j_1 0}(a_{j_1 j_1 s_2} - a_{j_1 j_1 s_1})(\Delta x_{j_1})^2.\end{aligned}$$

The flows at $x_{j_1}^* = a_{j_1 j_1 s_2}$ are upper-saddle and lower-saddle flows for $a_{j_1 j_1 0}(a_{j_1 j_1 s_2} - a_{j_1 j_1 s_1}) > 0$ and $a_{j_1 j_1 0}(a_{j_1 j_1 s_2} - a_{j_1 j_1 s_1}) < 0$ in the x_{j_1}-direction, and the flows are positive and negative in the x_{j_2}-direction for $a_{j_2 j_1 0}(a_{j_1 j_1 s_2} - a_{j_2 j_1 1}) > 0$ and $a_{j_2 j_1 0}(a_{j_1 j_1 s_2} - a_{j_2 j_1 1}) < 0$.

Therefore, the equilibrium of $x_{j_1}^* = a_{j_1 j_1 s_2}$ possesses the following properties as in Eqs. (2.27)–(2.30).

- For $a_{j_1 j_1 0}(a_{j_1 j_1 s_2} - a_{j_1 j_1 s_1}) > 0$ and $a_{j_2 j_1 0}(a_{j_1 j_1 s_2} - a_{j_2 j_1 1}) > 0$,

$$(a_{j_1 j_1 s_2}, \dot{x}_{j_2}) = \underbrace{(\text{US}, \text{pF})}_{\text{upper-saddle flow } (+)}.$$

  The flow at $x_{j_1}^* = a_{j_1 j_1 s_2}$ is a (US,pF)-positive upper-saddle flow.
- For $a_{j_1 j_1 0}(a_{j_1 j_1 s_2} - a_{j_1 j_1 s_1}) < 0$ and $a_{j_2 j_1 0}(a_{j_1 j_1 s_2} - a_{j_2 j_1 1}) > 0$,

$$(a_{j_1 j_1 s_2}, \dot{x}_{j_2}) = \underbrace{(\text{LS}, \text{pF})}_{\text{lower-saddle flow } (+)}.$$

  The flow at $x_{j_1}^* = a_{j_1 j_1 s_2}$ is an (LS,pF)-positive lower-saddle flow.
- For $a_{j_1 j_1 0}(a_{j_1 j_1 s_2} - a_{j_1 j_1 s_1}) > 0$ and $a_{j_2 j_1 0}(a_{j_1 j_1 s_2} - a_{j_2 j_1 1}) < 0$,

$$(a_{j_1 j_1 s_2}, \dot{x}_{j_2}) = \underbrace{(\text{US}, \text{nF})}_{\text{upper-saddle flow } (-)}.$$

  The flow at $x_{j_1}^* = a_{j_1 j_1 s_2}$ is a (US,nF)-negative upper-saddle flow.
- For $a_{j_1 j_1 0}(a_{j_1 j_1 s_2} - a_{j_1 j_1 s_1}) < 0$ and $a_{j_2 j_1 0}(a_{j_1 j_1 s_2} - a_{j_2 j_1 1}) < 0$,

$$(a_{j_1 j_1 s_2}, \dot{x}_{j_2}) = \underbrace{(\text{LS}, \text{nF})}_{\text{lower-saddle flow } (-)}.$$

  The flow at $x_{j_1}^* = a_{j_1 j_1 s_2}$ is an (LS,nF)-negative lower-saddle flow.

(ii$_{1c}$) At $x_{j_1}^* = a_{j_2 j_1 1}$ and $\bar{x}_{j_1} \neq a_{j_1 j_1 s_1} \neq a_{j_1 j_1 s_2}$, in phase space, we have

$$\frac{dx_{j_2}}{dx_{j_1}}\bigg|_{x_{j_1}^* = a_{j_2 j_1 1}} = \frac{a_{j_2 j_1 0}(x_{j_1} - a_{j_2 j_1 1})}{a_{j_1 j_1 0}(x_{j_1} - a_{j_1 j_1 s_1})(x_{j_1} - a_{j_1 j_1 s_2})^2}\bigg|_{x_{j_1}^* = a_{j_2 j_1 1}} = 0.$$

If

$$\frac{d^2 x_{j_2}}{dx_{j_1}^2}\bigg|_{x_{j_1}^* = a_{j_2 j_1 1}} = \frac{a_{j_2 j_1 0}}{a_{j_1 j_1 0}(a_{j_2 j_1 1} - a_{j_1 j_1 s_1})(a_{j_2 j_1 1} - a_{j_1 j_1 s_2})^2} > 0,$$

there is an up-parabola flow in the x_{j_2}-direction, and if

$$\frac{d^2 x_{j_2}}{dx_{j_1}^2}\bigg|_{x_{j_1}^*=a_{j_2 j_1 1}} = \frac{a_{j_2 j_1 0}}{a_{j_1 j_1 0}(a_{j_2 j_1 1} - a_{j_1 j_1 s_1})(a_{j_2 j_1 1} - a_{j_1 j_1 s_2})^2} < 0,$$

there is a down-parabola flow in the x_{j_2}-direction.

Because

$$\dot{x}_{j_1} = a_{j_1 j_1 0}(a_{j_2 j_1 1} - a_{j_1 j_1 s_1})(a_{j_2 j_1 1} - a_{j_1 j_1 s_2})^2,$$

the flows at $x_{j_1}^* = a_{j_2 j_1 1}$ are positive and negative in the x_{j_1}-direction for $a_{j_1 j_1 0}(a_{j_2 j_1 1} - a_{j_1 j_1 s_1}) > 0$ and $a_{j_1 j_1 0}(a_{j_2 j_1 1} - a_{j_1 j_1 s_1}) < 0$.

Therefore, the equilibrium of $x_{j_1}^* = a_{j_2 j_1 1}$ for the x_{j_2}-direction possesses the following properties as in Eqs. (2.31)–(2.34).

- For $a_{j_1 j_1 0}(a_{j_2 j_1 1} - a_{j_1 j_1 s_1})(a_{j_2 j_1 1} - a_{j_1 j_1 s_2})^2 > 0$ and $a_{j_2 j_1 0} > 0$,

$$(\dot{x}_{j_1}, a_{j_2 j_1 1}) = \underbrace{(\text{pF}, \text{UP})}_{\text{up-parabola flow }(+)}.$$

The flow at $x_{j_1}^* = a_{j_2 j_1 1}$ is a (pF,UP)-positive up-parabola flow.

- For $a_{j_1 j_1 0}(a_{j_2 j_1 1} - a_{j_1 j_1 s_1})(a_{j_2 j_1 1} - a_{j_1 j_1 s_2})^2 < 0$ and $a_{j_2 j_1 0} > 0$,

$$(\dot{x}_{j_1}, a_{j_2 j_1 1}) = \underbrace{(\text{nF}, \text{DP})}_{\text{down-parabola flow }(-)}.$$

The flow at $x_{j_1}^* = a_{j_2 j_1 1}$ is an (nF,DP)-negative down-parabola flow.

- For $a_{j_1 j_1 0}(a_{j_2 j_1 1} - a_{j_1 j_1 s_1})(a_{j_2 j_1 1} - a_{j_1 j_1 s_2})^2 > 0$ and $a_{j_2 j_1 0} < 0$,

$$(\dot{x}_{j_1}, a_{j_2 j_1 1}) = \underbrace{(\text{pF}, \text{DP})}_{\text{down-parabola flow }(+)}.$$

The flow at $x_{j_1}^* = a_{j_2 j_1 1}$ is a (pF,DP)-parabola down-parabola flow.

- For $a_{j_1 j_1 0}(a_{j_2 j_1 1} - a_{j_1 j_1 s_1})(a_{j_2 j_1 1} - a_{j_1 j_1 s_2})^2 < 0$ and $a_{j_2 j_1 0} < 0$,

$$(\dot{x}_{j_1}, a_{j_2 j_1 1}) = \underbrace{(\text{nF}, \text{UP})}_{\text{up-parabola flow }(-)}.$$

The flow at $x_{j_1}^* = a_{j_2 j_1 1}$ is an (nF,UP)-negative up-parabola flow.

(ii$_2$) For $a_{j_2j_11}=a_{j_1j_1s_1}\neq a_{j_1j_1s_2}(s_1,\, s_2\in\{1,2\})$,

$$\frac{dx_{j_2}}{dx_{j_1}}=\frac{a_{j_2j_10}}{a_{j_1j_10}(x_{j_1}-a_{j_1j_1s_2})^2}.$$

With an initial condition of (x_{j_10},x_{j_20}), the first integral manifold is

$$x_{j_2}-x_{j_20}=-\frac{a_{j_2j_10}}{a_{j_1j_10}}(\frac{1}{x_{j_1}-a_{j_1j_1s_2}}-\frac{1}{x_{j_10}-a_{j_1j_1s_2}}).$$

At $x_{j_1}^*=a_{j_1j_1s_1}=a_{j_2j_11}$, if

$$\left.\frac{dx_{j_2}}{dx_{j_1}}\right|_{x_{j_1}^*=a_{j_1j_1s_1}=a_{j_2j_11}}=\frac{a_{j_2j_10}}{a_{j_1j_10}(a_{j_1j_1s_1}-a_{j_1j_1s_2})^2}>0,$$

the flow at $x_{j_1}^*=a_{j_1j_1s_1}=a_{j_2j_11}$ is an increasing-inflection flow, and if

$$\left.\frac{dx_{j_2}}{dx_{j_1}}\right|_{x_{j_1}^*=a_{j_1j_1s_1}=a_{j_2j_11}}=\frac{a_{j_2j_10}}{a_{j_1j_10}(a_{j_1j_1s_1}-a_{j_1j_1s_2})^2}<0,$$

the flow at $x_{j_1}^*=a_{j_1j_1s_1}=a_{j_2j_11}$ is a decreasing-inflection flow.

At $x_{j_1}^*=a_{j_1j_1s_1}=a_{j_2j_11}$, the variational equations are

$$\begin{aligned}\Delta\dot{x}_{j_2}&=a_{j_2j_10}\Delta x_{j_1},\\ \Delta\dot{x}_{j_1}&=a_{j_1j_10}(a_{j_1j_1s_1}-a_{j_1j_1s_2})^2\Delta x_{j_1}.\end{aligned}$$

The flows at $x_{j_1}^*=a_{j_1j_1s_1}=a_{j_2j_11}$ are source and sink flows for $a_{j_1j_10}>0$ and $a_{j_1j_10}<0$ in the direction. From the first integral manifold, the flows for $a_{j_2j_10}>0$ and $a_{j_2j_10}<0$ are down-to-up and up-to-down flow in the x_{j_2}-direction. The singularity is of the first order in the x_{j_1} and x_{j_2}-directions. The flows are defined as the (1:1) increasing inflection and (1:1) decreasing inflection flows.

Therefore, the equilibrium of $x_{j_1}^*=a_{j_1j_1s_1}=a_{j_2j_11}$ with $\bar{x}_{j_2}\in(-\infty,\infty)$ has the following properties as in Eqs. (2.36)–(2.39).

- For $a_{j_1j_10}(a_{j_1j_1s_1}-a_{j_1j_1s_2})^2>0$ and $a_{j_2j_10}>0$,

$$(a_{j_1j_1s_1},\bar{x}_{j_2})=\underbrace{(\text{SO},\text{II})}_{\text{increasing-inflection source}}.$$

The equilibrium of $x_{j_1}^*=a_{j_1j_1s_1}=a_{j_2j_11}$ is an (SO,II)-increasing-inflection source infinite-equilibrium.

- For $a_{j_1j_10}(a_{j_1j_1s_1}-a_{j_1j_1s_2})^2<0$ and $a_{j_2j_10}>0$,

$$(a_{j_1j_1s_1},\bar{x}_{j_2})=\underbrace{(\text{SI, DI})}_{\text{decreasing-inflection sink}}.$$

The equilibrium of $x^*_{j_1}=a_{j_1j_1s_1}=a_{j_2j_11}$ is an (SI,DI)-decreasing-inflection sink infinite-equilibrium.

- For $a_{j_1j_10}(a_{j_1j_1s_1}-a_{j_1j_1s_2})^2>0$ and $a_{j_2j_10}<0$,

$$(a_{j_1j_1s_1},\bar{x}_{j_2})=\underbrace{(\text{SO, DI})}_{\text{decreasing-inflection source}}.$$

The equilibrium of $x^*_{j_1}=a_{j_1j_1s_1}=a_{j_2j_11}$ is an (SO,DI)-decreasing-inflection source infinite-equilibrium.

- For $a_{j_1j_10}(a_{j_1j_1s_1}-a_{j_1j_1s_2})^2<0$ and $a_{j_2j_10}<0$,

$$(a_{j_1j_1s_1},\bar{x}_{j_2})=\underbrace{(\text{SI, II})}_{\text{increasing-inflection sink}}.$$

The equilibrium of $x^*_{j_1}=a_{j_1j_1s_1}=a_{j_2j_11}$ is an (SI,II)-increasing-inflection sink infinite-equilibrium.

– The inflection infinite-equilibriums are the switching bifurcations of parabola flows with source (sink) flows.

(ii$_3$) For $a_{j_2j_11}=a_{j_1j_1s_2}\neq a_{j_1j_1s_1}$, in phase space,

$$\begin{aligned}\frac{dx_{j_2}}{dx_{j_1}}&=\frac{a_{j_2j_10}}{a_{j_1j_10}(x_{j_1}-a_{j_1j_1s_1})(x_{j_1}-a_{j_1j_1s_2})}\\&=\frac{a_{j_2j_10}}{a_{j_1j_10}}\frac{1}{a_{j_1j_1s_1}-a_{j_1j_1s_2}}\left(\frac{1}{x_{j_1}-a_{j_1j_1s_1}}-\frac{1}{x_{j_1}-a_{j_1j_1s_2}}\right).\end{aligned}$$

With an initial condition of (x_{j_10},x_{j_20}), the first integral manifold is

$$x_{j_2}-x_{j_20}=\frac{a_{j_2j_10}}{a_{j_1j_10}}\frac{1}{a_{j_1j_1s_1}-a_{j_1j_1s_2}}\left(\ln\frac{|x_{j_1}-a_{j_1j_1s_1}|}{|x_{j_10}-a_{j_1j_1s_1}|}-\ln\frac{|x_{j_1}-a_{j_1j_1s_2}|}{|x_{j_10}-a_{j_1j_1s_2}|}\right).$$

At $x^*_{j_1}=a_{j_1j_1s_2}=a_{j_2j_11}$, the variational equations are

$$\begin{aligned}\Delta\dot{x}_{j_2}&=a_{j_2j_10}\Delta x_{j_1},\\\Delta\dot{x}_{j_1}&=a_{j_1j_10}(a_{j_1j_1s_2}-a_{j_1j_1s_1})(\Delta x_{j_1})^2.\end{aligned}$$

The flows at $x^*_{j_1}=a_{j_1j_1s_2}=a_{j_2j_11}$ are upper-saddle and lower-saddle flows in the x_{j_1}-direction for $a_{j_1j_10}(a_{j_1j_1s_2}-a_{j_1j_1s_1})>0$ and $a_{j_1j_10}(a_{j_1j_1s_2}-a_{j_1j_1s_1})<0$. From the first integral manifold, the flows for $a_{j_2j_10} > 0$ and $a_{j_2j_10} < 0$ are down-up and up-down flow in the x_{j_2}-direction.

Therefore, the infinite-equilibrium of $x_{j_1}^* = a_{j_1j_1s_2} = a_{j_2j_11}$ with $\bar{x}_{j_2} \in (-\infty, \infty)$ has the following properties as in Eqs. (2.41)–(2.44).

- For $a_{j_1j_10}(a_{j_1j_1s_2} - a_{j_1j_1s_1}) > 0$ and $a_{j_2j_10} > 0$,

$$(a_{j_1j_1s_2}, \bar{x}_{j_2}) = \underbrace{(\mathrm{US}, \mathrm{DU})}_{\text{down-up asymptoic upper-saddle}}.$$

  The equilibrium of $x_{j_1}^* = a_{j_1j_1s_2} = a_{j_2j_11}$ is a (US,DU)-down-up asymptotic upper-saddle infinite-equilibrium.
- For $a_{j_1j_10}(a_{j_1j_1s_2} - a_{j_1j_1s_1}) < 0$ and $a_{j_2j_10} > 0$,

$$(a_{j_1j_1s_2}, \bar{x}_{j_2}) = \underbrace{(\mathrm{LS}, \mathrm{UD})}_{\text{up-down asymptotic lower-saddle}}.$$

  The equilibrium of $x_{j_1}^* = a_{j_1j_1s_2} = a_{j_2j_11}$ is an (LS,UD)-up-down asymptotic lower-saddle infinite-equilibrium.
- For $a_{j_1j_10}(a_{j_1j_1s_2} - a_{j_1j_1s_1}) > 0$ and $a_{j_2j_10} < 0$,

$$(a_{j_1j_1s_2}, \bar{x}_{j_2}) = \underbrace{(\mathrm{US}, \mathrm{UD})}_{\text{up-down asymptotic upper-saddle}}.$$

  The equilibrium of $x_{j_1}^* = a_{j_1j_1s_2} = a_{j_2j_11}$ is a (US,UD)-up-down asymptotic upper-saddle infinite-equilibrium.
- For $a_{j_1j_10}(a_{j_1j_1s_2} - a_{j_1j_1s_1}) < 0$ and $a_{j_2j_10} < 0$,

$$(a_{j_1j_1s_2}, \bar{x}_{j_2}) = \underbrace{(\mathrm{LS}, \mathrm{DU})}_{\text{down-up asymptotic lower-saddle}}.$$

  The equilibrium of $x_{j_1}^* = a_{j_1j_1s_2} = a_{j_2j_11}$ is an (LS,DU)-down-up asymptotic lower-saddle infinite-equilibrium.

The down-up and up-down saddle infinite-equilibriums are the switching bifurcations of parabola flows with upper-saddle (lower-saddle) flows.

(ii$_4$) For $a_{j_2j_11} \neq a_{j_1j_1s_1} = a_{j_1j_1s_2} \equiv a_{j_1j_11}(s_1, s_2 \in \{1, 2\})$, the two-dimensional dynamical system becomes

$$\begin{aligned}\dot{x}_{j_2} &= a_{j_2j_10}(x_{j_1} - a_{j_2j_11}),\\ \dot{x}_{j_1} &= a_{j_1j_10}(x_{j_1} - a_{j_1j_11})^3.\end{aligned}$$

In phase space,

$$\begin{aligned}\frac{dx_{j_2}}{dx_{j_1}} &= \frac{a_{j_2j_10}(x_{j_1}-a_{j_2j_11})}{a_{j_1j_10}(x_{j_1}-a_{j_1j_11})^3},\\ &= \frac{a_{j_2j_10}}{a_{j_1j_10}}\Big[\frac{1}{(x_{j_1}-a_{j_1j_11})^2}+\frac{(a_{j_1j_11}-a_{j_2j_11})}{(x_{j_1}-a_{j_1j_11})^3}\Big].\end{aligned}$$

With an initial condition of (x_{j_10}, x_{j_20}), the first integral manifold is

$$\begin{aligned}x_{j_2}-x_{j_20} = &-\frac{a_{j_2j_10}}{a_{j_1j_10}}\Big\{\Big(\frac{1}{x_{j_1}-a_{j_1j_11}}-\frac{1}{x_{j_10}-a_{j_1j_11}}\Big)\\ &+\frac{1}{2}(a_{j_1j_11}-a_{j_2j_11})\Big[\frac{1}{(x_{j_1}-a_{j_1j_11})^2}-\frac{1}{(x_{j_10}-a_{j_1j_11})^2}\Big]\Big\}.\end{aligned}$$

(ii$_{4a}$) At $x^*_{j_1}=a_{j_2j_11}\neq a_{j_1j_11}$, we have

$$\frac{dx_{j_2}}{dx_{j_1}}\Big|_{x^*_{j_1}=a_{j_2j_11}} = \frac{a_{j_2j_10}(x_{j_1}-a_{j_2j_11})}{a_{j_1j_10}(x_{j_1}-a_{j_1j_11})^3}\Big|_{x^*_{j_1}=a_{j_2j_11}}=0.$$

If

$$\frac{d^2x_{j_2}}{dx^2_{j_1}}\Big|_{x^*_{j_1}=a_{j_2j_11}} = \frac{a_{j_2j_10}}{a_{j_1j_10}(a_{j_2j_11}-a_{j_1j_11})^3}>0,$$

there is an up-parabola flow in x_{j_2}-direction, and if

$$\frac{d^2x_{j_2}}{dx^2_{j_1}}\Big|_{x^*_{j_1}=a_{j_2j_11}} = \frac{a_{j_2j_10}}{a_{j_1j_10}(a_{j_2j_11}-a_{j_1j_11})^3}<0,$$

there is a down-parabola flow in x_{j_2}-direction.

At $x^*_{j_1}=a_{j_2j_11}\neq a_{j_1j_11}$, the variational equation is

$$\begin{aligned}\Delta\dot{x}_{j_2} &= a_{j_2j_10}\Delta x_{j_1},\\ \dot{x}_{j_1} &= a_{j_1j_10}(a_{j_2j_11}-a_{j_1j_11})^3.\end{aligned}$$

The flows at $x^*_{j_1}=a_{j_2j_11}\neq a_{j_1j_11}$ are positive and negative flow for $\dot{x}_{j_1}>0$ and $\dot{x}_{j_1}<0$.

Therefore, the equilibrium of $x^*_{j_1}=a_{j_2j_11}\neq a_{j_1j_11}$ in the x_{j_2}-direction has the following properties as in Eqs. (2.47)–(2.50).

- For $a_{j_1j_10}(a_{j_2j_11}-a_{j_1j_11})^3>0$ and $a_{j_2j_10}>0$,

$$(\dot{x}_{j_1}, a_{j_2j_11}) = \underbrace{(\text{pF}, \text{UP})}_{\text{up-parabola flow } (+)}.$$

The flow at $x^*_{j_1}=a_{j_2j_11}\neq a_{j_1j_11}$ is a (pF,UP)-positive up-parabola flow.

- For $a_{j_1j_1 0}(a_{j_2j_1 1}-a_{j_1j_1 1})^3<0$ and $a_{j_2j_1 0}>0$,

$$(\dot{x}_{j_1},a_{j_2j_1 1})=\underbrace{(\text{nF},\text{DP})}_{\text{down-parabola flow }(-)}.$$

The flow at $x_{j_1}^*=a_{j_2j_1 1}\neq a_{j_1j_1 1}$ is an (nF,DP)-negative down-parabola flow.

- For $a_{j_1j_1 0}(a_{j_2j_1 1}-a_{j_1j_1 1})^3>0$ and $a_{j_2j_1 0}<0$,

$$(\dot{x}_{j_1},a_{j_2j_1 1})=\underbrace{(\text{pF},\text{DP})}_{\text{down-parabola flow }(+)}.$$

The flow at $x_{j_1}^*=a_{j_2j_1 1}\neq a_{j_1j_1 1}$ is a (pF,DP)-positive down-parabola flow.

- For $a_{j_1j_1 0}(a_{j_2j_1 1}-a_{j_1j_1 1})^3<0$ and $a_{j_2j_1 0}<0$,

$$(\dot{x}_{j_1},a_{j_2j_1 1})=\underbrace{(\text{nF},\text{UP})}_{\text{up-parabola flow }(-)}.$$

The flow at $x_{j_1}^*=a_{j_2j_1 1}\neq a_{j_1j_1 1}$ is an (nF,UP)-negative up-parabola flow.

($\text{ii}_{4\text{b}}$) At $x_{j_1}^*=a_{j_1j_1 1}\neq a_{j_2j_1 1}$, the variational equation is

$$\begin{aligned}\dot{x}_{j_2}&=a_{j_2j_1 0}(a_{j_1j_1 1}-a_{j_2j_1 1}),\\ \Delta\dot{x}_{j_1}&=a_{j_1j_1 0}(\Delta x_{j_1})^3.\end{aligned}$$

The flows at $x_{j_1}^*=a_{j_1j_1 1}\neq a_{j_2j_1 1}$ in the x_{j_1}-direction are the third-order source and sink flows for $a_{j_1j_1 0}>0$ and $a_{j_1j_1 0}<0$. In addition, the flows at $x_{j_1}^*=a_{j_1j_1 1}\neq a_{j_2j_1 1}$ in the x_{j_2}-direction are positive and negative for $a_{j_2j_1 0}(a_{j_1j_1 1}-a_{j_2j_1 1})>0$ and $a_{j_2j_1 0}\ (a_{j_1j_1 1}-a_{j_2j_1 1})<0$.

Thus, the equilibrium of $x_{j_1}^*=a_{j_1j_1 1}\neq a_{j_2j_1 1}$ in the x_{j_1}-direction has the following properties as in Eqs. (2.51)–(2.54).

- For $a_{j_1j_1 0}>0$ and $a_{j_2j_1 0}(a_{j_1j_1 1}-a_{j_2j_1 1})>0$,

$$(a_{j_1j_1 1},\dot{x}_{j_2})=\underbrace{(3^{\text{rd}}\text{SO},\text{pF})}_{\text{third-order source flow }(+)}.$$

The flow at $x_{j_1}^*=a_{j_1j_1 1}\neq a_{j_2j_1 1}$ is a (3^{rd}SO,pF)-positive third-order source flow.

- For $a_{j_1j_1 0}<0$ and $a_{j_2j_1 0}(a_{j_1j_1 1}-a_{j_2j_1 1})>0$,

$$(a_{j_1j_1 1},\dot{x}_{j_2})=\underbrace{(3^{\text{rd}}\text{SI},\text{pF})}_{\text{third-order sink flow }(+)}.$$

The flow at $x_{j_1}^*=a_{j_1j_1 1}\neq a_{j_2j_1 1}$ is a (3^{rd}SI,pF)-positive third-order sink flow.

- For $a_{j_1j_10} > 0$ and $a_{j_2j_10}(a_{j_1j_11} - a_{j_2j_11}) < 0$,

$$(a_{j_1j_11}, \dot{x}_{j_2}) = \underbrace{(3^{\text{rd}}\text{SO}, \text{nF})}_{\text{third-order source flow }(-)}.$$

The flow at $x_{j_1}^* = a_{j_1j_11} \neq a_{j_2j_11}$ is a (3rdSO,nF)-negative third-order source flow.

- For $a_{j_1j_10} < 0$ and $a_{j_2j_10}(a_{j_1j_11} - a_{j_2j_11}) < 0$,

$$(a_{j_1j_11}, \dot{x}_{j_2}) = \underbrace{(3^{\text{rd}}\text{SI}, \text{nF})}_{\text{third-order sink flow }(-)}.$$

The flow at $x_{j_1}^* = a_{j_1j_11} \neq a_{j_2j_11}$ is a (3rdSI,nF)-negative third-order sink flow.

(ii$_{4c}$) For $a_{j_2j_11} = a_{j_1j_11} \equiv a_{j_1j_1s_1} = a_{j_1j_1s_2}$, in phase space,

$$\frac{dx_{j_2}}{dx_{j_1}} = \frac{a_{j_2j_10}}{a_{j_1j_10}} \frac{1}{(x_{j_1} - a_{j_1j_11})^2}.$$

With an initial condition of (x_{j_10}, x_{j_20}), the first integral manifold is

$$x_{j_2} - x_{j_20} = -\frac{a_{j_2j_10}}{a_{j_1j_10}} (\frac{1}{x_{j_1} - a_{j_1j_11}} - \frac{1}{x_{j_10} - a_{j_1j_11}}).$$

At $x_{j_1}^* = a_{j_1j_11} = a_{j_2j_11}$, the variational equations are

$$\begin{aligned} \Delta\dot{x}_{j_2} &= a_{j_2j_10}\Delta x_{j_1}, \\ \Delta\dot{x}_{j_1} &= a_{j_1j_10}(\Delta x_{j_1})^3. \end{aligned}$$

The flows at $x_{j_1}^* = a_{j_1j_11} = a_{j_2j_11}$ are the third-order source and sink flows in the x_{j_1}-direction for $a_{j_1j_10} > 0$ and $a_{j_1j_10} < 0$. From the first integral manifold, the flows for $a_{j_2j_10} > 0$ and $a_{j_2j_10} < 0$ are down-to-up and up-to-down flow in the x_{j_2}-direction. The singularity is of the third and first order in the x_{j_1} and x_{j_2}-directions. The flows are defined as the (3:1) increasing-concave and (3:1) decreasing-concave asymptotic flows.

Therefore, the equilibrium of $x_{j_1}^* = a_{j_1j_11} = a_{j_2j_11}$ with $\bar{x}_{j_2} \in (-\infty, \infty)$ has the following properties as in Eqs. (2.56)–(2.59).

- For $a_{j_1j_10} > 0$ and $a_{j_2j_10} > 0$,

$$(a_{j_1j_11}, \bar{x}_{j_2}) = \underbrace{(3^{\text{rd}}\text{SO}, (3:1)\text{IC})}_{\text{third-order increasing-concave source}}.$$

The equilibrium of $x_{j_1}^* = a_{j_1j_1 1} = a_{j_2j_1 1}$ is an (3[rd]SO, (3:1)IC)-third-order increasing-concave source infinite-equilibrium.

- For $a_{j_1j_1 0} < 0$ and $a_{j_2j_1 0} > 0$,

$$(a_{j_1j_1 1}, \bar{x}_{j_2}) = \underbrace{(3^{\text{rd}}\text{SI}, (3:1)\text{DC})}_{\text{third-order decreasing-concave sink}}.$$

The equilibrium of $x_{j_1}^* = a_{j_1j_1 1} = a_{j_2j_1 1}$ is an (3[rd]SI, (3:1)DC)-third-order decreasing-concave sink infinite-equilibrium.

- For $a_{j_1j_1 0} > 0$ and $a_{j_2j_1 0} < 0$,

$$(a_{j_1j_1 1}, \bar{x}_{j_2}) = \underbrace{(3^{\text{rd}}\text{SO}, (3:1)\text{DC})}_{\text{third-order decreasing-concave source}}.$$

The equilibrium of $x_{j_1}^* = a_{j_1j_1 1} = a_{j_2j_1 1}$ is an (3[rd]SO, (3:1)DC)-third-order decreasing-concave source infinite-equilibrium.

- For $a_{j_1j_1 0} < 0$ and $a_{j_2j_1 0} < 0$,

$$(a_{j_1j_1 1}, \bar{x}_{j_2}) = \underbrace{(3^{\text{rd}}\text{SI}, (3:1)\text{IC})}_{\text{third-order increasing-concave sink}}.$$

The equilibrium of $x_{j_1}^* = a_{j_1j_1 1} = a_{j_2j_1 1}$ is a (3[rd]SI,(3:1)IC)-third-order increasing-concave sink infinite-equilibrium.

(iii) If

$$A_{j_2j_1 1} \neq 0, A_{j_2j_1 2} \neq 0 \text{ and } \Delta_{j_1j_1} > 0,$$

the two-dimensional dynamical system becomes

$$\begin{aligned}\dot{x}_{j_2} &= a_{j_2j_1 0}(x_{j_1} - a_{j_2j_1 1}),\\ \dot{x}_{j_1} &= a_{j_1j_1 0}(x_{j_1} - a_{j_1j_1 1})(x_{j_1} - a_{j_1j_1 2})(x_{j_1} - a_{j_1j_1 3})\end{aligned}$$

where

$$\begin{aligned}&\{a_{j_1j_1 1}, a_{j_1j_1 2}, a_{j_1j_1 3}\} = \text{sort}\{b_{j_1j_1 1}, b_{j_1j_1 2}, b_{j_1j_1 3}\}, a_{j_1j_1 1} < a_{j_1j_1 2} < a_{j_1j_1 3};\\ &b_{j_1j_1 2}, b_{j_1j_1 3} = -\frac{1}{2}(B_{j_1j_1 1} \pm \sqrt{\Delta_{j_1j_1}}), a_{j_1j_1 0} = A_{j_1j_1 0}.\end{aligned}$$

(iii$_1$) For $a_{j_1j_1s_1} \neq a_{j_2j_1 1}$ (s_1, $s_2 = 1, 2, 3$ and $s_1 \neq s_2$),

$$\begin{aligned}\frac{dx_{j_2}}{dx_{j_1}} &= \frac{a_{j_2j_1 0}(x_{j_1} - a_{j_2j_1 1})}{a_{j_1j_1 0}(x_{j_1} - a_{j_1j_1s_1})(x_{j_1} - a_{j_1j_1s_2})(x_{j_1} - a_{j_1j_1s_3})}\\ &= \frac{a_{j_2j_1 0}}{a_{j_1j_1 0}}\left[\frac{a_{j_1j_1s_1} - a_{j_2j_1 1}}{(a_{j_1j_1s_1} - a_{j_1j_1s_2})(a_{j_1j_1s_1} - a_{j_1j_1s_3})}\frac{1}{(x_{j_1} - a_{j_1j_1s_1})}\right.\\ &\quad + \frac{a_{j_1j_1s_2} - a_{j_2j_1 1}}{(a_{j_1j_1s_2} - a_{j_1j_1s_1})(a_{j_1j_1s_2} - a_{j_1j_1s_3})}\frac{1}{(x_{j_1} - a_{j_1j_1s_2})}\\ &\quad \left.+ \frac{a_{j_1j_1s_3} - a_{j_2j_1 1}}{(a_{j_1j_1s_3} - a_{j_1j_1s_1})(a_{j_1j_1s_3} - a_{j_1j_1s_2})}\frac{1}{(x_{j_1} - a_{j_1j_1s_3})}\right].\end{aligned}$$

With an initial condition of $(x_{j_1 0}, x_{j_2 0})$, the first integral manifold is

$$x_{j_2} - x_{j_2 0} = \frac{a_{j_2j_1 0}}{a_{j_1j_1 0}}\sum_{s_1=1}^{3}\frac{a_{j_1j_1s_1} - a_{j_2j_1 1}}{\prod_{s_2=1, s_2\neq s_1}^{3}(a_{j_1j_1s_1} - a_{j_1j_1s_{2'}})}\ln\frac{|x_{j_1} - a_{j_1j_1s_1}|}{|x_{j_1 0} - a_{j_1j_1s_1}|}.$$

The equilibriums for the x_{j_1} and x_{j_2}-directions are

$$\begin{aligned}&x_{j_1}^* = a_{j_1j_1 1}, a_{j_1j_1 2}, a_{j_1j_1 3} \text{ for the } x_{j_1}\text{-direction},\\ &x_{j_1}^* = a_{j_2j_1 1} \text{ for the } x_{j_2}\text{-direction}.\end{aligned}$$

(iii$_{1a}$) At $x_{j_1}^* = a_{j_1j_1s_1}$, the variational equation is

$$\begin{aligned}\dot{x}_{j_2} &= a_{j_2j_1 0}(a_{j_1j_1s_1} - a_{j_2j_1 1}),\\ \Delta\dot{x}_{j_1} &= a_{j_1j_1 0}\Pi_{s_2=1,s_2\neq s_1}^{3}(a_{j_1j_1s_1} - a_{j_1j_1s_{2'}})\Delta x_{j_1}\end{aligned}$$

The flows at $x_{j_1}^* = a_{j_1j_1s_1}$ (s_1, $s_2 \in \{1,2,3\}, s_1 \neq s_2$) are source and sink flows for $a_{j_1j_1 0}\Pi_{s_2=1,s_2\neq s_1}^{3}(a_{j_1j_1s_1} - a_{j_1j_1s_{2'}}) > 0$ and $a_{j_1j_1 0}\Pi_{s_2=1,s_2\neq s_1}^{3}(a_{j_1j_1s_1} - a_{j_1j_1s_{2'}}) < 0$ in the x_{j_1}-direction. For $a_{j_2j_1 0}(a_{j_1j_1s_1} - a_{j_2j_1 1}) > 0$ and $a_{j_2j_1 0}(a_{j_1j_1s_1} - a_{j_2j_1 1}) < 0$, the flows are positive and negative in the x_{j_2}-direction with $\dot{x}_{j_2} > 0$ and $\dot{x}_{j_2} < 0$.

Therefore, the equilibrium of $x_{j_1}^* = a_{j_1j_1s_1}$ (s_1, $s_2 \in \{1,2,3\}, s_1 \neq s_2$) and $a_{j_1j_1s_1} \neq a_{j_2j_1 1}$ for the x_{j_1}-direction has the following properties as in Eqs. (2.65)–(2.68).

- For $a_{j_1j_1 0}\prod_{s_2=1,s_2\neq s_1}^{3}(a_{j_1j_1s_1} - a_{j_1j_1s_2}) > 0$ and $a_{j_2j_1 0}(a_{j_1j_1s_1} - a_{j_2j_1 1}) > 0$,

$$(a_{j_1j_1s_1}, \dot{x}_{j_2}) = \underbrace{(\text{SO}, \text{pF})}_{\text{source flow }(+)}.$$

  The flow at $x_{j_1}^* = a_{j_1j_1s_1}$ is an (SO,pF)-positive source flow.

- For $a_{j_1 j_1 0}\prod_{s_2=1,s_2\neq s_1}^{3}(a_{j_1 j_1 s_1}-a_{j_1 j_1 s_2})<0$ and $a_{j_2 j_1 0}(a_{j_1 j_1 s_1}-a_{j_2 j_1 1})>0$,

$$(a_{j_1 j_1 s_1},\dot{x}_{j_2})=\underbrace{(\mathrm{SI},\mathrm{pF})}_{\text{sink flow }(+)}.$$

The flow at $x_{j_1}^*=a_{j_1 j_1 s_1}$ is an (SI,pF)-positive sink flow.

- For $a_{j_1 j_1 0}\prod_{s_2=1,s_2\neq s_1}^{3}(a_{j_1 j_1 s_1}-a_{j_1 j_1 s_2})>0$ and $a_{j_2 j_1 0}(a_{j_1 j_1 s_1}-a_{j_2 j_1 1})<0$,

$$(a_{j_1 j_1 s_1},\dot{x}_{j_2})=\underbrace{(\mathrm{SO},\mathrm{nF})}_{\text{source flow }(-)}.$$

The flow at $x_{j_1}^*=a_{j_1 j_1 s_1}$ is an (SO,nF)-negative source flow.

- For $a_{j_1 j_1 0}\prod_{s_2=1,s_2\neq s_1}^{3}(a_{j_1 j_1 s_1}-a_{j_1 j_1 s_2})<0$ and $a_{j_2 j_1 0}(a_{j_1 j_1 s_1}-a_{j_2 j_1 1})<0$,

$$(a_{j_1 j_1 s_1},\dot{x}_{j_2})=\underbrace{(\mathrm{SI},\mathrm{nF})}_{\text{sink flow }(-)}.$$

The flow at $x_{j_1}^*=a_{j_1 j_1 s_1}$ is an (SI,nF)-negative sink flow.

(iii$_{1b}$) In phase space, at $x_{j_1}^*=a_{j_2 j_1 1}$ and $a_{j_2 j_1 1}\neq a_{j_1 j_1 s_1}$ (s_1, $s_2\in\{1,2,3\}$,$s_1\neq s_2$),

$$\frac{dx_{j_2}}{dx_{j_1}}\Big|_{x_{j_1}^*=a_{j_2 j_1 1}}=\frac{a_{j_2 j_1 0}(x_{j_1}-a_{j_2 j_1 1})}{a_{j_1 j_1 0}(x_{j_1}-a_{j_1 j_1 s_1})(x_{j_1}-a_{j_1 j_1 s_2})(x_{j_1}-a_{j_1 j_1 s_3})}\Big|_{x_{j_1}^*=a_{j_2 j_1 1}}=0.$$

If

$$\frac{d^2x_{j_2}}{dx_{j_1}^2}\Big|_{x_{j_1}^*=a_{j_2 j_1 1}}=\frac{a_{j_2 j_1 0}(x_{j_1}-a_{j_2 j_1 1})}{a_{j_1 j_1 0}\prod_{s_1=1}^{3}(a_{j_2 j_1 1}-a_{j_1 j_1 s_1})}>0,$$

the flow at $x_{j_1}^*=a_{j_2 j_1 1}$ is an up-parabola flow in the x_{j_2}-direction, if

$$\frac{d^2x_{j_2}}{dx_{j_1}^2}\Big|_{x_{j_1}^*=a_{j_2 j_1 1}}=\frac{a_{j_2 j_1 0}(x_{j_1}-a_{j_2 j_1 1})}{a_{j_1 j_1 0}\prod_{s_1=1}^{3}(a_{j_2 j_1 1}-a_{j_1 j_1 s_1})}<0,$$

the flow at $x_{j_1}^*=a_{j_2 j_1 1}$ is a down-parabola flow in the x_{j_2}-direction.

Because of

$$\dot{x}_{j_1}=a_{j_1 j_1 0}\prod_{s_1=1}^{3}(a_{j_2 j_1 1}-a_{j_1 j_1 s_1}),$$

for $\dot{x}_{j_1}>0$ and $\dot{x}_{j_1}<0$, the flows at $x_{j_1}^*=a_{j_2 j_1 1}$ are positive and negative.

Therefore, the equilibrium of $x_{j_1}^* = a_{j_2 j_1 1}$ and $a_{j_2 j_1 1} \neq a_{j_1 j_1 s_1}$ (s_1, $s_2 \in \{1,2,3\}$, $s_1 \neq s_2$) for the x_{j_2}-direction has the following properties as in Eqs. (2.69)–(2.72).

- For $a_{j_1 j_1 0} \prod_{s_1=1}^{3} (a_{j_2 j_1 1} - a_{j_1 j_1 s_1}) > 0$ and $a_{j_2 j_1 0} > 0$,

$$(\dot{x}_{j_1}, a_{j_2 j_1 1}) = \underbrace{(\mathrm{pF}, \mathrm{UP})}_{\text{up-parabola flow } (+)}.$$

  The flow at $x_{j_1}^* = a_{j_2 j_1 1}$ is a (pF,UP)-positive up-parabola flow.
- For $a_{j_1 j_1 0} \prod_{s_1=1}^{3} (a_{j_2 j_1 1} - a_{j_1 j_1 s_1}) < 0$ and $a_{j_2 j_1 0} > 0$,

$$(\dot{x}_{j_1}, a_{j_2 j_1 1}) = \underbrace{(\mathrm{nF}, \mathrm{DP})}_{\text{down-parabola flow } (-)}.$$

  The flow at $x_{j_1}^* = a_{j_2 j_1 1}$ is an (nF,DP)-negative down-parabola flow.
- For $a_{j_1 j_1 0} \prod_{s_1=1}^{3} (a_{j_2 j_1 1} - a_{j_1 j_1 s_1}) > 0$ and $a_{j_2 j_1 0} < 0$,

$$(\dot{x}_{j_1}, a_{j_2 j_1 1}) = \underbrace{(\mathrm{pF}, \mathrm{DP})}_{\text{down-parabola flow } (+)}.$$

  The flow at $x_{j_1}^* = a_{j_2 j_1 1}$ is a (pF,DP)-positive down-parabola flow.
- For $a_{j_1 j_1 0} \prod_{s_1=1}^{3} (a_{j_2 j_1 1} - a_{j_1 j_1 s_1}) < 0$ and $a_{j_2 j_1 0} < 0$,

$$(\dot{x}_{j_1}, a_{j_2 j_1 1}) = \underbrace{(\mathrm{nF}, \mathrm{UP})}_{\text{up-parabola flow } (-)}.$$

  The flow at $x_{j_1}^* = a_{j_2 j_1 1}$ is an (nF,UP)-negative up-parabola flow.

(iii$_2$) For $a_{j_2 j_1 1} = a_{j_1 j_1 s_1}$ (s_1, s_2, $s_3 \in \{1,2,3\}$, $s_1 \neq s_2 \neq s_3$),

$$\begin{aligned}
\frac{dx_{j_2}}{dx_{j_1}} &= \frac{a_{j_2 j_1 0}}{a_{j_1 j_1 0}(x_{j_1} - a_{j_1 j_1 s_2})(x_{j_1} - a_{j_1 j_1 s_3})} \\
&= \frac{a_{j_2 j_1 0}}{a_{j_1 j_1 0}} \Big[\frac{1}{(a_{j_1 j_1 s_2} - a_{j_1 j_1 s_3})} \frac{1}{(x_{j_1} - a_{j_1 j_1 s_2})} + \frac{1}{(a_{j_1 j_1 s_3} - a_{j_1 j_1 s_2})} \frac{1}{(x_{j_1} - a_{j_1 j_1 s_3})} \Big].
\end{aligned}$$

With an initial condition of $(x_{j_1 0}, x_{j_2 0})$, the first integral manifold is

$$x_{j_2} - x_{j_2 0} = \frac{a_{j_2 j_1 0}}{a_{j_1 j_1 0}} \frac{1}{a_{j_1 j_1 s_2} - a_{j_1 j_1 s_3}} \Big(\ln \frac{|x_{j_1} - a_{j_1 j_1 s_2}|}{|x_{j_1 0} - a_{j_1 j_1 s_2}|} - \ln \frac{|x_{j_1} - a_{j_1 j_1 s_3}|}{|x_{j_1 0} - a_{j_1 j_1 s_3}|} \Big).$$

In phase space, at $x_{j_1}^* = a_{j_1 j_1 s_1} = a_{j_2 j_1 1}$, if

$$\left.\frac{dx_{j_2}}{dx_{j_1}}\right|_{x_{j_1}^*=a_{j_1j_1s_1}}=\frac{a_{j_2j_10}}{a_{j_1j_10}(a_{j_1j_1s_1}-a_{j_1j_1s_2})(a_{j_1j_1s_1}-a_{j_1j_1s_3})}>0,$$

the flow is an increasing-inflection flow, and if

$$\left.\frac{dx_{j_2}}{dx_{j_1}}\right|_{x_{j_1}^*=a_{j_1j_1s_1}}=\frac{a_{j_2j_10}}{a_{j_1j_10}(a_{j_1j_1s_1}-a_{j_1j_1s_2})(a_{j_1j_1s_1}-a_{j_1j_1s_3})}<0,$$

the flow is a decreasing-inflection flow.

In addition, at $x_{j_1}^*=a_{j_1j_1s_1}=a_{j_2j_11}$, the variational equations are

$$\begin{aligned}\Delta\dot{x}_{j_2}&=a_{j_2j_10}\Delta x_{j_1},\\ \Delta\dot{x}_{j_1}&=a_{j_1j_10}(a_{j_1j_1s_1}-a_{j_1j_1s_2})(a_{j_1j_1s_1}-a_{j_1j_1s_3})\Delta x_{j_1}.\end{aligned}$$

The flows at $x_{j_1}^*=a_{j_1j_1s_1}=a_{j_2j_11}$ in the x_{j_1}- direction are:

- a source flow for $a_{j_1j_10}(a_{j_1j_1s_1}-a_{j_1j_1s_2})(a_{j_1j_1s_1}-a_{j_1j_1s_3})>0$; and
- a sink flow for $a_{j_1j_10}(a_{j_1j_1s_1}-a_{j_1j_1s_2})(a_{j_1j_1s_1}-a_{j_1j_1s_3})>0$.

The flows at $x_{j_1}^*=a_{j_1j_1s_1}=a_{j_2j_11}$ in the x_{j_2}- direction are down-up and up-down flow for $a_{j_2j_10}>0$ and $a_{j_2j_10}<0$. From the singularity of the flow is of the first order, the flows are the (1:1)-increasing-inflection flow and (1:1)-decreasing inflection flow. However, from the first integral manifold, the increasing- and decreasing-inflections flows that are on the logarithmic curves are also called the up-logarithmic and down-logarithmic infinite-equilibriums. Such infinite-equilibriums are for switching bifurcations.

Therefore, the infinite-equilibrium of $x_{j_1}^*=a_{j_1j_1s_1}=a_{j_2j_11}$ (s_1, s_2, $s_3\in\{1,2,3\}$, $s_1\neq s_2\neq s_3$) with $\overline{x}_{j_2}\in(-\infty,\infty)$ possesses the following properties as in Eqs. (2.74)–(2.77).

- For $a_{j_1j_10}\prod_{s_2=1,s_2\neq s_1}^{3}(a_{j_1j_1s_1}-a_{j_1j_1s_2})>0$ and $a_{j_2j_10}>0$,

$$\left\{\begin{array}{c}\underbrace{(\text{nF},\text{DP})}_{\text{down-parabola flow }(-)}\\ \underbrace{(\text{SO},\text{pF})}_{\text{source flow }(+)}\end{array}\right\}\rightleftharpoons\underbrace{(\text{SO},\text{II})}_{\text{increasing-inflection source}}\rightleftharpoons\left\{\begin{array}{c}\underbrace{(\text{SO},\text{nF})}_{\text{source flow }(-)}\\ \underbrace{(\text{pF},\text{UP})}_{\text{up-parabola flow }(+)}\end{array}\right\}.$$

The equilibrium of $x_{j_1}^*=a_{j_1j_1s_1}=a_{j_2j_11}$ is an (SO,II)-increasing-inflection source infinite-equilibrium.

- For $a_{j_1j_10}\prod_{s_2=1,s_2\neq s_1}^{3}(a_{j_1j_1s_1}-a_{j_1j_1s_2})<0$ and $a_{j_2j_10}>0$,

$$\left\{\begin{array}{c}\underbrace{(\mathrm{nF},\mathrm{UP})}_{\text{up-parabola flow }(-)}\\ \underbrace{(\mathrm{SO},\mathrm{nF})}_{\text{source flow }(-)}\end{array}\right\}\rightleftharpoons\underbrace{(\mathrm{SI},\mathrm{DI})}_{\text{decreasing-inflection sink}}\rightleftharpoons\left\{\begin{array}{c}\underbrace{(\mathrm{SO},\mathrm{pF})}_{\text{source flow }(+)}\\ \underbrace{(\mathrm{pF},\mathrm{DP})}_{\text{down-parabola flow }(+)}\end{array}\right\}.$$

The equilibrium of $x_{j_1}^*=a_{j_1j_1s_1}=a_{j_2j_11}$ is an (SI,DI)-decreasing-inflection sink infinite-equilibrium.

- For $a_{j_1j_10}\prod_{s_2=1,s_2\neq s_1}^{3}(a_{j_1j_1s_1}-a_{j_1j_1s_2})>0$ and $a_{j_2j_10}<0$,

$$\left\{\begin{array}{c}\underbrace{(\mathrm{nF},\mathrm{DP})}_{\text{down-parabola flow }(-)}\\ \underbrace{(\mathrm{SO},\mathrm{pF})}_{\text{source flow }(+)}\end{array}\right\}\rightleftharpoons\underbrace{(\mathrm{SO},\mathrm{DI})}_{\text{decreasing-inflection source}}\rightleftharpoons\left\{\begin{array}{c}\underbrace{(\mathrm{SO},\mathrm{nF})}_{\text{source flow }(-)}\\ \underbrace{(\mathrm{pF},\mathrm{UP})}_{\text{up-parabola flow }(+)}\end{array}\right\}$$

The equilibrium of $x_{j_1}^*=a_{j_1j_1s_1}=a_{j_2j_11}$ is an (SO,DI)-decreasing-inflection source infinite-equilibrium.

- For $a_{j_1j_10}\prod_{s_2=1,s_2\neq s_1}^{3}(a_{j_1j_1s_1}-a_{j_1j_1s_2})<0$ and $a_{j_2j_10}<0$,

$$\left\{\begin{array}{c}\underbrace{(\mathrm{pF},\mathrm{DP})}_{\text{down-parabola flow }(+)}\\ \underbrace{(\mathrm{SI},\mathrm{nF})}_{\text{sink flow }(-)}\end{array}\right\}\rightleftharpoons\underbrace{(\mathrm{SO},\mathrm{II})}_{\text{increasing-inflection sink}}\rightleftharpoons\left\{\begin{array}{c}\underbrace{(\mathrm{SI},\mathrm{pF})}_{\text{sink flow }(+)}\\ \underbrace{(\mathrm{nF},\mathrm{DP})}_{\text{down-parabola flow }(-)}\end{array}\right\}.$$

The equilibrium of $x_{j_1}^*=a_{j_1j_1s_1}=a_{j_2j_11}$ is an (SO,II)-increasing-inflection sink infinite-equilibrium.

(iv) Under

$$A_{j_2j_11}\neq 0, A_{j_2j_12}\neq 0 \text{ and } \Delta_{j_1j_1}=0,$$

the two-dimensional dynamical system becomes

$$\begin{aligned}\dot{x}_{j_2}&=a_{j_2j_10}(x_{j_1}-a_{j_2j_11}),\\ \dot{x}_{j_1}&=a_{j_1j_10}(x_{j_1}-a_{j_1j_1s_1})(x_{j_1}-a_{j_1j_1s_2})^2.\end{aligned}$$

(iv_1) From case (ii_{1c}), the upper-saddle and lower-saddle flows are the upper-saddle and lower-saddle appearing bifurcations. Thus, for $a_{j_1j_1s_2}\neq a_{j_1j_1s_1}\neq a_{j_2j_11}$, the equilibrium of $x_{j_1}^*=a_{j_1j_1s_2}$ possesses the appearing bifurcations as in Eqs. (2.80)–(2.83).

- For $a_{j_1j_1 0}(a_{j_1j_1 s_2}-a_{j_1j_1 s_1})>0$ and $a_{j_1j_1 0}(a_{j_1j_1 s_2}-a_{j_2j_1 1})>0$,

$$(a_{j_1j_1 s_2},\dot{x}_{j_2})=\underbrace{(\text{US, pF})}_{\text{upper-saddle flow }(+)}\rightleftharpoons\left\{\begin{array}{c}\underbrace{(\text{SI, pF})}_{\text{sink flow }(+)}\\ \underbrace{(\text{SO, pF})}_{\text{source flow }(+)}\end{array}\right\}$$

The flow at $x_{j_1}^*=a_{j_1j_1 s_2}$ is called a (US,pF)-flow appearing bifurcation.

- For $a_{j_1j_1 0}(a_{j_1j_1 s_2}-a_{j_1j_1 s_1})<0$ and $a_{j_1j_1 0}(a_{j_1j_1 s_2}-a_{j_2j_1 1})>0$,

$$(a_{j_1j_1 s_2},\dot{x}_{j_2})=\underbrace{(\text{LS, pF})}_{\text{lower-saddle flow }(+)}\rightleftharpoons\left\{\begin{array}{c}\underbrace{(\text{SO, pF})}_{\text{source flow }(+)}\\ \underbrace{(\text{SI, pF})}_{\text{sink flow }(+)}\end{array}\right\}$$

The flow at $x_{j_1}^*=a_{j_1j_1 s_2}$ is called an (LS,pF)-flow appearing bifurcation.

- For $a_{j_1j_1 0}(a_{j_1j_1 s_2}-a_{j_1j_1 s_1})>0$ and $a_{j_1j_1 0}(a_{j_1j_1 1}-a_{j_2j_1 1})<0$,

$$(a_{j_1j_1 s_2},\dot{x}_{j_2})=\underbrace{(\text{US, nF})}_{\text{upper saddle flow }(-)}\rightleftharpoons\left\{\begin{array}{c}\underbrace{(\text{SI, nF})}_{\text{sink flow }(-)}\\ \underbrace{(\text{SO, nF})}_{\text{source flow }(-)}\end{array}\right\}$$

The flow at $x_{j_1}^*=a_{j_1j_1 s_2}$ is called a (US,pF)-flow appearing bifurcation.

- For $a_{j_1j_1 0}(a_{j_1j_1 s_2}-a_{j_1j_1 s_1})<0$ and $a_{j_1j_1 0}(a_{j_1j_1 1}-a_{j_2j_1 1})<0$,

$$(a_{j_1j_1 s_2},\dot{x}_{j_2})=\underbrace{(\text{LS, nF})}_{\text{lower-saddle flow }(-)}\rightleftharpoons\left\{\begin{array}{c}\underbrace{(\text{SO, pF})}_{\text{source flow }(-)}\\ \underbrace{(\text{SI, pF})}_{\text{sink flow }(-)}\end{array}\right\}$$

The flow at $x_{j_1}^*=a_{j_1j_1 s_2}$ is called an (LS,pF)-flow appearing bifurcation.

(iv$_2$) For $a_{j_1j_1 s_2}=a_{j_2j_1 1}$, the two-dimensional dynamical system becomes

$$\begin{aligned}\dot{x}_{j_2}&=a_{j_2j_1 0}(x_{j_1}-a_{j_2j_1 1}),\\ \dot{x}_{j_1}&=a_{j_1j_1 0}(x_{j_1}-a_{j_1j_1 s_1})(x_{j_1}-a_{j_1j_1 s_2})^2.\end{aligned}$$

(iv$_{2a}$) From case (ii$_3$) with cases (iii$_{1a}$) and (iii$_{1b}$), at $x_{j_1}^* = a_{j_1 j_1 s_2} = a_{j_2 j_1 1}$, the down-up and up-down asymptotic upper-saddle and lower-saddle infinite-equilibriums are the appearing bifurcations. Therefore, the equilibrium of $x_{j_1}^* = a_{j_1 j_1 s_2} = a_{j_2 j_1 1}$ possesses the appearing bifurcations as in Eqs. (2.85)–(2.88).

- For $a_{j_1 j_1 0}(a_{j_1 j_1 s_2} - a_{j_1 j_1 s_1}) > 0$ and $a_{j_2 j_1 0} > 0$,

$$\underbrace{(\text{pF}, \text{UP})}_{\text{up-parabola flow }(+)} \rightleftharpoons \underbrace{(\text{US}, \text{DU})}_{\substack{\text{down-up upper-saddle} \\ \text{appearing bifurcation}}} \rightleftharpoons \left\{ \begin{array}{c} \underbrace{(\text{SI}, \text{nF})}_{\text{sink flow }(-)} \\ \underbrace{(\text{nF}, \text{DP})}_{\text{down-parabola flow }(-)} \\ \underbrace{(\text{SO}, \text{pF})}_{\text{source flow }(+)} \end{array} \right\}.$$

The infinite-equilibrium is called a (US,DU)-appearing bifurcation.

- For $a_{j_1 j_1 0}(a_{j_1 j_1 s_2} - a_{j_1 j_1 s_1}) < 0$ and $a_{j_2 j_1 0} > 0$,

$$\underbrace{(\text{nF}, \text{DP})}_{\text{up-parabola flow }(-)} \rightleftharpoons \underbrace{(\text{LS}, \text{UD})}_{\substack{\text{up-down lower-saddle} \\ \text{appearing bifurcation}}} \rightleftharpoons \left\{ \begin{array}{c} \underbrace{(\text{SO}, \text{nF})}_{\text{source flow }(-)} \\ \underbrace{(\text{pF}, \text{UP})}_{\text{up-parabola flow }(+)} \\ \underbrace{(\text{SI}, \text{pF})}_{\text{sink flow }(+)} \end{array} \right\}.$$

The infinite-equilibrium is called an (LS,UD)-appearing bifurcation.

- For $a_{j_1 j_1 0}(a_{j_1 j_1 s_2} - a_{j_1 j_1 s_1}) > 0$ and $a_{j_2 j_1 0} < 0$,

$$\underbrace{(\text{pF}, \text{DP})}_{\text{down-parabola flow }(+)} \rightleftharpoons \underbrace{(\text{US}, \text{UD})}_{\substack{\text{infinite-equilibrium} \\ \text{apearing bifurcation}}} \rightleftharpoons \left\{ \begin{array}{c} \underbrace{(\text{SI}, \text{pF})}_{\text{sink flow }(+)} \\ \underbrace{(\text{nF}, \text{UP})}_{\text{up-parabola flow }(-)} \\ \underbrace{(\text{SO}, \text{nF})}_{\text{source flow }(-)} \end{array} \right\}.$$

The infinite-equilibrium is called a (US,UD)-appearing bifurcation.

- For $a_{j_1j_10}(a_{j_1j_1s_2}-a_{j_1j_1s_1})<0$ and $a_{j_2j_10}<0$,

$$\underbrace{(\text{nF},\text{UP})}_{\text{up-parabola flow }(-)} \rightleftharpoons \underbrace{(\text{US},\text{DU})}_{\substack{\text{down-up lower-saddle}\\ \text{appearing bifurcation}}} \rightleftharpoons \left\{\begin{array}{c}\underbrace{(\text{SO},\text{pF})}_{\text{source flow }(+)}\\ \underbrace{(\text{pF},\text{DP})}_{\text{down-parabola flow }(+)}\\ \underbrace{(\text{SI},\text{nF})}_{\text{sink flow }(-)}\end{array}\right\}.$$

The infinite-equilibrium is called an (LS,DU)-appearing bifurcation.

(iv$_{2b}$) From cases (ii$_{1b}$) and (ii$_{1c}$) with case (ii$_3$), the equilibrium of $x_{j_1}^*=a_{j_1j_1s_2}=a_{j_2j_11}$ possesses the switching bifurcations from up-parabola and down-parabola flows and up (down)-saddle flows as in Eqs. (2.89)–(2.92).

- For $a_{j_1j_10}(a_{j_1j_1s_2}-a_{j_1j_1s_1})>0$ and $a_{j_2j_10}>0$,

$$\left\{\begin{array}{c}\underbrace{(\text{US},\text{nF})}_{\text{upper-saddle flow }(-)}\\ \underbrace{(\text{pF},\text{UP})}_{\text{up-parabola flow }(+)}\end{array}\right\} \rightleftharpoons \underbrace{(\text{US},\text{DU})}_{\text{down-up upper-saddle}} \rightleftharpoons \left\{\begin{array}{c}\underbrace{(\text{pF},\text{UP})}_{\text{up-parabola flow }(+)}\\ \underbrace{(\text{US},\text{pF})}_{\text{upper-saddle flow }(+)}\end{array}\right\}.$$

The infinite-equilibrium is called a (US,DU)-switching bifurcation.

- For $a_{j_1j_10}(a_{j_1j_1s_2}-a_{j_1j_1s_1})<0$ and $a_{j_2j_10}>0$,

$$\left\{\begin{array}{c}\underbrace{(\text{LS},\text{nF})}_{\text{lower-saddle flow }(-)}\\ \underbrace{(\text{nF},\text{DP})}_{\text{down-parabola flow }(-)}\end{array}\right\} \rightleftharpoons \underbrace{(\text{LS},\text{UD})}_{\text{up-down lower-saddle}} \rightleftharpoons \left\{\begin{array}{c}\underbrace{(\text{nF},\text{DP})}_{\text{down-parabola flow }(-)}\\ \underbrace{(\text{LS},\text{pF})}_{\text{lower-saddle flow }(+)}\end{array}\right\}.$$

The infinite-equilibrium is called an (LS,UD)-switching bifurcation.

- For $a_{j_1j_10}(a_{j_1j_1s_2}-a_{j_1j_1s_1})>0$ and $a_{j_2j_10}<0$,

$$\left\{\begin{array}{c}\underbrace{(\text{US},\text{pF})}_{\text{upper-saddle flow }(+)}\\ \underbrace{(\text{pF},\text{DP})}_{\text{down-parabola flow }(+)}\end{array}\right\} \rightleftharpoons \underbrace{(\text{US},\text{UD})}_{\text{up-down upper-saddle}} \rightleftharpoons \left\{\begin{array}{c}\underbrace{(\text{pF},\text{DP})}_{\text{down-parabola flow }(+)}\\ \underbrace{(\text{US},\text{nF})}_{\text{upper-saddle flow }(-)}\end{array}\right\}.$$

The infinite-equilibrium is called a (US,UD)-switching bifurcation.

- For $a_{j_1j_10}(a_{j_1j_1s_2}-a_{j_1j_1s_1})<0$ and $a_{j_2j_10}<0$,

$$\left\{\begin{matrix}\underbrace{(\mathrm{LS},\mathrm{pF})}_{\text{lower-saddle flow }(+)}\\ \underbrace{(\mathrm{nF},\mathrm{UP})}_{\text{up-parabola flow }(-)}\end{matrix}\right\} \rightleftharpoons \underbrace{(\mathrm{LS},\mathrm{DU})}_{\text{down-up lower-saddle}} \rightleftharpoons \left\{\begin{matrix}\underbrace{(\mathrm{nF},\mathrm{UP})}_{\text{up-parabola flow }(-)}\\ \underbrace{(\mathrm{LS},\mathrm{nF})}_{\text{lower-saddle flow }(-)}\end{matrix}\right\}.$$

The infinite-equilibrium is called an (LS,DU)-switching bifurcation.

(v) Under

$$A_{j_2j_11}\neq 0, A_{j_2j_12}\neq 0 \text{ and } \Delta_{j_1j_1}=0 \text{ with } a_{j_1j_1s_1}=a_{j_1j_1s_2}\equiv a_{j_1j_11}$$

the two-dimensional dynamical system becomes

$$\begin{aligned}\dot{x}_{j_2} &= a_{j_2j_10}(x_{j_1}-a_{j_2j_11}),\\ \dot{x}_{j_1} &= a_{j_1j_10}(x_{j_1}-a_{j_1j_11})^3.\end{aligned}$$

(v_{1a}) From case (ii_{4b}), the third-order source and sink flows are the appearing bifurcations. Thus, the equilibrium of $a_{j_2j_11}\neq a_{j_1j_11}$ has the following bifurcations as in Eqs. (2.95)–(2.98).

- For $a_{j_1j_10}>0$ and $a_{j_2j_10}(a_{j_1j_11}-a_{j_2j_11})>0$,

$$\underbrace{(\mathrm{SO},\mathrm{pF})}_{\text{source flow }(+)} \rightleftharpoons \underbrace{(3^{\mathrm{rd}}\mathrm{SO},\mathrm{pF})}_{\text{third-order source flow }(+)} \rightleftharpoons \left\{\begin{matrix}\underbrace{(\mathrm{SO},\mathrm{pF})}_{\text{source flow }(+)}\\ \underbrace{(\mathrm{SI},\mathrm{pF})}_{\text{sink flow }(+)}\\ \underbrace{(\mathrm{SO},\mathrm{pF})}_{\text{source flow }(+)}\end{matrix}\right\}.$$

The (3^{rd}SO,pF)-positive third-order source flow of $x_{j_1}^*=a_{j_1j_11}\neq a_{j_2j_11}$ is a (3^{rd}SO, pF)-flow appearing bifurcation.

- For $a_{j_1j_10}<0$ and $a_{j_1j_11}(a_{j_1j_11}-a_{j_2j_11})>0$,

$$\underbrace{(\mathrm{SI},\mathrm{pF})}_{\text{sink flow }(+)} \rightleftharpoons \underbrace{(3^{\mathrm{rd}}\mathrm{SI},\mathrm{pF})}_{\text{third-order sink flow }(+)} \rightleftharpoons \left\{\begin{matrix}\underbrace{(\mathrm{SI},\mathrm{pF})}_{\text{sink flow }(+)}\\ \underbrace{(\mathrm{SO},\mathrm{pF})}_{\text{source flow }(+)}\\ \underbrace{(\mathrm{SI},\mathrm{pF})}_{\text{sink flow }(+)}\end{matrix}\right\}.$$

The (3[rd]SI,pF)-positive third-order sink flow of $x_{j_1}^* = a_{j_1j_1 1} \neq a_{j_2j_1 1}$ is a (3[rd]SI, pF)-flow appearing bifurcation.

- For $a_{j_1j_1 0} > 0$ and $a_{j_2j_1 0}(a_{j_1j_1 1} - a_{j_2j_1 1}) < 0$,

$$\underbrace{(\mathrm{SO,nF})}_{\text{source flow }(-)} \rightleftharpoons \underbrace{(3^{\mathrm{rd}}\mathrm{SO,nF})}_{\text{third-order source flow }(-)} \rightleftharpoons \left\{ \begin{array}{c} \underbrace{(\mathrm{SO,nF})}_{\text{source flow }(-)} \\ \underbrace{(\mathrm{SI,nF})}_{\text{sink flow }(-)} \\ \underbrace{(\mathrm{SO,nF})}_{\text{source flow }(-)} \end{array} \right\}.$$

The (3[rd]SO,nF)-negative third-order source flow of $x_{j_1}^* = a_{j_1j_1 1} \neq a_{j_2j_1 1}$ is a (3[rd]SO, nF)-flow appearing bifurcation.

- For $a_{j_1j_1 0} < 0$ and $a_{j_2j_1 0}(a_{j_1j_1 1} - a_{j_2j_1 1}) < 0$,

$$\underbrace{(\mathrm{SI,nF})}_{\text{sink flow }(-)} \rightleftharpoons \underbrace{(3^{\mathrm{rd}}\mathrm{SI,nF})}_{\text{third-order sink flow }(-)} \rightleftharpoons \left\{ \begin{array}{c} \underbrace{(\mathrm{SI,nF})}_{\text{sink flow }(-)} \\ \underbrace{(\mathrm{SO,nF})}_{\text{source flow }(-)} \\ \underbrace{(\mathrm{SI,nF})}_{\text{sink flow }(-)} \end{array} \right\}.$$

The (3[rd]SI,nF)-negative third-order sink flow of $x_{j_1}^* = a_{j_1j_1 1} \neq a_{j_2j_1 1}$ is a (3[rd]SI, nF)-flow appearing bifurcation.

(v_{1b}) From cases (ii_{4b}) and (iii), the equilibrium of $a_{j_2j_1 1} \neq a_{j_1j_1 1}$ has the following switching bifurcations.

- For $a_{j_1j_1 0} > 0$ and $a_{j_2j_1 0}(a_{j_1j_1 1} - a_{j_2j_1 1}) > 0$,

$$\left\{ \begin{array}{c} \underbrace{(\mathrm{SO,pF})}_{\text{source flow }(+)} \\ \underbrace{(\mathrm{SI,pF})}_{\text{sink flow }(+)} \\ \underbrace{(\mathrm{SO,pF})}_{\text{source flow }(+)} \end{array} \right\} \rightleftharpoons \underbrace{(3^{\mathrm{rd}}\mathrm{SO,pF})}_{\text{third-order source flow }(+)} \rightleftharpoons \left\{ \begin{array}{c} \underbrace{(\mathrm{SO,pF})}_{\text{source flow }(+)} \\ \underbrace{(\mathrm{SI,pF})}_{\text{sink flow }(+)} \\ \underbrace{(\mathrm{SO,pF})}_{\text{source flow }(+)} \end{array} \right\}.$$

The (3[rd]SO,pF)-positive third-order source flow of $x_{j_1}^* = a_{j_1j_1 1} \neq a_{j_2j_1 1}$ is a (3[rd]SO, pF)-flow switching bifurcation.

- For $a_{j_1j_1 0} < 0$ and $a_{j_2j_1 0}(a_{j_1j_1 1} - a_{j_2j_1 1}) > 0$,

$$\left\{\begin{array}{c}\underbrace{(\text{SI, pF})}_{\text{sink flow }(+)}\\ \underbrace{(\text{SO, pF})}_{\text{source flow }(+)}\\ \underbrace{(\text{SI, pF})}_{\text{sink flow }(+)}\end{array}\right\} \rightleftharpoons \underbrace{(3^{\text{rd}}\text{SI, pF})}_{\text{third-order sink flow }(+)} \rightleftharpoons \left\{\begin{array}{c}\underbrace{(\text{SI, pF})}_{\text{sink flow }(+)}\\ \underbrace{(\text{SO, pF})}_{\text{source flow }(+)}\\ \underbrace{(\text{SI, pF})}_{\text{sink flow }(+)}\end{array}\right\}.$$

  The (3^rd^SI,pF)-positive third-order sink flow of $x_{j_1}^* = a_{j_1j_1 1} \neq a_{j_2j_1 1}$ is a (3^rd^SI, pF)-flow switching bifurcation.
- For $a_{j_1j_1 0} > 0$ and $a_{j_2j_1 0}(a_{j_1j_1 1} - a_{j_2j_1 1}) < 0$,

$$\left\{\begin{array}{c}\underbrace{(\text{SO, nF})}_{\text{source flow }(-)}\\ \underbrace{(\text{SI, nF})}_{\text{sink flow }(-)}\\ \underbrace{(\text{SO, nF})}_{\text{source flow }(-)}\end{array}\right\} \rightleftharpoons \underbrace{(3^{\text{rd}}\text{SO, nF})}_{\text{third-order source flow }(-)} \rightleftharpoons \left\{\begin{array}{c}\underbrace{(\text{SO, nF})}_{\text{source flow }(-)}\\ \underbrace{(\text{SI, nF})}_{\text{sink flow }(-)}\\ \underbrace{(\text{SO, nF})}_{\text{source flow }(-)}\end{array}\right\}.$$

  The (3^{rd}SO,nF)-negative third-order source flow of $x_{j_1}^* = a_{j_1j_1 1} \neq a_{j_2j_1 1}$ is a (3^{rd}SO, nF)-flow switching bifurcation.
- For $a_{j_1j_1 0} < 0$ and $a_{j_2j_1 0}(a_{j_1j_1 1} - a_{j_2j_1 1}) < 0$,

$$\left\{\begin{array}{c}\underbrace{(\text{SI, nF})}_{\text{sink flow }(-)}\\ \underbrace{(\text{SO, nF})}_{\text{source flow }(-)}\\ \underbrace{(\text{SI, nF})}_{\text{sink flow }(-)}\end{array}\right\} \rightleftharpoons \underbrace{(3^{\text{rd}}\text{SI, nF})}_{\text{third-order sink flow }(-)} \rightleftharpoons \left\{\begin{array}{c}\underbrace{(\text{SI, nF})}_{\text{sink flow }(-)}\\ \underbrace{(\text{SO, nF})}_{\text{source flow }(-)}\\ \underbrace{(\text{SI, nF})}_{\text{sink flow }(-)}\end{array}\right\}.$$

  The (3^{rd}SI,nF)-negative third-order sink flow of $x_{j_1}^* = a_{j_1j_1 1} \neq a_{j_2j_1 1}$ is a (3^{rd}SI, nF)-flow switching bifurcation.

($\text{v}_{2\text{a}}$) From cases (ii_c) with cases (i) and (iii), the equilibrium of $x_{j_1}^* = a_{j_1j_1 1} = a_{j_2j_1 1}$ with $\bar{x}_{j_2} \in (-\infty, \infty)$ has the following appearing bifurcations as in Eqs. (2.103)–(2.110).

- For $a_{j_1j_10} > 0$ and $a_{j_2j_10} > 0$,

$$\left\{\begin{array}{c}\underbrace{(\mathrm{SO},\mathrm{nF})}_{\text{source flow }(-)}\\ \underbrace{(\mathrm{pF},\mathrm{UP})}_{\text{up-parabola flow }(+)}\end{array}\right\} \rightleftharpoons \underbrace{(3^{\mathrm{rd}}\mathrm{SO},(3:1)\mathrm{IC})}_{\text{third-order increasing-concave source}} \rightleftharpoons \left\{\begin{array}{c}\underbrace{(\mathrm{SO},\mathrm{nF})}_{\text{source flow }(-)}\\ \underbrace{(\mathrm{pF},\mathrm{UP})}_{\text{up-parabola flow }(+)}\\ \underbrace{(\mathrm{SI},\mathrm{pF})}_{\text{sink flow }(+)}\\ \underbrace{(\mathrm{SO},\mathrm{pF})}_{\text{source flow }(+)}\end{array}\right\}.$$

and

$$\left\{\begin{array}{c}\underbrace{(\mathrm{nF},\mathrm{DP})}_{\text{down-parabola flow }(-)}\\ \underbrace{(\mathrm{SO},\mathrm{pF})}_{\text{source flow }(+)}\end{array}\right\} \rightleftharpoons \underbrace{(3^{\mathrm{rd}}\mathrm{SO},(3:1)\mathrm{IC})}_{\text{third-order increasing-concave source}} \rightleftharpoons \left\{\begin{array}{c}\underbrace{(\mathrm{SO},\mathrm{nF})}_{\text{source flow }(-)}\\ \underbrace{(\mathrm{SI},\mathrm{nF})}_{\text{sink flow }(-)}\\ \underbrace{(\mathrm{nF},\mathrm{DP})}_{\text{down-parabola flow }(-)}\\ \underbrace{(\mathrm{SO},\mathrm{pF})}_{\text{source flow }(+)}\end{array}\right\}.$$

The $(3^{\mathrm{rd}}\mathrm{SO}$, (3:1)IC)-third-order increasing-concave source infinite-equilibrium of $x_{j_1}^* = a_{j_1j_11} = a_{j_2j_11}$ is a $(3^{\mathrm{rd}}\mathrm{SO}$,(3:1)IC)-appearing bifurcation from a parabola and source flows to source, sink, and source flows with a parabola flow.

- For $a_{j_1j_10} < 0$ and $a_{j_2j_10} > 0$,

$$\left\{\begin{array}{c}\underbrace{(\mathrm{SI},\mathrm{nF})}_{\text{sink flow }(-)}\\ \underbrace{(\mathrm{nF},\mathrm{DP})}_{\text{down-parabola flow }(-)}\end{array}\right\} \rightleftharpoons \underbrace{(3^{\mathrm{rd}}\mathrm{SI},(3:1)\mathrm{DC})}_{\text{third-order decreasing-concave sink}} \rightleftharpoons \left\{\begin{array}{c}\underbrace{(\mathrm{SI},\mathrm{nF})}_{\text{source flow }(-)}\\ \underbrace{(\mathrm{nF},\mathrm{DP})}_{\text{up-parabola flow }(-)}\\ \underbrace{(\mathrm{SO},\mathrm{pF})}_{\text{source flow }(+)}\\ \underbrace{(\mathrm{SI},\mathrm{pF})}_{\text{sink flow }(+)}\end{array}\right\}.$$

and

$$\left\{\begin{array}{c}\underbrace{(\text{pF, UP})}_{\text{up-parabola flow }(+)}\\ \underbrace{(\text{SI, pF})}_{\text{sink flow }(+)}\end{array}\right\}\rightleftharpoons\underbrace{(3^{\text{rd}}\text{SI}, (3{:}1)\text{DC})}_{\text{third-order decreasing-concave sink}}\rightleftharpoons\left\{\begin{array}{c}\underbrace{(\text{SI, nF})}_{\text{source flow }(-)}\\ \underbrace{(\text{SO, nF})}_{\text{source flow }(-)}\\ \underbrace{(\text{pF, UP})}_{\text{up-parabola flow }(+)}\\ \underbrace{(\text{SI, pF})}_{\text{sink flow }(+)}\end{array}\right\}.$$

The (3rdSI, (3:1)DC)-third-order decreasing-concave sink infinite-equilibrium of $x_{j_1}^* = a_{j_1 j_1 1} = a_{j_2 j_1 1}$ is a (3rdSI, (3:1)DC)-appearing bifurcation.

- For $a_{j_1 j_1 0} > 0$ and $a_{j_2 j_1 0} < 0$,

$$\left\{\begin{array}{c}\underbrace{(\text{SO, pF})}_{\text{sink flow }(+)}\\ \underbrace{(\text{pF, DP})}_{\text{down-parabola flow }(+)}\end{array}\right\}\rightleftharpoons\underbrace{(3^{\text{rd}}\text{SO}, (3:1)\text{DC})}_{\text{third-order decreasing-concave source}}\rightleftharpoons\left\{\begin{array}{c}\underbrace{(\text{SO, pF})}_{\text{source flow }(+)}\\ \underbrace{(\text{pF, DP})}_{\text{down-parabola flow }(+)}\\ \underbrace{(\text{SI, nF})}_{\text{sink flow }(-)}\\ \underbrace{(\text{SO, nF})}_{\text{source flow }(-)}\end{array}\right\}.$$

and

$$\left\{\begin{array}{c}\underbrace{(\text{nF, UP})}_{\text{up-parabola flow }(-)}\\ \underbrace{(\text{SO, nF})}_{\text{source flow }(-)}\end{array}\right\}\rightleftharpoons\underbrace{(3^{\text{rd}}\text{SO}, (3:1)\text{DC})}_{\text{third-order decreasing-concave source}}\rightleftharpoons\left\{\begin{array}{c}\underbrace{(\text{SO, pF})}_{\text{source flow }(+)}\\ \underbrace{(\text{SI, pF})}_{\text{sink flow }(+)}\\ \underbrace{(\text{nF, UP})}_{\text{up-parabola flow }(-)}\\ \underbrace{(\text{SO, nF})}_{\text{source flow }(-)}\end{array}\right\}.$$

The (3rdSO,(3:1)DC)-third-order decreasing-concave source infinite-equilibrium of $x_{j_1}^* = a_{j_1 j_1 1} = a_{j_2 j_1 1}$ is a (3rdSO, (3:1)DC)-appearing bifurcation.

- For $a_{j_1j_10} < 0$ and $a_{j_2j_10} < 0$,

$$\left\{\begin{array}{c}\underbrace{(\mathrm{SI},\mathrm{pF})}_{\text{sink flow }(+)}\\ \underbrace{(\mathrm{nF},\mathrm{UP})}_{\text{up-parabola flow }(-)}\end{array}\right\} \rightleftharpoons \underbrace{(3^{\mathrm{rd}}\mathrm{SI},(3:1)\mathrm{IC})}_{\text{third-order increasing-concave sink}} \rightleftharpoons \left\{\begin{array}{c}\underbrace{(\mathrm{SI},\mathrm{pF})}_{\text{sink flow }(+)}\\ \underbrace{(\mathrm{nF},\mathrm{UP})}_{\text{up-parabola flow }(-)}\\ \underbrace{(\mathrm{SO},\mathrm{nF})}_{\text{source flow }(-)}\\ \underbrace{(\mathrm{SI},\mathrm{nF})}_{\text{sink flow }(-)}\end{array}\right\}.$$

and

$$\left\{\begin{array}{c}\underbrace{(\mathrm{pF},\mathrm{DP})}_{\text{down-parabola flow }(+)}\\ \underbrace{(\mathrm{SI},\mathrm{nF})}_{\text{sink flow }(-)}\end{array}\right\} \rightleftharpoons \underbrace{(3^{\mathrm{rd}}\mathrm{SI},(3:1)\mathrm{IC})}_{\text{third-order increasing-concave sink}} \rightleftharpoons \left\{\begin{array}{c}\underbrace{(\mathrm{SI},\mathrm{pF})}_{\text{sink flow }(+)}\\ \underbrace{(\mathrm{SO},\mathrm{pF})}_{\text{source flow }(+)}\\ \underbrace{(\mathrm{pF},\mathrm{DP})}_{\text{down-parabola flow }(+)}\\ \underbrace{(\mathrm{SI},\mathrm{nF})}_{\text{sink flow }(-)}\end{array}\right\}.$$

The (3^{rd}SI, (3:1)IC)-third-order decreasing-concave sink infinite-equilibrium of $x_{j_1}^* = a_{j_1j_11} = a_{j_2j_11}$ is a (3^{rd}SI, (3:1)IC)-appearing bifurcation.

(v_{2b}) From cases (ii_{4a})-(ii_{4c}), the (3:1)-increasing-asymptotic and (3:1)-decreasing-asymptotic source and sink infinite-equilibriums are the switching bifurcation. Thus, the equilibrium of $x_{j_1}^* = a_{j_1j_11} = a_{j_2j_11}$ has the following switching bifurcations as in Eqs. (2.111)–(2.114).

- For $a_{j_1j_10} > 0$ and $a_{j_2j_10} > 0$,

$$\left\{\begin{array}{c}\underbrace{(3^{\mathrm{rd}}\mathrm{SO},\mathrm{nF})}_{\text{source flow }(-)}\\ \underbrace{(\mathrm{pF},\mathrm{UP})}_{\text{up-parabola flow }(+)}\end{array}\right\} \rightleftharpoons \underbrace{(3^{\mathrm{rd}}\mathrm{SO},(3:1)\mathrm{IC})}_{\text{third-order increasing-concave source}} \rightleftharpoons \left\{\begin{array}{c}\underbrace{(\mathrm{nF},\mathrm{DP})}_{\text{down-parabola flow }(-)}\\ \underbrace{(3^{\mathrm{rd}}\mathrm{SO},\mathrm{pF})}_{\text{source flow }(+)}\end{array}\right\}.$$

The (3^{rd}SO,(3:1)IC)-third-order increasing-concave source infinite-equilibrium of $x_{j_1}^* = a_{j_1j_11} = a_{j_2j_11}$ is a (3^{rd}SO,(3:1)IC)-switching bifurcation.

- For $a_{j_1j_10} < 0$ and $a_{j_2j_10} > 0$,

$$\left\{\begin{array}{c}\underbrace{(3^{\text{rd}}\text{SI, nF})}_{\text{sink flow }(-)}\\ \underbrace{(\text{nF, DP})}_{\text{down-parabola flow }(-)}\end{array}\right\} \rightleftharpoons \underbrace{(3^{\text{rd}}\text{SI}, (3:1)\text{DC})}_{\text{third-order decreasing-concave sink}} \rightleftharpoons \left\{\begin{array}{c}\underbrace{(\text{pF, UP})}_{\text{up-parabola flow }(+)}\\ \underbrace{(3^{\text{rd}}\text{SI, pF})}_{\text{sink flow }(+)}\end{array}\right\}.$$

The (3^{rd}SI,(3:1)DC)-third-order decreasing-concave sink infinite-equilibrium of $x_{j_1}^* = a_{j_1j_11} = a_{j_2j_11}$ is a (3^{rd}SI,(3:1)DC)-switching bifurcation.
- For $a_{j_1j_10} > 0$ and $a_{j_2j_10} < 0$,

$$\left\{\begin{array}{c}\underbrace{(3^{\text{rd}}\text{SO, nF})}_{\text{sink flow }(-)}\\ \underbrace{(\text{pF, DP})}_{\text{down-parabola flow }(+)}\end{array}\right\} \rightleftharpoons \underbrace{(3^{\text{rd}}\text{SO}, (3:1)\text{DC})}_{\text{third-order decreasing-concave source}} \rightleftharpoons \left\{\begin{array}{c}\underbrace{(\text{nF, UP})}_{\text{up-parabola flow }(-)}\\ \underbrace{(3^{\text{rd}}\text{SO, nF})}_{\text{ssource flow }(-)}\end{array}\right\}.$$

The (3^{rd}SO,(3:1)DC)-third-order decreasing-asymptotic source infinite-equilibrium of $x_{j_1}^* = a_{j_1j_11} = a_{j_2j_11}$ is a (3^{rd}SO,(3:1)DC)- switching bifurcation.
- For $a_{j_1j_10} < 0$ and $a_{j_2j_10} < 0$,

$$\left\{\begin{array}{c}\underbrace{(3^{\text{rd}}\text{SI, pF})}_{\text{sink flow }(+)}\\ \underbrace{(\text{nF, UP})}_{\text{up-parabola flow }(-)}\end{array}\right\} \rightleftharpoons \underbrace{(3^{\text{rd}}\text{SI}, (3:1)\text{IC})}_{\text{third-order increasing-concave sink}} \rightleftharpoons \left\{\begin{array}{c}\underbrace{(\text{pF, DP})}_{\text{down-parabola flow }(+)}\\ \underbrace{(3^{\text{rd}}\text{SI, nF})}_{\text{sink flow }(-)}\end{array}\right\}.$$

The (3^{rd}SI,(3:1)IC)-third-order decreasing-asymptotic source infinite-equilibrium of $x_{j_1}^* = a_{j_1j_11} = a_{j_2j_11}$ is a (3^{rd}SI,(3:1)IC)-increasing-asymptotic sink infinite-equilibrium.

Therefore, from the above local analysis, Theorem 2.1 is proved. ■

2.3 Flow Switching and Appearing Bifurcations

Consider a two-dimensional system as

$$\begin{aligned}\dot{x}_1 &= a_{110}(x_1 - a_{111})^3,\\ \dot{x}_2 &= a_{210}(x_1 - a_{211})\end{aligned} \tag{2.115}$$

where $a_{211} > a_{111}$, and the first integral manifold is

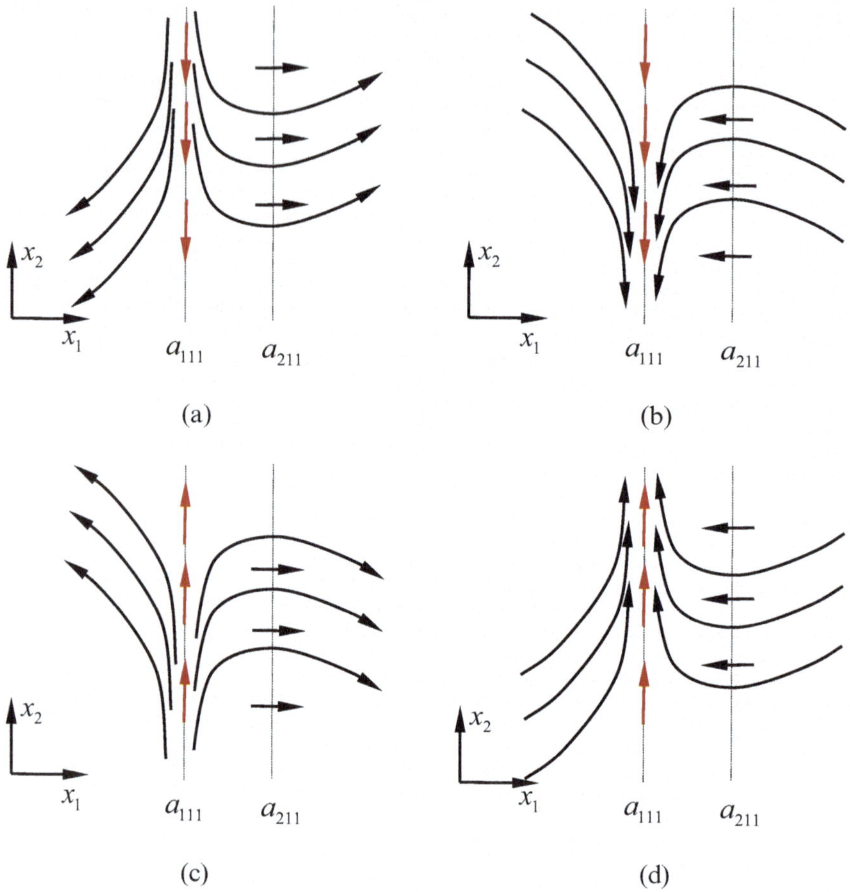

Fig. 2.1 Phase portraits for two-dimensional systems on the x_1-direction with $x_1^* = a_{111}, a_{211}$: (**a**) a (3^{rd}SO,nF)-negative third-order source flow and a (pF,UP)-positive up-parabola flow ($a_{110} > 0$, $a_{210} > 0$), (**b**) a (3^{rd}SI,nF)-negative third-order sink flow and an (nF,DP)-negative down-parabola flow ($a_{110} < 0$, $a_{210} > 0$), (**c**) a (3^{rd}SO,pF)-positive third-order source flow and a (DP,pF)-positive down-parabola flow ($a_{110} > 0$, $a_{210} < 0$), (**d**) a (3^{rd}SI,pF)-positive third-order sink flow and an (nF, UP)-negative up-parabola flow ($a_{110} < 0$, $a_{210} < 0$)

$$
\begin{aligned}
x_2 - x_{20} = -\frac{a_{210}}{a_{110}}\Big\{&(\frac{1}{x_1 - a_{111}} - \frac{1}{x_{10} - a_{111}}) \\
&+ \frac{(a_{111} - a_{211})}{2}[\frac{1}{(x_1 - a_{111})^2} - \frac{1}{(x_{10} - a_{111})^2}]\Big\}.
\end{aligned}
\tag{2.116}
$$

From the first integral manifold, the singular point is at $x_1^* = a_{111}$. Its phase portraits of third-order source and sink flows with parabola flows for a two-dimensional system on the x_1-direction are presented in Fig. 2.1 for $a_{111} < a_{211}$. For $x_1^* = a_{111}$, $\dot{x}_1 = 0$ and $\dot{x}_2 \neq 0$. Thus, for $x_1^* = a_{111}$, the flow is determined by $\dot{x}_2 \neq 0$. For $x_1^* = a_{211}$, $\dot{x}_1 \neq 0$ and $\dot{x}_2 = 0$. Thus for $x_1^* = a_{211}$, the flow is determined by $\dot{x}_1 \neq 0$.

In Fig. 2.1a, the phase portrait of the (3rdSO,nF)-negative third-order flow and the (UP,pF)-positive up-parabola flow for $a_{10} > 0$ and $a_{20} > 0$ is presented from the first integral manifold. At $x_1^* = a_{111}$, the flow of the (US,nF)-negative upper-saddle flow is toward the x_2-negative direction because of $\dot{x}_2 = a_{210}(a_{111} - a_{211}) < 0$. At $x_1^* = a_{211}$, the positive parabola flow is toward the x_1-positive direction because of $\dot{x}_1 = a_{110}(a_{211} - a_{111})^2 > 0$. Similarly, the phase portraits for a (3rdSI,nF)-negative third-order sink flow and a (DP,nF)-negative down-parabola flow($a_{110} < 0$, $a_{210} > 0$), a (3rdSO,pF)-positive third-order source flow and a (DP,pF)-positive down-parabola flow ($a_{110} > 0$, $a_{210} < 0$), and a (3rdSI,pF)-positive third-order sink flow and a (UP,nF)-negative up-parabola flow ($a_{110} < 0$, $a_{210} < 0$) are presented in Fig. 2.1b-d, respectively.

The third-order source and sink flows with parabola flows are presented in Fig. 2.1, and the third-order source and sink flows are the appearing bifurcations for sink and source flows. The third-order source and sink are asymptotic flows near $x_1^* = a_{111}$. The four cases are listed as follows.

$$\left\{\begin{array}{c}(a_{111},\dot{x}_2)\\(\dot{x}_1,a_{211})\end{array}\right\}=\left\{\begin{array}{c}\underbrace{(3^{rd}\text{SO, nF})}_{\text{third-order sink flow }(-)}\\\underbrace{(\text{pF, UP})}_{\text{up-parabola flow }(+)}\end{array}\right\},\text{ for } a_{110}>0, a_{210}>0; \tag{2.117}$$

$$\left\{\begin{array}{c}(a_{111},\dot{x}_2)\\(\dot{x}_1,a_{211})\end{array}\right\}=\left\{\begin{array}{c}\underbrace{(3^{rd}\text{SI, nF})}_{\text{third-order sink flow }(-)}\\\underbrace{(\text{nF, DP})}_{\text{down-parabola flow }(-)}\end{array}\right\},\text{ for } a_{110}<0, a_{210}>0; \tag{2.118}$$

$$\left\{\begin{array}{c}(a_{111},\dot{x}_2)\\(\dot{x}_1,a_{211})\end{array}\right\}=\left\{\begin{array}{c}\underbrace{(3^{rd}\text{SO, pF})}_{\text{third-order source flow }(+)}\\\underbrace{(\text{pF, DP})}_{\text{down-parabola flow }(+)}\end{array}\right\},\text{ for } a_{110}>0, a_{210}<0; \tag{2.119}$$

$$\left\{\begin{array}{c}(a_{111},\dot{x}_2)\\(\dot{x}_1,a_{211})\end{array}\right\}=\left\{\begin{array}{c}\underbrace{(3^{rd}\text{SI, pF})}_{\text{thid-order sink flow }(+)}\\\underbrace{(\text{nF, UP})}_{\text{up-parabola flow }(-)}\end{array}\right\},\text{ for } a_{110}<0, a_{210}<0. \tag{2.120}$$

Consider a two-dimensional system as

$$\begin{aligned}\dot{x}_1 &= a_{110}(x_1 - a_{111})^3,\\ \dot{x}_2 &= a_{210}(x_1 - a_{211})\end{aligned} \tag{2.121}$$

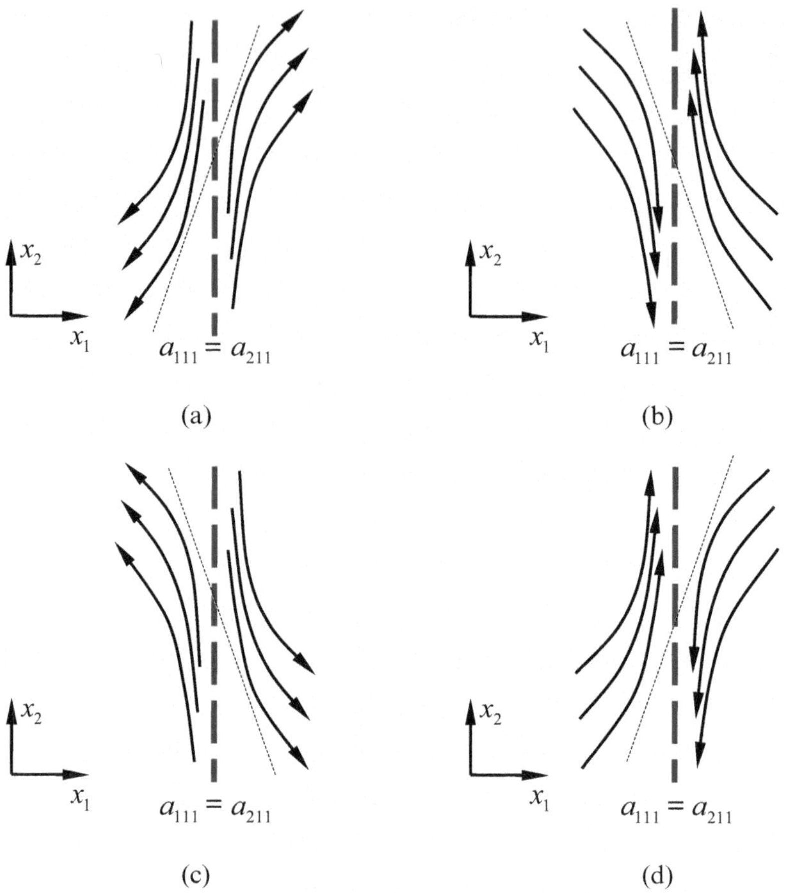

Fig. 2.2 Phase portraits for two-dimensional systems with an infinite-equilibrium line of $x_1^* = a_{111} = a_{211}$: (**a**) the (3[rd]SO,(3:1)IC)-infinite-equilibrium ($a_{110} > 0$, $a_{210} > 0$), (**b**) the (3[rd]SI, (3:1)DC)-infinite-equilibrium ($a_{110} < 0$, $a_{210} > 0$), (**c**) the (3[rd]SO,(3:1)DC)-infinite-equilibrium ($a_{110} > 0$, $a_{210} < 0$), (**d**) the (3[rd]SI,(3:1)IC)-infinite-equilibrium ($a_{110} < 0$, $a_{210} < 0$). IC: increasing-concave flows, DC: decreasing-concave flows. US: up-per-saddle, LS: Lower-saddle

with $a_{211} = a_{111}$. The first integral manifold is

$$x_2 - x_{20} = -\frac{a_{210}}{a_{110}}\left(\frac{1}{x_1 - a_{111}} - \frac{1}{x_{10} - a_{111}}\right). \quad (2.122)$$

The corresponding phase portraits for two-dimensional system with an infinite equilibrium line on the x_1-direction are presented in Fig. 2.2. The (3[rd]SO,(3:1)IC)-infinite-equilibrium ($a_{110} > 0$, $a_{210} > 0$), the (3[rd]SI,(3:1)DC)-infinite-equilibrium ($a_{110} < 0$, $a_{210} > 0$), the (3[rd]SO,(3:1)DC)-infinite-equilibrium ($a_{110} > 0$, $a_{210} < 0$), (3[rd]SI,(3:1)IC)-infinite-equilibrium ($a_{110} < 0$, $a_{210} < 0$) are presented in Fig. 2.2a-d, respectively. The infinite-equilibrium is depicted by the dashed line. The acronyms "IC" and "DC" are for increasing-concave and decreasing-concave flows at the infinite-equilibrium. SO and SI are for source and sink. Such third-order source flow bifurcations are from a source flow to source, sink, and source flows. Such third-order sink flow bifurcations are from a sink flow to sink, source, and sink flows. The infinite-equilibrium is asymptotic and not approachable. The infinite-equilibrium is an asymptote on both sides of the increasing-concave and decreasing-concave source or sink flows.

The appearing bifurcations of the increasing-concave and decreasing-concave sink and source-infinite equilibriums are from an up (down)-parabola flow to sink, source, and a down (up)-parabola flows. Such a flow switching bifurcation is similar to a bundle switching bifurcation, but they are different. The parabola flow will be invariant before and after bifurcation. The third-order source flow becomes source, sink, and source flows, and the third-order sink flow becomes sink, source, and sink flows. The four infinite-equilibriums of $x_1^* = a_{111} = a_{211}$ with $\bar{x}_2 \in (-\infty, \infty)$ are the appearing or switching bifurcations as follows:

$$
\begin{aligned}
(a_{111}, \bar{x}_2) &= \underbrace{(3^{\text{rd}}\text{SO}, (3:1)\text{IC})}_{\text{third-order increasing-concave source}}, \text{ for } a_{110} > 0, a_{210} > 0;\\
(a_{111}, \bar{x}_2) &= \underbrace{(3^{\text{rd}}\text{SI}, (3:1)\text{DC})}_{\text{third-order decreasing-concave sink}}, \text{ for } a_{110} < 0, a_{210} > 0;\\
(a_{111}, \bar{x}_2) &= \underbrace{(3^{\text{rd}}\text{SO}, (3:1)\text{DC})}_{\text{third-order decreasing-concave source}}, \text{ for } a_{110} > 0, a_{210} < 0;\\
(a_{111}, \bar{x}_2) &= \underbrace{(3^{\text{rd}}\text{SI}, (3:1)\text{IC})}_{\text{third-order increasing-concave sink}}, \text{ for } a_{110} < 0, a_{210} < 0.
\end{aligned}
\tag{2.123}
$$

Consider a two-dimensional system as

$$
\begin{aligned}
\dot{x}_1 &= a_{110}(x_1 - a_{111})(x_1 - a_{112})^2,\\
\dot{x}_2 &= a_{210}(x_1 - a_{211}).
\end{aligned}
\tag{2.124}
$$

For $a_{211} = a_{111}$, the first integral manifold is

$$
x_2 - x_{20} = -\frac{a_{210}}{a_{110}}\left(\frac{1}{x_1 - a_{112}} - \frac{1}{x_{10} - a_{112}}\right).
\tag{2.125}
$$

From the first integral manifold, the phase portraits for two-dimensional systems on the x_1-direction with $x_1^* = a_{111}, a_{112}$ and on the x_2-direction with $x_1^* = a_{211} = a_{111}$ are presented in Fig. 2.3. The solid and dashed lines are for sink and source infinite-equilibriums, respectively. The filled and hollow circular symbols are specific

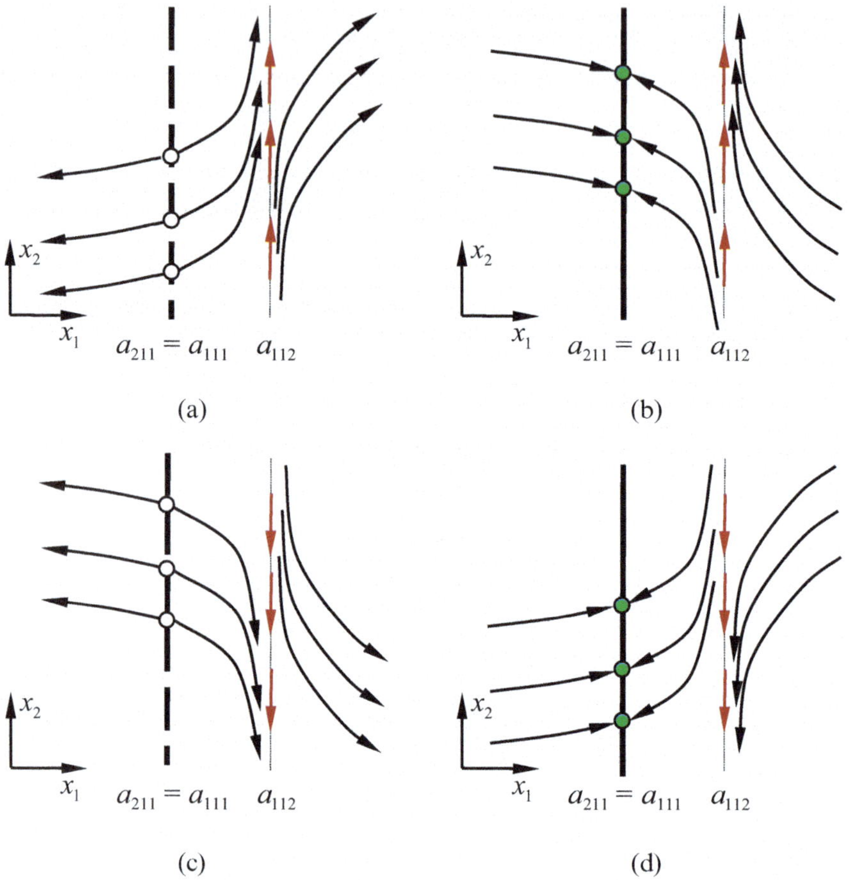

Fig. 2.3 Phase portraits for two-dimensional systems on the x_1-direction with $x_1^* = a_{111}, a_{112}$ and on the x_2-direction with $x_1^* = a_{211} = a_{111}$: (**a**) an (SO,II)-infinite-equilibrium and a (US,pF)-positive source flow ($a_{110} > 0$, $a_{210} > 0$), (**b**) an (SI,DI)-infinite-equilibrium and an (LS,pF)-positive sink flow ($a_{110} < 0$, $a_{210} > 0$), (**c**) an (SO,DI)-infinite-equilibrium and a (US,nF) negative source flow ($a_{110} > 0$, $a_{210} < 0$), (**d**) an (SI,II)-infinite-equilibrium and an (LS, nF) negative sink flow ($a_{110} < 0$, $a_{210} < 0$). The solid and dashed lines are for sink and source infinite-equilibriums, respectively. The filled and hollow circular symbols are specific equilibriums on the infinite-equilibriums, respectively

equilibriums on the infinite-equilibrium lines, respectively. The down- and up-logarithmic sink and source infinite-equilibriums of $x_1^* = a_{111} = a_{211}$ with $\bar{x}_2 \in (-\infty, \infty)$ are the switching bifurcations for up-parabola flows with sink and source flows, and down-parabola flow with sink and source flows, which are given as follows:

$$\begin{Bmatrix} (a_{111}, \bar{x}_2) \\ (a_{112}, \dot{x}_2) \end{Bmatrix} = \begin{Bmatrix} \underbrace{(\text{SO, II})}_{\text{increasing-inflection source}} \\ \underbrace{(\text{US, pF})}_{\text{upper-saddle flow }(+)} \end{Bmatrix}, \text{ for } a_{110} > 0, a_{210} > 0; \tag{2.126}$$

$$\begin{Bmatrix} (a_{111}, \bar{x}_2) \\ (a_{112}, \dot{x}_2) \end{Bmatrix} = \begin{Bmatrix} \underbrace{(\text{SI, DI})}_{\text{decreasing-inflection sink}} \\ \underbrace{(\text{LS, pF})}_{\text{lower-saddle flow }(+)} \end{Bmatrix}, \text{ for } a_{110} < 0, a_{210} > 0; \tag{2.127}$$

$$\begin{Bmatrix} (a_{111}, \bar{x}_2) \\ (a_{112}, \dot{x}_2) \end{Bmatrix} = \begin{Bmatrix} \underbrace{(\text{SO, DI})}_{\text{decreasing-inflection source}} \\ \underbrace{(\text{US, nF})}_{\text{upper-saddle flow }(-)} \end{Bmatrix}, \text{ for } a_{110} > 0, a_{210} < 0; \tag{2.128}$$

$$\begin{Bmatrix} (a_{111}, \bar{x}_2) \\ (a_{112}, \dot{x}_2) \end{Bmatrix} = \begin{Bmatrix} \underbrace{(\text{SI, II})}_{\text{increasing-inflection source}} \\ \underbrace{(\text{LS, nF})}_{\text{lower-saddle flow }(-)} \end{Bmatrix}, \text{ for } a_{110} < 0, a_{210} < 0. \tag{2.129}$$

In Fig. 2.3a, the phase portrait for $a_{110} > 0$ and $a_{210} > 0$ is presented from the first integral manifold. At $x_1^* = a_{111}$, on the x_2-direction is equilibrium because of $\dot{x}_1 = \dot{x}_2 = 0$. At $x_1^* = a_{112}$, the flow is toward the x_2-positive direction because of $\dot{x}_2 = a_{210}(a_{112} - a_{111}) > 0$. $x_1^* = a_{111}, a_{112}$ is an (SO,II)-increasing-inflection source infinite-equilibrium and a (US,pF)-positive upper-saddle flow ($a_{110} > 0$, $a_{210} > 0$). Similarly, the phase portraits for an (SI,DI)-decreasing-inflection sink infinite-equilibrium and an (LS,nF)-positive lower-saddle flow ($a_{110} < 0$, $a_{210} > 0$), an (SO,DI)-decreasing-inflection source infinite-equilibrium and a (US,nF)-negative source flow ($a_{110} > 0$, $a_{210} < 0$), an (SI,II)-increasing-inflection sink infinite-equilibrium and an (LS,nF)-negative lower-saddle flow ($a_{110} < 0$, $a_{210} < 0$) are presented in Fig. 2.3b-d, respectively.

For $a_{211} = a_{112}$, the first integral manifold is

$$x_2 - x_{20} = \frac{a_{210}}{a_{110}} \frac{1}{a_{112} - a_{111}} (\ln \frac{|x_1 - a_{112}|}{|x_{10} - a_{112}|} - \ln \frac{|x_1 - a_{111}|}{|x_{10} - a_{111}|}). \tag{2.130}$$

From the first integral manifold, the phase portraits for two-dimensional systems on the x_1-direction with $x_1^* = a_{111}, a_{112}$ and on the x_2-direction with $x_1^* = a_{211} = a_{111}$ are presented in Fig. 2.4. The solid and dashed lines are for sink and source infinite-equilibriums, respectively. The down- and up-logarithmic upper-saddle and lower-saddle infinite-equilibriums of $x_1^* = a_{112} = a_{211}$ with $\bar{x}_2 \in (-\infty, \infty)$ are the switching

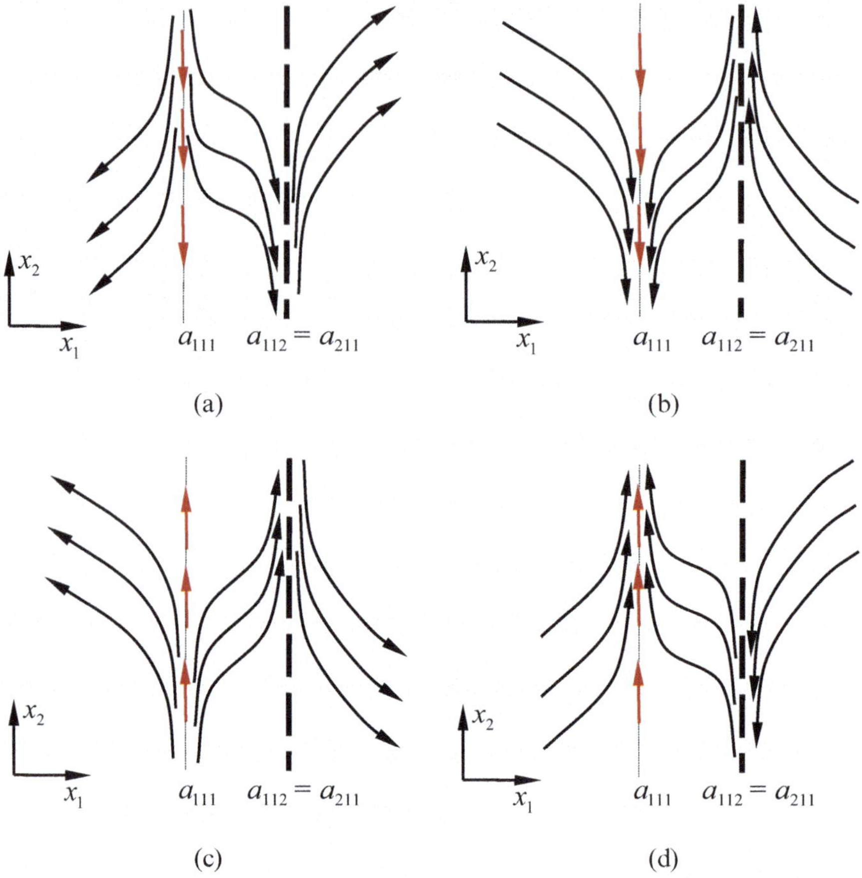

Fig. 2.4 Phase portraits for two-dimensional systems on the x_1-direction with $x_1^* = a_{111}, a_{112}$ and on the x_2-direction with $x_1^* = a_{211} = a_{111}$: (**a**) an (SO,nF)-negative source flow and a (US,DU)-down-up asymptotic upper-saddle ($a_{110} > 0, a_{210} > 0$), (**b**) an (SI,nF)-negative sink flow and an (LS,UD)-up-down asymptotic lower-saddle ($a_{110} < 0, a_{210} > 0$), (**c**) an (SO,pF)-positive source flow and a (US, UD)-up-down asymptotic upper-saddle ($a_{110} > 0, a_{210} < 0$), (**d**) an (SI,pF)-positive sink flow and an (LS,DU)-down-up asymptotic lower-saddle. ($a_{110} < 0, a_{210} < 0$). The dashed lines are for upper-saddle and lower-saddle infinite-equilibriums, respectively

bifurcations for up-parabola flows with sink and source flows, and down-parabola flow with sink and source flows, which are given as follows.

$$\begin{Bmatrix} (a_{111}, \dot{x}_2) \\ (a_{112}, \overline{x}_2) \end{Bmatrix} = \left\{ \begin{array}{c} \underbrace{(\mathrm{SO,\,nF})}_{\text{source flow }(-)} \\ \underbrace{(\mathrm{US,\,DU})}_{\text{down-up upper-saddle}} \end{array} \right\}, \text{for } a_{110} > 0, a_{210} > 0; \tag{2.131}$$

$$\left\{\begin{array}{l}(a_{111},\dot{x}_2)\\(a_{112},\overline{x}_2)\end{array}\right\}=\left\{\begin{array}{c}\underbrace{(\text{SI, nF})}_{\text{sink flow }(-)}\\ \underbrace{(\text{LS, UD})}_{\text{up-down lower-saddle}}\end{array}\right\},\text{ for } a_{110}<0, a_{210}>0; \tag{2.132}$$

$$\left\{\begin{array}{l}(a_{111},\dot{x}_2)\\(a_{112},\overline{x}_2)\end{array}\right\}=\left\{\begin{array}{c}\underbrace{(\text{SO, pF})}_{\text{source flow }(+)}\\ \underbrace{(\text{US, UD})}_{\text{up-down upper-saddle}}\end{array}\right\},\text{ for } a_{110}>0, a_{210}<0; \tag{2.133}$$

$$\left\{\begin{array}{l}(a_{111},\dot{x}_2)\\(a_{112},\overline{x}_2)\end{array}\right\}=\left\{\begin{array}{c}\underbrace{(\text{SI, pF})}_{\text{sink flow }(+)}\\ \underbrace{(\text{LS, DU})}_{\text{down-up lower-saddle}}\end{array}\right\},\text{ for } a_{110}<0, a_{210}<0. \tag{2.134}$$

In Fig. 2.4a, the phase portrait for $a_{110} > 0$ and $a_{210} > 0$ is presented from the first integral manifold. At $x_1^* = a_{111}$, on the x_2-direction is equilibrium because of $\dot{x}_1 = \dot{x}_2 = 0$. At $x_1^* = a_{112}$, the flow is toward the x_2-positive direction because of $\dot{x}_2 = a_{210}(a_{112} - a_{111}) > 0$. $x_1^* = a_{111}, a_{112}$ is an (SO,nF)-negative source flow and a (US, DU)-down-up asymptotic upper-saddle infinite-equilibrium ($a_{110} > 0$, $a_{210} > 0$). Similarly, the phase portraits for an (SI,pF)-negative flow and an (LS, UD)-up-down asymptotic lower-saddle infinite-equilibrium ($a_{110} < 0$, $a_{210} > 0$), an (SO,pF)-positive source flow and a (US,UD)-up-down upper-saddle infinite-equilibrium ($a_{110} > 0$, $a_{210} < 0$), an (SI,pF)-positive sink flow and an (LS, DU)-down-up asymptotic lower-saddle infinite-equilibrium ($a_{110} < 0$, $a_{210} < 0$) are presented in Fig. 2.4b-d, respectively.

Consider a dynamical system as

$$\begin{aligned}\dot{x}_2 &= a_{210}(x_1 - a_{211}),\\ \dot{x}_1 &= a_{110}(x_1 - a_{111})(x_1 - a_{112})(x_1 - a_{113}).\end{aligned} \tag{2.135}$$

The first integral manifold is

$$\begin{aligned}x_2 - x_{20} = \frac{a_{210}}{a_{110}}\bigg[&\frac{a_{111} - a_{211}}{(a_{111} - a_{112})(a_{111} - a_{113})}\ln\frac{|x_1 - a_{111}|}{|x_{10} - a_{111}|}\\ &+ \frac{a_{112} - a_{211}}{(a_{112} - a_{113})(a_{112} - a_{111})}\ln\frac{|x_1 - a_{112}|}{|x_{10} - a_{112}|}\\ &+ \frac{a_{113} - a_{211}}{(a_{113} - a_{111})(a_{113} - a_{112})}\ln\frac{|x_1 - a_{113}|}{|x_{10} - a_{113}|}\bigg].\end{aligned} \tag{2.136}$$

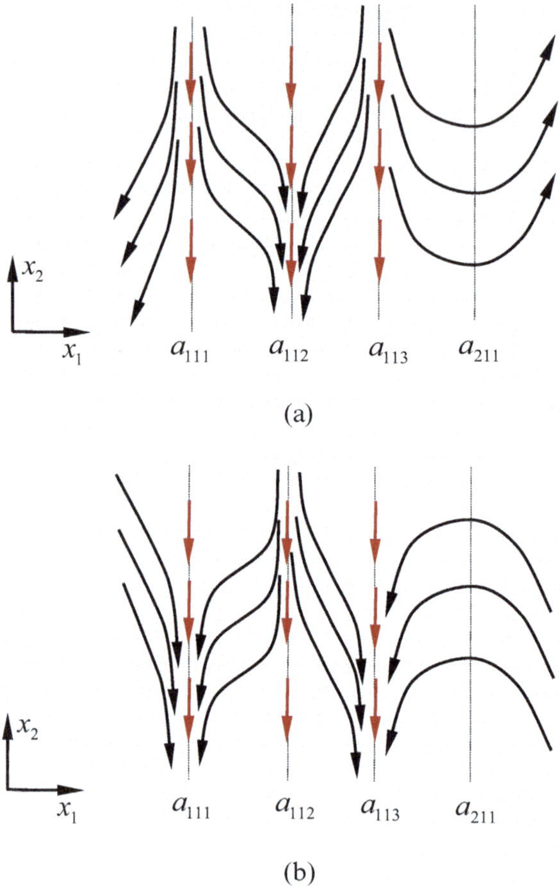

Fig. 2.5 Phase portraits for two-dimensional systems on the x_1-direction with $x_1^* = a_{111}, a_{112}, a_{113}$ and on the x_2-direction with $x_1^* = a_{211} > a_{113}$: (**a**) an (SO,nF)-negative source flow, an (SI,nF)-negative sink flow, an (SO,nF)-negative sink flow and a (UP,pF)-positive up-parabola flow ($a_{110} > 0$, $a_{210} > 0$), (**b**) an (SI,nF)-negative sink flow, an (SO,nF)-negative source flow, an (SI,nF)-negative sink flow and a (DP,nF)-negative down-parabola flow ($a_{110} < 0$, $a_{210} > 0$), (**c**) an (SO,pF)-positive source flow, an (SI,pF)-positive sink flow, an (SO,pF)-negative sink flow, and a (DP,pF)-positive down-parabola flow ($a_{110} > 0$, $a_{210} < 0$), (**d**) an (SI,pF)-positive sink flow, an (SO,pF)-positive source flow, an (SI,pF)-positive sink flow and a (UP,nF)-negative up-parabola flow ($a_{110} < 0$, $a_{210} < 0$)

For $a_{111} < a_{112} < a_{113} < a_{211}$, the sink, source and parabola flows are presented in Fig. 2.5, and the phase portraits for 2-dimensional systems on the x_1-direction with $x_1^* = a_{111}, a_{112}$ and on the x_2-direction with $x_1^* = a_{211} < a_{111}$ are sketched. The parabola flow lies on the left side of the sink and source flows. In Fig. 2.5a, an (SO,nF)-negative source flow, an (SI,nF)-negative sink flow, an (SO,nF)-negative source flow, and a (UP,pF)-positive up-parabola flow ($a_{110} > 0$, $a_{210} > 0$) are presented. The negative sink and source flows are logarithmic flows from the first

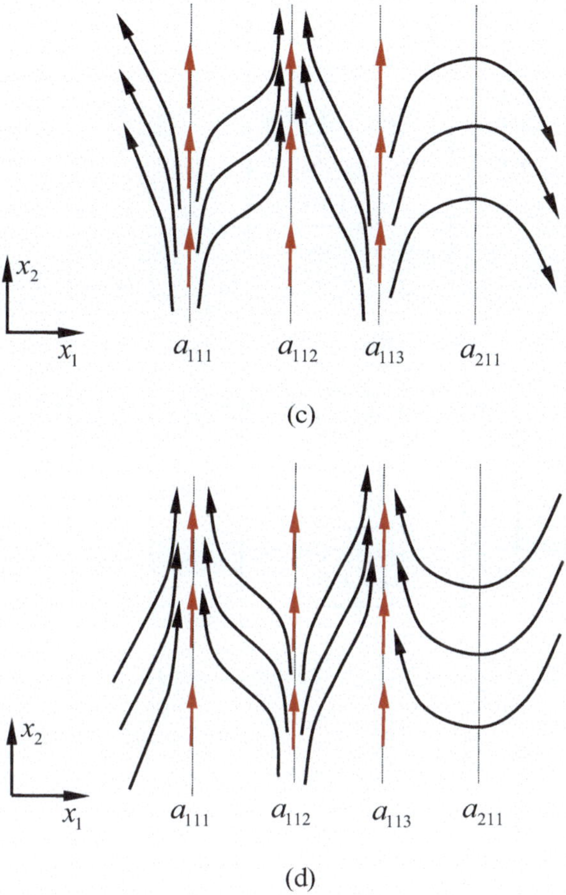

Fig. 2.5 (continued)

integral manifold, and they are asymptotic to the critical lines of $x_1^* = a_{111}, a_{112}, a_{113}$. For $x_1^* = a_{211}$, there is a positive up-parabola flow. Similarly, in Fig. 2.5b, an (SI,nF)-negative sink flow, an (SO,nF)-negative source flow, an (SI,nF)-negative sink flow, and a (DP,nF)-negative down-parabola flow are presented for $(a_{110} < 0, a_{210} > 0)$. In Fig. 2.5c, an (SO,pF)-positive source flow, an (SI,pF)-positive sink, an (SO,pF)-positive source flow, and a (DP,pF)-positive down-parabola flow are presented for $(a_{110} > 0,\ a_{210} < 0)$. Finally, an (SI,pF)-positive sink flow, an (SO,pF)-positive source flow, an (SI,pF)-positive sink flow, and a (UP,nF)-negative up-parabola flow are presented in Fig. 2.5d for $(a_{110} < 0, a_{210} < 0)$. The four cases of sink, source, and parabola flows are listed as follows for $x_1^* = a_{211} > a_{113}$.

$$\left\{\begin{array}{c}(a_{111},\dot{x}_2)\\(a_{112},\dot{x}_2)\\(a_{112},\dot{x}_2)\\(\dot{x}_1,a_{211})\end{array}\right\}=\left\{\begin{array}{c}\underbrace{(\text{SO},\text{nF})}_{\text{source flow }(-)}\\\underbrace{(\text{SI},\text{nF})}_{\text{sink flow }(-)}\\\underbrace{(\text{SO},\text{nF})}_{\text{source flow }(-)}\\\underbrace{(\text{pF},\text{UP})}_{\text{up-parabola flow }(+)}\end{array}\right\}\text{for } a_{110}>0, a_{210}>0 \tag{2.137}$$

$$\left\{\begin{array}{c}(a_{111},\dot{x}_2)\\(a_{112},\dot{x}_2)\\(a_{112},\dot{x}_2)\\(\dot{x}_1,a_{211})\end{array}\right\}=\left\{\begin{array}{c}\underbrace{(\text{SI},\text{nF})}_{\text{sink flow }(-)}\\\underbrace{(\text{SO},\text{nF})}_{\text{source flow }(-)}\\\underbrace{(\text{SI},\text{nF})}_{\text{sink flow }(-)}\\\underbrace{(\text{nF},\text{DP})}_{\text{down-parabola flow }(-)}\end{array}\right\}\text{for } a_{110}<0, a_{210}>0. \tag{2.138}$$

$$\left\{\begin{array}{c}(a_{111},\dot{x}_2)\\(a_{112},\dot{x}_2)\\(a_{112},\dot{x}_2)\\(\dot{x}_1,a_{211})\end{array}\right\}=\left\{\begin{array}{c}\underbrace{(\text{SO},\text{pF})}_{\text{source flow }(+)}\\\underbrace{(\text{SI},\text{pF})}_{\text{sink flow }(+)}\\\underbrace{(\text{SO},\text{pF})}_{\text{source flow }(+)}\\\underbrace{(\text{pF},\text{DP})}_{\text{down-parabola flow }(+)}\end{array}\right\}\text{for } a_{110}>0, a_{210}<0. \tag{2.139}$$

$$\left\{\begin{array}{c}(a_{111},\dot{x}_2)\\(a_{112},\dot{x}_2)\\(a_{112},\dot{x}_2)\\(\dot{x}_1,a_{211})\end{array}\right\}=\left\{\begin{array}{c}\underbrace{(\mathrm{SI},\mathrm{pF})}_{\text{sink flow }(+)}\\\underbrace{(\mathrm{SO},\mathrm{pF})}_{\text{source flow }(+)}\\\underbrace{(\mathrm{SI},\mathrm{pF})}_{\text{sink flow }(+)}\\\underbrace{(\mathrm{UP},\mathrm{nF})}_{\text{up-parabola flow }(-)}\end{array}\right\}\text{for } a_{110}<0, a_{210}<0. \tag{2.140}$$

Similarly, for $a_{211} < a_{111}$, $a_{211} \in (a_{111}, a_{112})$, and $a_{211} \in (a_{112}, a_{113})$, the phase portraits for two-dimensional systems can be sketched, which can be as an exercise for readers to complete.

Chapter 3
Crossing-Quadratic and Self-Cubic Vector Fields

In this chapter, consider two vector fields in single-variable dynamical systems to be quadratic crossing-variable and self-univariate cubic vector fields. The singularity, stability, and bifurcations of such dynamical systems are discussed. In addition to the first- and third-order sink and source flows with upper-saddle and lower-saddle flows, the parabola and inflection flows based on the quadratic crossing-variable vector fields are determined. The *up-parabola* sink (source) infinite-equilibriums are for the switching bifurcations of *increasing-inflection* flows and sink (source) flows. The *down-parabola* sink (source) infinite-equilibriums are for the switching bifurcations of *decreasing*-inflection flows and sink (source) flows. The down-up saddles are the appearing bifurcation from an up-parabola flow to source, down-parabola, and sink flows. The up-down saddles are the appearing bifurcation from a down-parabola flow to source, up-parabola, and sink flows. The (2:2)-inflection upper-saddle infinite-equilibriums are the switching bifurcations for sink and source flows with up-parabola and down-parabola flows. The third-order up and down-logarithmic *source* infinite-equilibriums are the switching bifurcations of the increasing- and decreasing-inflection flows with a third-order *source flow*. The third-order up and down-logarithmic *sink* infinite-equilibriums are the switching bifurcations of the increasing- and decreasing-inflection flows with a third-order *sink flow*.

3.1 Crossing-Quadratic and Self-Cubic Systems

In this section, consider two-dimensional dynamical systems with a variable-crossing quadratic-univariate vector field and a self-univariate cubic vector field. The corresponding stability and bifurcations of equilibriums will be presented through the following theorem.

Theorem 3.1 *Consider a single-variable system with a self-univariate cubic vector field and a crossing-quadratic vector field for* $A_{j_2 j_1 0} \neq 0$ $(j_1, j_2 \in \{1, 2\}, j_1 \neq j_2)$ *as*

A. C. J. Luo, *Two-dimensional Single-Variable Cubic Nonlinear Systems, Vol. I*,
https://doi.org/10.1007/978-3-031-48472-8_3

$$\begin{aligned}\dot{x}_{j_2} &= f_{j_2}(\mathbf{x},\mathbf{p}) = A_{j_2j_12} + A_{j_2j_11}x_{j_1} + A_{j_2j_10}x_{j_1}^2\\ &= a_{j_2j_10}(x_{j_1}^2 + B_{j_2j_11}x_{j_1} + C_{j_2j_11}),\\ \dot{x}_{j_1} &= f_{j_1}(\mathbf{x},\mathbf{p}) = A_{j_1j_13} + A_{j_1j_12}x_{j_1} + A_{j_1j_11}x_{j_1}^2 + A_{j_1j_10}x_{j_1}^3\\ &= a_{j_1j_10}(x_{j_1} - b_{j_1j_11})(x_{j_1}^2 + B_{j_1j_11}x_{j_1} + C_{j_1j_11}),\end{aligned} \tag{3.1}$$

with

$$\Delta_{jj_1} = (B_{jj_11})^2 - 4C_{jj_11} \;\; \text{for } j \in \{j_1, j_2\}, j_1, j_2 \in \{1,2\} j_1 \neq j_2. \tag{3.2}$$

(i) If

$$\Delta_{j_1j_1} < 0 \text{ and } \Delta_{j_2j_1} < 0, \tag{3.3}$$

then

$$\begin{aligned}&\dot{x}_{j_2} = a_{j_2j_10}\left[(x_{j_1} - a_{j_2j_1})^2 + b_{j_2j_1}\right],\\ &\dot{x}_{j_1} = a_{j_1j_10}(x_{j_1} - b_{j_1j_11})\left[(x_{j_1} - a_{j_1j_1})^2 + b_{j_1j_1}\right],\\ &\text{with } a_{sj_1} = -\frac{1}{2}B_{sj_11}, b_{sj_1} = \frac{1}{4}(-\Delta_{sj_1}), s = j_1, j_2;\end{aligned} \tag{3.4}$$

and the first integral manifold is

$$\begin{aligned}x_{j_2} - x_{j_20} = \frac{a_{j_2j_10}}{a_{j_1j_10}} &\left\{ \frac{1}{2}\ln\frac{|(x_{j_1} - a_{j_1j_1})^2 + b_{j_1j_1}|}{|(x_{j_10} - a_{j_1j_1})^2 + b_{j_1j_1}|} \right.\\ &+ \left[2(a_{j_1j_11} - a_{j_2j_1}) + a_{j_1j_1} - a_{j_1j_11}\right]\\ &\times \frac{1}{\sqrt{b_{j_1j_1}}}\left(\arctan\frac{x_{j_1} - a_{j_1j_1}}{\sqrt{b_{j_1j_1}}} - \arctan\frac{x_{j_10} - a_{j_1j_1}}{\sqrt{b_{j_1j_1}}}\right)\\ &+ \frac{(a_{j_1j_11} - a_{j_2j_1})^2 + b_{j_2j_1}}{(a_{j_1j_11} - a_{j_1j_1})^2 + b_{j_1j_1}}\left[\ln\frac{|x_{j_1} - a_{j_1j_11}|}{|x_{j_10} - a_{j_1j_11}|}\right.\\ &- \frac{1}{2}\ln\frac{|(x_{j_1} - a_{j_1j_1})^2 + b_{j_1j_1}|}{|(x_{j_10} - a_{j_1j_1})^2 + b_{j_1j_1}|}\\ &\left.\left. - \frac{a_{j_1j_11} - a_{j_1j_1}}{\sqrt{b_{j_1j_1}}}\left(\arctan\frac{x_{j_1} - a_{j_1j_1}}{\sqrt{b_{j_1j_1}}} - \arctan\frac{x_{j_10} - a_{j_1j_1}}{\sqrt{b_{j_1j_1}}}\right)\right]\right\},\\ &j_1, j_2 \in \{1,2\}, j_1 \neq j_2.\end{aligned} \tag{3.5}$$

The flow at $x_{j_1}^* = a_{j_1j_11}$ *for the system has the following properties.*

- *For $a_{j_1j_10} > 0$ and $a_{j_2j_10} > 0$,*

$$(a_{j_1j_11}, \dot{x}_{j_2}) = \underbrace{(\text{SO, pF})}_{\text{source flow }(+)}. \tag{3.6}$$

The (SO, pF)*-flow is a positive source flow at $x^*_{j_1} = a_{j_1j_11}$.*

- *For $a_{j_1j_10} < 0$ and $a_{j_2j_10} > 0$,*

$$(a_{j_1j_11}, \dot{x}_{j_2}) = \underbrace{(\text{SI, pF})}_{\text{sink flow}(+)}. \tag{3.7}$$

The (SI, pF)*-flow is a positive sink flow at $x^*_{j_1} = a_{j_1j_11}$.*

- *For $a_{j_1j_10} > 0$ and $a_{j_2j_10} < 0$,*

$$(a_{j_1j_11}, \dot{x}_{j_2}) = \underbrace{(\text{SO, nF})}_{\text{source flow }(-)}. \tag{3.8}$$

The (SO, nF)*-flow is a negative source flow at $x^*_{j_1} = a_{j_1j_11}$.*

- *For $a_{j_1j_10} < 0$ and $a_{j_2j_10} < 0$,*

$$(a_{j_1j_11}, \dot{x}_{j_2}) = \underbrace{(\text{SI, nF})}_{\text{sink flow }(-)}. \tag{3.9}$$

The (SI,nF)*-flow is a negative sink flow at $x^*_{j_1} = a_{j_1j_11}$.*

(ii) If

$$\Delta_{j_1j_1} = 0 \text{ and } \Delta_{j_2j_1} < 0, \tag{3.10}$$

the dynamical system becomes

$$\begin{aligned}
&\dot{x}_{j_2} = a_{j_2j_10}\left[(x_{j_1} - a_{j_2j_1})^2 + b_{j_2j_1}\right],\\
&\dot{x}_{j_1} = a_{j_1j_10}(x_{j_1} - a_{j_1j_2s_1})(x_{j_1} - a_{j_1j_2s_2})^2,\\
&\text{with } \{a_{j_1j_11}, a_{j_1j_12}\} = \text{sort}\{b_{j_1j_11}, a_{j_1j_1}\}, a_{j_1j_11} < a_{j_1j_12};\\
&a_{j_1j_1} = -\frac{1}{2}B_{j_1j_11}, b_{j_1j_1} = \frac{1}{4}(-\Delta_{j_1j_1}) = 0;\\
&a_{j_2j_1} = -\frac{1}{2}B_{j_2j_11}, b_{j_2j_1} = \frac{1}{4}(-\Delta_{j_2j_1});\\
&j_1, j_2 \in \{1,2\}; j_1 \neq j_2.
\end{aligned} \tag{3.11}$$

(ii₁) For $a_{j_1j_1s_1} \neq a_{j_1j_1s_2}$, the first integral manifold is

$$
\begin{aligned}
x_{j_2} - x_{j_20} = {} & \frac{a_{j_2j_10}}{a_{j_1j_10}} \Big\{ \ln \frac{|\, x_{j_1} - a_{j_1j_1s_2} \,|}{|\, x_{j_10} - a_{j_1j_1s_2} \,|} \\
& - (a_{j_1j_1} + a_{j_1j_1s_1} - 2a_{j_2j_1})(\frac{1}{x_{j_1} - a_{j_1j_1s_2}} - \frac{1}{x_{j_10} - a_{j_1j_1s_2}}) \\
& + \frac{(a_{j_1j_1s_1} - a_{j_2j_1})^2 + b_{j_2j_1}}{(a_{j_1j_1s_1} - a_{j_1j_1s_2})^2} \Big[\ln \frac{|\, x_{j_1} - a_{j_1j_1s_1} \,|}{|\, x_{j_10} - a_{j_1j_1s_1} \,|} - \ln \frac{|\, x_{j_1} - a_{j_1j_1s_2} \,|}{|\, x_{j_10} - a_{j_1j_1s_2} \,|} \\
& - (a_{j_1j_1s_2} - a_{j_1j_1s_1}) \Big(\frac{1}{x_{j_1} - a_{j_1j_1s_2}} - \frac{1}{x_{j_10} - a_{j_1j_1s_2}} \Big) \Big] \Big\}; \\
& j_1, j_2 \in \{1, 2\}, j_1 \neq j_2.
\end{aligned}
\tag{3.12}
$$

(ii₁ₐ) The equilibrium of $x^*_{j_1} = a_{j_1j_1s_1}$ for the x_{j_1}-direction has the following properties:

- *For $a_{j_1j_10}(a_{j_1j_1s_1} - a_{j_1j_1s_2})^2 > 0$ and $a_{j_2j_10} > 0$,*

$$
(a_{j_1j_1s_1}, \dot{x}_{j_2}) = \underbrace{(\mathrm{SO}, \mathrm{pF})}_{\text{source flow }(+)} . \tag{3.13}
$$

The flow at $x^*_{j_1} = a_{j_1j_1s_1}$ is an (SO,pF)*-positive source flow.*

- *For $a_{j_1j_10}(a_{j_1j_1s_1} - a_{j_1j_1s_2})^2 < 0$ and $a_{j_2j_10} > 0$,*

$$
(a_{j_1j_1s_1}, \dot{x}_{j_2}) = \underbrace{(\mathrm{SI}, \mathrm{pF})}_{\text{sink flow }(+)} . \tag{3.14}
$$

The flow at $x^*_{j_1} = a_{j_1j_1s_1}$ is an (SI,pF)*-positive sink flow.*

- *For $a_{j_1j_10}(a_{j_1j_1s_1} - a_{j_1j_1s_2})^2 > 0$ and $a_{j_2j_10} < 0$,*

$$
(a_{j_1j_1s_1}, \dot{x}_{j_2}) = \underbrace{(\mathrm{SO}, \mathrm{nF})}_{\text{source flow }(-)} . \tag{3.15}
$$

The flow at $x^*_{j_1} = a_{j_1j_1s_1}$ is an (SO,nF)*-negative source flow.*

- *For $a_{j_1j_10}(a_{j_1j_1s_1} - a_{j_1j_1s_2})^2 < 0$ and $a_{j_2j_10} < 0$,*

$$
(a_{j_1j_1s_1}, \dot{x}_{j_2}) = \underbrace{(\mathrm{SI}, \mathrm{nF})}_{\text{sink flow }(-)} . \tag{3.16}
$$

The flow at $x^*_{j_1} = a_{j_1j_1s_1}$ is an (SI,nF)*-negative sink flow.*

(ii_{1b}) The equilibrium of $x_{j_1}^* = a_{j_1j_1s_2}$ *for the* x_{j_1}*-direction has the following properties:*

- *For* $a_{j_1j_10}(a_{j_1j_1s_2} - a_{j_1j_1s_1}) > 0$ *and* $a_{j_2j_10} > 0$,

$$(a_{j_1j_1s_2}, \dot{x}_{j_2}) = \underbrace{(\text{US}, \text{pF})}_{\text{upper-saddle flow } (+)}. \tag{3.17}$$

  The flow at $x_{j_1}^* = a_{j_1j_1s_2}$ *is a* (US,pF)*-positive upper-saddle flow.*
- *For* $a_{j_1j_10}(a_{j_1j_1s_2} - a_{j_1j_1s_1}) < 0$ *and* $a_{j_2j_10} > 0$,

$$(a_{j_1j_1s_2}, \dot{x}_{j_2}) = \underbrace{(\text{LS}, \text{pF})}_{\text{lower-saddle flow } (+)}. \tag{3.18}$$

  The flow at $x_{j_1}^* = a_{j_1j_1s_2}$ *is an* (LS,pF)*-positive lower-saddle flow.*
- *For* $a_{j_1j_10}(a_{j_1j_1s_2} - a_{j_1j_1s_1}) > 0$ *and* $a_{j_2j_10} < 0$,

$$(a_{j_1j_1s_2}, \dot{x}_{j_2}) = \underbrace{(\text{US}, \text{nF})}_{\text{upper-saddle flow } (-)}. \tag{3.19}$$

  The flow at $x_{j_1}^* = a_{j_1j_1s_2}$ *is a* (US,nF)*-negative upper-saddle flow.*
- *For* $a_{j_1j_10}(a_{j_1j_1s_2} - a_{j_1j_1s_1}) < 0$ *and* $a_{j_2j_10} < 0$,

$$(a_{j_1j_1s_2}, \dot{x}_{j_2}) = \underbrace{(\text{LS}, \text{nF})}_{\text{lower-saddle flow } (-)}. \tag{3.20}$$

  The flow at $x_{j_1}^* = a_{j_1j_1s_2}$ *is an* (LS,nF)*-negative lower-saddle flow.*

– *The upper-saddle and lower-saddle flow bifurcations are for the sink and source flows.*

(ii_2) For $a_{j_1j_1s_2} = a_{j_1j_1s_1} \equiv a_{j_1j_11}$, *the dynamical system becomes*

$$\begin{aligned}
&\dot{x}_{j_1} = a_{j_1j_10}(x_{j_1} - a_{j_1j_11})^3,\\
&\dot{x}_{j_2} = a_{j_2j_10}\left[(x_{j_1} - a_{j_2j_1})^2 + b_{j_2j_1}\right],\\
&j_1, j_2 \in \{1,2\}; j_1 \neq j_2.
\end{aligned} \tag{3.21}$$

The first integral manifold is

$$
\begin{aligned}
x_{j_2} - x_{j_2 0} = & \frac{a_{j_2 j_1 0}}{a_{j_1 j_1 0}} \Big\{ \ln \frac{|x_{j_1} - a_{j_1 j_1 1}|}{|x_{j_1 0} - a_{j_1 j_1 1}|} \\
& - 2(a_{j_1 j_1 1} - a_{j_2 j_1})(\frac{1}{x_{j_1} - a_{j_1 j_1 1}} - \frac{1}{x_{j_1 0} - a_{j_1 j_1 1}}) \\
& - \frac{(a_{j_1 j_1 1} - a_{j_2 j_1})^2 + b_{j_2 j_1}}{2} \Big[\frac{1}{(x_{j_1} - a_{j_1 j_1 1})^2} - \frac{1}{(x_{j_1 0} - a_{j_1 j_1 1})^2} \Big] \Big\}, \\
& j_1, j_2 \in \{1,2\}, j_1 \neq j_2.
\end{aligned} \tag{3.22}
$$

The equilibrium of $x_{j_1}^* = a_{j_1 j_1 1}$ *for the* x_{j_1}*-direction has the following properties:*

- *For* $a_{j_1 j_1 0} > 0$ *and* $a_{j_2 j_1 0} > 0$,

$$
(a_{j_1 j_1 1}, \dot{x}_{j_2}) = \underbrace{(3^{\text{rd}}\text{SO, pF})}_{\text{third-order source flow } (+)} \tag{3.23}
$$

  The flow at $x_{j_1}^* = a_{j_1 j_1 1}$ *is a* (3^{rd}SO,pF)*-positive third-order source flow.*

- *For* $a_{j_1 j_1 0} < 0$ *and* $a_{j_2 j_1 0} > 0$,

$$
(a_{j_1 j_1 1}, \dot{x}_{j_2}) = \underbrace{(3^{\text{rd}}\text{SI, pF})}_{\text{third-order sink flow } (+)} \tag{3.24}
$$

  The flow at $x_{j_1}^* = a_{j_1 j_1 1}$ *is a* (3^{rd}SI,pF)*-positive third-order sink flow.*

- *For* $a_{j_1 j_1 0} > 0$ *and* $a_{j_2 j_1 0} < 0$,

$$
(a_{j_1 j_1 1}, \dot{x}_{j_2}) = \underbrace{(3^{\text{rd}}\text{SO, nF})}_{\text{third-order source flow } (-)} \tag{3.25}
$$

  The flow at $x_{j_1}^* = a_{j_1 j_1 1}$ *is a* (3^{rd}SO,nF)*-negative third-order source flow.*

- *For* $a_{j_1 j_1 0} < 0$ *and* $a_{j_2 j_1 0} < 0$,

$$
(a_{j_1 j_1 1}, \dot{x}_{j_2}) = \underbrace{(3^{\text{rd}}\text{SI, nF})}_{\text{third-order sink flow } (-)} \tag{3.26}
$$

  The flow at $x_{j_1}^* = a_{j_1 j_1 1}$ *is a* (3^{rd}SI,nF)*-negative third-order sink flow.*

– *The third-order source and sinks flow bifurcations are the appearing bifurcations for the source-sink-source and sink-source-sink flows.*

(iii) If

$$\Delta_{j_1j_1} > 0 \text{ and } \Delta_{j_2j_1} < 0, \tag{3.27}$$

then

$$\begin{aligned}
&\dot{x}_{j_1} = a_{j_1j_1 0}(x_{j_1} - a_{j_1j_1 1})(x_{j_1} - a_{j_1j_1 2})(x_{j_1} - a_{j_1j_1 3}),\\
&\dot{x}_{j_2} = a_{j_2j_1 0}\left[(x_{j_1} - a_{j_2j_1})^2 + b_{j_2j_1}\right],\\
&j_1, j_2 \in \{1,2\}; j_1 \neq j_2
\end{aligned} \tag{3.28}$$

where

$$\begin{aligned}
&\{a_{j_1j_1 1}, a_{j_1j_1 2}, a_{j_1j_1 3}\} = \text{sort}\{b_{j_1j_1 1}, b_{j_1j_1 2}, b_{j_1j_1 3}\}, a_{j_1j_1 1} < a_{j_1j_1 2} < a_{j_1j_1 3};\\
&b_{j_1j_1 2}, b_{j_1j_1 3} = -\frac{1}{2}(B_{j_1j_1 1} \pm \sqrt{\Delta_{j_1j_1}}); a_{j_2j_1} = -\frac{1}{2}B_{j_2j_1 1}.
\end{aligned} \tag{3.29}$$

The first integral manifold is

$$x_{j_2} - x_{j_2 0} = \sum_{s_1=1}^{3} \frac{(a_{j_1j_1 s_1} - a_{j_2j_1})^2 + b_{j_2j_1}}{\prod_{s_2=1, s_2 \neq s_1}^{3}(a_{j_1j_1 s_1} - a_{j_1j_1 s_2})} \ln \frac{|x_{j_1} - a_{j_1j_1 s_1}|}{|x_{j_1 0} - a_{j_1j_1 s_1}|}. \tag{3.30}$$

The equilibrium of $x_{j_1}^* = a_{j_1j_1 s_1}$ $(s_1, s_2 \in \{1,2,3\}, s_1 \neq s_2)$ in the x_{j_1}-direction possesses the following properties:

- *For $a_{j_1j_1 0}\prod_{s_2=1, s_2 \neq s_1}^{3}(a_{j_1j_1 s_1} - a_{j_1j_1 s_2}) > 0$ and $a_{j_2j_1 0} > 0$,*

$$(a_{j_1j_1 s_1}, \dot{x}_{j_2}) = \underbrace{(\text{SO}, \text{pF})}_{\text{source flow } (+)}. \tag{3.31}$$

The flow at $x_{j_1}^* = a_{j_1j_1 s_1}$ is an (SO,pF)*-positive source-flow.*

- *For $a_{j_1j_1 0}\prod_{s_2=1, s_2 \neq s_1}^{3}(a_{j_1j_1 s_1} - a_{j_1j_1 s_2}) < 0$ and $a_{j_2j_1 0} > 0$,*

$$(a_{j_1j_1 s_1}, \dot{x}_{j_2}) = \underbrace{(\text{SI}, \text{pF})}_{\text{sink flow } (+)}. \tag{3.32}$$

The flow at $x_{j_1}^* = a_{j_1j_1 s_1}$ is an (SI,pF)*-positive sink-flow.*

- *For $a_{j_1j_1 0}\prod_{s_2=1, s_2 \neq s_1}^{3}(a_{j_1j_1 s_1} - a_{j_1j_1 s_2}) > 0$ and $a_{j_2j_1 0} < 0$,*

$$(a_{j_1j_1s_1},\dot{x}_{j_2})=\underbrace{(\mathrm{SO},\mathrm{nF})}_{\text{source flow }(-)}. \tag{3.33}$$

The flow at $x_{j_1}^*=a_{j_1j_1s_1}$ *is an* (SO,nF)*-negative source-flow.*

- *For* $a_{j_1j_1 0}\prod_{s_2=1,s_2\neq s_1}^{3}(a_{j_1j_1s_1}-a_{j_1j_1s_2})<0$ *and* $a_{j_2j_1 0}<0$,

$$(a_{j_1j_1s_1},\dot{x}_{j_2})=\underbrace{(\mathrm{SI},\mathrm{nF})}_{\text{sink flow }(-)}. \tag{3.34}$$

The flow at $x_{j_1}^*=a_{j_1j_1s_1}$ *is an* (SI,nF)*-negative sink-flow.*

(iv) If

$$\Delta_{j_1j_1}<0 \text{ and } \Delta_{j_2j_1}=0, \tag{3.35}$$

then

$$\begin{aligned}
&\dot{x}_{j_2}=a_{j_2j_1 0}(x_{j_1}-a_{j_2j_1 1})^2,\\
&\dot{x}_{j_1}=a_{j_1j_1 0}(x_{j_1}-a_{j_1j_1 1})\left[(x_{j_1}-a_{j_1j_1})^2+b_{j_1j_1}\right],\\
&\text{with } a_{j_1j_1}=-\frac{1}{2}B_{j_1j_1 1},\, b_{j_1j_1}=\frac{1}{4}(-\Delta_{j_1j_1}),\, a_{j_1j_1 1}=b_{j_1j_1 1};\\
&a_{j_2j_1 1}=-\frac{1}{2}B_{j_2j_1 1},\, b_{j_2j_1}=\frac{1}{4}(-\Delta_{j_2j_1})=0.
\end{aligned} \tag{3.36}$$

(iv$_1$) For $a_{j_1j_1 1}\neq a_{j_2j_1 1}$, *the first integral manifold is*

$$\begin{aligned}
x_{j_2}-x_{j_2 0}=&\frac{a_{j_2j_1 0}}{a_{j_1j_1 0}}\left\{\frac{1}{2}\ln\frac{|(x_{j_1}-a_{j_1j_1})^2+b_{j_1j_1}|}{|(x_{j_1 0}-a_{j_1j_1})^2+b_{j_1j_1}|}\right.\\
&+\frac{(a_{j_1j_1 1}-2a_{j_2j_1 1}+a_{j_1j_1})}{\sqrt{b_{j_1j_1}}}(\arctan\frac{x_{j_1}-a_{j_1j_1}}{\sqrt{b_{j_1j_1}}}-\arctan\frac{x_{j_1 0}-a_{j_1j_1}}{\sqrt{b_{j_1j_1}}})\\
&+\frac{(a_{j_1j_1 1}-a_{j_2j_1 1})^2}{(a_{j_1j_1 1}-a_{j_1j_1})^2+b_{j_1j_1}}\left[\ln\frac{|x_{j_1}-a_{j_1j_1 1}|}{|x_{j_1 0}-a_{j_1j_1 1}|}\right.\\
&-\frac{1}{2}\ln\frac{|(x_{j_1}-a_{j_1j_1})^2+b_{j_1j_1}|}{|(x_{j_1 0}-a_{j_1j_1})^2+b_{j_1j_1}|}\\
&\left.\left.-\frac{a_{j_1j_1 1}-a_{j_1j_1}}{\sqrt{b_{j_1j_1}}}(\arctan\frac{x_{j_1}-a_{j_1j_1}}{\sqrt{b_{j_1j_1}}}-\arctan\frac{x_{j_1 0}-a_{j_1j_1}}{\sqrt{b_{j_1j_1}}})\right]\right\},\\
&j_1,j_2\in\{1,2\},j_1\neq j_2.
\end{aligned} \tag{3.37}$$

(iv_{1a}) The flow at $x_{j_1}^* = a_{j_1j_1 1} \neq a_{j_2j_1 1}$ for the system has the following properties.

- *For $a_{j_1j_1 0} > 0$ and $a_{j_2j_1 0}(a_{j_1j_1 1} - a_{j_2j_1 1})^2 > 0$,*

$$(a_{j_1j_1 1}, \dot{x}_{j_2}) = \underbrace{(\text{SO}, \text{pF})}_{\text{source flow }(+)}. \tag{3.38}$$

The (SO,pF)*-flow is a positive source flow at $x_{j_1}^* = a_{j_1j_1 1}$.*

- *For $a_{j_1j_1 0} < 0$ and $a_{j_2j_1 0}(a_{j_1j_1 1} - a_{j_2j_1 1})^2 > 0$,*

$$(a_{j_1j_1 1}, \dot{x}_{j_2}) = \underbrace{(\text{SI}, \text{pF})}_{\text{sink flow }(+)}. \tag{3.39}$$

The (SI, pF)*-flow is a positive sink flow at $x_{j_1}^* = a_{j_1j_1 1}$.*

- *For $a_{j_1j_1 0} > 0$ and $a_{j_2j_1 0}(a_{j_1j_1 1} - a_{j_2j_1 1})^2 < 0$,*

$$(a_{j_1j_1 1}, \dot{x}_{j_2}) = \underbrace{(\text{SO}, \text{nF})}_{\text{source flow }(-)}. \tag{3.40}$$

The (SO, nF)*-flow is a negative source flow at $x_{j_1}^* = a_{j_1j_1 1}$.*

- *For $a_{j_1j_1 0} < 0$ and $a_{j_2j_1 0}(a_{j_1j_1 1} - a_{j_2j_1 1})^2 < 0$,*

$$(a_{j_1j_1 1}, \dot{x}_{j_2}) = \underbrace{(\text{SI}, \text{nF})}_{\text{sink flow }(-)}. \tag{3.41}$$

The (SI,nF)*-flow is a negative sink flow at $x_{j_1}^* = a_{j_1j_1 1}$.*

(iv_{1b}) The flow at $x_{j_1}^* = a_{j_2j_1 1}$ for the system has the following properties.

- *For $a_{j_1j_1 0}(a_{j_2j_1 1} - a_{j_1j_1 1}) > 0$ and $a_{j_2j_1 0} > 0$,*

$$(\dot{x}_{j_1}, a_{j_2j_1 1}) = \underbrace{(\text{pF}, \text{II})}_{\text{increasing-inflection flow }(+)}. \tag{3.42}$$

The (pF,II)*-flow at $x_{j_1}^* = a_{j_2j_1 1}$ is a positive increasing-inflection flow.*

- *For $a_{j_1j_1 0}(a_{j_2j_1 1} - a_{j_1j_1 1}) < 0$ and $a_{j_2j_1 0} > 0$,*

$$(\dot{x}_{j_1}, a_{j_2j_1 1}) = \underbrace{(\text{nF}, \text{DI})}_{\text{decreasing-inflection flow }(-)}. \tag{3.43}$$

The (nF,DI)*-flow at $x_{j_1}^* = a_{j_2j_1 1}$ is a negative decreasing-inflection flow.*

- *For $a_{j_1j_1 0}(a_{j_2j_1 1} - a_{j_1j_1 1}) > 0$ and $a_{j_2j_1 0} < 0$,*

$$(\dot{x}_{j_1}, a_{j_2j_1 1}) = \underbrace{(\text{pF}, \text{DI})}_{\text{decreasing-inflection flow } (+)}. \tag{3.44}$$

The (pF, DI)*-flow at* $x^*_{j_1} = a_{j_2j_1 1}$ *is a positive decreasing-inflection flow.*

- *For* $a_{j_1j_1 0}(a_{j_2j_1 1} - a_{j_1j_1 1}) < 0$ *and* $a_{j_2j_1 0} < 0$,

$$(\dot{x}_{j_1}, a_{j_2j_1 1}) = \underbrace{(\text{nF}, \text{II})}_{\text{increasing-inflection flow } (-)}. \tag{3.45}$$

The (nF,II)*-flow at* $x^*_{j_1} = a_{j_2j_1 1}$ *is a negative increasing-inflection flow.*

– *The increasing-inflection and decreasing-inflection flows are the appearing (or switching) bifurcation for the up-parabola and down-parabola flows.*

(iv₂) For $a_{j_1j_1 1} = a_{j_2j_1 1}$, *the first integral manifold is*

$$\begin{aligned} x_{j_2} - x_{j_2 0} = {} & \frac{a_{j_2j_1 0}}{a_{j_1j_1 0}} \left[\frac{1}{2} \ln \frac{|(x_{j_1} - a_{j_1j_1})^2 + b_{j_1j_1}|}{|(x_{j_1 0} - a_{j_1j_1})^2 + b_{j_1j_1}|} \right. \\ & \left. + \frac{(a_{j_1j_1} - a_{j_1j_1 1})}{\sqrt{b_{j_1j_1}}} \left(\arctan \frac{x_{j_1} - a_{j_1j_1}}{\sqrt{b_{j_1j_1}}} - \arctan \frac{x_{j_1 0} - a_{j_1j_1}}{\sqrt{b_{j_1j_1}}} \right) \right]; \\ & j_1, j_2 \in \{1, 2\}, j_1 \neq j_2. \end{aligned} \tag{3.46}$$

The infinite-equilibrium of $x^*_{j_1} = a_{j_1j_1 1} = a_{j_2j_1 1}$ with $\bar{x}_{j_2} \in (-\infty, \infty)$ has the following properties:

- *For* $a_{j_1j_1 0} > 0$ *and* $a_{j_2j_1 0} > 0$,

$$(a_{j_1j_1 1}, \bar{x}_{j_2}) = \underbrace{(\text{SO}, \text{UP})}_{\text{up-parabola source}}. \tag{3.47}$$

The (SO,UP)*-equilibrium at* $x^*_{j_1} = a_{j_1j_1 1} = a_{j_2j_1 1}$ *is an up-parabola source infinite-equilibrium.*

- *For* $a_{j_1j_1 0} < 0$ *and* $a_{j_2j_1 0} > 0$,

$$(a_{j_1j_1 1}, \bar{x}_{j_2}) = \underbrace{(\text{SI}, \text{DP})}_{\text{down-parabola sink}}. \tag{3.48}$$

The (SI,DP)*-equilibrium at* $x^*_{j_1} = a_{j_1j_1 1} = a_{j_2j_1 1}$ *is a down-parabola sink infinite-equilibrium.*

- *For* $a_{j_1j_1 0} > 0$ *and* $a_{j_2j_1 0} < 0$,

$$(a_{j_1 j_1 1}, \overline{x}_{j_2}) = \underbrace{(\mathrm{SO}, \mathrm{DP})}_{\text{down-parabola source}} . \tag{3.49}$$

The (SO,DP)*-equilibrium at* $x_{j_1}^* = a_{j_1 j_1 1} = a_{j_2 j_1 1}$ *is a down-parabola source infinite-equilibrium.*

- *For* $a_{j_1 j_1 0} < 0$ *and* $a_{j_2 j_1 0} < 0$,

$$(a_{j_1 j_1 1}, \overline{x}_{j_2}) = \underbrace{(\mathrm{SI}, \mathrm{DP})}_{\text{down-parabola sink}} . \tag{3.50}$$

The (SI,UP)*-equilibrium at* $x_{j_1}^* = a_{j_1 j_1 1} = a_{j_2 j_1 1}$ *is an up-parabola sink infinite-equilibrium.*

- *The parabola source and sink infinite-equilibriums are the switching bifurcations of a source (sink) flow and an inflection flow.*
- *The up-parabola source flow is the appearing bifurcation from a positive source flow to negative up-parabola, negative source, positive up-parabola flows.*
- *The up-parabola sink flow is the appearing bifurcation from a negative sink flow to positive up-parabola, positive sink, negative up-parabola flows.*
- *The down-parabola source flow is the appearing bifurcation from a negative source flow to negative down-parabola, positive source, positive down-parabola flows.*
- *The down-parabola sink flow is the appearing bifurcation from a positive sink flow to positive down-parabola, negative sink, negative down-parabola flows.*

(v) If

$$\Delta_{j_1 j_1} = 0 \text{ and } \Delta_{j_2 j_1} = 0, \tag{3.51}$$

then

$$\begin{aligned} &\dot{x}_{j_1} = a_{j_1 j_1 0}(x_{j_1} - a_{j_1 j_1 s_1})(x_{j_1} - a_{j_1 j_1 s_2})^2, \\ &\dot{x}_{j_2} = a_{j_2 j_1 0}(x_{j_1} - a_{j_2 j_1 1})^2, \\ &j_1, j_2 \in \{1,2\}, j_1 \neq j_2; s_1, s_2 \in \{1,2\}, s_1 \neq s_2, \end{aligned} \tag{3.52}$$

where

$$\begin{aligned} &\{a_{j_1 j_1 1}, a_{j_1 j_1 2}\} = \mathrm{sort}\{b_{j_1 j_1 1}, b_{j_1 j_1 2}\}, a_{j_1 j_1 1} < a_{j_1 j_1 2} \\ &b_{j_1 j_1 2} = -\frac{1}{2} B_{j_1 j_1 1}, a_{j_2 j_1 1} = -\frac{1}{2} B_{j_2 j_1 1}; \\ &j_1, j_2 \in \{1,2\}, j_1 \neq j_2. \end{aligned} \tag{3.53}$$

(v_1) For $a_{j_1j_1s_1} \neq a_{j_2j_1 1} \neq a_{j_1j_1s_2}$, the first integral manifold is

$$\begin{aligned} x_{j_2} - x_{j_2 0} = \frac{a_{j_2j_1 0}}{a_{j_1j_1 0}} \Big\{ & \frac{(a_{j_1j_1s_1} - a_{j_2j_1 1})^2}{(a_{j_1j_1s_1} - a_{j_1j_1s_2})^2} \ln \frac{| x_{j_1} - a_{j_1j_1s_1} |}{| x_{j_1 0} - a_{j_1j_1s_1} |} \\ & - \frac{(a_{j_1j_1s_2} - a_{j_2j_1 1})^2}{(a_{j_1j_1s_2} - a_{j_1j_1s_1})} \Big(\frac{1}{x_{j_1} - a_{j_1j_1s_2}} - \frac{1}{x_{j_1 0} - a_{j_1j_1s_2}} \Big) \\ & + \Big[\frac{2(a_{j_1j_1s_2} - a_{j_2j_1 1})}{(a_{j_1j_1s_2} - a_{j_1j_1s_1})} - \frac{(a_{j_1j_1s_2} - a_{j_2j_1 1})^2}{(a_{j_1j_1s_2} - a_{j_1j_1s_1})^2} \Big] \ln \frac{| x_{j_1} - a_{j_1j_1s_2} |}{| x_{j_1 0} - a_{j_1j_1s_2} |} \Big\}, \\ & j_1, j_2 \in \{1,2\}, j_1 \neq j_2; s_1, s_2 \in \{1,2\}, s_1 \neq s_2. \end{aligned} \tag{3.54}$$

(v_{1a}) The equilibrium of $x_{j_1}^* = a_{j_1j_1s_1} \neq a_{j_2j_1 1}$ possesses the following properties.

- *For $a_{j_1j_1 0}(a_{j_1j_1s_1} - a_{j_1j_1s_2})^2 > 0$ and $a_{j_2j_1 0}(a_{j_1j_1s_1} - a_{j_2j_1 1})^2 > 0$,*

$$(a_{j_1j_1s_1}, \dot{x}_{j_2}) = \underbrace{(\text{SO, pF})}_{\text{source flow } (+)} . \tag{3.55}$$

The flow at $x_{j_1}^* = a_{j_1j_1s_1} \neq a_{j_2j_1 1}$ is an (SO,pF)-positive source flow.

- *For $a_{j_1j_1 0}(a_{j_1j_1s_1} - a_{j_1j_1s_2})^2 < 0$ and $a_{j_2j_1 0}(a_{j_1j_1s_1} - a_{j_2j_1 1})^2 > 0$,*

$$(a_{j_1j_1s_1}, \dot{x}_{j_2}) = \underbrace{(\text{SI, pF})}_{\text{sink flow } (+)} . \tag{3.56}$$

The flow at $x_{j_1}^* = a_{j_1j_1s_1} \neq a_{j_2j_1 1}$ is an (SI,pF)-positive sink flow.

- *For $a_{j_1j_1 0}(a_{j_1j_1s_1} - a_{j_1j_1s_2})^2 > 0$ and $a_{j_2j_1 0}(a_{j_1j_1s_1} - a_{j_2j_1 1})^2 < 0$,*

$$(a_{j_1j_1s_1}, \dot{x}_{j_2}) = \underbrace{(\text{SO, nF})}_{\text{source flow } (-)} . \tag{3.57}$$

The flow at $x_{j_1}^* = a_{j_1j_1s_1} \neq a_{j_2j_1 1}$ is an (SO,nF)-negative source flow.

- *For $a_{j_1j_1 0}(a_{j_1j_1s_1} - a_{j_1j_1s_2})^2 < 0$ and $a_{j_2j_1 0}(a_{j_1j_1s_1} - a_{j_2j_1 1})^2 < 0$,*

$$(a_{j_1j_1s_1}, \dot{x}_{j_2}) = \underbrace{(\text{SI, nF})}_{\text{sink flow } (-)} . \tag{3.58}$$

The flow at $x_{j_1}^* = a_{j_1j_1s_1} \neq a_{j_2j_1 1}$ is an (SI,pF)-negative sink flow.

(v_{1b}) *The equilibrium of* $x_{j_1}^* = a_{j_1j_1s_2} \neq a_{j_2j_1 1}$ *possesses the following properties.*

- *For* $a_{j_1j_1 0}(a_{j_1j_1s_2} - a_{j_1j_1s_1}) > 0$ *and* $a_{j_2j_1 0}(a_{j_1j_1s_2} - a_{j_2j_1 1})^2 > 0$,

$$(a_{j_1j_1s_2}, \dot{x}_{j_2}) = \underbrace{(\text{US}, \text{pF})}_{\text{upper-saddle flow }(+)}. \tag{3.59}$$

  The flow at $x_{j_1}^* = a_{j_1j_1s_2} \neq a_{j_2j_1 1}$ *is a* (US,pF)*-positive upper-saddle flow.*

- *For* $a_{j_1j_1 0}(a_{j_1j_1s_2} - a_{j_1j_1s_1}) < 0$ *and* $a_{j_2j_1 0}(a_{j_1j_1s_2} - a_{j_2j_1 1})^2 > 0$,

$$(a_{j_1j_1s_2}, \dot{x}_{j_2}) = \underbrace{(\text{LS}, \text{pF})}_{\text{lower-saddle flow }(+)}. \tag{3.60}$$

  The flow at $x_{j_1}^* = a_{j_1j_1s_2} \neq a_{j_2j_1 1}$ *is an* (LS,pF)*-positive lower-saddle flow.*

- *For* $a_{j_1j_1 0}(a_{j_1j_1s_2} - a_{j_1j_1s_1}) > 0$ *and* $a_{j_2j_1 0}(a_{j_1j_1s_2} - a_{j_2j_1 1})^2 < 0$,

$$(a_{j_1j_1s_2}, \dot{x}_{j_2}) = \underbrace{(\text{US}, \text{nF})}_{\text{upper-saddle flow }(-)}. \tag{3.61}$$

  The flow at $x_{j_1}^* = a_{j_1j_1s_2} \neq a_{j_2j_1 1}$ *is a* (US,nF)*-negative upper-saddle flow.*

- *For* $a_{j_1j_1 0}(a_{j_1j_1s_2} - a_{j_1j_1s_1}) < 0$ *and* $a_{j_2j_1 0}(a_{j_1j_1s_2} - a_{j_2j_1 1})^2 < 0$,

$$(a_{j_1j_1s_2}, \dot{x}_{j_2}) = \underbrace{(\text{LS}, \text{nF})}_{\text{lower-saddle flow }(-)}. \tag{3.62}$$

  The flow at $x_{j_1}^* = a_{j_1j_1s_2} \neq a_{j_2j_1 1}$ *is an* (LS,nF)*-negative lower-saddle flow.*

– *The lower-saddle (LS) and upper-saddle (US) flows are the appearing-switching bifurcation for the sink and source flows.*

(v_{1c}) *The equilibrium of* $x_{j_1}^* = a_{j_2j_1 1} \neq a_{j_1j_1s_1}, a_{j_1j_1s_2}$ *possesses the following properties.*

- *For* $a_{j_1j_1 0}(a_{j_2j_1 1} - a_{j_1j_1s_1})(a_{j_2j_1 1} - a_{j_1j_1s_2})^2 > 0$ *and* $a_{j_2j_1 0} > 0$,

$$(\dot{x}_{j_1}, a_{j_2j_1 1}) = \underbrace{(\text{pF}, \text{II})}_{\text{increasing-inflection flow }(+)}. \tag{3.63}$$

  The flow at $x_{j_1}^* = a_{j_2j_1 1}$ *is a* (pF,II)*-positive increasing-inflection flow.*

- *For* $a_{j_1j_1 0}(a_{j_2j_1 1} - a_{j_1j_1s_1})(a_{j_2j_1 1} - a_{j_1j_1s_2})^2 < 0$ *and* $a_{j_2j_1 0} > 0$,

$$(\dot{x}_{j_1}, a_{j_2 j_1 1}) = \underbrace{(\text{nF}, \text{DI})}_{\text{decreasing-inflection flow }(-)}. \tag{3.64}$$

The flow at $x_{j_1}^* = a_{j_2 j_1 1}$ *is an* (nF,DI)*-negative decreasing-inflection flow.*

- *For* $a_{j_1 j_1 0}(a_{j_2 j_1 1} - a_{j_1 j_1 s_1})(a_{j_2 j_1 1} - a_{j_1 j_1 s_2})^2 > 0$ *and* $a_{j_2 j_1 0} < 0$,

$$(\dot{x}_{j_1}, a_{j_2 j_1 1}) = \underbrace{(\text{pF}, \text{DI})}_{\text{decreasing-inflection flow }(+)}. \tag{3.65}$$

The flow at $x_{j_1}^* = a_{j_2 j_1 1}$ *is a* (pF,DI)*-positive increasing-inflection flow.*

- *For* $a_{j_1 j_1 0}(a_{j_2 j_1 1} - a_{j_1 j_1 s_1})(a_{j_2 j_1 1} - a_{j_1 j_1 s_2})^2 < 0$ *and* $a_{j_2 j_1 0} < 0$,

$$(\dot{x}_{j_1}, a_{j_2 j_1 1}) = \underbrace{(\text{nF}, \text{II})}_{\text{increasing-inflection flow }(-)}. \tag{3.66}$$

The flow at $x_{j_1}^* = a_{j_2 j_1 1}$ *is an* (nF,II)*-positive increasing-inflection flow.*

– *The increasing-inflection* (II) *and decreasing-inflection* (DI) *flows are the appearing-switching bifurcations for up-parabola and down-parabola flows.*

(v_2*) For* $a_{j_1 j_1 s_2} = a_{j_2 j_1 1}$,

$$\begin{aligned}
&\dot{x}_{j_1} = a_{j_1 j_1 0}(x_{j_1} - a_{j_1 j_1 s_1})(x_{j_1} - a_{j_1 j_1 s_2})^2, \\
&\dot{x}_{j_2} = a_{j_2 j_1 0}(x_{j_1} - a_{j_2 j_1 1})^2, \\
&j_1, j_2 \in \{1, 2\}, j_1 \neq j_2.
\end{aligned} \tag{3.67}$$

The first integral manifold is

$$\begin{aligned}
&x_{j_2} - x_{j_2 0} = \frac{a_{j_2 j_1 0}}{a_{j_1 j_1 0}} \ln \frac{|\, x_{j_1} - a_{j_1 j_1 s_1} \,|}{|\, x_{j_1 0} - a_{j_1 j_1 s_1} \,|}, \\
&j_1, j_2 \in \{1, 2\}, j_1 \neq j_2.
\end{aligned} \tag{3.68}$$

(v_{2a}*) The infinite-equilibrium of* $x_{j_1}^* = a_{j_1 j_1 s_2} = a_{j_2 j_1 1}$ with $\bar{x}_{j_2} \in (-\infty, \infty)$ *has the following properties.*

- *For* $a_{j_1 j_1 0}(a_{j_1 j_1 s_2} - a_{j_1 j_1 s_1}) > 0$ *and* $a_{j_2 j_1 0} > 0$,

$$(a_{j_1 j_1 s_2}, \bar{x}_{j_2}) = \underbrace{(\text{US}, (2{:}2)\text{II})}_{\text{increasing-inflection upper-saddle}}. \tag{3.69}$$

The equilibrium at $x_{j_1}^* = a_{j_1 j_1 s_2} = a_{j_2 j_1 1}$ *is a* (US,(2:2)II)*-increasing-inflection upper-saddle infinite-equilibrium.*

- *For* $a_{j_1j_10}(a_{j_1j_1s_2} - a_{j_1j_1s_1}) < 0$ *and* $a_{j_2j_10} > 0$,

$$(a_{j_1j_1s_2}, \overline{x}_{j_2}) = \underbrace{(\text{LS}, (2{:}2)\text{DI})}_{\text{decreasing-inflection lower-saddle}} . \tag{3.70}$$

The equilibrium at $x_{j_1}^* = a_{j_1j_1s_2} = a_{j_2j_11}$ *is an* (LS,(2:2)DI)-*decreasing-inflection lower-saddle infinite-equilibrium.*

- *For* $a_{j_1j_10}(a_{j_1j_1s_2} - a_{j_1j_1s_1}) > 0$ *and* $a_{j_2j_10} < 0$,

$$(a_{j_1j_1s_2}, \overline{x}_{j_2}) = \underbrace{(\text{US}, (2{:}2)\text{DI})}_{\text{decreasing-inflection upper-saddle}} . \tag{3.71}$$

The equilibrium at $x_{j_1}^* = a_{j_1j_1s_2} = a_{j_2j_11}$ *is* a (US,(2:2)DI)-*increasing-inflection upper-saddle infinite-equilibrium.*

- *For* $a_{j_1j_10}(a_{j_1j_1s_2} - a_{j_1j_1s_1}) < 0$ *and* $a_{j_2j_10} < 0$,

$$(a_{j_1j_1s_2}, \overline{x}_{j_2}) = \underbrace{(\text{LS}, (2{:}2)\text{II})}_{\text{increasing-inflection lower-saddle}} . \tag{3.72}$$

The equilibrium at $x_{j_1}^* = a_{j_1j_1s_2} = a_{j_2j_11}$ *is an* (LS,(2:2)II)-*increasing-inflection lower-saddle infinite-equilibrium.*

– *The infinite-equilibriums are the switching bifurcations for the increasing-inflection and decreasing-inflection flows with the upper-saddle and lower-saddle flows.*

(v_{2b}) *The infinite-equilibriums of* $x_{j_1}^* = a_{j_1j_1s_2} = a_{j_2j_11}$ with $\overline{x}_{j_2} \in (-\infty, \infty)$ *are the appearing bifurcations for the sink and source flows and the up-parabola and down-parabola flows with* $\Delta_{j_1j_1}^{+} = \Delta_{j_1j_1} + 0$ and $\Delta_{j_2j_1}^{+} = \Delta_{j_2j_1} + 0$.

- *For* $a_{j_1j_10}(a_{j_1j_1s_2} - a_{j_1j_1s_1}) > 0$ *and* $a_{j_2j_10} > 0$,

$$\underbrace{(\text{US}, (2{:}2)\text{II})}_{\text{appearing bifurcation}} \rightleftharpoons \left\{ \begin{array}{cc} \underbrace{(\text{SI}, \text{pF})}_{\text{sink flow }(+)}, & \underbrace{(\text{nF}, \text{UP})}_{\text{up-parabola flow }(-)}, \\ \underbrace{(\text{nF}, \text{DP})}_{\text{down-parabola flow }(-)}, & \underbrace{(\text{SO}, \text{pF})}_{\text{source flow }(+)} \end{array} \right\} \tag{3.73}$$

for $\Delta_{j_1j_1}^{+} > \Delta_{j_2j_1}^{+}$;

$$\underbrace{(\text{US},(2{:}2)\text{II})}_{\text{appearing bifurcation}} \rightleftharpoons \left\{\begin{array}{cc} \underbrace{(\text{pF},\text{DP})}_{\text{down-parabola flow }(+)}, & \underbrace{(\text{SI},\text{nF})}_{\text{sink flow }(-)}, \\ \underbrace{(\text{SO},\text{nF})}_{\text{source flow }(-)}, & \underbrace{(\text{nF},\text{UP})}_{\text{up-parabola flow }(+)} \end{array}\right\} \tag{3.74}$$

for $\Delta^+_{j_1j_1} < \Delta^+_{j_2j_1}$.

- *For* $a_{j_1j_10}(a_{j_1j_1s_2} - a_{j_1j_1s_1}) < 0$ *and* $a_{j_2j_10} > 0$,

$$\underbrace{(\text{LS},(2{:}2)\text{DI})}_{\text{appearing bifurcation}} \rightleftharpoons \left\{\begin{array}{cc} \underbrace{(\text{SO},\text{pF})}_{\text{source flow }(+)}, & \underbrace{(\text{pF},\text{DP})}_{\text{down-parabola flow }(+)}, \\ \underbrace{(\text{pF},\text{UP})}_{\text{up-parabola flow }(+)}, & \underbrace{(\text{SI},\text{pF})}_{\text{sink flow }(+)} \end{array}\right\} \tag{3.75}$$

for $\Delta^+_{j_1j_1} > \Delta^+_{j_2j_1}$;

$$\underbrace{(\text{LS},(2{:}2)\text{DI})}_{\text{appearing bifurcation}} \rightleftharpoons \left\{\begin{array}{cc} \underbrace{(\text{nF},\text{UP})}_{\text{up-parabola flow }(-)}, & \underbrace{(\text{SO},\text{nF})}_{\text{source flow }(-)}, \\ \underbrace{(\text{SI},\text{nF})}_{\text{sink flow }(-)}, & \underbrace{(\text{nF},\text{DP})}_{\text{down-parabola flow }(-)} \end{array}\right\} \tag{3.76}$$

for$\Delta^+_{j_1j_1} < \Delta^+_{j_2j_1}$.

- *For* $a_{j_1j_10}(a_{j_1j_1s_2} - a_{j_1j_1s_1}) > 0$ *and* $a_{j_2j_10} < 0$,

$$\underbrace{(\text{US},(2{:}2)\text{DI})}_{\text{appearing bifurcation}} \rightleftharpoons \left\{\begin{array}{cc} \underbrace{(\text{SI},\text{nF})}_{\text{sink flow }(-)}, & \underbrace{(\text{nF},\text{DP})}_{\text{down-parabola flow }(-)}, \\ \underbrace{(\text{nF},\text{UP})}_{\text{up-parabola flow }(-)}, & \underbrace{(\text{SO},\text{nF})}_{\text{source flow }(-)} \end{array}\right\} \tag{3.77}$$

for$\Delta^+_{j_1j_1} > \Delta^+_{j_2j_1}$;

$$\underbrace{(\text{US},(2{:}2)\text{DI})}_{\text{appearing bifurcation}} \rightleftharpoons \left\{\begin{array}{cc} \underbrace{(\text{pF},\text{UP})}_{\text{up-parabola flow }(+)}, & \underbrace{(\text{SI},\text{pF})}_{\text{sink flow }(+)}, \\ \underbrace{(\text{SO},\text{pF})}_{\text{source flow }(+)}, & \underbrace{(\text{pF},\text{DP})}_{\text{down-parabola flow }(+)} \end{array}\right\} \tag{3.78}$$

for $\Delta^+_{j_1j_1} > \Delta^+_{j_2j_1}$.

- *For* $a_{j_1j_10}(a_{j_1j_1s_2}-a_{j_1j_1s_1})<0$ *and* $a_{j_2j_10}<0$,

$$\underbrace{(\mathrm{LS},(2{:}2)\mathrm{II})}_{\text{appearing bifurcation}} \rightleftharpoons \left\{\begin{array}{cc} \underbrace{(\mathrm{SO},\mathrm{nF})}_{\text{source flow }(-)}, & \underbrace{(\mathrm{pF},\mathrm{UP})}_{\text{up-parabola flow }(+)}, \\ \underbrace{(\mathrm{pF},\mathrm{DP})}_{\text{down-parabola flow }(+)}, & \underbrace{(\mathrm{SI},\mathrm{nF})}_{\text{sink flow }(-)} \end{array}\right\} \tag{3.79}$$

for $\Delta^+_{j_1j_1}>\Delta^+_{j_2j_1}$;

$$\underbrace{(\mathrm{LS},(2{:}2)\mathrm{II})}_{\text{appearing bifurcation}} \rightleftharpoons \left\{\begin{array}{cc} \underbrace{(\mathrm{nF},\mathrm{DP})}_{\text{down-parabola flow }(-)}, & \underbrace{(\mathrm{SO},\mathrm{pF})}_{\text{source flow }(+)}, \\ \underbrace{(\mathrm{SI},\mathrm{pF})}_{\text{sink flow }(+)}, & \underbrace{(\mathrm{nF},\mathrm{UP})}_{\text{up-parabola flow }(-)} \end{array}\right\} \tag{3.80}$$

for $\Delta^+_{j_1j_1}<\Delta^+_{j_2j_1}$.

(v_3) For $a_{j_1j_1s_1}=a_{j_2j_11}$,

$$\begin{aligned} \dot{x}_{j_1} &= a_{j_1j_10}(x_{j_1}-a_{j_1j_1s_1})(x_{j_1}-a_{j_1j_1s_2})^2, \\ \dot{x}_{j_2} &= a_{j_2j_10}(x_{j_1}-a_{j_2j_11})^2, \\ j_1,j_2 &\in \{1,2\}, j_1\neq j_2. \end{aligned} \tag{3.81}$$

The first integral manifold is

$$\begin{aligned} x_{j_2}-x_{j_20} = \frac{a_{j_2j_10}}{a_{j_1j_10}}\Big[&\ln\frac{|\,x_{j_1}-a_{j_1j_1s_2}\,|}{|\,x_{j_10}-a_{j_1j_1s_2}\,|} \\ &-(a_{j_1j_1s_2}-a_{j_2j_11})\Big(\frac{1}{x_{j_1}-a_{j_1j_1s_2}}-\frac{1}{x_{j_10}-a_{j_1j_1s_2}}\Big)\Big], \\ j_1,j_2 \in \{1,2\}, j_1\neq j_2.& \end{aligned} \tag{3.82}$$

(v_{3a}) The infinite-equilibrium of $x^*_{j_1}=a_{j_1j_1s_1}=a_{j_2j_11}$ with $\bar{x}_{j_2}\in(-\infty,\infty)$ *has the following properties.*

- *For* $a_{j_1j_10}(a_{j_1j_1s_1}-a_{j_1j_1s_2})^2>0$ *and* $a_{j_2j_10}>0$,

$$(a_{j_1j_1s_1},\bar{x}_{j_2}) = \underbrace{(\mathrm{SO},\mathrm{UP})}_{\text{up-parabola source}}. \tag{3.83}$$

  The equilibrium at $x^*_{j_1}=a_{j_1j_1s_1}=a_{j_2j_11}$ *is an* (SO,UP)*-up-parabola source infinite-equilibrium.*

- *For* $a_{j_1j_10}(a_{j_1j_1s_1}-a_{j_1j_1s_2})^2<0$ *and* $a_{j_2j_10}>0$,

$$(a_{j_1j_1s_1}, \overline{x}_{j_2}) = \underbrace{(\text{SI}, \text{DP})}_{\text{down-parabola sink}}. \tag{3.84}$$

The equilibrium at $x_{j_1}^* = a_{j_1j_1s_1} = a_{j_2j_11}$ *is an* (SI,DP)*-down-parabola sink infinite-equilibrium.*

- *For* $a_{j_1j_10}(a_{j_1j_1s_1} - a_{j_1j_1s_2})^2 > 0$ *and* $a_{j_2j_10} < 0$,

$$(a_{j_1j_1s_1}, \overline{x}_{j_2}) = \underbrace{(\text{SO}, \text{DP})}_{\text{down-parabola source}}. \tag{3.85}$$

The equilibrium at $x_{j_1}^* = a_{j_1j_1s_1} = a_{j_2j_11}$ *is an* (SO,DP)*-down-parabola source infinite-equilibrium.*

- *For* $a_{j_1j_10}(a_{j_1j_1s_1} - a_{j_1j_1s_2})^2 < 0$ *and* $a_{j_2j_10} < 0$,

$$(a_{j_1j_1s_1}, \overline{x}_{j_2}) = \underbrace{(\text{SI}, \text{UP})}_{\text{up-parabola sink}}. \tag{3.86}$$

The equilibrium at $x_{j_1}^* = a_{j_1j_1s_1} = a_{j_2j_11}$ *is an* (SI,UP)*-up-parabola sink infinite-equilibrium.*

(v_{3b}) The infinite-equilibriums are the switching bifurcation for the sink and source flows with increasing-inflection and decreasing-inflection flows.

- *For* $a_{j_1j_10}(a_{j_1j_1s_1} - a_{j_1j_1s_2})^2 > 0$ *and* $a_{j_2j_10} > 0$,

$$\left\{\begin{array}{c}\underbrace{(\text{nF}, \text{DI})}_{\text{decreasing-inflection flow }(-)}\\ \underbrace{(\text{SO}, \text{pF})}_{\text{source flow }(+)}\end{array}\right\} \rightleftharpoons \underbrace{(\text{SO}, \text{UP})}_{\text{switching bifurcation}} \rightleftharpoons \left\{\begin{array}{c}\underbrace{(\text{SO}, \text{pF})}_{\text{source flow }(+)}\\ \underbrace{(\text{pF}, \text{II})}_{\text{increasing-inflection flow }(+)}\end{array}\right\}. \tag{3.87}$$

The (SO,UP)*-up-parabola source infinite-equilibrium is called an* (SO,UP)*-switching bifurcation.*

- *For* $a_{j_1j_10}(a_{j_1j_1s_1} - a_{j_1j_1s_2})^2 < 0$ *and* $a_{j_2j_10} > 0$,

$$\left\{\begin{array}{c}\underbrace{(\text{pF}, \text{II})}_{\text{increasing-inflection flow }(+)}\\ \underbrace{(\text{SI}, \text{pF})}_{\text{sink flow }(+)}\end{array}\right\} \rightleftharpoons \underbrace{(\text{SI}, \text{DP})}_{\text{switching bifurcation}} \rightleftharpoons \left\{\begin{array}{c}\underbrace{(\text{SI}, \text{pF})}_{\text{sink flow }(+)}\\ \underbrace{(\text{nF}, \text{DI})}_{\text{decreasing-inflection flow }(-)}\end{array}\right\}. \tag{3.88}$$

The (SI,DP)*-down-parabola sink infinite-equilibrium is called an (SI,DP)-switching bifurcation.*

- *For $a_{j_1j_10}(a_{j_1j_1s_1}-a_{j_1j_1s_2})^2>0$ and $a_{j_2j_10}<0$,*

$$\left\{\begin{array}{c}\underbrace{(\text{nF},\text{II})}_{\text{increasing-inflection flow }(-)}\\ \underbrace{(\text{SO},\text{nF})}_{\text{source flow }(-)}\end{array}\right\}\rightleftharpoons\underbrace{(\text{SO},\text{DP})}_{\text{switching bifurcation}}\rightleftharpoons\left\{\begin{array}{c}\underbrace{(\text{SO},\text{nF})}_{\text{source flow }(-)}\\ \underbrace{(\text{pF},\text{DI})}_{\text{decreasing-inflection flow }(+)}\end{array}\right\}. \tag{3.89}$$

The (SO,DP)-*down-parabola source infinite-equilibrium is called an* (SO, DP)-*switching bifurcation.*

- *For $a_{j_1j_10}(a_{j_1j_1s_1}-a_{j_1j_1s_2})^2<0$ and $a_{j_2j_10}<0$,*

$$\left\{\begin{array}{c}\underbrace{(\text{pF},\text{DI})}_{\text{decreasing-inflection flow }(+)}\\ \underbrace{(\text{SI},\text{nF})}_{\text{sink flow }(-)}\end{array}\right\}\rightleftharpoons\underbrace{(\text{SI},\text{UP})}_{\text{switching bifurcation}}\rightleftharpoons\left\{\begin{array}{c}\underbrace{(\text{SI},\text{nF})}_{\text{sink flow }(-)}\\ \underbrace{(\text{nF},\text{II})}_{\text{increasing-inflection flow }(-)}\end{array}\right\}. \tag{3.90}$$

The (SI,UP)-*up-parabola sink infinite-equilibrium is called an* (SI,UP)-*switching bifurcation.*

(vi) If

$$\begin{aligned}&\Delta_{j_1j_1}=0\text{ and }\Delta_{j_2j_1}=0,\\&a_{j_1j_1s_1}=a_{j_1j_1s_2}\equiv a_{j_1j_11}\end{aligned} \tag{3.91}$$

then

$$\begin{aligned}&\dot{x}_{j_1}=a_{j_1j_10}(x_{j_1}-a_{j_1j_11})^3,\\&\dot{x}_{j_2}=a_{j_2j_10}(x_{j_1}-a_{j_2j_11})^2,\\&j_1,j_2\in\{1,2\},j_1\neq j_2.\end{aligned} \tag{3.92}$$

(vi_1) For $a_{j_1j_11}\neq a_{j_2j_11}$, the first integral manifold is

$$\begin{aligned}x_{j_2}-x_{j_20}=\frac{a_{j_2j_10}}{a_{j_1j_10}}\Big\{&\ln\frac{|\,x_{j_1}-a_{j_1j_11}\,|}{|\,x_{j_10}-a_{j_1j_11}\,|}\\&-2(a_{j_1j_11}-a_{j_2j_11})(\frac{1}{x_{j_1}-a_{j_1j_11}}-\frac{1}{x_{j_10}-a_{j_1j_11}})\\&-\frac{1}{2}(a_{j_1j_11}-a_{j_2j_11})^2\Big[\frac{1}{(x_{j_1}-a_{j_1j_11})^2}-\frac{1}{(x_{j_10}-a_{j_1j_11})^2}\Big]\Big\},\\ j_1,j_2\in\{1,2\},j_1\neq j_2.&\end{aligned} \tag{3.93}$$

(vi$_{1a}$) The equilibrium of $x_{j_1}^* = a_{j_1j_1 1} \neq a_{j_2j_1 1}$ has the following properties.

- *For $a_{j_1j_1 0} > 0$ and $a_{j_2j_1 0}(a_{j_1j_1 1} - a_{j_2j_1 1})^2 > 0$,*

$$(a_{j_1j_1 1}, \dot{x}_{j_2}) = \underbrace{(3^{\text{rd}}\text{SO}, \text{pF})}_{\text{third-order source flow }(+)}. \tag{3.94}$$

The flow at $x_{j_1}^* = a_{j_1j_1 1}$ is a (3rdSO,pF)*-positive third-order source flow.*

- *For $a_{j_1j_1 0} < 0$ and $a_{j_2j_1 0}(a_{j_1j_1 1} - a_{j_2j_1 1})^2 > 0$,*

$$(a_{j_1j_1 1}, \dot{x}_{j_2}) = \underbrace{(3^{\text{rd}}\text{SI}, \text{pF})}_{\text{third-order sink flow }(+)}. \tag{3.95}$$

The flow at $x_{j_1}^* = a_{j_1j_1 1}$ is a (3rdSI,pF)*-positive third-order sink flow.*

- *For $a_{j_1j_1 0} > 0$ and $a_{j_2j_1 0}(a_{j_1j_1 1} - a_{j_2j_1 1})^2 < 0$,*

$$(a_{j_1j_1 1}, \dot{x}_{j_2}) = \underbrace{(3^{\text{rd}}\text{SO}, \text{nF})}_{\text{third-order source flow }(-)}. \tag{3.96}$$

The flow at $x_{j_1}^* = a_{j_1j_1 1}$ is a (3rdSO,nF)*-negative third-order source flow.*

- *For $a_{j_1j_1 0} < 0$ and $a_{j_2j_1 0}(a_{j_1j_1 1} - a_{j_2j_1 1})^2 < 0$,*

$$(a_{j_1j_1 1}, \dot{x}_{j_2}) = \underbrace{(3^{\text{rd}}\text{SI}, \text{nF})}_{\text{third-order sink flow }(-)}. \tag{3.97}$$

The flow at $x_{j_1}^* = a_{j_1j_1 1}$ is a (3rdSI,pF)*-negative third-order sink flow.*

– *The third-order source and sink flows bifurcations are the appearing bifurcations for source-sink-source and sink-source-sink flows.*

(vi$_{1b}$) The equilibrium of $x_{j_1}^* = a_{j_2j_1 1} \neq a_{j_1j_1 1}$ has the following properties.

- *For $a_{j_1j_1 0}(a_{j_2j_1 1} - a_{j_1j_1 1})^3 > 0$ and $a_{j_2j_1 0} > 0$,*

$$(\dot{x}_{j_1}, a_{j_2j_1 1}) = \underbrace{(\text{pF}, \text{II})}_{\text{increasing-inflection flow }(+)}. \tag{3.98}$$

The flow at $x_{j_1}^* = a_{j_2j_1 1}$ is a (pF,II)*-positive increasing-inflection flow.*

- *For $a_{j_1j_1 0}(a_{j_2j_1 1} - a_{j_1j_1 1})^3 < 0$ and $a_{j_2j_1 0} > 0$,*

$$(\dot{x}_{j_1}, a_{j_2j_1 1}) = \underbrace{(\text{nF, DI})}_{\text{decreasing-inflection flow }(-)}. \tag{3.99}$$

The flow at $x^*_{j_1} = a_{j_2j_1 1}$ *is an* (nF,DI)*-negative decreasing-inflection flow.*

- *For* $a_{j_1j_1 0}(a_{j_2j_1 1} - a_{j_1j_1 1})^3 > 0$ *and* $a_{j_2j_1 0} < 0$,

$$(\dot{x}_{j_1}, a_{j_2j_1 1}) = \underbrace{(\text{pF, DI})}_{\text{decreasing-inflection flow }(+)}. \tag{3.100}$$

The flow at $x^*_{j_1} = a_{j_2j_1 1}$ *is a* (pF,DI)*-positive decreasing-inflection flow.*

- *For* $a_{j_1j_1 0}(a_{j_2j_1 1} - a_{j_1j_1 1})^3 < 0$ *and* $a_{j_2j_1 0} < 0$,

$$(\dot{x}_{j_1}, a_{j_2j_1 1}) = \underbrace{(\text{nF, II})}_{\text{increasing-inflection flow }(-)}. \tag{3.101}$$

The flow at $x^*_{j_1} = a_{j_2j_1 1}$ *is an* (nF,II)*-negative increasing-inflection flow.*

– *The increasing-inflection and decreasing-inflection flow bifurcations are the appearing bifurcations for the up-parabola and down-parabola flows.*

(vi$_2$) For $a_{j_2j_1 1} = a_{j_1j_1 1}$,

$$\begin{aligned} &\dot{x}_{j_1} = a_{j_1j_1 0}(x_{j_1} - a_{j_1j_1 1})^3, \\ &\dot{x}_{j_2} = a_{j_2j_1 0}(x_{j_1} - a_{j_2j_1 1})^2, \\ &j_1, j_2 \in \{1,2\}, j_1 \neq j_2. \end{aligned} \tag{3.102}$$

The first integral manifold is

$$\begin{aligned} &x_{j_2} - x_{j_2 0} = \frac{a_{j_2j_1 0}}{a_{j_1j_1 0}} \ln \frac{|\, x_{j_1} - a_{j_1j_1 1} \,|}{|\, x_{j_1 0} - a_{j_1j_1 1} \,|}, \\ &j_1, j_2 \in \{1,2\}, j_1 \neq j_2. \end{aligned} \tag{3.103}$$

(vi$_{2a}$) The infinite-equilibrium of $x^*_{j_1} = a_{j_2j_1 1} = a_{j_1j_1 1}$ with $\overline{x}_{j_2} \in (-\infty, \infty)$ *has the following properties.*

- *For* $a_{j_1j_1 0} > 0$ *and* $a_{j_2j_1 0} > 0$,

$$(a_{j_1j_1 1}, \overline{x}_{j_2}) = \underbrace{(3^{\text{rd}}\text{SO}, (3{:}2)\text{DL})}_{\text{third-order down-logarithmic source}}. \tag{3.104}$$

The equilibrium at $x^*_{j_1} = a_{j_2j_1 1} = a_{j_1j_1 1}$ *is a* (3rdSO,(3:2)DL)*-third-order down-logarithmic source infinite-equilibrium.*

- *For* $a_{j_1j_10} < 0$ *and* $a_{j_2j_10} > 0$,

$$(a_{j_1j_11}, \bar{x}_{j_2}) = \underbrace{(3^{\text{rd}}\text{SI}, (3{:}2)\text{UL})}_{\text{third-order up-logarithmic sink}}. \tag{3.105}$$

The equilibrium at $x_{j_1}^* = a_{j_2j_11} = a_{j_1j_11}$ *is a* $(3^{\text{rd}}\text{SI},(3{:}2)\text{UL})$*-third-order up-logarithmic sink infinite-equilibrium.*

- *For* $a_{j_1j_10} > 0$ *and* $a_{j_2j_10} < 0$,

$$(a_{j_1j_11}, \bar{x}_{j_2}) = \underbrace{(3^{\text{rd}}\text{SO}, (3{:}2)\text{UL})}_{\text{third-order up-logarithmic source}}. \tag{3.106}$$

The equilibrium at $x_{j_1}^* = a_{j_2j_11} = a_{j_1j_11}$ *is a* $(3^{\text{rd}}\text{SO},(3{:}2)\text{UL})$*-third-order up-logarithmic source infinite-equilibrium.*

- *For* $a_{j_1j_10} < 0$ *and* $a_{j_2j_10} < 0$,

$$(a_{j_1j_11}, \bar{x}_{j_2}) = \underbrace{(3^{\text{rd}}\text{SI}, (3{:}2)\text{DL})}_{\text{third-order down-logarithmic sink}}. \tag{3.107}$$

The equilibrium at $x_{j_1}^* = a_{j_2j_11} = a_{j_1j_11}$ *is a* $(3^{\text{rd}}\text{SI},(3{:}2)\text{DL})$*-third-order down-logarithmic sink infinite-equilibrium.*

(vi_{2b}) The infinite-equilibriums of $x_{j_1}^* = a_{j_2j_11} = a_{j_1j_11}$ with $\bar{x}_{j_2} \in (-\infty, \infty)$ *are the appearing bifurcation for the sink and source flows and the up-parabola and down-parabola flows with* $\Delta_{j_1j_1}^+ = \Delta_{j_1j_1} + 0$ and $\Delta_{j_2j_1}^+ = \Delta_{j_2j_1} + 0$.

- *For* $a_{j_1j_10} > 0$ *and* $a_{j_2j_10} > 0$,

$$\underbrace{(\text{SO}, \text{pF})}_{\text{source flow }(+)} \rightleftharpoons \underbrace{(3^{\text{rd}}\text{SO}, (3{:}2)\text{DL})}_{\text{appearing bifurcation}} \rightleftharpoons \left\{ \begin{array}{c} \underbrace{(\text{SO}, \text{pF})}_{\text{sink flow }(+)}, \underbrace{(\text{UP}, \text{nF})}_{\text{up-parabola flow }(-)}, \\ \underbrace{(\text{SI}, \text{pF})}_{\text{source flow }(+)}, \\ \underbrace{(\text{DP}, \text{nF})}_{\text{down-parabola flow }(-)}, \underbrace{(\text{SO}, \text{pF})}_{\text{source flow }(+)} \end{array} \right\} \tag{3.108}$$

for $\Delta_{j_1j_1}^+ > \Delta_{j_2j_1}^+$;

$$\underbrace{(\text{SO},\text{pF})}_{\text{source flow }(+)} \rightleftharpoons \underbrace{(3^{\text{rd}}\text{SO},(3{:}2)\text{DL})}_{\text{appearing bifurcation}} \rightleftharpoons \left\{\begin{array}{c} \underbrace{(\text{DP},\text{pF})}_{\text{down-parabola flow }(+)}, \underbrace{(\text{SO},\text{nF})}_{\text{sink flow }(-)}, \\ \underbrace{(\text{SI},\text{nF})}_{\text{sink flow }(-)}, \\ \underbrace{(\text{SO},\text{nF})}_{\text{source flow }(-)}, \underbrace{(\text{UP},\text{nF})}_{\text{up-parabola flow }(+)} \end{array}\right\} \tag{3.109}$$

for $\Delta^{+}_{j_1j_1} < \Delta^{+}_{j_2j_1}$.

- *For* $a_{j_1j_10} < 0$ *and* $a_{j_2j_10} > 0$,

$$\underbrace{(\text{SI},\text{pF})}_{\text{sink flow }(+)} \rightleftharpoons \underbrace{(3^{\text{rd}}\text{SI},(3{:}2)\text{UL})}_{\text{appearing bifurcation}} \rightleftharpoons \left\{\begin{array}{c} \underbrace{(\text{SI},\text{pF})}_{\text{source flow }(+)}, \underbrace{(\text{pF},\text{DP})}_{\text{down-parabola flow }(+)}, \\ \underbrace{(\text{SO},\text{pF})}_{\text{source flow }(+)}, \\ \underbrace{(\text{pF},\text{UP})}_{\text{up-parabola flow }(+)}, \underbrace{(\text{SI},\text{pF})}_{\text{sink flow }(+)} \end{array}\right\} \tag{3.110}$$

for $\Delta^{+}_{j_1j_1} > \Delta^{+}_{j_2j_1}$;

$$\underbrace{(\text{SI},\text{pF})}_{\text{sink flow }(+)} \rightleftharpoons \underbrace{(3^{\text{rd}}\text{SI},(3{:}2)\text{UL})}_{\text{appearing bifurcation}} \rightleftharpoons \left\{\begin{array}{c} \underbrace{(\text{nF},\text{UP})}_{\text{up-parabola flow }(-)}, \underbrace{(\text{SI},\text{nF})}_{\text{source flow }(-)}, \\ \underbrace{(\text{SO},\text{nF})}_{\text{source flow }(+)}, \\ \underbrace{(\text{SI},\text{nF})}_{\text{sink flow }(-)}, \underbrace{(\text{nF},\text{DP})}_{\text{down-parabola flow }(-)} \end{array}\right\} \tag{3.111}$$

for$\Delta^{+}_{j_1j_1} < \Delta^{+}_{j_2j_1}$.

- *For* $a_{j_1j_10} > 0$ *and* $a_{j_2j_10} < 0$,

$$\underbrace{(\text{SO},\text{nF})}_{\text{source flow }(-)} \rightleftharpoons \underbrace{(3^{\text{rd}}\text{SO},(3{:}2)\text{UL})}_{\text{appearing bifurcation}} \rightleftharpoons \left\{\begin{array}{c} \underbrace{(\text{SO},\text{nF})}_{\text{sink flow }(-)}, \underbrace{(\text{nF},\text{DP})}_{\text{down-parabola flow }(-)}, \\ \underbrace{(\text{SI},\text{nF})}_{\text{sink flow }(-)}, \\ \underbrace{(\text{nF},\text{UP})}_{\text{up-parabola flow }(-)}, \underbrace{(\text{SO},\text{nF})}_{\text{source flow }(-)} \end{array}\right\} \tag{3.112}$$

for $\Delta^{+}_{j_1j_1} > \Delta^{+}_{j_2j_1}$;

$$\underbrace{(\mathrm{SO},\mathrm{nF})}_{\text{source flow }(-)} \rightleftharpoons \underbrace{(3^{\mathrm{rd}}\mathrm{SO},(3:2)\mathrm{UL})}_{\text{appearing bifurcation}} \rightleftharpoons \left\{ \begin{array}{c} \underbrace{(\mathrm{pF},\mathrm{UP})}_{\text{up-parabola flow }(+)}, \underbrace{(\mathrm{SO},\mathrm{pF})}_{\text{source flow }(+)}, \\ \underbrace{(\mathrm{SI},\mathrm{pF})}_{\text{sink flow }(+)}, \\ \underbrace{(\mathrm{SO},\mathrm{pF})}_{\text{source flow }(+)}, \underbrace{(\mathrm{pF},\mathrm{DP})}_{\text{down-parabola flow }(+)} \end{array} \right\} \tag{3.113}$$

for $\Delta^{+}_{j_1j_1} < \Delta^{+}_{j_2j_1}$.

- *For* $a_{j_1j_10} < 0$ *and* $a_{j_2j_10} < 0$,

$$\underbrace{(\mathrm{SI},\mathrm{nF})}_{\text{sink flow }(-)} \rightleftharpoons \underbrace{(3^{\mathrm{rd}}\mathrm{SI},(3:2)\mathrm{DL})}_{\text{appearing bifurcation}} \rightleftharpoons \left\{ \begin{array}{c} \underbrace{(\mathrm{SI},\mathrm{nF})}_{\text{source flow }(-)}, \underbrace{(\mathrm{nF},\mathrm{DP})}_{\text{down-parabola flow }(-)}, \\ \underbrace{(\mathrm{SO},\mathrm{pF})}_{\text{source flow }(+)}, \\ \underbrace{(\mathrm{pF},\mathrm{DP})}_{\text{down-parabola flow }(+)}, \underbrace{(\mathrm{SI},\mathrm{nF})}_{\text{sink flow }(-)} \end{array} \right\} \tag{3.114}$$

for $\Delta^{+}_{j_1j_1} > \Delta^{+}_{j_2j_1}$;

$$\underbrace{(\mathrm{SI},\mathrm{nF})}_{\text{sink flow }(-)} \rightleftharpoons \underbrace{(3^{\mathrm{rd}}\mathrm{SI},(3:2)\mathrm{DL})}_{\text{appearing bifurcation}} \rightleftharpoons \left\{ \begin{array}{c} \underbrace{(\mathrm{DP},\mathrm{nF})}_{\text{down-parabola flow }(-)}, \underbrace{(\mathrm{SI},\mathrm{pF})}_{\text{source flow }(+)}, \\ \underbrace{(\mathrm{SO},\mathrm{pF})}_{\text{source flow }(+)}, \\ \underbrace{(\mathrm{SI},\mathrm{pF})}_{\text{sink flow }(+)}, \underbrace{(\mathrm{UP},\mathrm{nF})}_{\text{up-parabola flow }(-)} \end{array} \right\} \tag{3.115}$$

for $\Delta^{+}_{j_1j_1} < \Delta^{+}_{j_2j_1}$.

(vi_{2c}) The infinite-equilibriums of $x^{*}_{j_1} = a_{j_2j_11} = a_{j_1j_11}$ *are the switching bifurcation for the third-order sink and source flows and the increasing-inflection and decreasing-inflection flows.*

- *For* $a_{j_1j_10}>0$ *and* $a_{j_2j_10}>0$,

$$\left\{\begin{array}{c}\underbrace{(\mathrm{nF},\mathrm{DI})}_{\text{decreasing-inflection flow }(-)}\\ \underbrace{(3^{\mathrm{rd}}\mathrm{SO},\mathrm{pF})}_{\text{third order source flow }(+)}\end{array}\right\}\rightleftharpoons\underbrace{(3^{\mathrm{rd}}\mathrm{SO},(3{:}2)\mathrm{DL})}_{\text{switching bifurcation}}\rightleftharpoons\left\{\begin{array}{c}\underbrace{(3^{\mathrm{rd}}\mathrm{SO},\mathrm{pF})}_{\text{third order source flow }(+)}\\ \underbrace{(\mathrm{pF},\mathrm{II})}_{\text{increasing-inflection flow }(+)}\end{array}\right\}. \tag{3.116}$$

- *For* $a_{j_1j_10}<0$ *and* $a_{j_2j_10}>0$,

$$\left\{\begin{array}{c}\underbrace{(\mathrm{pF},\mathrm{II})}_{\text{increasing-inflection flow }(+)}\\ \underbrace{(3^{\mathrm{rd}}\mathrm{SI},\mathrm{pF})}_{\text{third order sink flow }(+)}\end{array}\right\}\rightleftharpoons\underbrace{(3^{\mathrm{rd}}\mathrm{SI},(3{:}2)\mathrm{UL})}_{\text{switching bifurcation}}\rightleftharpoons\left\{\begin{array}{c}\underbrace{(3^{\mathrm{rd}}\mathrm{SI},\mathrm{pF})}_{\text{third order sink flow }(+)}\\ \underbrace{(\mathrm{nF},\mathrm{DI})}_{\text{decreasing-inflection flow }(-)}\end{array}\right\}. \tag{3.117}$$

- *For* $a_{j_1j_10}>0$ *and* $a_{j_2j_10}<0$,

$$\left\{\begin{array}{c}\underbrace{(\mathrm{nF},\mathrm{II})}_{\text{increasing-inflection flow }(-)}\\ \underbrace{(3^{\mathrm{rd}}\mathrm{SO},\mathrm{nF})}_{\text{third order source flow }(-)}\end{array}\right\}\rightleftharpoons\underbrace{(3^{\mathrm{rd}}\mathrm{SO},(3{:}2)\mathrm{UL})}_{\text{switching bifurcation}}\rightleftharpoons\left\{\begin{array}{c}\underbrace{(3^{\mathrm{rd}}\mathrm{SO},\mathrm{nF})}_{\text{third order source flow }(-)}\\ \underbrace{(\mathrm{pF},\mathrm{DI})}_{\text{decreasing-inflection flow }(+)}\end{array}\right\}. \tag{3.118}$$

- *For* $a_{j_1j_10}<0$ *and* $a_{j_2j_10}<0$,

$$\left\{\begin{array}{c}\underbrace{(\mathrm{nF},\mathrm{DI})}_{\text{decreasing-inflection flow }(-)}\\ \underbrace{(3^{\mathrm{rd}}\mathrm{SI},\mathrm{nF})}_{\text{third order sink flow }(-)}\end{array}\right\}\rightleftharpoons\underbrace{(3^{\mathrm{rd}}\mathrm{SI},(3{:}2)\mathrm{DL})}_{\text{switching bifurcation}}\rightleftharpoons\left\{\begin{array}{c}\underbrace{(3^{\mathrm{rd}}\mathrm{SI},\mathrm{nF})}_{\text{third order sink flow }(-)}\\ \underbrace{(\mathrm{nF},\mathrm{II})}_{\text{increasing-inflection flow }(-)}\end{array}\right\}. \tag{3.119}$$

(vii) If

$$\Delta_{j_1j_1}>0 \text{ and } \Delta_{j_2j_1}=0, \tag{3.120}$$

then

$$\begin{aligned}\dot{x}_{j_1} &= a_{j_1j_10}(x_{j_1} - a_{j_1j_11})(x_{j_1} - a_{j_1j_12})(x_{j_1} - a_{j_1j_13}),\\ \dot{x}_{j_2} &= a_{j_2j_10}(x_{j_1} - a_{j_2j_11})^2,\end{aligned} \tag{3.121}$$

where

$$\begin{aligned}&\{a_{j_1j_11}, a_{j_1j_12}, a_{j_1j_13}\} = \mathrm{sort}\{b_{j_1j_11}, b_{j_1j_12}, b_{j_1j_13}\}, a_{j_1j_11} < a_{j_1j_12} < a_{j_1j_13};\\ &b_{j_1j_12}, b_{j_1j_13} = -\frac{1}{2}(B_{j_1j_11} \pm \sqrt{\Delta_{j_1j_1}}), a_{j_2j_11} = -\frac{1}{2}B_{j_2j_11};\\ &j_1, j_2 \in \{1,2\}; j_1 \neq j_2.\end{aligned} \tag{3.122}$$

(vii₁) For $x_{j_1}^* = a_{j_1j_1s_1} \neq a_{j_2j_11}$ (s_1, $s_2 \in \{1,2,3\}$; $s_1 \neq s_2$), the first integral manifold is

$$x_{j_2} - x_{j_20} = \frac{a_{j_2j_10}}{a_{j_1j_10}} \sum_{s_1=1}^{3} \frac{(a_{j_1j_1s_1} - a_{j_2j_11})^2}{\prod_{s_2=1, s_2\neq s_1}^{3} (a_{j_1j_1s_1} - a_{j_1j_1s_2})} \ln \frac{|x_{j_1} - a_{j_1j_1s_1}|}{|x_{j_10} - a_{j_1j_1s_1}|}. \tag{3.123}$$

(vii_{1a}) The equilibrium of $x_{j_1}^* = a_{j_1j_1s_1} \neq a_{j_2j_11}$ (s_1, $s_2 \in \{1,2,3\}$; $s_1 \neq s_2$) in the x_{j_1}-direction possesses the following properties.

- *For $a_{j_1j_10}\prod_{s_2=1,s_2\neq s_1}^{3}(a_{j_1j_1s_1} - a_{j_1j_1s_2}) > 0$ and $a_{j_2j_10}(a_{j_1j_1s_1} - a_{j_2j_11})^2 > 0$,*

$$(a_{j_1j_1s_1}, \dot{x}_{j_2}) = \underbrace{(\mathrm{SO}, \mathrm{pF})}_{\text{source flow }(+)}. \tag{3.124}$$

  The flow at $x_{j_1}^* = a_{j_1j_1s_1}$ is an (SO,pF)*-positive source-flow.*

- *For $a_{j_1j_10}\prod_{s_2=1,s_2\neq s_1}^{3}(a_{j_1j_1s_1} - a_{j_1j_1s_2}) < 0$ and $a_{j_2j_10}(a_{j_1j_1s_1} - a_{j_2j_11})^2 > 0$,*

$$(a_{j_1j_1s_1}, \dot{x}_{j_2}) = \underbrace{(\mathrm{SI}, \mathrm{pF})}_{\text{sink flow }(+)}. \tag{3.125}$$

  The flow at $x_{j_1}^* = a_{j_1j_1s_1}$ is an (SI,pF)*-positive sink-flow.*

- *For $a_{j_1j_10}\prod_{s_2=1,s_2\neq s_1}^{3}(a_{j_1j_1s_1} - a_{j_1j_1s_2}) > 0$ and $a_{j_2j_10}(a_{j_1j_1s_1} - a_{j_2j_11})^2 < 0$,*

$$(a_{j_1j_1s_1}, \dot{x}_{j_2}) = \underbrace{(\mathrm{SO}, \mathrm{nF})}_{\text{source flow }(-)}. \tag{3.126}$$

  The flow at $x_{j_1}^* = a_{j_1j_1s_1}$ is an (SO,nF)*-negative source-flow.*

- *For* $a_{j_1j_10}\prod_{s_2=1,s_2\neq s_1}^{3}(a_{j_1j_1s_1}-a_{j_1j_1s_2})<0$ *and* $a_{j_2j_10}(a_{j_1j_1s_1}-a_{j_2j_11})^2<0$,

$$(a_{j_1j_1s_1},\dot{x}_{j_2})=\underbrace{(\text{SI},\text{nF})}_{\text{sink flow }(-)}.\tag{3.127}$$

The flow at $x_{j_1}^*=a_{j_1j_1s_1}$ *is an* (SI,nF)*-negative sink-flow.*

(vii$_{Ib}$*) The equilibrium of* $x_{j_1}^*=a_{j_2j_11}\neq a_{j_1j_1s_1},a_{j_1j_1s_2}$ (s_1, $s_2\in\{1,2,3\}$; $s_1\neq s_2$) *in the* x_{j_2} *-direction possesses the following properties:*

- *For* $a_{j_1j_10}\prod_{s_1=1}^{3}(a_{j_2j_11}-a_{j_1j_1s_1})>0$ *and* $a_{j_2j_10}>0$,

$$(\dot{x}_{j_1},a_{j_2j_11})=\underbrace{(\text{pF},\text{II})}_{\text{increasing-inflection flow }(+)}.\tag{3.128}$$

The flow at $x_{j_1}^*=a_{j_2j_11}$ *is a* (pF II)*-positive increasing-inflection flow.*

- *For* $a_{j_1j_10}\prod_{s_1=1}^{3}(a_{j_2j_11}-a_{j_1j_1s_1})<0$ *and* $a_{j_2j_10}>0$,

$$(\dot{x}_{j_1},a_{j_2j_11})=\underbrace{(\text{nF},\text{DI})}_{\text{decreasing-inflection flow }(-)}.\tag{3.129}$$

The flow at $x_{j_1}^*=a_{j_2j_11}$ *is an* (nF, DI)*-negative decreasing-inflection flow.*

- *For* $a_{j_1j_10}\prod_{s_1=1}^{3}(a_{j_2j_11}-a_{j_1j_1s_1})>0$ *and* $a_{j_2j_10}<0$,

$$(\dot{x}_{j_1},a_{j_2j_11})=\underbrace{(\text{pF},\text{DI})}_{\text{decreasing-inflection flow }(+)}.\tag{3.130}$$

The flow at $x_{j_1}^*=a_{j_2j_11}$ *is a* (pF, DI)*-positive decreasing-inflection flow.*

- *For* $a_{j_1j_10}\prod_{s_1=1}^{3}(a_{j_2j_11}-a_{j_1j_1s_1})<0$ *and* $a_{j_2j_10}<0$,

$$(\dot{x}_{j_1},a_{j_2j_11})=\underbrace{(\text{nF},\text{II})}_{\text{increasing-inflection flow }(-)}.\tag{3.131}$$

The flow at $x_{j_1}^*=a_{j_2j_11}$ *is an (nF,II)-negative increasing-inflection flow.*

– *The decreasing-inflection and increasing-inflection flows are the appearing bifurcations of the up-parabola and down-parabola flows.*

(vii$_2$) For $a_{j_2j_11}=a_{j_1j_1s_1}$ (s_1, s_2, $s_3 \in \{1,2,3\}$; $s_1 \neq s_2 \neq s_3$).

$$\begin{aligned}
&\dot{x}_{j_1}=a_{j_1j_10}(x_{j_1}-a_{j_1j_11})(x_{j_1}-a_{j_1j_12})(x_{j_1}-a_{j_1j_13}),\\
&\dot{x}_{j_2}=a_{j_2j_10}(x_{j_1}-a_{j_2j_11})^2,\\
&j_1,j_2 \in \{1,2\}, j_1 \neq j_2;
\end{aligned} \tag{3.132}$$

and the first integral manifold is

$$x_{j_2}-x_{j_20}=\frac{a_{j_2j_10}}{a_{j_1j_10}}\sum_{s_2=1,s_2\neq s_1}^{3}\frac{(a_{j_1j_1s_2}-a_{j_2j_11})}{\prod_{s_3=1,s_3\neq s_1,s_2}^{3}(a_{j_1j_1s_2}-a_{j_1j_1s_3})}\ln\frac{|x_{j_1}-a_{j_1j_1s_2}|}{|x_{j_10}-a_{j_1j_1s_2}|}. \tag{3.133}$$

(vii$_{2a}$) The infinite equilibrium of $x_{j_1}^*=a_{j_1j_1s_1}=a_{j_2j_11}$ ($s_1 \in \{1,2,3\}$) with $\overline{x}_{j_2} \in (-\infty,\infty)$ possesses the following cases.

- *For $a_{j_1j_10}\prod_{s_2=1,s_2\neq s_1}^{3}(a_{j_1j_1s_1}-a_{j_1j_1s_2})>0$ and $a_{j_2j_10}>0$,*

$$(a_{j_1j_1s_1},\overline{x}_{j_2})=\underbrace{(\mathrm{SO},\mathrm{UP})}_{\text{up-parabola source}}. \tag{3.134}$$

  The equilibrium at $x_{j_1}^*=a_{j_1j_1s_1}=a_{j_2j_11}$ is an (SO,UP)*-up-parabola source infinite-equilibrium.*

- *For $a_{j_1j_10}\prod_{s_2=1,s_2\neq s_1}^{3}(a_{j_1j_1s_1}-a_{j_1j_1s_2})<0$ and $a_{j_2j_10}>0$,*

$$(a_{j_1j_1s_1},\overline{x}_{j_2})=\underbrace{(\mathrm{SI},\mathrm{DP})}_{\text{down-parabola sink}}. \tag{3.135}$$

  The equilibrium at $x_{j_1}^*=a_{j_1j_1s_1}=a_{j_2j_11}$ is an (SI,DP)*-down-parabola sink infinite-equilibrium.*

- *For $a_{j_1j_10}\prod_{s_2=1,s_2\neq s_1}^{3}(a_{j_1j_1s_1}-a_{j_1j_1s_2})>0$ and $a_{j_2j_10}<0$,*

$$(a_{j_1j_1s_1},\overline{x}_{j_2})=\underbrace{(\mathrm{SO},\mathrm{DP})}_{\text{down-parabola source}}. \tag{3.136}$$

  The equilibrium at $x_{j_1}^*=a_{j_1j_1s_1}=a_{j_2j_11}$ is an (SO,DP)*-down-parabola source infinite-equilibrium.*

- *For $a_{j_1j_10}\prod_{s_2=1,s_2\neq s_1}^{3}(a_{j_1j_1s_1}-a_{j_1j_1s_2})<0$ and $a_{j_2j_10}<0$,*

$$(a_{j_1j_1s_1}, \bar{x}_{j_2}) = \underbrace{(\mathrm{SI}, \mathrm{UP})}_{\text{up-parabola sink}}. \tag{3.137}$$

The equilibrium at $x_{j_1}^* = a_{j_1j_1s_1} = a_{j_2j_1 1}$ *is an* (SI,UP)-*up-parabola sink infinite-equilibrium.*

– *The up-parabola and down-parabola sink, and source infinite-equilibriums are the appearing and switching bifurcations for the up-parabola and down-parabola flows with sink or source flows.*

(vii$_{2b}$) The equilibrium of $x_{j_1}^* = a_{j_1j_1s_2} \neq a_{j_2j_1 1}$ $(s_1, s_2 \in \{1,2,3\}; s_1 \neq s_2)$ *in the* x_{j_1}*-direction possesses the following properties:*

- *For* $a_{j_1j_1 0}\prod_{s_1=1, s_1\neq s_2}^{3}(a_{j_1j_1s_2} - a_{j_1j_1s_1}) > 0$ *and* $a_{j_2j_1 0}(a_{j_1j_1s_2} - a_{j_2j_1 1})^2 > 0$,

$$(a_{j_1j_1s_2}, \dot{x}_{j_2}) = \underbrace{(\mathrm{SO}, \mathrm{pF})}_{\text{source flow }(+)}. \tag{3.138}$$

  The flow at $x_{j_1}^* = a_{j_1j_1s_2}$ *is an* (SO,pF)-*positive source-flow.*
- *For* $a_{j_1j_1 0}\prod_{s_1=1, s_1\neq s_2}^{3}(a_{j_1j_1s_2} - a_{j_1j_1s_1}) < 0$ *and* $a_{j_2j_1 0}(a_{j_1j_1s_2} - a_{j_2j_1 1})^2 > 0$,

$$(a_{j_1j_1s_2}, \dot{x}_{j_2}) = \underbrace{(\mathrm{SI}, \mathrm{pF})}_{\text{sink flow }(+)}. \tag{3.139}$$

  The flow at $x_{j_1}^* = a_{j_1j_1s_2}$ *is an* (SI,pF)-*positive sink-flow.*
- *For* $a_{j_1j_1 0}\prod_{s_1=1, s_1\neq s_2}^{3}(a_{j_1j_1s_2} - a_{j_1j_1s_1}) > 0$ *and* $a_{j_2j_1 0}(a_{j_1j_1s_2} - a_{j_2j_1 1})^2 < 0$,

$$(a_{j_1j_1s_2}, \dot{x}_{j_2}) = \underbrace{(\mathrm{SO}, \mathrm{nF})}_{\text{source flow }(-)}. \tag{3.140}$$

  The flow at $x_{j_1}^* = a_{j_1j_1s_2}$ *is an* (SO,nF)-*negative source-flow.*
- *For* $a_{j_1j_1 0}\prod_{s_1=1, s_1\neq s_2}^{3}(a_{j_1j_1s_2} - a_{j_1j_1s_1}) < 0$ *and* $a_{j_2j_1 0}(a_{j_1j_1s_2} - a_{j_2j_1 1})^2 < 0$,

$$(a_{j_1j_1s_2}, \dot{x}_{j_2}) = \underbrace{(\mathrm{SI}, \mathrm{nF})}_{\text{sink flow }(-)}. \tag{3.141}$$

  The flow at $x_{j_1}^* = a_{j_1j_1s_2}$ *is an* (SI,nF)-*negative sink-flow.*

(viii) If

$$\Delta_{j_1j_1} > 0 \text{ and } \Delta_{j_2j_1} > 0, \tag{3.142}$$

then

$$\begin{aligned}\dot{x}_{j_1} &= a_{j_1j_10}(x_{j_1} - a_{j_1j_11})(x_{j_1} - a_{j_1j_12})(x_{j_1} - a_{j_1j_13}),\\ \dot{x}_{j_2} &= a_{j_2j_10}(x_{j_1} - a_{j_2j_11})(x_{j_1} - a_{j_2j_12}),\end{aligned} \tag{3.143}$$

where

$$\begin{aligned}&\{a_{j_1j_11}, a_{j_1j_12}, a_{j_1j_13}\} = \text{sort}\{b_{j_1j_11}, b_{j_1j_12}, b_{j_1j_13}\}, a_{j_1j_11} < a_{j_1j_12} < a_{j_1j_13};\\ &\{a_{j_2j_11}, a_{j_2j_12}) = \text{sort}\{b_{j_2j_11}, b_{j_2j_12}\}, a_{j_2j_11} < a_{j_2j_12};\\ &b_{j_1j_12}, b_{j_1j_13} = -\frac{1}{2}(B_{j_1j_11} \pm \sqrt{\Delta_{j_1j_1}}),\\ &b_{j_2j_11}, b_{j_2j_12} = -\frac{1}{2}(B_{j_2j_11} \pm \sqrt{\Delta_{j_2j_1}});\\ &j_1, j_2 \in \{1,2\}; j_1 \neq j_2.\end{aligned} \tag{3.144}$$

(viii$_1$) For $a_{j_1j_1s_1} \neq a_{j_2j_1l_1}(s_1, s_2 \in \{1,2,3\}; s_1 \neq s_2)$ and $(l_1, l_2 \in \{1,2\}; l_1 \neq l_2)$, *the first integral manifold is*

$$x_{j_2} - x_{j_20} = \frac{a_{j_2j_10}}{a_{j_1j_10}} \sum_{s_1=1}^{3} \frac{\prod_{l_1=1}^{2}(a_{j_1j_1s_1} - a_{j_2j_1l_1})}{\prod_{s_2=1,s_2\neq s_1}^{3}(a_{j_1j_1s_1} - a_{j_1j_1s_2})} \ln \frac{|x_{j_1} - a_{j_1j_1s_1}|}{|x_{j_10} - a_{j_1j_1s_1}|}. \tag{3.145}$$

(viii$_{1a}$) The equilibriums of $x_{j_1}^* = a_{j_1j_1s_1} \neq a_{j_2j_1l_1}, a_{j_2j_1l_2}$ $(s_1, s_2 \in \{1,2,3\}; s_1 \neq s_2)$ *and* $(l_1, l_2 \in \{1,2\}; l_1 \neq l_2)$ *are the following cases.*

- *For* $a_{j_1j_10}\prod_{s_2=1,s_2\neq s_1}^{3}(a_{j_1j_1s_1} - a_{j_1j_1s_2}) > 0$ *and* $a_{j_2j_10}\prod_{l_1=1}^{2}(a_{j_1j_1s_1} - a_{j_2j_1l_1}) > 0$,

$$(a_{j_1j_1s_1}, \dot{x}_{j_2}) = \underbrace{(\text{SO}, \text{pF})}_{\text{source flow }(+)}. \tag{3.146}$$

The flow at $x_{j_1}^* = a_{j_1j_1s_1}$ *is an* (SO,pF)*-positive source-flow.*

- *For* $a_{j_1j_10}\prod_{s_2=1,s_2\neq s_1}^{3}(a_{j_1j_1s_1} - a_{j_1j_1s_2}) < 0$ *and* $a_{j_2j_10}\prod_{l_1=1}^{2}(a_{j_1j_1s_1} - a_{j_2j_1l_1}) > 0$,

$$(a_{j_1j_1s_1}, \dot{x}_{j_2}) = \underbrace{(\text{SI}, \text{pF})}_{\text{sink flow }(+)}. \tag{3.147}$$

The flow at $x_{j_1}^* = a_{j_1j_1s_1}$ *is an* (SI,pF)*-positive sink-flow.*

- *For* $a_{j_1j_10}\prod_{s_2=1,s_2\neq s_1}^{3}(a_{j_1j_1s_1}-a_{j_1j_1s_2})>0$ *and* $a_{j_2j_10}\prod_{l_1=1}^{2}(a_{j_1j_1s_1}-a_{j_2j_1l_1})<0$,

$$(a_{j_1j_1s_1},\dot{x}_{j_2})=\underbrace{(\mathrm{SO},\mathrm{nF})}_{\text{source flow }(-)}. \tag{3.148}$$

The flow at $x_{j_1}^*=a_{j_1j_1s_1}$ *is an* (SO,nF)*-negative source-flow.*

- *For* $a_{j_1j_10}\prod_{s_2=1,s_2\neq s_1}^{3}(a_{j_1j_1s_1}-a_{j_1j_1s_2})<0$ *and* $a_{j_2j_10}\prod_{l_1=1}^{2}(a_{j_1j_1s_1}-a_{j_2j_1l_1})<0$,

$$(a_{j_1j_1s_1},\dot{x}_{j_2})=\underbrace{(\mathrm{SI},\mathrm{nF})}_{\text{sink flow }(-)}. \tag{3.149}$$

The flow at $x_{j_1}^*=a_{j_1j_1s_1}$ *is an* (SI,nF)*-negative sink-flow.*

(viii$_{1b}$*) The equilibrium of* $x_{j_1}^*=a_{j_1j_1l_1}\neq a_{j_2j_1s_1},a_{j_2j_1s_2},a_{j_2j_1s_3}$ *(*$l_1,\ l_2\in\{1,2\};\ l_1\neq l_2$*) and (*$s_1,\ s_2,\ s_3\in\{1,2,3\};\ s_1\neq s_2\neq s_2$*) has the following properties.*

- *For* $a_{j_1j_10}\prod_{s_1=1}^{3}(a_{j_2j_1l_1}-a_{j_1j_1s_1})>0$ *and* $a_{j_2j_10}(a_{j_2j_1l_1}-a_{j_2j_1l_2})>0$,

$$(\dot{x}_{j_1},a_{j_1j_1l_1})=\underbrace{(\mathrm{pF},\mathrm{UP})}_{\text{up-parabola flow }(+)}. \tag{3.150}$$

The flow at $x_{j_1}^*=a_{j_1j_1l_1}$ *is a* (pF,UP)*-positive up-parabola flow.*

- *For* $a_{j_1j_10}\prod_{s_1=1}^{3}(a_{j_2j_1l_1}-a_{j_1j_1s_1})<0$ *and* $a_{j_2j_10}(a_{j_2j_1l_1}-a_{j_2j_1l_2})>0$,

$$(\dot{x}_{j_1},a_{j_1j_1l_1})=\underbrace{(\mathrm{nF},\mathrm{DP})}_{\text{down-parabola flow }(-)}. \tag{3.151}$$

The flow at $x_{j_1}^*=a_{j_1j_1l_1}$ *is an* (nF,DP)*-negative down-parabola flow.*

- *For* $a_{j_1j_10}\prod_{s_1=1}^{3}(a_{j_2j_1l_1}-a_{j_1j_1s_1})>0$ *and* $a_{j_2j_10}(a_{j_2j_1l_1}-a_{j_2j_1l_2})<0$,

$$(\dot{x}_{j_1},a_{j_1j_1l_1})=\underbrace{(\mathrm{pF},\mathrm{DP})}_{\text{down-parabola flow }(+)}. \tag{3.152}$$

The flow at $x_{j_1}^*=a_{j_1j_1l_1}$ *is a* (pF,DP)*-positive down-parabola flow.*

- *For* $a_{j_1j_10}\prod_{s_1=1}^{3}(a_{j_2j_1l_1}-a_{j_1j_1s_1})<0$ *and* $a_{j_2j_10}(a_{j_2j_1l_1}-a_{j_2j_1l_2})<0$,

$$(\dot{x}_{j_1},a_{j_1j_1l_1})=\underbrace{(\mathrm{nF},\mathrm{UP})}_{\text{up-parabola flow }(-)}. \tag{3.153}$$

The flow at $x_{j_1}^*=a_{j_1j_1l_1}$ *is an* (nF,UP)*-negative up-parabola flow.*

(viii$_2$) For $a_{j_1j_1s_1}=a_{j_2j_1l_1}(s_1, s_2, s_3 \in \{1,2,3\}; s_1 \neq s_2 \neq s_3)$ and $(l_1, l_2 \in \{1,2\}; l_1 \neq l_2)$,

$$\begin{aligned}\dot{x}_{j_1} &= a_{j_1j_10}(x_{j_1}-a_{j_1j_11})(x_{j_1}-a_{j_1j_12})(x_{j_1}-a_{j_1j_13}),\\ \dot{x}_{j_2} &= a_{j_2j_10}(x_{j_1}-a_{j_2j_11})(x_{j_1}-a_{j_2j_12}).\end{aligned} \tag{3.154}$$

The first integral manifold is

$$x_{j_2}-x_{j_20}=\frac{a_{j_2j_10}}{a_{j_1j_10}}\sum_{s_2=1,s_2\neq s_1}^{3}\frac{\prod_{l_2=1,l_2\neq l_1}^{2}(a_{j_1j_1s_2}-a_{j_2j_1l_2})}{\prod_{s_3=1,s_3\neq s_1,s_2}^{3}(a_{j_1j_1s_2}-a_{j_1j_1s_3})}\ln\frac{|x_{j_1}-a_{j_1j_1s_2}|}{|x_{j_10}-a_{j_1j_1s_2}|}. \tag{3.155}$$

The infinite-equilibrium of $x_{j_1}^*=a_{j_1j_1s_1}=a_{j_2j_1l_1}$ with $\bar{x}_{j_2}\in(-\infty,\infty)$ has the following properties.

- *For* $a_{j_1j_10}\prod_{s_2=1,s_2\neq s_1}^{3}(a_{j_1j_1s_1}-a_{j_1j_1s_2})>0$ *and* $a_{j_2j_10}\prod_{l_2=1,l_2\neq l_1}^{2}(a_{j_2j_1l_1}-a_{j_2j_1l_2})>0$,

$$(a_{j_1j_1s_1},\bar{x}_{j_2})=\underbrace{(\text{SO, II})}_{\text{increasing-inflection source}}. \tag{3.156}$$

  The equilibrium at $x_{j_1}^*=a_{j_1j_1s_1}=a_{j_2j_1l_1}$ *is an* (SO,II)*-increasing-inflection source infinite-equilibrium.*

- *For* $a_{j_1j_10}\prod_{s_2=1,s_2\neq s_1}^{3}(a_{j_1j_1s_1}-a_{j_1j_1s_2})<0$ *and* $a_{j_2j_10}\prod_{l_2=1,l_2\neq l_1}^{2}(a_{j_2j_1l_1}-a_{j_2j_1l_2})>0$,

$$(a_{j_1j_1s_1},\bar{x}_{j_2})=\underbrace{(\text{SI, DI})}_{\text{decreasing-inflection sink}}. \tag{3.157}$$

  The equilibrium at $x_{j_1}^*=a_{j_1j_1s_1}=a_{j_2j_1l_1}$ *is an* (SI,DI)*-decreasing-inflection sink infinite-equilibrium.*

- *For* $a_{j_1j_10}\prod_{s_2=1,s_2\neq s_1}^{3}(a_{j_1j_1s_1}-a_{j_1j_1s_2})>0$ *and* $a_{j_2j_10}\prod_{l_2=1,l_2\neq l_1}^{2}(a_{j_2j_1l_1}-a_{j_2j_1l_2})<0$,

$$(a_{j_1j_1s_1},\bar{x}_{j_2})=\underbrace{(\text{SO, DI})}_{\text{decreasing-inflection source}}. \tag{3.158}$$

  The equilibrium at $x_{j_1}^*=a_{j_1j_1s_1}=a_{j_2j_1l_1}$ *is an* (SO,DI)*-increasing-inflection source infinite-equilibrium.*

- *For* $a_{j_1j_10}\prod_{s_2=1,s_2\neq s_1}^{3}(a_{j_1j_1s_1}-a_{j_1j_1s_2})<0$ *and* $a_{j_2j_10}\prod_{l_2=1,l_2\neq l_1}^{2}(a_{j_2j_1l_1}-a_{j_2j_1l_2})<0$,

$$(a_{j_1j_1s_1},\bar{x}_{j_2}) = \underbrace{(\mathrm{SI},\mathrm{II})}_{\text{increasing-inflection sink}}. \quad (3.159)$$

The equilibrium at $x_{j_1}^* = a_{j_1j_1s_1} = a_{j_2j_1l_1}$ *is an* (SI,II)*-increasing-inflection sink infinite-equilibrium.*

– *The infinite-equilibriums are the switching bifurcations for the sink and source flows with the sink and source flows.*

(ix) If

$$\Delta_{j_1j_1} < 0 \text{ and } \Delta_{j_2j_1} > 0, \quad (3.160)$$

then

$$\begin{aligned}
&\dot{x}_{j_2} = a_{j_2j_1 0}(x_{j_1} - a_{j_2j_1 1})(x_{j_1} - a_{j_2j_1 2}),\\
&\dot{x}_{j_1} = a_{j_1j_1 0}(x_{j_1} - a_{j_1j_1 1})\big[(x_{j_1} - a_{j_1j_1})^2 + b_{j_1j_1}\big],\\
&\text{with } a_{j_1j_1} = -\frac{1}{2}B_{j_1j_1 1}, b_{j_1j_1} = \frac{1}{4}(-\Delta_{j_1j_1}), a_{j_1j_1 1} = b_{j_1j_1 1};\\
&a_{j_2j_1 1}, a_{j_2j_1 2} = \mathrm{sort}\{b_{j_2j_1 1}, b_{j_2j_1 2}\}, a_{j_2j_1 1} < a_{j_2j_1 2};\\
&b_{j_2j_1 1}, b_{j_2j_1 2} = -\frac{1}{2}(B_{j_2j_1 1} \pm \sqrt{\Delta_{j_2j_1}}).
\end{aligned} \quad (3.161)$$

(ix₁) For $a_{j_1j_1 1} \neq a_{j_2j_1 1}$*, the first integral manifold is*

$$\begin{aligned}
x_{j_2} - x_{j_2 0} = &\frac{a_{j_2j_1 0}}{a_{j_1j_1 0}}\Big\{\frac{1}{2}\ln\frac{|(x_{j_1} - a_{j_1j_1})^2 + b_{j_1j_1}|}{|(x_{j_1 0} - a_{j_1j_1})^2 + b_{j_1j_1}|}\\
&+(a_{j_1j_1 1} + a_{j_1j_1} - a_{j_2j_1 1} - a_{j_2j_1 2})\\
&\times\frac{1}{\sqrt{b_{j_1j_1}}}(\arctan\frac{x_{j_1} - a_{j_1j_1}}{\sqrt{b_{j_1j_1}}} - \arctan\frac{x_{j_1 0} - a_{j_1j_1}}{\sqrt{b_{j_1j_1}}})\\
&+\frac{(a_{j_1j_1 1} - a_{j_2j_1 1})(a_{j_1j_1 1} - a_{j_2j_1 2})}{(a_{j_1j_1 1} - a_{j_1j_1})^2 + b_{j_1j_1}}\Big[\ln\frac{|x_{j_1} - a_{j_1j_1 1}|}{|x_{j_1 0} - a_{j_1j_1 1}|}\\
&-\frac{1}{2}\ln\frac{|(x_{j_1} - a_{j_1j_1})^2 + b_{j_1j_1}|}{|(x_{j_1 0} - a_{j_1j_1})^2 + b_{j_1j_1}|}\\
&-\frac{a_{j_1j_1 1} - a_{j_1j_1}}{\sqrt{b_{j_1j_1}}}(\arctan\frac{x_{j_1} - a_{j_1j_1}}{\sqrt{b_{j_1j_1}}} - \arctan\frac{x_{j_1 0} - a_{j_1j_1}}{\sqrt{b_{j_1j_1}}})\Big]\Big\},\\
&j_1, j_2 \in \{1,2\}, j_1 \neq j_2.
\end{aligned} \quad (3.162)$$

(ix₁ₐ) The flow at $x_{j_1}^* = a_{j_1j_1 1} \neq a_{j_2j_1 1}$ *for the system has the following properties.*

- *For* $a_{j_1j_1 0} > 0$ *and* $a_{j_2j_1 0}(a_{j_1j_1 1} - a_{j_2j_1 1})(a_{j_1j_1 1} - a_{j_2j_1 2}) > 0$,

$$(a_{j_1j_1 1}, \dot{x}_{j_2}) = \underbrace{(\text{SO}, \text{pF})}_{\text{source flow }(+)}. \tag{3.163}$$

The (SO,pF)*-flow is a positive source flow at* $x_{j_1}^* = a_{j_1j_1 1}$.

- *For* $a_{j_1j_1 0} < 0$ *and* $a_{j_2j_1 0}(a_{j_1j_1 1} - a_{j_2j_1 1})(a_{j_1j_1 1} - a_{j_2j_1 2}) > 0$,

$$(a_{j_1j_1 1}, \dot{x}_{j_2}) = \underbrace{(\text{SI}, \text{pF})}_{\text{sink flow }(+)}. \tag{3.164}$$

The (SI, pF)*-flow is a positive sink flow at* $x_{j_1}^* = a_{j_1j_1 1}$.

- *For* $a_{j_1j_1 0} > 0$ *and* $a_{j_2j_1 0}(a_{j_1j_1 1} - a_{j_2j_1 1})(a_{j_1j_1 1} - a_{j_2j_1 2}) < 0$,

$$(a_{j_1j_1 1}, \dot{x}_{j_2}) = \underbrace{(\text{SO}, \text{nF})}_{\text{source flow }(-)}. \tag{3.165}$$

The (SO, nF)*-flow is a negative source flow at* $x_{j_1}^* = a_{j_1j_1 1}$.

- *For* $a_{j_1j_1 0} < 0$ *and* $a_{j_2j_1 0}(a_{j_1j_1 1} - a_{j_2j_1 1})(a_{j_1j_1 1} - a_{j_2j_1 2}) < 0$,

$$(a_{j_1j_1 1}, \dot{x}_{j_2}) = \underbrace{(\text{SI}, \text{nF})}_{\text{sink flow }(-)}. \tag{3.166}$$

The (SI,nF)*-flow is a negative sink flow at* $x_{j_1}^* = a_{j_1j_1 1}$.

(ix$_{1b}$) The flow at $x_{j_1}^* = a_{j_2j_1 l_1}$ $(l_1, l_2 \in \{1, 2\};\ l_1 \neq l_2)$ *for the system has the following properties.*

- *For* $a_{j_1j_1 0}(a_{j_2j_1 l_1} - a_{j_1j_1 1}) > 0$ *and* $a_{j_2j_1 0}(a_{j_2j_1 l_1} - a_{j_2j_1 l_2}) > 0$,

$$(\dot{x}_{j_1}, a_{j_2j_1 l_1}) = \underbrace{(\text{pF}, \text{UP})}_{\text{up-parabola flow }(+)}. \tag{3.167}$$

The (pF,UP)*-flow at* $x_{j_1}^* = a_{j_2j_1 l_1}$ *is a positive up-parabola flow.*

- *For* $a_{j_1j_1 0}(a_{j_2j_1 l_1} - a_{j_1j_1 1}) < 0$ *and* $a_{j_2j_1 0}(a_{j_2j_1 l_1} - a_{j_2j_1 l_2}) > 0$,

$$(\dot{x}_{j_1}, a_{j_2j_1 l_1}) = \underbrace{(\text{nF}, \text{DP})}_{\text{down-parabola flow }(-)}. \tag{3.168}$$

The (nF,DP)*-flow at* $x_{j_1}^* = a_{j_2j_1 l_1}$ *is a negative down-parabola flow.*

- *For* $a_{j_1j_1 0}(a_{j_2j_1 l_1} - a_{j_1j_1 l_1}) > 0$ *and* $a_{j_2j_1 0}(a_{j_2j_1 l_1} - a_{j_2j_1 l_2}) < 0$,

$$(\dot{x}_{j_1}, a_{j_2 j_1 l_1}) = \underbrace{(\mathrm{pF}, \mathrm{DP})}_{\text{down-parabola flow }(+)}. \tag{3.169}$$

The (pF, DP)*-flow at* $x^*_{j_1} = a_{j_2 j_1 l_1}$ *is a positive down-parabola flow.*

- *For* $a_{j_1 j_1 0}(a_{j_2 j_1 l_1} - a_{j_1 j_1 1}) < 0$ *and* $a_{j_2 j_1 0}(a_{j_2 j_1 l_1} - a_{j_2 j_1 l_2}) < 0$,

$$(\dot{x}_{j_1}, a_{j_2 j_1 l_1}) = \underbrace{(\mathrm{nF}, \mathrm{UP})}_{\text{up-parabola flow }(-)}. \tag{3.170}$$

The (nF,UP)*-flow at* $x^*_{j_1} = a_{j_2 j_1 l_1}$ *is a negative up-parabola flow.*

(ix$_2$) For $a_{j_1 j_1 1} = a_{j_2 j_1 l_1}$, *the first integral manifold is*

$$\begin{aligned} x_{j_2} - x_{j_2 0} &= \frac{a_{j_2 j_1 0}}{a_{j_1 j_1 0}} \left[\frac{1}{2} \ln \frac{|(x_{j_1} - a_{j_1 j_1})^2 + b_{j_1 j_1}|}{|(x_{j_1 0} - a_{j_1 j_1})^2 + b_{j_1 j_1}|} \right. \\ &\quad \left. + \frac{(a_{j_1 j_1} - a_{j_1 j_1 l_2})}{\sqrt{b_{j_1 j_1}}} \left(\arctan \frac{x_{j_1} - a_{j_1 j_1}}{\sqrt{b_{j_1 j_1}}} - \arctan \frac{x_{j_1 0} - a_{j_1 j_1}}{\sqrt{b_{j_1 j_1}}} \right) \right]; \\ &\quad j_1, j_2 \in \{1,2\}, j_1 \neq j_2. \end{aligned} \tag{3.171}$$

The infinite-equilibrium of $x^*_{j_1} = a_{j_1 j_1 1} = a_{j_2 j_1 l_1}$ with $\bar{x}_{j_2} \in (-\infty, \infty)$ has the following properties:

- *For* $a_{j_1 j_1 0} > 0$ *and* $a_{j_2 j_1 0}(a_{j_2 j_1 l_1} - a_{j_2 j_1 l_2}) > 0$,

$$(a_{j_1 j_1 1}, \bar{x}_{j_2}) = \underbrace{(\mathrm{SO}, \mathrm{II})}_{\text{increasing-inflection source}}. \tag{3.172}$$

The (SO,II)*-equilibrium at* $x^*_{j_1} = a_{j_1 j_1 1} = a_{j_2 j_1 l_1}$ *is an increasing-inflection source infinite-equilibrium.*

- *For* $a_{j_1 j_1 0} < 0$ *and* $a_{j_2 j_1 0}(a_{j_2 j_1 l_1} - a_{j_2 j_1 l_2}) > 0$,

$$(a_{j_1 j_1 1}, \bar{x}_{j_2}) = \underbrace{(\mathrm{SI}, \mathrm{DI})}_{\text{decreasing-inflection sink}}. \tag{3.173}$$

The (SI,DI)*-equilibrium at* $x^*_{j_1} = a_{j_1 j_1 1} = a_{j_2 j_1 l_1}$ *is a decreasing-inflection sink infinite-equilibrium.*

- *For* $a_{j_1 j_1 0} > 0$ *and* $a_{j_2 j_1 0}(a_{j_2 j_1 l_1} - a_{j_2 j_1 l_2}) < 0$,

$$(a_{j_1 j_1 1}, \bar{x}_{j_2}) = \underbrace{(\mathrm{SO}, \mathrm{DI})}_{\text{decreasing-inflection source}}. \tag{3.174}$$

The (SO,DI)*-equilibrium at* $x^*_{j_1} = a_{j_1j_1 1} = a_{j_2j_1 l_1}$ *is an increasing-inflection source infinite-equilibrium.*

- *For* $a_{j_1j_1 0} < 0$ *and* $a_{j_2j_1 0}(a_{j_2j_1 l_1} - a_{j_2j_1 l_2}) < 0$,

$$(a_{j_1j_1 1}, \bar{x}_{j_2}) = \underbrace{(\text{SI, II})}_{\text{increasing-inflection sink}}. \tag{3.175}$$

The (SI,II)*-equilibrium at* $x^*_{j_1} = a_{j_1j_1 1} = a_{j_2j_1 l_1}$ *is an increasing-inflection sink infinite-equilibrium.*

– *The inflection source and sink infinite-equilibriums are the switching bifurcations of a source (sink) flow and a parabola flow.*

(x) If

$$\Delta_{j_1j_1} = 0 \text{ and } \Delta_{j_2j_1} > 0, \tag{3.176}$$

then

$$\begin{aligned}
&\dot{x}_{j_1} = a_{j_1j_1 0}(x_{j_1} - a_{j_1j_1 s_1})(x_{j_1} - a_{j_1j_1 s_2})^2,\\
&\dot{x}_{j_2} = a_{j_2j_1 0}(x_{j_1} - a_{j_2j_1 1})(x_{j_1} - a_{j_2j_1 2}),\\
&j_1, j_2 \in \{1,2\}, j_1 \neq j_2; s_1, s_2 \in \{1,2\}, s_1 \neq s_2,
\end{aligned} \tag{3.177}$$

where

$$\begin{aligned}
&\{a_{j_1j_1 1}, a_{j_1j_1 2}\} = \text{sort}\{b_{j_1j_1 1}, b_{j_1j_1 2}\}, a_{j_1j_1 1} < a_{j_1j_1 2};\\
&\{a_{j_2j_1 1}, a_{j_2j_1 2}\} = \text{sort}\{b_{j_2j_1 1}, b_{j_2j_1 2}\}, a_{j_2j_1 1} < a_{j_2j_1 2};\\
&b_{j_1j_1 2} = -\frac{1}{2}B_{j_1j_1 1}; b_{j_2j_1 1}, b_{j_2j_1 2} = -\frac{1}{2}(B_{j_2j_1 1} \pm \sqrt{\Delta_{j_2j_1}});\\
&j_1, j_2 \in \{1,2\}, j_1 \neq j_2.
\end{aligned} \tag{3.178}$$

(x₁) For $a_{j_1j_1 s_1} \neq a_{j_2j_1 l_1} \neq a_{j_1j_1 s_2}$ $(l_1, l_2 \in \{1,2\}; l_1 \neq l_2)$, *the first integral manifold is*

$$\begin{aligned}
x_{j_2} - x_{j_2 0} = \frac{a_{j_2j_1 0}}{a_{j_1j_1 0}}\Big\{&\frac{\prod_{l_1=1}^{2}(a_{j_1j_1 s_1} - a_{j_2j_1 l_1})}{(a_{j_1j_1 s_1} - a_{j_1j_1 s_2})^2}\ln\frac{|x_{j_1} - a_{j_1j_1 s_1}|}{|x_{j_1 0} - a_{j_1j_1 s_1}|}\\
&-\frac{\prod_{l_1=1}^{2}(a_{j_1j_1 s_1} - a_{j_2j_1 l_1})}{(a_{j_1j_1 s_2} - a_{j_1j_1 s_1})}\Big(\frac{1}{x_{j_1} - a_{j_1j_1 s_2}} - \frac{1}{x_{j_1 0} - a_{j_1j_1 s_2}}\Big)\\
&+\Big[\frac{\sum_{l_1=1}^{2}(a_{j_1j_1 s_2} - a_{j_2j_1 l_1})}{(a_{j_1j_1 s_2} - a_{j_1j_1 s_1})} - \frac{\prod_{l_1=1}^{2}(a_{j_1j_1 s_2} - a_{j_2j_1 l_1})}{(a_{j_1j_1 s_2} - a_{j_1j_1 s_1})^2}\Big]\ln\frac{|x_{j_1} - a_{j_1j_1 s_2}|}{|x_{j_1 0} - a_{j_1j_1 s_2}|}\Big\},\\
&j_1, j_2 \in \{1,2\}, j_1 \neq j_2; s_1, s_2 \in \{1,2\}, s_1 \neq s_2.
\end{aligned} \tag{3.179}$$

(x₁ₐ) The equilibrium of $x^*_{j_1} = a_{j_1j_1 s_1} \neq a_{j_2j_1 l_1}$ $(l_1 \in \{1,2\})$ *possesses the following properties.*

- *For* $a_{j_1j_10}(a_{j_1j_1s_1}-a_{j_1j_1s_2})^2>0$ *and* $a_{j_2j_10}\prod_{l_1=1}^{2}(a_{j_1j_1s_1}-a_{j_2j_1l_1})>0$,

$$(a_{j_1j_1s_1},\dot{x}_{j_2})=\underbrace{(\text{SO},\text{pF})}_{\text{source flow }(+)}. \tag{3.180}$$

The flow at $x_{j_1}^*=a_{j_1j_1s_1}\neq a_{j_2j_1l_1}$ *is an* (SO,pF)*-positive source flow.*

- *For* $a_{j_1j_10}(a_{j_1j_1s_1}-a_{j_1j_1s_2})^2<0$ *and* $a_{j_2j_10}\prod_{l_1=1}^{2}(a_{j_1j_1s_1}-a_{j_2j_1l_1})>0$,

$$(a_{j_1j_1s_1},\dot{x}_{j_2})=\underbrace{(\text{SI},\text{pF})}_{\text{sink flow }(+)}. \tag{3.181}$$

The flow at $x_{j_1}^*=a_{j_1j_1s_1}\neq a_{j_2j_1l_1}$ *is an (SI,pF)-positive sink flow.*

- *For* $a_{j_1j_10}(a_{j_1j_1s_1}-a_{j_1j_1s_2})^2>0$ *and* $a_{j_2j_10}\prod_{l_1=1}^{2}(a_{j_1j_1s_1}-a_{j_2j_1l_1})<0$,

$$(a_{j_1j_1s_1},\dot{x}_{j_2})=\underbrace{(\text{SO},\text{nF})}_{\text{source flow }(-)}. \tag{3.182}$$

The flow at $x_{j_1}^*=a_{j_1j_1s_1}\neq a_{j_2j_1l_1}$ *is an* (SO,nF)*-negative source flow.*

- *For* $a_{j_1j_10}(a_{j_1j_1s_1}-a_{j_1j_1s_2})^2<0$ *and* $a_{j_2j_10}\prod_{l_1=1}^{2}(a_{j_1j_1s_1}-a_{j_2j_1l_1})<0$,

$$(a_{j_1j_1s_1},\dot{x}_{j_2})=\underbrace{(\text{SI},\text{nF})}_{\text{sink flow }(-)}. \tag{3.183}$$

The flow at $x_{j_1}^*=a_{j_1j_1s_1}\neq a_{j_2j_1l_1}$ *is an* (SI,pF)*-negative sink flow.*

(x_{1b}) The equilibrium of $x_{j_1}^*=a_{j_1j_1s_2}\neq a_{j_2j_1l_1}$ $(l_1\in\{1,2\})$ *possesses the following properties.*

- *For* $a_{j_1j_10}(a_{j_1j_1s_2}-a_{j_1j_1s_1})>0$ *and* $a_{j_2j_10}\prod_{l_1=1}^{2}(a_{j_1j_1s_2}-a_{j_2j_1l_1})>0$,

$$(a_{j_1j_1s_2},\dot{x}_{j_2})=\underbrace{(\text{US},\text{pF})}_{\text{upper-saddle flow }(+)}. \tag{3.184}$$

The flow at $x_{j_1}^*=a_{j_1j_1s_2}\neq a_{j_2j_1l_1}$ *is a* (US,pF)*-positive upper-saddle flow.*

- *For* $a_{j_1j_10}(a_{j_1j_1s_2}-a_{j_1j_1s_1})<0$ *and* $a_{j_2j_10}\prod_{l_1=1}^{2}(a_{j_1j_1s_2}-a_{j_2j_1l_1})>0$,

$$(a_{j_1j_1s_2}, \dot{x}_{j_2}) = \underbrace{(\text{LS}, \text{pF})}_{\text{lower-saddle flow } (+)}. \tag{3.185}$$

The flow at $x^*_{j_1} = a_{j_1j_1s_2} \neq a_{j_2j_1l_1}$ *is an* (LS,pF)*-positive lower-saddle flow.*

- *For* $a_{j_1j_10}(a_{j_1j_1s_2} - a_{j_1j_1s_1}) > 0$ *and* $a_{j_2j_10}\prod_{l_1=1}^{2}(a_{j_1j_1s_2} - a_{j_2j_1l_1}) < 0$,

$$(a_{j_1j_1s_2}, \dot{x}_{j_2}) = \underbrace{(\text{US}, \text{nF})}_{\text{upper-saddle flow } (-)}. \tag{3.186}$$

The flow at $x^*_{j_1} = a_{j_1j_1s_2} \neq a_{j_2j_1l_1}$ *is a* (US,nF)*-negative upper-saddle flow.*

- *For* $a_{j_1j_10}(a_{j_1j_1s_2} - a_{j_1j_1s_1}) < 0$ *and* $a_{j_2j_10}\prod_{l_1=1}^{2}(a_{j_1j_1s_2} - a_{j_2j_1l_1}) < 0$,

$$(a_{j_1j_1s_2}, \dot{x}_{j_2}) = \underbrace{(\text{LS}, \text{nF})}_{\text{lower-saddle flow } (-)}. \tag{3.187}$$

The flow at $x^*_{j_1} = a_{j_1j_1s_2} \neq a_{j_2j_1l_1}$ *is an* (LS,nF)*-negative lower-saddle flow.*

– *The lower-saddle (LS) and upper-saddle (US) flows are the appearing-switching bifurcation for the sink and source flows.*

(x_{Ic}) *The equilibrium of* $x^*_{j_1} = a_{j_2j_1l_1} \neq a_{j_1j_1s_1}, a_{j_1j_1s_2}$ *possesses the following properties.*

- *For* $a_{j_1j_10}(a_{j_2j_1l_1} - a_{j_1j_1s_1})(a_{j_2j_1l_1} - a_{j_1j_1s_2})^2 > 0$ *and* $a_{j_2j_10}(a_{j_2j_1l_1} - a_{j_2j_1l_2}) > 0$,

$$(\dot{x}_{j_1}, a_{j_2j_1l_1}) = \underbrace{(\text{pF}, \text{UP})}_{\text{up-parabola flow } (+)}. \tag{3.188}$$

The flow at $x^*_{j_1} = a_{j_2j_1l_1}$ *is a* (pF,II)*-positive up-parabola flow.*

- *For* $a_{j_1j_10}(a_{j_2j_1l_1} - a_{j_1j_1s_1})(a_{j_2j_1l_1} - a_{j_1j_1s_2})^2 < 0$ *and* $a_{j_2j_10}(a_{j_2j_1l_1} - a_{j_2j_1l_2}) > 0$,

$$(\dot{x}_{j_1}, a_{j_2j_1l_1}) = \underbrace{(\text{nF}, \text{DP})}_{\text{down-parabola flow } (-)}. \tag{3.189}$$

The flow at $x^*_{j_1} = a_{j_2j_1l_1}$ *is an* (nF,DP)*-negative down-parabola flow.*

- *For* $a_{j_1j_10}(a_{j_2j_1l_1} - a_{j_1j_1s_1})(a_{j_2j_1l_1} - a_{j_1j_1s_2})^2 > 0$ *and* $a_{j_2j_10}(a_{j_2j_1l_1} - a_{j_2j_1l_2}) < 0$,

$$(\dot{x}_{j_1}, a_{j_2j_1l_1}) = \underbrace{(\text{pF}, \text{DP})}_{\text{down-parabola flow } (+)}. \tag{3.190}$$

The flow at $x^*_{j_1} = a_{j_2j_1l_1}$ *is a* (pF,DI)*-positive down-parabola flow.*

- *For* $a_{j_1j_10}(a_{j_2j_1l_1} - a_{j_1j_1s_1})(a_{j_2j_1l_1} - a_{j_1j_1s_2})^2 < 0$ *and* $a_{j_2j_10}(a_{j_2j_1l_1} - a_{j_2j_1l_2}) < 0$,

$$(\dot{x}_{j_1}, a_{j_2j_1l_1}) = \underbrace{(\text{pF}, \text{UP})}_{\text{up-parabola flow }(-)}. \tag{3.191}$$

The flow at $x_{j_1}^* = a_{j_2j_1l_1}$ *is an* (nF,UP)*-positive up-parabola flow.*

(x₂) For $a_{j_1j_1s_2} = a_{j_2j_1l_1}$,

$$\begin{aligned} \dot{x}_{j_1} &= a_{j_1j_1 0}(x_{j_1} - a_{j_1j_1s_1})(x_{j_1} - a_{j_1j_1s_2})^2, \\ \dot{x}_{j_2} &= a_{j_2j_1 0}(x_{j_1} - a_{j_2j_1l_1})(x_{j_1} - a_{j_2j_1l_2}), \\ & j_1, j_2 \in \{1,2\}, j_1 \neq j_2. \end{aligned} \tag{3.192}$$

The first integral manifold is

$$\begin{aligned} x_{j_2} - x_{j_2 0} = & \frac{a_{j_2j_1 0}}{a_{j_1j_1 0}} \Big(\frac{a_{j_1j_1s_1} - a_{j_2j_1l_2}}{a_{j_1j_1s_1} - a_{j_1j_1s_2}} \ln \frac{|\, x_{j_1} - a_{j_1j_1s_1} \,|}{|\, x_{j_1 0} - a_{j_1j_1s_1} \,|} \\ & + \frac{a_{j_1j_1s_2} - a_{j_2j_1l_2}}{a_{j_1j_1s_2} - a_{j_1j_1s_1}} \ln \frac{|\, x_{j_1} - a_{j_1j_1s_2} \,|}{|\, x_{j_1 0} - a_{j_1j_1s_2} \,|} \Big); \\ & j_1, j_2 \in \{1,2\}, j_1 \neq j_2. \end{aligned} \tag{3.193}$$

(x_{2a}) The infinite-equilibrium of $x_{j_1}^* = a_{j_1j_1s_2} = a_{j_2j_1l_1}$ with $\bar{x}_{j_2} \in (-\infty, \infty)$ *has the following properties.*

- *For* $a_{j_1j_1 0}(a_{j_1j_1s_2} - a_{j_1j_1s_1}) > 0$ *and* $a_{j_2j_1 0}(a_{j_2j_1l_1} - a_{j_2j_1l_2}) > 0$,

$$(a_{j_1j_1s_2}, \bar{x}_{j_2}) = \underbrace{(\text{US}, \text{DU})}_{\text{down-up upper-saddle}}. \tag{3.194}$$

  The equilibrium at $x_{j_1}^* = a_{j_1j_1s_2} = a_{j_2j_1l_1}$ *is a* (US,DU)*-down-up upper-saddle infinite-equilibrium.*

- *For* $a_{j_1j_1 0}(a_{j_1j_1s_2} - a_{j_1j_1s_1}) < 0$ *and* $a_{j_2j_1 0}(a_{j_2j_1l_1} - a_{j_2j_1l_2}) > 0$,

$$(a_{j_1j_1s_2}, \bar{x}_{j_2}) = \underbrace{(\text{LS}, \text{UD})}_{\text{up-down lower-saddle}}. \tag{3.195}$$

  The equilibrium at $x_{j_1}^* = a_{j_1j_1s_2} = a_{j_2j_1l_1}$ *is an* (LS,UD)*-up-down lower-saddle infinite-equilibrium.*

- *For* $a_{j_1j_10}(a_{j_1j_1s_2}-a_{j_1j_1s_1})>0$ *and* $a_{j_2j_10}(a_{j_2j_1l_1}-a_{j_2j_1l_2})<0$,

$$(a_{j_1j_1s_2},\bar{x}_{j_2})=\underbrace{(\mathrm{US},\mathrm{UD})}_{\text{up-down upper-saddle}}. \tag{3.196}$$

The equilibrium at $x_{j_1}^*=a_{j_1j_1s_2}=a_{j_2j_1l_1}$ *is a* (US,UD)*-up-down upper-saddle infinite-equilibrium.*

- *For* $a_{j_1j_10}(a_{j_1j_1s_2}-a_{j_1j_1s_1})<0$ *and* $a_{j_2j_10}(a_{j_2j_1l_1}-a_{j_2j_1l_2})<0$,

$$(a_{j_1j_1s_2},\bar{x}_{j_2})=\underbrace{(\mathrm{LS},\mathrm{DU})}_{\text{down-up lower-saddle}}. \tag{3.197}$$

The equilibrium at $x_{j_1}^*=a_{j_1j_1s_2}=a_{j_2j_1l_1}$ *is an* (LS,DU)*-down-up lower-saddle infinite-equilibrium.*

- *The down-up upper-saddle (lower-saddle) infinite-equilibriums are the switching bifurcations for an up-parabola flow with an upper-saddle (lower-saddle) flow.*
- *The up-down upper-saddle (lower-saddle) infinite-equilibriums are the switching bifurcations for a down-parabola flow with an upper-saddle (lower-saddle) flow.*

(x_3) For $a_{j_1j_1s_1}=a_{j_2j_1l_1}$,

$$\begin{aligned}&\dot{x}_{j_1}=a_{j_1j_10}(x_{j_1}-a_{j_1j_1s_1})(x_{j_1}-a_{j_1j_1s_2})^2,\\&\dot{x}_{j_2}=a_{j_2j_10}(x_{j_1}-a_{j_2j_1l_1})(x_{j_1}-a_{j_2j_1l_2}),\\&j_1,j_2\in\{1,2\},j_1\neq j_2.\end{aligned} \tag{3.198}$$

The first integral manifold is

$$\begin{aligned}x_{j_2}-x_{j_20}=&\frac{a_{j_2j_10}}{a_{j_1j_10}}\Big[\ln\frac{|x_{j_1}-a_{j_1j_1s_2}|}{|x_{j_10}-a_{j_1j_1s_2}|}\\&-(a_{j_1j_1s_2}-a_{j_2j_1l_2})(\frac{1}{x_{j_1}-a_{j_1j_1s_2}}-\frac{1}{x_{j_10}-a_{j_1j_1s_2}})\Big],\\&j_1,j_2\in\{1,2\},j_1\neq j_2.\end{aligned} \tag{3.199}$$

(x_{3a}) The infinite-equilibrium of $x_{j_1}^*=a_{j_1j_1s_1}=a_{j_2j_1l_1}$ with $\bar{x}_{j_2}\in(-\infty,\infty)$ *has the following properties.*

- *For* $a_{j_1j_10}(a_{j_1j_1s_1}-a_{j_1j_1s_2})^2>0$ *and* $a_{j_2j_10}(a_{j_2j_1l_1}-a_{j_1j_1l_2})>0$,

$$(a_{j_1j_1s_1},\bar{x}_{j_2})=\underbrace{(\mathrm{SO},\mathrm{II})}_{\text{increasing-inflection source}}. \tag{3.200}$$

The equilibrium at $x_{j_1}^* = a_{j_1j_1s_1} = a_{j_2j_1l_1}$ *is an* (SO,II)*-increasing-inflection source infinite-equilibrium.*

- *For* $a_{j_1j_10}(a_{j_1j_1s_1} - a_{j_1j_1s_2})^2 < 0$ *and* $a_{j_2j_10}(a_{j_2j_1l_1} - a_{j_1j_1l_2}) > 0$,

$$(a_{j_1j_1s_1}, \bar{x}_{j_2}) = \underbrace{(\text{SI, DI})}_{\text{decreasing-inflection sink}}. \tag{3.201}$$

The equilibrium at $x_{j_1}^* = a_{j_1j_1s_1} = a_{j_2j_1l_1}$ *is an* (SI,DI)*-decreasing-inflection sink infinite-equilibrium.*

- *For* $a_{j_1j_10}(a_{j_1j_1s_1} - a_{j_1j_1s_2})^2 > 0$ *and* $a_{j_2j_10}(a_{j_2j_1l_1} - a_{j_1j_1l_2}) < 0$,

$$(a_{j_1j_1s_1}, \bar{x}_{j_2}) = \underbrace{(\text{SO, DI})}_{\text{decreasing-inflection source}}. \tag{3.202}$$

The equilibrium at $x_{j_1}^* = a_{j_1j_1s_1} = a_{j_2j_1l_1}$ *is an* (SO,DI)*-decreasing-inflection source infinite-equilibrium.*

- *For* $a_{j_1j_10}(a_{j_1j_1s_1} - a_{j_1j_1s_2})^2 < 0$ *and* $a_{j_2j_10}(a_{j_2j_1l_1} - a_{j_1j_1l_2}) < 0$,

$$(a_{j_1j_1s_1}, \bar{x}_{j_2}) = \underbrace{(\text{SI, II})}_{\text{increasing-inflection sink}}. \tag{3.203}$$

The equilibrium at $x_{j_1}^* = a_{j_1j_1s_1} = a_{j_2j_1l_1}$ *is an* (SI,II)*-increasing-inflection sink infinite-equilibrium.*

– *The inflection source (sink) infinite-equilibriums are the switching bifurcation for a source (sink) flow with a parabola flow.*

(xi) If

$$\begin{gathered} \Delta_{j_1j_1} = 0 \text{ and } \Delta_{j_2j_1} > 0, \\ a_{j_1j_1s_1} = a_{j_1j_1s_2} \equiv a_{j_1j_11} \end{gathered} \tag{3.204}$$

then

$$\begin{aligned} &\dot{x}_{j_1} = a_{j_1j_10}(x_{j_1} - a_{j_1j_11})^3, \\ &\dot{x}_{j_2} = a_{j_2j_10}(x_{j_1} - a_{j_2j_1l_1})(x_{j_1} - a_{j_2j_1l_2}), \\ &j_1, j_2 \in \{1, 2\}, j_1 \neq j_2. \end{aligned} \tag{3.205}$$

(xi_1) For $a_{j_1j_1 1} \neq a_{j_2j_1 1}$, the first integral manifold is

$$\begin{aligned} x_{j_2} - x_{j_2 0} = \frac{a_{j_2j_1 0}}{a_{j_1j_1 0}} \Big\{ \ln \frac{| x_{j_1} - a_{j_1j_1 1} |}{| x_{j_1 0} - a_{j_1j_1 1} |} \\ - \sum\nolimits_{l_1=1}^{2} (a_{j_1j_1 1} - a_{j_2j_1 l_1}) \Big(\frac{1}{x_{j_1} - a_{j_1j_1 1}} - \frac{1}{x_{j_1 0} - a_{j_1j_1 1}} \Big) \\ - \frac{1}{2} \prod\nolimits_{l_1=1}^{2} (a_{j_1j_1 1} - a_{j_2j_1 l_1}) \Big[\frac{1}{(x_{j_1} - a_{j_1j_1 1})^2} - \frac{1}{(x_{j_1 0} - a_{j_1j_1 1})^2} \Big] \Big\}, \\ j_1, j_2 \in \{1,2\}, j_1 \neq j_2. \end{aligned} \tag{3.206}$$

(xi_{1a}) The equilibrium of $x^*_{j_1} = a_{j_1j_1 1} \neq a_{j_2j_1 l_1}$ has the following properties.

- *For $a_{j_1j_1 0} > 0$ and $a_{j_2j_1 0} \prod_{l_1=1}^{2} (a_{j_1j_1 1} - a_{j_2j_1 l_1}) > 0$,*

$$(a_{j_1j_1 1}, \dot{x}_{j_2}) = \underbrace{(3^{\text{rd}}\text{SO, pF})}_{\text{third-order source flow }(+)}. \tag{3.207}$$

The flow at $x^*_{j_1} = a_{j_1j_1 1}$ is a (3^{rd}SO,pF)-positive third-order source flow.

- *For $a_{j_1j_1 0} < 0$ and $a_{j_2j_1 0} \prod_{l_1=1}^{2} (a_{j_1j_1 1} - a_{j_2j_1 l_1}) > 0$,*

$$(a_{j_1j_1 1}, \dot{x}_{j_2}) = \underbrace{(3^{\text{rd}}\text{SI, pF})}_{\text{third-order sink flow }(+)}. \tag{3.208}$$

The flow at $x^*_{j_1} = a_{j_1j_1 1}$ is a (3^{rd}SI,pF)-positive third-order sink flow.

- *For $a_{j_1j_1 0} > 0$ and $a_{j_2j_1 0} \prod_{l_1=1}^{2} (a_{j_1j_1 1} - a_{j_2j_1 l_1}) < 0$,*

$$(a_{j_1j_1 1}, \dot{x}_{j_2}) = \underbrace{(3^{\text{rd}}\text{SO, nF})}_{\text{third-order source flow }(-)}. \tag{3.209}$$

The flow at $x^*_{j_1} = a_{j_1j_1 1}$ is a (3^{rd}SO,nF)-negative third-order source flow.

- *For $a_{j_1j_1 0} < 0$ and $a_{j_2j_1 0} \prod_{l_1=1}^{2} (a_{j_1j_1 1} - a_{j_2j_1 l_1}) < 0$,*

$$(a_{j_1j_1 1}, \dot{x}_{j_2}) = \underbrace{(3^{\text{rd}}\text{SI, nF})}_{\text{third-order sink flow }(-)}. \tag{3.210}$$

The flow at $x^*_{j_1} = a_{j_1j_1 1}$ is a (3^{rd}SI,pF)-negative third-order sink flow.

– *The third-order source and sink flows bifurcations are the appearing bifurcations for source-sink-source and sink-source-sink flows.*

(xi_{1b}) The equilibrium of $x_{j_1}^* = a_{j_2 j_1 l_1} \neq a_{j_1 j_1 1}$ *has the following properties.*

- *For* $a_{j_1 j_1 0}(a_{j_2 j_1 l_1} - a_{j_1 j_1 1})^3 > 0$ *and* $a_{j_2 j_1 0}(a_{j_2 j_1 l_1} - a_{j_2 j_1 l_2}) > 0$,

$$(\dot{x}_{j_1}, a_{j_2 j_1 l_1}) = \underbrace{(\text{pF}, \text{UP})}_{\text{up-parabola flow } (+)}. \tag{3.211}$$

The flow at $x_{j_1}^* = a_{j_2 j_1 l_1}$ *is a* (pF,UP)*-positive up-parabola flow.*

- *For* $a_{j_1 j_1 0}(a_{j_2 j_1 l_1} - a_{j_1 j_1 1})^3 < 0$ *and* $a_{j_2 j_1 0}(a_{j_2 j_1 l_1} - a_{j_2 j_1 l_2}) > 0$,

$$(\dot{x}_{j_1}, a_{j_2 j_1 l_1}) = \underbrace{(\text{nF}, \text{DP})}_{\text{down-parabola flow } (-)}. \tag{3.212}$$

The flow at $x_{j_1}^* = a_{j_2 j_1 l_1}$ *is an* (nF,DP)*-negative down-parabola flow.*

- *For* $a_{j_1 j_1 0}(a_{j_2 j_1 l_1} - a_{j_1 j_1 1})^3 > 0$ *and* $a_{j_2 j_1 0}(a_{j_2 j_1 l_1} - a_{j_2 j_1 l_2}) < 0$,

$$(\dot{x}_{j_1}, a_{j_2 j_1 l_1}) = \underbrace{(\text{pF}, \text{DP})}_{\text{down-parabola flow } (+)}. \tag{3.213}$$

The flow at $x_{j_1}^* = a_{j_2 j_1 l_1}$ *is a* (pF,DP)*-positive down-parabola flow.*

- *For* $a_{j_1 j_1 0}(a_{j_2 j_1 l_1} - a_{j_1 j_1 1})^3 < 0$ *and* $a_{j_2 j_1 0}(a_{j_2 j_1 l_1} - a_{j_2 j_1 l_2}) < 0$,

$$(\dot{x}_{j_1}, a_{j_2 j_1 l_1}) = \underbrace{(\text{nF}, \text{UP})}_{\text{up-parabola flow } (-)}. \tag{3.214}$$

The flow at $x_{j_1}^* = a_{j_2 j_1 l_1}$ *is an* (nF,UP)*-negative up-parabola flow.*

(xi_2) For $a_{j_2 j_1 l_1} = a_{j_1 j_1 1}$,

$$\begin{aligned}
&\dot{x}_{j_1} = a_{j_1 j_1 0}(x_{j_1} - a_{j_1 j_1 1})^3,\\
&\dot{x}_{j_2} = a_{j_2 j_1 0}(x_{j_1} - a_{j_2 j_1 l_1})(x_{j_1} - a_{j_2 j_1 l_2}),\\
&j_1, j_2 \in \{1, 2\}, j_1 \neq j_2.
\end{aligned} \tag{3.215}$$

The first integral manifold is

$$\begin{aligned} x_{j_2} - x_{j_2 0} = \frac{a_{j_2 j_1 0}}{a_{j_1 j_1 0}} \Big[& \ln \frac{| x_{j_1} - a_{j_1 j_1 1} |}{| x_{j_1 0} - a_{j_1 j_1 1} |} \\ & - (a_{j_2 j_1 l_1} - a_{j_2 j_1 l_2})(\frac{1}{x_{j_1} - a_{j_1 j_1 1}} - \frac{1}{x_{j_1 0} - a_{j_1 j_1 1}}) \Big], \\ & j_1, j_2 \in \{1,2\}, j_1 \neq j_2. \end{aligned} \tag{3.216}$$

The infinite-equilibrium of $x_{j_1}^* = a_{j_2 j_1 l_1} = a_{j_1 j_1 1}$ with $\bar{x}_{j_2} \in (-\infty, \infty)$ *has the following properties.*

- *For* $a_{j_1 j_1 0} > 0$ *and* $a_{j_2 j_1 0}(a_{j_2 j_1 l_1} - a_{j_2 j_1 l_2}) > 0$,

$$(a_{j_1 j_1 1}, \bar{x}_{j_2}) = \underbrace{(3^{\text{rd}}\text{SO}, (3{:}1)\text{IC})}_{\text{third-order increasing-concave source}}. \tag{3.217}$$

  The equilibrium at $x_{j_1}^* = a_{j_2 j_1 l_1} = a_{j_1 j_1 1}$ *is a* (3^{rd}SO,(3:1)IC)*-third-order increasing-concave source infinite-equilibrium.*

- *For* $a_{j_1 j_1 0} < 0$ *and* $a_{j_2 j_1 0}(a_{j_2 j_1 l_1} - a_{j_2 j_1 l_2}) > 0$,

$$(a_{j_1 j_1 1}, \bar{x}_{j_2}) = \underbrace{(3^{\text{rd}}\text{SO}, (3{:}1)\text{DC})}_{\text{third-order decreasing-concave source}}. \tag{3.218}$$

  The equilibrium at $x_{j_1}^* = a_{j_2 j_1 l_1} = a_{j_1 j_1 1}$ *is a* (3^{rd}SI,(3:1)DC)*-third-order decreasing-concave sink infinite-equilibrium.*

- *For* $a_{j_1 j_1 0} > 0$ *and* $a_{j_2 j_1 0}(a_{j_2 j_1 l_1} - a_{j_2 j_1 l_2}) < 0$,

$$(a_{j_1 j_1 1}, \bar{x}_{j_2}) = \underbrace{(3^{\text{rd}}\text{SO}, (3{:}1)\text{DC})}_{\text{third-order decreasing-concave source}}. \tag{3.219}$$

  The equilibrium at $x_{j_1}^* = a_{j_2 j_1 l_1} = a_{j_1 j_1 1}$ *is a* (3^{rd}SO,(3:1)DC)*-third-order decreasing-concave source infinite-equilibrium.*

- *For* $a_{j_1 j_1 0} < 0$ *and* $a_{j_2 j_1 0}(a_{j_2 j_1 l_1} - a_{j_2 j_1 l_2}) < 0$,

$$(a_{j_1 j_1 1}, \bar{x}_{j_2}) = \underbrace{(3^{\text{rd}}\text{SI}, (3{:}1)\text{IC})}_{\text{third-order increasing-concave sink}}. \tag{3.220}$$

  The equilibrium at $x_{j_1}^* = a_{j_2 j_1 l_1} = a_{j_1 j_1 1}$ *is a* (3^{rd}SI,(3:1)IC)*-third-order increasing-concave sink infinite-equilibrium.*

– *The third-order concave source (sink) infinite-equilibriums are the switching bifurcations of a third-order source (sink) flow with a parabola flow.*

3.2 Proof of Theorem 3.1

Consider

$$\begin{aligned}\dot{x}_{j_2} &= a_{j_2j_1 0}(x_{j_1}^2 + B_{j_2j_1 1}x_{j_1} + C_{j_2j_1 1})\\ \dot{x}_{j_1} &= a_{j_1j_1 0}(x_{j_1} - b_{j_1j_1 1})(x_{j_1}^2 + B_{j_1j_1 1}x_{j_1} + C_{j_1j_1 1})\end{aligned}$$

with

$$\Delta_{jj_1} = (B_{jj_1 1})^2 - 4C_{jj_1 1} \ \text{ for } j \in \{j_1, j_2\}, j_1, j_2 \in \{1,2\} j_1 \neq j_2.$$

(i) If

$$\Delta_{j_1j_1} < 0 \text{ and } \Delta_{j_2j_1} < 0,$$

then

$$\begin{aligned}&\dot{x}_{j_2} = a_{j_2j_1 0}\left[(x_{j_1} - a_{j_2j_1})^2 + b_{j_2j_1}\right],\\ &\dot{x}_{j_1} = a_{j_1j_1 0}(x_{j_1} - b_{j_1j_1 1})\left[(x_{j_1} - a_{j_1j_1})^2 + b_{j_1j_1}\right],\\ &\text{with } a_{sj_1} = -\frac{1}{2}B_{sj_1 1}, b_{sj_1} = \frac{1}{4}(-\Delta_{sj_1}), s = j_1, j_2.\end{aligned}$$

In phase space, we have

$$\begin{aligned}\frac{dx_{j_2}}{dx_{j_1}} &= \frac{a_{j_2j_1 0}}{a_{j_1j_1 0}} \frac{(x_{j_1} - a_{j_2j_1})^2 + b_{j_2j_1}}{(x_{j_1} - a_{j_1j_1 1})\left[(x_{j_1} - a_{j_1j_1})^2 + b_{j_1j_1}\right]}\\ &= \frac{a_{j_2j_1 0}}{a_{j_1j_1 0}}\left\{\frac{(x_{j_1} - a_{j_1j_1}) + (a_{j_1j_1} - a_{j_1j_1 1})}{(x_{j_1} - a_{j_1j_1})^2 + b_{j_1j_1}} + \frac{2(a_{j_1j_1 1} - a_{j_2j_1})}{(x_{j_1} - a_{j_1j_1})^2 + b_{j_1j_1}}\right.\\ &\quad \left. + \frac{(a_{j_1j_1 1} - a_{j_2j_1})^2 + b_{j_2j_1}}{(x_{j_1} - a_{j_1j_1 1})\left[(x_{j_1} - a_{j_1j_1})^2 + b_{j_1j_1}\right]}\right\}\\ &= \frac{a_{j_2j_1 0}}{a_{j_1j_1 0}}\left\{\frac{(x_{j_1} - a_{j_1j_1})}{(x_{j_1} - a_{j_1j_1})^2 + b_{j_1j_1}} + \frac{a_{j_1j_1} + a_{j_1j_1 1} - 2a_{j_2j_1}}{(x_{j_1} - a_{j_1j_1})^2 + b_{j_1j_1}}\right.\\ &\quad + \frac{(a_{j_1j_1 1} - a_{j_2j_1})^2 + b_{j_2j_1}}{(a_{j_1j_1 1} - a_{j_1j_1})^2 + b_{j_1j_1}}\left[\frac{1}{x_{j_1} - a_{j_1j_1 1}} - \frac{(x_{j_1} - a_{j_1j_1})}{(x_{j_1} - a_{j_1j_1})^2 + b_{j_1j_1}}\right.\\ &\quad \left.\left. - \frac{a_{j_1j_1 1} - a_{j_1j_1}}{(x_{j_1} - a_{j_1j_1})^2 + b_{j_1j_1}}\right]\right\}.\end{aligned}$$

With an initial condition $(x_{j_1 0}, x_{j_2 0})$, the integration of the foregoing equation gives the first integral manifold as

$$
\begin{aligned}
x_{j_2} - x_{j_2 0} = \frac{a_{j_2 j_1 0}}{a_{j_1 j_1 0}} \Big\{ & \frac{1}{2} \ln \frac{|(x_{j_1} - a_{j_1 j_1})^2 + b_{j_1 j_1}|}{|(x_{j_1 0} - a_{j_1 j_1})^2 + b_{j_1 j_1}|} \\
& + \frac{a_{j_1 j_1 1} + a_{j_1 j_1} - 2a_{j_2 j_1}}{\sqrt{b_{j_1 j_1}}} (\arctan \frac{x_{j_1} - a_{j_1 j_1}}{\sqrt{b_{j_1 j_1}}} - \arctan \frac{x_{j_1 0} - a_{j_1 j_1}}{\sqrt{b_{j_1 j_1}}}) \\
& + \frac{(a_{j_1 j_1 1} - a_{j_2 j_1})^2 + b_{j_2 j_1}}{(a_{j_1 j_1 1} - a_{j_1 j_1})^2 + b_{j_1 j_1}} \Big[\ln \frac{|x_{j_1} - a_{j_1 j_1 1}|}{|x_{j_1 0} - a_{j_1 j_1 1}|} \\
& - \frac{1}{2} \ln \frac{|(x_{j_1} - a_{j_1 j_1})^2 + b_{j_1 j_1}|}{|(x_{j_1 0} - a_{j_1 j_1})^2 + b_{j_1 j_1}|} \\
& - \frac{a_{j_1 j_1 1} - a_{j_1 j_1}}{\sqrt{b_{j_1 j_1}}} (\arctan \frac{x_{j_1} - a_{j_1 j_1}}{\sqrt{b_{j_1 j_1}}} - \arctan \frac{x_{j_1 0} - a_{j_1 j_1}}{\sqrt{b_{j_1 j_1}}}) \Big] \Big\}, \\
& j_1, j_2 \in \{1, 2\}, j_1 \neq j_2.
\end{aligned}
$$

For $x_{j_1}^* = a_{j_1 j_1 1}$, the variational equation is

$$
\begin{aligned}
\dot{x}_{j_2} &= a_{j_2 j_1 0} \left[(a_{j_1 j_1 1} - a_{j_2 j_1})^2 + b_{j_2 j_1} \right], \\
\Delta \dot{x}_{j_1} &= a_{j_1 j_1 0} \left[(a_{j_1 j_1 1} - a_{j_1 j_1})^2 + b_{j_1 j_1} \right] \Delta x_{j_1}.
\end{aligned}
$$

Due to

$$
(a_{j_1 j_1 1} - a_{j_2 j_1})^2 + b_{j_2 j_1} > 0, \text{and } (a_{j_1 j_1 1} - a_{j_1 j_1})^2 + b_{j_1 j_1} > 0,
$$

the flows at $x_{j_1}^* = a_{j_1 j_1 1}$ are source and sink flows for $a_{j_1 j_1 0} > 0$ and $a_{j_1 j_1 0} < 0$ in the x_{j_1}-direction, and the flows at $x_{j_1}^* = a_{j_1 j_1 1}$ are positive and negative for $a_{j_2 j_1 0} > 0$ and $a_{j_2 j_1 0} < 0$ in the x_{j_2}-direction.

Therefore, the flow at $x_{j_1}^* = a_{j_1 j_1 1}$ for the system has the following properties as in Eqs. (3.6–3.9).

- For $a_{j_1 j_1 0} > 0$ and $a_{j_2 j_1 0} > 0$,

$$
(a_{j_1 j_1 1}, \dot{x}_{j_2}) = \underbrace{(\text{SO}, \text{pF})}_{\text{source flow } (+)}.
$$

The(SO,pF)-flow is a positive source flow at $x_{j_1}^* = a_{j_1 j_1 1}$.

- For $a_{j_1 j_1 0} < 0$ and $a_{j_2 j_1 0} > 0$,

$$
(a_{j_1 j_1 1}, \dot{x}_{j_2}) = \underbrace{(\text{SI}, \text{pF})}_{\text{sink flow}(+)}.
$$

The (SI, pF)-flow is a positive sink flow at $x_{j_1}^* = a_{j_1 j_1 1}$.

- For $a_{j_1j_10} > 0$ and $a_{j_2j_10} < 0$,

$$(a_{j_1j_11}, \dot{x}_{j_2}) = \underbrace{(\text{SO}, \text{nF})}_{\text{source flow }(-)}.$$

The (SO, nF)-flow is a negative source flow at $x^*_{j_1} = a_{j_1j_11}$.

- For $a_{j_1j_10} < 0$ and $a_{j_2j_10} < 0$,

$$(a_{j_1j_11}, \dot{x}_{j_2}) = \underbrace{(\text{SI}, \text{nF})}_{\text{sink flow }(-)}.$$

The (SI,nF)-flow is a negative sink flow at $x^*_{j_1} = a_{j_1j_11}$.

(ii) If

$$\Delta_{j_1j_1} = 0 \text{ and } \Delta_{j_2j_1} < 0,$$

the dynamical system becomes

$$\begin{aligned}
\dot{x}_{j_1} &= a_{j_1j_10}(x_{j_1} - a_{j_1j_1s_1})(x_{j_1} - a_{j_1j_1s_2})^2, \\
\dot{x}_{j_2} &= a_{j_2j_10}\left[(x_{j_1} - a_{j_2j_1})^2 + b_{j_2j_1}\right],
\end{aligned}$$

with

$$\begin{aligned}
&\{a_{j_1j_11}, a_{j_1j_12}\} = \text{sort}\{b_{j_1j_11}, a_{j_1j_1}\}, a_{j_1j_11} < a_{j_1j_12}; \\
&a_{j_1j_1} = -\frac{1}{2}B_{j_1j_11}, b_{j_1j_1} = \frac{1}{4}(-\Delta_{j_1j_1}) = 0; \\
&a_{j_2j_1} = -\frac{1}{2}B_{j_2j_11}, b_{j_2j_1} = \frac{1}{4}(-\Delta_{j_2j_1}); \\
&j_1, j_2 \in \{1, 2\}; j_1 \neq j_2.
\end{aligned}$$

(ii$_1$) For $a_{j_1j_1s_1} \neq a_{j_1j_1s_2}$, in phase space, we have

$$\begin{aligned}
\frac{dx_{j_2}}{dx_{j_1}} &= \frac{a_{j_2j_10}}{a_{j_1j_10}} \frac{(x_{j_1} - a_{j_2j_1})^2 + b_{j_2j_1}}{(x_{j_1} - a_{j_1j_1s_1})(x_{j_1} - a_{j_1j_1s_2})^2} \\
&= \frac{a_{j_2j_10}}{a_{j_1j_10}} \left[\frac{(x_{j_1} - a_{j_1j_1s_2}) + (a_{j_1j_1s_2} - a_{j_1j_1s_1})}{(x_{j_1} - a_{j_1j_1s_2})^2} \right. \\
&\quad \left. + \frac{2(a_{j_1j_1s_1} - a_{j_2j_1})}{(x_{j_1} - a_{j_1j_1s_2})^2} + \frac{(a_{j_1j_1s_1} - a_{j_2j_1})^2 + b_{j_2j_1}}{(x_{j_1} - a_{j_1j_1s_1})(x_{j_1} - a_{j_1j_1s_2})^2} \right],
\end{aligned}$$

Furthermore,

$$\frac{dx_{j_2}}{dx_{j_1}} = \frac{a_{j_2j_10}}{a_{j_1j_10}}\left\{\frac{(x_{j_1}-a_{j_1j_1s_2})}{(x_{j_1}-a_{j_1j_1s_2})^2} + \frac{a_{j_1j_1}+a_{j_1j_1s_1}-2a_{j_2j_1}}{(x_{j_1}-a_{j_1j_1s_2})^2}\right.$$
$$\left.+\frac{(a_{j_1j_1s_1}-a_{j_2j_1})^2+b_{j_2j_1}}{(a_{j_1j_1s_1}-a_{j_1j_1s_2})^2}\left[\frac{1}{x_{j_1}-a_{j_1j_1s_1}} - \frac{(x_{j_1}-a_{j_1j_1s_2})}{(x_{j_1}-a_{j_1j_1s_2})^2} + \frac{a_{j_1j_1s_2}-a_{j_1j_1s_1}}{(x_{j_1}-a_{j_1j_1s_2})^2}\right]\right\}.$$

With an initial condition (x_{j_10}, x_{j_20}), the integration of the foregoing equation gives the first integral manifold as

$$x_{j_2} - x_{j_20} = \frac{a_{j_2j_10}}{a_{j_1j_10}}\left\{\ln\frac{|x_{j_1}-a_{j_1j_1s_2}|}{|x_{j_10}-a_{j_1j_1s_2}|}\right.$$
$$-(a_{j_1j_1}+a_{j_1j_1s_1}-2a_{j_2j_1})(\frac{1}{x_{j_1}-a_{j_1j_1s_2}} - \frac{1}{x_{j_10}-a_{j_1j_1s_2}})$$
$$+\frac{(a_{j_1j_1s_1}-a_{j_2j_1})^2+b_{j_2j_1}}{(a_{j_1j_1s_1}-a_{j_1j_1s_2})^2}\left[\ln\frac{|x_{j_1}-a_{j_1j_1s_1}|}{|x_{j_10}-a_{j_1j_1s_1}|} - \ln\frac{|x_{j_1}-a_{j_1j_1s_2}|}{|x_{j_10}-a_{j_1j_1s_2}|}\right.$$
$$\left.\left.-(a_{j_1j_1s_2}-a_{j_1j_1s_1})(\frac{1}{x_{j_1}-a_{j_1j_1s_2}} - \frac{1}{x_{j_10}-a_{j_1j_1s_2}})\right]\right\}.$$

(ii$_{1a}$) For $x_{j_1}^* = a_{j_1j_1s_1}$, the variational equation is

$$\dot{x}_{j_2} = a_{j_2j_10}\left[(a_{j_1j_1s_1}-a_{j_2j_1})^2+b_{j_2j_1}\right],$$
$$\Delta\dot{x}_{j_1} = a_{j_1j_10}(a_{j_1j_1s_1}-a_{j_1j_1s_2})^2\Delta x_{j_1}.$$

For $a_{j_1j_1s_1} \neq a_{j_1j_1s_2}$, due to

$$(a_{j_1j_1s_1}-a_{j_2j_1})^2+b_{j_2j_1} > 0 \text{ and } (a_{j_1j_1s_1}-a_{j_1j_1s_2})^2 > 0,$$

the flows at $x_{j_1}^* = a_{j_1j_1s_1}$ are source and sink flows for $a_{j_1j_10} > 0$ and $a_{j_1j_10} < 0$ in the x_{j_1}-direction, and the flows at $x_{j_1}^* = a_{j_1j_1s_1}$ are positive and negative for $a_{j_2j_10} > 0$ and $a_{j_2j_10} < 0$ in the x_{j_2}-direction.

Therefore, the equilibrium of $x_{j_1}^* = a_{j_1j_1s_1}$ for the x_{j_1}-direction has the following properties as in Eqs. (3.13)–(3.16):

- For $a_{j_1j_10}(a_{j_1j_1s_1}-a_{j_1j_1s_2})^2 > 0$ and $a_{j_2j_10} > 0$,

$$(a_{j_1j_1s_1}, \dot{x}_{j_2}) = \underbrace{(\text{SO}, \text{pF})}_{\text{source flow }(+)}.$$

The flow at $x_{j_1}^* = a_{j_1j_1s_1}$ is an (SO,pF)-positive source flow.

- For $a_{j_1j_10}(a_{j_1j_1s_1} - a_{j_1j_1s_2})^2 < 0$ and $a_{j_2j_10} > 0$,

$$(a_{j_1j_1s_1}, \dot{x}_{j_2}) = \underbrace{(\mathrm{SI}, \mathrm{pF})}_{\text{sink flow } (+)}.$$

The flow at $x_{j_1}^* = a_{j_1j_1s_1}$ is an (SI,pF)-positive sink flow.

- For $a_{j_1j_10}(a_{j_1j_1s_1} - a_{j_1j_1s_2})^2 > 0$ and $a_{j_2j_10} < 0$,

$$(a_{j_1j_1s_1}, \dot{x}_{j_2}) = \underbrace{(\mathrm{SO}, \mathrm{nF})}_{\text{source flow } (-)}.$$

The flow at $x_{j_1}^* = a_{j_1j_1s_1}$ is an (SO,nF)-negative source flow.

- For $a_{j_1j_10}(a_{j_1j_1s_1} - a_{j_1j_1s_2})^2 < 0$ and $a_{j_2j_10} < 0$,

$$(a_{j_1j_1s_1}, \dot{x}_{j_2}) = \underbrace{(\mathrm{SI}, \mathrm{nF})}_{\text{sink flow } (-)}.$$

The flow at $x_{j_1}^* = a_{j_1j_1s_1}$ is an (SI,nF)-negative sink flow.

(ii_{1b}) For $x_{j_1}^* = a_{j_1j_1s_1}$, the variational equation is

$$\begin{aligned} \dot{x}_{j_2} &= a_{j_2j_10}\left[(a_{j_1j_1s_2} - a_{j_2j_1})^2 + b_{j_2j_1}\right], \\ \Delta\dot{x}_{j_1} &= a_{j_1j_10}(a_{j_1j_1s_2} - a_{j_1j_1s_1})(\Delta x_{j_1})^2. \end{aligned}$$

Due to, for $a_{j_1j_1s_1} \neq a_{j_1j_1s_2}$,

$$(a_{j_1j_1s_2} - a_{j_2j_1})^2 + b_{j_2j_1} > 0,$$

the flows at $x_{j_1}^* = a_{j_1j_1s_2}$ in the x_{j_1}-direction are upper-saddle and lower-saddle flows for $a_{j_1j_10}(a_{j_1j_1s_2} - a_{j_1j_1s_1}) > 0$ and $a_{j_1j_10}(a_{j_1j_1s_2} - a_{j_1j_1s_1}) < 0$, and the flows at $x_{j_1}^* = a_{j_1j_1s_2}$ are positive and negative for $a_{j_2j_10} > 0$ and $a_{j_2j_10} < 0$ in the x_{j_2}-direction.

Therefore, the equilibrium of $x_{j_1}^* = a_{j_1j_1s_2}$ for the x_{j_1}-direction has the following properties as in Eqs. (3.17)–(3.20):

- For $a_{j_1j_10}(a_{j_1j_1s_2} - a_{j_1j_1s_1}) > 0$ and $a_{j_2j_10} > 0$,

$$(a_{j_1j_1s_2}, \dot{x}_{j_2}) = \underbrace{(\mathrm{US}, \mathrm{pF})}_{\text{upper-saddle flow } (+)}.$$

The flow at $x_{j_1}^* = a_{j_1j_1s_2}$ is a (US,pF)-positive upper-saddle flow.

- For $a_{j_1 j_1 0}(a_{j_1 j_1 s_2} - a_{j_1 j_1 s_1}) < 0$ and $a_{j_2 j_1 0} > 0$,

$$(a_{j_1 j_1 s_2}, \dot{x}_{j_2}) = \underbrace{(\text{LS}, \text{pF})}_{\text{lower-saddle flow }(+)}.$$

  The flow at $x_{j_1}^* = a_{j_1 j_1 s_2}$ is an (LS,pF)-positive lower-saddle flow.
- For $a_{j_1 j_1 0}(a_{j_1 j_1 s_2} - a_{j_1 j_1 s_1}) > 0$ and $a_{j_2 j_1 0} < 0$,

$$(a_{j_1 j_1 s_2}, \dot{x}_{j_2}) = \underbrace{(\text{US}, \text{nF})}_{\text{upper-saddle flow }(-)}.$$

  The flow at $x_{j_1}^* = a_{j_1 j_1 s_2}$ is a (US,nF)-negative upper-saddle flow.
- For $a_{j_1 j_1 0}(a_{j_1 j_1 s_2} - a_{j_1 j_1 s_1}) < 0$ and $a_{j_2 j_1 0} < 0$,

$$(a_{j_1 j_1 s_2}, \dot{x}_{j_2}) = \underbrace{(\text{LS}, \text{nF})}_{\text{lower-saddle flow }(-)}.$$

  The flow at $x_{j_1}^* = a_{j_1 j_1 s_2}$ is an (LS,nF)-negative lower-saddle flow.

– The upper-saddle and lower-saddle flow bifurcations are for the sink and source flows.

(ii$_2$) For $a_{j_1 j_1 s_2} = a_{j_1 j_1 s_1} \equiv a_{j_1 j_1 1}$,

$$\begin{aligned} &\dot{x}_{j_1} = a_{j_1 j_1 0}(x_{j_1} - a_{j_1 j_1 1})^3 \\ &\dot{x}_{j_2} = a_{j_2 j_1 0}\left[(x_{j_1} - a_{j_2 j_1})^2 + b_{j_2 j_1}\right], \\ &j_1, j_2 \in \{1,2\}; j_1 \neq j_2. \end{aligned}$$

In phase space, we have

$$\begin{aligned} \frac{dx_{j_2}}{dx_{j_1}} &= \frac{a_{j_2 j_1 0}}{a_{j_1 j_1 0}} \frac{(x_{j_1} - a_{j_2 j_1})^2 + b_{j_2 j_1}}{(x_{j_1} - a_{j_1 j_1 1})^3} \\ &= \frac{a_{j_2 j_1 0}}{a_{j_1 j_1 0}} \left[\frac{1}{(x_{j_1} - a_{j_1 j_1 1})} + \frac{2(a_{j_1 j_1 1} - a_{j_2 j_1})}{(x_{j_1} - a_{j_1 j_1 1})^2} + \frac{(a_{j_1 j_1 1} - a_{j_2 j_1})^2 + b_{j_2 j_1}}{(x_{j_1} - a_{j_1 j_1 1})^3}\right]. \end{aligned}$$

With an initial condition $(x_{j_1 0}, x_{j_2 0})$, the first integral manifold is

$$x_{j_2} - x_{j_2 0} = \frac{a_{j_2 j_1 0}}{a_{j_1 j_1 0}} \Big\{ \ln \frac{| x_{j_1} - a_{j_1 j_1 1} |}{| x_{j_1 0} - a_{j_1 j_1 1} |} \\ - 2(a_{j_1 j_1 1} - a_{j_2 j_1})(\frac{1}{x_{j_1} - a_{j_1 j_1 1}} - \frac{1}{x_{j_1 0} - a_{j_1 j_1 1}}) \\ - \frac{(a_{j_1 j_1 1} - a_{j_2 j_1})^2 + b_{j_2 j_1}}{2} \Big[\frac{1}{(x_{j_1} - a_{j_1 j_1 1})^2} - \frac{1}{(x_{j_1 0} - a_{j_1 j_1 1})^2} \Big] \Big\}, \\ j_1, j_2 \in \{1,2\}, j_1 \neq j_2.$$

At $x^*_{j_1} = a_{j_1 j_1 1}$, the variational equation is

$$\Delta \dot{x}_{j_1} = a_{j_1 j_1 0}(\Delta x_{j_1})^3 \\ \dot{x}_{j_2} = a_{j_2 j_1 0}\left[(a_{j_1 j_1 1} - a_{j_2 j_1})^2 + b_{j_2 j_1}\right].$$

Due to

$$(a_{j_1 j_1 1} - a_{j_2 j_1})^2 + b_{j_2 j_1} > 0,$$

the flows at $x^*_{j_1} = a_{j_1 j_1 1}$ are the third-order source and sink flows for $a_{j_1 j_1 0} > 0$ and $a_{j_1 j_1 0} < 0$ in the x_{j_1}-direction, and the flows at $x^*_{j_1} = a_{j_1 j_1 1}$ are positive and negative for $a_{j_2 j_1 0} > 0$ and $a_{j_2 j_1 0} < 0$ in the x_{j_2}-direction.

Therefore, the equilibrium of $x^*_{j_1} = a_{j_1 j_1 1}$ for the x_{j_1}-direction has the following properties as in Eqs. (3.23)–(3.26).

- For $a_{j_1 j_1 0} > 0$ and $a_{j_2 j_1 0} > 0$,

$$(a_{j_1 j_1 1}, \dot{x}_{j_2}) = \underbrace{(3^{\text{rd}}\text{SO, pF})}_{\text{third-order source flow }(+)},$$

The flow at $x^*_{j_1} = a_{j_1 j_1 1}$ is a (3rdSO,pF)-positive third-order source flow.

- For $a_{j_1 j_1 0} < 0$ and $a_{j_2 j_1 0} > 0$,

$$(a_{j_1 j_1 1}, \dot{x}_{j_2}) = \underbrace{(3^{\text{rd}}\text{SI, pF})}_{\text{third-order sink flow }(+)}.$$

The flow at $x^*_{j_1} = a_{j_1 j_1 1}$ is a (3rdSI,pF)-positive third-order sink flow.

- For $a_{j_1 j_1 0} > 0$ and $a_{j_2 j_1 0} < 0$,

$$(a_{j_1 j_1 1}, \dot{x}_{j_2}) = \underbrace{(3^{\text{rd}}\text{SO, nF})}_{\text{third-order source flow }(-)}.$$

The flow at $x^*_{j_1} = a_{j_1 j_1 1}$ is a (3rdSO,nF)-negative third-order source flow.

- For $a_{j_1j_10}<0$ and $a_{j_2j_10}<0$,

$$(a_{j_1j_1 1},\dot{x}_{j_2})=\underbrace{(3^{\text{rd}}\text{SI, nF})}_{\text{third-order sink flow }(-)}.$$

The flow at $x^*_{j_1}=a_{j_1j_1 1}$ is a (3rdSI,nF)-negative third-order sink flow.

– The third-order source and sinks flow bifurcations are the appearing bifurcations for the source-sink-source and sink-source-sink flows.

(iii) If

$$\Delta_{j_1j_1}>0 \text{ and } \Delta_{j_2j_1}<0,$$

then

$$\begin{aligned}
&\dot{x}_{j_1}=a_{j_1j_10}(x_{j_1}-a_{j_1j_11})(x_{j_1}-a_{j_1j_12})(x_{j_1}-a_{j_1j_13}),\\
&\dot{x}_{j_2}=a_{j_2j_10}\left[(x_{j_1}-a_{j_2j_1})^2+b_{j_2j_1}\right],\\
&j_1,j_2\in\{1,2\};j_1\neq j_2
\end{aligned}$$

where

$$\begin{aligned}
&\{a_{j_1j_11},a_{j_1j_12},a_{j_1j_13}\}=\text{sort}\{b_{j_1j_11},b_{j_1j_12},b_{j_1j_13}\},a_{j_1j_11}<a_{j_1j_12}<a_{j_1j_13};\\
&b_{j_1j_12},b_{j_1j_13}=-\frac{1}{2}(B_{j_1j_11}\pm\sqrt{\Delta_{j_1j_1}});a_{j_2j_1}=-\frac{1}{2}B_{j_2j_11}.
\end{aligned}$$

In phase space, we have

$$\begin{aligned}
\frac{dx_{j_2}}{dx_{j_1}}&=\frac{a_{j_2j_10}}{a_{j_1j_10}}\frac{(x_{j_1}-a_{j_2j_1})^2+b_{j_2j_1}}{(x_{j_1}-a_{j_1j_11})(x_{j_1}-a_{j_1j_12})(x_{j_1}-a_{j_1j_13})}\\
&=\frac{a_{j_2j_10}}{a_{j_1j_10}}\sum_{s_1=1}^{3}\frac{(a_{j_1j_1s_1}-a_{j_2j_1})^2+b_{j_2j_1}}{\prod_{s_2=1,s_2\neq s_1}^{3}(a_{j_1j_1s_1}-a_{j_1j_1s_2})}\frac{1}{(x_{j_1}-a_{j_1j_1s_1})}
\end{aligned}$$

With an initial condition (x_{j_10},x_{j_20}), the first integral manifold is

$$x_{j_2}-x_{j_20}=\sum_{s_1=1}^{3}\frac{(a_{j_1j_1s_1}-a_{j_2j_1})^2+b_{j_2j_1}}{\prod_{s_2=1,s_2\neq s_1}^{3}(a_{j_1j_1s_1}-a_{j_1j_1s_2})}\ln\frac{|x_{j_1}-a_{j_1j_1s_1}|}{|x_{j_10}-a_{j_1j_1s_1}|}.$$

At $x_{j_1}^* = a_{j_1 j_1 s_1}$ (s_1, $s_2 \in \{1, 2, 3\}$, $s_1 \neq s_2$), the variational equation is

$$\Delta\dot{x}_{j_1} = a_{j_1 j_1 0}\prod\nolimits_{s_2=1, s_2 \neq s_1}^{3} (a_{j_1 j_1 s_1} - a_{j_1 j_1 s_2})\Delta x_{j_1}$$
$$\dot{x}_{j_2} = a_{j_2 j_1 0}\left[(a_{j_1 j_1 s_1} - a_{j_2 j_1})^2 + b_{j_2 j_1}\right].$$

Due to

$$(a_{j_1 j_1 s_1} - a_{j_2 j_1})^2 + b_{j_2 j_1} > 0,$$

the flows at $x_{j_1}^* = a_{j_1 j_1 s_1}$ in the x_{j_1}-direction are source and sink flows for $a_{j_1 j_1 0}\prod\nolimits_{s_2=1, s_2 \neq s_1}^{3} (a_{j_1 j_1 s_1} - a_{j_1 j_1 s_2}) > 0$ and $a_{j_1 j_1 0}\prod\nolimits_{s_2=1, s_2 \neq s_1}^{3} (a_{j_1 j_1 s_1} - a_{j_1 j_1 s_2}) < 0$, respectively, and the flows at $x_{j_1}^* = a_{j_1 j_1 s_1}$ are positive and negative for $a_{j_2 j_1 0} > 0$ and $a_{j_2 j_1 0} < 0$ in the x_{j_2}-direction, respectively.

Therefore, the equilibrium of $x_{j_1}^* = a_{j_1 j_1 s_1}$ (s_1, $s_2 \in \{1, 2, 3\}$, $s_1 \neq s_2$) in the x_{j_1}-direction possesses the following properties as in Eqs. (3.31)–(3.34):

- For $a_{j_1 j_1 0}\prod\nolimits_{s_2=1, s_2 \neq s_1}^{3} (a_{j_1 j_1 s_1} - a_{j_1 j_1 s_2}) > 0$ and $a_{j_2 j_1 0} > 0$,

$$(a_{j_1 j_1 s_1}, \dot{x}_{j_2}) = \underbrace{(\text{SO}, \text{pF})}_{\text{source flow }(+)}.$$

  The flow at $x_{j_1}^* = a_{j_1 j_1 s_1}$ is an (SO,pF)-positive source-flow.
- For $a_{j_1 j_1 0}\prod\nolimits_{s_2=1, s_2 \neq s_1}^{3} (a_{j_1 j_1 s_1} - a_{j_1 j_1 s_2}) < 0$ and $a_{j_2 j_1 0} > 0$,

$$(a_{j_1 j_1 s_1}, \dot{x}_{j_2}) = \underbrace{(\text{SI}, \text{pF})}_{\text{sink flow }(+)}.$$

  The flow at $x_{j_1}^* = a_{j_1 j_1 s_1}$ is an (SI,pF)-positive sink-flow.
- For $a_{j_1 j_1 0}\prod\nolimits_{s_2=1, s_2 \neq s_1}^{3} (a_{j_1 j_1 s_1} - a_{j_1 j_1 s_2}) > 0$ and $a_{j_2 j_1 0} < 0$,

$$(a_{j_1 j_1 s_1}, \dot{x}_{j_2}) = \underbrace{(\text{SO}, \text{nF})}_{\text{source flow }(-)}.$$

  The flow at $x_{j_1}^* = a_{j_1 j_1 s_1}$ is an (SO,nF)-negative source-flow.
- For $a_{j_1 j_1 0}\prod\nolimits_{s_2=1, s_2 \neq s_1}^{3} (a_{j_1 j_1 s_1} - a_{j_1 j_1 s_2}) < 0$ and $a_{j_2 j_1 0} < 0$,

$$(a_{j_1j_1s_1}, \dot{x}_{j_2}) = \underbrace{(\mathrm{SI}, \mathrm{nF})}_{\text{sink flow }(-)}.$$

The flow at $x^*_{j_1} = a_{j_1j_1s_1}$ is an (SI,nF)-negative sink-flow.

(iv) If

$$\Delta_{j_1j_1} < 0 \text{ and } \Delta_{j_2j_1} = 0,$$

then

$$\begin{aligned}
\dot{x}_{j_2} &= a_{j_2j_10}(x_{j_1} - a_{j_2j_11})^2,\\
\dot{x}_{j_1} &= a_{j_1j_10}(x_{j_1} - a_{j_1j_11})\big[(x_{j_1} - a_{j_1j_1})^2 + b_{j_1j_1}\big],
\end{aligned}$$

with

$$\begin{aligned}
a_{j_1j_1} &= -\frac{1}{2}B_{j_1j_11}, b_{j_1j_1} = \frac{1}{4}(-\Delta_{j_1j_1}), a_{j_1j_11} = b_{j_1j_11};\\
a_{j_2j_11} &= -\frac{1}{2}B_{j_2j_11}, b_{j_2j_1} = \frac{1}{4}(-\Delta_{j_2j_1}) = 0.
\end{aligned}$$

(iv$_1$) For $a_{j_1j_11} \neq a_{j_2j_11}$, we have

$$\begin{aligned}
\frac{dx_{j_2}}{dx_{j_1}} &= \frac{a_{j_2j_10}}{a_{j_1j_10}} \frac{(x_{j_1} - a_{j_2j_11})^2}{(x_{j_1} - a_{j_1j_11})\big[(x_{j_1} - a_{j_1j_1})^2 + b_{j_1j_1}\big]}\\
&= \frac{a_{j_2j_10}}{a_{j_1j_10}} \left\{ \frac{(x_{j_1} - a_{j_1j_1})}{(x_{j_1} - a_{j_1j_1})^2 + b_{j_1j_1}} + \frac{a_{j_1j_1} + a_{j_1j_11} - 2a_{j_2j_11}}{(x_{j_1} - a_{j_1j_1})^2 + b_{j_1j_1}} \right.\\
&\quad + \frac{(a_{j_1j_11} - a_{j_2j_11})^2}{(a_{j_1j_11} - a_{j_1j_1})^2 + b_{j_1j_1}} \left[\frac{1}{x_{j_1} - a_{j_1j_11}} - \frac{(x_{j_1} - a_{j_1j_1})}{(x_{j_1} - a_{j_1j_1})^2 + b_{j_1j_1}} \right.\\
&\quad \left.\left. - \frac{a_{j_1j_11} - a_{j_1j_1}}{(x_{j_1} - a_{j_1j_1})^2 + b_{j_1j_1}} \right] \right\}.
\end{aligned}$$

With an initial condition (x_{j_10}, x_{j_20}), the first integral manifold is

$$
\begin{aligned}
x_{j_2} - x_{j_2 0} = {} & \frac{a_{j_2 j_1 0}}{a_{j_1 j_1 0}} \Big\{ \frac{1}{2} \ln \frac{|(x_{j_1} - a_{j_1 j_1})^2 + b_{j_1 j_1}|}{|(x_{j_1 0} - a_{j_1 j_1})^2 + b_{j_1 j_1}|} \\
& + \frac{(a_{j_1 j_1 1} - 2a_{j_2 j_1 1} + a_{j_1 j_1})}{\sqrt{b_{j_1 j_1}}} (\arctan \frac{x_{j_1} - a_{j_1 j_1}}{\sqrt{b_{j_1 j_1}}} - \arctan \frac{x_{j_1 0} - a_{j_1 j_1}}{\sqrt{b_{j_1 j_1}}}) \\
& + \frac{(a_{j_1 j_1 1} - a_{j_2 j_1 1})^2}{(a_{j_1 j_1 1} - a_{j_1 j_1})^2 + b_{j_1 j_1}} \Big[\ln \frac{|x_{j_1} - a_{j_1 j_1 1}|}{|x_{j_1 0} - a_{j_1 j_1 1}|} \\
& - \frac{1}{2} \ln \frac{|(x_{j_1} - a_{j_1 j_1})^2 + b_{j_1 j_1}|}{|(x_{j_1 0} - a_{j_1 j_1})^2 + b_{j_1 j_1}|} \\
& - \frac{a_{j_1 j_1 1} - a_{j_1 j_1}}{\sqrt{b_{j_1 j_1}}} (\arctan \frac{x_{j_1} - a_{j_1 j_1}}{\sqrt{b_{j_1 j_1}}} - \arctan \frac{x_{j_1 0} - a_{j_1 j_1}}{\sqrt{b_{j_1 j_1}}}) \Big] \Big\}, \\
& j_1, j_2 \in \{1, 2\}, j_1 \neq j_2.
\end{aligned}
$$

(iv_{1a}) For $x_{j_1}^* = a_{j_1 j_1 1} \neq a_{j_2 j_1 1}$, the variational equation is

$$
\begin{aligned}
& \dot{x}_{j_2} = a_{j_2 j_1 0} (a_{j_1 j_1 1} - a_{j_2 j_1 1})^2, \\
& \Delta \dot{x}_{j_1} = a_{j_1 j_1 0} \left[(a_{j_1 j_1 1} - a_{j_1 j_1})^2 + b_{j_1 j_1} \right] \Delta x_{j_1}.
\end{aligned}
$$

Due to

$$
(a_{j_1 j_1 1} - a_{j_2 j_1 1})^2 > 0 \text{ and } (a_{j_1 j_1 1} - a_{j_1 j_1})^2 + b_{j_1 j_1} > 0,
$$

the flows at $x_{j_1}^* = a_{j_1 j_1 1}$ are source and sink flows for $a_{j_1 j_1 0} > 0$ and $a_{j_1 j_1 0} < 0$ in the x_{j_1}-direction; and the flows at $x_{j_1}^* = a_{j_1 j_1 1}$ are positive and negative for $a_{j_2 j_1 0} > 0$ and $a_{j_2 j_1 0} < 0$ in the x_{j_2}-direction.

Therefore, the flow at $x_{j_1}^* = a_{j_1 j_1 1}$ for the system has the following properties as in Eqs. (3.38)–(3.41).

- For $a_{j_1 j_1 0} > 0$ and $a_{j_2 j_1 0} (a_{j_1 j_1 1} - a_{j_2 j_1 1})^2 > 0$,

$$
(a_{j_1 j_1 1}, \dot{x}_{j_2}) = \underbrace{(\mathrm{SO}, \mathrm{pF})}_{\text{source flow } (+)}.
$$

The (SO,pF)-flow is a positive source flow at $x_{j_1}^* = a_{j_1 j_1 1}$.

- For $a_{j_1 j_1 0} < 0$ and $a_{j_2 j_1 0} (a_{j_1 j_1 1} - a_{j_2 j_1 1})^2 > 0$,

$$
(a_{j_1 j_1 1}, \dot{x}_{j_2}) = \underbrace{(\mathrm{SI}, \mathrm{pF})}_{\text{sink flow } (+)}.
$$

The (SI, pF)-flow is a positive sink flow at $x_{j_1}^* = a_{j_1 j_1 1}$.

- For $a_{j_1j_1 0}>0$ and $a_{j_2j_1 0}(a_{j_1j_1 1}-a_{j_2j_1 1})^2<0$,

$$(a_{j_1j_1 1},\dot{x}_{j_2})=\underbrace{(\text{SO},\text{nF})}_{\text{source flow }(-)}.$$

The (SO, nF)-flow is a negative source flow at $x_{j_1}^*=a_{j_1j_1 1}$.

- For $a_{j_1j_1 0}<0$ and $a_{j_2j_1 0}(a_{j_1j_1 1}-a_{j_2j_1 1})^2<0$,

$$(a_{j_1j_1 1},\dot{x}_{j_2})=\underbrace{(\text{SI},\text{nF})}_{\text{sink flow }(-)}.$$

The (SI,nF)-flow is a negative sink flow at $x_{j_1}^*=a_{j_1j_1 1}$.

(iv$_{1b}$) For $x_{j_1}^*=a_{j_2j_1 1}\neq a_{j_1j_1 1}$, we have

$$\frac{dx_{j_2}}{dx_{j_1}}\bigg|_{x_{j_1}^*=a_{j_2j_1 1}}=\frac{a_{j_2j_1 0}}{a_{j_1j_1 0}}\frac{(x_{j_1}-a_{j_2j_1 1})^2}{(x_{j_1}-a_{j_1j_1 1})\left[(x_{j_1}-a_{j_1j_1})^2+b_{j_1j_1}\right]}\bigg|_{x_{j_1}^*=a_{j_2j_1 1}}=0,$$

$$\frac{d^2x_{j_2}}{dx_{j_1}^2}\bigg|_{x_{j_1}^*=a_{j_2j_1 1}}=0.$$

If

$$\frac{d^3x_{j_2}}{dx_{j_1}^3}\bigg|_{x_{j_1}^*=a_{j_2j_1 1}}=\frac{2a_{j_2j_1 0}}{a_{j_1j_1 0}(a_{j_2j_1 1}-a_{j_1j_1 1})\left[(a_{j_2j_1 1}-a_{j_1j_1})^2+b_{j_1j_1}\right]}>0,$$

the flow at $x_{j_1}^*=a_{j_2j_1 1}\neq a_{j_1j_1 1}$ is an increasing-inflection flow in the x_{j_2}-direction, and if

$$\frac{d^3x_{j_2}}{dx_{j_1}^3}\bigg|_{x_{j_1}^*=a_{j_2j_1 1}}=\frac{2a_{j_2j_1 0}}{a_{j_1j_1 0}(a_{j_2j_1 1}-a_{j_1j_1 1})\left[(a_{j_2j_1 1}-a_{j_1j_1})^2+b_{j_1j_1}\right]}<0,$$

the flow at $x_{j_1}^*=a_{j_2j_1 1}\neq a_{j_1j_1 1}$ is a deceasing-inflection flow in the x_{j_2}-direction.

Let

$$\dot{x}_{j_1}=a_{j_1j_1 0}(a_{j_2j_1 1}-a_{j_1j_1 1})\left[(a_{j_2j_1 1}-a_{j_1j_1})^2+b_{j_1j_1}\right].$$

Due to

$$(a_{j_1j_1 1}-a_{j_1j_1})^2+b_{j_1j_1}>0,$$

the flows at $x_{j_1}^*=a_{j_2j_1 1}$ are positive and negative for $a_{j_1j_1 0}(a_{j_2j_1 1}-a_{j_1j_1 1})>0$ and $a_{j_1j_1 0}(a_{j_2j_1 1}-a_{j_1j_1 1})<0$ in the x_{j_1}-direction.

Therefore, the flow at $x_{j_1}^* = a_{j_2 j_1 1}$ for the system has the following properties as in Eqs. (3.42)–(3.45).

- For $a_{j_1 j_1 0}(a_{j_2 j_1 1} - a_{j_1 j_1 1}) > 0$ and $a_{j_2 j_1 0} > 0$,

$$(\dot{x}_{j_1}, a_{j_2 j_1 1}) = \underbrace{(\text{pF, II})}_{\text{increasing-inflection flow } (+)}.$$

  The (pF,II)-flow at $x_{j_1}^* = a_{j_2 j_1 1}$ is a positive increasing-inflection flow.
- For $a_{j_1 j_1 0}(a_{j_2 j_1 1} - a_{j_1 j_1 1}) < 0$ and $a_{j_2 j_1 0} > 0$,

$$(\dot{x}_{j_1}, a_{j_2 j_1 1}) = \underbrace{(\text{nF, DI})}_{\text{decreasing-inflection flow } (-)}.$$

  The (nF,DI)-flow at $x_{j_1}^* = a_{j_2 j_1 1}$ is a negative decreasing-inflection flow.
- For $a_{j_1 j_1 0}(a_{j_2 j_1 1} - a_{j_1 j_1 1}) > 0$ and $a_{j_2 j_1 0} < 0$,

$$(\dot{x}_{j_1}, a_{j_2 j_1 1}) = \underbrace{(\text{pF, DI})}_{\text{decreasing-inflection flow } (+)}.$$

  The (pF, DI)-flow at $x_{j_1}^* = a_{j_2 j_1 1}$ is a positive decreasing-inflection flow.
- For $a_{j_1 j_1 0}(a_{j_2 j_1 1} - a_{j_1 j_1 1}) < 0$ and $a_{j_2 j_1 0} < 0$,

$$(\dot{x}_{j_1}, a_{j_2 j_1 1}) = \underbrace{(\text{nF, II})}_{\text{increasing-inflection flow } (-)}.$$

  The (nF,II)-flow at $x_{j_1}^* = a_{j_2 j_1 1}$ is a negative increasing-inflection flow.

– The increasing-inflection and decreasing-inflection flows are the appearing (or switching) bifurcation for the up-parabola and down-parabola flows.

(iv$_2$) For $a_{j_1 j_1 1} = a_{j_2 j_1 1}$, we have

$$\begin{aligned}
\frac{dx_{j_2}}{dx_{j_1}} &= \frac{a_{j_2 j_1 0}}{a_{j_1 j_1 0}} \frac{(x_{j_1} - a_{j_2 j_1 1})^2}{(x_{j_1} - a_{j_1 j_1 1})\left[(x_{j_1} - a_{j_1 j_1})^2 + b_{j_1 j_1}\right]} \\
&= \frac{a_{j_2 j_1 0}}{a_{j_1 j_1 0}} \left[\frac{(x_{j_1} - a_{j_1 j_1})}{(x_{j_1} - a_{j_1 j_1})^2 + b_{j_1 j_1}} + \frac{a_{j_1 j_1} - a_{j_1 j_1 1}}{(x_{j_1} - a_{j_1 j_1})^2 + b_{j_1 j_1}}\right].
\end{aligned}$$

With an initial condition $(x_{j_1 0}, x_{j_2 0})$, the first integral manifold is

$$\begin{aligned}x_{j_2}-x_{j_2 0}=\frac{a_{j_2 j_1 0}}{a_{j_1 j_1 0}}\Big[\frac{1}{2}\ln\frac{|(x_{j_1}-a_{j_1 j_1})^2+b_{j_1 j_1}|}{|(x_{j_1 0}-a_{j_1 j_1})^2+b_{j_1 j_1}|}\\+\frac{(a_{j_1 j_1}-a_{j_1 j_1 1})}{\sqrt{b_{j_1 j_1}}}(\arctan\frac{x_{j_1}-a_{j_1 j_1}}{\sqrt{b_{j_1 j_1}}}-\arctan\frac{x_{j_1 0}-a_{j_1 j_1}}{\sqrt{b_{j_1 j_1}}})\Big];\\ j_1,j_2\in\{1,2\},j_1\neq j_2.\end{aligned}$$

In phase space, for $x_{j_1}^*=a_{j_1 j_1 1}=a_{j_2 j_1 1}$, we have

$$\frac{dx_{j_2}}{dx_{j_1}}\Big|_{x_{j_1}^*=a_{j_1 j_1 1}=a_{j_2 j_1 1}}=\frac{a_{j_2 j_1 0}}{a_{j_1 j_1 0}}\frac{(x_{j_1}-a_{j_2 j_1 1})}{[(x_{j_1}-a_{j_1 j_1})^2+b_{j_1 j_1}]}=0.$$

If

$$\frac{d^2x_{j_2}}{dx_{j_1}^2}\Big|_{x_{j_1}^*=a_{j_1 j_1 1}=a_{j_2 j_1 1}}=\frac{a_{j_2 j_1 0}}{a_{j_1 j_1 0}[(a_{j_1 j_1 1}-a_{j_1 j_1})^2+b_{j_1 j_1}]}>0,$$

the flow at $x_{j_1}^*=a_{j_1 j_1 1}=a_{j_2 j_1 1}$ is an up-parabola flow in the x_{j_2}-direction, and if

$$\frac{d^2x_{j_2}}{dx_{j_1}^2}\Big|_{x_{j_1}^*=a_{j_1 j_1 1}=a_{j_2 j_1 1}}=\frac{a_{j_2 j_1 0}}{a_{j_1 j_1 0}[(a_{j_1 j_1 1}-a_{j_1 j_1})^2+b_{j_1 j_1}]}<0,$$

the flow at $x_{j_1}^*=a_{j_1 j_1 1}=a_{j_2 j_1 1}$ is a down-parabola flow in the x_{j_2}-direction.

At $x_{j_1}^*=a_{j_1 j_1 1}=a_{j_2 j_1 1}$, the variational equations are

$$\begin{aligned}\Delta\dot{x}_{j_2}&=a_{j_2 j_1 0}(\Delta x_{j_1})^2,\\ \Delta\dot{x}_{j_1}&=a_{j_1 j_1 0}[(a_{j_2 j_1 1}-a_{j_1 j_1})^2+b_{j_1 j_1}]\Delta x_{j_1}.\end{aligned}$$

Due to

$$(a_{j_1 j_1 1}-a_{j_1 j_1})^2+b_{j_1 j_1}>0,$$

The flows at $x_{j_1}^*=a_{j_1 j_1 1}=a_{j_2 j_1 1}$ are source and sink for $a_{j_1 j_1 0}>0$ and $a_{j_1 j_1 0}<0$ in the x_{j_1}-direction, respectively.

Thus, the infinite-equilibrium of $x_{j_1}^*=a_{j_1 j_1 1}=a_{j_2 j_1 1}$ with $\overline{x}_{j_2}\in(-\infty,\infty)$ has the following properties as in Eqs. (3.47)–(3.50).

- For $a_{j_1 j_1 0}>0$ and $a_{j_2 j_1 0}>0$,

$$(a_{j_1 j_1 1},\overline{x}_{j_2})=\underbrace{(\text{SO},\text{UP})}_{\text{up-parabola source}}.$$

The (SO,UP)-equilibrium at $x_{j_1}^* = a_{j_1 j_1 1} = a_{j_2 j_1 1}$ is an up-parabola source infinite-equilibrium.

- For $a_{j_1 j_1 0} < 0$ and $a_{j_2 j_1 0} > 0$,

$$(a_{j_1 j_1 1}, \bar{x}_{j_2}) = \underbrace{(\text{SI}, \text{DP})}_{\text{down-parabola sink}}.$$

The (SI,DP)-equilibrium at $x_{j_1}^* = a_{j_1 j_1 1} = a_{j_2 j_1 1}$ is a down-parabola sink infinite-equilibrium.

- For $a_{j_1 j_1 0} > 0$ and $a_{j_2 j_1 0} < 0$,

$$(a_{j_1 j_1 1}, \bar{x}_{j_2}) = \underbrace{(\text{SO}, \text{DP})}_{\text{down-parabola source}}.$$

The (SO,DP)-equilibrium at $x_{j_1}^* = a_{j_1 j_1 1} = a_{j_2 j_1 1}$ is a down-parabola source infinite-equilibrium.

- For $a_{j_1 j_1 0} < 0$ and $a_{j_2 j_1 0} < 0$,

$$(a_{j_1 j_1 1}, \bar{x}_{j_2}) = \underbrace{(\text{SI}, \text{DP})}_{\text{down-parabola sink}}.$$

The (SI,UP)-equilibrium at $x_{j_1}^* = a_{j_1 j_1 1} = a_{j_2 j_1 1}$ is an up-parabola sink infinite-equilibrium.

- The parabola source and sink infinite-equilibriums are the switching bifurcations of a source (sink) flow and an inflection flow.
- The up-parabola source flow is the appearing bifurcation from a positive source flow to negative up-parabola, negative source, positive up-parabola flows.
- The up-parabola sink flow is the appearing bifurcation from a negative sink flow to positive up-parabola, positive sink, negative up-parabola flows.
- The down-parabola source flow is the appearing bifurcation from a negative source flow to negative down-parabola, positive source, positive down-parabola flows.
- The down-parabola sink flow is the appearing bifurcation from a positive sink flow to positive down-parabola, negative sink, negative down-parabola flows.

(v) If

$$\Delta_{j_1 j_1} = 0, \Delta_{j_2 j_1} = 0,$$

then

$$\begin{aligned}
&\dot{x}_{j_1}=a_{j_1j_10}(x_{j_1}-a_{j_1j_1s_1})(x_{j_1}-a_{j_1j_1s_2})^2,\\
&\dot{x}_{j_2}=a_{j_2j_10}(x_{j_1}-a_{j_2j_11})^2,\\
&j_1,j_2\in\{1,2\},j_1\neq j_2;s_1,s_2\in\{1,2\},s_1\neq s_2
\end{aligned}$$

where

$$\begin{aligned}
&\{a_{j_1j_11},a_{j_1j_12}\}=\mathrm{sort}\{b_{j_1j_11},b_{j_1j_12}\},a_{j_1j_11}<a_{j_1j_12}\\
&b_{j_1j_12}=-\frac{1}{2}B_{j_1j_11},a_{j_2j_11}=-\frac{1}{2}B_{j_2j_11};\\
&j_1,j_2\in\{1,2\},j_1\neq j_2.
\end{aligned}$$

(v_1) For $a_{j_1j_1s_1}\neq a_{j_2j_11}\neq a_{j_1j_1s_2}$, we have

$$\begin{aligned}
\frac{dx_{j_2}}{dx_{j_1}}&=\frac{a_{j_2j_10}}{a_{j_1j_10}}\frac{(x_{j_1}-a_{j_2j_11})^2}{(x_{j_1}-a_{j_1j_1s_1})(x_{j_1}-a_{j_1j_1s_2})^2}\\
&=\frac{a_{j_2j_10}}{a_{j_1j_10}}\left\{\frac{(a_{j_1j_1s_1}-a_{j_2j_11})^2}{(a_{j_1j_1s_1}-a_{j_1j_1s_2})^2}\frac{1}{x_{j_1}-a_{j_1j_11}}\right.\\
&\quad+\frac{(a_{j_1j_1s_2}-a_{j_2j_11})^2}{(a_{j_1j_1s_2}-a_{j_1j_1s_1})}\frac{1}{(x_{j_1}-a_{j_1j_1s_2})^2}\\
&\quad\left.+\left[\frac{2(a_{j_1j_1s_2}-a_{j_2j_11})}{(a_{j_1j_1s_2}-a_{j_1j_1s_1})}-\frac{(a_{j_1j_1s_2}-a_{j_2j_11})^2}{(a_{j_1j_1s_2}-a_{j_1j_1s_1})^2}\right]\frac{1}{x_{j_1}-a_{j_1j_1s_2}}\right\}.
\end{aligned}$$

With an initial condition (x_{j_10},x_{j_20}), the first integral manifold is

$$\begin{aligned}
x_{j_2}-x_{j_20}=&\frac{a_{j_2j_10}}{a_{j_1j_10}}\left\{\frac{(a_{j_1j_1s_1}-a_{j_2j_11})^2}{(a_{j_1j_1s_1}-a_{j_1j_1s_2})^2}\ln\frac{|x_{j_1}-a_{j_1j_1s_1}|}{|x_{j_10}-a_{j_1j_1s_1}|}\right.\\
&-\frac{(a_{j_1j_1s_2}-a_{j_2j_11})^2}{(a_{j_1j_1s_2}-a_{j_1j_1s_1})}\left(\frac{1}{x_{j_1}-a_{j_1j_1s_2}}-\frac{1}{x_{j_10}-a_{j_1j_1s_2}}\right)\\
&\left.+\left[\frac{2(a_{j_1j_1s_2}-a_{j_2j_11})}{(a_{j_1j_1s_2}-a_{j_1j_1s_1})}-\frac{(a_{j_1j_1s_2}-a_{j_2j_11})^2}{(a_{j_1j_1s_2}-a_{j_1j_1s_1})^2}\right]\ln\frac{|x_{j_1}-a_{j_1j_1s_2}|}{|x_{j_10}-a_{j_1j_1s_2}|}\right\},\\
&j_1,j_2\in\{1,2\},j_1\neq j_2;s_1,s_2\in\{1,2\},s_1\neq s_2.
\end{aligned}$$

(v_{1a}) For $x_{j_1}^*=a_{j_1j_1s_1}\neq a_{j_2j_11}$, the variational equation is

$$\begin{aligned}
&\dot{x}_{j_2}=a_{j_2j_10}(a_{j_1j_1s_1}-a_{j_2j_11})^2,\\
&\Delta\dot{x}_{j_1}=a_{j_1j_10}(a_{j_1j_1s_1}-a_{j_1j_1s_2})^2\Delta x_{j_1}.
\end{aligned}$$

Due to

$$(a_{j_1j_1s_1} - a_{j_2j_11})^2 > 0 \text{ and } (a_{j_1j_1s_1} - a_{j_1j_1s_2})^2 > 0,$$

the flows at $x_{j_1}^* = a_{j_1j_1s_1} \neq a_{j_2j_11}$ are source and sink flows for $a_{j_1j_10} > 0$ and $a_{j_1j_10} < 0$ in the x_{j_1}-direction, and the flows at $x_{j_1}^* = a_{j_1j_1s_1} \neq a_{j_2j_11}$ are positive and negative for $a_{j_2j_10} > 0$ and $a_{j_2j_10} < 0$ in the x_{j_2}-direction.

Therefore, the equilibrium of $x_{j_1}^* = a_{j_1j_1s_1} \neq a_{j_2j_11}$ possesses the following properties as in Eqs. (3.55)–(3.58).

- For $a_{j_1j_10}(a_{j_1j_1s_1} - a_{j_1j_1s_2})^2 > 0$ and $a_{j_2j_10}(a_{j_1j_1s_1} - a_{j_2j_11})^2 > 0$,

$$(a_{j_1j_1s_1}, \dot{x}_{j_2}) = \underbrace{(\mathrm{SO}, \mathrm{pF})}_{\text{source flow } (+)}.$$

The flow at $x_{j_1}^* = a_{j_1j_1s_1} \neq a_{j_2j_11}$ is an (SO,pF)-positive source flow.

- For $a_{j_1j_10}(a_{j_1j_1s_1} - a_{j_1j_1s_2})^2 < 0$ and $a_{j_2j_10}(a_{j_1j_1s_1} - a_{j_2j_11})^2 > 0$,

$$(a_{j_1j_1s_1}, \dot{x}_{j_2}) = \underbrace{(\mathrm{SI}, \mathrm{pF})}_{\text{sink flow } (+)}.$$

The flow at $x_{j_1}^* = a_{j_1j_1s_1} \neq a_{j_2j_11}$ is an (SI,pF)-positive sink flow.

- For $a_{j_1j_10}(a_{j_1j_1s_1} - a_{j_1j_1s_2})^2 > 0$ and $a_{j_2j_10}(a_{j_1j_1s_1} - a_{j_2j_11})^2 < 0$,

$$(a_{j_1j_1s_1}, \dot{x}_{j_2}) = \underbrace{(\mathrm{SO}, \mathrm{nF})}_{\text{source flow } (-)}.$$

The flow at $x_{j_1}^* = a_{j_1j_1s_1} \neq a_{j_2j_11}$ is an (SO,nF)-negative source flow.

- For $a_{j_1j_10}(a_{j_1j_1s_1} - a_{j_1j_1s_2})^2 < 0$ and $a_{j_2j_10}(a_{j_1j_1s_1} - a_{j_2j_11})^2 < 0$,

$$(a_{j_1j_1s_1}, \dot{x}_{j_2}) = \underbrace{(\mathrm{SI}, \mathrm{nF})}_{\text{sink flow } (-)}.$$

The flow at $x_{j_1}^* = a_{j_1j_1s_1} \neq a_{j_2j_11}$ is an (SI,pF)-negative sink flow.

($\mathrm{v}_{1\mathrm{b}}$) For $x_{j_1}^* = a_{j_1j_1s_2} \neq a_{j_2j_11}$, the variational equation is

$$\begin{aligned}\dot{x}_{j_2} &= a_{j_2j_10}(a_{j_1j_1s_2} - a_{j_2j_11})^2,\\ \Delta\dot{x}_{j_1} &= a_{j_1j_10}(a_{j_1j_1s_2} - a_{j_1j_1s_1})(\Delta x_{j_1})^2.\end{aligned}$$

The flows at $x_{j_1}^* = a_{j_1j_1s_2} \neq a_{j_2j_11}$ are upper-saddle and lower-saddle flows for $a_{j_1j_10}(a_{j_1j_1s_2} - a_{j_1j_1s_1}) > 0$ and $a_{j_1j_10}(a_{j_1j_1s_2} - a_{j_1j_1s_1}) < 0$ in the x_{j_1}-direction, and the flows at $x_{j_1}^* = a_{j_1j_1s_2} \neq a_{j_2j_11}$ are positive and negative for $a_{j_2j_10} > 0$ and $a_{j_2j_10} < 0$ in the x_{j_2}-direction.

Thus, the equilibrium of $x_{j_1}^* = a_{j_1 j_1 s_2} \neq a_{j_2 j_1 1}$ possesses the following properties as in Eqs. (3.59)–(3.62).

- For $a_{j_1 j_1 0}(a_{j_1 j_1 s_2} - a_{j_1 j_1 s_1}) > 0$ and $a_{j_2 j_1 0}(a_{j_1 j_1 s_2} - a_{j_2 j_1 1})^2 > 0$,

$$(a_{j_1 j_1 s_2}, \dot{x}_{j_2}) = \underbrace{(\text{US}, \text{pF})}_{\text{upper-saddle flow } (+)}.$$

  The flow at $x_{j_1}^* = a_{j_1 j_1 s_2} \neq a_{j_2 j_1 1}$ is a (US,pF)-positive upper-saddle flow.
- For $a_{j_1 j_1 0}(a_{j_1 j_1 s_2} - a_{j_1 j_1 s_1}) < 0$ and $a_{j_2 j_1 0}(a_{j_1 j_1 s_2} - a_{j_2 j_1 1})^2 > 0$,

$$(a_{j_1 j_1 s_2}, \dot{x}_{j_2}) = \underbrace{(\text{LS}, \text{pF})}_{\text{lower-saddle flow } (+)}.$$

  The flow at $x_{j_1}^* = a_{j_1 j_1 s_2} \neq a_{j_2 j_1 1}$ is an (LS,pF)-positive lower-saddle flow.
- For $a_{j_1 j_1 0}(a_{j_1 j_1 s_2} - a_{j_1 j_1 s_1}) > 0$ and $a_{j_2 j_1 0}(a_{j_1 j_1 s_2} - a_{j_2 j_1 1})^2 < 0$,

$$(a_{j_1 j_1 s_2}, \dot{x}_{j_2}) = \underbrace{(\text{US}, \text{nF})}_{\text{upper-saddle flow } (-)}.$$

  The flow at $x_{j_1}^* = a_{j_1 j_1 s_2} \neq a_{j_2 j_1 1}$ is a (US,nF)-negative upper-saddle flow.
- For $a_{j_1 j_1 0}(a_{j_1 j_1 s_2} - a_{j_1 j_1 s_1}) < 0$ and $a_{j_2 j_1 0}(a_{j_1 j_1 s_2} - a_{j_2 j_1 1})^2 < 0$,

$$(a_{j_1 j_1 s_2}, \dot{x}_{j_2}) = \underbrace{(\text{LS}, \text{nF})}_{\text{lower-saddle flow } (-)}.$$

  The flow at $x_{j_1}^* = a_{j_1 j_1 s_2} \neq a_{j_2 j_1 1}$ is an (LS,nF)-negative lower-saddle flow.

– The lower-saddle and upper-saddle flows are the appearing-switching bifurcation for the sink and source flows.

($\text{v}_{1\text{c}}$) In phase space, for $x_{j_1}^* = a_{j_2 j_1 1} \neq a_{j_1 j_1 s_1}, a_{j_1 j_1 s_2}$, we have

$$\begin{aligned}
&\left.\frac{dx_{j_2}}{dx_{j_1}}\right|_{x_{j_1}^* = a_{j_2 j_1 1}} = \frac{a_{j_2 j_1 0}}{a_{j_1 j_1 0}} \frac{(x_{j_1} - a_{j_2 j_1 1})^2}{(x_{j_1} - a_{j_1 j_1 s_1})(x_{j_1} - a_{j_1 j_1 s_2})^2}\right|_{x_{j_1}^* = a_{j_2 j_1 1}} = 0,\\
&\left.\frac{d^2 x_{j_2}}{dx_{j_1}^2}\right|_{x_{j_1}^* = a_{j_2 j_1 1}} = 0.
\end{aligned}$$

If

$$\frac{d^3 x_{j_2}}{dx_{j_1}^3}\bigg|_{x_{j_1}^*=a_{j_2j_11}} = \frac{2a_{j_2j_10}}{a_{j_1j_10}(a_{j_2j_11}-a_{j_1j_1s_1})(a_{j_2j_11}-a_{j_1j_1s_2})^2} > 0,$$

the flow at $x_{j_1}^* = a_{j_2j_11} \neq a_{j_1j_1s_1}, a_{j_1j_1s_2}$ is an increasing-inflection flow in the x_{j_2}-direction, and if

$$\frac{d^3 x_{j_2}}{dx_{j_1}^3}\bigg|_{x_{j_1}^*=a_{j_2j_11}} = \frac{2a_{j_2j_10}}{a_{j_1j_10}(a_{j_2j_11}-a_{j_1j_1s_1})(a_{j_2j_11}-a_{j_1j_1s_2})^2} < 0,$$

the flow at $x_{j_1}^* = a_{j_2j_11} \neq a_{j_1j_1s_1}, a_{j_1j_1s_2}$ is a decreasing-inflection flow in the x_{j_2}-direction.

At $x_{j_1}^* = a_{j_2j_11} \neq a_{j_1j_1s_1}, a_{j_1j_1s_2}$, let

$$\dot{x}_{j_1} = a_{j_1j_10}(a_{j_2j_11}-a_{j_1j_1s_1})(a_{j_2j_11}-a_{j_1j_1s_2})^2.$$

The flows at $x_{j_1}^* = a_{j_2j_11} \neq a_{j_1j_1s_1}, a_{j_1j_1s_2}$ are positive and negative for $\dot{x}_{j_1} > 0$ and $\dot{x}_{j_1} < 0$ in the x_{j_1}-direction.

Therefore, the equilibrium of $x_{j_1}^* = a_{j_2j_11} \neq a_{j_1j_1s_1}, a_{j_1j_1s_2}$ possesses the following properties as in Eqs. (3.63)–(3.66).

- For $a_{j_1j_10}(a_{j_2j_11}-a_{j_1j_1s_1})(a_{j_2j_11}-a_{j_1j_1s_2})^2 > 0$ and $a_{j_2j_10} > 0$,

$$(\dot{x}_{j_1}, a_{j_2j_11}) = \underbrace{(\text{pF}, \text{II})}_{\text{increasing-inflection flow } (+)}.$$

The flow at $x_{j_1}^* = a_{j_2j_11}$ is a (pF,II)-positive increasing-inflection flow.

- For $a_{j_1j_10}(a_{j_2j_11}-a_{j_1j_1s_1})(a_{j_2j_11}-a_{j_1j_1s_2})^2 < 0$ and $a_{j_2j_10} > 0$,

$$(\dot{x}_{j_1}, a_{j_2j_11}) = \underbrace{(\text{nF}, \text{DI})}_{\text{decreasing-inflection flow } (-)}.$$

The flow at $x_{j_1}^* = a_{j_2j_11}$ is an (nF,DI)-negative decreasing-inflection flow.

- For $a_{j_1j_10}(a_{j_2j_11}-a_{j_1j_1s_1})(a_{j_2j_11}-a_{j_1j_1s_2})^2 > 0$ and $a_{j_2j_10} < 0$,

$$(\dot{x}_{j_1}, a_{j_2j_11}) = \underbrace{(\text{pF}, \text{DI})}_{\text{decreasing-inflection flow } (+)}.$$

The flow at $x_{j_1}^* = a_{j_2j_11}$ is a (pF,DI)-positive increasing-inflection flow.

- For $a_{j_1j_10}(a_{j_2j_11}-a_{j_1j_1s_1})(a_{j_2j_11}-a_{j_1j_1s_2})^2<0$ and $a_{j_2j_10}<0$,

$$(\dot{x}_{j_1},a_{j_2j_11})=\underbrace{(\text{nF},\text{II})}_{\text{increasing-inflection flow }(-)}.$$

The flow at $x_{j_1}^*=a_{j_2j_11}$ is an (nF,II)-positive increasing-inflection flow.

– The increasing-inflection and decreasing-inflection flows are the appearing-switching bifurcations for up-parabola and down-parabola flows.

(v_2) For $a_{j_1j_1s_2}=a_{j_2j_11}$,

$$\begin{aligned}&\dot{x}_{j_1}=a_{j_1j_10}(x_{j_1}-a_{j_1j_1s_1})(x_{j_1}-a_{j_1j_1s_2})^2,\\&\dot{x}_{j_2}=a_{j_2j_10}(x_{j_1}-a_{j_2j_11})^2,\\&j_1,j_2\in\{1,2\},j_1\neq j_2.\end{aligned}$$

In phase space, we have

$$\frac{dx_{j_2}}{dx_{j_1}}=\frac{a_{j_2j_10}}{a_{j_1j_10}}\frac{1}{(x_{j_1}-a_{j_1j_1s_1})}.$$

With an initial condition (x_{j_10},x_{j_20}),the first integral manifold is

$$\begin{aligned}x_{j_2}-x_{j_20}&=\frac{a_{j_2j_10}}{a_{j_1j_10}}\ln\frac{|x_{j_1}-a_{j_1j_1s_1}|}{|x_{j_10}-a_{j_1j_1s_1}|},\\&j_1,j_2\in\{1,2\},j_1\neq j_2.\end{aligned}$$

(v_{2a}) In phase space, for $x_{j_1}^*=a_{j_1j_1s_2}=a_{j_2j_11}$, if

$$\left.\frac{dx_{j_2}}{dx_{j_1}}\right|_{x_{j_1}^*=a_{j_2j_11}}=\frac{a_{j_2j_10}}{a_{j_1j_10}}\frac{1}{(a_{j_1j_1s_2}-a_{j_1j_1s_1})}>0,$$

the flow at $x_{j_1}^*=a_{j_1j_1s_2}=a_{j_2j_11}$ is an increasing-inflection flow in the x_{j_2}-direction, and if

$$\left.\frac{dx_{j_2}}{dx_{j_1}}\right|_{x_{j_1}^*=a_{j_2j_11}}=\frac{a_{j_2j_10}}{a_{j_1j_10}}\frac{1}{(a_{j_1j_1s_2}-a_{j_1j_1s_1})}<0,$$

the flow at $x_{j_1}^*=a_{j_1j_1s_2}=a_{j_2j_11}$ is a decreasing-inflection flow in the x_{j_2}-direction.

At $x_{j_1}^*=a_{j_1j_1s_2}=a_{j_2j_11}$, the variational equation is

$$\Delta\dot{x}_{j_2} = a_{j_2 j_1 0}(\Delta x_{j_1})^2,$$
$$\Delta\dot{x}_{j_1} = a_{j_1 j_1 0}(a_{j_2 j_1 1} - a_{j_1 j_1 s_1})(\Delta x_{j_1})^2.$$

The flows at $x^*_{j_1} = a_{j_1 j_1 s_2} = a_{j_2 j_1 1}$ are upper-saddle and lower-saddle flows for $a_{j_1 j_1 0}(a_{j_2 j_1 1} - a_{j_1 j_1 s_1}) > 0$ and $a_{j_1 j_1 0}(a_{j_2 j_1 1} - a_{j_1 j_1 s_1}) < 0$ in the x_{j_1}-direction. On the both direction, the second-order singularity exist; thus, (2:2)II and (2:2)DI are used for increasing-inflection and decreasing-inflection.

Therefore, the infinite-equilibrium of $x^*_{j_1} = a_{j_1 j_1 s_2} = a_{j_2 j_1 1}$ with $\overline{x}_{j_2} \in (-\infty, \infty)$ has the following properties as in Eqs. (3.69)–(3.72).

- For $a_{j_1 j_1 0}(a_{j_1 j_1 s_2} - a_{j_1 j_1 s_1}) > 0$ and $a_{j_2 j_1 0} > 0$,

$$(a_{j_1 j_1 s_2}, \overline{x}_{j_2}) = \underbrace{(\text{US}, (2{:}2)\text{II})}_{\text{increasing-inflection upper-saddle}}.$$

  The equilibrium at $x^*_{j_1} = a_{j_1 j_1 s_2} = a_{j_2 j_1 1}$ is a (US,(2:2)II)-increasing-inflection upper-saddle infinite-equilibrium.
- For $a_{j_1 j_1 0}(a_{j_1 j_1 s_2} - a_{j_1 j_1 s_1}) < 0$ and $a_{j_2 j_1 0} > 0$,

$$(a_{j_1 j_1 s_2}, \overline{x}_{j_2}) = \underbrace{(\text{LS}, (2{:}2)\text{DI})}_{\text{decreasing-inflection lower-saddle}}.$$

  The equilibrium at $x^*_{j_1} = a_{j_1 j_1 s_2} = a_{j_2 j_1 1}$ is an (LS,(2:2)DI)-decreasing-inflection lower-saddle infinite-equilibrium.
- For $a_{j_1 j_1 0}(a_{j_1 j_1 s_2} - a_{j_1 j_1 s_1}) > 0$ and $a_{j_2 j_1 0} < 0$,

$$(a_{j_1 j_1 s_2}, \overline{x}_{j_2}) = \underbrace{(\text{US}, (2{:}2)\text{DI})}_{\text{decreasing-inflection upper-saddle}}.$$

  The equilibrium at $x^*_{j_1} = a_{j_1 j_1 s_2} = a_{j_2 j_1 1}$ is a (US,(2:2)DI)-increasing-inflection upper-saddle infinite-equilibrium.
- For $a_{j_1 j_1 0}(a_{j_1 j_1 s_2} - a_{j_1 j_1 s_1}) < 0$ and $a_{j_2 j_1 0} < 0$,

$$(a_{j_1 j_1 s_2}, \overline{x}_{j_2}) = \underbrace{(\text{LS}, (2{:}2)\text{II})}_{\text{increasing-inflection lower-saddle}}.$$

  The equilibrium at $x^*_{j_1} = a_{j_1 j_1 s_2} = a_{j_2 j_1 1}$ is an (LS,(2:2)II)-increasing-inflection lower-saddle infinite-equilibrium.

– The infinite-equilibriums are the switching bifurcations for the increasing-inflection and decreasing-inflection flows with the upper-saddle and lower-saddle flows.

($\mathrm{v}_{2\mathrm{b}}$) Let $\Delta^{+}_{j_1j_1} = \Delta_{j_1j_1} + 0$ and $\Delta^{+}_{j_2j_1} = \Delta_{j_2j_1} + 0$ varying with parameters, the infinite-equilibriums can be the appearing bifurcations for the sink and source flows and the up-parabola and down-parabola flows. Thus, the infinite-equilibriums of $x^{*}_{j_1} = a_{j_1j_1s_2} = a_{j_2j_1 1}$ have the following appearing bifurcations as in Eqs. (3.73)–(3.80).

- For $a_{j_1j_1 0}(a_{j_1j_1s_2} - a_{j_1j_1s_1}) > 0$ and $a_{j_2j_1 0} > 0$,

$$\underbrace{(\text{US}, (2{:}2)\text{II})}_{\text{appearing bifurcation}} \rightleftharpoons \left\{ \begin{array}{ll} \underbrace{(\text{SI}, \text{pF})}_{\text{sink flow } (+)}, & \underbrace{(\text{anF}, \text{UP})}_{\text{up-parabola flow } (-)}, \\ \underbrace{(\text{nF}, \text{DP})}_{\text{down-parabola flow } (-)}, & \underbrace{(\text{SO}, \text{pF})}_{\text{source flow } (+)} \end{array} \right\}$$

for $\Delta^{+}_{j_1j_1} > \Delta^{+}_{j_2j_1}$;

$$\underbrace{(\text{US}, (2{:}2)\text{II})}_{\text{appearing bifurcation}} \rightleftharpoons \left\{ \begin{array}{ll} \underbrace{(\text{pF}, \text{DP})}_{\text{down-parabola flow } (+)}, & \underbrace{(\text{SI}, \text{nF})}_{\text{sink flow } (-)}, \\ \underbrace{(\text{SO}, \text{nF})}_{\text{source flow } (-)}, & \underbrace{(\text{nF}, \text{UP})}_{\text{up-parabola flow } (+)} \end{array} \right\}$$

for $\Delta^{+}_{j_1j_1} < \Delta^{+}_{j_2j_1}$.

- For $a_{j_1j_1 0}(a_{j_1j_1s_2} - a_{j_1j_1s_1}) < 0$ and $a_{j_2j_1 0} > 0$,

$$\underbrace{(\text{LS}, (2{:}2)\text{DI})}_{\text{appearing bifurcation}} \rightleftharpoons \left\{ \begin{array}{ll} \underbrace{(\text{SO}, \text{pF})}_{\text{source flow } (+)}, & \underbrace{(\text{pF}, \text{DP})}_{\text{down-parabola flow } (+)}, \\ \underbrace{(\text{pF}, \text{UP})}_{\text{up-parabola flow } (+)}, & \underbrace{(\text{SI}, \text{pF})}_{\text{sink flow } (+)} \end{array} \right\}$$

for $\Delta^{+}_{j_1j_1} > \Delta^{+}_{j_2j_1}$;

$$\underbrace{(\text{LS}, (2{:}2)\text{DI})}_{\text{appearing bifurcation}} \rightleftharpoons \left\{ \begin{array}{ll} \underbrace{(\text{nF}, \text{UP})}_{\text{up-parabola flow } (-)}, & \underbrace{(\text{SO}, \text{nF})}_{\text{source flow } (-)}, \\ \underbrace{(\text{SI}, \text{nF})}_{\text{sink flow } (-)}, & \underbrace{(\text{nF}, \text{DP})}_{\text{down-parabola flow } (-)} \end{array} \right\}$$

for$\Delta^{+}_{j_1j_1} < \Delta^{+}_{j_2j_1}$.

- For $a_{j_1j_1 0}(a_{j_1j_1s_2}-a_{j_1j_1s_1})>0$ and $a_{j_2j_1 0}<0$,

$$\underbrace{(\text{US},(2{:}2)\text{DI})}_{\text{appearing bifurcation}} \rightleftharpoons \left\{\begin{array}{ll}\underbrace{(\text{SI},\text{nF})}_{\text{sink flow }(-)}, & \underbrace{(\text{nF},\text{DP})}_{\text{down-parabola flow }(-)}, \\ \underbrace{(\text{nF},\text{UP})}_{\text{up-parabola flow }(-)}, & \underbrace{(\text{SO},\text{nF})}_{\text{source flow }(-)}\end{array}\right\}$$

for$\Delta^{+}_{j_1j_1}>\Delta^{+}_{j_2j_1}$;

$$\underbrace{(\text{US},(2{:}2)\text{DI})}_{\text{appearing bifurcation}} \rightleftharpoons \left\{\begin{array}{ll}\underbrace{(\text{pF},\text{UP})}_{\text{up-parabola flow }(+)}, & \underbrace{(\text{SI},\text{pF})}_{\text{sink flow }(+)}, \\ \underbrace{(\text{SO},\text{pF})}_{\text{source flow }(+)}, & \underbrace{(\text{pF},\text{DP})}_{\text{down-parabola flow }(+)}\end{array}\right\}$$

for $\Delta^{+}_{j_1j_1}>\Delta^{+}_{j_2j_1}$.

- For $a_{j_1j_1 0}(a_{j_1j_1s_2}-a_{j_1j_1s_1})<0$ and $a_{j_2j_1 0}<0$,

$$\underbrace{(\text{LS},(2{:}2)\text{II})}_{\text{appearing bifurcation}} \rightleftharpoons \left\{\begin{array}{ll}\underbrace{(\text{SO},\text{nF})}_{\text{source flow }(-)}, & \underbrace{(\text{pF},\text{UP})}_{\text{up-parabola flow }(+)}, \\ \underbrace{(\text{pF},\text{DP})}_{\text{down-parabola flow }(+)}, & \underbrace{(\text{SI},\text{nF})}_{\text{sink flow }(-)}\end{array}\right\}$$

for$\Delta^{+}_{j_1j_1}>\Delta^{+}_{j_2j_1}$;

$$\underbrace{(\text{LS},(2{:}2)\text{II})}_{\text{appearing bifurcation}} \rightleftharpoons \left\{\begin{array}{ll}\underbrace{(\text{nF},\text{DP})}_{\text{down-parabola flow }(-)}, & \underbrace{(\text{SO},\text{pF})}_{\text{source flow }(+)}, \\ \underbrace{(\text{SI},\text{pF})}_{\text{sink flow }(+)}, & \underbrace{(\text{nF},\text{UP})}_{\text{up-parabola flow }(-)}\end{array}\right\}$$

for$\Delta^{+}_{j_1j_1}<\Delta^{+}_{j_2j_1}$.

(v_3) For $a_{j_1j_1s_1}=a_{j_2j_1 1}$,

$$\begin{aligned}&\dot{x}_{j_1}=a_{j_1j_1 0}(x_{j_1}-a_{j_1j_1s_1})(x_{j_1}-a_{j_1j_1s_2})^2,\\&\dot{x}_{j_2}=a_{j_2j_1 0}(x_{j_1}-a_{j_2j_1 1})^2,\\&j_1,j_2\in\{1,2\},j_1\neq j_2.\end{aligned}$$

In phase space, we have

$$\begin{aligned}\frac{dx_{j_2}}{dx_{j_1}} &= \frac{a_{j_2j_10}}{a_{j_1j_10}}\frac{(x_{j_1}-a_{j_2j_11})}{(x_{j_1}-a_{j_1j_1s_2})^2}\\ &= \frac{a_{j_2j_10}}{a_{j_1j_10}}\left[\frac{1}{x_{j_1}-a_{j_1j_1s_2}}+\frac{(a_{j_1j_1s_2}-a_{j_2j_11})}{(x_{j_1}-a_{j_1j_1s_2})^2}\right].\end{aligned}$$

With an initial condition (x_{j_10}, x_{j_20}),the first integral manifold is

$$\begin{aligned}x_{j_2}-x_{j_20} = \frac{a_{j_2j_10}}{a_{j_1j_10}}&\left[\ln\frac{|x_{j_1}-a_{j_1j_1s_2}|}{|x_{j_10}-a_{j_1j_1s_2}|}\right.\\ &\left.-(a_{j_1j_1s_2}-a_{j_2j_11})(\frac{1}{x_{j_1}-a_{j_1j_1s_2}}-\frac{1}{x_{j_10}-a_{j_1j_1s_2}})\right],\\ &j_1, j_2\in\{1,2\}, j_1\neq j_2.\end{aligned}$$

(v_{3a}) For $x^*_{j_1}=a_{j_1j_1s_1}=a_{j_2j_11}$, we have

$$\frac{dx_{j_2}}{dx_{j_1}}\Big|_{x^*_{j_1}=a_{j_2j_11}} = \frac{a_{j_2j_10}}{a_{j_1j_10}}\frac{(x_{j_1}-a_{j_2j_11})}{(x_{j_1}-a_{j_1j_1s_2})^2}\Big|_{x^*_{j_1}=a_{j_2j_11}}=0.$$

If

$$\frac{d^2x_{j_2}}{dx_{j_1}^2}\Big|_{x^*_{j_1}=a_{j_2j_11}} = \frac{a_{j_2j_10}}{a_{j_1j_10}}\frac{1}{(a_{j_2j_11}-a_{j_1j_1s_2})^2}\Big|_{x^*_{j_1}=a_{j_2j_11}}>0,$$

the flow at $x^*_{j_1}=a_{j_1j_1s_1}=a_{j_2j_11}$ is an up-parabola flow in the x_{j_2}-direction, and if

$$\frac{d^2x_{j_2}}{dx_{j_1}^2}\Big|_{x^*_{j_1}=a_{j_2j_11}} = \frac{a_{j_2j_10}}{a_{j_1j_10}}\frac{1}{(a_{j_2j_11}-a_{j_1j_1s_2})^2}\Big|_{x^*_{j_1}=a_{j_2j_11}}<0,$$

the flow at $x^*_{j_1}=a_{j_1j_1s_1}=a_{j_2j_11}$ is a down-parabola flow in the x_{j_2}-direction.

At $x^*_{j_1}=a_{j_1j_1s_1}=a_{j_2j_11}$, the variational equations are

$$\begin{aligned}\Delta\dot{x}_{j_2} &= a_{j_2j_10}(\Delta x_{j_1})^2,\\ \Delta\dot{x}_{j_1} &= a_{j_1j_10}(a_{j_1j_1s_1}-a_{j_1j_1s_2})^2\Delta x_{j_1}.\end{aligned}$$

The flows at $x^*_{j_1}=a_{j_1j_1s_1}=a_{j_2j_11}$ are source and sink flows for $a_{j_1j_10}>0$ and $a_{j_1j_10}<0$ in the x_{j_1}-direction.

Therefore, the infinite-equilibrium of $x^*_{j_1}=a_{j_1j_1s_1}=a_{j_2j_11}$ with $\bar{x}_{j_2}\in(-\infty,\infty)$ has the following properties as in Eqs. (3.83)–(3.86).

- For $a_{j_1j_10}(a_{j_1j_1s_1}-a_{j_1j_1s_2})^2>0$ and $a_{j_2j_10}>0$,

$$(a_{j_1j_1s_1}, \bar{x}_{j_2}) = \underbrace{(\mathrm{SO}, \mathrm{UP})}_{\text{up-parabola source}}.$$

The equilibrium at $x_{j_1}^* = a_{j_1j_1s_1} = a_{j_2j_11}$ is an (SO,UP)-up-parabola source infinite-equilibrium.

- For $a_{j_1j_10}(a_{j_1j_1s_1} - a_{j_1j_1s_2})^2 < 0$ and $a_{j_2j_10} > 0$,

$$(a_{j_1j_1s_1}, \bar{x}_{j_2}) = \underbrace{(\mathrm{SI}, \mathrm{DP})}_{\text{down-parabola sink}}.$$

The equilibrium at $x_{j_1}^* = a_{j_1j_1s_1} = a_{j_2j_11}$ is an (SI,DP)-down-parabola sink infinite-equilibrium.

- For $a_{j_1j_10}(a_{j_1j_1s_1} - a_{j_1j_1s_2})^2 > 0$ and $a_{j_2j_10} < 0$,

$$(a_{j_1j_1s_1}, \bar{x}_{j_2}) = \underbrace{(\mathrm{SO}, \mathrm{DP})}_{\text{down-parabola source}}.$$

The equilibrium at $x_{j_1}^* = a_{j_1j_1s_1} = a_{j_2j_11}$ is an (SO,DP)-down-parabola source infinite-equilibrium.

- For $a_{j_1j_10}(a_{j_1j_1s_1} - a_{j_1j_1s_2})^2 < 0$ and $a_{j_2j_10} < 0$,

$$(a_{j_1j_1s_1}, \bar{x}_{j_2}) = \underbrace{(\mathrm{SI}, \mathrm{UP})}_{\text{up-parabola sink}}.$$

The equilibrium at $x_{j_1}^* = a_{j_1j_1s_1} = a_{j_2j_11}$ is an (SI,UP)-up-parabola sink infinite-equilibrium.

(v_{3b}) The infinite-equilibriums of $x_{j_1}^* = a_{j_1j_1s_1} = a_{j_2j_11}$ are the switching bifurcation for the sink and source flows with increasing-inflection and decreasing-inflection flows as in Eqs. (3.87)–(3.90).

- For $a_{j_1j_10}(a_{j_1j_1s_1} - a_{j_1j_1s_2})^2 > 0$ and $a_{j_2j_10} > 0$,

$$\left\{\begin{array}{c}\underbrace{(\mathrm{nF}, \mathrm{DI})}_{\text{decreasing-inflection flow }(-)}\\ \underbrace{(\mathrm{SO}, \mathrm{pF})}_{\text{source flow }(+)}\end{array}\right\} \rightleftharpoons \underbrace{(\mathrm{SO}, \mathrm{UP})}_{\text{switching bifurcation}} \rightleftharpoons \left\{\begin{array}{c}\underbrace{(\mathrm{SO}, \mathrm{pF})}_{\text{source flow }(+)}\\ \underbrace{(\mathrm{pF}, \mathrm{II})}_{\text{increasing-inflection flow }(+)}\end{array}\right\}.$$

The (SO,UP)-up-parabola source infinite-equilibrium is called an (SO,UP)-switching bifurcation.

- For $a_{j_1j_10}(a_{j_1j_1s_1} - a_{j_1j_1s_2})^2 < 0$ and $a_{j_2j_10} > 0$,

$$\left\{\begin{array}{c}\underbrace{(\text{pF},\text{II})}_{\text{increasing-inflection flow }(+)}\\\underbrace{(\text{SI},\text{pF})}_{\text{sink flow }(+)}\end{array}\right\}\rightleftharpoons\underbrace{(\text{SI},\text{DP})}_{\text{switching bifurcation}}\rightleftharpoons\left\{\begin{array}{c}\underbrace{(\text{SI},\text{pF})}_{\text{sink flow }(+)}\\\underbrace{(\text{nF},\text{DI})}_{\text{decreasing-inflection flow }(-)}\end{array}\right\}.$$

The (SI,DP)-down-parabola sink infinite-equilibrium is called an (SI,DP)-switching bifurcation.

- For $a_{j_1j_10}(a_{j_1j_1s_1}-a_{j_1j_1s_2})^2>0$ and $a_{j_2j_10}<0$,

$$\left\{\begin{array}{c}\underbrace{(\text{nF},\text{II})}_{\text{increasing-inflection flow }(-)}\\\underbrace{(\text{SO},\text{nF})}_{\text{source flow }(-)}\end{array}\right\}\rightleftharpoons\underbrace{(\text{SO},\text{DP})}_{\text{switching bifurcation}}\rightleftharpoons\left\{\begin{array}{c}\underbrace{(\text{SO},\text{nF})}_{\text{source flow }(-)}\\\underbrace{(\text{pF},\text{DI})}_{\text{decreasing-inflection flow }(+)}\end{array}\right\}.$$

The (SO,DP)-down-parabola source infinite-equilibrium is called an (SO, DP)-switching bifurcation.

- For $a_{j_1j_10}(a_{j_1j_1s_1}-a_{j_1j_1s_2})^2<0$ and $a_{j_2j_10}<0$,

$$\left\{\begin{array}{c}\underbrace{(\text{pF},\text{DI})}_{\text{decreasing-inflection flow }(+)}\\\underbrace{(\text{SI},\text{nF})}_{\text{sink flow }(-)}\end{array}\right\}\rightleftharpoons\underbrace{(\text{SI},\text{UP})}_{\text{switching bifurcation}}\rightleftharpoons\left\{\begin{array}{c}\underbrace{(\text{SI},\text{nF})}_{\text{sink flow }(-)}\\\underbrace{(\text{nF},\text{II})}_{\text{increasing-inflection flow }(-)}\end{array}\right\}.$$

The (SI,UP)-up-parabola sink infinite-equilibrium is called an (SI,UP)-switching bifurcation.

(vi) If

$$\begin{aligned}&\Delta_{j_1j_1}=0\text{ and }\Delta_{j_2j_1}=0,\\&a_{j_1j_1s_1}=a_{j_1j_1s_2}\equiv a_{j_1j_11}\end{aligned}$$

then

$$\begin{aligned}&\dot{x}_{j_1}=a_{j_1j_10}(x_{j_1}-a_{j_1j_11})^3,\\&\dot{x}_{j_2}=a_{j_2j_10}(x_{j_1}-a_{j_2j_11})^2,\\&j_1,j_2\in\{1,2\},j_1\neq j_2.\end{aligned}$$

(vi$_1$) For $a_{j_1j_11}\neq a_{j_2j_11}$, we have

$$\begin{aligned}\frac{dx_{j_2}}{dx_{j_1}} &= \frac{a_{j_2j_10}\,(x_{j_1}-a_{j_2j_11})^2}{a_{j_1j_10}\,(x_{j_1}-a_{j_1j_11})^3}\\ &= \frac{a_{j_2j_10}}{a_{j_1j_10}}\Big[\frac{1}{x_{j_1}-a_{j_1j_11}}+\frac{2(a_{j_1j_11}-a_{j_2j_11})}{(x_{j_1}-a_{j_1j_1s_2})^2}+\frac{(a_{j_1j_11}-a_{j_2j_11})^2}{(x_{j_1}-a_{j_1j_11})^3}\Big].\end{aligned}$$

With an initial condition (x_{j_10}, x_{j_20}), the first integral manifold is

$$\begin{aligned}x_{j_2}-x_{j_20} = &\frac{a_{j_2j_10}}{a_{j_1j_10}}\Big\{\ln\frac{|\,x_{j_1}-a_{j_1j_11}\,|}{|\,x_{j_10}-a_{j_1j_11}\,|}\\ &-2(a_{j_1j_11}-a_{j_2j_11})(\frac{1}{x_{j_1}-a_{j_1j_11}}-\frac{1}{x_{j_10}-a_{j_1j_11}})\\ &-\frac{1}{2}(a_{j_1j_11}-a_{j_2j_11})^2\Big[\frac{1}{(x_{j_1}-a_{j_1j_11})^2}-\frac{1}{(x_{j_10}-a_{j_1j_11})^2}\Big]\Big\},\\ &j_1,j_2\in\{1,2\}, j_1\neq j_2.\end{aligned}$$

(vi_{1a}) For $x_{j_1}^* = a_{j_1j_11} \neq a_{j_2j_11}$, the variational equation is

$$\begin{aligned}\dot{x}_{j_2} &= a_{j_2j_10}(a_{j_1j_11}-a_{j_2j_11})^2,\\ \Delta\dot{x}_{j_1} &= a_{j_1j_10}(\Delta x_{j_1})^3.\end{aligned}$$

The flows at $x_{j_1}^* = a_{j_1j_11} \neq a_{j_2j_11}$ are the third-order source and sink flows for $a_{j_1j_10} > 0$ and $a_{j_1j_10} < 0$ in the x_{j_1}-direction. The flows at $x_{j_1}^* = a_{j_1j_11} \neq a_{j_2j_11}$ are positive and negative for $a_{j_2j_10} > 0$ and $a_{j_2j_10} < 0$ in the x_{j_2}-direction.

Therefore, the equilibrium of $x_{j_1}^* = a_{j_1j_11} \neq a_{j_2j_11}$ has the following properties as in Eqs. (3.94–3.97).

- For $a_{j_1j_10} > 0$ and $a_{j_2j_10}(a_{j_1j_11}-a_{j_2j_11})^2 > 0$,

$$(a_{j_1j_11}, \dot{x}_{j_2}) = \underbrace{(3^{\mathrm{rd}}\mathrm{SO}, \mathrm{pF})}_{\text{third-order source flow }(+)}.$$

The flow at $x_{j_1}^* = a_{j_1j_11}$ is a (3^{rd}SO,pF)-positive third-order source flow.

- For $a_{j_1j_10} < 0$ and $a_{j_2j_10}(a_{j_1j_11}-a_{j_2j_11})^2 > 0$,

$$(a_{j_1j_11}, \dot{x}_{j_2}) = \underbrace{(3^{\mathrm{rd}}\mathrm{SI}, \mathrm{pF})}_{\text{third-order sink flow }(+)}.$$

The flow at $x_{j_1}^* = a_{j_1j_11}$ is a (3^{rd}SI,pF)-positive third-order sink flow.

- For $a_{j_1j_1 0} > 0$ and $a_{j_2j_1 0}(a_{j_1j_1 1} - a_{j_2j_1 1})^2 < 0$,

$$(a_{j_1j_1 1}, \dot{x}_{j_2}) = \underbrace{(3^{\text{rd}}\text{SO}, \text{nF})}_{\text{third-order source flow } (-)}.$$

  The flow at $x^*_{j_1} = a_{j_1j_1 1}$ is a (3rdSO,nF)-negative third-order source flow.

- For $a_{j_1j_1 0} < 0$ and $a_{j_2j_1 0}(a_{j_1j_1 1} - a_{j_2j_1 1})^2 < 0$,

$$(a_{j_1j_1 1}, \dot{x}_{j_2}) = \underbrace{(3^{\text{rd}}\text{SI}, \text{nF})}_{\text{third-order sink flow } (-)}.$$

  The flow at $x^*_{j_1} = a_{j_1j_1 1}$ is a (3rdSI,pF)-negative third-order sink flow.

– The third-order source and sink flows bifurcations are the appearing bifurcations for source-sink-source and sink-source-sink flows.

(vi$_{1b}$) For $x^*_{j_1} = a_{j_2j_1 1} \neq a_{j_1j_1 1}$, we have

$$\frac{dx_{j_2}}{dx_{j_1}}\Big|_{x^*_{j_1} = a_{j_2j_1 1}} = \frac{a_{j_2j_1 0}\,(x_{j_1} - a_{j_2j_1 1})^2}{a_{j_1j_1 0}\,(x_{j_1} - a_{j_1j_1 1})^3}\Big|_{x^*_{j_1} = a_{j_2j_1 1}} = 0,$$
$$\frac{d^2x_{j_2}}{dx^2_{j_1}}\Big|_{x^*_{j_1} = a_{j_2j_1 1}} = 0.$$

If

$$\frac{d^3x_{j_2}}{dx^3_{j_1}}\Big|_{x^*_{j_1} = a_{j_2j_1 1}} = \frac{a_{j_2j_1 0}}{a_{j_1j_1 0}}\,\frac{1}{(a_{j_2j_1 1} - a_{j_1j_1 1})^3} > 0,$$

the flow at $x^*_{j_1} = a_{j_1j_1 s_1} = a_{j_2j_1 1}$ is an increasing-inflection flow in the x_{j_2}-direction, and if

$$\frac{d^3x_{j_2}}{dx^3_{j_1}}\Big|_{x^*_{j_1} = a_{j_2j_1 1}} = \frac{a_{j_2j_1 0}}{a_{j_1j_1 0}}\,\frac{1}{(a_{j_2j_1 1} - a_{j_1j_1 1})^3} < 0,$$

the flow at $x^*_{j_1} = a_{j_1j_1 s_1} = a_{j_2j_1 1}$ is a decreasing-inflection flow in the x_{j_2}-direction.

At $x^*_{j_1} = a_{j_1j_1 s_1} = a_{j_2j_1 1}$, the variational equation is

$$\Delta\dot{x}_{j_2} = a_{j_2j_1 0}(\Delta x_{j_1})^2,$$
$$\dot{x}_{j_1} = a_{j_1j_1 0}(a_{j_2j_1 1} - a_{j_1j_1 1})^3.$$

The flows at $x^*_{j_1} = a_{j_1j_1 s_1} = a_{j_2j_1 1}$ are positive and negative for $\dot{x}_{j_1} > 0$ and $\dot{x}_{j_1} < 0$ in the x_{j_1}-direction.

Therefore, the equilibrium of $x^*_{j_1} = a_{j_2j_1 1} \neq a_{j_1j_1 1}$ has the following properties as in Eqs. (3.98)–(3.101).

- For $a_{j_1j_1 0}(a_{j_2j_1 1} - a_{j_1j_1 1})^3 > 0$ and $a_{j_2j_1 0} > 0$,

$$(\dot{x}_{j_1}, a_{j_2j_1 1}) = \underbrace{(\text{pF}, \text{II})}_{\text{increasing-inflection flow } (+)}.$$

  The flow at $x^*_{j_1} = a_{j_2j_1 1}$ is a (pF,II)-positive increasing-inflection flow.
- For $a_{j_1j_1 0}(a_{j_2j_1 1} - a_{j_1j_1 1})^3 < 0$ and $a_{j_2j_1 0} > 0$,

$$(\dot{x}_{j_1}, a_{j_2j_1 1}) = \underbrace{(\text{nF}, \text{DI})}_{\text{decreasing-inflection flow } (-)}.$$

  The flow at $x^*_{j_1} = a_{j_2j_1 1}$ is an (nF,DI)-negative decreasing-inflection flow.
- For $a_{j_1j_1 0}(a_{j_2j_1 1} - a_{j_1j_1 1})^3 > 0$ and $a_{j_2j_1 0} < 0$,

$$(\dot{x}_{j_1}, a_{j_2j_1 1}) = \underbrace{(\text{pF}, \text{DI})}_{\text{decreasing-inflection flow } (+)}.$$

  The flow at $x^*_{j_1} = a_{j_2j_1 1}$ is a (pF,DI)-positive decreasing-inflection flow.
- For $a_{j_1j_1 0}(a_{j_2j_1 1} - a_{j_1j_1 1})^3 < 0$ and $a_{j_2j_1 0} < 0$,

$$(\dot{x}_{j_1}, a_{j_2j_1 1}) = \underbrace{(\text{nF}, \text{II})}_{\text{increasing-inflection flow } (-)}.$$

  The flow at $x^*_{j_1} = a_{j_2j_1 1}$ is an (nF,II)-negative increasing-inflection flow.

– The increasing-inflection and decreasing-inflection flow bifurcations are the appearing bifurcations for the up-parabola and down-parabola flows.

(vi$_2$) For $a_{j_2j_1 1} = a_{j_1j_1 1}$,

$$\begin{aligned}
&\dot{x}_{j_1} = a_{j_1j_1 0}(x_{j_1} - a_{j_1j_1 1})^3, \\
&\dot{x}_{j_2} = a_{j_2j_1 0}(x_{j_1} - a_{j_2j_1 1})^2, \\
&j_1, j_2 \in \{1, 2\}, j_1 \neq j_2.
\end{aligned}$$

we have

$$\frac{dx_{j_2}}{dx_{j_1}} = \frac{a_{j_2j_1 0}}{a_{j_1j_1 0}} \frac{1}{(x_{j_1} - a_{j_1j_1 1})}.$$

With an initial condition $(x_{j_1 0}, x_{j_2 0})$, the first integral manifold is

$$x_{j_2} - x_{j_2 0} = \frac{a_{j_2 j_1 0}}{a_{j_1 j_1 0}} \ln \frac{|x_{j_1} - a_{j_1 j_1 1}|}{|x_{j_1 0} - a_{j_1 j_1 1}|},$$
$$j_1, j_2 \in \{1,2\}, j_1 \neq j_2.$$

(vi$_{2a}$) At $x_{j_1}^* = a_{j_2 j_1 1} = a_{j_1 j_1 1}$, the variational equations are

$$\Delta\dot{x}_{j_1} = a_{j_1 j_1 0}(\Delta x_{j_1})^3,$$
$$\Delta\dot{x}_{j_2} = a_{j_2 j_1 0}(\Delta x_{j_1})^2.$$

From the first integral manifold, the flow at the equilibrium at $x_{j_1}^* = a_{j_2 j_1 1} = a_{j_1 j_1 1}$ is a logarithmic flow. The flow $x_{j_1}^* = a_{j_2 j_1 1} = a_{j_1 j_1 1}$ is the third-order source and sink flows in the x_{j_1} direction. The singularity at the equilibrium is of the second-order and third-order in the x_{j_2}-direction and the x_{j_1}-direction. Thus, (3:2) is used to identify the singularity of the logarithmic flows.

Therefore, the infinite-equilibrium of $x_{j_1}^* = a_{j_2 j_1 1} = a_{j_1 j_1 1}$ with $\bar{x}_{j_2} \in (-\infty, \infty)$ has the following properties as in Eqs. (3.104)–(3.107).

- For $a_{j_1 j_1 0} > 0$ and $a_{j_2 j_1 0} > 0$,

$$(a_{j_1 j_1 1}, \bar{x}_{j_2}) = \underbrace{(3^{\text{rd}}\text{SO}, (3:2)\text{DL})}_{\text{third-order down-logarithmic source}}.$$

  The equilibrium at $x_{j_1}^* = a_{j_2 j_1 1} = a_{j_1 j_1 1}$ is a (3^{rd}SO,(3:2)DL)-third-order down-logarithmic source infinite-equilibrium.
- For $a_{j_1 j_1 0} < 0$ and $a_{j_2 j_1 0} > 0$,

$$(a_{j_1 j_1 1}, \bar{x}_{j_2}) = \underbrace{(3^{\text{rd}}\text{SI}, (3:2)\text{UL})}_{\text{third-order up-logarithmic sink}}.$$

  The equilibrium at $x_{j_1}^* = a_{j_2 j_1 1} = a_{j_1 j_1 1}$ is a (3^{rd}SI,(3:2)UL)-third-order up-logarithmic sink infinite-equilibrium.
- For $a_{j_1 j_1 0} > 0$ and $a_{j_2 j_1 0} < 0$,

$$(a_{j_1 j_1 1}, \bar{x}_{j_2}) = \underbrace{(3^{\text{rd}}\text{SO}, (3:2)\text{UL})}_{\text{third-order up-logarithmic source}}.$$

  The equilibrium at $x_{j_1}^* = a_{j_2 j_1 1} = a_{j_1 j_1 1}$ is a (3^{rd}SO,(3:2)UL)-third-order up-logarithmic source infinite-equilibrium.
- For $a_{j_1 j_1 0} < 0$ and $a_{j_2 j_1 0} < 0$,

$$(a_{j_1j_1 1}, \bar{x}_{j_2}) = \underbrace{(3^{\rm rd}\mathrm{SI}, (3{:}2)\mathrm{DL})}_{\text{third-order down-logarithmic sink}}.$$

The equilibrium at $x_{j_1}^* = a_{j_2j_1 1} = a_{j_1j_1 1}$ is a (3$^{\rm rd}$SI,(3:2)DL)-third-order down-logarithmic sink infinite-equilibrium.

(vi$_{2b}$) Let $\Delta_{j_1j_1}^+ = \Delta_{j_1j_1} + 0$ and $\Delta_{j_2j_1}^+ = \Delta_{j_2j_1} + 0$. The infinite-equilibriums of $x_{j_1}^* = a_{j_2j_1 1} = a_{j_1j_1 1}$ are the appearing bifurcation for the sink and source flows and the up-parabola and down-parabola flows as in Eqs. (3.108)–(3.115).

- For $a_{j_1j_1 0} > 0$ and $a_{j_2j_1 0} > 0$,

$$\underbrace{(\mathrm{SO}, \mathrm{pF})}_{\text{source flow }(+)} \rightleftharpoons \underbrace{(3^{\rm rd}\mathrm{SO}, (3{:}2)\mathrm{DL})}_{\text{appearing bifurcation}} \rightleftharpoons \left\{ \begin{array}{c} \underbrace{(\mathrm{SO}, \mathrm{pF})}_{\text{sink flow }(+)}, \underbrace{(\mathrm{UP}, \mathrm{nF})}_{\text{up-parabola flow }(-)}, \\ \underbrace{(\mathrm{SI}, \mathrm{pF})}_{\text{source flow }(+)}, \\ \underbrace{(\mathrm{DP}, \mathrm{nF})}_{\text{down-parabola flow }(-)}, \underbrace{(\mathrm{SO}, \mathrm{pF})}_{\text{source flow }(+)} \end{array} \right\}$$

for $\Delta_{j_1j_1}^+ > \Delta_{j_2j_1}^+$;

$$\underbrace{(\mathrm{SO}, \mathrm{pF})}_{\text{source flow }(+)} \rightleftharpoons \underbrace{(3^{\rm rd}\mathrm{SO}, (3{:}2)\mathrm{DL})}_{\text{appearing bifurcation}} \rightleftharpoons \left\{ \begin{array}{c} \underbrace{(\mathrm{DP}, \mathrm{pF})}_{\text{down-parabola flow }(+)}, \underbrace{(\mathrm{SO}, \mathrm{nF})}_{\text{sink flow }(-)}, \\ \underbrace{(\mathrm{SI}, \mathrm{nF})}_{\text{sink flow }(-)}, \\ \underbrace{(\mathrm{SO}, \mathrm{nF})}_{\text{source flow }(-)}, \underbrace{(\mathrm{UP}, \mathrm{nF})}_{\text{up-parabola flow }(+)} \end{array} \right\}$$

for $\Delta_{j_1j_1}^+ < \Delta_{j_2j_1}^+$.

- For $a_{j_1j_1 0} < 0$ and $a_{j_2j_1 0} > 0$,

$$\underbrace{(\mathrm{SI}, \mathrm{pF})}_{\text{sink flow }(+)} \rightleftharpoons \underbrace{(3^{\rm rd}\mathrm{SI}, (3{:}2)\mathrm{UL})}_{\text{appearing bifurcation}} \rightleftharpoons \left\{ \begin{array}{c} \underbrace{(\mathrm{SI}, \mathrm{pF})}_{\text{source flow }(+)}, \underbrace{(\mathrm{pF}, \mathrm{DP})}_{\text{down-parabola flow }(+)}, \\ \underbrace{(\mathrm{SO}, \mathrm{pF})}_{\text{source flow }(+)}, \\ \underbrace{(\mathrm{pF}, \mathrm{UP})}_{\text{up-parabola flow }(+)}, \underbrace{(\mathrm{SI}, \mathrm{pF})}_{\text{sink flow }(+)} \end{array} \right\}$$

for $\Delta_{j_1j_1}^+ > \Delta_{j_2j_1}^+$;

$$\underbrace{(\mathrm{SI},\mathrm{pF})}_{\text{sink flow }(+)} \rightleftharpoons \underbrace{(3^{\mathrm{rd}}\mathrm{SI},(3:2)\mathrm{UL})}_{\text{appearing bifurcation}} \rightleftharpoons \left\{\begin{array}{c} \underbrace{(\mathrm{nF},\mathrm{UP})}_{\text{up-parabola flow }(-)}, \underbrace{(\mathrm{SI},\mathrm{nF})}_{\text{source flow }(-)}, \\ \underbrace{(\mathrm{SO},\mathrm{nF})}_{\text{source flow }(+)}, \\ \underbrace{(\mathrm{SI},\mathrm{nF})}_{\text{sink flow }(-)}, \underbrace{(\mathrm{nF},\mathrm{DP})}_{\text{down-parabola flow }(-)} \end{array}\right\}$$

for$\Delta^{+}_{j_1 j_1} < \Delta^{+}_{j_2 j_1}$.

- For $a_{j_1 j_1 0} > 0$ and $a_{j_2 j_1 0} < 0$,

$$\underbrace{(\mathrm{SO},\mathrm{nF})}_{\text{source flow }(-)} \rightleftharpoons \underbrace{(3^{\mathrm{rd}}\mathrm{SO},(3:2)\mathrm{UL})}_{\text{appearing bifurcation}} \rightleftharpoons \left\{\begin{array}{c} \underbrace{(\mathrm{SO},\mathrm{nF})}_{\text{sink flow }(-)}, \underbrace{(\mathrm{nF},\mathrm{DP})}_{\text{down-parabola flow }(-)}, \\ \underbrace{(\mathrm{SI},\mathrm{nF})}_{\text{sink flow }(-)}, \\ \underbrace{(\mathrm{nF},\mathrm{UP})}_{\text{up-parabola flow }(-)}, \underbrace{(\mathrm{SO},\mathrm{nF})}_{\text{source flow }(-)} \end{array}\right\}$$

for $\Delta^{+}_{j_1 j_1} > \Delta^{+}_{j_2 j_1}$;

$$\underbrace{(\mathrm{SO},\mathrm{nF})}_{\text{source flow }(-)} \rightleftharpoons \underbrace{(3^{\mathrm{rd}}\mathrm{SO},(3:2)\mathrm{UL})}_{\text{appearing bifurcation}} \rightleftharpoons \left\{\begin{array}{c} \underbrace{(\mathrm{pF},\mathrm{UP})}_{\text{up-parabola flow }(+)}, \underbrace{(\mathrm{SO},\mathrm{pF})}_{\text{source flow }(+)}, \\ \underbrace{(\mathrm{SI},\mathrm{pF})}_{\text{sink flow }(+)}, \\ \underbrace{(\mathrm{SO},\mathrm{pF})}_{\text{source flow }(+)}, \underbrace{(\mathrm{pF},\mathrm{DP})}_{\text{down-parabola flow }(+)} \end{array}\right\}$$

for $\Delta^{+}_{j_1 j_1} < \Delta^{+}_{j_2 j_1}$.

- For $a_{j_1 j_1 0} < 0$ and $a_{j_2 j_1 0} < 0$,

$$\underbrace{(\mathrm{SI},\mathrm{nF})}_{\text{sink flow }(-)} \rightleftharpoons \underbrace{(3^{\mathrm{rd}}\mathrm{SI},(3:2)\mathrm{DL})}_{\text{appearing bifurcation}} \rightleftharpoons \left\{\begin{array}{c} \underbrace{(\mathrm{SI},\mathrm{nF})}_{\text{source flow }(-)}, \underbrace{(\mathrm{nF},\mathrm{DP})}_{\text{down-parabola flow }(-)}, \\ \underbrace{(\mathrm{SO},\mathrm{pF})}_{\text{source flow }(+)}, \\ \underbrace{(\mathrm{pF},\mathrm{DP})}_{\text{down-parabola flow }(+)}, \underbrace{(\mathrm{SI},\mathrm{nF})}_{\text{sink flow }(-)} \end{array}\right\}$$

for $\Delta^{+}_{j_1 j_1} > \Delta^{+}_{j_2 j_1}$;

$$\underbrace{(\mathrm{SI},\mathrm{nF})}_{\text{sink flow }(-)} \rightleftharpoons \underbrace{(3^{\mathrm{rd}}\mathrm{SI},(3{:}2)\mathrm{DL})}_{\text{appearing bifurcation}} \rightleftharpoons \left\{\begin{array}{c} \underbrace{(\mathrm{DP},\mathrm{nF})}_{\text{down-parabola flow }(-)}, \underbrace{(\mathrm{SI},\mathrm{pF})}_{\text{source flow }(+)}, \\ \underbrace{(\mathrm{SO},\mathrm{pF})}_{\text{source flow }(+)}, \\ \underbrace{(\mathrm{SI},\mathrm{pF})}_{\text{sink flow }(+)}, \underbrace{(\mathrm{UP},\mathrm{nF})}_{\text{up-parabola flow }(-)} \end{array}\right\}$$

for $\Delta^{+}_{j_1 j_1} < \Delta^{+}_{j_2 j_1}$.

(vi_{2c}) The infinite-equilibriums of $x^{*}_{j_1} = a_{j_2 j_1 1} = a_{j_1 j_1 1}$ are the switching bifurcation for the third-order sink and source flows and the increasing-inflection and decreasing-inflection flows as in Eqs. (3.116)–(3.119).

- For $a_{j_1 j_1 0} > 0$ and $a_{j_2 j_1 0} > 0$,

$$\left\{\begin{array}{c} \underbrace{(\mathrm{nF},\mathrm{DI})}_{\text{decreasing-inflection flow }(-)} \\ \underbrace{(3^{\mathrm{rd}}\mathrm{SO},\mathrm{pF})}_{\text{third order source flow }(+)} \end{array}\right\} \rightleftharpoons \underbrace{(3^{\mathrm{rd}}\mathrm{SO},(3{:}2)\mathrm{DL})}_{\text{switching bifurcation}} \rightleftharpoons \left\{\begin{array}{c} \underbrace{(3^{\mathrm{rd}}\mathrm{SO},\mathrm{pF})}_{\text{third order source flow }(+)} \\ \underbrace{(\mathrm{pF},\mathrm{II})}_{\text{increasing-inflection flow }(+)} \end{array}\right\}.$$

- For $a_{j_1 j_1 0} < 0$ and $a_{j_2 j_1 0} > 0$,

$$\left\{\begin{array}{c} \underbrace{(\mathrm{pF},\mathrm{II})}_{\text{increasing-inflection flow }(+)} \\ \underbrace{(3^{\mathrm{rd}}\mathrm{SI},\mathrm{pF})}_{\text{third order sink flow }(+)} \end{array}\right\} \rightleftharpoons \underbrace{(3^{\mathrm{rd}}\mathrm{SI},(3{:}2)\mathrm{UL})}_{\text{switching bifurcation}} \rightleftharpoons \left\{\begin{array}{c} \underbrace{(3^{\mathrm{rd}}\mathrm{SI},\mathrm{pF})}_{\text{third order sink flow }(+)} \\ \underbrace{(\mathrm{nF},\mathrm{DI})}_{\text{decreasing-inflection flow }(-)} \end{array}\right\}.$$

- For $a_{j_1 j_1 0} > 0$ and $a_{j_2 j_1 0} < 0$,

$$\left\{\begin{array}{c} \underbrace{(\mathrm{nF},\mathrm{II})}_{\text{increasing-inflection flow }(-)} \\ \underbrace{(3^{\mathrm{rd}}\mathrm{SO},\mathrm{nF})}_{\text{third order source flow }(-)} \end{array}\right\} \rightleftharpoons \underbrace{(3^{\mathrm{rd}}\mathrm{SO},(3{:}2)\mathrm{UL})}_{\text{switching bifurcation}} \rightleftharpoons \left\{\begin{array}{c} \underbrace{(3^{\mathrm{rd}}\mathrm{SO},\mathrm{nF})}_{\text{third order source flow }(-)} \\ \underbrace{(\mathrm{nF},\mathrm{DI})}_{\text{decreasing-inflection flow }(+)} \end{array}\right\}.$$

- For $a_{j_1 j_1 0} < 0$ and $a_{j_2 j_1 0} < 0$,

$$\left\{\begin{array}{c}\underbrace{(\text{nF},\text{DI})}_{\text{decreasing-inflection flow }(-)}\\ \underbrace{(3^{\text{rd}}\text{SI},\text{nF})}_{\text{third order sink flow }(-)}\end{array}\right\}\rightleftharpoons\underbrace{(3^{\text{rd}}\text{SI},(3{:}2)\text{DL})}_{\text{switching bifurcation}}\rightleftharpoons\left\{\begin{array}{c}\underbrace{(3^{\text{rd}}\text{SI},\text{nF})}_{\text{third order sink flow }(-)}\\ \underbrace{(\text{nF},\text{II})}_{\text{increasing-inflection flow }(-)}\end{array}\right\}.$$

(vii) If

$$\Delta_{j_1j_1}>0 \text{ and } \Delta_{j_2j_1}=0,$$

then

$$\begin{aligned}\dot{x}_{j_1}&=a_{j_1j_10}(x_{j_1}-a_{j_1j_11})(x_{j_1}-a_{j_1j_12})(x_{j_1}-a_{j_1j_13}),\\ \dot{x}_{j_2}&=a_{j_2j_10}(x_{j_1}-a_{j_2j_11})^2,\end{aligned}$$

where

$$\begin{aligned}&\{a_{j_1j_11},a_{j_1j_12},a_{j_1j_13}\}=\text{sort}\{b_{j_1j_11},b_{j_1j_12},b_{j_1j_13}\},a_{j_1j_11}<a_{j_1j_12}<a_{j_1j_13};\\ &b_{j_1j_12},b_{j_1j_13}=-\frac{1}{2}(B_{j_1j_11}\pm\sqrt{\Delta_{j_1j_1}}),a_{j_2j_11}=-\frac{1}{2}B_{j_2j_11};\\ &j_1,j_2\in\{1,2\};j_1\neq j_2.\end{aligned}$$

(vii$_1$) For $x^*_{j_1}=a_{j_1j_1s_1}\neq a_{j_2j_11}$ (s_1, $s_2\in\{1,2,3\}$; $s_1\neq s_2$), we have

$$\begin{aligned}\frac{dx_{j_2}}{dx_{j_1}}&=\frac{a_{j_2j_10}}{a_{j_1j_10}}\frac{(x_{j_1}-a_{j_2j_11})^2}{(x_{j_1}-a_{j_1j_1s_1})(x_{j_1}-a_{j_1j_1s_2})(x_{j_1}-a_{j_1j_1s_3})}\\ &=\frac{a_{j_2j_10}}{a_{j_1j_10}}\sum_{s_1=1}^{3}\frac{(a_{j_1j_1s_1}-a_{j_2j_11})^2}{\prod_{s_2=1,s_2\neq s_1}^{3}(a_{j_1j_1s_1}-a_{j_1j_1s_2})}\frac{1}{x_{j_1}-a_{j_1j_1s_1}}.\end{aligned}$$

With an initial condition (x_{j_10},x_{j_20}),the first integral manifold is

$$x_{j_2}-x_{j_20}=\frac{a_{j_2j_10}}{a_{j_1j_10}}\sum_{s_1=1}^{3}\frac{(a_{j_1j_1s_1}-a_{j_2j_11})^2}{\prod_{s_2=1,s_2\neq s_1}^{3}(a_{j_1j_1s_1}-a_{j_1j_1s_2})}\ln\frac{|x_{j_1}-a_{j_1j_1s_1}|}{|x_{j_10}-a_{j_1j_1s_1}|}.$$

(vii$_{1a}$) At $x^*_{j_1}=a_{j_1j_1s_1}\neq a_{j_2j_11}$, the variational equation is

$$\begin{aligned}\Delta\dot{x}_{j_1}&=a_{j_1j_10}\prod_{s_2=1,s_2\neq s_1}^{3}(a_{j_1j_1s_1}-a_{j_1j_1s_2})\Delta x_{j_1},\\ \dot{x}_{j_2}&=a_{j_2j_10}(a_{j_1j_1s_1}-a_{j_2j_11})^2.\end{aligned}$$

The flows at $x_{j_1}^* = a_{j_1 j_1 s_1} \neq a_{j_2 j_1 1}$ are source and sink flows in the x_{j_1}-direction for $a_{j_1 j_1 0}\prod_{s_2=1, s_2 \neq s_1}^{3}(a_{j_1 j_1 s_1} - a_{j_1 j_1 s_2}) > 0$ and $a_{j_1 j_1 0}\prod_{s_2=1, s_2 \neq s_1}^{3}(a_{j_1 j_1 s_1} - a_{j_1 j_1 s_2}) < 0$, and the flows at $x_{j_1}^* = a_{j_1 j_1 s_1} \neq a_{j_2 j_1 1}$ are positive and negative in the x_{j_2}-direction for $a_{j_2 j_1 0} > 0$ and $a_{j_2 j_1 0} < 0$.

Therefore, the equilibrium of $x_{j_1}^* = a_{j_1 j_1 s_1} \neq a_{j_2 j_1 1}$ $(s_1, s_2 \in \{1,2,3\};\ s_1 \neq s_2)$ in the x_{j_1}-direction possesses the following properties as in Eqs. (3.124)–(3.127).

- For $a_{j_1 j_1 0}\prod_{s_2=1, s_2 \neq s_1}^{3}(a_{j_1 j_1 s_1} - a_{j_1 j_1 s_2}) > 0$ and $a_{j_2 j_1 0}(a_{j_1 j_1 s_1} - a_{j_2 j_1 1})^2 > 0$,

$$(a_{j_1 j_1 s_1}, \dot{x}_{j_2}) = \underbrace{(\mathrm{SO}, \mathrm{pF})}_{\text{source flow } (+)}.$$

The flow at $x_{j_1}^* = a_{j_1 j_1 s_1}$ is an (SO,pF)-positive source-flow.

- For $a_{j_1 j_1 0}\prod_{s_2=1, s_2 \neq s_1}^{3}(a_{j_1 j_1 s_1} - a_{j_1 j_1 s_2}) < 0$ and $a_{j_2 j_1 0}(a_{j_1 j_1 s_1} - a_{j_2 j_1 1})^2 > 0$,

$$(a_{j_1 j_1 s_1}, \dot{x}_{j_2}) = \underbrace{(\mathrm{SI}, \mathrm{pF})}_{\text{sink flow } (+)}.$$

The flow at $x_{j_1}^* = a_{j_1 j_1 s_1}$ is an (SI,pF)-positive sink-flow.

- For $a_{j_1 j_1 0}\prod_{s_2=1, s_2 \neq s_1}^{3}(a_{j_1 j_1 s_1} - a_{j_1 j_1 s_2}) > 0$ and $a_{j_2 j_1 0}(a_{j_1 j_1 s_1} - a_{j_2 j_1 1})^2 < 0$,

$$(a_{j_1 j_1 s_1}, \dot{x}_{j_2}) = \underbrace{(\mathrm{SO}, \mathrm{nF})}_{\text{source flow } (-)}.$$

The flow at $x_{j_1}^* = a_{j_1 j_1 s_1}$ is an (SO,nF)-negative source-flow.

- For $a_{j_1 j_1 0}\prod_{s_2=1, s_2 \neq s_1}^{3}(a_{j_1 j_1 s_1} - a_{j_1 j_1 s_2}) < 0$ and $a_{j_2 j_1 0}(a_{j_1 j_1 s_1} - a_{j_2 j_1 1})^2 < 0$,

$$(a_{j_1 j_1 s_1}, \dot{x}_{j_2}) = \underbrace{(\mathrm{SI}, \mathrm{nF})}_{\text{sink flow } (-)}.$$

The flow at $x_{j_1}^* = a_{j_1 j_1 s_1}$ is an (SI,nF)-negative sink-flow.

(vii_{1b}) For $x_{j_1}^* = a_{j_2 j_1 1} \neq a_{j_1 j_1 s_1}$ $(s_1 = 1, 2, 3)$, we have

$$\frac{dx_{j_2}}{dx_{j_1}}\bigg|_{x_{j_1}^* = a_{j_2 j_1 1}} = \frac{a_{j_2 j_1 0}}{a_{j_1 j_1 0}} \frac{(x_{j_1} - a_{j_2 j_1 1})^2}{\prod_{s_1=1}^{3}(x_{j_1} - a_{j_1 j_1 s_1})}\bigg|_{x_{j_1}^* = a_{j_2 j_1 1}} = 0,$$

$$\frac{d^2 x_{j_2}}{dx_{j_1}^2}\bigg|_{x_{j_1}^* = a_{j_2 j_1 1}} = 0.$$

If

$$\frac{d^3x_{j_2}}{dx_{j_1}^3}\bigg|_{x_{j_1}^*=a_{j_2j_1 1}} = \frac{2a_{j_2j_1 0}}{a_{j_1j_1 0}\prod_{s_1=1}^{3}(a_{j_2j_1 1}-a_{j_1j_1 s_1})} > 0,$$

the flow at $x_{j_1}^* = a_{j_2j_1 1} \neq a_{j_1j_1 s_1}$ ($s_1 = 1, 2, 3$) is an increasing-inflection flow, and if

$$\frac{d^3x_{j_2}}{dx_{j_1}^3}\bigg|_{x_{j_1}^*=a_{j_2j_1 1}} = \frac{2a_{j_2j_1 0}}{a_{j_1j_1 0}\prod_{s_1=1}^{3}(a_{j_1j_1 1}-a_{j_1j_1 s_1})} < 0,$$

the flow at $x_{j_1}^* = a_{j_2j_1 1} \neq a_{j_1j_1 s_1}$ ($s_1 = 1, 2, 3$) is a decreasing-inflection flow. Let

$$\dot{x}_{j_1} = a_{j_1j_1 0}\prod_{s_1=1}^{3}(a_{j_1j_1 1}-a_{j_1j_1 s_1}).$$

The flows at $x_{j_1}^* = a_{j_2j_1 1} \neq a_{j_1j_1 s_1}, a_{j_1j_1 s_2}$ ($s_1 = 1, 2, 3$) are positive and negative for $\dot{x}_{j_1} > 0$ and $\dot{x}_{j_1} < 0$ in the x_{j_1}-direction.

Therefore, the equilibrium of $x_{j_1}^* = a_{j_2j_1 1} \neq a_{j_1j_1 s_1}$ ($s_1 = 1, 2, 3$) in the x_{j_2}-direction possesses the following properties as in (3.128)–(3.131).

- For $a_{j_1j_1 0}\prod_{s_1=1}^{3}(a_{j_2j_1 1}-a_{j_1j_1 s_1}) > 0$ and $a_{j_2j_1 0} > 0$,

$$(\dot{x}_{j_1}, a_{j_2j_1 1}) = \underbrace{(\text{pF}, \text{II})}_{\text{increasing-inflection flow }(+)}.$$

  The flow at $x_{j_1}^* = a_{j_2j_1 1}$ is a (pF II)-positive increasing-inflection flow.
- For $a_{j_1j_1 0}\prod_{s_1=1}^{3}(a_{j_2j_1 1}-a_{j_1j_1 s_1}) < 0$ and $a_{j_2j_1 0} > 0$,

$$(\dot{x}_{j_1}, a_{j_2j_1 1}) = \underbrace{(\text{nF}, \text{DI})}_{\text{decreasing-inflection flow }(-)}.$$

  The flow at $x_{j_1}^* = a_{j_2j_1 1}$ is an (nF, DI)-negative decreasing-inflection flow.
- For $a_{j_1j_1 0}\prod_{s_1=1}^{3}(a_{j_2j_1 1}-a_{j_1j_1 s_1}) > 0$ and $a_{j_2j_1 0} < 0$,

$$(\dot{x}_{j_1}, a_{j_2j_1 1}) = \underbrace{(\text{pF}, \text{DI})}_{\text{decreasing-inflection flow }(+)}.$$

  The flow at $x_{j_1}^* = a_{j_2j_1 1}$ is a (pF, DI)-positive decreasing-inflection flow.
- For $a_{j_1j_1 0}\prod_{s_1=1}^{3}(a_{j_2j_1 1}-a_{j_1j_1 s_1}) < 0$ and $a_{j_2j_1 0} < 0$,

$$(\dot{x}_{j_1}, a_{j_2 j_1 1}) = \underbrace{(\text{nF, II})}_{\text{increasing-inflection flow } (-)}.$$

The flow at $x^*_{j_1} = a_{j_2 j_1 1}$ is an (nF,II)-negative increasing-inflection flow.

- The decreasing-inflection and increasing-inflection flows are the appearing bifurcations of the up-parabola and down-parabola flows.

(vii$_2$) For $a_{j_2 j_1 1} = a_{j_1 j_1 s_1}$ (s_1, s_2, $s_3 \in \{1, 2, 3\}$; $s_1 \neq s_2 \neq s_3$)

$$\begin{aligned}
&\dot{x}_{j_1} = a_{j_1 j_1 0}(x_{j_1} - a_{j_1 j_1 1})(x_{j_1} - a_{j_1 j_1 2})(x_{j_1} - a_{j_1 j_1 3}),\\
&\dot{x}_{j_2} = a_{j_2 j_1 0}(x_{j_1} - a_{j_2 j_1 1})^2,\\
&j_1, j_2 \in \{1, 2\}, j_1 \neq j_2.
\end{aligned}$$

In phase space,

$$\begin{aligned}
\frac{dx_{j_2}}{dx_{j_1}} &= \frac{a_{j_2 j_1 0}}{a_{j_1 j_1 0}} \frac{(x_{j_1} - a_{j_2 j_1 1})}{(x_{j_1} - a_{j_1 j_1 s_2})(x_{j_1} - a_{j_1 j_1 s_3})}\\
&= \frac{a_{j_2 j_1 0}}{a_{j_1 j_1 0}} \sum_{s_2 = 1, s_2 \neq s_1}^{3} \frac{(a_{j_1 j_1 s_2} - a_{j_2 j_1 1})}{\prod_{s_3 = 1, s_3 \neq s_1, s_2}^{3} (a_{j_1 j_1 s_2} - a_{j_1 j_1 s_3})} \frac{1}{x_{j_1} - a_{j_1 j_1 s_2}}.
\end{aligned}$$

With an initial condition $(x_{j_1 0}, x_{j_2 0})$, the first integral manifold is

$$x_{j_2} - x_{j_2 0} = \frac{a_{j_2 j_1 0}}{a_{j_1 j_1 0}} \sum_{s_2 = 1, s_2 \neq s_1}^{3} \frac{(a_{j_1 j_1 s_2} - a_{j_2 j_1 1})}{\prod_{s_3 = 1, s_3 \neq s_1, s_2}^{3} (a_{j_1 j_1 s_2} - a_{j_1 j_1 s_3})} \ln \frac{|\, x_{j_1} - a_{j_1 j_1 s_2} \,|}{|\, x_{j_1 0} - a_{j_1 j_1 s_2} \,|}.$$

(vii$_{2a}$) For $x^*_{j_1} = a_{j_1 j_1 s_1} = a_{j_2 j_1 1}$ ($s_1 \in \{1, 2, 3\}$), we have

$$\left. \frac{dx_{j_2}}{dx_{j_1}} \right|_{x^*_{j_1} = a_{j_2 j_1 1}} = \frac{a_{j_2 j_1 0}}{a_{j_1 j_1 0}} \left. \frac{(x_{j_1} - a_{j_2 j_1 1})}{\prod_{s_2 = 1, s_2 \neq s_1}^{3} (x_{j_1} - a_{j_1 j_1 s_2})} \right|_{x^*_{j_1} = a_{j_2 j_1 1}} = 0,$$

If

$$\left. \frac{d^2 x_{j_2}}{dx^2_{j_1}} \right|_{x^*_{j_1} = a_{j_2 j_1 1}} = \frac{a_{j_2 j_1 0}}{a_{j_1 j_1 0} \prod_{s_2 = 1, s_2 \neq s_1}^{3} (a_{j_2 j_1 1} - a_{j_1 j_1 s_2})} > 0,$$

the flow at $x^*_{j_1} = a_{j_2 j_1 1} = a_{j_1 j_1 s_1}$ ($s_1 \in \{1, 2, 3\}$) is an up-parabola flow, and if

$$\left. \frac{d^2 x_{j_2}}{dx^2_{j_1}} \right|_{x^*_{j_1} = a_{j_2 j_1 1}} = \frac{a_{j_2 j_1 0}}{a_{j_1 j_1 0} \prod_{s_2 = 1, s_2 \neq s_1}^{3} (a_{j_2 j_1 1} - a_{j_1 j_1 s_2})} < 0,$$

the flow at $x^*_{j_1} = a_{j_2 j_1 1} = a_{j_1 j_1 s_1}$ ($s_1 \in \{1, 2, 3\}$) is a down-parabola flow.

At $x_{j_1}^* = a_{j_2 j_1 1} = a_{j_1 j_1 s_1} (s_1 \in \{1, 2, 3\})$, the variational equations are

$$\begin{aligned}
\Delta \dot{x}_{j_2} &= a_{j_2 j_1 0} (\Delta x_{j_1})^2, \\
\Delta \dot{x}_{j_1} &= a_{j_1 j_1 0} \prod\nolimits_{s_2 = 1, s_2 \neq s_1}^{3} (a_{j_1 j_1 s_1} - a_{j_1 j_1 s_2}) \Delta x_{j_1}.
\end{aligned}$$

The flows at $x_{j_1}^* = a_{j_2 j_1 1} = a_{j_1 j_1 s_1}$ in the x_{j_1}-direction are source and sink flows for $a_{j_1 j_1 0} \prod\nolimits_{s_2 = 1, s_2 \neq s_1}^{3} (a_{j_1 j_1 s_1} - a_{j_1 j_1 s_2}) > 0$ and $a_{j_1 j_1 0} \prod\nolimits_{s_2 = 1, s_2 \neq s_1}^{3} (a_{j_1 j_1 s_1} - a_{j_1 j_1 s_2}) < 0$.

Therefore, the infinite equilibrium of $x_{j_1}^* = a_{j_1 j_1 s_1} = a_{j_2 j_1 1}$ $(s_1 \in \{1, 2, 3\})$ with $\overline{x}_{j_2} \in (-\infty, \infty)$ possesses the following cases as in Eqs. (3.134)–(3.137).

- For $a_{j_1 j_1 0} \prod\nolimits_{s_2 = 1, s_2 \neq s_1}^{3} (a_{j_1 j_1 s_1} - a_{j_1 j_1 s_2}) > 0$ and $a_{j_2 j_1 0} > 0$,

$$(a_{j_1 j_1 s_1}, \overline{x}_{j_2}) = \underbrace{(\text{SO}, \text{UP})}_{\text{up-parabola source}}.$$

  The equilibrium at $x_{j_1}^* = a_{j_1 j_1 s_1} = a_{j_2 j_1 1}$ is an (SO,UP)-up-parabola source infinite-equilibrium.

- For $a_{j_1 j_1 0} \prod\nolimits_{s_2 = 1, s_2 \neq s_1}^{3} (a_{j_1 j_1 s_1} - a_{j_1 j_1 s_2}) < 0$ and $a_{j_2 j_1 0} > 0$,

$$(a_{j_1 j_1 s_1}, \overline{x}_{j_2}) = \underbrace{(\text{SI}, \text{DP})}_{\text{down-parabola sink}}.$$

  The equilibrium at $x_{j_1}^* = a_{j_1 j_1 s_1} = a_{j_2 j_1 1}$ is an (SI,DP)-down-parabola sink infinite-equilibrium.

- For $a_{j_1 j_1 0} \prod\nolimits_{s_2 = 1, s_2 \neq s_1}^{3} (a_{j_1 j_1 s_1} - a_{j_1 j_1 s_2}) > 0$ and $a_{j_2 j_1 0} < 0$,

$$(a_{j_1 j_1 s_1}, \overline{x}_{j_2}) = \underbrace{(\text{SO}, \text{DP})}_{\text{down-parabola source}}.$$

  The equilibrium at $x_{j_1}^* = a_{j_1 j_1 s_1} = a_{j_2 j_1 1}$ is an (SO,DP)-down-parabola source infinite-equilibrium.

- For $a_{j_1 j_1 0} \prod\nolimits_{s_2 = 1, s_2 \neq s_1}^{3} (a_{j_1 j_1 s_1} - a_{j_1 j_1 s_2}) < 0$ and $a_{j_2 j_1 0} < 0$,

$$(a_{j_1 j_1 s_1}, \overline{x}_{j_2}) = \underbrace{(\text{SI}, \text{UP})}_{\text{up-parabola sink}}.$$

  The equilibrium at $x_{j_1}^* = a_{j_1 j_1 s_1} = a_{j_2 j_1 1}$ is an (SI,UP)-up-parabola sink infinite-equilibrium.

– The up-parabola and down-parabola sink, and source infinite-equilibriums are the appearing and switching bifurcations for the up-parabola and down-parabola flows with sink or source flows.

(vii$_{2b}$) For $x_{j_1}^* = a_{j_1j_1s_2} \neq a_{j_2j_11}$ ($s_1, s_2 \in \{1,2,3\}$; $s_1 \neq s_2$), the variational equation is

$$\begin{aligned}\dot{x}_{j_2} &= a_{j_2j_10}(a_{j_1j_1s_2} - a_{j_2j_11})^2,\\ \Delta\dot{x}_{j_1} &= a_{j_1j_10}\prod\nolimits_{s_1=1,s_1\neq s_2}^{3}(a_{j_1j_1s_2} - a_{j_1j_1s_1})\Delta x_{j_1}.\end{aligned}$$

The flows at $x_{j_1}^* = a_{j_1j_1s_2} \neq a_{j_2j_11}$ in the x_{j_1}-direction are source and sink flows for $a_{j_1j_10}\prod_{s_1=1,s_1\neq s_2}^{3}(a_{j_1j_1s_2} - a_{j_1j_1s_1}) > 0$ and $a_{j_1j_10}\prod_{s_1=1,s_1\neq s_2}^{3}(a_{j_1j_1s_2} - a_{j_1j_1s_1}) < 0$, and the flows are positive and negative for $a_{j_2j_10} > 0$ and $a_{j_2j_10} < 0$ in the x_{j_1}-direction.

Therefore, the equilibrium of $x_{j_1}^* = a_{j_1j_1s_2} \neq a_{j_2j_11}$ ($s_1, s_2 \in \{1,2,3\}$; $s_1 \neq s_2$) in the x_{j_1}-direction possesses the following properties as in Eqs. (3.138)–(3.141).

- For $a_{j_1j_10}\prod_{s_1=1,s_1\neq s_2}^{3}(a_{j_1j_1s_2} - a_{j_1j_1s_1}) > 0$ and $a_{j_2j_10}(a_{j_1j_1s_2} - a_{j_2j_11})^2 > 0$,

$$(a_{j_1j_1s_2}, \dot{x}_{j_2}) = \underbrace{(\mathrm{SO}, \mathrm{pF})}_{\text{source flow }(+)}.$$

  The flow at $x_{j_1}^* = a_{j_1j_1s_2}$ is an (SO,pF)-positive source-flow.
- For $a_{j_1j_10}\prod_{s_1=1,s_1\neq s_2}^{3}(a_{j_1j_1s_2} - a_{j_1j_1s_1}) < 0$ and $a_{j_2j_10}(a_{j_1j_1s_2} - a_{j_2j_11})^2 > 0$,

$$(a_{j_1j_1s_2}, \dot{x}_{j_2}) = \underbrace{(\mathrm{SI}, \mathrm{pF})}_{\text{sink flow }(+)}.$$

  The flow at $x_{j_1}^* = a_{j_1j_1s_2}$ is an (SI,pF)-positive sink-flow.
- For $a_{j_1j_10}\prod_{s_1=1,s_1\neq s_2}^{3}(a_{j_1j_1s_2} - a_{j_1j_1s_1}) > 0$ and $a_{j_2j_10}(a_{j_1j_1s_2} - a_{j_2j_11})^2 < 0$,

$$(a_{j_1j_1s_2}, \dot{x}_{j_2}) = \underbrace{(\mathrm{SO}, \mathrm{nF})}_{\text{source flow }(-)}.$$

  The flow at $x_{j_1}^* = a_{j_1j_1s_2}$ is an (SO,nF)-negative source-flow.
- For $a_{j_1j_10}\prod_{s_1=1,s_1\neq s_2}^{3}(a_{j_1j_1s_2} - a_{j_1j_1s_1}) < 0$ and $a_{j_2j_10}(a_{j_1j_1s_2} - a_{j_2j_11})^2 < 0$,

$$(a_{j_1j_1s_2}, \dot{x}_{j_2}) = \underbrace{(\mathrm{SI}, \mathrm{nF})}_{\text{sink flow }(-)}.$$

  The flow at $x_{j_1}^* = a_{j_1j_1s_2}$ is an (SI,nF)-negative sink-flow.

(viii) If

$$\Delta_{j_1j_1} > 0 \text{ and } \Delta_{j_2j_1} > 0,$$

then

$$\begin{aligned}\dot{x}_{j_1} &= a_{j_1j_1 0}(x_{j_1} - a_{j_1j_1 1})(x_{j_1} - a_{j_1j_1 2})(x_{j_1} - a_{j_1j_1 3}),\\ \dot{x}_{j_2} &= a_{j_2j_1 0}(x_{j_1} - a_{j_2j_1 1})(x_{j_1} - a_{j_2j_1 2}),\end{aligned}$$

where

$$\begin{aligned}&\{a_{j_1j_1 1}, a_{j_1j_1 2}, a_{j_1j_1 3}\} = \text{sort}\{b_{j_1j_1 1}, b_{j_1j_1 2}, b_{j_1j_1 3}\}, a_{j_1j_1 1} < a_{j_1j_1 2} < a_{j_1j_1 3};\\ &\{a_{j_2j_1 1}, a_{j_2j_1 2}) = \text{sort}\{b_{j_2j_1 1}, b_{j_2j_1 2}\}, a_{j_2j_1 1} < a_{j_2j_1 2};\\ &b_{j_1j_1 2}, b_{j_1j_1 3} = -\frac{1}{2}(B_{j_1j_1 1} \pm \sqrt{\Delta_{j_1j_1}}),\\ &b_{j_2j_1 1}, b_{j_2j_1 2} = -\frac{1}{2}(B_{j_2j_1 1} \pm \sqrt{\Delta_{j_2j_1}});\\ &j_1, j_2 \in \{1, 2\}; j_1 \neq j_2.\end{aligned}$$

(viii$_1$) For $a_{j_1j_1s_1} \neq a_{j_2j_1l_1}$ (s_1, $s_2 \in \{1, 2, 3\}$; $s_1 \neq s_2$) and (l_1, $l_2 \in \{1, 2\}$; $l_1 \neq l_2$), in phase space,

$$\frac{dx_{j_2}}{dx_{j_1}} = \frac{a_{j_2j_1 0}}{a_{j_1j_1 0}} \sum_{s_1=1}^{3} \frac{\prod_{l_1=1}^{2}(a_{j_1j_1s_1} - a_{j_2j_1l_1})}{\prod_{s_2=1, s_2\neq s_1}^{3}(a_{j_1j_1s_1} - a_{j_1j_1s_2})} \frac{1}{x_{j_1} - a_{j_1j_1s_1}},$$

With an initial condition $(x_{j_1 0}, x_{j_2 0})$, the first integral manifold is

$$x_{j_2} - x_{j_2 0} = \frac{a_{j_2j_1 0}}{a_{j_1j_1 0}} \sum_{s_1=1}^{3} \frac{\prod_{l_1=1}^{2}(a_{j_1j_1s_1} - a_{j_2j_1l_1})}{\prod_{s_2=1, s_2\neq s_1}^{3}(a_{j_1j_1s_1} - a_{j_1j_1s_2})} \ln \frac{|x_{j_1} - a_{j_1j_1s_1}|}{|x_{j_1 0} - a_{j_1j_1s_1}|}.$$

(viii$_{1a}$) For the equilibriums of $x_{j_1}^* = a_{j_1j_1s_1} \neq a_{j_2j_1l_1}, a_{j_2j_1l_2}$ (s_1, $s_2 \in \{1, 2, 3\}$; $s_1 \neq s_2$) and (l_1, $l_2 \in \{1, 2\}$; $l_1 \neq l_2$), the variational equation is

$$\begin{aligned}\Delta\dot{x}_{j_1} &= a_{j_1j_1 0}\prod\nolimits_{s_2=1, s_2\neq s_1}^{3}(a_{j_1j_1s_1} - a_{j_1j_1s_2})\Delta x_{j_1},\\ \dot{x}_{j_2} &= a_{j_2j_1 0}\prod\nolimits_{l_1=1}^{2}(a_{j_1j_1s_1} - a_{j_2j_1l_1}).\end{aligned}$$

The flows at $x_{j_1}^* = a_{j_1j_1s_1}$ are source and sink flows in the x_{j_1}-direction for $a_{j_1j_1 0}\prod_{s_2=1, s_2\neq s_1}^{3}(a_{j_1j_1s_1} - a_{j_1j_1s_2}) > 0$ and $a_{j_1j_1 0}\prod_{s_2=1, s_2\neq s_1}^{3}(a_{j_1j_1s_1} - a_{j_1j_1s_2}) < 0$. The flow at $x_{j_1}^* = a_{j_1j_1s_1}$ are positive and negative in the x_{j_2}-direction for $\dot{x}_{j_2} > 0$ and $\dot{x}_{j_2} < 0$.

Therefore, the equilibriums of $x_{j_1}^* = a_{j_1 j_1 s_1} \neq a_{j_2 j_1 l_1}, a_{j_2 j_1 l_2}$ $(s_1, s_2 \in \{1,2,3\}; s_1 \neq s_2)$ and $(l_1, l_2 \in \{1,2\}; l_1 \neq l_2)$ are the following properties as in Eqs. (3.146)–(3.149).

- For $a_{j_1 j_1 0}\prod_{s_2=1, s_2 \neq s_1}^{3}(a_{j_1 j_1 s_1} - a_{j_1 j_1 s_2}) > 0$ and $a_{j_2 j_1 0}\prod_{l_1=1}^{2}(a_{j_1 j_1 s_1} - a_{j_2 j_1 l_1}) > 0$,

$$(a_{j_1 j_1 s_1}, \dot{x}_{j_2}) = \underbrace{(\text{SO}, \text{pF})}_{\text{source flow } (+)} .$$

The flow at $x_{j_1}^* = a_{j_1 j_1 s_1}$ is an (SO,pF)-positive source-flow.

- For $a_{j_1 j_1 0}\prod_{s_2=1, s_2 \neq s_1}^{3}(a_{j_1 j_1 s_1} - a_{j_1 j_1 s_2}) < 0$ and $a_{j_2 j_1 0}\prod_{l_1=1}^{2}(a_{j_1 j_1 s_1} - a_{j_2 j_1 l_1}) > 0$,

$$(a_{j_1 j_1 s_1}, \dot{x}_{j_2}) = \underbrace{(\text{SI}, \text{pF})}_{\text{sink flow } (+)} .$$

The flow at $x_{j_1}^* = a_{j_1 j_1 s_1}$ is an (SI,pF)-positive sink-flow.

- For $a_{j_1 j_1 0}\prod_{s_2=1, s_2 \neq s_1}^{3}(a_{j_1 j_1 s_1} - a_{j_1 j_1 s_2}) > 0$ and $a_{j_2 j_1 0}\prod_{l_1=1}^{2}(a_{j_1 j_1 s_1} - a_{j_2 j_1 l_1}) < 0$,

$$(a_{j_1 j_1 s_1}, \dot{x}_{j_2}) = \underbrace{(\text{SO}, \text{nF})}_{\text{source flow } (-)} .$$

The flow at $x_{j_1}^* = a_{j_1 j_1 s_1}$ is an (SO,nF)-negative source-flow.

- For $a_{j_1 j_1 0}\prod_{s_2=1, s_2 \neq s_1}^{3}(a_{j_1 j_1 s_1} - a_{j_1 j_1 s_2}) < 0$ and $a_{j_2 j_1 0}\prod_{l_1=1}^{2}(a_{j_1 j_1 s_1} - a_{j_2 j_1 l_1}) < 0$,

$$(a_{j_1 j_1 s_1}, \dot{x}_{j_2}) = \underbrace{(\text{SI}, \text{nF})}_{\text{sink flow } (-)} .$$

The flow at $x_{j_1}^* = a_{j_1 j_1 s_1}$ is an (SI,nF)-negative sink-flow.

(viii$_{1b}$) For the equilibrium of $x_{j_1}^* = a_{j_1 j_1 l_1} \neq a_{j_2 j_1 s_1}, a_{j_2 j_1 s_2}, a_{j_2 j_1 s_3} (l_1, l_2 \in \{1,2\}; l_1 \neq l_2)$ and $(s_1, s_2, s_3 \in \{1,2,3\}; s_1 \neq s_2 \neq s_2)$, we have

$$\frac{dx_{j_2}}{dx_{j_1}}\bigg|_{x_{j_1}^* = a_{j_1 j_1 l_1}} = \frac{a_{j_2 j_1 0}}{a_{j_1 j_1 0}} \frac{(x_{j_1} - a_{j_2 j_1 l_1})(x_{j_1} - a_{j_2 j_1 l_2})}{\prod_{s_1=1}^{3}(x_{j_1} - a_{j_1 j_1 s_1})}\bigg|_{x_{j_1}^* = a_{j_1 j_1 l_1}} = 0.$$

If

$$\frac{d^2 x_{j_2}}{dx_{j_1}^2}\bigg|_{x_{j_1}^* = a_{j_1 j_1 l_1}} = \frac{a_{j_2 j_1 0}}{a_{j_1 j_1 0}} \frac{(a_{j_2 j_1 l_1} - a_{j_2 j_1 l_2})}{\prod_{s_1=1}^{3}(a_{j_2 j_1 l_1} - a_{j_1 j_1 s_1})} > 0,$$

the flow at $x_{j_1}^* = a_{j_1 j_1 l_1}$ is an up-parabola flow, and if

$$\frac{d^2 x_{j_2}}{dx_{j_1}^2}\bigg|_{x_{j_1}^*=a_{j_1 j_1 l_1}} = \frac{a_{j_2 j_1 0}}{a_{j_1 j_1 0}} \frac{(a_{j_2 j_1 l_1} - a_{j_2 j_1 l_2})}{\prod_{s_1=1}^{3}(a_{j_2 j_1 l_1} - a_{j_1 j_1 s_1})} < 0,$$

the flow at $x_{j_1}^* = a_{j_1 j_1 l_1}$ is a down-parabola flow. Let

$$\dot{x}_{j_1} = a_{j_1 j_1 0} \prod_{s_1=1,}^{3} (a_{j_2 j_1 l_1} - a_{j_1 j_1 s_1}).$$

The flows at $x_{j_1}^* = a_{j_1 j_1 l_1}$ are positive and negative for $\dot{x}_{j_1} > 0$ and $\dot{x}_{j_1} < 0$.

Therefore, the equilibrium of $x_{j_1}^* = a_{j_1 j_1 l_1} \neq a_{j_2 j_1 s_1}, a_{j_2 j_1 s_2}, a_{j_2 j_1 s_3}$ (l_1, $l_2 \in \{1,2\}$; $l_1 \neq l_2$) and (s_1, s_2, $s_3 \in \{1,2,3\}$; $s_1 \neq s_2 \neq s_2$) has the following properties as in Eqs. (3.150)–(3.153).

- For $a_{j_1 j_1 0} \prod_{s_1=1}^{3} (a_{j_2 j_1 l_1} - a_{j_1 j_1 s_1}) > 0$ and $a_{j_2 j_1 0}(a_{j_2 j_1 l_1} - a_{j_2 j_1 l_2}) > 0$,

$$(\dot{x}_{j_1}, a_{j_1 j_1 l_1}) = \underbrace{(\text{pF}, \text{UP})}_{\text{up-parabola flow } (+)}.$$

  The flow at $x_{j_1}^* = a_{j_1 j_1 l_1}$ is a (pF,UP)-positive up-parabola flow.

- For $a_{j_1 j_1 0} \prod_{s_1=1}^{3} (a_{j_2 j_1 l_1} - a_{j_1 j_1 s_1}) < 0$ and $a_{j_2 j_1 0}(a_{j_2 j_1 l_1} - a_{j_2 j_1 l_2}) > 0$,

$$(\dot{x}_{j_1}, a_{j_1 j_1 l_1}) = \underbrace{(\text{nF}, \text{DP})}_{\text{down-parabola flow } (-)}.$$

  The flow at $x_{j_1}^* = a_{j_1 j_1 l_1}$ is an (nF,DP)-negative down-parabola flow.

- For $a_{j_1 j_1 0} \prod_{s_1=1}^{3} (a_{j_2 j_1 l_1} - a_{j_1 j_1 s_1}) > 0$ and $a_{j_2 j_1 0}(a_{j_2 j_1 l_1} - a_{j_2 j_1 l_2}) < 0$,

$$(\dot{x}_{j_1}, a_{j_1 j_1 l_1}) = \underbrace{(\text{pF}, \text{DP})}_{\text{down-parabola flow } (+)}.$$

  The flow at $x_{j_1}^* = a_{j_1 j_1 l_1}$ is a (pF,DP)-positive down-parabola flow.

- For $a_{j_1 j_1 0} \prod_{s_1=1}^{3} (a_{j_2 j_1 l_1} - a_{j_1 j_1 s_1}) < 0$ and $a_{j_2 j_1 0}(a_{j_2 j_1 l_1} - a_{j_2 j_1 l_2}) < 0$,

$$(\dot{x}_{j_1}, a_{j_1 j_1 l_1}) = \underbrace{(\text{nF}, \text{UP})}_{\text{up-parabola flow } (-)}.$$

  The flow at $x_{j_1}^* = a_{j_1 j_1 l_1}$ is an (nF,UP)-negative up-parabola flow.

(viii$_2$) For $a_{j_1j_1s_1} = a_{j_2j_1l_1}$ $(s_1, s_2, s_3 \in \{1,2,3\}; s_1 \neq s_2 \neq s_3)$ and $(l_1, l_2 \in \{1,2\}; l_1 \neq l_2)$,

$$\begin{aligned}\dot{x}_{j_1} &= a_{j_1j_10}(x_{j_1} - a_{j_1j_11})(x_{j_1} - a_{j_1j_12})(x_{j_1} - a_{j_1j_13}),\\ \dot{x}_{j_2} &= a_{j_2j_10}(x_{j_1} - a_{j_2j_11})(x_{j_1} - a_{j_2j_12}).\end{aligned}$$

In phase space, we have,

$$\frac{dx_{j_2}}{dx_{j_1}} = \frac{a_{j_2j_10}}{a_{j_1j_10}} \sum_{s_2=1,\, s_2 \neq s_1}^{3} \frac{\prod_{l_2=1, l_2 \neq l_1}^{2} (a_{j_1j_1s_2} - a_{j_2j_1l_2})}{\prod_{s_3=1, s_3 \neq s_1, s_2}^{3} (a_{j_1j_1s_2} - a_{j_1j_1s_3})} \frac{1}{x_{j_1} - a_{j_1j_1s_2}},$$

With an initial condition $(x_{j_1 0}, x_{j_2 0})$, the first integral manifold is

$$x_{j_2} - x_{j_20} = \frac{a_{j_2j_10}}{a_{j_1j_10}} \sum_{s_2=1,\, s_2 \neq s_1}^{3} \frac{\prod_{l_2=1, l_2 \neq l_1}^{2} (a_{j_1j_1s_2} - a_{j_2j_1l_2})}{\prod_{s_3=1, s_3 \neq s_1, s_2}^{3} (a_{j_1j_1s_2} - a_{j_1j_1s_3})} \ln \frac{|\, x_{j_1} - a_{j_1j_1s_2} \,|}{|\, x_{j_1 0} - a_{j_1j_1s_2} \,|}.$$

At $x_{j_1}^* = a_{j_1j_1s_1} = a_{j_2j_1l_1}$, if

$$\left.\frac{dx_{j_2}}{dx_{j_1}}\right|_{x_{j_1}^* = a_{j_1j_1s_1} = a_{j_2j_1l_1}} = \frac{a_{j_2j_10}}{a_{j_1j_10}} \frac{\prod_{l_2=1, l_2 \neq l_1}^{2} (a_{j_2j_1l_1} - a_{j_2j_1l_2})}{\prod_{s_2=1, s_2 \neq s_1}^{3} (a_{j_1j_1s_1} - a_{j_1j_1s_2})} > 0,$$

the flow is an increasing-inflection flow, and if

$$\left.\frac{dx_{j_2}}{dx_{j_1}}\right|_{x_{j_1}^* = a_{j_1j_1s_1} = a_{j_2j_1l_1}} = \frac{a_{j_2j_10}}{a_{j_1j_10}} \frac{\prod_{l_2=1, l_2 \neq l_1}^{2} (a_{j_2j_1l_1} - a_{j_2j_1l_2})}{\prod_{s_2=1, s_2 \neq s_1}^{3} (a_{j_1j_1s_1} - a_{j_1j_1s_2})} < 0,$$

the flow is a decreasing-inflection flow.

At $x_{j_1}^* = a_{j_1j_1s_1} = a_{j_2j_1l_1}$, the variational equation is

$$\begin{aligned}\Delta\dot{x}_{j_1} &= a_{j_1j_10} \prod\nolimits_{s_2=1, s_2 \neq s_1}^{3} (a_{j_1j_1s_1} - a_{j_1j_1s_2}) \Delta x_{j_1},\\ \Delta\dot{x}_{j_2} &= a_{j_2j_10} \prod\nolimits_{l_2=1, l_2 \neq l_1}^{2} (a_{j_1j_1l_1} - a_{j_2j_1l_2}) \Delta x_{j_1}.\end{aligned}$$

the flows at $x_{j_1}^* = a_{j_1j_1s_1} = a_{j_2j_1l_1}$ are source and sink flows in the x_{j_1} - direction for $a_{j_1j_10} \prod_{s_2=1, s_2 \neq s_1}^{3} (a_{j_1j_1s_1} - a_{j_1j_1s_2}) > 0$ and $a_{j_1j_10} \prod_{s_2=1, s_2 \neq s_1}^{3} (a_{j_1j_1s_1} - a_{j_1j_1s_2}) < 0$.

Therefore, the infinite-equilibrium of $x_{j_1}^* = a_{j_1j_1s_1} = a_{j_2j_1l_1}$ with $\overline{x}_{j_2} \in (-\infty, \infty)$ has the following properties as in Eqs. (3.156)–(3.159).

- For $a_{j_1j_10} \prod_{s_2=1, s_2 \neq s_1}^{3} (a_{j_1j_1s_1} - a_{j_1j_1s_2}) > 0$ and $a_{j_2j_10} \prod_{l_2=1, l_2 \neq l_1}^{2} (a_{j_2j_1l_1} - a_{j_2j_1l_2}) > 0$,

$$(a_{j_1j_1s_1}, \bar{x}_{j_2}) = \underbrace{(\text{SO}, \text{II})}_{\text{increasing-inflection source}}.$$

The equilibrium at $x_{j_1}^* = a_{j_1j_1s_1} = a_{j_2j_1l_1}$ is an (SO,II)-increasing-inflection source infinite-equilibrium.

- For $a_{j_1j_10}\prod_{s_2=1,s_2\neq s_1}^{3}(a_{j_1j_1s_1} - a_{j_1j_1s_2}) < 0$ and $a_{j_2j_10}\prod_{l_2=1,l_2\neq l_1}^{2}(a_{j_2j_1l_1} - a_{j_2j_1l_2}) > 0$,

$$(a_{j_1j_1s_1}, \bar{x}_{j_2}) = \underbrace{(\text{SI}, \text{DI})}_{\text{decreasing-inflection sink}}.$$

The equilibrium at $x_{j_1}^* = a_{j_1j_1s_1} = a_{j_2j_1l_1}$ is an (SI,DI)-decreasing-inflection sink infinite-equilibrium.

- For $a_{j_1j_10}\prod_{s_2=1,s_2\neq s_1}^{3}(a_{j_1j_1s_1} - a_{j_1j_1s_2}) > 0$ and $a_{j_2j_10}\prod_{l_2=1,l_2\neq l_1}^{2}(a_{j_2j_1l_1} - a_{j_2j_1l_2}) < 0$,

$$(a_{j_1j_1s_1}, \bar{x}_{j_2}) = \underbrace{(\text{SO}, \text{DI})}_{\text{decreasing-inflection source}}.$$

The equilibrium at $x_{j_1}^* = a_{j_1j_1s_1} = a_{j_2j_1l_1}$ is an (SO,DI)-increasing-inflection source infinite-equilibrium.

- For $a_{j_1j_10}\prod_{s_2=1,s_2\neq s_1}^{3}(a_{j_1j_1s_1} - a_{j_1j_1s_2}) < 0$ and $a_{j_2j_10}\prod_{l_2=1,l_2\neq l_1}^{2}(a_{j_2j_1l_1} - a_{j_2j_1l_2}) < 0$,

$$(a_{j_1j_1s_1}, \bar{x}_{j_2}) = \underbrace{(\text{SI}, \text{II})}_{\text{increasing-inflection sink}}.$$

The equilibrium at $x_{j_1}^* = a_{j_1j_1s_1} = a_{j_2j_1l_1}$ is an (SI,II)-increasing-inflection sink infinite-equilibrium.

(ix) If

$$\Delta_{j_1j_1} < 0 \text{ and } \Delta_{j_2j_1} > 0,$$

then

$$\begin{aligned}
\dot{x}_{j_2} &= a_{j_2j_10}(x_{j_1} - a_{j_2j_11})(x_{j_1} - a_{j_2j_12}),\\
\dot{x}_{j_1} &= a_{j_1j_10}(x_{j_1} - a_{j_1j_11})\left[(x_{j_1} - a_{j_1j_1})^2 + b_{j_1j_1}\right],
\end{aligned}$$

with

$$
\begin{aligned}
&a_{j_1j_1} = -\frac{1}{2}B_{j_1j_1 1}, b_{j_1j_1} = \frac{1}{4}(-\Delta_{j_1j_1}), a_{j_1j_1 1} = b_{j_1j_1 1};\\
&a_{j_2j_1 1}, a_{j_2j_1 2} = \text{sort}\{b_{j_2j_1 1}, b_{j_2j_1 2}\}, a_{j_2j_1 1} < a_{j_2j_1 2};\\
&b_{j_2j_1 1}, b_{j_2j_1 2} = -\frac{1}{2}(B_{j_2j_1 1} \pm \sqrt{\Delta_{j_2j_1}}).
\end{aligned}
$$

(ix_1) For $a_{j_1j_1 1} \neq a_{j_2j_1 1}$, we have

$$
\begin{aligned}
\frac{dx_{j_2}}{dx_{j_1}} &= \frac{a_{j_2j_1 0}}{a_{j_1j_1 0}} \frac{(x_{j_1} - a_{j_2j_1 1})(x_{j_1} - a_{j_2j_1 2})}{(x_{j_1} - a_{j_1j_1 1})\left[(x_{j_1} - a_{j_1j_1})^2 + b_{j_1j_1}\right]}\\
&= \frac{a_{j_2j_1 0}}{a_{j_1j_1 0}} \left[\frac{(x_{j_1} - a_{j_1j_1})}{(x_{j_1} - a_{j_1j_1})^2 + b_{j_1j_1}} + \frac{a_{j_1j_1} + a_{j_1j_1 1} - a_{j_2j_1 1} - a_{j_2j_1 2}}{(x_{j_1} - a_{j_1j_1})^2 + b_{j_1j_1}} \right.\\
&\quad + \frac{(a_{j_1j_1 1} - a_{j_2j_1 1})(a_{j_1j_1 1} - a_{j_2j_1 2})}{(a_{j_1j_1 1} - a_{j_1j_1})^2 + b_{j_1j_1}} \left[\frac{1}{x_{j_1} - a_{j_1j_1 1}} - \frac{(x_{j_1} - a_{j_1j_1})}{(x_{j_1} - a_{j_1j_1})^2 + b_{j_1j_1}} \right.\\
&\quad \left.\left. - \frac{a_{j_1j_1 1} - a_{j_1j_1}}{(x_{j_1} - a_{j_1j_1})^2 + b_{j_1j_1}} \right]\right].
\end{aligned}
$$

With an initial condition $(x_{j_1 0}, x_{j_2 0})$, the first integral manifold is

$$
\begin{aligned}
x_{j_2} - x_{j_2 0} = &\frac{a_{j_2j_1 0}}{a_{j_1j_1 0}} \left\{ \frac{1}{2} \ln \frac{|(x_{j_1} - a_{j_1j_1})^2 + b_{j_1j_1}|}{|(x_{j_1 0} - a_{j_1j_1})^2 + b_{j_1j_1}|} \right.\\
&+ (a_{j_1j_1 1} + a_{j_1j_1} - a_{j_2j_1 1} - a_{j_2j_1 2})\\
&\times \frac{1}{\sqrt{b_{j_1j_1}}} (\arctan \frac{x_{j_1} - a_{j_1j_1}}{\sqrt{b_{j_1j_1}}} - \arctan \frac{x_{j_1 0} - a_{j_1j_1}}{\sqrt{b_{j_1j_1}}})\\
&+ \frac{(a_{j_1j_1 1} - a_{j_2j_1 1})(a_{j_1j_1 1} - a_{j_2j_1 2})}{(a_{j_1j_1 1} - a_{j_1j_1})^2 + b_{j_1j_1}} \left[\ln \frac{|x_{j_1} - a_{j_1j_1 1}|}{|x_{j_1 0} - a_{j_1j_1 1}|} \right.\\
&- \frac{1}{2} \ln \frac{|(x_{j_1} - a_{j_1j_1})^2 + b_{j_1j_1}|}{|(x_{j_1 0} - a_{j_1j_1})^2 + b_{j_1j_1}|}\\
&\left.\left. - \frac{a_{j_1j_1 1} - a_{j_1j_1}}{\sqrt{b_{j_1j_1}}} (\arctan \frac{x_{j_1} - a_{j_1j_1}}{\sqrt{b_{j_1j_1}}} - \arctan \frac{x_{j_1 0} - a_{j_1j_1}}{\sqrt{b_{j_1j_1}}}) \right] \right\},\\
&j_1, j_2 \in \{1, 2\}, j_1 \neq j_2.
\end{aligned}
$$

(ix_{1a}) For $x_{j_1}^* = a_{j_1j_1 1} \neq a_{j_2j_1 1}$, the variational equation is

$$
\begin{aligned}
\dot{x}_{j_2} &= a_{j_2j_1 0}(a_{j_1j_1 1} - a_{j_2j_1 1})(a_{j_1j_1 1} - a_{j_2j_1 2}),\\
\Delta\dot{x}_{j_1} &= a_{j_1j_1 0}\left[(a_{j_1j_1 1} - a_{j_1j_1})^2 + b_{j_1j_1}\right]\Delta x_{j_1}.
\end{aligned}
$$

Due to

$$(a_{j_1j_1 1}-a_{j_1j_1})^2+b_{j_1j_1}>0,$$

the flows at $x_{j_1}^*=a_{j_1j_1 1}$ are source and sink flows for $a_{j_1j_1 0}>0$ and $a_{j_1j_1 0}<0$ in the x_{j_1}-direction, and the flows at $x_{j_1}^*=a_{j_1j_1 1}$ are positive and negative for $a_{j_2j_1 0}>0$ and $a_{j_2j_1 0}<0$ in the x_{j_2}-direction.

Therefore, the flow at $x_{j_1}^*=a_{j_1j_1 1}\neq a_{j_2j_1 1}$ for the system has the following properties as in Eqs. (3.163)–(3.166) .

- For $a_{j_1j_1 0}>0$ and $a_{j_2j_1 0}(a_{j_1j_1 1}-a_{j_2j_1 1})(a_{j_1j_1 1}-a_{j_2j_1 2})>0$,

$$(a_{j_1j_1 1},\dot{x}_{j_2})=\underbrace{(\text{SO},\text{pF})}_{\text{source flow }(+)}.$$

The (SO,pF)-flow is a positive source flow at $x_{j_1}^*=a_{j_1j_1 1}$.

- For $a_{j_1j_1 0}<0$ and $a_{j_2j_1 0}(a_{j_1j_1 1}-a_{j_2j_1 1})(a_{j_1j_1 1}-a_{j_2j_1 2})>0$,

$$(a_{j_1j_1 1},\dot{x}_{j_2})=\underbrace{(\text{SI},\text{pF})}_{\text{sink flow }(+)}.$$

The (SI, pF)-flow is a positive sink flow at $x_{j_1}^*=a_{j_1j_1 1}$.

- For $a_{j_1j_1 0}>0$ and $a_{j_2j_1 0}(a_{j_1j_1 1}-a_{j_2j_1 1})(a_{j_1j_1 1}-a_{j_2j_1 2})<0$,

$$(a_{j_1j_1 1},\dot{x}_{j_2})=\underbrace{(\text{SO},\text{nF})}_{\text{source flow }(-)}.$$

The (SO, nF)-flow is a negative source flow at $x_{j_1}^*=a_{j_1j_1 1}$.

- For $a_{j_1j_1 0}<0$ and $a_{j_2j_1 0}(a_{j_1j_1 1}-a_{j_2j_1 1})(a_{j_1j_1 1}-a_{j_2j_1 2})<0$,

$$(a_{j_1j_1 1},\dot{x}_{j_2})=\underbrace{(\text{SI},\text{nF})}_{\text{sink flow }(-)}.$$

The (SI,nF)-flow is a negative sink flow at $x_{j_1}^*=a_{j_1j_1 1}$.

(ix$_{1b}$) For $x_{j_1}^*=a_{j_2j_1 l_1}$, we have

$$\frac{dx_{j_2}}{dx_{j_1}}\bigg|_{x_{j_1}^*=a_{j_2j_1 l_1}}=\frac{a_{j_2j_1 0}}{a_{j_1j_1 0}}\frac{(x_{j_1}-a_{j_2j_1 l_1})(x_{j_1}-a_{j_2j_1 l_2})}{(x_{j_1}-a_{j_1j_1 1})\left[(x_{j_1}-a_{j_1j_1})^2+b_{j_1j_1}\right]}\bigg|_{x_{j_1}^*=a_{j_2j_1 l_1}}=0.$$

If

$$\frac{d^2x_{j_2}}{dx_{j_1}^2}\bigg|_{x_{j_1}^*=a_{j_2j_1l_1}} = \frac{a_{j_2j_1 0}(a_{j_2j_1l_1}-a_{j_2j_1l_2})}{a_{j_1j_1 0}(a_{j_2j_1l_1}-a_{j_1j_1 1})\left[(a_{j_2j_1l_1}-a_{j_1j_1})^2+b_{j_1j_1}\right]} > 0,$$

the flow at $x_{j_1}^*=a_{j_2j_1l_1}$ is an up-parabola flow in the x_{j_2}-direction, and if

$$\frac{d^2x_{j_2}}{dx_{j_1}^2}\bigg|_{x_{j_1}^*=a_{j_2j_1l_1}} = \frac{a_{j_2j_1 0}(a_{j_2j_1l_1}-a_{j_2j_1l_2})}{a_{j_1j_1 0}(a_{j_2j_1l_1}-a_{j_1j_1 1})\left[(a_{j_2j_1l_1}-a_{j_1j_1})^2+b_{j_1j_1}\right]} < 0,$$

the flow at $x_{j_1}^*=a_{j_2j_1l_1}$ is a down-parabola flow in the x_{j_2}-direction.

Let

$$\dot{x}_{j_1} = a_{j_1j_1 0}(a_{j_2j_1l_1}-a_{j_1j_1 1})\left[(a_{j_2j_1l_1}-a_{j_1j_1})^2+b_{j_1j_1}\right].$$

Due to

$$(a_{j_2j_1l_1}-a_{j_1j_1})^2+b_{j_1j_1} > 0,$$

the flows at $x_{j_1}^*=a_{j_2j_1l_1}$ are positive and negative for $a_{j_1j_1 0}(a_{j_2j_1l_1}-a_{j_1j_1 1})>0$ and $a_{j_1j_1 0}(a_{j_2j_1l_1}-a_{j_1j_1 1})<0$ in the x_{j_1}-direction.

Therefore, the flow at $x_{j_1}^*=a_{j_2j_1l_1}$ ($l_1, l_2 \in \{1,2\}$; $l_1 \neq l_2$) for the system has the following properties as in Eqs. (3.167)–(3.170)

- For $a_{j_1j_1 0}(a_{j_2j_1l_1}-a_{j_1j_1 1})>0$ and $a_{j_2j_1 0}(a_{j_2j_1l_1}-a_{j_2j_1l_2})>0$,

$$(\dot{x}_{j_1}, a_{j_2j_1l_1}) = \underbrace{(\text{pF, UP})}_{\text{up-parabola flow }(+)}.$$

  The (pF,UP)-flow at $x_{j_1}^*=a_{j_2j_1l_1}$ is a positive up-parabola flow.
- For $a_{j_1j_1 0}(a_{j_2j_1l_1}-a_{j_1j_1 1})<0$ and $a_{j_2j_1 0}(a_{j_2j_1l_1}-a_{j_2j_1l_2})>0$,

$$(\dot{x}_{j_1}, a_{j_2j_1l_1}) = \underbrace{(\text{nF, DP})}_{\text{down-parabola flow }(-)}.$$

  The (nF,DP)-flow at $x_{j_1}^*=a_{j_2j_1l_1}$ is a negative down-parabola flow.
- For $a_{j_1j_1 0}(a_{j_2j_1l_1}-a_{j_1j_1 1})>0$ and $a_{j_2j_1 0}(a_{j_2j_1l_1}\quad a_{j_2j_1l_2})<0$,

$$(\dot{x}_{j_1}, a_{j_2j_1l_1}) = \underbrace{(\text{pF, DP})}_{\text{down-parabola flow }(+)}.$$

  The (pF, DP)-flow at $x_{j_1}^*=a_{j_2j_1l_1}$ is a positive down-parabola flow.
- For $a_{j_1j_1 0}(a_{j_2j_1l_1}-a_{j_1j_1 1})<0$ and $a_{j_2j_1 0}(a_{j_2j_1l_1}-a_{j_2j_1l_2})<0$,

$$(\dot{x}_{j_1}, a_{j_2 j_1 l_1}) = \underbrace{(\text{nF}, \text{UP})}_{\text{up-parabola flow }(-)}.$$

The (nF,UP)-flow at $x_{j_1}^* = a_{j_2 j_1 l_1}$ is a negative up-parabola flow.

(ix$_2$) For $a_{j_1 j_1 1} = a_{j_2 j_1 l_1}$, we have

$$\begin{aligned}\frac{dx_{j_2}}{dx_{j_1}} &= \frac{a_{j_2 j_1 0}}{a_{j_1 j_1 0}} \frac{(x_{j_1} - a_{j_2 j_1 l_1})(x_{j_1} - a_{j_2 j_1 l_2})}{(x_{j_1} - a_{j_1 j_1 1})\left[(x_{j_1} - a_{j_1 j_1})^2 + b_{j_1 j_1}\right]} \\ &= \frac{a_{j_2 j_1 0}}{a_{j_1 j_1 0}} \left[\frac{(x_{j_1} - a_{j_1 j_1})}{(x_{j_1} - a_{j_1 j_1})^2 + b_{j_1 j_1}} + \frac{a_{j_1 j_1} - a_{j_1 j_1 l_2}}{(x_{j_1} - a_{j_1 j_1})^2 + b_{j_1 j_1}}\right].\end{aligned}$$

With an initial condition $(x_{j_1 0}, x_{j_2 0})$, the first integral manifold is

$$\begin{aligned}x_{j_2} - x_{j_2 0} = {} & \frac{a_{j_2 j_1 0}}{a_{j_1 j_1 0}} \left[\frac{1}{2} \ln \frac{|(x_{j_1} - a_{j_1 j_1})^2 + b_{j_1 j_1}|}{|(x_{j_1 0} - a_{j_1 j_1})^2 + b_{j_1 j_1}|}\right. \\ & \left. + \frac{(a_{j_1 j_1} - a_{j_1 j_1 l_2})}{\sqrt{b_{j_1 j_1}}} \left(\arctan \frac{x_{j_1} - a_{j_1 j_1}}{\sqrt{b_{j_1 j_1}}} - \arctan \frac{x_{j_1 0} - a_{j_1 j_1}}{\sqrt{b_{j_1 j_1}}}\right)\right]; \\ & j_1, j_2 \in \{1, 2\}, j_1 \neq j_2.\end{aligned}$$

In phase space, for $x_{j_1}^* = a_{j_1 j_1 1} = a_{j_2 j_1 l_1}$, if

$$\left.\frac{dx_{j_2}}{dx_{j_1}}\right|_{x_{j_1}^* = a_{j_1 j_1 1} = a_{j_2 j_1 l_1}} = \frac{a_{j_2 j_1 0}}{a_{j_1 j_1 0}} \frac{(a_{j_2 j_1 l_1} - a_{j_2 j_1 l_2})}{\left[(x_{j_1} - a_{j_1 j_1})^2 + b_{j_1 j_1}\right]} > 0,$$

the flow at $x_{j_1}^* = a_{j_1 j_1 1} = a_{j_2 j_1 l_1}$ is an increasing-flow flow in the x_{j_2}-direction, and if

$$\left.\frac{dx_{j_2}}{dx_{j_1}}\right|_{x_{j_1}^* = a_{j_1 j_1 1} = a_{j_2 j_1 l_1}} = \frac{a_{j_2 j_1 0}}{a_{j_1 j_1 0}} \frac{(a_{j_2 j_1 l_1} - a_{j_2 j_1 l_2})}{\left[(x_{j_1} - a_{j_1 j_1})^2 + b_{j_1 j_1}\right]} > 0,$$

the flow at $x_{j_1}^* = a_{j_1 j_1 1} = a_{j_2 j_1 l_1}$ is a decreasing flow in the x_{j_2}-direction.

At $x_{j_1}^* = a_{j_1 j_1 1} = a_{j_2 j_1 l_1}$, the variational equation is

$$\begin{aligned}\Delta \dot{x}_{j_2} &= a_{j_2 j_1 0}(a_{j_2 j_1 l_1} - a_{j_1 j_1 l_2}) \Delta x_{j_1}, \\ \Delta \dot{x}_{j_1} &= a_{j_1 j_1 0}\left[(a_{j_2 j_1 l_1} - a_{j_1 j_1})^2 + b_{j_1 j_1}\right] \Delta x_{j_1}.\end{aligned}$$

Due to

$$(a_{j_2j_1l_1} - a_{j_1j_1})^2 + b_{j_1j_1} > 0,$$

The flows at $x^*_{j_1} = a_{j_1j_1 1} = a_{j_2j_1l_1}$ are source and sink for $a_{j_1j_1 0} > 0$ and $a_{j_1j_1 0} < 0$ in the x_{j_1}-direction, respectively.

Thus, the infinite-equilibrium of $x^*_{j_1} = a_{j_1j_1 1} = a_{j_2j_1l_1}$ with $\bar{x}_{j_2} \in (-\infty, \infty)$ has the following properties as in Eqs. (3.172)–(3.175).

- For $a_{j_1j_1 0} > 0$ and $a_{j_2j_1 0}(a_{j_2j_1l_1} - a_{j_2j_1l_2}) > 0$,

$$(a_{j_1j_1 1}, \bar{x}_{j_2}) = \underbrace{(\text{SO, II})}_{\text{increasing-inflection source}}.$$

  The (SO,II)-equilibrium at $x^*_{j_1} = a_{j_1j_1 1} = a_{j_2j_1l_1}$ is an increasing-inflection source infinite-equilibrium.
- For $a_{j_1j_1 0} < 0$ and $a_{j_2j_1 0}(a_{j_2j_1l_1} - a_{j_2j_1l_2}) > 0$,

$$(a_{j_1j_1 1}, \bar{x}_{j_2}) = \underbrace{(\text{SI, DI})}_{\text{decreasing-inflection sink}}.$$

  The (SI,DI)-equilibrium at $x^*_{j_1} = a_{j_1j_1 1} = a_{j_2j_1l_1}$ is a decreasing-inflection sink infinite-equilibrium.
- For $a_{j_1j_1 0} > 0$ and $a_{j_2j_1 0}(a_{j_2j_1l_1} - a_{j_2j_1l_2}) < 0$,

$$(a_{j_1j_1 1}, \bar{x}_{j_2}) = \underbrace{(\text{SO, DI})}_{\text{decreasing-inflection source}}.$$

  The (SO,DI)-equilibrium at $x^*_{j_1} = a_{j_1j_1 1} = a_{j_2j_1l_1}$ is an decreasing-inflection source infinite-equilibrium.
- For $a_{j_1j_1 0} < 0$ and $a_{j_2j_1 0}(a_{j_2j_1l_1} - a_{j_2j_1l_2}) < 0$,

$$(a_{j_1j_1 1}, \bar{x}_{j_2}) = \underbrace{(\text{SI, II})}_{\text{increasing-inflection sink}}.$$

  The (SI,II)-equilibrium at $x^*_{j_1} = a_{j_1j_1 1} = a_{j_2j_1l_1}$ is an increasing-inflection sink infinite-equilibrium.

– The inflection source and sink infinite-equilibriums are the switching bifurcations of a source (sink) flow and a parabola flow.

(x) If

$$\Delta_{j_1j_1} = 0, \Delta_{j_2j_1} = 0,$$

then

$$\begin{aligned}
&\dot{x}_{j_1} = a_{j_1j_10}(x_{j_1} - a_{j_1j_1s_1})(x_{j_1} - a_{j_1j_1s_2})^2,\\
&\dot{x}_{j_2} = a_{j_2j_10}(x_{j_1} - a_{j_2j_11})(x_{j_1} - a_{j_2j_12}),\\
&j_1, j_2 \in \{1,2\}, j_1 \neq j_2; s_1, s_2 \in \{1,2\}, s_1 \neq s_2,
\end{aligned}$$

where

$$\begin{aligned}
&\{a_{j_1j_11}, a_{j_1j_12}\} = \mathrm{sort}\{b_{j_1j_11}, b_{j_1j_12}\}, a_{j_1j_11} < a_{j_1j_12};\\
&\{a_{j_2j_11}, a_{j_2j_12}\} = \mathrm{sort}\{b_{j_2j_11}, b_{j_2j_12}\}, a_{j_2j_11} < a_{j_2j_12}\\
&b_{j_1j_12} = -\frac{1}{2}B_{j_1j_11}; b_{j_2j_11}, b_{j_2j_12} = -\frac{1}{2}(B_{j_2j_11} \pm \sqrt{\Delta_{j_2j_1}});\\
&j_1, j_2 \in \{1,2\}, j_1 \neq j_2.
\end{aligned}$$

(x$_1$) For $a_{j_1j_1s_1} \neq a_{j_2j_11} \neq a_{j_1j_1s_2}$, we have

$$\begin{aligned}
\frac{dx_{j_2}}{dx_{j_1}} &= \frac{a_{j_2j_10}}{a_{j_1j_10}} \frac{(x_{j_1} - a_{j_2j_11})(x_{j_1} - a_{j_2j_12})}{(x_{j_1} - a_{j_1j_1s_1})(x_{j_1} - a_{j_1j_1s_2})^2}\\
&= \frac{a_{j_2j_10}}{a_{j_1j_10}} \Big\{ \frac{\prod_{l_1=1}^{2}(a_{j_1j_1s_1} - a_{j_2j_1l_1})}{(a_{j_1j_1s_1} - a_{j_1j_1s_2})^2} \frac{1}{x_{j_1} - a_{j_1j_11}}\\
&\quad + \frac{\prod_{l_1=1}^{2}(a_{j_1j_1s_2} - a_{j_2j_1l_1})}{(a_{j_1j_1s_2} - a_{j_1j_1s_1})} \frac{1}{(x_{j_1} - a_{j_1j_1s_2})^2}\\
&\quad + \Big[\frac{\sum_{l_1=1}^{2}(a_{j_1j_1s_2} - a_{j_2j_1l_1})}{(a_{j_1j_1s_2} - a_{j_1j_1s_1})} - \frac{\prod_{l_1=1}^{2}(a_{j_1j_1s_2} - a_{j_2j_1l_1})}{(a_{j_1j_1s_2} - a_{j_1j_1s_1})^2}\Big] \frac{1}{x_{j_1} - a_{j_1j_1s_2}} \Big\}.
\end{aligned}$$

With an initial condition (x_{j_10}, x_{j_20}), the first integral manifold is

$$\begin{aligned}
x_{j_2} - x_{j_20} &= \frac{a_{j_2j_10}}{a_{j_1j_10}} \Big\{ \frac{\prod_{l_1=1}^{2}(a_{j_1j_1s_1} - a_{j_2j_1l_1})}{(a_{j_1j_1s_1} - a_{j_1j_1s_2})^2} \ln \frac{|x_{j_1} - a_{j_1j_1s_1}|}{|x_{j_10} - a_{j_1j_1s_1}|}\\
&\quad - \frac{\prod_{l_1=1}^{2}(a_{j_1j_1s_2} - a_{j_2j_1l_1})}{(a_{j_1j_1s_2} - a_{j_1j_1s_1})} \Big(\frac{1}{x_{j_1} - a_{j_1j_1s_2}} - \frac{1}{x_{j_10} - a_{j_1j_1s_2}}\Big)\\
&\quad + \Big[\frac{\sum_{l_1=1}^{2}(a_{j_1j_1s_2} - a_{j_2j_1l_1})}{(a_{j_1j_1s_2} - a_{j_1j_1s_1})} - \frac{\prod_{l_1=1}^{2}(a_{j_1j_1s_2} - a_{j_2j_1l_1})}{(a_{j_1j_1s_2} - a_{j_1j_1s_1})^2}\Big] \ln \frac{|x_{j_1} - a_{j_1j_1s_2}|}{|x_{j_10} - a_{j_1j_1s_2}|} \Big\},\\
&j_1, j_2 \in \{1,2\}, j_1 \neq j_2; s_1, s_2 \in \{1,2\}, s_1 \neq s_2.
\end{aligned}$$

(x_{1a}) For $x^*_{j_1} = a_{j_1 j_1 s_1} \neq a_{j_2 j_1 1}$, the variational equation is

$$\dot{x}_{j_2} = a_{j_2 j_1 0} \prod\nolimits_{l_1=1}^{2} (a_{j_1 j_1 s_1} - a_{j_2 j_1 l_1}),$$
$$\Delta \dot{x}_{j_1} = a_{j_1 j_1 0} (a_{j_1 j_1 s_1} - a_{j_1 j_1 s_2})^2 \Delta x_{j_1}.$$

Due to

$$(a_{j_1 j_1 s_1} - a_{j_1 j_1 s_2})^2 > 0,$$

the flows at $x^*_{j_1} = a_{j_1 j_1 s_1} \neq a_{j_2 j_1 1}$ are source and sink flows for $a_{j_1 j_1 0} > 0$ and $a_{j_1 j_1 0} < 0$ in the x_{j_1}-direction, and the flows at $x^*_{j_1} = a_{j_1 j_1 s_1} \neq a_{j_2 j_1 1}$ are positive and negative for $a_{j_2 j_1 0} > 0$ and $a_{j_2 j_1 0} < 0$ in the x_{j_2}-direction.

Therefore, the equilibrium of $x^*_{j_1} = a_{j_1 j_1 s_1} \neq a_{j_2 j_1 l_1}$ ($l_1 \in \{1, 2\}$) possesses the following properties as in Eqs. (3.180)–(3.183).

- For $a_{j_1 j_1 0} (a_{j_1 j_1 s_1} - a_{j_1 j_1 s_2})^2 > 0$ and $a_{j_2 j_1 0} \prod_{l_1=1}^{2} (a_{j_1 j_1 s_1} - a_{j_2 j_1 l_1}) > 0$,

$$(a_{j_1 j_1 s_1}, \dot{x}_{j_2}) = \underbrace{(\text{SO}, \text{pF})}_{\text{source flow } (+)}.$$

  The flow at $x^*_{j_1} = a_{j_1 j_1 s_1} \neq a_{j_2 j_1 l_1}$ is an (SO,pF)-positive source flow.
- For $a_{j_1 j_1 0} (a_{j_1 j_1 s_1} - a_{j_1 j_1 s_2})^2 < 0$ and $a_{j_2 j_1 0} \prod_{l_1=1}^{2} (a_{j_1 j_1 s_1} - a_{j_2 j_1 l_1}) > 0$,

$$(a_{j_1 j_1 s_1}, \dot{x}_{j_2}) = \underbrace{(\text{SI}, \text{pF})}_{\text{sink flow } (+)}.$$

  The flow at $x^*_{j_1} = a_{j_1 j_1 s_1} \neq a_{j_2 j_1 l_1}$ is an (SI,pF)-positive sink flow.
- For $a_{j_1 j_1 0} (a_{j_1 j_1 s_1} - a_{j_1 j_1 s_2})^2 > 0$ and $a_{j_2 j_1 0} \prod_{l_1=1}^{2} (a_{j_1 j_1 s_1} - a_{j_2 j_1 l_1}) < 0$,

$$(a_{j_1 j_1 s_1}, \dot{x}_{j_2}) = \underbrace{(\text{SO}, \text{nF})}_{\text{source flow } (-)}.$$

  The flow at $x^*_{j_1} = a_{j_1 j_1 s_1} \neq a_{j_2 j_1 l_1}$ is an (SO,nF)-negative source flow.
- For $a_{j_1 j_1 0} (a_{j_1 j_1 s_1} - a_{j_1 j_1 s_2})^2 < 0$ and $a_{j_2 j_1 0} \prod_{l_1=1}^{2} (a_{j_1 j_1 s_1} - a_{j_2 j_1 l_1}) < 0$,

$$(a_{j_1 j_1 s_1}, \dot{x}_{j_2}) = \underbrace{(\text{SI}, \text{nF})}_{\text{sink flow } (-)}.$$

  The flow at $x^*_{j_1} = a_{j_1 j_1 s_1} \neq a_{j_2 j_1 l_1}$ is an (SI,pF)-negative sink flow.

($\mathrm{x}_{1\mathrm{b}}$) For $x_{j_1}^* = a_{j_1 j_1 s_2} \neq a_{j_2 j_1 l_1}$ ($l_1 \in \{1,2\}$, the variational equation is

$$\dot{x}_{j_2} = a_{j_2 j_1 0} \prod\nolimits_{l_1=1}^{2} (a_{j_1 j_1 s_2} - a_{j_2 j_1 l_1}),$$
$$\Delta \dot{x}_{j_1} = a_{j_1 j_1 0} (a_{j_1 j_1 s_2} - a_{j_1 j_1 s_1}) (\Delta x_{j_1})^2.$$

The flows at $x_{j_1}^* = a_{j_1 j_1 s_2} \neq a_{j_2 j_1 l_1}$ ($l_1 \in \{1,2\}$ are upper-saddle and lower-saddle flows for $a_{j_1 j_1 0}(a_{j_1 j_1 s_2} - a_{j_1 j_1 s_1}) > 0$ and $a_{j_1 j_1 0}(a_{j_1 j_1 s_2} - a_{j_1 j_1 s_1}) < 0$ in the x_{j_1}-direction; and the flows at $x_{j_1}^* = a_{j_1 j_1 s_2} \neq a_{j_2 j_1 l_1}$ are positive and negative for $\dot{x}_{j_2} > 0$ and $\dot{x}_{j_2} < 0$ in the x_{j_2}-direction.

Thus, the equilibrium of $x_{j_1}^* = a_{j_1 j_1 s_2} \neq a_{j_2 j_1 l_1}$ ($l_1 \in \{1,2\}$) possesses the following properties as in Eqs. (3.184)–(3.187).

- For $a_{j_1 j_1 0}(a_{j_1 j_1 s_2} - a_{j_1 j_1 s_1}) > 0$ and $a_{j_2 j_1 0} \prod_{l_1=1}^{2} (a_{j_1 j_1 s_2} - a_{j_2 j_1 l_1}) > 0$,

$$(a_{j_1 j_1 s_2}, \dot{x}_{j_2}) = \underbrace{(\mathrm{US}, \mathrm{pF})}_{\text{upper-saddle flow } (+)}.$$

  The flow at $x_{j_1}^* = a_{j_1 j_1 s_2} \neq a_{j_2 j_1 l_1}$ is a (US,pF)-positive upper-saddle flow.
- For $a_{j_1 j_1 0}(a_{j_1 j_1 s_2} - a_{j_1 j_1 s_1}) < 0$ and $a_{j_2 j_1 0} \prod_{l_1=1}^{2} (a_{j_1 j_1 s_2} - a_{j_2 j_1 l_1}) > 0$,

$$(a_{j_1 j_1 s_2}, \dot{x}_{j_2}) = \underbrace{(\mathrm{LS}, \mathrm{pF})}_{\text{lower-saddle flow } (+)}.$$

  The flow at $x_{j_1}^* = a_{j_1 j_1 s_2} \neq a_{j_2 j_1 l_1}$ is an (LS,pF)-positive lower-saddle flow.
- For $a_{j_1 j_1 0}(a_{j_1 j_1 s_2} - a_{j_1 j_1 s_1}) > 0$ and $a_{j_2 j_1 0} \prod_{l_1=1}^{2} (a_{j_1 j_1 s_2} - a_{j_2 j_1 l_1}) < 0$,

$$(a_{j_1 j_1 s_2}, \dot{x}_{j_2}) = \underbrace{(\mathrm{US}, \mathrm{nF})}_{\text{upper-saddle flow } (-)}.$$

  The flow at $x_{j_1}^* = a_{j_1 j_1 s_2} \neq a_{j_2 j_1 l_1}$ is a (US,nF)-negative upper-saddle flow.
- For $a_{j_1 j_1 0}(a_{j_1 j_1 s_2} - a_{j_1 j_1 s_1}) < 0$ and $a_{j_2 j_1 0} \prod_{l_1=1}^{2} (a_{j_1 j_1 s_2} - a_{j_2 j_1 l_1}) < 0$,

$$(a_{j_1 j_1 s_2}, \dot{x}_{j_2}) = \underbrace{(\mathrm{LS}, \mathrm{nF})}_{\text{lower-saddle flow } (-)}.$$

  The flow at $x_{j_1}^* = a_{j_1 j_1 s_2} \neq a_{j_2 j_1 l_1}$ is an (LS,nF)-negative lower-saddle flow.

– The lower-saddle and upper-saddle flows are the appearing-switching bifurcation for the sink and source flows.

(x_{1c}) In phase space, for $x_{j_1}^* = a_{j_2 j_1 l_1} \neq a_{j_1 j_1 s_1}, a_{j_1 j_1 s_2}$ ($l_1 \in \{1,2\}$), we have

$$\frac{dx_{j_2}}{dx_{j_1}}\Big|_{x_{j_1}^* = a_{j_2 j_1 l_1}} = \frac{a_{j_2 j_1 0}}{a_{j_1 j_1 0}} \frac{(x_{j_1} - a_{j_2 j_1 l_1})(x_{j_1} - a_{j_2 j_1 l_2})}{(x_{j_1} - a_{j_1 j_1 s_1})(x_{j_1} - a_{j_1 j_1 s_2})^2}\Big|_{x_{j_1}^* = a_{j_2 j_1 l_1}} = 0.$$

If

$$\frac{d^2 x_{j_2}}{dx_{j_1}^2}\Big|_{x_{j_1}^* = a_{j_2 j_1 l_1}} = \frac{a_{j_2 j_1 0}}{a_{j_1 j_1 0}} \frac{(a_{j_2 j_1 l_1} - a_{j_2 j_1 l_2})}{(a_{j_2 j_1 l_1} - a_{j_1 j_1 s_1})(a_{j_2 j_1 l_1} - a_{j_1 j_1 s_2})^2} > 0,$$

the flow at $x_{j_1}^* = a_{j_2 j_1 l_1} \neq a_{j_1 j_1 s_1}, a_{j_1 j_1 s_2}$ is an up-parabola flow in the x_{j_2}-direction, and if

$$\frac{d^2 x_{j_2}}{dx_{j_1}^2}\Big|_{x_{j_1}^* = a_{j_2 j_1 l_1}} = \frac{a_{j_2 j_1 0}}{a_{j_1 j_1 0}} \frac{(a_{j_2 j_1 l_1} - a_{j_2 j_1 l_2})}{(a_{j_2 j_1 l_1} - a_{j_1 j_1 s_1})(a_{j_2 j_1 l_1} - a_{j_1 j_1 s_2})^2} < 0,$$

the flow at $x_{j_1}^* = a_{j_2 j_1 l_1} \neq a_{j_1 j_1 s_1}, a_{j_1 j_1 s_2}$ is a down-parabola flow in the x_{j_2}-direction.

At $x_{j_1}^* = a_{j_2 j_1 l_1} \neq a_{j_1 j_1 s_1}, a_{j_1 j_1 s_2}$, let

$$\dot{x}_{j_1} = a_{j_1 j_1 0}(a_{j_2 j_1 l_1} - a_{j_1 j_1 s_1})(a_{j_2 j_1 l_1} - a_{j_1 j_1 s_2})^2.$$

The flows at $x_{j_1}^* = a_{j_2 j_1 l_1} \neq a_{j_1 j_1 s_1}, a_{j_1 j_1 s_2}$ are positive and negative for $\dot{x}_{j_1} > 0$ and $\dot{x}_{j_1} < 0$ in the x_{j_1}-direction.

Therefore, the equilibrium of $x_{j_1}^* = a_{j_2 j_1 l_1} \neq a_{j_1 j_1 s_1}, a_{j_1 j_1 s_2}$ possesses the following properties as in Eqs. (3.188)–(3.191).

- For $a_{j_1 j_1 0}(a_{j_2 j_1 l_1} - a_{j_1 j_1 s_1})(a_{j_2 j_1 l_1} - a_{j_1 j_1 s_2})^2 > 0$ and $a_{j_2 j_1 0}(a_{j_2 j_1 l_1} - a_{j_2 j_1 l_2}) > 0$,

$$(\dot{x}_{j_1}, a_{j_2 j_1 l_1}) = \underbrace{(\text{pF}, \text{UP})}_{\text{up-parabola flow }(+)}.$$

  The flow at $x_{j_1}^* = a_{j_2 j_1 l_1}$ is a (pF,UP)-positive up-parabola flow.
- For $a_{j_1 j_1 0}(a_{j_2 j_1 l_1} - a_{j_1 j_1 s_1})(a_{j_2 j_1 l_1} - a_{j_1 j_1 s_2})^2 < 0$ and $a_{j_2 j_1 0}(a_{j_2 j_1 l_1} - a_{j_2 j_1 l_2}) > 0$,

$$(\dot{x}_{j_1}, a_{j_2 j_1 l_1}) = \underbrace{(\text{nF}, \text{DP})}_{\text{down-parabola flow }(-)}.$$

  The flow at $x_{j_1}^* = a_{j_2 j_1 l_1}$ is an (nF,DP)-negative down-parabola flow.
- For $a_{j_1 j_1 0}(a_{j_2 j_1 l_1} - a_{j_1 j_1 s_1})(a_{j_2 j_1 l_1} - a_{j_1 j_1 s_2})^2 > 0$ and $a_{j_2 j_1 0}(a_{j_2 j_1 l_1} - a_{j_2 j_1 l_2}) < 0$,

$$(\dot{x}_{j_1}, a_{j_2 j_1 l_1}) = \underbrace{(\mathrm{pF}, \mathrm{DP})}_{\text{down-parabola flow }(+)}.$$

The flow at $x_{j_1}^* = a_{j_2 j_1 l_1}$ is a (pF,DP)-positive down-parabola flow.

- For $a_{j_1 j_1 0}(a_{j_2 j_1 1} - a_{j_1 j_1 s_1})(a_{j_2 j_1 1} - a_{j_1 j_1 s_2})^2 < 0$ and $a_{j_2 j_1 0}(a_{j_2 j_1 l_1} - a_{j_2 j_1 l_2}) < 0$,

$$(\dot{x}_{j_1}, a_{j_2 j_1 l_1}) = \underbrace{(\mathrm{nF}, \mathrm{UP})}_{\text{up-parabola flow }(-)}.$$

The flow at $x_{j_1}^* = a_{j_2 j_1 l_1}$ is an (nF, UP)-positive up-parabola flow.

(x_2) For $a_{j_1 j_1 s_2} = a_{j_2 j_1 l_1} (l_1 \in \{1, 2\})$,

$$\begin{aligned} &\dot{x}_{j_1} = a_{j_1 j_1 0}(x_{j_1} - a_{j_1 j_1 s_1})(x_{j_1} - a_{j_1 j_1 s_2})^2, \\ &\dot{x}_{j_2} = a_{j_2 j_1 0}(x_{j_1} - a_{j_2 j_1 l_1})(x_{j_1} - a_{j_2 j_1 l_2}), \\ &j_1, j_2 \in \{1, 2\}, j_1 \neq j_2. \end{aligned}$$

In phase space, we have

$$\begin{aligned} \frac{dx_{j_2}}{dx_{j_1}} &= \frac{a_{j_2 j_1 0}}{a_{j_1 j_1 0}} \frac{(x_{j_1} - a_{j_2 j_1 l_2})}{(x_{j_1} - a_{j_1 j_1 s_1})(x_{j_1} - a_{j_1 j_1 s_2})} \\ &= \frac{a_{j_2 j_1 0}}{a_{j_1 j_1 0}} \Big(\frac{a_{j_1 j_1 s_1} - a_{j_2 j_1 l_2}}{a_{j_1 j_1 s_1} - a_{j_1 j_1 s_2}} \frac{1}{x_{j_1} - a_{j_1 j_1 s_1}} + \frac{a_{j_1 j_1 s_2} - a_{j_2 j_1 l_2}}{a_{j_1 j_1 s_2} - a_{j_1 j_1 s_1}} \frac{1}{x_{j_1} - a_{j_1 j_1 s_1}} \Big). \end{aligned}$$

With an initial condition $(x_{j_1 0}, x_{j_2 0})$, the first integral manifold is

$$\begin{aligned} x_{j_2} - x_{j_2 0} = &\frac{a_{j_2 j_1 0}}{a_{j_1 j_1 0}} \Big(\frac{a_{j_1 j_1 s_1} - a_{j_2 j_1 l_2}}{a_{j_1 j_1 s_1} - a_{j_1 j_1 s_2}} \ln \frac{|x_{j_1} - a_{j_1 j_1 s_1}|}{|x_{j_1 0} - a_{j_1 j_1 s_1}|}. \\ &+ \frac{a_{j_1 j_1 s_2} - a_{j_2 j_1 l_2}}{a_{j_1 j_1 s_2} - a_{j_1 j_1 s_1}} \ln \frac{|x_{j_1} - a_{j_1 j_1 s_1}|}{|x_{j_1 0} - a_{j_1 j_1 s_1}|} \Big), \\ &j_1, j_2 \in \{1, 2\}, j_1 \neq j_2. \end{aligned}$$

In phase space, for $x_{j_1}^* = a_{j_1 j_1 s_2} = a_{j_2 j_1 l_1}$,

$$\frac{dx_{j_2}}{dx_{j_1}} \Big|_{x_{j_1}^* = a_{j_2 j_1 l_1} \pm \varepsilon} = \pm \frac{a_{j_2 j_1 0}}{a_{j_1 j_1 0}} \frac{(a_{j_2 j_1 l_1} - a_{j_2 j_1 l_2})}{(x_{j_1} - a_{j_1 j_1 s_1}) \varepsilon}.$$

If

$$\frac{a_{j_2j_10}}{a_{j_1j_10}}\frac{(a_{j_2j_1l_1}-a_{j_2j_1l_2})}{(x_{j_1}-a_{j_1j_1s_1})}>0,$$

the flow at $x^*_{j_1}=a_{j_1j_1s_2}=a_{j_2j_1l_1}$ is a down-up asymptotic flow in the x_{j_2}-direction, and if

$$\frac{a_{j_2j_10}}{a_{j_1j_10}}\frac{(a_{j_2j_1l_1}-a_{j_2j_1l_2})}{(x_{j_1}-a_{j_1j_1s_1})}<0,$$

the flow at $x^*_{j_1}=a_{j_1j_1s_2}=a_{j_2j_1l_1}$ is a up-down asymptotic flow in the x_{j_2}-direction.

At $x^*_{j_1}=a_{j_1j_1s_2}=a_{j_2j_1l_1}$, the variational equations are

$$\begin{aligned}\Delta\dot{x}_{j_2}&=a_{j_2j_10}(a_{j_2j_1l_1}-a_{j_2j_1l_2})\Delta x_{j_1},\\ \Delta\dot{x}_{j_1}&=a_{j_1j_10}(a_{j_2j_1s_2}-a_{j_1j_1s_1})(\Delta x_{j_1})^2.\end{aligned}$$

The flows at $x^*_{j_1}=a_{j_1j_1s_2}=a_{j_2j_1l_1}$ are upper-saddle and lower-saddle flows in the x_{j_1}-direction for $a_{j_1j_10}(a_{j_2j_1s_2}-a_{j_1j_1s_1})>0$ and $a_{j_1j_10}(a_{j_2j_1s_2}-a_{j_1j_1s_1})<0$.

Therefore, the infinite-equilibrium of $x^*_{j_1}=a_{j_1j_1s_2}=a_{j_2j_1l_1}$ with $\bar{x}_{j_2}\in(-\infty,\infty)$ has the following properties as in Eqs. (3.194)–(3.197).

- For $a_{j_1j_10}(a_{j_1j_1s_2}-a_{j_1j_1s_1})>0$ and $a_{j_2j_10}(a_{j_2j_1l_1}-a_{j_2j_1l_2})>0$,

$$(a_{j_1j_1s_2},\bar{x}_{j_2})=\underbrace{(\mathrm{US},\mathrm{DU})}_{\text{down-up upper-saddle}}.$$

The equilibrium at $x^*_{j_1}=a_{j_1j_1s_2}=a_{j_2j_1l_1}$ is a (US,DU)-down-up upper-saddle infinite-equilibrium.

- For $a_{j_1j_10}(a_{j_1j_1s_2}-a_{j_1j_1s_1})<0$ and $a_{j_2j_10}(a_{j_2j_1l_1}-a_{j_2j_1l_2})>0$,

$$(a_{j_1j_1s_2},\bar{x}_{j_2})=\underbrace{(\mathrm{LS},\mathrm{UD})}_{\text{up-down lower-saddle}}.$$

The equilibrium at $x^*_{j_1}=a_{j_1j_1s_2}=a_{j_2j_1l_1}$ is an (LS,UD)-up-down lower-saddle infinite-equilibrium.

- For $a_{j_1j_10}(a_{j_1j_1s_2}-a_{j_1j_1s_1})>0$ and $a_{j_2j_10}(a_{j_2j_1l_1}-a_{j_2j_1l_2})<0$,

$$(a_{j_1j_1s_2},\bar{x}_{j_2})=\underbrace{(\mathrm{US},\mathrm{UD})}_{\text{up-down upper-saddle}}.$$

The equilibrium at $x^*_{j_1}=a_{j_1j_1s_2}=a_{j_2j_1l_1}$ is a (US,UD)-up-down upper-saddle infinite-equilibrium.

- For $a_{j_1j_10}(a_{j_1j_1s_2}-a_{j_1j_1s_1})<0$ and $a_{j_2j_10}<0$,

$$(a_{j_1j_1s_2},\bar{x}_{j_2})=\underbrace{(\mathrm{LS},\mathrm{DU})}_{\text{down-up lower-saddle}}.$$

The equilibrium at $x_{j_1}^*=a_{j_1j_1s_2}=a_{j_2j_1l_1}$ is an (LS,DU)-down-up lower-saddle infinite-equilibrium.

– The down-up upper-saddle (lower-saddle) infinite-equilibriums are the switching bifurcations for an up-parabola flow with an upper-saddle (lower-saddle) flow.
– The up-down upper-saddle (lower-saddle) infinite-equilibriums are the switching bifurcations for a down-parabola flow with an upper-saddle (lower-saddle) flow.

(x$_3$) For $a_{j_1j_1s_1}=a_{j_2j_1l_1}$,

$$\begin{aligned}&\dot{x}_{j_1}=a_{j_1j_10}(x_{j_1}-a_{j_1j_1s_1})(x_{j_1}-a_{j_1j_1s_2})^2,\\&\dot{x}_{j_2}=a_{j_2j_10}(x_{j_1}-a_{j_2j_1l_1})(x_{j_1}-a_{j_2j_1l_2}),\\&j_1,j_2\in\{1,2\},j_1\neq j_2.\end{aligned}$$

In phase space, we have

$$\begin{aligned}\frac{dx_{j_2}}{dx_{j_1}}&=\frac{a_{j_2j_10}}{a_{j_1j_10}}\frac{(x_{j_1}-a_{j_2j_1l_2})}{(x_{j_1}-a_{j_1j_1s_2})^2}\\&=\frac{a_{j_2j_10}}{a_{j_1j_10}}\left[\frac{1}{x_{j_1}-a_{j_1j_1s_2}}+\frac{(a_{j_1j_1s_2}-a_{j_2j_1l_2})}{(x_{j_1}-a_{j_1j_1s_2})^2}\right].\end{aligned}$$

With an initial condition (x_{j_10},x_{j_20}),the first integral manifold is

$$\begin{aligned}x_{j_2}-x_{j_20}=&\frac{a_{j_2j_10}}{a_{j_1j_10}}\left[\ln\frac{|x_{j_1}-a_{j_1j_1s_2}|}{|x_{j_10}-a_{j_1j_1s_2}|}\right.\\&\left.-(a_{j_1j_1s_2}-a_{j_2j_1l_2})(\frac{1}{x_{j_1}-a_{j_1j_1s_2}}-\frac{1}{x_{j_10}-a_{j_1j_1s_2}})\right],\\&j_1,j_2\in\{1,2\},j_1\neq j_2.\end{aligned}$$

(x$_{3a}$) For $x_{j_1}^*=a_{j_1j_1s_1}=a_{j_2j_1l_1}$, if

$$\left.\frac{dx_{j_2}}{dx_{j_1}}\right|_{x_{j_1}^*=a_{j_2j_1l_1}}=\frac{a_{j_2j_10}}{a_{j_1j_10}}\frac{(a_{j_2j_1l_2}-a_{j_2j_1l_2})}{(a_{j_1j_1s_1}-a_{j_1j_1s_2})^2}>0,$$

the flow at $x_{j_1}^*=a_{j_1j_1s_1}=a_{j_2j_1l_1}$ is an increasing-inflection flow in the x_{j_2}-direction, and if

$$\frac{dx_{j_2}}{dx_{j_1}}\bigg|_{x^*_{j_1}=a_{j_2j_1l_1}} = \frac{a_{j_2j_10}}{a_{j_1j_10}} \frac{(a_{j_2j_1l_2}-a_{j_2j_1l_2})}{(a_{j_1j_1s_1}-a_{j_1j_1s_2})^2} < 0,$$

the flow at $x^*_{j_1}=a_{j_1j_1s_1}=a_{j_2j_1l_1}$ is a decreasing-inflection flow in the x_{j_2}-direction.

At $x^*_{j_1}=a_{j_1j_1s_1}=a_{j_2j_1l_1}$, the variational equations are

$$\Delta\dot{x}_{j_2} = a_{j_2j_10}(a_{j_2j_1l_1}-a_{j_1j_1l_2})\Delta x_{j_1},$$
$$\Delta\dot{x}_{j_1} = a_{j_1j_10}(a_{j_2j_1l_1}-a_{j_1j_1s_2})^2\Delta x_{j_1}.$$

The flows at $x^*_{j_1}=a_{j_1j_1s_1}=a_{j_2j_1l_1}$ are source and sink flows for $a_{j_1j_10}>0$ and $a_{j_1j_10}<0$ in the x_{j_1}-direction.

Therefore, the infinite-equilibrium of $x^*_{j_1}=a_{j_1j_1s_1}=a_{j_2j_1l_1}$ with $\bar{x}_{j_2}\in(-\infty,\infty)$ has the following properties as in Eqs. (3.200)–(3.203).

- For $a_{j_1j_10}(a_{j_1j_1s_1}-a_{j_1j_1s_2})^2>0$ and $a_{j_2j_10}(a_{j_2j_1l_1}-a_{j_1j_1l_2})>0$,

$$(a_{j_1j_1s_1},\bar{x}_{j_2}) = \underbrace{(\text{SO, II})}_{\text{increasing-inflection source}}.$$

  The equilibrium at $x^*_{j_1}=a_{j_1j_1s_1}=a_{j_2j_1l_1}$ is an (SO,II)-increasing-inflection source infinite-equilibrium.

- For $a_{j_1j_10}(a_{j_1j_1s_1}-a_{j_1j_1s_2})^2<0$ and $a_{j_2j_10}(a_{j_2j_1l_1}-a_{j_1j_1l_2})>0$,

$$(a_{j_1j_1s_1},\bar{x}_{j_2}) = \underbrace{(\text{SI, DI})}_{\text{decreasing-inflection sink}}.$$

  The equilibrium at $x^*_{j_1}=a_{j_1j_1s_1}=a_{j_2j_1l_1}$ is an (SI,DI)-decreasing-inflection sink infinite-equilibrium.

- For $a_{j_1j_10}(a_{j_1j_1s_1}-a_{j_1j_1s_2})^2>0$ and $a_{j_2j_10}(a_{j_2j_1l_1}-a_{j_1j_1l_2})<0$,

$$(a_{j_1j_1s_1},\bar{x}_{j_2}) = \underbrace{(\text{SO, DI})}_{\text{decreasing-inflection source}}.$$

  The equilibrium at $x^*_{j_1}=a_{j_1j_1s_1}=a_{j_2j_1l_1}$ is an (SO,DI)-decreasing-inflection source infinite-equilibrium.

- For $a_{j_1j_10}(a_{j_1j_1s_1}-a_{j_1j_1s_2})^2<0$ and $a_{j_2j_10}(a_{j_2j_1l_1}-a_{j_1j_1l_2})<0$,

$$(a_{j_1j_1s_1},\bar{x}_{j_2}) = \underbrace{(\text{SI, II})}_{\text{increasing-inflection sink}}.$$

  The equilibrium at $x^*_{j_1}=a_{j_1j_1s_1}=a_{j_2j_1l_1}$ is an (SI,II)-increasing-inflection sink infinite-equilibrium.

- The inflection source (sink) infinite-equilibriums are the switching bifurcation for a source (sink) flow with a parabola flow.

(xi) If

$$\Delta_{j_1j_1}=0 \text{ and } \Delta_{j_2j_1}>0,$$
$$a_{j_1j_1s_1}=a_{j_1j_1s_2}\equiv a_{j_1j_1 1}$$

then

$$\dot{x}_{j_1}=a_{j_1j_1 0}(x_{j_1}-a_{j_1j_1 1})^3,$$
$$\dot{x}_{j_2}=a_{j_2j_1 0}(x_{j_1}-a_{j_2j_1 1})(x_{j_1}-a_{j_2j_1 2}),$$
$$j_1,j_2\in\{1,2\},j_1\neq j_2.$$

(xi$_1$) For $a_{j_1j_1 1}\neq a_{j_2j_1 1}$, we have

$$\frac{dx_{j_2}}{dx_{j_1}}=\frac{a_{j_2j_1 0}}{a_{j_1j_1 0}}\frac{(x_{j_1}-a_{j_2j_1 1})(x_{j_1}-a_{j_2j_1 2})}{(x_{j_1}-a_{j_1j_1 1})^3}$$
$$=\frac{a_{j_2j_1 0}}{a_{j_1j_1 0}}\left[\frac{1}{x_{j_1}-a_{j_1j_1 1}}+\frac{\sum_{l_1=1}^{2}(a_{j_1j_1 1}-a_{j_2j_1l_1})}{(x_{j_1}-a_{j_1j_1s_2})^2}+\frac{\prod_{l_1=1}^{2}(a_{j_1j_1 1}-a_{j_2j_1l_1})}{(x_{j_1}-a_{j_1j_1 1})^3}\right].$$

With an initial condition $(x_{j_1 0},x_{j_2 0})$,the first integral manifold is

$$x_{j_2}-x_{j_2 0}=\frac{a_{j_2j_1 0}}{a_{j_1j_1 0}}\left\{\ln\frac{|x_{j_1}-a_{j_1j_1 1}|}{|x_{j_1 0}-a_{j_1j_1 1}|}\right.$$
$$-\sum_{l_1=1}^{2}(a_{j_1j_1 1}-a_{j_2j_1l_1})\left(\frac{1}{x_{j_1}-a_{j_1j_1 1}}-\frac{1}{x_{j_1 0}-a_{j_1j_1 1}}\right)$$
$$\left.-\frac{1}{2}\prod_{l_1=1}^{2}(a_{j_1j_1 1}-a_{j_2j_1l_1})\left[\frac{1}{(x_{j_1}-a_{j_1j_1 1})^2}-\frac{1}{(x_{j_1 0}-a_{j_1j_1 1})^2}\right]\right\},$$
$$j_1,j_2\in\{1,2\},j_1\neq j_2.$$

(xi$_{1a}$) For $x_{j_1}^*=a_{j_1j_1 1}\neq a_{j_2j_1l_1}(l_1\in\{1,2\})$, the variational equation is

$$\dot{x}_{j_2}=a_{j_2j_1 0}\prod_{l_1=1}^{2}(a_{j_1j_1 1}-a_{j_2j_1l_1}),$$
$$\Delta\dot{x}_{j_1}=a_{j_1j_1 0}(\Delta x_{j_1})^3.$$

The flows at $x_{j_1}^*=a_{j_1j_1 1}\neq a_{j_2j_1 1}$ are the third-order source and sink flows for $a_{j_1j_1 0}>0$ and $a_{j_1j_1 0}<0$ in the x_{j_1}-direction. The flows at $x_{j_1}^*=a_{j_1j_1 1}\neq a_{j_2j_1 1}$ are positive and negative for $a_{j_2j_1 0}>0$ and $a_{j_2j_1 0}<0$ in the x_{j_2}-direction.

Therefore, the equilibrium of $x_{j_1}^* = a_{j_1j_1 1} \neq a_{j_2j_1 1}$ has the following properties as in Eqs. (3.207)–(3.210).

- For $a_{j_1j_1 0} > 0$ and $a_{j_2j_1 0}\prod_{l_1=1}^{2}(a_{j_1j_1 1} - a_{j_2j_1 l_1}) > 0$,

$$(a_{j_1j_1 1}, \dot{x}_{j_2}) = \underbrace{(3^{\text{rd}}\text{SO, pF})}_{\text{third-order source flow } (+)}.$$

The flow at $x_{j_1}^* = a_{j_1j_1 1}$ is a (3^{rd}SO,pF)-positive third-order source flow.

- For $a_{j_1j_1 0} < 0$ and $a_{j_2j_1 0}\prod_{l_1=1}^{2}(a_{j_1j_1 1} - a_{j_2j_1 l_1}) > 0$,

$$(a_{j_1j_1 1}, \dot{x}_{j_2}) = \underbrace{(3^{\text{rd}}\text{SI, pF})}_{\text{third-order sink flow } (+)}.$$

The flow at $x_{j_1}^* = a_{j_1j_1 1}$ is a (3^{rd}SI,pF)-positive third-order sink flow.

- For $a_{j_1j_1 0} > 0$ and $a_{j_2j_1 0}\prod_{l_1=1}^{2}(a_{j_1j_1 1} - a_{j_2j_1 l_1}) < 0$,

$$(a_{j_1j_1 1}, \dot{x}_{j_2}) = \underbrace{(3^{\text{rd}}\text{SO, nF})}_{\text{third-order source flow } (-)}.$$

The flow at $x_{j_1}^* = a_{j_1j_1 1}$ is a (3^{rd}SO,nF)-negative third-order source flow.

- For $a_{j_1j_1 0} < 0$ and $a_{j_2j_1 0}\prod_{l_1=1}^{2}(a_{j_1j_1 1} - a_{j_2j_1 l_1}) < 0$,

$$(a_{j_1j_1 1}, \dot{x}_{j_2}) = \underbrace{(3^{\text{rd}}\text{SI, nF})}_{\text{third-order sink flow } (-)}.$$

The flow at $x_{j_1}^* = a_{j_1j_1 1}$ is a (3^{rd}SI,pF)-negative third-order sink flow.

– The third-order source and sink flows bifurcations are the appearing bifurcations for source-sink-source and sink-source-sink flows.

($\text{xi}_{1\text{b}}$) For $x_{j_1}^* = a_{j_2j_1 l_1} \neq a_{j_1j_1 1}$, we have

$$\frac{dx_{j_2}}{dx_{j_1}}\bigg|_{x_{j_1}^* = a_{j_2j_1 l_1}} = \frac{a_{j_2j_1 0}}{a_{j_1j_1 0}} \frac{(x_{j_1} - a_{j_2j_1 l_1})(x_{j_1} - a_{j_2j_1 l_2})}{(x_{j_1} - a_{j_1j_1 1})^3}\bigg|_{x_{j_1}^* = a_{j_2j_1 l_1}} = 0.$$

If

$$\frac{d^2x_{j_2}}{dx_{j_1}^2}\bigg|_{x_{j_1}^* = a_{j_2j_1 l_1}} = \frac{a_{j_2j_1 0}}{a_{j_1j_1 0}} \frac{(a_{j_2j_1 l_1} - a_{j_2j_1 l_2})}{(a_{j_2j_1 l_1} - a_{j_1j_1 1})^3} > 0,$$

the flow at $x_{j_1}^* = a_{j_2j_1 l_1} \neq a_{j_1j_1 1}$ is an up-parabola flow in the x_{j_2}-direction, and if

$$\frac{d^2 x_{j_2}}{dx_{j_1}^2}\bigg|_{x_{j_1}^*=a_{j_2j_1l_1}} = \frac{a_{j_2j_10}}{a_{j_1j_10}} \frac{(a_{j_2j_1l_1} - a_{j_2j_1l_2})}{(a_{j_2j_1l_1} - a_{j_1j_11})^3} < 0,$$

the flow at $x_{j_1}^* = a_{j_2j_1l_1} \neq a_{j_1j_11}$ is a down-parabola flow in the x_{j_2}-direction.

At $x_{j_1}^* = a_{j_1j_1s_1} = a_{j_2j_11}$, the variational equation is

$$\Delta\dot{x}_{j_2} = a_{j_2j_10}(a_{j_2j_1l_1} - a_{j_2j_1l_2})\Delta x_{j_1},$$
$$\dot{x}_{j_1} = a_{j_1j_10}(a_{j_2j_1l_1} - a_{j_1j_11})^3.$$

The flows at $x_{j_1}^* = a_{j_1j_1s_1} = a_{j_2j_11}$ are positive and negative for $\dot{x}_{j_1} > 0$ and $\dot{x}_{j_1} < 0$ in the x_{j_1}-direction.

Therefore, the equilibrium of $x_{j_1}^* = a_{j_2j_1l_1} \neq a_{j_1j_11}$ has the following properties as in Eqs. (3.211)–(3.214).

- For $a_{j_1j_10}(a_{j_2j_1l_1} - a_{j_1j_11})^3 > 0$ and $a_{j_2j_10}(a_{j_2j_1l_1} - a_{j_2j_1l_2}) > 0$,

$$(\dot{x}_{j_1}, a_{j_2j_1l_1}) = \underbrace{(\text{pF, UP})}_{\text{up-parabola flow }(+)}.$$

  The flow at $x_{j_1}^* = a_{j_2j_1l_1}$ is a (pF,UP)-positive up-parabola flow.
- For $a_{j_1j_10}(a_{j_2j_1l_1} - a_{j_1j_11})^3 < 0$ and $a_{j_2j_10}(a_{j_2j_1l_1} - a_{j_2j_1l_2}) > 0$,

$$(\dot{x}_{j_1}, a_{j_2j_1l_1}) = \underbrace{(\text{nF, DP})}_{\text{down-parabola flow }(-)}.$$

  The flow at $x_{j_1}^* = a_{j_2j_1l_1}$ is an (nF,DI)-negative down-parabola flow.
- For $a_{j_1j_10}(a_{j_2j_1l_1} - a_{j_1j_11})^3 > 0$ and $a_{j_2j_10}(a_{j_2j_1l_1} - a_{j_2j_1l_2}) < 0$,

$$(\dot{x}_{j_1}, a_{j_2j_1l_1}) = \underbrace{(\text{pF, DP})}_{\text{down-parabola flow }(+)}.$$

  The flow at $x_{j_1}^* = a_{j_2j_1l_1}$ is a (pF,DP)-positive down-parabola flow.
- For $a_{j_1j_10}(a_{j_2j_1l_1} - a_{j_1j_11})^3 < 0$ and $a_{j_2j_10}(a_{j_2j_1l_1} - a_{j_2j_1l_2}) < 0$,

$$(\dot{x}_{j_1}, a_{j_2j_1l_1}) = \underbrace{(\text{nF, UP})}_{\text{up-parabola flow }(-)}.$$

  The flow at $x_{j_1}^* = a_{j_2j_1l_1}$ is an (nF,UP)-negative up-parabola flow.

(xi$_2$) For $a_{j_2j_1l_1}=a_{j_1j_1 1}$,

$$\begin{aligned}&\dot{x}_{j_1}=a_{j_1j_10}(x_{j_1}-a_{j_1j_11})^3,\\&\dot{x}_{j_2}=a_{j_2j_10}(x_{j_1}-a_{j_2j_1l_1})(x_{j_1}-a_{j_2j_1l_2}),\\&j_1,j_2\in\{1,2\},j_1\neq j_2.\end{aligned}$$

we have

$$\frac{dx_{j_2}}{dx_{j_1}}=\frac{a_{j_2j_10}}{a_{j_1j_10}}\frac{(x_{j_1}-a_{j_2j_1l_2})}{(x_{j_1}-a_{j_1j_11})^2}.$$

With an initial condition (x_{j_10},x_{j_20}), the first integral manifold is

$$\begin{aligned}x_{j_2}-x_{j_20}=&\frac{a_{j_2j_10}}{a_{j_1j_10}}\Bigg[\ln\frac{|x_{j_1}-a_{j_1j_11}|}{|x_{j_10}-a_{j_1j_11}|}\\&-(a_{j_2j_1l_1}-a_{j_2j_1l_2})(\frac{1}{x_{j_1}-a_{j_2j_11}}-\frac{1}{x_{j_10}-a_{j_2j_11}})\Bigg],\\&j_1,j_2\in\{1,2\},j_1\neq j_2.\end{aligned}$$

At $x_{j_1}^*=a_{j_2j_1l_1}=a_{j_1j_11}$, the variational equation is

$$\begin{aligned}&\Delta\dot{x}_{j_1}=a_{j_1j_10}(\Delta x_{j_1})^3,\\&\Delta\dot{x}_{j_2}=a_{j_2j_10}(a_{j_2j_1l_1}-a_{j_2j_1l_2})\Delta x_{j_1}.\end{aligned}$$

From the first integral manifold, the flow at the equilibrium at $x_{j_1}^*=a_{j_2j_1l_1}=a_{j_1j_11}$ is a logarithmic flow. The flow $x_{j_1}^*=a_{j_2j_1l_1}=a_{j_1j_11}$ is the third-order source and sink flows in the x_{j_1} direction. The singularity at the equilibrium is of the first-order and third-order in the x_{j_2}-direction and the x_{j_1}-direction. Thus, (3:1) is used to identify the singularity of the concave asymptotic flows.

Therefore, the infinite-equilibrium of $x_{j_1}^*=a_{j_2j_1l_1}=a_{j_1j_11}$ with $\bar{x}_{j_2}\in(-\infty,\infty)$ has the following properties as in Eqs. (3.217)–(3.220).

- For $a_{j_1j_10}>0$ and $a_{j_2j_10}(a_{j_2j_1l_1}-a_{j_2j_1l_2})>0$,

$$(a_{j_1j_11},\bar{x}_{j_2})=\underbrace{(3^{\text{rd}}\text{SO},(3{:}1)\text{IC})}_{\text{third-order increasing-concave source}}.$$

The equilibrium at $x_{j_1}^*=a_{j_2j_1l_1}=a_{j_1j_11}$ is a (3rdSO,(3:1)IC)-third-order increasing-concave source infinite-equilibrium.

- For $a_{j_1j_10}<0$ and $a_{j_2j_10}(a_{j_2j_1l_1}-a_{j_2j_1l_2})>0$,

$$(a_{j_1j_1 1},\bar{x}_{j_2}) = \underbrace{(3^{\text{rd}}\text{SI}, (3:1)\text{DC})}_{\text{third-order decreasing-concave sink}}.$$

The equilibrium at $x_{j_1}^* = a_{j_2j_1l_1} = a_{j_1j_1 1}$ is a (3^{rd}SI,(3:1)DC)-third-order decreasing-concave sink infinite-equilibrium.

- For $a_{j_1j_1 0} > 0$ and $a_{j_2j_1 0}(a_{j_2j_1l_1} - a_{j_2j_1l_2}) > 0$,

$$(a_{j_1j_1 1},\bar{x}_{j_2}) = \underbrace{(3^{\text{rd}}\text{SO}, (3:1)\text{DC})}_{\text{third-order decreasing-concave source}}.$$

The equilibrium at $x_{j_1}^* = a_{j_2j_1l_1} = a_{j_1j_1 1}$ is a (3^{rd}SO,(3:1)DC)-third-order decreasing-concave source infinite-equilibrium.

- For $a_{j_1j_1 0} < 0$ and $a_{j_2j_1 0}(a_{j_2j_1l_1} - a_{j_2j_1l_2}) < 0$,

$$(a_{j_1j_1 1},\bar{x}_{j_2}) = \underbrace{(3^{\text{rd}}\text{SI}, (3:1)\text{IC})}_{\text{third-order increasing-inflection sink}}.$$

The equilibrium at $x_{j_1}^* = a_{j_2j_1l_1} = a_{j_1j_1 1}$ is a (3^{rd}SI,(3:1)IC)-third-order increasing-concave sink infinite-equilibrium.

– The third-order concave source (sink) infinite-equilibriums are the switching bifurcations of a third-order source (sink) flow with a parabola flow.

In the end, from the above local analysis and the first integral of such a dynamical system, this theorem is proved. ■

3.3 Singular Dynamics and Bifurcations

In this section, dynamical systems with cubic and quadratic univariate vector fields with a single variable will be discussed. The corresponding stability and bifurcations of equilibriums will be presented.

Consider

$$\begin{aligned}\dot{x}_1 &= f_1(x_1,\mathbf{p}) = a_{110}(x_1 - a_{111})(x_1 - a_{112})(x_1 - a_{113}),\\ \dot{x}_2 &= f_2(x_1,\mathbf{p}) = a_{210}(x_1 - a_{211})^2\end{aligned} \tag{3.221}$$

where $a_{111} < a_{112}$ with $a_{211} > a_{112}$. The first integral manifield is

$$x_2 - x_{20} = \frac{a_{210}}{a_{110}} \sum_{s_1=1}^{3} \frac{(a_{11s_1} - a_{211})^2}{\prod_{s_2=1,s_2\neq s_1}^{3}(a_{11s_1} - a_{11s_2})} \ln \frac{|\, x_1 - a_{11s_1} \,|}{|\, x_{10} - a_{11s_1} \,|}. \tag{3.222}$$

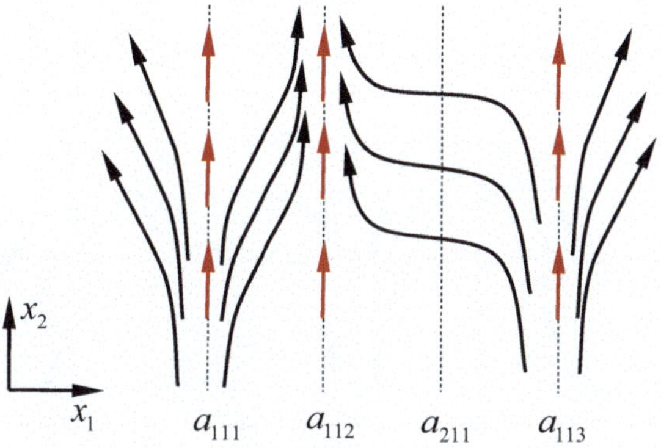

(a)

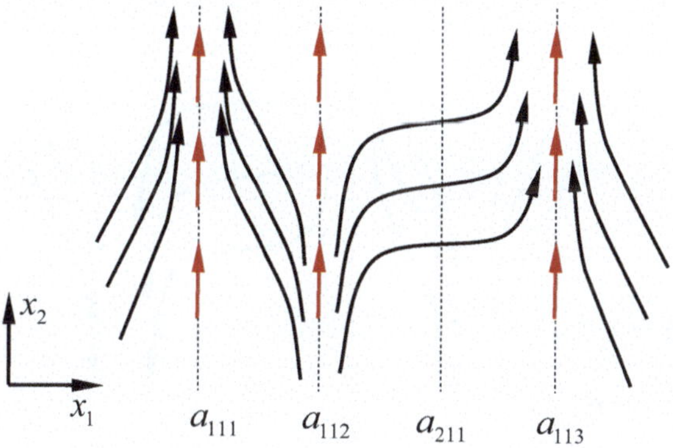

(b)

Fig. 3.1 Phase portraits for two-dimensional systems on the x_1-direction with $x_1^* = a_{111}, a_{112}, a_{113}$ and on the x_2-direction with $x_1^* = a_{211} \in (a_{112}, a_{113})$: (**a**) an (SO,pF)-positive source flow, an (SI, pF)-positive sink flow, a (pF, DI)-positive decreasing-inflection flow, and an (SO,pF)-positive source flow ($a_{110} > 0$, $a_{210} > 0$), (**b**) an (SI,pF)-positive sink flow, an (SO,pF)-positive source flow, a (pF, II) positive-increasing inflection flow, and an (SI,pF)-positive sink flow ($a_{110} < 0$, $a_{210} > 0$), (**c**) an (SO,nF)-negative source flow, an (SI,nF)-negative sink flow, an (nF,II)-negative increasing-inflection flow, and an (SO, nF)-negative flow ($a_{110} > 0$, $a_{210} < 0$), (**d**) an (SI,nF)-negative sink flow, an (SO,nF)-negative source flow, an (nF,DI)-negative decreasing-inflection flow, and an (SI,nF)-negative sink flow, ($a_{110} < 0$, $a_{210} < 0$)

The corresponding phase portraits for two-dimensional systems on the x_1-direction with $x_1^* = a_{111}, a_{112}, a_{113}$ and on the x_2-direction with $x_1^* = a_{211} \in (a_{112}, a_{113})$ are presented in Fig. 3.1. In Fig. 3.1a, the positive source, sink and source flows at $x_1^* = a_{111}, a_{112}, a_{113}$, and a positive decreasing-inflection flow at $x_1^* = a_{211} \in (a_{112}, a_{113})$ for $a_{110} > 0$ and $a_{210} > 0$ are presented from the first integral manifold. At $x_1^* = a_{111}, a_{113}$, the flows are positive source flows because of $df_1/dx_1|_{x_1^* = a_{111}, a_{113}} > 0$

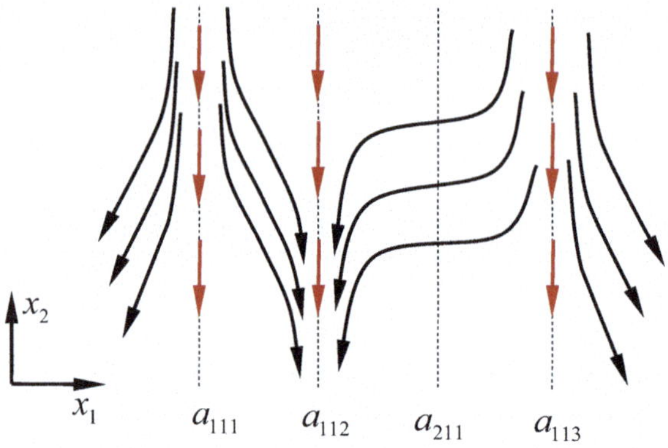

(c)

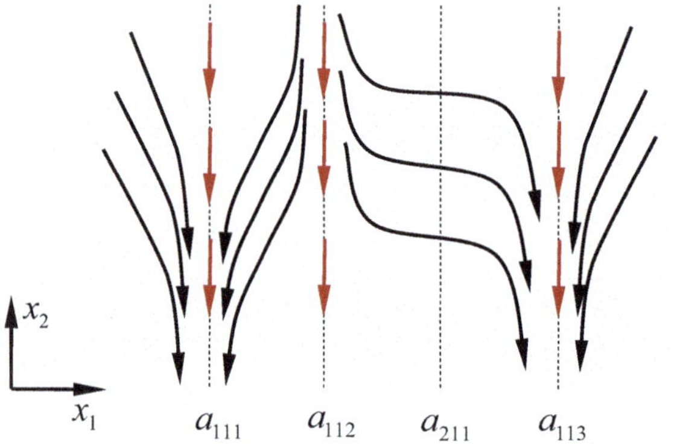

(d)

Fig. 3.1 (continued)

and $\dot{x}_2 > 0$. At $x_1^* = a_{112}$, the flow is a positive sink flow due to $df_1/dx_1|_{x_1^*=a_{112}} < 0$ and $\dot{x}_2 > 0$. $x_1^* = a_{211}$ is a positive decreasing-inflection flow due to $d^2x_2/dx_1^2|_{x_1^*=a_{211}} = a_{210}/\prod_{s_1=1}^{3}(a_{211} - a_{11s_1}) < 0$ and $\dot{x}_1 > 0$. Similarly, the positive sink, source and sink flows, and a positive increasing-inflection flow ($a_{110} < 0$, $a_{210} > 0$) are sketched in Fig. 3.1b for $x_1^* = a_{111}, a_{112}, a_{113}$ on the x_1-direction and $x_1^* = a_{211} \in (a_{112}, a_{113})$ on the x_2-direction. For $a_{110} > 0$ and $a_{210} < 0$, $x_1^* = a_{111}, a_{112}, a_{113}$ on the x_1-direction and $x_1^* = a_{211} \in (a_{112}, a_{113})$ on the x_2-direction are a negative source, sink and source flows, and a negative increasing-inflection flow, as shown in Fig. 3.1c. $x_1^* = a_{111}, a_{112}, a_{113}$ on the x_1-direction and $x_1^* = a_{211} \in (a_{112}, a_{113})$ on the x_2-direction are the negative sink, source and sink flows and a negative decreasing-inflection flow ($a_{110} < 0$, $a_{210} < 0$) is presented in Fig. 3.1d. The flows at equilibriums are listed as follows.

$$\left\{\begin{array}{l}(a_{111}, \dot{x}_2)\\(a_{112}, \dot{x}_2)\\(\dot{x}_1, a_{211})\\(a_{113}, \dot{x}_2)\end{array}\right\} = \left\{\begin{array}{l}(\mathrm{SO}, \mathrm{pF})\\(\mathrm{SI}, \mathrm{pF})\\(\mathrm{nF}, \mathrm{DI})\\(\mathrm{SO}, \mathrm{pF})\end{array}\right\} \text{ for } a_{110} > 0, a_{210} > 0; \tag{3.223}$$

$$\left\{\begin{array}{l}(a_{111}, \dot{x}_2)\\(a_{112}, \dot{x}_2)\\(\dot{x}_1, a_{211})\\(a_{113}, \dot{x}_2)\end{array}\right\} = \left\{\begin{array}{l}(\mathrm{SI}, \mathrm{pF})\\(\mathrm{SO}, \mathrm{pF})\\(\mathrm{pF}, \mathrm{II})\\(\mathrm{SI}, \mathrm{pF})\end{array}\right\} \text{ for } a_{110} < 0, a_{210} > 0; \tag{3.224}$$

$$\left\{\begin{array}{l}(a_{111}, \dot{x}_2)\\(a_{112}, \dot{x}_2)\\(\dot{x}_1, a_{211})\\(a_{113}, \dot{x}_2)\end{array}\right\} = \left\{\begin{array}{l}(\mathrm{SO}, \mathrm{nF})\\(\mathrm{SI}, \mathrm{nF})\\(\mathrm{nF}, \mathrm{II})\\(\mathrm{SO}, \mathrm{nF})\end{array}\right\} \text{ for } a_{110} > 0, a_{210} < 0; \tag{3.225}$$

$$\left\{\begin{array}{l}(a_{111}, \dot{x}_2)\\(a_{112}, \dot{x}_2)\\(\dot{x}_1, a_{211})\\(a_{113}, \dot{x}_2)\end{array}\right\} = \left\{\begin{array}{l}(\mathrm{SI}, \mathrm{nF})\\(\mathrm{SO}, \mathrm{nF})\\(\mathrm{pF}, \mathrm{DI})\\(\mathrm{SI}, \mathrm{nF})\end{array}\right\} \text{ for } a_{110} < 0, a_{210} < 0. \tag{3.226}$$

Similarly, the phase portraits of two-dimensional systems for $x_1^* = a_{111}, a_{112}, a_{113}$ on the x_1-direction with three cases of $a_{211} \in (a_{111}, a_{112})$, $a_{211} < a_{111}$, and $a_{211} > a_{113}$ on the x_2-direction can be presented. Readers can finish them as an exercise.

Consider a dynamical system as

$$\begin{aligned}\dot{x}_1 &= a_{110}(x_1 - a_{111})(x_1 - a_{112})(x_1 - a_{113}),\\ \dot{x}_2 &= a_{210}(x_1 - a_{211})^2\end{aligned} \tag{3.227}$$

with $a_{211} = a_{11s_1}(s_1 \in \{1, 2, 3\})$. The first integral manifold is

$$x_2 - x_{20} = \frac{a_{210}}{a_{110}} \sum_{s_2=1, s_2 \neq s_1}^{3} \frac{(a_{11s_2} - a_{211})}{\prod_{s_3=1, s_3 \neq s_1, s_2}^{3} (a_{11s_1} - a_{11s_2})} \ln \frac{|x_1 - a_{11s_2}|}{|x_{10} - a_{11s_2}|}. \tag{3.228}$$

For $a_{211} = a_{112}$, the phase portraits with equilibriums of $x_1^* = a_{111}, a_{112}, a_{113}$ on the x_1-direction and $x_1^* = a_{211} = a_{112}$ on the x_2-direction are presented in Fig. 3.2. In Fig. 3.2a, there are a down-parabola sink-infinite-equilibrium (i.e., (SI,DP)-infinite-equilibrium) and two positive source flows for $a_{110} > 0$ and $a_{210} > 0$. At $x_1^* = a_{112} = a_{211}$, the equilibrium is a down-parabola sink infinite-equilibrium due to sink on the x_1-direction and $\dot{x}_2 = a_{210}(x_1 - a_{211})^2 \geq 0$. At $x_1^* = a_{111}, a_{113}$, the flows are a positive source flow due to $\dot{x}_2 > 0$. An up-parabola source-equilibrium (i.e., (SO,UP)-infinite-equilibrium) and two positive sink flows for $a_{110} < 0$ and

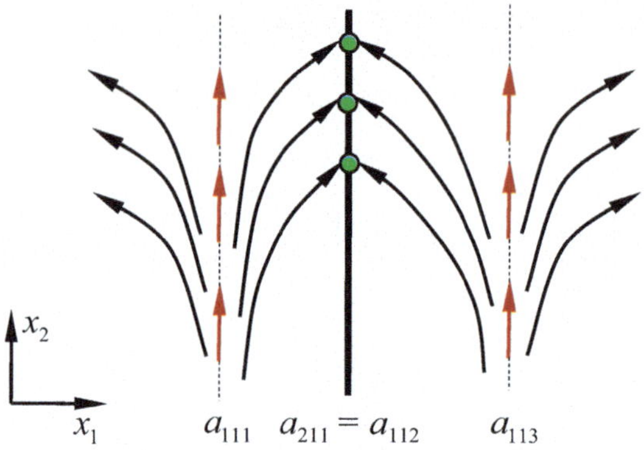

(a)

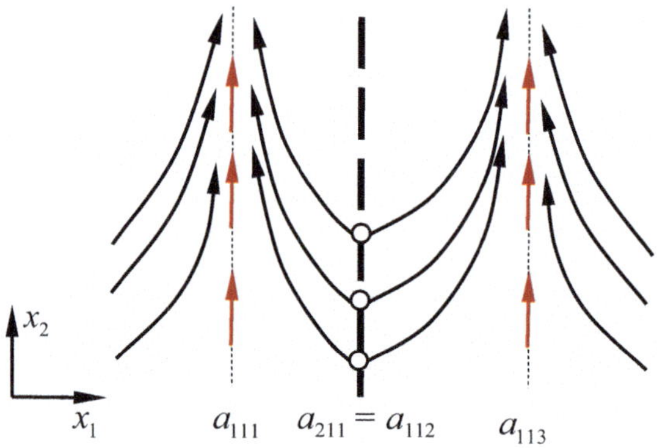

(b)

Fig. 3.2 Phase portraits for two-dimensional systems with equilibriums of $x_1^* = a_{111}, a_{112}, a_{113}$ on the x_1-direction and $x_1^* = a_{211} = a_{112}$ on the x_2-direction: (**a**) an (SO,pF)-positive source flow, an (SI, DP)-infinite-equilibrium, and an (SO,pF)-positive source flow ($a_{110} > 0$, $a_{210} > 0$), (**b**) an (SI,pF)-positive sink flow, an (SO,UP)-infinite-equilibrium, and an (SI,pF)-positive sink flow ($a_{110} < 0$, $a_{210} > 0$), (**c**) an (SO,pF)-negative source flow, an (SI,UP)-infinite-equilibrium, and an (SO,nF)-negative source flow ($a_{110} > 0$, $a_{210} < 0$), (**d**) an (SI,nF)-negative sink flow, an (SO,DP)-infinite-equilibrium, and an (SI,nF)-negative sink flow ($a_{110} < 0$, $a_{210} < 0$). The solid and dashed lines are for sink and source infinite-equilibriums, respectively. The filled and hollow circular symbols are specific equilibriums on the infinite-equilibrium lines, respectively

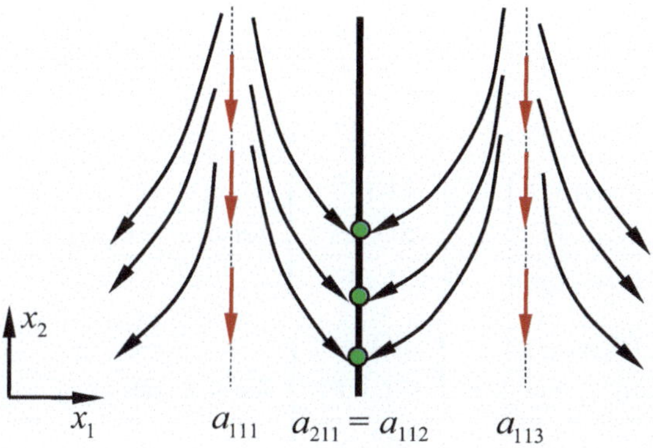

(c)

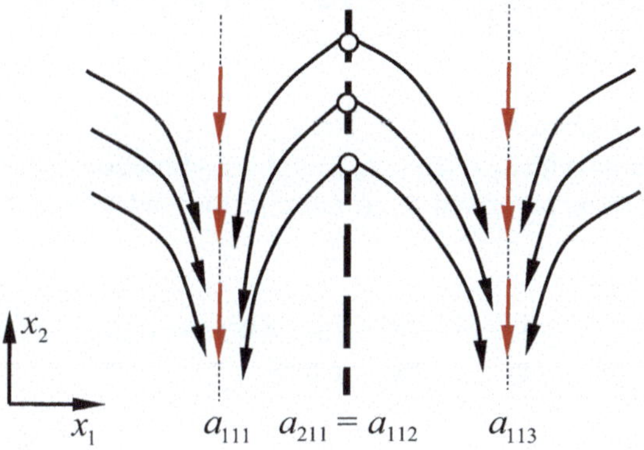

(d)

Fig. 3.2 (continued)

$a_{210} > 0$ at $x_1^* = a_{112}, a_{111}, a_{113}$ are presented in Fig. 3.2b. $x_1^* = a_{211} = a_{112}$ and $x_1^* = a_{111}, a_{113}$ are an up-parabola sink-infinite-equilibrium (i.e., (SI,UP)-infinite-equilibrium) and two negative source flows for $a_{110} > 0$ and $a_{210} < 0$, as shown in Fig. 3.2c. A down-parabola source-infinite equilibrium (i.e., (SO,DP)-infinite-equilibrium) and two negative sink flow at $x_1^* = a_{112} = a_{211}$ and $x_1^* = a_{111}, a_{113}$ for $a_{110} < 0$ and $a_{210} < 0$ are presented in Fig. 3.2d. The solid and dashed lines are for sink and source infinite-equilibriums, respectively. The filled and hollow circular symbols are specific equilibriums on the infinite-equilibrium lines, respectively. The flows at equilibriums are presented as follows.

$$\left\{\begin{array}{l}(a_{111},\dot{x}_2)\\(a_{112},\overline{x}_2)\\(a_{113},\dot{x}_2)\end{array}\right\}=\left\{\begin{array}{l}(\mathrm{SO,pF})\\(\mathrm{SI,DP})\\(\mathrm{SO,pF})\end{array}\right\}\text{ for } a_{110}>0, a_{210}>0; \tag{3.229}$$

$$\left\{\begin{array}{l}(a_{111},\dot{x}_2)\\(a_{112},\overline{x}_2)\\(a_{113},\dot{x}_2)\end{array}\right\}=\left\{\begin{array}{l}(\mathrm{SI,pF})\\(\mathrm{SO,UP})\\(\mathrm{SI,pF})\end{array}\right\}\text{ for } a_{110}<0, a_{210}>0; \tag{3.230}$$

$$\left\{\begin{array}{l}(a_{111},\dot{x}_2)\\(a_{112},\overline{x}_2)\\(a_{113},\dot{x}_2)\end{array}\right\}=\left\{\begin{array}{l}(\mathrm{SO,nF})\\(\mathrm{SI,UP})\\(\mathrm{SO,nF})\end{array}\right\}\text{ for } a_{110}>0, a_{210}<0; \tag{3.231}$$

$$\left\{\begin{array}{l}(a_{111},\dot{x}_2)\\(a_{112},\overline{x}_2)\\(a_{113},\dot{x}_2)\end{array}\right\}=\left\{\begin{array}{l}(\mathrm{SO,nF})\\(\mathrm{SI,DP})\\(\mathrm{SO,nF})\end{array}\right\}\text{ for } a_{110}<0, a_{210}<0. \tag{3.232}$$

Similarly, for cases of $a_{211}=a_{111}$ and $a_{211}=a_{113}$, the phase portraits for two-dimensional systems with $x_1^*=a_{111},a_{112}$ on the x_1-direction and $x_1^*=a_{211}=a_{112}$ on the x_2-direction can be presented as an exercise for readers to complete.

Consider

$$\begin{aligned}\dot{x}_1&=a_{110}(x_1-a_{11s_1})(x_1-a_{11s_2})^2,\\\dot{x}_2&=a_{210}(x_1-a_{211})^2.\end{aligned} \tag{3.233}$$

The first integral manifold is

$$\begin{aligned}x_2-x_{20}=&\frac{a_{210}}{a_{110}}\left\{\frac{(a_{11s_1}-a_{211})^2}{(a_{11s_1}-a_{11s_2})^2}\ln\frac{|\,x_1-a_{11s_1}\,|}{|\,x_{10}-a_{11s_1}\,|}\right.\\&-\frac{(a_{11s_2}-a_{211})^2}{(a_{11s_2}-a_{11s_1})}\left(\frac{1}{(x_1-a_{11s_2})}-\frac{1}{(x_{10}-a_{11s_2})}\right)\\&\left.+\left[\frac{2(a_{11s_2}-a_{211})}{(a_{11s_2}-a_{11s_1})}-\frac{(a_{11s_2}-a_{211})^2}{(a_{11s_2}-a_{11s_1})^2}\right]\ln\frac{|\,x_1-a_{11s_2}\,|}{|\,x_{10}-a_{11s_2}\,|}\right\}.\end{aligned} \tag{3.234}$$

For $a_{211}\in(a_{111},a_{112})$ with $a_{11s_1}=a_{112}$ and $a_{11s_2}=a_{111}$, the phase portraits for two-dimensional systems on the x_1-direction are presented in Fig. 3.3. For $x_1^*=a_{111}$, $\dot{x}_1=0$ and $\dot{x}_2>0$. Thus, for $x_1^*=a_{11}$, the flow is determined by $\dot{x}_2\neq0$. For $x_1^*=a_{211}$, $\dot{x}_1>0$ and $\dot{x}_2=0$. Thus, for $x_1^*=a_{211}$, the flow is determined by $\dot{x}_1\neq0$. In Fig. 3.3a, the phase portrait for $a_{110}>0$ and $a_{210}>0$ is presented from the first integral manifold. At $x_1^*=a_{111}$, the lower-saddle flow is toward the x_2-positive direction due to $\dot{x}_2=a_{210}(a_{111}-a_{211})^2>0$. At $x_1^*=a_{211}$, the decreasing-inflection flow is toward the x_1-negative direction because of $\dot{x}_1=a_{110}(a_{211}-a_{111})^2(a_{211}-a_{112})<0$. At

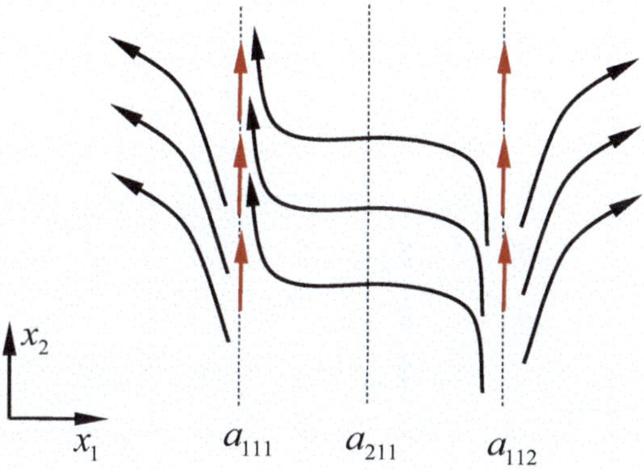

(a)

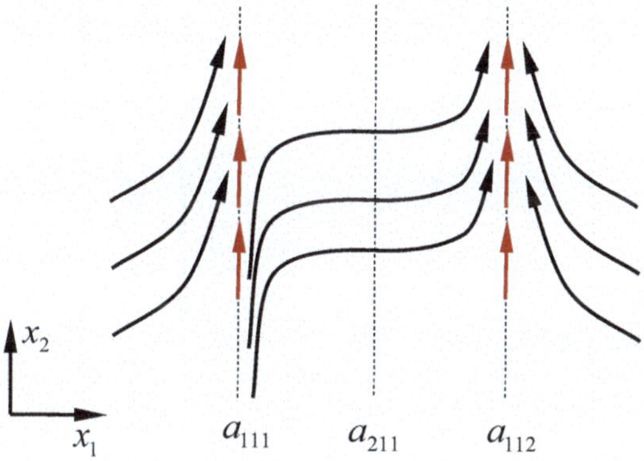

(b)

Fig. 3.3 Phase portraits for two-dimensional systems on the x_2-direction with $x_1^* = a_{111}, a_{112}$ and on the x_1-direction with $x_1^* = a_{211} \in (a_{111}, a_{112})$: (**a**) an (LS,pF)-positive upper-saddle flow, an (nF,DI) positive increasing-inflection flow, and an (SO,pF)-positive source flow ($a_{110} > 0$, $a_{210} > 0$), (**b**) a (US,pF)-positive lower-saddle flow, a (pF,II)-positive decreasing-inflection flow, and an (SI,pF)-positive sink flow ($a_{110} < 0$, $a_{210} > 0$), (**c**) an (LS,nF)-negative upper-saddle flow, an (nF,II)-negative increasing-inflection flow, an (SO,nF)-negative sink flow ($a_{110} > 0$, $a_{210} < 0$), (**d**) a (US, nF)-negative upper-saddle flow, a (pF,DI)-negative increasing-inflection flow, and an (SI,nF)-negative source flow ($a_{110} < 0$, $a_{210} < 0$)

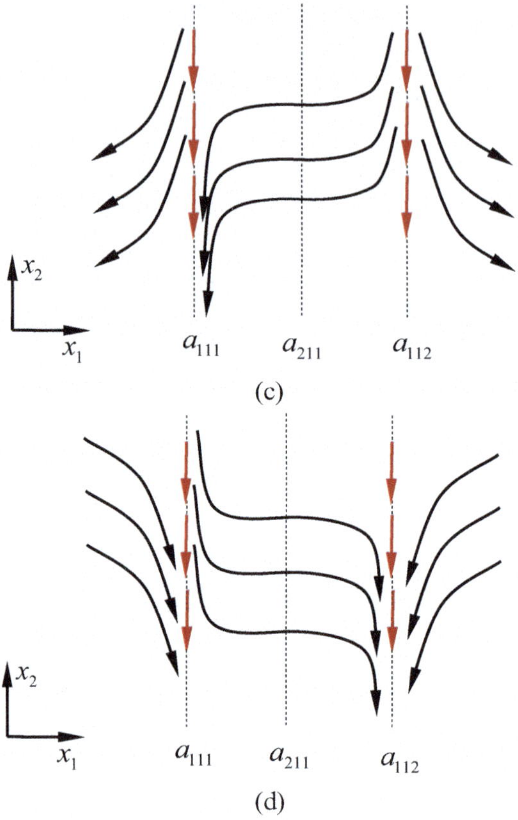

Fig. 3.3 (continued)

$x_1^* = a_{112}$, the positive source flow is toward the x_2-negative direction due to $\dot{x}_2 = a_{210}(a_{111} - a_{211})^2 > 0$. Similarly, the phase portraits for $(a_{110} < 0,\ a_{210} > 0)$, $(a_{110} > 0,\ a_{210} < 0)$, and $(a_{110} < 0,\ a_{210} < 0)$ are presented in Fig. 3.3b-d, respectively. The flows at the equilibriums are listed as follows.

$$\left\{ \begin{array}{l} (a_{111}, \dot{x}_2) \\ (\dot{x}_1, a_{211}) \\ (a_{112}, \dot{x}_2) \end{array} \right\} = \left\{ \begin{array}{l} (\text{LS}, \text{pF}) \\ (\text{nF}, \text{DI}) \\ (\text{SO}, \text{pF}) \end{array} \right\} \text{ for } a_{110} > 0, a_{210} > 0; \tag{3.235}$$

$$\left\{ \begin{array}{l} (a_{111}, \dot{x}_2) \\ (\dot{x}_2, a_{211}) \\ (a_{112}, \dot{x}_2) \end{array} \right\} = \left\{ \begin{array}{l} (\text{US}, \text{pF}) \\ (\text{pF}, \text{II}) \\ (\text{SI}, \text{pF}) \end{array} \right\} \text{ for } a_{110} < 0, a_{210} > 0; \tag{3.236}$$

$$\left\{\begin{array}{c}(a_{111},\dot{x}_2)\\(\dot{x}_1,a_{211})\\(a_{112},\dot{x}_2)\end{array}\right\}=\left\{\begin{array}{c}(\text{LS},\text{nF})\\(\text{nF},\text{II})\\(\text{SO},\text{nF})\end{array}\right\}\text{ for } a_{110}>0, a_{210}<0; \tag{3.237}$$

$$\left\{\begin{array}{c}(a_{111},\dot{x}_2)\\(\dot{x}_1,a_{211})\\(a_{113},\dot{x}_2)\end{array}\right\}=\left\{\begin{array}{c}(\text{US},\text{nF})\\(\text{pF},\text{DI})\\(\text{SI},\text{nF})\end{array}\right\}\text{ for } a_{110}<0, a_{210}<0. \tag{3.238}$$

For $a_{211} < a_{111}$ and $a_{211} > a_{112}$, the phase portraits for two-dimensional systems on the x_1-direction can be sketched for $a_{11s_1} = a_{112}$ and $a_{11s_2} = a_{111}$. The dynamical behaviors for $a_{211} < a_{111}$ and $a_{211} > a_{112}$ are similar. For $a_{11s_1} = a_{111}$ and $a_{11s_2} = a_{112}$, the corresponding phase portraits can be sketched. The upper-saddle and lower-saddle flows are the appearing bifurcation for the sink and source flows. The increasing-inflection and decreasing-inflection flows are the appearing bifurcations for the up-parabola and down-parabola flows.

Consider

$$\begin{aligned}\dot{x}_1 &= a_{110}(x_1 - a_{11s_1})(x_1 - a_{11s_2})^2,\\ \dot{x}_2 &= a_{210}(x_1 - a_{211})^2.\end{aligned} \tag{3.239}$$

For $a_{211} = a_{11s_1}$ (s_1, $s_2 \in \{1,2\}$, $s_1 \neq s_2$), the first integral manifold is

$$x_2 - x_{20} = \frac{a_{210}}{a_{110}}\left[\ln\frac{|x_1 - a_{11s_2}|}{|x_{10} - a_{11s_2}|} - (a_{11s_2} - a_{211})\left(\frac{1}{x_1 - a_{11s_2}} - \frac{1}{x_{10} - a_{11s_2}}\right)\right]. \tag{3.240}$$

For $a_{211} = a_{112}$ with $a_{11s_1} = a_{112}$ and $a_{11s_2} = a_{111}$, the phase portraits for two-dimensional systems on the x_1-direction are presented in Fig. 3.4. In Fig. 3.4a, the phase portrait for $a_{110} > 0$ and $a_{210} > 0$ is presented from the first integral manifold. At $x_1^* = a_{111}$, the lower-saddle flow is toward the x_2-positive direction due to $\dot{x}_2 = a_{210}(a_{111} - a_{211})^2 > 0$. At $x_1^* = a_{211} = a_{112}$, the up-parabola source exists. Similarly, the phase portraits for ($a_{110} < 0$, $a_{210} > 0$), ($a_{110} > 0$, $a_{210} < 0$), and ($a_{110} < 0$, $a_{210} < 0$) are presented in Fig. 3.4b-d, respectively. The flows at the equilibriums of $x_1^* = a_{211} = a_{112}$ with $\bar{x}_2 \in (-\infty, \infty)$ are listed as follows.

$$\left\{\begin{array}{c}(a_{111},\dot{x}_2)\\(a_{112},\bar{x}_2)\end{array}\right\}=\left\{\begin{array}{c}(\text{LS},\text{pF})\\(\text{SO},\text{UP})\end{array}\right\}\text{ for } a_{110}>0, a_{210}>0; \tag{3.241}$$

$$\left\{\begin{array}{c}(a_{111},\dot{x}_2)\\(a_{112},\bar{x}_2)\end{array}\right\}=\left\{\begin{array}{c}(\text{US},\text{pF})\\(\text{SI},\text{DP})\end{array}\right\}\text{ for } a_{110}<0, a_{210}>0; \tag{3.242}$$

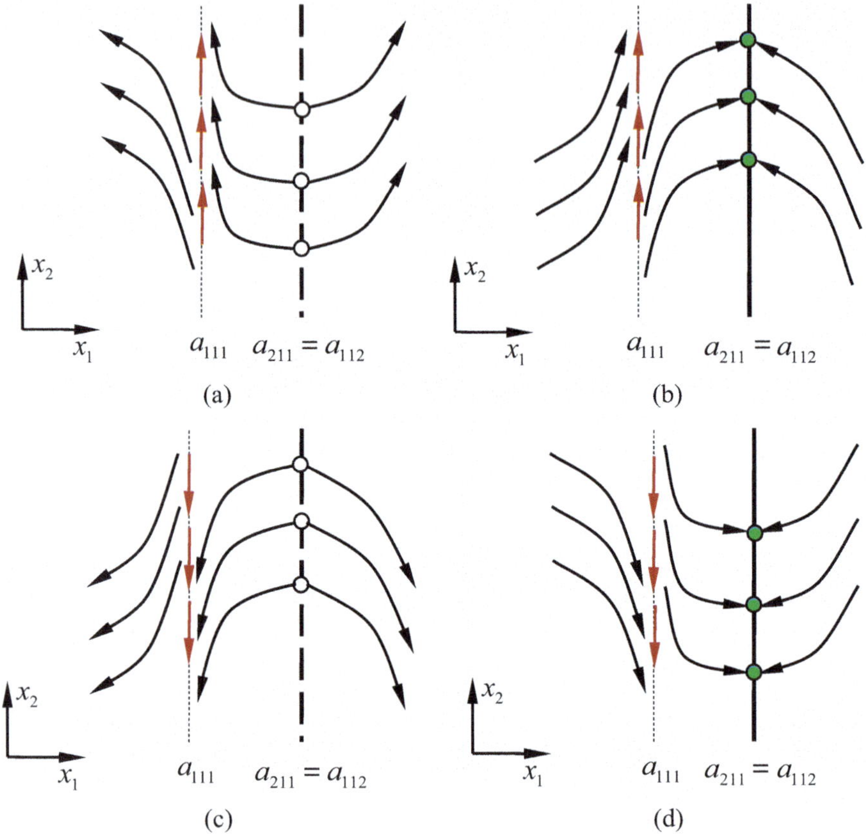

Fig. 3.4 Phase portraits for two-dimensional systems on the x_2-direction with $x_1^* = a_{111}, a_{112}$ and on the x_1-direction with $x_1^* = a_{211} = a_{112}$: (**a**) an (LS,pF)-positive upper-saddle flow and an (SO,UP)-up-parabola source, ($a_{110} > 0$, $a_{210} > 0$), (**b**) a (US,pF)-positive lower-saddle flow and an (SI,DP)-down-parabola sink ($a_{110} < 0$, $a_{210} > 0$), (**c**) an (LS,nF)-negative upper-saddle flow and an (SO, DP)-down-parabola source ($a_{110} > 0$, $a_{210} < 0$), (**d**) a (US,nF)-negative upper-saddle flow and an (SI,UP)-up-parabola sink ($a_{110} < 0$, $a_{210} < 0$)

$$\left\{ \begin{array}{c} (a_{111}, \dot{x}_2) \\ (a_{112}, \bar{x}_2) \end{array} \right\} = \left\{ \begin{array}{c} (\text{LS, nF}) \\ (\text{SO, DP}) \end{array} \right\} \text{ for } a_{110} > 0, a_{210} < 0; \tag{3.243}$$

$$\left\{ \begin{array}{c} (a_{111}, \dot{x}_2) \\ (a_{112}, \bar{x}_2) \end{array} \right\} = \left\{ \begin{array}{c} (\text{US, nF}) \\ (\text{SI, UP}) \end{array} \right\} \text{ for } a_{110} < 0, a_{210} < 0. \tag{3.244}$$

For $a_{211} = a_{111}$, the phase portraits for two-dimensional systems on the x_1-direction can be sketched for $a_{11s_1} = a_{112}$ and $a_{11s_2} = a_{111}$. The up-parabola and down-parabola sink and source are for switching bifurcations for the increasing-inflection and decreasing-inflection flows with sink and source flows.

Consider

$$\begin{aligned}\dot{x}_1 &= a_{110}(x_1 - a_{11s_1})(x_1 - a_{11s_2})^2,\\ \dot{x}_2 &= a_{210}(x_1 - a_{211})^2.\end{aligned} \tag{3.245}$$

For $a_{211} = a_{11s_2}$ (s_1, $s_2 \in \{1,2\}$, $s_1 \neq s_2$), the first integral manifold is

$$x_2 - x_{20} = \frac{a_{210}}{a_{110}} \ln \frac{|\, x_1 - a_{11s_1} \,|}{|\, x_{10} - a_{11s_1} \,|}. \tag{3.246}$$

For $a_{211} = a_{11s_2} = a_{112}$ with $a_{11s_1} = a_{111}$, the phase portraits for two-dimensional systems on the x_1-direction are presented in Fig. 3.5. In Fig. 3.5a, the phase portrait for $a_{110} > 0$ and $a_{210} > 0$ is presented from the first integral manifold. At $x_1^* = a_{111}$, the positive source flow exists due to $\dot{x}_2 = a_{210}(a_{111} - a_{211})^2 > 0$. At $x_1^* = a_{211} = a_{112}$, the increasing-inflection upper-saddle exists. Similarly, the phase portraits for ($a_{110} < 0$, $a_{210} > 0$), ($a_{110} > 0$, $a_{210} < 0$), and ($a_{110} < 0$, $a_{210} < 0$) are presented in Fig. 3.5b-d, respectively. The flows at the equilibriums with $\bar{x}_2 \in (-\infty, \infty)$ are listed as follows.

$$\begin{Bmatrix} (a_{111}, \dot{x}_2) \\ (a_{112}, \bar{x}_2) \end{Bmatrix} = \begin{Bmatrix} (\text{SO, pF}) \\ (\text{US, (2:2)II}) \end{Bmatrix} \text{ for } a_{110} > 0, a_{210} > 0; \tag{3.247}$$

$$\begin{Bmatrix} (a_{111}, \dot{x}_2) \\ (a_{112}, \bar{x}_2) \end{Bmatrix} = \begin{Bmatrix} (\text{SI, pF}) \\ (\text{LS, (2:2)DI}) \end{Bmatrix} \text{ for } a_{110} < 0, a_{210} > 0; \tag{3.248}$$

$$\begin{Bmatrix} (a_{111}, \dot{x}_2) \\ (a_{112}, \bar{x}_2) \end{Bmatrix} = \begin{Bmatrix} (\text{SO, nF}) \\ (\text{US, (2:2)DI}) \end{Bmatrix} \text{ for } a_{110} > 0, a_{210} < 0; \tag{3.249}$$

$$\begin{Bmatrix} (a_{111}, \dot{x}_2) \\ (a_{112}, \bar{x}_2) \end{Bmatrix} = \begin{Bmatrix} (\text{SI, nF}) \\ (\text{LS, (2:2)II}) \end{Bmatrix} \text{ for } a_{110} < 0, a_{210} < 0. \tag{3.250}$$

Similarly, for $a_{211} = a_{11s_2} = a_{111}$ with $a_{11s_1} = a_{112}$, the phase portraits for two-dimensional systems on the x_1-direction can be sketched. The increasing-inflection and decreasing-inflection upper-saddle and lower-saddle are switching bifurcations for the up and decreasing-inflection flows with sink and source flows.

Consider

$$\begin{aligned}\dot{x}_1 &= a_{110}(x_1 - a_{111})^3,\\ \dot{x}_2 &= a_{210}(x_1 - a_{211})^2\end{aligned} \tag{3.251}$$

with $a_{111} \neq a_{211}$ and the first integral manifold is

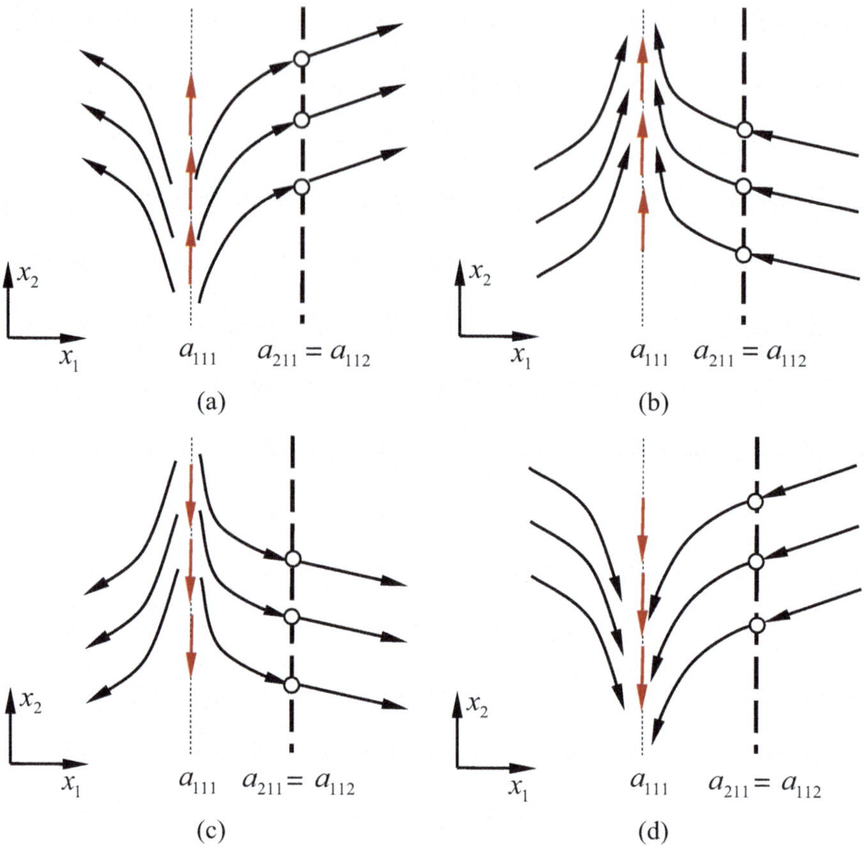

Fig. 3.5 Phase portraits for two-dimensional systems on the x_2-direction with $x_1^* = a_{111}, a_{112}$ and on the x_1-direction with $x_1^* = a_{211} = a_{112}$: (**a**) an (SO,pF)-positive source flow and a (US,(2:2)II)-increasing-inflection upper-saddle, ($a_{110} > 0$, $a_{210} > 0$), (**b**) an (SI,pF)-positive sink flow and an (LS,(2:2)DI)-decreasing-inflection lower-saddle ($a_{110} < 0$, $a_{210} > 0$), (**c**) an (SO,nF)-negative source flow and a (US,(2:2)DI)-decreasing-inflection upper-saddle ($a_{110} > 0$, $a_{210} < 0$), (**d**) an (SI,nF)-negative sink flow and an (LS,(2:2)II)-increasing-inflection lower-saddle ($a_{110} < 0$, $a_{210} < 0$)

$$
\begin{aligned}
x_2 - x_{20} = \frac{a_{210}}{a_{110}} \Big\{ \ln \frac{|x_1 - a_{111}|}{|x_{10} - a_{111}|} \\
&- 2(a_{111} - a_{211}) \left[\frac{1}{(x_1 - a_{111})} - \frac{1}{(x_{10} - a_{111})} \right] \\
&- \frac{1}{2}(a_{111} - a_{211})^2 \left[\frac{1}{(x_1 - a_{111})^2} - \frac{1}{(x_{10} - a_{111})^2} \right] \Big\}.
\end{aligned}
\tag{3.252}
$$

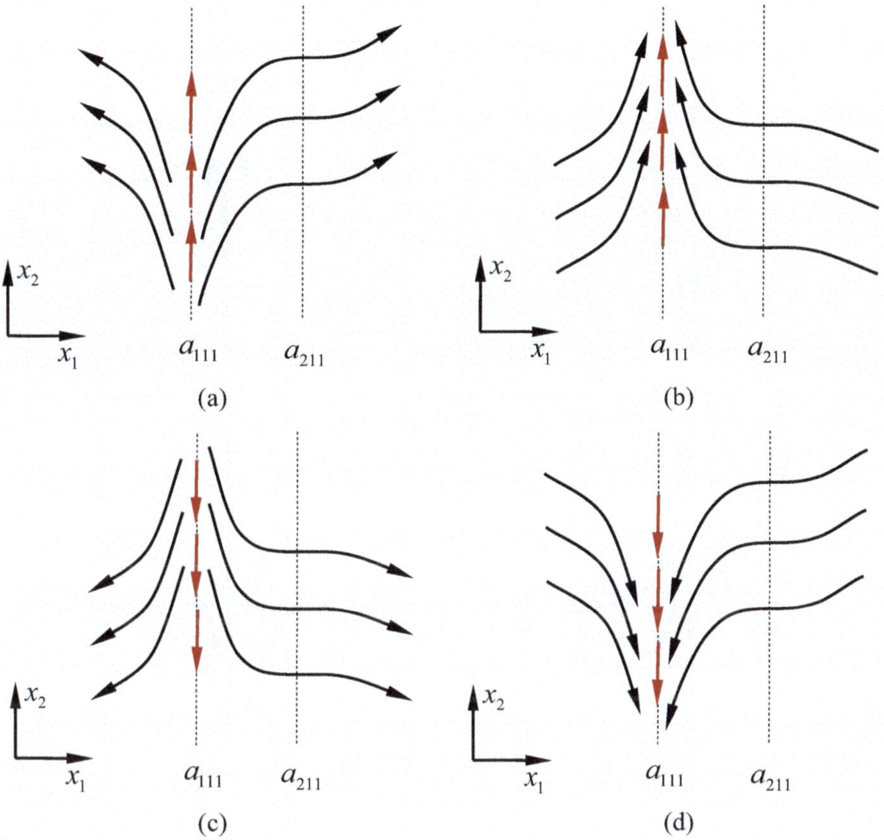

Fig. 3.6 Phase portraits for two-dimensional systems on the x_2-direction with $x_1^* = a_{111}$ and on the x_1-direction with $x_1^* = a_{211} > a_{111}$: (**a**) a ($3^{rd}$SO,pF)-positive third-order source flow and a (pF,II)-positive increasing-inflection flow, ($a_{110} > 0$, $a_{210} > 0$), (**b**) a (3^{rd}SI,pF)-positive third-sink flow and an (nF,DI)-negative decreasing-inflection flow ($a_{110} < 0$, $a_{210} > 0$), (**c**) a (3^{rd}SO,nF)-negative third-order source flow and a (pF,DI)-positive decreasing-inflection flow ($a_{110} > 0$, $a_{210} < 0$), (**d**) a (3^{rd}SI,nF)-negative third-order sink flow and an (nF,II)-negative increasing-inflection flow ($a_{110} < 0$, $a_{210} < 0$)

Its phase portraits for two-dimensional systems on the x_1-direction are presented in Fig. 3.6 for $a_{211} > a_{111}$. In Fig. 3.6a, the phase portrait for $a_{110} > 0$ and $a_{210} > 0$ is presented from the first integral manifold. At $x_1^* = a_{111}$, the positive third-order source flow exists because $\dot{x}_2 > 0$, and the positive increasing-inflection flow exists at $x_1^* = a_{211}$. Similarly, in Fig. 3.6b-d, at $x_1^* = a_{111}, a_{211}$, a positive third-sink flow and a negative decreasing-inflection flow exist for ($a_{110} < 0$, $a_{210} > 0$), a negative third-order source flow and a positive decreasing-inflection flow at $x_1^* = a_{111}, a_{211}$ exist for ($a_{110} > 0$, $a_{210} < 0$), a negative third-order sink flow and a negative increasing-

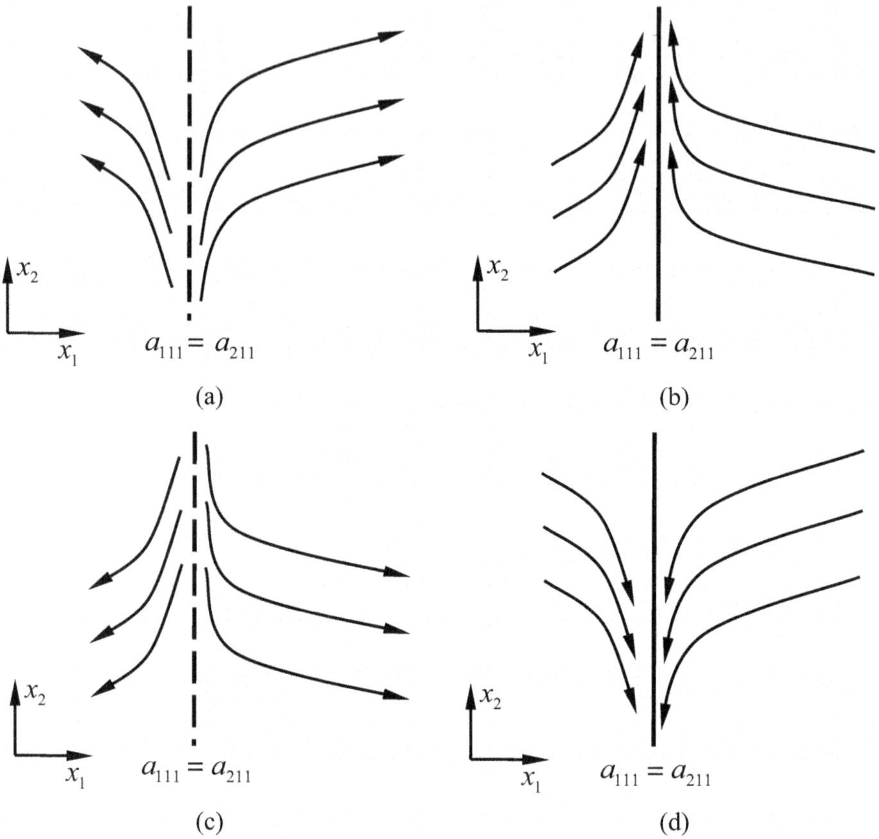

Fig. 3.7 Phase portraits for two-dimensional systems on the x_2-direction with $x_1^* = a_{111}$ and on the x_1-direction with $x_1^* = a_{211} > a_{111}$: (**a**) a (3[rd]SO,(3:2)DL)-third-order down-logarithmic source ($a_{110} > 0$, $a_{210} > 0$), (**b**) a (3[rd]SI,(3:2)UL)-third-order up-logarithmic sink ($a_{110} < 0$, $a_{210} > 0$), (**c**) a (3[rd]SO,(3:2)UL)-third-order up-logarithmic source ($a_{110} > 0$, $a_{210} < 0$), (**d**) a (3[rd]SI,(3:2)DL)-third-order down-logarithmic sink ($a_{110} < 0$, $a_{210} < 0$)

inflection flow at $x_1^* = a_{111}, a_{211}$ exist for ($a_{110} < 0$, $a_{210} < 0$). The flows at the equilibriums are listed as follows.

$$\left\{\begin{array}{c}(a_{111}, \dot{x}_2)\\(a_{112}, a_{211})\end{array}\right\} = \left\{\begin{array}{c}(3^{\text{rd}}\text{SO, pF})\\(\text{pF, II})\end{array}\right\} \text{ for } a_{110} > 0, a_{210} > 0; \tag{3.253}$$

$$\left\{\begin{array}{c}(a_{111}, \dot{x}_2)\\(a_{112}, a_{211})\end{array}\right\} = \left\{\begin{array}{c}(3^{\text{rd}}\text{SI, pF})\\(\text{nF, DI})\end{array}\right\} \text{ for } a_{110} < 0, a_{210} > 0; \tag{3.254}$$

$$\left\{\begin{array}{l}(a_{111},\dot{x}_2)\\(a_{112},a_{211})\end{array}\right\}=\left\{\begin{array}{c}(3^{\rm rd}\mathrm{SO},\mathrm{nF})\\(\mathrm{pF},\mathrm{DI})\end{array}\right\}\text{ for } a_{110}>0, a_{210}<0; \tag{3.255}$$

$$\left\{\begin{array}{l}(a_{111},\dot{x}_2)\\(a_{112},a_{211})\end{array}\right\}=\left\{\begin{array}{c}(3^{\rm rd}\mathrm{SI},\mathrm{nF})\\(\mathrm{nF},\mathrm{II})\end{array}\right\}\text{ for } a_{110}<0, a_{210}<0. \tag{3.256}$$

Similarly, for $a_{211} < a_{111}$, the phase portraits for two-dimensional systems on the x_1-direction can be sketched. Readers can complete them as an exercise.

Consider

$$\begin{aligned}\dot{x}_1&=a_{110}(x_1-a_{111})^3,\\ \dot{x}_2&=a_{210}(x_1-a_{211})^2\end{aligned} \tag{3.257}$$

with $a_{111}=a_{211}$ and the first integral manifold is

$$x_2-x_{20}=\frac{a_{210}}{a_{110}}\ln\frac{|x_1-a_{111}|}{|x_{10}-a_{111}|}. \tag{3.258}$$

The phase portraits for two-dimensional systems on the x_1-direction are presented in Fig. 3.7 for $a_{211}=a_{111}$. In Fig. 3.7a, the phase portrait for $a_{110}>0$ and $a_{210}>0$ is presented from the first integral manifold. At $x_1^*=a_{111}=a_{211}$, the increasing-inflection source exists, which is called a ($3^{\rm rd}$SO,(3:2)DL)-third-order down-logarithmic source. Similarly, a ($3^{\rm rd}$SI,(3:2)UL)-third-order up-logarithmic sink ($a_{110}<0$, $a_{210}>0$), a ($3^{\rm rd}$SO,(3:2)UL)-third-order up-logarithmic source ($a_{110}>0$, $a_{210}<0$), a ($3^{\rm rd}$SI,(3:2)DL)-third-order down-logarithmic sink ($a_{110}<0$, $a_{210}<0$) are presented in Fig. 3.7b-d. The infinite-equilibrium at $x_1^*=a_{111}=a_{211}$ with $\bar{x}_2\in(-\infty,\infty)$ is listed as follows.

$$(a_{111},\bar{x}_2)=\begin{cases}(3^{\rm rd}\mathrm{SO},(3{:}2)\mathrm{DL}) & \text{for } a_{110}>0, a_{210}>0;\\(3^{\rm rd}\mathrm{SI},(3{:}2)\mathrm{UL}) & \text{for } a_{110}<0, a_{210}>0;\\(3^{\rm rd}\mathrm{SO},(3{:}2)\mathrm{UL}) & \text{for } a_{110}>0, a_{210}<0;\\(3^{\rm rd}\mathrm{SI},(3{:}2)\mathrm{DL}) & \text{for } a_{110}<0, a_{210}<0.\end{cases} \tag{3.259}$$

Chapter 4
Two Single-Variable Cubic Vector Fields

In this chapter, two vector fields in dynamical systems to be two cubic univariate vector fields with a single variable are considered. The singularity, stability, and bifurcations of such dynamical systems will be discussed. In addition to the first- and third-order sink and source flows with upper-saddle and lower-saddle flows, the first- and third-order parabola and inflection flows based on the cubic crossing-variable vector fields are determined. The (1:3)-*up-parabola* sink (source) infinite-equilibriums, the down-up and up-down saddles, the (2:2)-inflection upper-saddle infinite-equilibriums, the third-order increasing and decreasing-concave *source and sink* infinite-equilibriums, the third-order up and down-logarithmic *source and sink* infinite-equilibriums discussed are presented in Chapter 3. The third-order inflection source and sink infinite-equilibrium are presented for the appearing and switching bifurcations of the source, sink, source flows and the sink, source, and sink flows with up-parabola, down-parabola, and up-parabola flows and the down-parabola, up-parabola, and down-parabola flows. The third-order inflection source and sink infinite-equilibrium are also for the switching bifurcations of the source and sink flows and the upper-saddle and lower-saddle flows with the up-parabola and down-parabola flows and the increasing- and decreasing-inflection flows.

4.1 Two Single-Variable Cubic Systems

In this section, consider two-dimensional dynamical systems with a variable-crossing univariate cubic vector field and a self-univariate cubic vector field. The corresponding stability and bifurcations of equilibriums will be presented through the following theorem.

Theorem 4.1 *Consider a single-variable system with a self-univariate cubic vector field and a crossing-variable cubic vector field for* $A_{j_2 j_1 0} \neq 0\,(j_1, j_2 \in \{1, 2\}, j_1 \neq j_2)$ *as*

A. C. J. Luo, *Two-dimensional Single-Variable Cubic Nonlinear Systems, Vol. I*,
https://doi.org/10.1007/978-3-031-48472-8_4

$$\begin{aligned}\dot{x}_{j_2} &= f_{j_2}(\mathbf{x},\mathbf{p}) = A_{j_2j_13} + A_{j_2j_12}x_{j_1} + A_{j_2j_11}x_{j_1}^2 + A_{j_2j_10}x_{j_1}^3\\ &= a_{j_2j_10}(x_{j_1} - b_{j_2j_11})(x_{j_1}^2 + B_{j_2j_11}x_{j_1} + C_{j_2j_11}),\\ \dot{x}_{j_1} &= f_{j_1}(\mathbf{x},\mathbf{p}) = A_{j_1j_13} + A_{j_1j_12}x_{j_1} + A_{j_1j_11}x_{j_1}^2 + A_{j_1j_10}x_{j_1}^3\\ &= a_{j_1j_10}(x_{j_1} - b_{j_1j_11})(x_{j_1}^2 + B_{j_1j_11}x_{j_1} + C_{j_1j_11}),\end{aligned} \tag{4.1}$$

with

$$\Delta_{jj_1} = (B_{jj_11})^2 - 4C_{jj_11} \ \text{ for } j \in \{j_1, j_2\}, j_1, j_2 \in \{1,2\} j_1 \neq j_2. \tag{4.2}$$

(i) If

$$\Delta_{j_1j_1} < 0 \text{ and } \Delta_{j_2j_1} < 0, \tag{4.3}$$

then

$$\begin{aligned}&\dot{x}_{j_2} = a_{j_2j_10}(x_{j_1} - b_{j_2j_11})\left[(x_{j_1} - a_{j_2j_1})^2 + b_{j_2j_1}\right],\\ &\dot{x}_{j_1} = a_{j_1j_10}(x_{j_1} - b_{j_1j_11})\left[(x_{j_1} - a_{j_1j_1})^2 + b_{j_1j_1}\right],\\ &\text{with } a_{sj_1} = -\frac{1}{2}B_{sj_11}, b_{sj_1} = \frac{1}{4}(-\Delta_{sj_1}), s = j_1, j_2.\end{aligned} \tag{4.4}$$

(i_1) For $a_{j_2j_11} \neq a_{j_1j_11}$, the first integral manifold is

$$\begin{aligned}x_{j_2} - x_{j_20} = &\frac{a_{j_2j_10}}{a_{j_1j_10}}\bigg(\bigg[x_{j_2} - x_{j_20} + (a_{j_1j_1} - a_{j_2j_1})\ln\frac{|(x_{j_1} - a_{j_1j_1})^2 + b_{j_1j_1}|}{|(x_{j_10} - a_{j_1j_1})^2 + b_{j_1j_1}|}\\ &+\left[(a_{j_1j_1} - a_{j_2j_1})^2 + b_{j_2j_1} - b_{j_1j_1}\right]\frac{1}{\sqrt{b_{j_1j_1}}}(\arctan\frac{x_{j_1} - a_{j_1j_1}}{\sqrt{b_{j_1j_1}}}\\ &- \arctan\frac{x_{j_10} - a_{j_1j_1}}{\sqrt{b_{j_1j_1}}})\bigg] + (a_{j_1j_11} - a_{j_2j_11})\bigg\{\frac{1}{2}\ln\frac{|(x_{j_1} - a_{j_1j_1})^2 + b_{j_1j_1}|}{|(x_{j_10} - a_{j_1j_1})^2 + b_{j_1j_1}|}\\ &+\left[2(a_{j_1j_11} - a_{j_2j_1}) + a_{j_1j_1} - a_{j_1j_11}\right]\frac{1}{\sqrt{b_{j_1j_1}}}(\arctan\frac{x_{j_1} - a_{j_1j_1}}{\sqrt{b_{j_1j_1}}}\\ &- \arctan\frac{x_{j_10} - a_{j_1j_1}}{\sqrt{b_{j_1j_1}}}) + \frac{(a_{j_1j_11} - a_{j_2j_1})^2 + b_{j_2j_1}}{(a_{j_1j_11} - a_{j_1j_1})^2 + b_{j_1j_1}}\bigg[\ln\frac{|x_{j_1} - a_{j_1j_11}|}{|x_{j_10} - a_{j_1j_11}|}\\ &-\frac{1}{2}\ln\frac{|(x_{j_1} - a_{j_1j_1})^2 + b_{j_1j_1}|}{|(x_{j_10} - a_{j_1j_1})^2 + b_{j_1j_1}|} - \frac{a_{j_1j_11} - a_{j_1j_1}}{\sqrt{b_{j_1j_1}}}(\arctan\frac{x_{j_1} - a_{j_1j_1}}{\sqrt{b_{j_1j_1}}}\\ &- \arctan\frac{x_{j_10} - a_{j_1j_1}}{\sqrt{b_{j_1j_1}}})\bigg]\bigg\}\bigg), \quad j_1, j_2 \in \{1,2\}, j_1 \neq j_2.\end{aligned} \tag{4.5}$$

(i_{1a}) The flow at $x^*_{j_1} = a_{j_1j_1}1 \neq a_{j_2j_1}1$ for the system has the following properties.

- *For $a_{j_1j_1}0 > 0$ and $a_{j_2j_1}0(a_{j_1j_1}1 - a_{j_2j_1}1) > 0$,*

$$(a_{j_1j_1}1, \dot{x}_{j_2}) = \underbrace{(\text{SO}, \text{pF})}_{\text{source flow } (+)}. \tag{4.6}$$

The flow of $x^*_{j_1} = a_{j_1j_1}1$ is an (SO,pF)*-positive source flow.*

- *For $a_{j_1j_1}0 < 0$ and $a_{j_2j_1}0(a_{j_1j_1}1 - a_{j_2j_1}1) > 0$,*

$$(a_{j_1j_1}1, \dot{x}_{j_2}) = \underbrace{(\text{SI}, \text{pF})}_{\text{sink flow } (+)}. \tag{4.7}$$

The flow of $x^*_{j_1} = a_{j_1j_1}1$ is an (SI, pF)*-positive sink flow.*

- *For $a_{j_1j_1}0 > 0$ and $a_{j_2j_1}0(a_{j_1j_1}1 - a_{j_2j_1}1) < 0$,*

$$(a_{j_1j_1}1, \dot{x}_{j_2}) = \underbrace{(\text{SO}, \text{nF})}_{\text{source flow } (-)}. \tag{4.8}$$

The flow of $x^*_{j_1} = a_{j_1j_1}1$ is an (SO, nF)*-negative source flow.*

- *For $a_{j_1j_1}0 < 0$ and $a_{j_2j_1}0(a_{j_1j_1}1 - a_{j_2j_1}1) < 0$,*

$$(a_{j_1j_1}1, \dot{x}_{j_2}) = \underbrace{(\text{SI}, \text{nF})}_{\text{sink flow } (-)}. \tag{4.9}$$

The flow of $x^*_{j_1} = a_{j_1j_1}1$ is an (SI,nF*)-negative sink flow.*

(i_{1b}) The flow at $x^*_{j_1} = a_{j_2j_1}1 \neq a_{j_1j_1}1$ for the system has the following properties.

- *For $a_{j_1j_1}0(a_{j_2j_1}1 - a_{j_1j_1}1) > 0$ and $a_{j_2j_1}0 > 0$,*

$$(\dot{x}_{j_1}, a_{j_2j_1}1) = \underbrace{(\text{pF}, \text{UP})}_{\text{up-parabola flow } (+)}. \tag{4.10}$$

The flow of $x^*_{j_1} = a_{j_2j_1}1$ is a (pF,UP)*-positive up-parabola flow.*

- *For $a_{j_1j_1}0(a_{j_2j_1}1 - a_{j_1j_1}1) < 0$ and $a_{j_2j_1}0 > 0$,*

$$(\dot{x}_{j_1}, a_{j_2j_1}1) = \underbrace{(\text{nF}, \text{DP})}_{\text{down-parabola flow } (-)}. \tag{4.11}$$

The flow of $x^*_{j_1} = a_{j_2j_1}1$ is an (nF,DP*)-negative down-parabola flow.*

- *For* $a_{j_1j_10}(a_{j_2j_11}-a_{j_1j_11})>0$ *and* $a_{j_2j_10}<0$,

$$(\dot{x}_{j_1}, a_{j_2j_11}) = \underbrace{(\text{pF}, \text{DP})}_{\text{down-parabola flow }(+)}. \tag{4.12}$$

The flow of $x_{j_1}^* = a_{j_2j_11}$ *is a* (pF,DP)-*positive down-parabola flow.*

- *For* $a_{j_1j_10}(a_{j_2j_11}-a_{j_1j_11})<0$ *and* $a_{j_2j_10}<0$,

$$(\dot{x}_{j_1}, a_{j_2j_11}) = \underbrace{(\text{nF}, \text{UP})}_{\text{up-parabola flow }(-)}. \tag{4.13}$$

The flow of $x_{j_1}^* = a_{j_2j_11}$ *is an* (nF,UP)-*negative up-parabola flow.*

(i_2) *For* $a_{j_2j_11}=a_{j_1j_11}$, *the first integral manifold is*

$$\begin{aligned} x_{j_2} - x_{j_20} = & \frac{a_{j_2j_10}}{a_{j_1j_10}}\Big[x_{j_1} - x_{j_10} + (a_{j_1j_1} - a_{j_2j_1}) \ln \frac{|(x_{j_1} - a_{j_1j_1})^2 + b_{j_1j_1}|}{|(x_{j_10} - a_{j_1j_1})^2 + b_{j_1j_1}|} \\ & + \big[(a_{j_1j_1} - a_{j_2j_1})^2 + b_{j_2j_1} - b_{j_1j_1}\big] \\ & \times \frac{1}{\sqrt{b_{j_1j_1}}}(\arctan \frac{x_{j_1} - a_{j_1j_1}}{\sqrt{b_{j_1j_1}}} - \arctan \frac{x_{j_10} - a_{j_1j_1}}{\sqrt{b_{j_1j_1}}})\Big], \\ & j_1, j_2 \in \{1,2\}, j_1 \neq j_2. \end{aligned} \tag{4.14}$$

The equilibrium of $x_{j_1}^* = a_{j_2j_11} = a_{j_1j_11}$ with $\overline{x}_{j_2} \in (-\infty, \infty)$ *for the system has the following properties.*

- *For* $a_{j_1j_10}>0$ *and* $a_{j_2j_10}>0$,

$$(a_{j_1j_11}, \overline{x}_{j_2}) = \underbrace{(\text{SO}, \text{II})}_{\text{increasing-inflection source}}. \tag{4.15}$$

The equilibrium of $x_{j_1}^* = a_{j_1j_11} = a_{j_2j_11}$ *is an* (SO,II)-*increasing-inflection source infinite-equilibrium.*

- *For* $a_{j_1j_10}<0$ *and* $a_{j_2j_10}>0$,

$$(a_{j_1j_11}, \overline{x}_{j_2}) = \underbrace{(\text{SI}, \text{DI})}_{\text{decreasing-inflection sink}}. \tag{4.16}$$

The equilibrium of $x_{j_1}^* = a_{j_1j_11} = a_{j_2j_11}$ *is an* (SI,DI)-*decreasing-inflection sink infinite-equilibrium.*

- *For $a_{j_1j_1 0} > 0$ and $a_{j_2j_1 0} < 0$,*

$$(a_{j_1j_1 1}, \bar{x}_{j_2}) = \underbrace{(\text{SO}, \text{DI})}_{\text{decreasing-inflection source}}. \tag{4.17}$$

The equilibrium of $x^*_{j_1} = a_{j_1j_1 1} = a_{j_2j_1 1}$ is an (SO,DI)*-decreasing-inflection source infinite-equilibrium.*

- *For $a_{j_1j_1 0} < 0$ and $a_{j_2j_1 0} < 0$,*

$$(a_{j_1j_1 1}, \bar{x}_{j_2}) = \underbrace{(\text{SI}, \text{II})}_{\text{increasing-inflection sink}}. \tag{4.18}$$

The equilibrium at $x^*_{j_1} = a_{j_1j_1 1} = a_{j_2j_1 1}$ is an (SI,II)*-increasing-inflection sink infinite-equilibrium.*

– *The inflection infinite-equilibriums are for the switching bifurcation of a parabola flow with a source (sink) flow.*

(ii) If

$$\Delta_{j_1j_1} = 0 \text{ and } \Delta_{j_2j_1} < 0, \tag{4.19}$$

the dynamical system becomes

$$\begin{aligned}
&\dot{x}_{j_1} = a_{j_1j_1 0}(x_{j_1} - a_{j_1j_1 s_1})(x_{j_1} - a_{j_1j_1 s_2})^2,\\
&\dot{x}_{j_2} = a_{j_2j_1 0}(x_{j_1} - a_{j_2j_1 1})\left[(x_{j_1} - a_{j_2j_1})^2 + b_{j_2j_1}\right],\\
&\text{with } \{a_{j_1j_1 1}, a_{j_1j_1 2}\} = \text{sort}\{b_{j_1j_1 1}, a_{j_1j_1}\}, a_{j_1j_1 1} < a_{j_1j_1 2};\\
&a_{j_2j_1 1} = b_{j_2j_1 1}; a_{j_1j_1} = -\frac{1}{2}B_{j_1j_1 1}, b_{j_1j_1} = \frac{1}{4}(-\Delta_{j_1j_1}) = 0;\\
&a_{j_2j_1} = -\frac{1}{2}B_{j_2j_1 1}, b_{j_2j_1} = \frac{1}{4}(-\Delta_{j_2j_1});\\
&j_1, j_2 \in \{1, 2\}; j_1 \neq j_2.
\end{aligned} \tag{4.20}$$

(ii$_1$) The first integral manifold is

$$
\begin{aligned}
x_{j_2} - x_{j_2 0} = {} & \frac{a_{j_2 j_1 0}}{a_{j_1 j_1 0}} \Big(\Big\{ x_{j_1} - x_{j_1 0} + 2(a_{j_1 j_1 s_2} - a_{j_2 j_1}) \ln \frac{| x_{j_1} - a_{j_1 j_1 s_2} |}{| x_{j_1 0} - a_{j_1 j_1 s_2} |} \\
& - \big[(a_{j_1 j_1 s_2} - a_{j_2 j_1})^2 + b_{j_2 j_1}\big]\Big(\frac{1}{x_{j_1} - a_{j_1 j_1 s_2}} - \frac{1}{x_{j_1 0} - a_{j_1 j_1 s_2}}\Big)\Big\} \\
& + (a_{j_1 j_1 s_1} - a_{j_2 j_1 1}) \Big\{ \ln \frac{| x_{j_1} - a_{j_1 j_1 s_2} |}{| x_{j_1 0} - a_{j_1 j_1 s_2} |} \\
& - (a_{j_1 j_1} + a_{j_1 j_1 s_1} - 2a_{j_2 j_1})\Big(\frac{1}{x_{j_1} - a_{j_1 j_1 s_2}} - \frac{1}{x_{j_1 0} - a_{j_1 j_1 s_2}}\Big) \\
& + \frac{(a_{j_1 j_1 s_1} - a_{j_2 j_1})^2 + b_{j_2 j_1}}{(a_{j_1 j_1 s_1} - a_{j_1 j_1 s_2})^2} \Big[\ln \frac{| x_{j_1} - a_{j_1 j_1 s_1} |}{| x_{j_1 0} - a_{j_1 j_1 s_1} |} - \ln \frac{| x_{j_1} - a_{j_1 j_1 s_2} |}{| x_{j_1 0} - a_{j_1 j_1 s_2} |} \\
& - (a_{j_1 j_1 s_2} - a_{j_1 j_1 s_1})\Big(\frac{1}{x_{j_1} - a_{j_1 j_1 s_2}} - \frac{1}{x_{j_1 0} - a_{j_1 j_1 s_2}}\Big)\Big]\Big\}\Big); \\
& j_1, j_2 \in \{1, 2\}, j_1 \neq j_2.
\end{aligned}
\tag{4.21}
$$

(ii$_{1a}$) The equilibrium of $x_{j_1}^* = a_{j_1 j_1 s_1}$ for the x_{j_1}-direction has the following properties:

- *For $a_{j_1 j_1 0}(a_{j_1 j_1 s_1} - a_{j_1 j_1 s_2})^2 > 0$ and $a_{j_2 j_1 0}(a_{j_1 j_1 s_1} - a_{j_2 j_1 1}) > 0$,*

$$
(a_{j_1 j_1 s_1}, \dot{x}_{j_2}) = \underbrace{(\text{SO}, \text{pF})}_{\text{source flow } (+)} . \tag{4.22}
$$

The flow at $x_{j_1}^* = a_{j_1 j_1 s_1}$ is an (SO,pF)*-positive source flow.*

- *For $a_{j_1 j_1 0}(a_{j_1 j_1 s_1} - a_{j_1 j_1 s_2})^2 < 0$ and $a_{j_2 j_1 0}(a_{j_1 j_1 s_1} - a_{j_2 j_1 1}) > 0$,*

$$
(a_{j_1 j_1 s_1}, \dot{x}_{j_2}) = \underbrace{(\text{SI}, \text{pF})}_{\text{sink flow } (+)} . \tag{4.23}
$$

The flow at $x_{j_1}^* = a_{j_1 j_1 s_1}$ is an (SI,pF)*-positive sink flow.*

- *For $a_{j_1 j_1 0}(a_{j_1 j_1 s_1} - a_{j_1 j_1 s_2})^2 < 0$ and $a_{j_2 j_1 0}(a_{j_1 j_1 s_1} - a_{j_2 j_1 1}) > 0$,*

$$
(a_{j_1 j_1 s_1}, \dot{x}_{j_2}) = \underbrace{(\text{SO}, \text{nF})}_{\text{source flow } (-)} . \tag{4.24}
$$

The flow at $x_{j_1}^* = a_{j_1 j_1 s_1}$ is an (SO,nF)*-negative source flow.*

- *For* $a_{j_1j_1 0}(a_{j_1j_1s_1} - a_{j_1j_1s_2})^2 < 0$ *and* $a_{j_2j_1 0}(a_{j_1j_1s_1} - a_{j_2j_1 1}) < 0$,

$$(a_{j_1j_1s_1}, \dot{x}_{j_2}) = \underbrace{(\mathrm{SI}, \mathrm{nF})}_{\text{sink flow }(-)} . \tag{4.25}$$

The flow at $x^*_{j_1} = a_{j_1j_1s_1}$ *is an* (SI,nF)-*negative sink flow.*

(ii_{1b}) *The equilibrium of* $x^*_{j_1} = a_{j_1j_1s_2}$ *for the* x_{j_1}*-direction has the following properties:*

- *For* $a_{j_1j_1 0}(a_{j_1j_1s_2} - a_{j_1j_1s_1}) > 0$ *and* $a_{j_2j_1 0}(a_{j_1j_1s_2} - a_{j_2j_1 1}) > 0$,

$$(a_{j_1j_1s_2}, \dot{x}_{j_2}) = \underbrace{(\mathrm{US}, \mathrm{pF})}_{\text{up-parabola flow }(+)} . \tag{4.26}$$

The flow at $x^*_{j_1} = a_{j_1j_1s_2}$ *is a* (US,pF)-*positive upper-saddle flow.*

- *For* $a_{j_1j_1 0}(a_{j_1j_1s_2} - a_{j_1j_1s_1}) < 0$ *and* $a_{j_2j_1 0}(a_{j_1j_1s_2} - a_{j_2j_1 1}) > 0$,

$$(a_{j_1j_1s_2}, \dot{x}_{j_2}) = \underbrace{(\mathrm{LS}, \mathrm{pF})}_{\text{lower-saddle flow }(+)} . \tag{4.27}$$

The flow at $x^*_{j_1} = a_{j_1j_1s_2}$ *is an* (LS,pF)-*positive lower-saddle flow.*

- *For* $a_{j_1j_1 0}(a_{j_1j_1s_2} - a_{j_1j_1s_1}) > 0$ *and* $a_{j_2j_1 0}(a_{j_1j_1s_2} - a_{j_2j_1 1}) < 0$,

$$(a_{j_1j_1s_2}, \dot{x}_{j_2}) = \underbrace{(\mathrm{US}, \mathrm{nF})}_{\text{upper-saddle flow }(-)} . \tag{4.28}$$

The flow at $x^*_{j_1} = a_{j_1j_1s_2}$ *is a* (US,nF)-*negative upper-saddle flow.*

- *For* $a_{j_1j_1 0}(a_{j_1j_1s_2} - a_{j_1j_1s_1}) < 0$ *and* $a_{j_2j_1 0}(a_{j_1j_1s_2} - a_{j_2j_1 1}) < 0$,

$$(a_{j_1j_1s_2}, \dot{x}_{j_2}) = \underbrace{(\mathrm{LS}, \mathrm{nF})}_{\text{lower-saddle flow }(-)} . \tag{4.29}$$

The flow at $x^*_{j_1} = a_{j_1j_1s_2}$ *is an* (LS,nF)-*negative lower-saddle flow.*

– *The upper-saddle and lower-saddle flow bifurcations are for the sink and source flows.*

(ii_{1c}) *The flow at* $x^*_{j_1} = a_{j_2j_1 1}$ *for the system has the following properties.*

- *For* $a_{j_1j_1 0}(a_{j_2j_1 1} - a_{j_1j_1s_1})(a_{j_2j_1 1} - a_{j_1j_1s_2})^2 > 0$ *and* $a_{j_2j_1 0} > 0$,

$$(\dot{x}_{j_1}, a_{j_2j_1 1}) = \underbrace{(\mathrm{pF}, \mathrm{UP})}_{\text{up-parabola flow }(+)} . \tag{4.30}$$

The flow of $x^*_{j_1} = a_{j_2j_1 1}$ *is a* (pF,UP)-*positive up-parabola flow.*

- *For* $a_{j_1j_10}(a_{j_2j_11}-a_{j_1j_1s_1})(a_{j_2j_11}-a_{j_1j_1s_2})^2<0$ *and* $a_{j_2j_10}>0$,

$$(\dot{x}_{j_1},a_{j_2j_11})=\underbrace{(\mathrm{nF},\mathrm{DP})}_{\text{down-parabola flow }(-)}. \tag{4.31}$$

The flow of $x^*_{j_1}=a_{j_2j_11}$ *is an* (nF,DP)*-negative down-parabola flow.*

- *For* $a_{j_1j_10}(a_{j_2j_11}-a_{j_1j_1s_1})(a_{j_2j_11}-a_{j_1j_1s_2})^2>0$ *and* $a_{j_2j_10}<0$,

$$(\dot{x}_{j_1},a_{j_2j_11})=\underbrace{(\mathrm{pF},\mathrm{DP})}_{\text{down-parabola flow }(+)}. \tag{4.32}$$

The flow of $x^*_{j_1}=a_{j_2j_11}$ *is a* (pF,DP)*-positive down-parabola flow.*

- *For* $a_{j_1j_10}(a_{j_2j_11}-a_{j_1j_1s_1})(a_{j_2j_11}-a_{j_1j_1s_2})^2<0$ *and* $a_{j_2j_10}<0$,

$$(\dot{x}_{j_1},a_{j_2j_11})=\underbrace{(\mathrm{nF},\mathrm{UP})}_{\text{up-parabola flow }(-)}. \tag{4.33}$$

The flow of $x^*_{j_1}=a_{j_2j_11}$ *is an* (nF,UP)*-negative up-parabola flow.*

(ii$_2$) For $a_{j_1j_1s_1}=a_{j_2j_11}$, *the first integral manifold is*

$$\begin{aligned}x_{j_2}-x_{j_20}=\frac{a_{j_2j_10}}{a_{j_1j_10}}\Big\{&x_{j_1}-x_{j_10}+2(a_{j_1j_1s_2}-a_{j_2j_1})\ln\frac{|x_{j_1}-a_{j_1j_1s_2}|}{|x_{j_10}-a_{j_1j_1s_2}|}\\&-\big[(a_{j_1j_1s_2}-a_{j_2j_1})^2+b_{j_2j_1}\big]\Big(\frac{1}{x_{j_1}-a_{j_1j_1s_2}}-\frac{1}{x_{j_10}-a_{j_1j_1s_2}}\Big)\Big\}.\end{aligned} \tag{4.34}$$

The infinite-equilibrium of $x^*_{j_1}=a_{j_1j_1s_1}=a_{j_2j_11}$ *with* $\bar{x}_{j_2}\in(-\infty,\infty)$ *for the* x_{j_1} *-direction has the following properties.*

- *For* $a_{j_1j_10}(a_{j_1j_1s_1}-a_{j_1j_1s_2})^2>0$ *and* $a_{j_2j_10}>0$,

$$(a_{j_1j_1s_1},\bar{x}_{j_2})=\underbrace{(\mathrm{SO},\mathrm{II})}_{\text{increasing-infelction source}}. \tag{4.35}$$

The infinite-equilibrium of $x^*_{j_1}=a_{j_1j_1s_1}=a_{j_2j_11}$ *is an* (SO,II)*-increasing-inflection source.*

- *For* $a_{j_1j_10}(a_{j_1j_1s_1}-a_{j_1j_1s_2})^2<0$ *and* $a_{j_2j_10}>0$,

$$(a_{j_1j_1s_1},\bar{x}_{j_2})=\underbrace{(\mathrm{SI},\mathrm{DI})}_{\text{decreasing-infelction sink}}. \tag{4.36}$$

The infinite-equilibrium of $x^*_{j_1}=a_{j_1j_1s_1}=a_{j_2j_11}$ *is an* (SI,DI)*-decreasing-inflection sink.*

- *For* $a_{j_1j_10}(a_{j_1j_1s_1}-a_{j_1j_1s_2})^2>0$ *and* $a_{j_2j_10}<0$,

$$(a_{j_1j_1s_1},\bar{x}_{j_2})=\underbrace{(\mathrm{SO},\mathrm{DI})}_{\text{decreasing-infelction source}}. \tag{4.37}$$

The infinite-equilibrium of $x^*_{j_1}=a_{j_1j_1s_1}=a_{j_2j_11}$ *is an* (SO,DI)-*decreasing-inflection source.*

- *For* $a_{j_1j_10}(a_{j_1j_1s_1}-a_{j_1j_1s_2})^2<0$ *and* $a_{j_2j_10}<0$,

$$(a_{j_1j_1s_1},\bar{x}_{j_2})=\underbrace{(\mathrm{SI},\mathrm{II})}_{\text{increasing-infelction sink}}. \tag{4.38}$$

The infinite-equilibrium of $x^*_{j_1}=a_{j_1j_1s_1}=a_{j_2j_11}$ *is an* (SI,II)-*increasing-inflection sink.*

– *The inflection source and sink are the switching bifurcations of sink and source flows with parabola flows.*

(ii$_3$*) For* $a_{j_1j_1s_2}=a_{j_2j_11}$, *the first integral manifold is*

$$\begin{aligned}x_{j_2}-x_{j_20}=\frac{a_{j_2j_10}}{a_{j_1j_10}}\Bigg[&x_{j_1}-x_{j_10}+\frac{(a_{j_1j_1s_1}-a_{j_2j_1})^2+b_{j_2j_1}}{(a_{j_1j_1s_1}-a_{j_1j_1s_2})}\ln\frac{|\,x_{j_1}-a_{j_1j_1s_1}\,|}{|\,x_{j_10}-a_{j_1j_1s_1}\,|}\\&+\frac{(a_{j_1j_1s_2}-a_{j_2j_1})^2+b_{j_2j_1}}{(a_{j_1j_1s_2}-a_{j_1j_1s_1})}\ln\frac{|\,x_{j_1}-a_{j_1j_1s_2}\,|}{|\,x_{j_10}-a_{j_1j_1s_2}\,|}\Bigg].\end{aligned} \tag{4.39}$$

The finite-equilibrium of $x^*_{j_1}=a_{j_1j_1s_2}=a_{j_2j_11}$ *with* $\bar{x}_{j_2}\in(-\infty,\infty)$ *for the* x_{j_1}-*direction has the following properties.*

- *For* $a_{j_1j_10}(a_{j_1j_1s_2}-a_{j_1j_1s_1})>0$ *and* $a_{j_2j_10}>0$,

$$(a_{j_1j_1s_2},\bar{x}_{j_2})=\underbrace{(\mathrm{US},\mathrm{DU})}_{\text{down-up upper-saddle}}. \tag{4.40}$$

The finite-equilibrium of $x^*_{j_1}=a_{j_1j_1s_2}=a_{j_2j_11}$ *is a* (US,DU)-*down-up upper-saddle flow.*

- *For* $a_{j_1j_10}(a_{j_1j_1s_2}-a_{j_1j_1s_1})<0$ *and* $a_{j_2j_10}>0$,

$$(a_{j_1j_1s_2},\bar{x}_{j_2})=\underbrace{(\mathrm{LS},\mathrm{UD})}_{\text{up-down lower-saddle}}. \tag{4.41}$$

The finite-equilibrium of $x^*_{j_1}=a_{j_1j_1s_2}=a_{j_2j_11}$ *is an* (LS,UD)-*up-down lower-saddle flow.*

- *For* $a_{j_1j_10}(a_{j_1j_1s_2}-a_{j_1j_1s_1})>0$ *and* $a_{j_2j_10}<0$,

$$(a_{j_1j_1s_2},\bar{x}_{j_2})=\underbrace{(\mathrm{US},\mathrm{UD})}_{\text{up-down upper-saddle}}. \tag{4.42}$$

The finite-equilibrium of $x^*_{j_1}=a_{j_1j_1s_2}=a_{j_2j_11}$ *is a* (US,UD)-*up-down upper-saddle flow.*

- *For* $a_{j_1j_10}(a_{j_1j_1s_2}-a_{j_1j_1s_1})<0$ *and* $a_{j_2j_10}<0$,

$$(a_{j_1j_1s_2},\bar{x}_{j_2})=\underbrace{(\mathrm{LS},\mathrm{DU})}_{\text{down-up lower-saddle}}. \tag{4.43}$$

The finite-equilibrium of $x^*_{j_1}=a_{j_1j_1s_2}=a_{j_2j_11}$ *is a* (LS,DU)-*down-up upper-saddle flow.*

– *The up-down and down-up saddle bifurcations are for switching bifurcations of saddle flow with parabola flows.*

(iii) For $a_{j_1j_1s_2}=a_{j_1j_1s_1}\equiv a_{j_1j_11}$, *the dynamical system becomes*

$$\begin{aligned}&\dot{x}_{j_1}=a_{j_1j_10}(x_{j_1}-a_{j_1j_11})^3\\&\dot{x}_{j_2}=a_{j_2j_10}(x_{j_1}-a_{j_2j_11})\left[(x_{j_1}-a_{j_2j_1})^2+b_{j_2j_1}\right],\\&j_1,j_2\in\{1,2\};j_1\neq j_2.\end{aligned} \tag{4.44}$$

The first integral manifold is

$$\begin{aligned}x_{j_2}-x_{j_20}=&\frac{a_{j_2j_10}}{a_{j_1j_10}}\Big(\Big\{x_{j_1}-x_{j_10}+2(a_{j_1j_11}-a_{j_2j_1})\ln\frac{|x_{j_1}-a_{j_1j_11}|}{|x_{j_10}-a_{j_1j_11}|}\\&-\left[(a_{j_1j_11}-a_{j_2j_1})^2+b_{j_2j_1}\right]\left(\frac{1}{x_{j_1}-a_{j_1j_11}}-\frac{1}{x_{j_10}-a_{j_1j_11}}\right)\Big\}\\&+(a_{j_1j_11}-a_{j_2j_11})\Big\{\ln\frac{|x_{j_1}-a_{j_1j_11}|}{|x_{j_10}-a_{j_1j_11}|}\\&-2(a_{j_1j_11}-a_{j_2j_1})\left(\frac{1}{x_{j_1}-a_{j_1j_11}}-\frac{1}{x_{j_10}-a_{j_1j_11}}\right)\\&-\frac{(a_{j_1j_11}-a_{j_2j_1})^2+b_{j_2j_1}}{2}\left[\frac{1}{(x_{j_1}-a_{j_1j_11})^2}-\frac{1}{(x_{j_10}-a_{j_1j_11})^2}\right]\Big\}\Big),\\&j_1,j_2\in\{1,2\},j_1\neq j_2.\end{aligned} \tag{4.45}$$

(iii$_{1a}$*) The equilibrium of* $x^*_{j_1}=a_{j_1j_11}$ *for the* x_{j_1}*-direction has the following properties:*

- *For* $a_{j_1j_10}>0$ *and* $a_{j_2j_10}(a_{j_1j_11}-a_{j_2j_11})>0$,

$$(a_{j_1j_1 1}, \dot{x}_{j_2}) = \underbrace{(3^{\text{rd}}\text{SO}, \text{pF})}_{\text{third-order source flow }(+)}. \tag{4.46}$$

The flow of $x_{j_1}^* = a_{j_1j_1 1}$ *is a* (3rdSO,pF)*-positive third-order source flow.*

- *For* $a_{j_1j_1 0} < 0$ *and* $a_{j_2j_1 0}(a_{j_1j_1 1} - a_{j_2j_1 1}) > 0$,

$$(a_{j_1j_1 1}, \dot{x}_{j_2}) = \underbrace{(3^{\text{rd}}\text{SI}, \text{pF})}_{\text{third-order sink flow }(+)}. \tag{4.47}$$

The flow of $x_{j_1}^* = a_{j_1j_1 1}$ *is a* (3rdSI,pF)*-positive third-order sink flow.*

- *For* $a_{j_1j_1 0} > 0$ *and* $a_{j_2j_1 0}(a_{j_1j_1 1} - a_{j_2j_1 1}) < 0$,

$$(a_{j_1j_1 1}, \dot{x}_{j_2}) = \underbrace{(3^{\text{rd}}\text{SO}, \text{nF})}_{\text{third-order source flow }(-)}. \tag{4.48}$$

The flow of $x_{j_1}^* = a_{j_1j_1 1}$ *is a* (3rdSO,nF)*-negative third-order source flow.*

- *For* $a_{j_1j_1 0} < 0$ *and* $a_{j_2j_1 0}(a_{j_1j_1 1} - a_{j_2j_1 1}) < 0$,

$$(a_{j_1j_1 1}, \dot{x}_{j_2}) = \underbrace{(3^{\text{rd}}\text{SI}, \text{nF})}_{\text{third-order sink flow }(+)}. \tag{4.49}$$

The flow of $x_{j_1}^* = a_{j_1j_1 1}$ *is a* (3rdSI,nF)*-negative third-order sink flow.*

– *The third-order source and sink flow bifurcations are the appearing and switching bifurcations for the source-sink-source and sink-source-sink flows.*

(*iii*$_{1b}$) *The flow at* $x_{j_1}^* = a_{j_2j_1 1}$ *for the system has the following properties.*

- *For* $a_{j_1j_1 0}(a_{j_2j_1 1} - a_{j_1j_1 1})^3 > 0$ *and* $a_{j_2j_1 0} > 0$,

$$(\dot{x}_{j_1}, a_{j_2j_1 1}) = \underbrace{(\text{pF}, \text{UP})}_{\text{up-parabola flow }(+)}. \tag{4.50}$$

The flow of $x_{j_1}^* = a_{j_2j_1 1}$ *is a* (pF,UP)*-positive up-parabola flow.*

- *For* $a_{j_1j_1 0}(a_{j_2j_1 1} - a_{j_1j_1 1})^3 < 0$ *and* $a_{j_2j_1 0} > 0$,

$$(\dot{x}_{j_1}, a_{j_2j_1 1}) = \underbrace{(\text{nF}, \text{DP})}_{\text{down-parabola flow }(-)}. \tag{4.51}$$

The flow of $x_{j_1}^* = a_{j_2j_1 1}$ *is an* (nF,DP)*-negative down-parabola flow.*

- *For* $a_{j_1j_10}(a_{j_2j_11} - a_{j_1j_11})^3 > 0$ *and* $a_{j_2j_10} < 0$,

$$(\dot{x}_{j_1}, a_{j_2j_11}) = \underbrace{(\text{pF}, \text{DP})}_{\text{down-parabola flow } (+)}. \tag{4.52}$$

The flow of $x_{j_1}^* = a_{j_2j_11}$ *is a* (pF,DP)*-positive down-parabola flow.*

- *For* $a_{j_1j_10}(a_{j_2j_11} - a_{j_1j_11})^3 < 0$ *and* $a_{j_2j_10} < 0$,

$$(\dot{x}_{j_1}, a_{j_2j_11}) = \underbrace{(\text{nF}, \text{UP})}_{\text{up-parabola flow } (-)}. \tag{4.53}$$

The flow of $x_{j_1}^* = a_{j_2j_11}$ *is an* (nF,UP)*-negative up-parabola flow.*

(iii₂) For $a_{j_1j_11} = a_{j_2j_11}$, *the first integral manifold is*

$$\begin{aligned} x_{j_2} - x_{j_20} = & \frac{a_{j_2j_10}}{a_{j_1j_10}} \Big\{ x_{j_1} - x_{j_10} + 2(a_{j_1j_11} - a_{j_2j_1}) \ln \frac{|x_{j_1} - a_{j_1j_11}|}{|x_{j_10} - a_{j_1j_11}|} \\ & - [(a_{j_1j_11} - a_{j_2j_1})^2 + b_{j_2j_1}] (\frac{1}{x_{j_1} - a_{j_1j_11}} - \frac{1}{x_{j_10} - a_{j_1j_11}}) \Big\}; \\ & j_1, j_2 \in \{1, 2\}, j_1 \neq j_2. \end{aligned} \tag{4.54}$$

The infinite equilibrium of $x_{j_1}^* = a_{j_1j_11} = a_{j_2j_11}$ *with* $\bar{x}_{j_2} \in (-\infty, \infty)$ *for the* x_{j_1}*-direction has the following properties:*

- *For* $a_{j_1j_10} > 0$ *and* $a_{j_2j_10} > 0$,

$$(a_{j_1j_11}, \bar{x}_{j_2}) = \underbrace{(3^{\text{rd}}\text{SO}, (3:1)\text{IC})}_{\text{third-order increasing-concave source}}. \tag{4.55}$$

The infinite-equilibrium at $x_{j_1}^* = a_{j_1j_11} = a_{j_2j_11}$ *is a* (3^{rd}SO,(3:1)IC)*-third-order increasing-concave source.*

- *For* $a_{j_1j_10} < 0$ *and* $a_{j_2j_10} > 0$,

$$(a_{j_1j_11}, \bar{x}_{j_2}) = \underbrace{(3^{\text{rd}}\text{SI}, (3:1)\text{DC})}_{\text{third-order decreasing-inflection sink}}. \tag{4.56}$$

The infinite-equilibrium at $x_{j_1}^* = a_{j_1j_11} = a_{j_2j_11}$ *is a* (3^{rd}SI,(3:1)DC)*-third-order decreasing-concave sink.*

- *For* $a_{j_1j_10} > 0$ *and* $a_{j_2j_10} < 0$,

$$(a_{j_1j_11}, \bar{x}_{j_2}) = \underbrace{(3^{\text{rd}}\text{SO}, (3:1)\text{DC})}_{\text{third-order decreasing-inflection source}}. \tag{4.57}$$

The infinite-equilibrium at $x_{j_1}^* = a_{j_1 j_1 1} = a_{j_2 j_1 1}$ *is a* ($3^{\rm rd}$SO,(3:1)DC)-*third-order decreasing-concave source.*

- *For* $a_{j_1 j_1 0} < 0$ *and* $a_{j_2 j_1 0} < 0$,

$$(a_{j_1 j_1 1}, \bar{x}_{j_2}) = \underbrace{(3^{\rm rd}{\rm SI}, (3:1){\rm IC})}_{\text{third-order increasing-concave sink}}. \tag{4.58}$$

The infinite-equilibrium at $x_{j_1}^* = a_{j_1 j_1 1} = a_{j_2 j_1 1}$ *is a* ($3^{\rm rd}$SI,(3:1)IC)-*third-order increasing-concave sink.*

- *The third-order increasing and decreasing concave source (sink) infinite-equilibriums are the switching bifurcations of third-order source (sink) flows with a parabola flow.*
- *The third-order increasing and decreasing concave source (sink) infinite-equilibriums are the appearing bifurcations from source (sink) and parabola flows to source (sink), sink (source), and source (sink) flows with a parabola flow.*

(iv) If

$$\Delta_{j_1 j_1} > 0 \text{ and } \Delta_{j_2 j_1} < 0, \tag{4.59}$$

then

$$\begin{aligned}
&\dot{x}_{j_1} = a_{j_1 j_1 0}(x_{j_1} - a_{j_1 j_1 1})(x_{j_1} - a_{j_1 j_1 2})(x_{j_1} - a_{j_1 j_1 3}),\\
&\dot{x}_{j_2} = a_{j_2 j_1 0}(x_{j_1} - a_{j_2 j_1 1})\left[(x_{j_1} - a_{j_2 j_1})^2 + b_{j_2 j_1}\right],\\
&j_1, j_2 \in \{1, 2\}; j_1 \neq j_2.
\end{aligned} \tag{4.60}$$

where

$$\begin{aligned}
&\{a_{j_1 j_1 1}, a_{j_1 j_1 2}, a_{j_1 j_1 3}\} = \text{sort}\{b_{j_1 j_1 1}, b_{j_1 j_1 2}, b_{j_1 j_1 3}\}, a_{j_1 j_1 1} < a_{j_1 j_1 2} < a_{j_1 j_1 3};\\
&a_{211} = b_{211}; b_{j_1 j_1 2}, b_{j_1 j_1 3} = -\frac{1}{2}(B_{j_1 j_1 1} \pm \sqrt{\Delta_{j_1 j_1}}); a_{j_2 j_1} = -\frac{1}{2} B_{j_2 j_1 1}.
\end{aligned} \tag{4.61}$$

(iv_1) The first integral manifold is

$$\begin{aligned}
x_{j_2} - x_{j_2 0} = \frac{a_{j_1 j_1 0}}{a_{j_1 j_1 0}} \Bigg\{ \Bigg[& x_{j_1} - x_{j_1 0} + \frac{(a_{j_1 j_1 s_1} - a_{j_2 j_1})^2 + b_{j_2 j_1}}{(a_{j_1 j_1 s_1} - a_{j_1 j_1 s_2})} \ln \frac{| x_{j_1} - a_{j_1 j_1 s_1} |}{| x_{j_1 0} - a_{j_1 j_1 s_1} |}\\
& + \frac{(a_{j_1 j_1 s_2} - a_{j_2 j_1})^2 + b_{j_2 j_1}}{(a_{j_1 j_1 s_2} - a_{j_1 j_1 s_1})} \ln \frac{| x_{j_1} - a_{j_1 j_1 s_2} |}{| x_{j_1 0} - a_{j_1 j_1 s_2} |}\Bigg]\\
& + (a_{j_1 j_1 s_3} - a_{j_2 j_1 1}) \Bigg[\sum_{s_1 = 1}^{3} \frac{(a_{j_1 j_1 s_1} - a_{j_2 j_1 1})^2 + b_{j_2 j_1}}{\prod_{s_2 = 1, s_2 \neq s_1}^{3} (a_{j_1 j_1 s_1} - a_{j_1 j_1 s_2})} \ln \frac{| x_{j_1} - a_{j_1 j_1 s_1} |}{| x_{j_1 0} - a_{j_1 j_1 s_1} |}\Bigg]\Bigg\}.
\end{aligned} \tag{4.62}$$

(iv_{1a}) The equilibrium of $x^*_{j_1} = a_{j_1 j_1 s_1}$ (s_1, $s_2 \in \{1,2,3\}$, $s_1 \neq s_2$) in the x_{j_1}-direction possesses the following properties:

- *For $a_{j_1 j_1 0}\prod_{s_2=1, s_2\neq s_1}^{3}(a_{j_1 j_1 s_1} - a_{j_1 j_1 s_2}) > 0$ and $a_{j_2 j_1 0}(a_{j_1 j_1 s_1} - a_{j_2 j_1 1}) > 0$,*

$$(a_{j_1 j_1 s_1}, \dot{x}_{j_2}) = \underbrace{(\text{SO}, \text{pF})}_{\text{source flow } (+)} . \tag{4.63}$$

The flow of $x^*_{j_1} = a_{j_1 j_1 s_1}$ is an (SO,pF)*-positive source-flow.*

- *For $a_{j_1 j_1 0}\prod_{s_2=1, s_2\neq s_1}^{3}(a_{j_1 j_1 s_1} - a_{j_1 j_1 s_2}) < 0$ and $a_{j_2 j_1 0}(a_{j_1 j_1 s_1} - a_{j_2 j_1 1}) > 0$,*

$$(a_{j_1 j_1 s_1}, \dot{x}_{j_2}) = \underbrace{(\text{SI}, \text{pF})}_{\text{sink flow } (+)} . \tag{4.64}$$

The flow of $x^*_{j_1} = a_{j_1 j_1 s_1}$ is an (SI,pF)*-positive sink-flow.*

- *For $a_{j_1 j_1 0}\prod_{s_2=1, s_2\neq s_1}^{3}(a_{j_1 j_1 s_1} - a_{j_1 j_1 s_2}) > 0$ and $a_{j_2 j_1 0}(a_{j_1 j_1 s_1} - a_{j_2 j_1 1}) < 0$,*

$$(a_{j_1 j_1 s_1}, \dot{x}_{j_2}) = \underbrace{(\text{SO}, \text{nF})}_{\text{source flow } (-)} . \tag{4.65}$$

The flow of $x^*_{j_1} = a_{j_1 j_1 s_1}$ is an (SO,nF)*-negative source-flow.*

- *For $a_{j_1 j_1 0}\prod_{s_2=1, s_2\neq s_1}^{3}(a_{j_1 j_1 s_1} - a_{j_1 j_1 s_2}) < 0$ and $a_{j_2 j_1 0}(a_{j_1 j_1 s_1} - a_{j_2 j_1 1}) < 0$,*

$$(a_{j_1 j_1 s_1}, \dot{x}_{j_2}) = \underbrace{(\text{SI}, \text{nF})}_{\text{sink flow } (-)} . \tag{4.66}$$

The flow of $x^*_{j_1} = a_{j_1 j_1 s_1}$ is an (SI,nF)*-negative sink-flow.*

(iv_{1b}) The equilibrium of $x^*_{j_1} = a_{j_2 j_1 1} \neq a_{j_1 j_1 s_1}$ ($s_1 = 1, 2, 3$) in the x_{j_2}-direction possesses the following properties.

- *For $a_{j_1 j_1 0}\prod_{s_1=1}^{3}(a_{j_2 j_1 1} - a_{j_1 j_1 s_1}) > 0$ and $a_{j_2 j_1 0} > 0$,*

$$(\dot{x}_{j_1}, a_{j_2 j_1 1}) = \underbrace{(\text{pF}, \text{UP})}_{\text{up-parabola flow } (+)} . \tag{4.67}$$

The flow of $x^*_{j_1} = a_{j_2 j_1 1}$ is a (pF,UP)*-positive up-parabola flow.*

- *For $a_{j_1 j_1 0}\prod_{s_1=1}^{3}(a_{j_2 j_1 1} - a_{j_1 j_1 s_1}) < 0$ and $a_{j_2 j_1 0} > 0$,*

$$(\dot{x}_{j_1}, a_{j_2 j_1 1}) = \underbrace{(\text{nF}, \text{DP})}_{\text{down-parabola flow } (-)} . \tag{4.68}$$

The flow of $x^*_{j_1} = a_{j_2 j_1 1}$ is an (nF,DP)*-negative down-parabola flow.*

- *For* $a_{j_1 j_1 0}\prod_{s_1=1}^{3}(a_{j_2 j_1 1}-a_{j_1 j_1 s_1})>0$ *and* $a_{j_2 j_1 0}<0$,

$$(\dot{x}_{j_1}, a_{j_2 j_1 1}) = \underbrace{(\text{pF}, \text{DP})}_{\text{down-parabola flow } (+)}. \tag{4.69}$$

The flow of $x_{j_1}^* = a_{j_2 j_1 1}$ *is a* (pF,DP)-*positive down-parabola flow.*

- *For* $a_{j_1 j_1 0}\prod_{s_1=1}^{3}(a_{j_2 j_1 1}-a_{j_1 j_1 s_1})<0$ *and* $a_{j_2 j_1 0}<0$,

$$(\dot{x}_{j_1}, a_{j_2 j_1 1}) = \underbrace{(\text{nF}, \text{UP})}_{\text{up-parabola flow } (-)}. \tag{4.70}$$

The flow of $x_{j_1}^* = a_{j_2 j_1 1}$ *is an* (pF,UP)-*negative up-parabola flow.*

(iv$_2$) For $a_{j_2 j_1 1} = a_{11 s_1}$ $(s_1 = 1, 2, 3)$, *the first integral manifold is*

$$\begin{aligned} x_{j_2} - x_{j_2 0} = \frac{a_{j_2 j_2 0}}{a_{j_1 j_1 0}} \Bigg[& x_{j_1} - x_{j_1 0} + \frac{(a_{j_1 j_1 s_2} - a_{j_2 j_1})^2 + b_{j_2 j_1}}{a_{j_1 j_1 s_2} - a_{j_1 j_1 s_3}} \ln \frac{| x_{j_1} - a_{j_1 j_1 s_2} |}{| x_{j_1 0} - a_{j_1 j_1 s_2} |} \\ & + \frac{(a_{j_1 j_1 s_3} - a_{j_2 j_1})^2 + b_{j_2 j_1}}{a_{j_1 j_1 s_3} - a_{j_1 j_1 s_2}} \ln \frac{| x_{j_1} - a_{j_1 j_1 s_3} |}{| x_{j_1 0} - a_{j_1 j_1 s_3} |} \Bigg]. \end{aligned} \tag{4.71}$$

The infinite-equilibrium at $x_{j_1}^* = a_{j_2 j_1 1} = a_{11 s_1}$ *with* $\overline{x}_{j_2} \in (-\infty, \infty)$ has the following properties.

- *For* $a_{j_1 j_1 0}\prod_{s_2=1, s_2 \neq s_1}^{3}(a_{j_1 j_1 s_1}-a_{j_1 j_1 s_2})>0$ *and* $a_{j_2 j_1 0}>0$,

$$(a_{j_1 j_1 s_1}, \overline{x}_{j_2}) = \underbrace{(\text{SO}, \text{II})}_{\text{increasing-inflecton source}}. \tag{4.72}$$

The infinite-equilibrium at $x_{j_1}^* = a_{j_1 j_1 s_1}$ *is an* (SO,II)-*increasing-inflection source.*

- *For* $a_{j_1 j_1 0}\prod_{s_2=1, s_2 \neq s_1}^{3}(a_{j_1 j_1 s_1}-a_{j_1 j_1 s_2})<0$ *and* $a_{j_2 j_1 0}>0$,

$$(a_{j_1 j_1 s_1}, \overline{x}_{j_2}) = \underbrace{(\text{SI}, \text{DI})}_{\text{decreasing-inflecton sink}}. \tag{4.73}$$

The infinite-equilibrium at $x_{j_1}^* = a_{j_1 j_1 s_1}$ *is an* (SI,DI)-*decreasing-inflection sink.*

- *For* $a_{j_1 j_1 0}\prod_{s_2=1, s_2 \neq s_1}^{3}(a_{j_1 j_1 s_1}-a_{j_1 j_1 s_2})>0$ *and* $a_{j_2 j_1 0}<0$,

$$(a_{j_1 j_1 s_1}, \overline{x}_{j_2}) = \underbrace{(\text{SO}, \text{DI})}_{\text{decreasing-inflecton source}}. \tag{4.74}$$

The infinite-equilibrium at $x_{j_1}^* = a_{j_1 j_1 s_1}$ *is an* (SO,DI)-*decreasing-inflection source.*

- *For* $a_{j_1j_10}\prod_{s_2=1,s_2\neq s_1}^{3}(a_{j_1j_1s_1}-a_{j_1j_1s_2})<0$ *and* $a_{j_2j_10}<0$,

$$(a_{j_1j_1s_1},\bar{x}_{j_2})=\underbrace{(\mathrm{SI,II})}_{\text{increasing-inflecton sink}}\quad. \tag{4.75}$$

The infinite-equilibrium at $x_{j_1}^{*}=a_{j_1j_1s_1}$ *is an* (SI,II)*-increasing-inflection sink.*

– *The inflection-source (sink) infinite-equilibriums are the switching bifurcations of a source (sink) flow with a parabola flow.*

(v) If

$$\Delta_{j_1j_1}<0 \text{ and } \Delta_{j_2j_1}=0, \tag{4.76}$$

then

$$\begin{aligned}
&\dot{x}_{j_2}=a_{j_2j_10}(x_{j_1}-a_{j_2j_1l_1})(x_{j_1}-a_{j_2j_1l_2})^2,\\
&\dot{x}_{j_1}=a_{j_1j_10}(x_{j_1}-a_{j_1j_11})\big[(x_{j_1}-a_{j_1j_1})^2+b_{j_1j_1}\big],\\
&\text{with } a_{j_1j_1}=-\frac{1}{2}B_{j_1j_11},b_{j_1j_1}=\frac{1}{4}(-\Delta_{j_1j_1}),a_{j_1j_11}=b_{j_1j_11};\\
&a_{j_2j_1l_1}=b_{j_2j_11};a_{j_2j_1l_2}=-\frac{1}{2}B_{j_2j_11},b_{j_2j_1}=\frac{1}{4}(-\Delta_{j_2j_1})=0.
\end{aligned} \tag{4.77}$$

(v₁) For $a_{j_1j_11}\neq a_{j_2j_1l_1},a_{j_2j_1l_2}$, *the first integral manifold is*

$$\begin{aligned}
x_{j_2}-x_{j_20}=&\frac{a_{j_2j_10}}{a_{j_1j_10}}\Big(\Big\{x_{j_1}-x_{j_10}+(a_{j_1j_1}-a_{j_2j_1l_2})\ln\frac{|(x_{j_1}-a_{j_1j_1})^2+b_{j_1j_1}|}{|(x_{j_10}-a_{j_1j_1})^2+b_{j_1j_1}|}\\
&+\big[(a_{j_1j_1}-a_{j_2j_1l_2})^2-b_{j_1j_1}\big]\\
&\times\frac{1}{\sqrt{b_{j_1j_1}}}(\arctan\frac{x_{j_1}-a_{j_1j_1}}{\sqrt{b_{j_1j_1}}}-\arctan\frac{x_{j_10}-a_{j_1j_1}}{\sqrt{b_{j_1j_1}}})\Big\}\\
&+(a_{j_1j_11}-a_{j_2j_1l_1})\Big\{\frac{1}{2}\ln\frac{|(x_{j_1}-a_{j_1j_1})^2+b_{j_1j_1}|}{|(x_{j_10}-a_{j_1j_1})^2+b_{j_1j_1}|}\\
&+\big[2(a_{j_1j_11}-a_{j_2j_11})+a_{j_1j_1}-a_{j_1j_11}\big]\\
&\times\frac{1}{\sqrt{b_{j_1j_1}}}(\arctan\frac{x_{j_1}-a_{j_1j_1}}{\sqrt{b_{j_1j_1}}}-\arctan\frac{x_{j_10}-a_{j_1j_1}}{\sqrt{b_{j_1j_1}}})\\
&+\frac{(a_{j_1j_11}-a_{j_2j_11})^2}{(a_{j_1j_11}-a_{j_1j_1})^2+b_{j_1j_1}}\Big[\ln\frac{|x_{j_1}-a_{j_1j_11}|}{|x_{j_10}-a_{j_1j_11}|}\\
&-\frac{1}{2}\ln\frac{|(x_{j_1}-a_{j_1j_1})^2+b_{j_1j_1}|}{|(x_{j_10}-a_{j_1j_1})^2+b_{j_1j_1}|}\\
&-\frac{a_{j_1j_11}-a_{j_1j_1}}{\sqrt{b_{j_1j_1}}}(\arctan\frac{x_{j_1}-a_{j_1j_1}}{\sqrt{b_{j_1j_1}}}-\arctan\frac{x_{j_10}-a_{j_1j_1}}{\sqrt{b_{j_1j_1}}})\Big]\Big\}\Big),\\
&j_1,j_2\in\{1,2\},j_1\neq j_2.
\end{aligned} \tag{4.78}$$

(v_{1a}) The flow at $x_{j_1}^* = a_{j_1 j_1 1}$ for the system has the following properties.

- *For $a_{j_1 j_1 0} > 0$ and $a_{j_2 j_1 0}(a_{j_1 j_1 1} - a_{j_2 j_1 l_1})(a_{j_1 j_1 1} - a_{j_2 j_1 l_2})^2 > 0$,*

$$(a_{j_1 j_1 1}, \dot{x}_{j_2}) = \underbrace{(\mathrm{SO}, \mathrm{pF})}_{\text{source flow }(+)} . \tag{4.79}$$

The flow of $x_{j_1}^* = a_{j_1 j_1 1}$ is an (SO,pF)*-positive source flow.*

- *For $a_{j_1 j_1 0} < 0$ and $a_{j_2 j_1 0}(a_{j_1 j_1 1} - a_{j_2 j_1 l_1})(a_{j_1 j_1 1} - a_{j_2 j_1 l_2})^2 > 0$,*

$$(a_{j_1 j_1 1}, \dot{x}_{j_2}) = \underbrace{(\mathrm{SI}, \mathrm{pF})}_{\text{sink flow }(+)} . \tag{4.80}$$

The flow of $x_{j_1}^* = a_{j_1 j_1 1}$ is an (SI, pF)*-positive sink flow.*

- *For $a_{j_1 j_1 0} > 0$ and $a_{j_2 j_1 0}(a_{j_1 j_1 1} - a_{j_2 j_1 l_1})(a_{j_1 j_1 1} - a_{j_2 j_1 l_2})^2 < 0$,*

$$(a_{j_1 j_1 1}, \dot{x}_{j_2}) = \underbrace{(\mathrm{SO}, \mathrm{nF})}_{\text{source flow }(-)} . \tag{4.81}$$

The flow of $x_{j_1}^* = a_{j_1 j_1 1}$ is an (SO, nF)*-negative source flow.*

- *For $a_{j_1 j_1 0} < 0$ and $a_{j_2 j_1 0}(a_{j_1 j_1 1} - a_{j_2 j_1 l_1})(a_{j_1 j_1 1} - a_{j_2 j_1 l_2})^2 < 0$,*

$$(a_{j_1 j_1 1}, \dot{x}_{j_2}) = \underbrace{(\mathrm{SI}, \mathrm{nF})}_{\text{sink flow }(-)} . \tag{4.82}$$

The flow of $x_{j_1}^* = a_{j_1 j_1 1}$ is an (SI,nF)*-negative sink flow.*

(v_{1b}) The flow at $x_{j_1}^* = a_{j_2 j_1 l_1}$ for the system has the following properties.

- *For $a_{j_1 j_1 0}(a_{j_2 j_1 l_1} - a_{j_1 j_1 1}) > 0$ and $a_{j_2 j_1 0}(a_{j_2 j_1 l_1} - a_{j_2 j_1 l_2})^2 > 0$,*

$$(\dot{x}_{j_1}, a_{j_2 j_1 l_1}) = \underbrace{(\mathrm{pF}, \mathrm{UP})}_{\text{up-parabola flow }(+)} . \tag{4.83}$$

The flow of $x_{j_1}^* = a_{j_2 j_1 l_1}$ is a (pF,UP)*-positive up-parabola flow.*

- *For $a_{j_1 j_1 0}(a_{j_2 j_1 l_1} - a_{j_1 j_1 1}) < 0$ and $a_{j_2 j_1 0}(a_{j_2 j_1 l_1} - a_{j_2 j_1 l_2})^2 > 0$,*

$$(\dot{x}_{j_1}, a_{j_2 j_1 l_1}) = \underbrace{(\mathrm{nF}, \mathrm{DP})}_{\text{down-parabola flow }(-)} . \tag{4.84}$$

The flow of $x_{j_1}^* = a_{j_2 j_1 l_1}$ is an (nF,DP)*-negative down-parabola flow.*

- *For $a_{j_1j_10}(a_{j_2j_1l_1} - a_{j_1j_11}) > 0$ and $a_{j_2j_10}(a_{j_2j_1l_1} - a_{j_2j_1l_2})^2 < 0$,*

$$(\dot{x}_{j_1}, a_{j_2j_1l_1}) = \underbrace{(\text{pF}, \text{DP})}_{\text{down-parabola flow } (+)}. \tag{4.85}$$

The flow of $x^*_{j_1} = a_{j_2j_1l_1}$ is a (pF, DP)*-positive down-parabola flow.*

- *For $a_{j_1j_10}(a_{j_2j_1l_1} - a_{j_1j_11}) < 0$ and $a_{j_2j_10}(a_{j_2j_1l_1} - a_{j_2j_1l_2})^2 < 0$,*

$$(\dot{x}_{j_1}, a_{j_2j_1l_1}) = \underbrace{(\text{nF}, \text{UP})}_{\text{up-parabola flow } (-)}. \tag{4.86}$$

The flow of $x^*_{j_1} = a_{j_2j_1l_1}$ is an (nF,UP)*-negative up-parabola flow.*

(v_{1c}) *The flow at $x^*_{j_1} = a_{j_2j_1l_2}$ for the system has the following properties.*

- *For $a_{j_1j_10}(a_{j_2j_1l_2} - a_{j_1j_11}) > 0$ and $a_{j_2j_10}(a_{j_2j_1l_2} - a_{j_2j_1l_1}) > 0$,*

$$(\dot{x}_{j_1}, a_{j_2j_1l_2}) = \underbrace{(\text{pF}, \text{II})}_{\text{increasing-inflection flow } (+)}. \tag{4.87}$$

The flow of $x^*_{j_1} = a_{j_2j_1l_2}$ is a (pF,II)*-positive increasing-inflection flow.*

- *For $a_{j_1j_10}(a_{j_2j_1l_2} - a_{j_1j_11}) < 0$ and $a_{j_2j_10}(a_{j_2j_1l_2} - a_{j_2j_1l_1}) > 0$,*

$$(\dot{x}_{j_1}, a_{j_2j_1l_2}) = \underbrace{(\text{nF}, \text{DI})}_{\text{decreasing-inflection flow } (-)}. \tag{4.88}$$

The flow of $x^*_{j_1} = a_{j_2j_1l_2}$ is an (nF,DI)*-negative decreasing-inflection flow.*

- *For $a_{j_1j_10}(a_{j_2j_1l_2} - a_{j_1j_11}) > 0$ and $a_{j_2j_10}(a_{j_2j_1l_2} - a_{j_2j_1l_1}) < 0$,*

$$(\dot{x}_{j_1}, a_{j_2j_1l_2}) = \underbrace{(\text{pF}, \text{DI})}_{\text{decreasing-inflection flow } (+)}. \tag{4.89}$$

The flow of $x^*_{j_1} = a_{j_2j_1l_2}$ is a (pF, DI)*-positive decreasing-inflection flow.*

- *For $a_{j_1j_10}(a_{j_2j_1l_2} - a_{j_1j_11}) < 0$ and $a_{j_2j_10}(a_{j_2j_1l_2} - a_{j_2j_1l_1}) < 0$,*

$$(\dot{x}_{j_1}, a_{j_2j_1l_2}) = \underbrace{(\text{nF}, \text{II})}_{\text{increasing-inflection flow } (-)}. \tag{4.90}$$

The flow of $x^*_{j_1} = a_{j_2j_1l_2}$ is an (nF,II)*-negative increasing-inflection flow.*

– *The increasing-inflection and decreasing-inflection flows are the appearing (or switching) bifurcation for the up-parabola and down-parabola flows.*

(v_2) For $a_{j_1j_11}=a_{j_2j_1l_1}$, the first integral manifold is

$$
\begin{aligned}
x_{j_2}-x_{j_20}=\frac{a_{j_2j_10}}{a_{j_1j_10}}\Big\{&x_{j_1}-x_{j_10}+(a_{j_1j_1}-a_{j_2j_1l_2})\ln\frac{|\,(x_{j_1}-a_{j_1j_1})^2+b_{j_1j_1}\,|}{|\,(x_{j_10}-a_{j_1j_1})^2+b_{j_1j_1}\,|}\\
&+\big[(a_{j_1j_1}-a_{j_2j_1l_2})^2-b_{j_1j_1}\big]\\
&\times\frac{1}{\sqrt{b_{j_1j_1}}}(\arctan\frac{x_{j_1}-a_{j_1j_1}}{\sqrt{b_{j_1j_1}}}-\arctan\frac{x_{j_10}-a_{j_1j_1}}{\sqrt{b_{j_1j_1}}})\Big\},\\
&j_1,j_2\in\{1,2\},j_1\neq j_2.
\end{aligned}
\tag{4.91}
$$

The infinite-equilibrium of $x^*_{j_1}=a_{j_1j_11}=a_{j_2j_1l_1}$ with $\bar{x}_{j_2}\in(-\infty,\infty)$ has the following properties:

- *For $a_{j_1j_10}>0$ and $a_{j_2j_10}(a_{j_2j_1l_1}-a_{j_2j_1l_2})^2>0$,*

$$
(a_{j_1j_11},\bar{x}_{j_2})=\underbrace{(\text{SO, II})}_{\text{increasing-inflection source}}. \tag{4.92}
$$

  The equilibrium of $x^*_{j_1}=a_{j_1j_11}=a_{j_2j_1l_1}$ is an (SO,II)*-increasing-inflection source infinite-equilibrium.*

- *For $a_{j_1j_10}<0$ and $a_{j_2j_10}(a_{j_2j_1l_1}-a_{j_2j_1l_2})^2>0$,*

$$
(a_{j_1j_11},\bar{x}_{j_2})=\underbrace{(\text{SI, DI})}_{\text{decreasing-inflection sink}}. \tag{4.93}
$$

  The equilibrium of $x^*_{j_1}=a_{j_1j_11}=a_{j_2j_1l_1}$ is an (SI,DI)*-deceasing-inflection sink infinite-equilibrium.*

- *For $a_{j_1j_10}>0$ and $a_{j_2j_10}(a_{j_2j_1l_1}-a_{j_2j_1l_2})^2<0$,*

$$
(a_{j_1j_11},\bar{x}_{j_2})=\underbrace{(\text{SO, DI})}_{\text{decreasing-inflection source}}. \tag{4.94}
$$

  The equilibrium of $x^*_{j_1}=a_{j_1j_11}=a_{j_2j_1l_1}$ is an (SO,DI)*-deceasing-inflection source infinite-equilibrium.*

- *For $a_{j_1j_10}<0$ and $a_{j_2j_10}(a_{j_2j_1l_1}-a_{j_2j_1l_2})^2<0$,*

$$
(a_{j_1j_11},\bar{x}_{j_2})=\underbrace{(\text{SI, II})}_{\text{increasing-inflection sink}}. \tag{4.95}
$$

  The equilibrium of $x^*_{j_1}=a_{j_1j_11}=a_{j_2j_1l_1}$ is an (SI,II)*-increasing-inflection sink infinite-equilibrium.*

– *The (1:1)-increasing-inflection and decreasing-inflection source and sink are the switching bifurcations for the sink and source flows with up-parabola and down-parabola flows.*

(v_3) For $a_{j_1j_1 1}=a_{j_2j_1l_2}$, the first integral manifold is

$$
\begin{aligned}
x_{j_2}-x_{j_2 0}=&\frac{a_{j_2j_1 0}}{a_{j_1j_1 0}}\Big[x_{j_1}-x_{j_1 0}+\frac{1}{2}(2a_{j_1j_1}-a_{j_2j_1l_1}-a_{j_2j_1l_2})\\
&\times\ln\frac{|\,(x_{j_1}-a_{j_1j_1})^2+b_{j_1j_1}\,|}{|\,(x_{j_1 0}-a_{j_1j_1})^2+b_{j_1j_1}\,|}\\
&+\frac{(a_{j_1j_1}-a_{j_2j_1l_1})(a_{j_1j_1}-a_{j_2j_1l_2})-b_{j_1j_1}}{\sqrt{b_{j_1j_1}}}\\
&\times(\arctan\frac{x_{j_1}-a_{j_1j_1}}{\sqrt{b_{j_1j_1}}}-\arctan\frac{x_{j_1 0}-a_{j_1j_1}}{\sqrt{b_{j_1j_1}}})\Big];\\
&j_1,j_2\in\{1,2\},j_1\neq j_2.
\end{aligned}
\tag{4.96}
$$

The infinite-equilibrium of $x^*_{j_1}=a_{j_1j_1 1}=a_{j_2j_1l_2}$ with $\bar{x}_{j_2}\in(-\infty,\infty)$ has the following properties.

- *For $a_{j_1j_1 0}>0$ and $a_{j_2j_1 0}(a_{j_2j_1l_2}-a_{j_2j_1l_1})>0$,*

$$
(a_{j_1j_1 1},\bar{x}_{j_2})=\underbrace{(\mathrm{SO},\mathrm{UP})}_{\text{up-parabola source}}. \tag{4.97}
$$

  The equilibrium of $x^*_{j_1}=a_{j_1j_1 1}=a_{j_2j_1l_2}$ is an (SO,UP)*-up-parabola source infinite-equilibrium.*

- *For $a_{j_1j_1 0}<0$ and $a_{j_2j_1 0}(a_{j_2j_1l_2}-a_{j_2j_1l_1})>0$,*

$$
(a_{j_1j_1 1},\bar{x}_{j_2})=\underbrace{(\mathrm{SI},\mathrm{DP})}_{\text{down-parabola sink}}. \tag{4.98}
$$

  The equilibrium of $x^*_{j_1}=a_{j_1j_1 1}=a_{j_2j_1l_2}$ is an (SI,DP)*-down-parabola sink infinite-equilibrium.*

- *For $a_{j_1j_1 0}>0$ and $a_{j_2j_1 0}(a_{j_2j_1l_2}-a_{j_2j_1l_1})<0$,*

$$
(a_{j_1j_1 1},\bar{x}_{j_2})=\underbrace{(\mathrm{SO},\mathrm{DP})}_{\text{down-parabola source}}. \tag{4.99}
$$

  The equilibrium of $x^*_{j_1}=a_{j_1j_1 1}=a_{j_2j_1l_2}$ is an (SO,DP)*-down-parabola source infinite-equilibrium.*

- *For* $a_{j_1j_10} < 0$ *and* $a_{j_2j_10}(a_{j_2j_1l_2} - a_{j_2j_1l_1}) < 0$,

$$(a_{j_1j_11}, \bar{x}_{j_2}) = \underbrace{(\text{SI}, \text{UP})}_{\text{up-parabola sink}}. \tag{4.100}$$

The equilibrium of $x_{j_1}^* = a_{j_1j_11} = a_{j_2j_1l_2}$ *is an* (SI,UP)*-up-parabola sink infinite-equilibrium.*

– *The up-parabola and down-parabola source and sink are the switching bifurcations for the sink and source flows with increasing-inflection and decreasing-inflection flows.*

(vi) If

$$\begin{aligned} &\Delta_{j_1j_1} < 0 \text{ and } \Delta_{j_2j_1} = 0,\\ &a_{j_2j_1l_2} = a_{j_2j_1l_1} = a_{j_2j_11}, \end{aligned} \tag{4.101}$$

then the dynamical system becomes

$$\begin{aligned} \dot{x}_{j_2} &= a_{j_2j_10}(x_{j_1} - a_{j_2j_11})^3,\\ \dot{x}_{j_1} &= a_{j_1j_10}(x_{j_1} - a_{j_1j_11})\left[(x_{j_1} - a_{j_1j_1})^2 + b_{j_1j_1}\right]. \end{aligned} \tag{4.102}$$

(vi$_1$*) The first integral manifold is*

$$\begin{aligned} x_{j_2} - x_{j_20} = &\frac{a_{j_2j_10}}{a_{j_1j_10}}\Big\{x_{j_1} - x_{j_10} + \left[\frac{3}{2}(a_{j_1j_11} - a_{j_2j_11}) + (a_{j_1j_1} - a_{j_1j_11})\right]\\ &\times \ln\frac{|(x_{j_1} - a_{j_1j_1})^2 + b_{j_1j_1}|}{|(x_{j_10} - a_{j_1j_1})^2 + b_{j_1j_1}|} + \left[3(a_{j_1j_11} - a_{j_2j_11})(a_{j_1j_1} - a_{j_2j_11})\right.\\ &\left.+(a_{j_1j_1} - a_{j_1j_11})^2 - b_{j_1j_1}\right]\\ &\times \frac{1}{\sqrt{b_{j_1j_1}}}(\arctan\frac{x_{j_1} - a_{j_1j_1}}{\sqrt{b_{j_1j_1}}} - \arctan\frac{x_{j_10} - a_{j_1j_1}}{\sqrt{b_{j_1j_1}}})\\ &+\frac{(a_{j_1j_11} - a_{j_2j_11})^3}{(a_{j_1j_11} - a_{j_1j_1})^2 + b_{j_1j_1}}\Big[\ln\frac{|x_{j_1} - a_{j_1j_11}|}{|x_{j_10} - a_{j_1j_11}|}\\ &-\frac{1}{2}\ln\frac{|(x_{j_1} - a_{j_1j_1})^2 + b_{j_1j_1}|}{|(x_{j_10} - a_{j_1j_1})^2 + b_{j_1j_1}|}\\ &+\frac{a_{j_1j_1} - a_{j_1j_11}}{\sqrt{b_{j_1j_1}}}(\arctan\frac{x_{j_1} - a_{j_1j_1}}{\sqrt{b_{j_1j_1}}} - \arctan\frac{x_{j_10} - a_{j_1j_1}}{\sqrt{b_{j_1j_1}}})\Big]\Big\},\\ &j_1, j_2 \in \{1,2\}, j_1 \neq j_2. \end{aligned} \tag{4.103}$$

(vi_{1a}) The flow of $x^*_{j_1} = a_{j_1 j_1 1}$ for the system has the following properties.

- *For $a_{j_1 j_1 0} > 0$ and $a_{j_2 j_1 0}(a_{j_1 j_1 1} - a_{j_2 j_1 1})^3 > 0$,*

$$(a_{j_1 j_1 1}, \dot{x}_{j_2}) = \underbrace{(\mathrm{SO}, \mathrm{pF})}_{\text{source flow } (+)} . \quad (4.104)$$

The flow of $x^*_{j_1} = a_{j_1 j_1 1}$ is an (SO,pF)*-positive source flow.*

- *For $a_{j_1 j_1 0} < 0$ and $a_{j_2 j_1 0}(a_{j_1 j_1 1} - a_{j_2 j_1 1})^3 > 0$,*

$$(a_{j_1 j_1 1}, \dot{x}_{j_2}) = \underbrace{(\mathrm{SI}, \mathrm{pF})}_{\text{sink flow } (+)} . \quad (4.105)$$

The flow of $x^*_{j_1} = a_{j_1 j_1 1}$ is an (SI, pF)*-positive sink flow.*

- *For $a_{j_1 j_1 0} > 0$ and $a_{j_2 j_1 0}(a_{j_1 j_1 1} - a_{j_2 j_1 1})^3 < 0$,*

$$(a_{j_1 j_1 1}, \dot{x}_{j_2}) = \underbrace{(\mathrm{SO}, \mathrm{nF})}_{\text{source flow } (-)} . \quad (4.106)$$

The flow of $x^*_{j_1} = a_{j_1 j_1 1}$ is an (SO, nF)*-negative source flow.*

- *For $a_{j_1 j_1 0} < 0$ and $a_{j_2 j_1 0}(a_{j_1 j_1 1} - a_{j_2 j_1 1})^3 < 0$,*

$$(a_{j_1 j_1 1}, \dot{x}_{j_2}) = \underbrace{(\mathrm{SI}, \mathrm{nF})}_{\text{sink flow } (-)} . \quad (4.107)$$

The flow of $x^*_{j_1} = a_{j_1 j_1 1}$ is an (SI,nF)*-negative sink flow.*

(vi_{1b}) The flow at $x^*_{j_1} = a_{j_2 j_1 1}$ for the system has the following properties.

- *For $a_{j_1 j_1 0}(a_{j_2 j_1 1} - a_{j_1 j_1 1}) > 0$ and $a_{j_2 j_1 0} > 0$,*

$$(\dot{x}_{j_1}, a_{j_2 j_1 1}) = \underbrace{(\mathrm{pF}, 3^{\mathrm{rd}}\mathrm{UP})}_{\text{third-order up-parabola flow } (+)} . \quad (4.108)$$

The flow of $x^*_{j_1} = a_{j_2 j_1 1}$ is a (pF,3^rd^UP)*-positive third-order up-parabola flow.*

- *For $a_{j_1 j_1 0}(a_{j_2 j_1 1} - a_{j_1 j_1 1}) < 0$ and $a_{j_2 j_1 0} > 0$,*

$$(\dot{x}_{j_1}, a_{j_2 j_1 1}) = \underbrace{(\mathrm{nF}, 3^{\mathrm{rd}}\mathrm{DP})}_{\text{third-order down-parabola flow } (-)} . \quad (4.109)$$

The flow of $x^*_{j_1} = a_{j_2 j_1 1}$ is an (nF,3^{rd}DP)*-negative third-order down-parabola flow.*

- *For* $a_{j_1j_10}(a_{j_2j_11}-a_{j_1j_11})>0$ *and* $a_{j_2j_10}<0$,

$$(\dot{x}_{j_1},a_{j_2j_11})=\underbrace{(\text{pF},3^{\text{rd}}\text{DP})}_{\text{third-order down-parabola flow }(+)}. \quad (4.110)$$

The flow of $x^*_{j_1}=a_{j_2j_11}$ *is a* (pF,3rdDP)-*positive third-order down-parabola flow.*

- *For* $a_{j_1j_10}(a_{j_2j_11}-a_{j_1j_11})<0$ *and* $a_{j_2j_10}<0$,

$$(\dot{x}_{j_1},a_{j_2j_11})=\underbrace{(\text{nF},3^{\text{rd}}\text{UP})}_{\text{third-order up-parabola flow }(-)}. \quad (4.111)$$

The flow of $x^*_{j_1}=a_{j_2j_11}$ *is an* (nF,3rdUP)-*negative third-order up-parabola flow.*

– *The third-order up-parabola and down-parabola flows are the appearing (or switching) bifurcations for the up-parabola, down-parabola, and down-parabola flows and down-parabola, up-parabola, and down-parabola flows.*

(vi₂) For $a_{j_1j_11}=a_{j_2j_11}$, *the first integral manifold is*

$$\begin{aligned}x_{j_2}-x_{j_20}=&\frac{a_{j_2j_10}}{a_{j_1j_10}}\left[x_{j_1}-x_{j_10}+(a_{j_1j_1}-a_{j_2j_11})\ln\frac{|(x_{j_1}-a_{j_1j_1})^2+b_{j_1j_1}|}{|(x_{j_10}-a_{j_1j_1})^2+b_{j_1j_1}|}\right.\\&\left.+\frac{(a_{j_1j_1}-a_{j_1j_11})^2-b_{j_1j_1}}{\sqrt{b_{j_1j_1}}}(\arctan\frac{x_{j_1}-a_{j_1j_1}}{\sqrt{b_{j_1j_1}}}-\arctan\frac{x_{j_10}-a_{j_1j_1}}{\sqrt{b_{j_1j_1}}})\right],\\&j_1,j_2\in\{1,2\},j_1\neq j_2.\end{aligned} \quad (4.112)$$

The infinite-equilibrium of $x^*_{j_1}=a_{j_1j_11}=a_{j_2j_11}$ with $\bar{x}_{j_2}\in(-\infty,\infty)$ *for the system has the following properties.*

- *For* $a_{j_1j_10}(a_{j_2j_11}-a_{j_1j_11})>0$ *and* $a_{j_2j_10}>0$,

$$(a_{j_1j_11},\bar{x}_{j_2})=\underbrace{(\text{SO},(1:3)\text{II})}_{(1:3)\text{-increasing-inflection source}}. \quad (4.113)$$

The (SO,(1:3)II)-*infinite-equilibrium at* $x^*_{j_1}=a_{j_2j_11}$ *is a* (1:3)-*increasing-inflection source.*

- *For* $a_{j_1j_10}(a_{j_2j_11}-a_{j_1j_11})<0$ *and* $a_{j_2j_10}>0$,

$$(a_{j_1j_11},\bar{x}_{j_2})=\underbrace{(\text{SI},(1:3)\text{DI})}_{(1:3)\text{-decreasing-inflection sink}}. \quad (4.114)$$

The (SI,(1:3)DI)-*infinite-equilibrium at* $x^*_{j_1} = a_{j_2 j_1 1}$ *is a* (1:3)-*decreasing-inflection sink.*

- *For* $a_{j_1 j_1 0}(a_{j_2 j_1 1} - a_{j_1 j_1 1}) > 0$ *and* $a_{j_2 j_1 0} < 0$,

$$(a_{j_1 j_1 1}, \overline{x}_{j_2}) = \underbrace{(\text{SO}, (1{:}3)\text{DI})}_{(1:3)\text{-decreasing-inflection source}}. \tag{4.115}$$

The (SO,(1:3)DI)-*infinite-equilibrium at* $x^*_{j_1} = a_{j_2 j_1 1}$ *is a* (1:3)-*decreasing-inflection source*

- *For* $a_{j_1 j_1 0}(a_{j_2 j_1 1} - a_{j_1 j_1 1}) < 0$ *and* $a_{j_2 j_1 0} < 0$,

$$(a_{j_1 j_1 1}, \overline{x}_{j_2}) = \underbrace{(\text{SI}, (1{:}3)\text{II})}_{(1:3)\text{-increasing-inflection sink}}. \tag{4.116}$$

The (SI,(1:3)II)-*infinite-equilibrium at* $x^*_{j_1} = a_{j_2 j_1 1}$ *is a* (1:3)-*increasing-inflection sink.*

– *The* (1:3)-*increasing-inflection and (1:3)-decreasing-inflection source and sink are the appearing (or switching) bifurcations for the third-order increasing-inflection and decreasing-inflection flows with source and sink flows.*

(vii) If

$$\Delta_{j_1 j_1} = 0 \text{ and } \Delta_{j_2 j_1} = 0, \tag{4.117}$$

then

$$\begin{aligned} &\dot{x}_{j_1} = a_{j_1 j_1 0}(x_{j_1} - a_{j_1 j_1 s_1})(x_{j_1} - a_{j_1 j_1 s_2})^2, \\ &\dot{x}_{j_2} = a_{j_2 j_1 0}(x_{j_1} - a_{j_2 j_1 l_1})(x_{j_1} - a_{j_2 j_1 l_2})^2, \\ &j_1, j_2 \in \{1, 2\}, j_1 \neq j_2; s_1, s_2 \in \{1, 2\}, s_1 \neq s_2, \end{aligned} \tag{4.118}$$

where

$$\begin{aligned} &\{a_{j_1 j_1 1}, a_{j_1 j_1 2}\} = \text{sort}\{b_{j_1 j_1 1}, b_{j_1 j_1 2}\}, a_{j_1 j_1 1} < a_{j_1 j_1 2} \\ &\{a_{j_2 j_1 1}, a_{j_2 j_1 2}\} = \text{sort}\{b_{j_2 j_1 1}, b_{j_2 j_1 2}\}, a_{j_2 j_1 1} < a_{j_2 j_1 2} \\ &b_{j_1 j_1 2} = -\frac{1}{2}B_{j_1 j_1 1}, b_{j_2 j_1 2} = -\frac{1}{2}B_{j_2 j_1 1}; \\ &j_1, j_2 \in \{1, 2\}, j_1 \neq j_2. \end{aligned} \tag{4.119}$$

(vii$_1$) For $a_{j_1 j_1 s_1} \neq a_{j_2 j_1 l_1} \neq a_{j_1 j_1 s_2} \neq a_{j_2 j_1 l_2}$, *the first integral manifold is*

$$
\begin{aligned}
x_{j_2} - x_{j_20} = \frac{a_{j_2j_10}}{a_{j_1j_10}} \Big(\Big\{ & (x_{j_1} - x_{j_10}) + 2(a_{j_1j_1s_2} - a_{j_1j_1l_2}) \ln \frac{|x_{j_1} - a_{j_1j_1s_2}|}{|x_{j_10} - a_{j_1j_1s_2}|} \\
& + (a_{j_1j_1s_2} - a_{j_1j_1l_2})^2 \Big(\frac{1}{x_{j_1} - a_{j_1j_1s_2}} - \frac{1}{x_{j_10} - a_{j_1j_1s_2}} \Big) \Big\} \\
& + (a_{j_1j_1s_1} - a_{j_1j_1l_1}) \Big\{ \frac{(a_{j_1j_1s_1} - a_{j_2j_11})^2}{(a_{j_1j_1s_1} - a_{j_1j_1s_2})^2} \ln \frac{|x_{j_1} - a_{j_1j_1s_1}|}{|x_{j_10} - a_{j_1j_1s_1}|} \\
& - \frac{(a_{j_1j_1s_2} - a_{j_2j_11})^2}{(a_{j_1j_1s_2} - a_{j_1j_1s_1})} \Big(\frac{1}{x_{j_1} - a_{j_1j_1s_2}} - \frac{1}{x_{j_10} - a_{j_1j_1s_2}} \Big) \\
& + \Big[\frac{2(a_{j_1j_1s_2} - a_{j_2j_11})}{(a_{j_1j_1s_2} - a_{j_1j_1s_1})} - \frac{(a_{j_1j_1s_2} - a_{j_2j_11})^2}{(a_{j_1j_1s_2} - a_{j_1j_1s_1})^2} \Big] \ln \frac{|x_{j_1} - a_{j_1j_1s_2}|}{|x_{j_10} - a_{j_1j_1s_2}|} \Big\} \Big), \\
& j_1, j_2 \in \{1,2\}, j_1 \neq j_2; s_1, s_2 \in \{1,2\}, s_1 \neq s_2.
\end{aligned}
\tag{4.120}
$$

(vii_{1a}) The equilibrium of $x_{j_1}^* = a_{j_1j_1s_1} \neq a_{j_2j_1l_1}, a_{j_2j_1l_2}, a_{j_1j_1s_2}$ possesses the following properties.

- *For $a_{j_1j_10}(a_{j_1j_1s_1} - a_{j_1j_1s_2})^2 > 0$ and $a_{j_2j_10}(a_{j_1j_1s_1} - a_{j_2j_1l_1})(a_{j_1j_1s_1} - a_{j_2j_1l_2})^2 > 0$,*

$$
(a_{j_1j_1s_1}, \dot{x}_{j_2}) = \underbrace{(\text{SO}, \text{pF})}_{\text{source flow }(+)} . \tag{4.121}
$$

The flow of $x_{j_1}^* = a_{j_1j_1s_1}$ is an (SO,pF)*-positive source flow.*

- *For $a_{j_1j_10}(a_{j_1j_1s_1} - a_{j_1j_1s_2})^2 < 0$ and $a_{j_2j_10}(a_{j_1j_1s_1} - a_{j_2j_1l_1})(a_{j_1j_1s_1} - a_{j_2j_1l_2})^2 > 0$,*

$$
(a_{j_1j_1s_1}, \dot{x}_{j_2}) = \underbrace{(\text{SI}, \text{pF})}_{\text{sink flow }(+)} . \tag{4.122}
$$

The flow of $x_{j_1}^* = a_{j_1j_1s_1}$ is an (SI,pF)*-positive sink flow.*

- *For $a_{j_1j_10}(a_{j_1j_1s_1} - a_{j_1j_1s_2})^2 > 0$ and $a_{j_2j_10}(a_{j_1j_1s_1} - a_{j_2j_1l_1})(a_{j_1j_1s_1} - a_{j_2j_1l_2})^2 < 0$,*

$$
(a_{j_1j_1s_1}, \dot{x}_{j_2}) = \underbrace{(\text{SO}, \text{nF})}_{\text{source flow }(-)} . \tag{4.123}
$$

The flow of $x_{j_1}^* = a_{j_1j_1s_1}$ is an (SO,nF)*-negative source flow.*

- *For $a_{j_1j_10}(a_{j_1j_1s_1} - a_{j_1j_1s_2})^2 < 0$ and $a_{j_2j_10}(a_{j_1j_1s_1} - a_{j_2j_1l_1})(a_{j_1j_1s_1} - a_{j_2j_1l_2})^2 < 0$,*

$$
(a_{j_1j_1s_1}, \dot{x}_{j_2}) = \underbrace{(\text{SI}, \text{nF})}_{\text{sink flow }(-)} . \tag{4.124}
$$

The flow of $x_{j_1}^* = a_{j_1j_1s_1}$ is an (SI,pF)*-negative sink flow.*

(vii$_{1b}$) The equilibrium of $x^*_{j_1} = a_{j_1j_1s_2} \neq a_{j_1j_1s_1}, a_{j_2j_1l_1}, a_{j_2j_1l_2}$ *possesses the following properties.*

- *For* $a_{j_1j_10}(a_{j_1j_1s_2} - a_{j_1j_1s_1}) > 0$ *and* $a_{j_2j_10}(a_{j_1j_1s_2} - a_{j_2j_1l_1})(a_{j_1j_1s_2} - a_{j_2j_1l_2})^2 > 0$,

$$(a_{j_1j_1s_2}, \dot{x}_{j_2}) = \underbrace{(\mathrm{US}, \mathrm{pF})}_{\text{upper-saddle flow }(+)}. \tag{4.125}$$

The flow of $x^*_{j_1} = a_{j_1j_1s_2}$ *is a* (US,pF)*-positive upper-saddle flow.*

- *For* $a_{j_1j_10}(a_{j_1j_1s_2} - a_{j_1j_1s_1}) < 0$ *and* $a_{j_2j_10}(a_{j_1j_1s_2} - a_{j_2j_1l_1})(a_{j_1j_1s_2} - a_{j_2j_1l_2})^2 > 0$,

$$(a_{j_1j_1s_2}, \dot{x}_{j_2}) = \underbrace{(\mathrm{LS}, \mathrm{pF})}_{\text{lower-saddle flow }(+)}. \tag{4.126}$$

The flow of $x^*_{j_1} = a_{j_1j_1s_2}$ *is an* (LS,pF)*-positive lower-saddle flow.*

- *For* $a_{j_1j_10}(a_{j_1j_1s_2} - a_{j_1j_1s_1}) > 0$ *and* $a_{j_2j_10}(a_{j_1j_1s_2} - a_{j_2j_1l_1})(a_{j_1j_1s_2} - a_{j_2j_1l_2})^2 < 0$,

$$(a_{j_1j_1s_2}, \dot{x}_{j_2}) = \underbrace{(\mathrm{US}, \mathrm{nF})}_{\text{upper-saddle flow }(-)}. \tag{4.127}$$

The flow of $x^*_{j_1} = a_{j_1j_1s_2}$ *is a* (US,nF)*-negative upper-saddle flow.*

- *For* $a_{j_1j_10}(a_{j_1j_1s_2} - a_{j_1j_1s_1}) < 0$ *and* $a_{j_2j_10}(a_{j_1j_1s_2} - a_{j_2j_1l_1})(a_{j_1j_1s_2} - a_{j_2j_1l_2})^2 < 0$,

$$(a_{j_1j_1s_2}, \dot{x}_{j_2}) = \underbrace{(\mathrm{LS}, \mathrm{nF})}_{\text{lower-saddle flow }(-)}. \tag{4.128}$$

The flow of $x^*_{j_1} = a_{j_1j_1s_2}$ *is an* (LS,nF)*-negative lower-saddle flow.*

– *The lower-saddle (LS) and upper-saddle (US) flows are the appearing-switching bifurcation for the sink and source flows.*

(vii$_{1c}$) The equilibrium of $x^*_{j_1} = a_{j_2j_1l_1} \neq a_{j_1j_1s_1}, a_{j_1j_1s_2}, a_{j_2j_1l_2}$ *possesses the following properties.*

- *For* $a_{j_1j_10}(a_{j_2j_1l_1} - a_{j_1j_1s_1})(a_{j_2j_1l_1} - a_{j_1j_1s_2})^2 > 0$ *and* $a_{j_2j_10}(a_{j_2j_1l_1} - a_{j_2j_1l_2})^2 > 0$,

$$(\dot{x}_{j_1}, a_{j_2j_1l_1}) = \underbrace{(\mathrm{pF}, \mathrm{UP})}_{\text{up-parabola flow }(+)}. \tag{4.129}$$

The flow of $x^*_{j_1} = a_{j_2j_1l_1}$ *is a* (pF,UP)*-positive up-parabola flow.*

- *For* $a_{j_1j_10}(a_{j_2j_1l_1}-a_{j_1j_1s_1})(a_{j_2j_1l_1}-a_{j_1j_1s_2})^2<0$ *and* $a_{j_2j_10}(a_{j_2j_1l_1}-a_{j_2j_1l_2})^2>0$,

$$(\dot{x}_{j_1},a_{j_2j_1l_1})=\underbrace{(\text{nF},\text{DP})}_{\text{deown-parabola }(-)}. \tag{4.130}$$

The flow of $x^*_{j_1}=a_{j_2j_1l_1}$ *is an* (nF,DP)*-negative down-parabola flow.*

- *For* $a_{j_1j_10}(a_{j_2j_1l_1}-a_{j_1j_1s_1})(a_{j_2j_1l_1}-a_{j_1j_1s_2})^2>0$ *and* $a_{j_2j_10}(a_{j_2j_1l_1}-a_{j_2j_1l_2})^2<0$,

$$(\dot{x}_{j_1},a_{j_2j_1l_1})=\underbrace{(\text{pF},\text{DP})}_{\text{down-parabola flow }(+)}. \tag{4.131}$$

The flow of $x^*_{j_1}=a_{j_2j_1l_1}$ *is a* (pF,DP)*-positive down-parabola flow.*

- *For* $a_{j_1j_10}(a_{j_2j_1l_1}-a_{j_1j_1s_1})(a_{j_2j_1l_1}-a_{j_1j_1s_2})^2<0$ *and* $a_{j_2j_10}(a_{j_2j_1l_1}-a_{j_2j_1l_2})^2<0$,

$$(\dot{x}_{j_1},a_{j_2j_1l_1})=\underbrace{(\text{nF},\text{UP})}_{\text{iup-parabola flow }(-)}. \tag{4.132}$$

The flow of $x^*_{j_1}=a_{j_2j_1l_1}$ *is an* (nF,UP)*-positive increasing-inflection flow.*

(vii$_{1d}$) The equilibrium of $x^*_{j_1}=a_{j_2j_1l_2}\neq a_{j_1j_1s_1},a_{j_1j_1s_2},a_{j_2j_1l_1}$ *possesses the following properties.*

- *For* $a_{j_1j_10}(a_{j_2j_1l_2}-a_{j_1j_1s_1})(a_{j_2j_1l_2}-a_{j_1j_1s_2})^2>0$ *and* $a_{j_2j_10}(a_{j_2j_1l_2}-a_{j_2j_1l_1})>0$,

$$(\dot{x}_{j_1},a_{j_2j_1l_2})=\underbrace{(\text{pF},\text{II})}_{\text{increasing-inflection flow }(+)}. \tag{4.133}$$

The flow of $x^*_{j_1}=a_{j_2j_1l_2}$ *is a* (pF,II)*-positive increasing-inflection flow.*

- *For* $a_{j_1j_10}(a_{j_2j_1l_2}-a_{j_1j_1s_1})(a_{j_2j_1l_2}-a_{j_1j_1s_2})^2<0$ *and* $a_{j_2j_10}(a_{j_2j_1l_2}-a_{j_2j_1l_1})>0$,

$$(\dot{x}_{j_1},a_{j_2j_1l_2})=\underbrace{(\text{nF},\text{DI})}_{\text{decreasing-inflection flow }(-)}. \tag{4.134}$$

The flow of $x^*_{j_1}=a_{j_2j_1l_2}$ *is an* (nF,DI)*-negative decreasing-inflection flow.*

- *For* $a_{j_1j_10}(a_{j_2j_1l_2}-a_{j_1j_1s_1})(a_{j_2j_1l_2}-a_{j_1j_1s_2})^2>0$ *and* $a_{j_2j_10}(a_{j_2j_1l_2}-a_{j_2j_1l_1})<0$,

$$(\dot{x}_{j_1},a_{j_2j_1l_2})=\underbrace{(\text{pF},\text{DI})}_{\text{decreasing-inflection flow }(+)}. \tag{4.135}$$

The flow of $x^*_{j_1}=a_{j_2j_1l_2}$ *is a* (pF,DI)*-positive increasing-inflection flow.*

- *For* $a_{j_1j_10}(a_{j_2j_1l_2}-a_{j_1j_1s_1})(a_{j_2j_1l_2}-a_{j_1j_1s_2})^2<0$ *and* $a_{j_2j_10}(a_{j_2j_1l_2}-a_{j_2j_1l_1})<0$,

$$(\dot{x}_{j_1}, a_{j_2j_1l_2}) = \underbrace{(\text{nF, II})}_{\text{increasing-inflection flow }(-)}. \tag{4.136}$$

The flow of $x^*_{j_1}=a_{j_2j_1l_2}$ *is an* (nF,II)-*positive increasing-inflection flow.*

– *The increasing-inflection* (II) *and decreasing-inflection* (DI) *flows are the appearing-switching bifurcations for up-parabola and down-parabola flows.*

(vii$_2$) For $a_{j_1j_1s_2}=a_{j_2j_1l_2}$,

$$\begin{aligned}
&\dot{x}_{j_1}=a_{j_1j_10}(x_{j_1}-a_{j_1j_1s_1})(x_{j_1}-a_{j_1j_1s_2})^2,\\
&\dot{x}_{j_2}=a_{j_2j_10}(x_{j_1}-a_{j_2j_1l_1})(x_{j_1}-a_{j_2j_1l_2})^2,\\
&j_1,j_2\in\{1,2\},j_1\neq j_2.
\end{aligned} \tag{4.137}$$

The first integral manifold is

$$\begin{aligned}
x_{j_2}-x_{j_20}&=\frac{a_{j_2j_10}}{a_{j_1j_10}}\left[x_{j_1}-x_{j_10}+(a_{j_1j_1s_1}-a_{j_2j_1l_1})\ln\frac{|x_{j_1}-a_{j_1j_1s_1}|}{|x_{j_10}-a_{j_1j_1s_1}|}\right],\\
&j_1,j_2\in\{1,2\},j_1\neq j_2.
\end{aligned} \tag{4.138}$$

The infinite-equilibrium of $x^*_{j_1}=a_{j_1j_1s_2}=a_{j_2j_1l_2}$ with $\bar{x}_{j_2}\in(-\infty,\infty)$ *has the following properties.*

- *For* $a_{j_1j_10}(a_{j_1j_1s_2}-a_{j_1j_1s_1})>0$ *and* $a_{j_2j_10}(a_{j_2j_1l_2}-a_{j_2j_1l_1})>0$,

$$(a_{j_1j_1s_2},\bar{x}_{j_2}) = \underbrace{(\text{US}, (2:2)\text{II})}_{\text{increasing-inflection upper-saddle}}. \tag{4.139}$$

The equilibrium at $x^*_{j_1}=a_{j_1j_1s_2}=a_{j_2j_1l_2}$ *is a* (US,(2:2)II)-*increasing-inflection upper-saddle infinite-equilibrium.*

- *For* $a_{j_1j_10}(a_{j_1j_1s_2}-a_{j_1j_1s_1})<0$ *and* $a_{j_2j_10}(a_{j_2j_1l_2}-a_{j_2j_1l_1})>0$,

$$(a_{j_1j_1s_2},\bar{x}_{j_2}) = \underbrace{(\text{LS}, (2:2)\text{DI})}_{\text{decreasing-inflection lower-saddle}}. \tag{4.140}$$

The equilibrium at $x^*_{j_1}=a_{j_1j_1s_2}=a_{j_2j_1l_2}$ *is an* (LS,(2:2)DI)-*decreasing-inflection lower-saddle infinite-equilibrium.*

- *For* $a_{j_1j_10}(a_{j_1j_1s_2}-a_{j_1j_1s_1})>0$ *and* $a_{j_2j_10}(a_{j_2j_1l_2}-a_{j_2j_1l_1})<0$,

$$(a_{j_1j_1s_2},\bar{x}_{j_2}) = \underbrace{(\text{US}, (2:2)\text{DI})}_{\text{decreasing-inflection upper-saddle}}. \tag{4.141}$$

The equilibrium at $x_{j_1}^{*}=a_{j_1j_1s_2}=a_{j_2j_1l_2}$ *is a* (US,(2:2)DI)-*increasing-inflection upper-saddle infinite-equilibrium.*

- *For* $a_{j_1j_1 0}(a_{j_1j_1s_2}-a_{j_1j_1s_1})<0$ *and* $a_{j_2j_1 0}(a_{j_2j_1l_2}-a_{j_2j_1l_1})<0$,

$$(a_{j_1j_1s_2},\bar{x}_{j_2})=\underbrace{(\text{LS},(2:2)\text{II})}_{\text{increasing-inflection lower-saddle}}. \tag{4.142}$$

The equilibrium at $x_{j_1}^{*}=a_{j_1j_1s_2}=a_{j_2j_1l_2}$ *is an* (LS,(2:2)II)-*increasing-inflection lower-saddle infinite-equilibrium.*

- *The infinite-equilibriums are the switching bifurcations for the increasing-inflection and decreasing-inflection flows with the upper-saddle and lower-saddle flows.*
- *The infinite-equilibriums are the appearing bifurcations for the sink and source flows and the up-parabola and down-parabola flows.*

(vii$_3$*) For* $a_{j_1j_1s_1}=a_{j_2j_1l_2}$,

$$\begin{aligned}&\dot{x}_{j_1}=a_{j_1j_1 0}(x_{j_1}-a_{j_1j_1s_1})(x_{j_1}-a_{j_1j_1s_2})^2,\\&\dot{x}_{j_2}=a_{j_2j_1 0}(x_{j_1}-a_{j_2j_1l_1})(x_{j_1}-a_{j_2j_1l_2})^2,\\&j_1,j_2\in\{1,2\},j_1\neq j_2.\end{aligned} \tag{4.143}$$

The first integral manifold is

$$\begin{aligned}x_{j_2}-x_{j_2 0}=&\frac{a_{j_2j_1 0}}{a_{j_1j_1 0}}\Big\{x_{j_1}-x_{j_1 0}+(a_{j_1j_1s_2}-a_{j_2j_1l_2})\ln\frac{|x_{j_1}-a_{j_1j_1s_2}|}{|x_{j_1 0}-a_{j_1j_1s_2}|}\\&+(a_{j_1j_1s_2}-a_{j_2j_1l_1})\Big[\ln\frac{|x_{j_1}-a_{j_1j_1s_2}|}{|x_{j_1 0}-a_{j_1j_1s_2}|}\\&-(a_{j_1j_1s_2}-a_{j_2j_1l_2})\Big(\frac{1}{x_{j_1}-a_{j_1j_1s_2}}-\frac{1}{x_{j_1 0}-a_{j_1j_1s_2}}\Big)\Big]\Big\},\\&j_1,j_2\in\{1,2\},j_1\neq j_2.\end{aligned} \tag{4.144}$$

The infinite-equilibrium of $x_{j_1}^{*}=a_{j_1j_1s_1}=a_{j_2j_1l_2}$ with $\bar{x}_{j_2}\in(-\infty,\infty)$ *has the following properties.*

- *For* $a_{j_1j_1 0}(a_{j_1j_1s_1}-a_{j_1j_1s_2})^2>0$ *and* $a_{j_2j_1 0}(a_{j_2j_1l_2}-a_{j_2j_1l_1})>0$,

$$(a_{j_1j_1s_1},\bar{x}_{j_2})=\underbrace{(\text{SO},\text{UP})}_{\text{up-parabola source}}. \tag{4.145}$$

The equilibrium at $x_{j_1}^{*}=a_{j_1j_1s_1}=a_{j_2j_1l_2}$ *is an* (SO,UP)-*up-parabola source infinite-equilibrium.*

- *For* $a_{j_1j_10}(a_{j_1j_1s_1} - a_{j_1j_1s_2})^2 < 0$ *and* $a_{j_2j_10}(a_{j_2j_1l_2} - a_{j_2j_1l_1}) > 0$,

$$(a_{j_1j_1s_1}, \bar{x}_{j_2}) = \underbrace{(\text{SI}, \text{DP})}_{\text{down-parabola sink}} . \tag{4.146}$$

The equilibrium at $x^*_{j_1} = a_{j_1j_1s_1} = a_{j_2j_1l_2}$ *is an* (SI,DP)*-down-parabola sink infinite-equilibrium.*

- *For* $a_{j_1j_10}(a_{j_1j_1s_1} - a_{j_1j_1s_2})^2 > 0$ *and* $a_{j_2j_10}(a_{j_2j_1l_2} - a_{j_2j_1l_1}) < 0$,

$$(a_{j_1j_1s_1}, \bar{x}_{j_2}) = \underbrace{(\text{SO}, \text{DP})}_{\text{down-parabola source}} . \tag{4.147}$$

The equilibrium at $x^*_{j_1} = a_{j_1j_1s_1} = a_{j_2j_1l_2}$ *is an* (SO,DP)*-down-parabola source infinite-equilibrium.*

- *For* $a_{j_1j_10}(a_{j_1j_1s_1} - a_{j_1j_1s_2})^2 < 0$ *and* $a_{j_2j_10}(a_{j_2j_1l_2} - a_{j_2j_1l_1}) < 0$,

$$(a_{j_1j_1s_1}, \bar{x}_{j_2}) = \underbrace{(\text{SI}, \text{UP})}_{\text{up-parabola sink}} . \tag{4.148}$$

The equilibrium at $x^*_{j_1} = a_{j_1j_1s_1} = a_{j_2j_1l_2}$ *is an* (SI,UP)*-up-parabola sink infinite-equilibrium.*

– *The infinite-equilibriums are the switching bifurcation for sink and source flows with increasing-inflection and decreasing-inflection flows.*

(vi$_4$*) For* $a_{j_1j_1s_1} = a_{j_2j_1l_1}$,

$$\begin{aligned} &\dot{x}_{j_1} = a_{j_1j_10}(x_{j_1} - a_{j_1j_1s_1})(x_{j_1} - a_{j_1j_1s_2})^2, \\ &\dot{x}_{j_2} = a_{j_2j_10}(x_{j_1} - a_{j_2j_1l_1})(x_{j_1} - a_{j_2j_1l_2})^2, \\ &j_1, j_2 \in \{1,2\}, j_1 \neq j_2. \end{aligned} \tag{4.149}$$

The first integral manifold is

$$\begin{aligned} x_{j_2} - x_{j_20} = &\frac{a_{j_2j_10}}{a_{j_1j_10}} \Big[x_{j_1} - x_{j_10} + 2(a_{j_1j_1s_2} - a_{j_2j_1l_2}) \ln \frac{|x_{j_1} - a_{j_1j_1s_2}|}{|x_{j_10} - a_{j_1j_1s_2}|} \\ &- (a_{j_1j_1s_2} - a_{j_2j_1l_2})^2 \Big(\frac{1}{x_{j_1} - a_{j_1j_1s_2}} - \frac{1}{x_{j_10} - a_{j_1j_1s_2}}\Big)\Big], \\ &j_1, j_2 \in \{1,2\}, j_1 \neq j_2. \end{aligned} \tag{4.150}$$

The infinite-equilibrium of $x^*_{j_1} = a_{j_1j_1s_1} = a_{j_2j_1l_1}$ with $\bar{x}_{j_2} \in (-\infty, \infty)$ *has the following properties.*

- *For* $a_{j_1j_10}(a_{j_1j_1s_1}-a_{j_1j_1s_2})^2>0$ *and* $a_{j_2j_10}(a_{j_2j_1l_1}-a_{j_2j_1l_2})^2>0$,

$$(a_{j_1j_1s_1},\bar{x}_{j_2})=\underbrace{(\text{SO},\text{II})}_{\text{increasing-inflection source}}. \tag{4.151}$$

The equilibrium at $x^*_{j_1}=a_{j_1j_1s_1}=a_{j_2j_1l_1}$ *is an* (SO,II)-*increasing-inflection source infinite-equilibrium.*

- *For* $a_{j_1j_10}(a_{j_1j_1s_1}-a_{j_1j_1s_2})^2<0$ *and* $a_{j_2j_10}(a_{j_2j_1l_1}-a_{j_2j_1l_2})^2>0$,

$$(a_{j_1j_1s_1},\bar{x}_{j_2})=\underbrace{(\text{SI},\text{DI})}_{\text{decreasing-inflection sink}}. \tag{4.152}$$

The equilibrium at $x^*_{j_1}=a_{j_1j_1s_1}=a_{j_2j_1l_1}$ *is an* (SI,DI)-*decreasing-inflection sink infinite-equilibrium.*

- *For* $a_{j_1j_10}(a_{j_1j_1s_1}-a_{j_1j_1s_2})^2>0$ *and* $a_{j_2j_10}(a_{j_2j_1l_1}-a_{j_2j_1l_2})^2<0$,

$$(a_{j_1j_1s_1},\bar{x}_{j_2})=\underbrace{(\text{SO},\text{DI})}_{\text{decreasing-inflection source}}. \tag{4.153}$$

The equilibrium at $x^*_{j_1}=a_{j_1j_1s_1}=a_{j_2j_1l_1}$ *is an* (SO,DI)-*increasing-inflection source infinite-equilibrium.*

- *For* $a_{j_1j_10}(a_{j_1j_1s_1}-a_{j_1j_1s_2})^2<0$ *and* $a_{j_2j_10}(a_{j_2j_1l_1}-a_{j_2j_1l_2})^2<0$,

$$(a_{j_1j_1s_1},\bar{x}_{j_2})=\underbrace{(\text{SI},\text{II})}_{\text{increasing-inflection sink}}. \tag{4.154}$$

The equilibrium at $x^*_{j_1}=a_{j_1j_1s_1}=a_{j_2j_1l_1}$ *is an* (SI,II)-*increasing-inflection sink infinite-equilibrium.*

– *The infinite-equilibriums are the switching bifurcations for the up-parabola and down-parabola flows with the source and sink flows.*

(vii$_5$*) For* $a_{j_1j_1s_2}=a_{j_2j_1l_1}$,

$$\begin{aligned}
&\dot{x}_{j_1}=a_{j_1j_10}(x_{j_1}-a_{j_1j_1s_1})(x_{j_1}-a_{j_1j_1s_2})^2,\\
&\dot{x}_{j_2}=a_{j_2j_10}(x_{j_1}-a_{j_2j_1l_1})(x_{j_1}-a_{j_2j_1l_2})^2,\\
&j_1,j_2\in\{1,2\},j_1\neq j_2.
\end{aligned} \tag{4.155}$$

The first integral manifold is

$$
\begin{aligned}
x_{j_2} - x_{j_2 0} = & \frac{a_{j_2 j_2 0}}{a_{j_1 j_1 0}} \Big[x_{j_1} - x_{j_1 0} + (a_{j_1 j_1 s_2} + a_{j_1 j_1 s_1} - 2a_{j_2 j_1 l_2}) \ln \frac{| x_{j_1} - a_{j_1 j_1 s_2} |}{| x_{j_1 0} - a_{j_1 j_1 s_2} |} \\
& + \frac{(a_{j_1 j_1 s_1} - a_{j_2 j_1 l_2})^2}{a_{j_1 j_1 s_1} - a_{j_1 j_1 s_2}} \Big(\ln \frac{| x_{j_1} - a_{j_1 j_1 s_1} |}{| x_{j_1 0} - a_{j_1 j_1 s_1} |} - \ln \frac{| x_{j_1} - a_{j_1 j_1 s_2} |}{| x_{j_1 0} - a_{j_1 j_1 s_2} |} \Big) \Big], \\
& j_1, j_2 \in \{1,2\}, j_1 \neq j_2.
\end{aligned} \tag{4.156}
$$

The infinite-equilibrium of $x_{j_1}^* = a_{j_1 j_1 s_2} = a_{j_2 j_1 l_1}$ with $\bar{x}_{j_2} \in (-\infty, \infty)$ *has the following properties.*

- *For* $a_{j_1 j_1 0}(a_{j_1 j_1 s_2} - a_{j_1 j_1 s_1}) > 0$ *and* $a_{j_2 j_1 0}(a_{j_2 j_1 l_1} - a_{j_2 j_1 l_2})^2 > 0$,

$$
(a_{j_1 j_1 s_2}, \bar{x}_{j_2}) = \underbrace{(\text{US}, \text{DU})}_{\text{down-up upper-saddle}}. \tag{4.157}
$$

  The equilibrium at $x_{j_1}^* = a_{j_1 j_1 s_2} = a_{j_2 j_1 l_1}$ *is a* (US,DU)*-down-up asymptotic upper-saddle infinite-equilibrium.*

- *For* $a_{j_1 j_1 0}(a_{j_1 j_1 s_2} - a_{j_1 j_1 s_1}) < 0$ *and* $a_{j_2 j_1 0}(a_{j_2 j_1 l_1} - a_{j_2 j_1 l_2})^2 > 0$,

$$
(a_{j_1 j_1 s_2}, \bar{x}_{j_2}) = \underbrace{(\text{LS}, \text{UD})}_{\text{up-down lower-saddle}}. \tag{4.158}
$$

  The equilibrium at $x_{j_1}^* = a_{j_1 j_1 s_2} = a_{j_2 j_1 l_1}$ *is an* (LS,UD)*-up-down asymptotic lower-saddle infinite-equilibrium.*

- *For* $a_{j_1 j_1 0}(a_{j_1 j_1 s_2} - a_{j_1 j_1 s_1}) > 0$ *and* $a_{j_2 j_1 0}(a_{j_2 j_1 l_1} - a_{j_2 j_1 l_2})^2 < 0$,

$$
(a_{j_1 j_1 s_2}, \bar{x}_{j_2}) = \underbrace{(\text{US}, \text{UD})}_{\text{up-down upper-saddle}}. \tag{4.159}
$$

  The equilibrium at $x_{j_1}^* = a_{j_1 j_1 s_2} = a_{j_2 j_1 l_1}$ *is a* (US,UD)*-up-down asymptotic upper-saddle infinite-equilibrium.*

- *For* $a_{j_1 j_1 0}(a_{j_1 j_1 s_2} - a_{j_1 j_1 s_1}) < 0$ *and* $a_{j_2 j_1 0}(a_{j_2 j_1 l_1} - a_{j_2 j_1 l_2})^2 < 0$,

$$
(a_{j_1 j_1 s_2}, a_{j_2 j_1 l_1}) = \underbrace{(\text{LS}, \text{DU})}_{\text{down-up lower-saddle}}. \tag{4.160}
$$

  The equilibrium at $x_{j_1}^* = a_{j_1 j_1 s_2} = a_{j_2 j_1 l_1}$ *is an* (LS,DU)*-down-up asymptotic lower-saddle infinite-equilibrium.*

– *The up-down and down-up asymptotic upper-saddle and lower-saddle are for switching bifurcations of sink and source flows with up-parabola or down-parabola flow.*

(viii) If

$$\begin{aligned} &\Delta_{j_1j_1}=0 \text{ and } \Delta_{j_2j_1}=0,\\ &a_{j_1j_1s_1}=a_{j_1j_1s_2}\equiv a_{j_1j_1 1} \end{aligned} \tag{4.161}$$

then

$$\begin{aligned} &\dot{x}_{j_1}=a_{j_1j_1 0}(x_{j_1}-a_{j_1j_1 1})^3,\\ &\dot{x}_{j_2}=a_{j_2j_1 0}(x_{j_1}-a_{j_2j_1 l_1})(x_{j_1}-a_{j_2j_1 l_2})^2,\\ &j_1,j_2\in\{1,2\},j_1\neq j_2. \end{aligned} \tag{4.162}$$

(viii$_1$) For $a_{j_1j_1 1}\neq a_{j_2j_1 l_1},a_{j_2j_1 l_2}$, the first integral manifold is

$$\begin{aligned} x_{j_2}-x_{j_2 0}=&\frac{a_{j_2j_1 0}}{a_{j_1j_1 0}}\Big(\Big[(x_{j_1}-x_{j_1 0})+2(a_{j_1j_1 1}-a_{j_2j_1 l_2})\ln\frac{|x_{j_1}-a_{j_1j_1 1}|}{|x_{j_1 0}-a_{j_1j_1 1}|}\\ &-(a_{j_1j_1 1}-a_{j_2j_1 l_2})^2(\frac{1}{x_{j_1}-a_{j_1j_1 1}}-\frac{1}{x_{j_1 0}-a_{j_1j_1 1}})\Big]\\ &+(a_{j_1j_1 1}-a_{j_2j_1 l_1})\Big\{\ln\frac{|x_{j_1}-a_{j_1j_1 1}|}{|x_{j_1 0}-a_{j_1j_1 1}|}\\ &-2(a_{j_1j_1 1}-a_{j_2j_1 l_2})(\frac{1}{x_{j_1}-a_{j_1j_1 1}}-\frac{1}{x_{j_1 0}-a_{j_1j_1 1}})\\ &-\frac{1}{2}(a_{j_1j_1 1}-a_{j_2j_1 l_2})^2\Big[\frac{1}{(x_{j_1}-a_{j_1j_1 1})^2}-\frac{1}{(x_{j_1 0}-a_{j_1j_1 1})^2}\Big]\Big\}\Big),\\ &j_1,j_2\in\{1,2\},j_1\neq j_2. \end{aligned} \tag{4.163}$$

(viii$_{1a}$) The equilibrium of $x^*_{j_1}=a_{j_1j_1 1}\neq a_{j_2j_1 l_1},a_{j_2j_1 l_2}$ has the following properties.

- *For $a_{j_1j_1 0}>0$ and $a_{j_2j_1 0}(a_{j_1j_1 1}-a_{j_2j_1 l_1})(a_{j_1j_1 1}-a_{j_2j_1 l_2})^2>0$,*

$$(a_{j_1j_1 1},\dot{x}_{j_2})=\underbrace{(3^{\text{rd}}\text{SO, pF})}_{\text{third-order source flow }(+)}. \tag{4.164}$$

The flow at $x^*_{j_1}=a_{j_1j_1 1}$ is a (3^{rd}SO,pF)*-positive third-order source flow.*

- *For $a_{j_1j_1 0}<0$ and $a_{j_2j_1 0}(a_{j_1j_1 1}-a_{j_2j_1 l_1})(a_{j_1j_1 1}-a_{j_2j_1 l_2})^2>0$,*

$$(a_{j_1j_1 1},\dot{x}_{j_2})=\underbrace{(3^{\text{rd}}\text{SI, pF})}_{\text{third-order sink flow }(+)}. \tag{4.165}$$

The flow at $x^*_{j_1}=a_{j_1j_1 1}$ is a (3^{rd}SI,pF)*-positive third-order sink flow.*

- *For* $a_{j_1j_1 0} > 0$ *and* $a_{j_2j_1 0}(a_{j_1j_1 1} - a_{j_2j_1 l_1})(a_{j_1j_1 1} - a_{j_2j_1 l_2})^2 < 0$,

$$(a_{j_1j_1 1}, \dot{x}_{j_2}) = \underbrace{(3^{\text{rd}}\text{SO, nF})}_{\text{third-order source flow } (-)} . \tag{4.166}$$

The flow at $x^*_{j_1} = a_{j_1j_1 1}$ *is a* (3^rd^SO,nF)*-negative third-order source flow.*

- *For* $a_{j_1j_1 0} < 0$ *and* $a_{j_2j_1 0}(a_{j_1j_1 1} - a_{j_2j_1 l_1})(a_{j_1j_1 1} - a_{j_2j_1 l_2})^2 < 0$,

$$(a_{j_1j_1 1}, \dot{x}_{j_2}) = \underbrace{(3^{\text{rd}}\text{SI, nF})}_{\text{third-order sink flow } (-)} . \tag{4.167}$$

The flow at $x^*_{j_1} = a_{j_1j_1 1}$ *is a* (3^{rd}SI,pF)*-negative third-order sink flow.*

– *The third-order source and sink flows bifurcations are the appearing bifurcations for source-sink-source and sink-source-sink flows.*

($viii_{1b}$) *The equilibrium of* $x^*_{j_1} = a_{j_2j_1 l_2} \neq a_{j_1j_1 1}$ *has the following properties.*

- *For* $a_{j_1j_1 0}(a_{j_2j_1 l_2} - a_{j_1j_1 1})^3 > 0$ *and* $a_{j_2j_1 0}(a_{j_2j_1 l_2} - a_{j_2j_1 l_1}) > 0$,

$$(\dot{x}_{j_1}, a_{j_2j_1 l_2}) = \underbrace{(\text{pF, II})}_{\text{increasing-inflection flow } (+)} . \tag{4.168}$$

The flow at $x^*_{j_1} = a_{j_2j_1 l_2}$ *is a* (pF,II)*-positive increasing-inflection flow.*

- *For* $a_{j_1j_1 0}(a_{j_2j_1 l_2} - a_{j_1j_1 1})^3 < 0$ *and* $a_{j_2j_1 0}(a_{j_2j_1 l_2} - a_{j_2j_1 l_1}) > 0$,

$$(\dot{x}_{j_1}, a_{j_2j_1 l_2}) = \underbrace{(\text{nF, DI})}_{\text{decreasing-inflection flow } (-)} . \tag{4.169}$$

The flow at $x^*_{j_1} = a_{j_2j_1 l_2}$ *is a* (pF,DI)*-positive decreasing-inflection flow.*

- *For* $a_{j_1j_1 0}(a_{j_2j_1 l_2} - a_{j_1j_1 1})^3 > 0$ *and* $a_{j_2j_1 0}(a_{j_2j_1 l_2} - a_{j_2j_1 l_1}) < 0$,

$$(\dot{x}_{j_1}, a_{j_2j_1 l_2}) = \underbrace{(\text{pF, DI})}_{\text{decreasing-inflection flow } (+)} . \tag{4.170}$$

The flow at $x^*_{j_1} = a_{j_2j_1 l_2}$ *is an* (nF,DI)*-negative decreasing-inflection flow.*

- *For* $a_{j_1j_1 0}(a_{j_2j_1 l_2} - a_{j_1j_1 1})^3 < 0$ *and* $a_{j_2j_1 0}(a_{j_2j_1 l_2} - a_{j_2j_1 l_1}) < 0$,

$$(\dot{x}_{j_1}, a_{j_2j_1 l_2}) = \underbrace{(\text{nF, II})}_{\text{increasing-inflection flow } (-)} . \tag{4.171}$$

The flow at $x^*_{j_1} = a_{j_2j_1 l_2}$ *is an* (nF,II)*-negative increasing-inflection flow.*

– *The increasing-inflection and decreasing-inflection flow bifurcations are the appearing bifurcations for the up-parabola and down-parabola flows.*

(viii$_{1c}$) The equilibrium of $x_{j_1}^* = a_{j_2j_1l_1} \neq a_{j_1j_1 1}$ *has the following properties.*

- *For* $a_{j_1j_1 0}(a_{j_2j_1l_1} - a_{j_1j_1 1})^3 > 0$ *and* $a_{j_2j_1 0}(a_{j_2j_1l_1} - a_{j_2j_1l_2})^2 > 0$,

$$(\dot{x}_{j_1}, a_{j_2j_1l_1}) = \underbrace{(\text{pF}, \text{UP})}_{\text{up-parabola flow } (+)}. \tag{4.172}$$

The flow at $x_{j_1}^* = a_{j_2j_1l_1}$ *is a* (pF,UP)*-positive up-parabola flow.*

- *For* $a_{j_1j_1 0}(a_{j_2j_1l_1} - a_{j_1j_1 1})^3 < 0$ *and* $a_{j_2j_1 0}(a_{j_2j_1l_1} - a_{j_2j_1l_2})^2 > 0$,

$$(\dot{x}_{j_1}, a_{j_2j_1l_1}) = \underbrace{(\text{nF}, \text{DP})}_{\text{down-parabola flow } (-)}. \tag{4.173}$$

The flow at $x_{j_1}^* = a_{j_2j_1l_1}$ *is an* (nF,DP)*-negative down-parabola flow.*

- *For* $a_{j_1j_1 0}(a_{j_2j_1l_1} - a_{j_1j_1 1})^3 > 0$ *and* $a_{j_2j_1 0}(a_{j_2j_1l_1} - a_{j_2j_1l_2})^2 < 0$,

$$(\dot{x}_{j_1}, a_{j_2j_1l_1}) = \underbrace{(\text{pF}, \text{DP})}_{\text{down-parabola flow } (+)}. \tag{4.174}$$

The flow at $x_{j_1}^* = a_{j_2j_1l_1}$ *is a* (pF,DP)*-positive down-parabola flow.*

- *For* $a_{j_1j_1 0}(a_{j_2j_1l_1} - a_{j_1j_1 1})^3 < 0$ *and* $a_{j_2j_1 0}(a_{j_2j_1l_1} - a_{j_2j_1l_2})^2 < 0$,

$$(\dot{x}_{j_1}, a_{j_2j_1l_1}) = \underbrace{(\text{nF}, \text{UP})}_{\text{up-parabola flow } (-)}. \tag{4.175}$$

The flow at $x_{j_1}^* = a_{j_2j_1l_1}$ *is an* (nF,UP)*-negative up-parabola flow.*

(viii$_2$) For $a_{j_2j_1l_2} = a_{j_1j_1 1}$,

$$\begin{aligned}
&\dot{x}_{j_1} = a_{j_1j_1 0}(x_{j_1} - a_{j_1j_1 1})^3,\\
&\dot{x}_{j_2} = a_{j_2j_1 0}(x_{j_1} - a_{j_2j_1l_1})(x_{j_1} - a_{j_2j_1l_2})^2,\\
&j_1, j_2 \in \{1,2\}, j_1 \neq j_2.
\end{aligned} \tag{4.176}$$

The first integral manifold is

$$\begin{aligned}
x_{j_2} - x_{j_2 0} = &\frac{a_{j_2j_1 0}}{a_{j_1j_1 0}}\left[(x_{j_1} - x_{j_1 0}) + (a_{j_1j_1 1} - a_{j_2j_1l_1})\ln\frac{|x_{j_1} - a_{j_1j_1 1}|}{|x_{j_1 0} - a_{j_1j_1 1}|}\right],\\
&j_1, j_2 \in \{1,2\}, j_1 \neq j_2.
\end{aligned} \tag{4.177}$$

The infinite-equilibrium of $x_{j_1}^* = a_{j_2j_1l_2} = a_{j_1j_1 1}$ with $\bar{x}_{j_2} \in (-\infty, \infty)$ *has the following properties.*

- *For* $a_{j_1 j_1 0} > 0$ *and* $a_{j_2 j_1 0}(a_{j_2 j_1 l_2} - a_{j_2 j_1 l_1}) > 0$,

$$(a_{j_1 j_1 1}, \bar{x}_{j_2}) = \underbrace{(3^{\text{rd}}\text{SO}, (3:2)\text{DL})}_{\text{third-order down-logarithmic source}}. \tag{4.178}$$

The equilibrium at $x_{j_1}^* = a_{j_2 j_1 l_2} = a_{j_1 j_1 1}$ *is a* (3^{rd}SO,(3:2)DL)*-third-order down-logarithmic source infinite-equilibrium.*

- *For* $a_{j_1 j_1 0} < 0$ *and* $a_{j_2 j_1 0}(a_{j_2 j_1 l_2} - a_{j_2 j_1 l_1}) > 0$,

$$(a_{j_1 j_1 1}, \bar{x}_{j_2}) = \underbrace{(3^{\text{rd}}\text{SI}, (3:2)\text{UL})}_{\text{third-order up-logarithmic sink}}. \tag{4.179}$$

The equilibrium at $x_{j_1}^* = a_{j_2 j_1 l_2} = a_{j_1 j_1 1}$ *is a* (3^{rd}SI,(3:2)UL)*-third-order up-logarithmic sink infinite-equilibrium.*

- *For* $a_{j_1 j_1 0} > 0$ *and* $a_{j_2 j_1 0}(a_{j_2 j_1 l_2} - a_{j_2 j_1 l_1}) < 0$,

$$(a_{j_1 j_1 1}, \bar{x}_{j_2}) = \underbrace{(3^{\text{rd}}\text{SO}, (3:2)\text{UL})}_{\text{third-order up-logarithmic source}}. \tag{4.180}$$

The equilibrium at $x_{j_1}^* = a_{j_2 j_1 l_2} = a_{j_1 j_1 1}$ *is a* (3^{rd}SO,(3:2)UL)*-third-order up-logarithmic source infinite-equilibrium.*

- *For* $a_{j_1 j_1 0} < 0$ *and* $a_{j_2 j_1 0}(a_{j_2 j_1 l_2} - a_{j_2 j_1 l_1}) < 0$,

$$(a_{j_1 j_1 1}, \bar{x}_{j_2}) = \underbrace{(3^{\text{rd}}\text{SI}, (3:2)\text{DL})}_{\text{third-order down-logarithmic sink}}. \tag{4.181}$$

The equilibrium at $x_{j_1}^* = a_{j_2 j_1 l_2} = a_{j_1 j_1 1}$ *is a* (3^{rd}SI,(3:2)DL)*-third-order down-logarithmic sink infinite-equilibrium.*

- *The third-order up-logarithmic and down-logarithmic source and sink infinite-equilibriums of* $x_{j_1}^* = a_{j_2 j_1 l_2} = a_{j_1 j_1 1}$ *are the appearing bifurcation for the sink and source flows and the up-parabola and down-parabola flows.*
- *The third-order up-logarithmic and down-logarithmic source and sink infinite-equilibriums of* $x_{j_1}^* = a_{j_2 j_1 l_2} = a_{j_1 j_1 1}$ *are the switching bifurcation for the third-order sink and source flows and the increasing-inflection and decreasing-inflection flows.*

(viii$_3$) For $a_{j_2 j_1 l_1} = a_{j_1 j_1 1}$,

$$\begin{aligned}
&\dot{x}_{j_1} = a_{j_1 j_1 0}(x_{j_1} - a_{j_1 j_1 1})^3, \\
&\dot{x}_{j_2} = a_{j_2 j_1 0}(x_{j_1} - a_{j_2 j_1 l_1})(x_{j_1} - a_{j_2 j_1 l_2})^2, \\
&j_1, j_2 \in \{1, 2\}, j_1 \neq j_2.
\end{aligned} \tag{4.182}$$

The first integral manifold is

$$\begin{aligned}x_{j_2}-x_{j_20}=&\frac{a_{j_2j_10}}{a_{j_1j_10}}\Big[(x_{j_1}-x_{j_10})+2(a_{j_1j_11}-a_{j_2j_1l_2})\ln\frac{|\,x_{j_1}-a_{j_1j_11}\,|}{|\,x_{j_10}-a_{j_1j_11}\,|}\\&-(a_{j_1j_11}-a_{j_2j_1l_2})^2(\frac{1}{x_{j_1}-a_{j_1j_11}}-\frac{1}{x_{j_10}-a_{j_1j_11}})\Big],\\&j_1,j_2\in\{1,2\},j_1\neq j_2.\end{aligned}\tag{4.183}$$

The infinite-equilibrium of $x_{j_1}^*=a_{j_2j_1l_1}=a_{j_1j_11}$ *with* $\overline{x}_{j_2}\in(-\infty,\infty)$ *has the following properties.*

- *For* $a_{j_1j_10}>0$ *and* $a_{j_2j_10}(a_{j_2j_1l_1}-a_{j_2j_1l_2})^2>0$*,*

$$(a_{j_1j_11},\overline{x}_{j_2})=\underbrace{(3^{\mathrm{rd}}\mathrm{SO},(3:1)\mathrm{IC})}_{\text{third-order increasing-concave source}}.\tag{4.184}$$

  The equilibrium at $x_{j_1}^*=a_{j_2j_1l_1}=a_{j_1j_11}$ *is a* (3^rd^SO,(3:1)IC)*-third-order increasing-concave source infinite-equilibrium.*
- *For* $a_{j_1j_10}<0$ *and* $a_{j_2j_10}(a_{j_2j_1l_1}-a_{j_2j_1l_2})^2>0$*,*

$$(a_{j_1j_11},\overline{x}_{j_2})=\underbrace{(3^{\mathrm{rd}}\mathrm{SI},(3:1)\mathrm{DC})}_{\text{third-order decreasing-concave sink}}.\tag{4.185}$$

  The equilibrium at $x_{j_1}^*=a_{j_2j_1l_1}=a_{j_1j_11}$ *is a* (3^{rd}SI,(3:1)DC)*-third-order decreasing-concave sink infinite-equilibrium.*
- *For* $a_{j_1j_10}>0$ *and* $a_{j_2j_10}(a_{j_2j_1l_1}-a_{j_2j_1l_2})^2<0$*,*

$$(a_{j_1j_11},a_{j_2j_1l_2})=\underbrace{(3^{\mathrm{rd}}\mathrm{SO},(3:1)\mathrm{DC})}_{\text{third-order decreasing-concave source}}.\tag{4.186}$$

  The equilibrium at $x_{j_1}^*=a_{j_2j_1l_1}=a_{j_1j_11}$ *is a* (3^{rd}SO,(3:1)DC)*-third-order decreasing-concave source infinite-equilibrium.*
- *For* $a_{j_1j_10}<0$ *and* $a_{j_2j_10}(a_{j_2j_1l_1}-a_{j_2j_1l_2})^2<0$*,*

$$(a_{j_1j_11},\overline{x}_{j_2})=\underbrace{(3^{\mathrm{rd}}\mathrm{SI},(3:1)\mathrm{IC})}_{\text{third-order increasing-concave sink}}.\tag{4.187}$$

  The equilibrium at $x_{j_1}^*=a_{j_2j_1l_1}=a_{j_1j_11}$ *is a* (3^{rd}SI,(3:1)IC)*-third-order increasing-concave source infinite-equilibrium.*

- *The third-order increasing-concave and decreasing-concave source and sink infinite-equilibriums of* $x_{j_1}^* = a_{j_2j_1l_1} = a_{j_1j_1 1}$ *are the appearing bifurcation for the sink and source flows and the up-parabola and down-parabola flows.*
- *The third-order increasing-concave and decreasing-concave source and sink infinite-equilibriums of* $x_{j_1}^* = a_{j_2j_1l_1} = a_{j_1j_1 1}$ *are the switching bifurcation for the third-order sink and source flows and the up-parabola and down-parabola flows.*

(ix) If

$$\begin{array}{l}\Delta_{j_1j_1} = 0 \text{ and } \Delta_{j_2j_1} = 0,\\ a_{j_2j_1l_1} = a_{j_2j_1l_2} \equiv a_{j_2j_1 1}, a_{j_1j_1s_1} \neq a_{j_1j_1s_2},\end{array} \tag{4.188}$$

then

$$\begin{array}{l}\dot{x}_{j_1} = a_{j_1j_1 0}(x_{j_1} - a_{j_1j_1s_1})(x_{j_1} - a_{j_1j_1s_2})^2,\\ \dot{x}_{j_2} = a_{j_2j_1 0}(x_{j_1} - a_{j_2j_1 1})^3,\\ j_1, j_2 \in \{1,2\}, j_1 \neq j_2.\end{array} \tag{4.189}$$

(ix$_1$) For $a_{j_1j_1 1} \neq a_{j_2j_1l_1}, a_{j_2j_1l_2}$, *the first integral manifold is*

$$\begin{aligned}x_{j_2} - x_{j_2 0} = & \frac{a_{j_2j_1 0}}{a_{j_1j_1 0}}\Big\{(x_{j_1} - x_{j_1 0}) + \frac{(a_{j_1j_1s_1} - a_{j_2j_1 1})^3}{(a_{j_1j_1s_1} - a_{j_1j_1s_2})^2}\ln\frac{|x_{j_1} - a_{j_1j_1s_1}|}{|x_{j_1 0} - a_{j_1j_1s_1}|}\\ & - \frac{(a_{j_1j_1s_2} - a_{j_2j_1 1})^3}{(a_{j_1j_1s_1} - a_{j_1j_1s_2})}\Big(\frac{1}{x_{j_1} - a_{j_1j_1s_2}} - \frac{1}{x_{j_1 0} - a_{j_1j_1s_2}}\Big)\\ & + \Big[\frac{3(a_{j_1j_1s_2} - a_{j_2j_1 1})^2}{a_{j_1j_1s_2} - a_{j_1j_1s_1}} - \frac{(a_{j_1j_1s_2} - a_{j_2j_1 1})^3}{(a_{j_1j_1s_2} - a_{j_1j_1s_1})^2}\Big]\ln\frac{|x_{j_1} - a_{j_1j_1s_2}|}{|x_{j_1 0} - a_{j_1j_1s_2}|}\Big\},\\ & s_1, s_2 \in \{1,2\}, s_1 \neq s_2.\end{aligned} \tag{4.190}$$

(ix$_{1a}$) The equilibrium of $x_{j_1}^* = a_{j_2j_1 1} \neq a_{j_1j_1s_1}, a_{j_2j_1s_2}$ *has the following properties.*

- *For* $a_{j_1j_1 0}(a_{j_2j_1 1} - a_{j_1j_1s_1})(a_{j_2j_1 1} - a_{j_1j_1s_2})^2 > 0$ *and* $a_{j_2j_1 0} > 0$,

$$(\dot{x}_{j_1}, a_{j_2j_1 1}) = \underbrace{(\text{pF}, 3^{\text{rd}}\text{UP})}_{\text{third-order up-parabola flow }(+)}. \tag{4.191}$$

The flow at $x_{j_1}^* = a_{j_2j_1 1}$ *is a* (pF, 3rdUP)*-positive third-order up-parabola flow.*

- *For* $a_{j_1j_1 0}(a_{j_2j_1 1} - a_{j_1j_1s_1})(a_{j_2j_1 1} - a_{j_1j_1s_2})^2 < 0$ *and* $a_{j_2j_1 0} > 0$,

$$(\dot{x}_{j_1}, a_{j_2j_1 1}) = \underbrace{(\text{nF}, 3^{\text{rd}}\text{DP})}_{\text{third-order down-parabola flow }(-)}. \tag{4.192}$$

The flow at $x_{j_1}^* = a_{j_2 j_1 1}$ *is an* (nF,3$^{\rm rd}$DP)-*negative third-order down-parabola flow.*

- *For* $a_{j_1 j_1 0}(a_{j_2 j_1 1} - a_{j_1 j_1 s_1})(a_{j_2 j_1 1} - a_{j_1 j_1 s_2})^2 > 0$ *and* $a_{j_2 j_1 0} < 0$,

$$(\dot{x}_{j_1}, a_{j_2 j_1 1}) = \underbrace{(\mathrm{pF}, 3^{\mathrm{rd}}\mathrm{DP})}_{\text{third-order down-parabola flow } (+)}. \tag{4.193}$$

The flow at $x_{j_1}^* = a_{j_2 j_1 1}$ *is a* (pF,3$^{\rm rd}$DP)-*negative third-order down-parabola flow.*

- *For* $a_{j_1 j_1 0}(a_{j_2 j_1 1} - a_{j_1 j_1 s_1})(a_{j_2 j_1 1} - a_{j_1 j_1 s_2})^2 < 0$ *and* $a_{j_2 j_1 0} < 0$,

$$(\dot{x}_{j_1}, a_{j_2 j_1 1}) = \underbrace{(\mathrm{nF}, 3^{\mathrm{rd}}\mathrm{UP})}_{\text{third-order up-parabola flow } (-)}. \tag{4.194}$$

The flow at $x_{j_1}^* = a_{j_2 j_1 1}$ *is an* (nF,3$^{\rm rd}$UP)-*negative third-order up-parabola flow.*

– *The third-order up-parabola and down-parabola flows bifurcations are the appearing bifurcations for up-parabola, down-parabola, and up-parabola flows and down-parabola, up-parabola, and down-parabola flows.*

(ix_{1b}) *The equilibrium of* $x_{j_1}^* = a_{j_1 j_1 s_1} \neq a_{j_2 j_1 1}$ *has the following properties.*

- *For* $a_{j_1 j_1 0}(a_{j_1 j_1 s_1} - a_{j_1 j_1 s_2})^2 > 0$ *and* $a_{j_2 j_1 0}(a_{j_1 j_1 s_1} - a_{j_2 j_1 1})^3 > 0$,

$$(a_{j_1 j_1 s_1}, \dot{x}_{j_2}) = \underbrace{(\mathrm{SO}, \mathrm{pF})}_{\text{source flow } (+)}. \tag{4.195}$$

The flow at $x_{j_1}^* = a_{j_1 j_1 s_1}$ *is an* (SO,pF)-*positive source flow.*

- *For* $a_{j_1 j_1 0}(a_{j_1 j_1 s_1} - a_{j_1 j_1 s_2})^2 < 0$ *and* $a_{j_2 j_1 0}(a_{j_1 j_1 s_1} - a_{j_2 j_1 1})^3 > 0$,

$$(a_{j_1 j_1 s_1}, \dot{x}_{j_2}) = \underbrace{(\mathrm{SI}, \mathrm{pF})}_{\text{sink flow } (+)}. \tag{4.196}$$

The flow at $x_{j_1}^* = a_{j_1 j_1 s_1}$ *is an* (SI,pF)-*positive sink flow.*

- *For* $a_{j_1 j_1 0}(a_{j_1 j_1 s_1} - a_{j_1 j_1 s_2})^2 > 0$ *and* $a_{j_2 j_1 0}(a_{j_1 j_1 s_1} - a_{j_2 j_1 1})^3 < 0$,

$$(a_{j_1 j_1 s_1}, \dot{x}_{j_2}) = \underbrace{(\mathrm{SO}, \mathrm{nF})}_{\text{source flow } (-)}. \tag{4.197}$$

The flow at $x_{j_1}^* = a_{j_1 j_1 s_1}$ *is an* (SO,nF)-*negative source flow.*

- *For* $a_{j_1j_10}(a_{j_1j_1s_1}-a_{j_1j_1s_2})^2<0$ *and* $a_{j_2j_10}(a_{j_1j_1s_1}-a_{j_2j_11})^3<0$,

$$(a_{j_1j_1s_1},\dot{x}_{j_2})=\underbrace{(\text{SI},\text{nF})}_{\text{sink flow }(-)}. \tag{4.198}$$

The flow at $x^*_{j_1}=a_{j_1j_1s_1}$ *is an* (SI,nF)*-negative sink flow.*

(ix$_{1c}$*) The equilibrium of* $x^*_{j_1}=a_{j_1j_1s_2}\neq a_{j_2j_11}$ *has the following properties.*

- *For* $a_{j_1j_10}(a_{j_1j_1s_2}-a_{j_1j_1s_1})>0$ *and* $a_{j_2j_10}(a_{j_1j_1s_2}-a_{j_2j_11})^3>0$,

$$(a_{j_1j_1s_2},\dot{x}_{j_2})=\underbrace{(\text{US},\text{pF})}_{\text{upper-saddle flow }(+)}. \tag{4.199}$$

The flow at $x^*_{j_1}=a_{j_1j_1s_2}$ *is a* (US,pF)*-positive upper-saddle flow.*

- *For* $a_{j_1j_10}(a_{j_1j_1s_2}-a_{j_1j_1s_1})<0$ *and* $a_{j_2j_10}(a_{j_1j_1s_2}-a_{j_2j_11})^3>0$,

$$(a_{j_1j_1s_2},\dot{x}_{j_2})=\underbrace{(\text{LS},\text{pF})}_{\text{lower-saddle flow }(+)}. \tag{4.200}$$

The flow at $x^*_{j_1}=a_{j_1j_1s_2}$ *is an* (LS,pF)*-positive lower-saddle flow.*

- *For* $a_{j_1j_10}(a_{j_1j_1s_2}-a_{j_1j_1s_1})>0$ *and* $a_{j_2j_10}(a_{j_1j_1s_2}-a_{j_2j_11})^3<0$,

$$(a_{j_1j_1s_2},\dot{x}_{j_2})=\underbrace{(\text{US},\text{nF})}_{\text{upper-saddle flow }(-)}. \tag{4.201}$$

The flow at $x^*_{j_1}=a_{j_1j_1s_2}$ *is a* (US,nF)*-negative upper-saddle flow.*

- *For* $a_{j_1j_10}(a_{j_1j_1s_2}-a_{j_1j_1s_1})<0$ *and* $a_{j_2j_10}(a_{j_1j_1s_2}-a_{j_2j_11})^3<0$,

$$(a_{j_1j_1s_2},\dot{x}_{j_2})=\underbrace{(\text{LS},\text{nF})}_{\text{lower-saddle flow }(-)}. \tag{4.202}$$

The flow at $x^*_{j_1}=a_{j_1j_1s_2}$ *is an* (LS,pF)*-negative lower-saddle flow.*

(ix$_2$*) For* $a_{j_2j_11}=a_{j_1j_1s_2}$,

$$\begin{aligned}
&\dot{x}_{j_1}=a_{j_1j_10}(x_{j_1}-a_{j_1j_1s_1})(x_{j_1}-a_{j_1j_1s_2})^2,\\
&\dot{x}_{j_2}=a_{j_2j_10}(x_{j_1}-a_{j_2j_11})^3,\\
&j_1,j_2\in\{1,2\},j_1\neq j_2.
\end{aligned} \tag{4.203}$$

The first integral manifold is

$$x_{j_2}-x_{j_2 0}=\frac{a_{j_2 j_1 0}}{a_{j_1 j_1 0}}\left[(x_{j_1}-x_{j_1 0})+(a_{j_1 j_1 s_1}-a_{j_2 j_1 1})\ln\frac{|x_{j_1}-a_{j_1 j_1 s_1}|}{|x_{j_1 0}-a_{j_1 j_1 s_1}|}\right],\qquad j_1,j_2\in\{1,2\},j_1\neq j_2. \tag{4.204}$$

The infinite-equilibrium of $x_{j_1}^*=a_{j_1 j_1 s_2}=a_{j_2 j_1 1}$ *with* $\bar{x}_{j_2}\in(-\infty,\infty)$ *has the following properties.*

- *For* $a_{j_1 j_1 0}(a_{j_1 j_1 s_2}-a_{j_1 j_1 s_1})>0$ *and* $a_{j_2 j_1 0}>0$,

$$(a_{j_1 j_1 s_2},\bar{x}_{j_2})=\underbrace{(\mathrm{US},(2:3)\mathrm{UP})}_{\text{up-parabola upper-saddle}}. \tag{4.205}$$

  The equilibrium at $x_{j_1}^*=a_{j_1 j_1 s_2}=a_{j_2 j_1 1}$ *is a* (US,(2:3)UP)-*up-parabola upper-saddle infinite-equilibrium.*
- *For* $a_{j_1 j_1 0}(a_{j_1 j_1 s_2}-a_{j_1 j_1 s_1})<0$ *and* $a_{j_2 j_1 0}>0$,

$$(a_{j_1 j_1 s_2},\bar{x}_{j_2})=\underbrace{(\mathrm{LS},(2:3)\mathrm{DP})}_{\text{down-parabola lower-saddle}}. \tag{4.206}$$

  The equilibrium at $x_{j_1}^*=a_{j_1 j_1 s_2}=a_{j_2 j_1 1}$ *is an* (LS,(2:3)DP)-*down-parabola lower-saddle infinite-equilibrium.*
- *For* $a_{j_1 j_1 0}(a_{j_1 j_1 s_2}-a_{j_1 j_1 s_1})>0$ *and* $a_{j_2 j_1 0}<0$,

$$(a_{j_1 j_1 s_2},\bar{x}_{j_2})=\underbrace{(\mathrm{US},(2:3)\mathrm{DP})}_{\text{down-parabola upper-saddle}}. \tag{4.207}$$

  The equilibrium at $x_{j_1}^*=a_{j_1 j_1 s_2}=a_{j_2 j_1 1}$ *is a* (US,(2:3)DP)-*up-logarithmic source infinite-equilibrium.*
- *For* $a_{j_1 j_1 0}(a_{j_1 j_1 s_2}-a_{j_1 j_1 s_1})<0$ *and* $a_{j_2 j_1 0}<0$,

$$(a_{j_1 j_1 s_2},\bar{x}_{j_2})=\underbrace{(\mathrm{LS},(2:3)\mathrm{UP})}_{\text{up-parabola lower-saddle}}. \tag{4.208}$$

  The equilibrium at $x_{j_1}^*=a_{j_1 j_1 s_2}=a_{j_2 j_1 1}$ *is an* (LS,(2:3)UP)-*up-parabola lower-saddle infinite-equilibrium.*

– *The (2:3)-up-parabola and down-parabola, upper-saddle and lower-saddle infinite-equilibriums of* $x_{j_1}^*=a_{j_1 j_1 s_2}=a_{j_2 j_1 1}$ *are the appearing bifurcation for the sink and source flows and the up-parabola and down-parabola flows.*

– *The (2:3)-up-parabola and down-parabola, upper-saddle and lower-saddle infinite-equilibriums of* $x_{j_1}^* = a_{j_1 j_1 s_2} = a_{j_2 j_1 1}$ *are the switching bifurcation for the third-order up-parabola and down-parabola flows and the upper-saddle and lower-saddle flows.*

(ix₃) For $a_{j_1 j_1 s_1} = a_{j_2 j_1 1}$,

$$\begin{aligned} \dot{x}_{j_1} &= a_{j_1 j_1 0}(x_{j_1} - a_{j_1 j_1 s_1})(x_{j_1} - a_{j_1 j_1 s_2})^2, \\ \dot{x}_{j_2} &= a_{j_2 j_1 0}(x_{j_1} - a_{j_2 j_1 1})^3, \\ & j_1, j_2 \in \{1,2\}, j_1 \neq j_2. \end{aligned} \tag{4.209}$$

The first integral manifold is

$$\begin{aligned} x_{j_2} - x_{j_2 0} = &\frac{a_{j_2 j_1 0}}{a_{j_1 j_1 0}} \Big[(x_{j_1} - x_{j_1 0}) + 2(a_{j_1 j_1 s_2} - a_{j_2 j_1 1}) \ln \frac{|x_{j_1} - a_{j_1 j_1 s_2}|}{|x_{j_1 0} - a_{j_1 j_1 s_2}|} \\ & - (a_{j_1 j_1 s_2} - a_{j_2 j_1 1})^2 (\frac{1}{x_{j_1} - a_{j_1 j_1 s_2}} - \frac{1}{x_{j_1 0} - a_{j_1 j_1 s_2}}) \Big], \\ & s_1, s_2 \in \{1,2\}, s_1 \neq s_2. \end{aligned} \tag{4.210}$$

The infinite-equilibrium of $x_{j_1}^* = a_{j_1 j_1 s_1} = a_{j_2 j_1 1}$ *with* $\bar{x}_{j_2} \in (-\infty, \infty)$ *has the following properties.*

- *For* $a_{j_1 j_1 0}(a_{j_1 j_1 s_1} - a_{j_1 j_1 s_2})^2 > 0$ *and* $a_{j_2 j_1 0} > 0$,

$$(a_{j_1 j_1 s_1}, \bar{x}_{j_2}) = \underbrace{(\text{SO}, (1{:}3)\text{II})}_{(1:3)\text{-increasing-inflection source}}. \tag{4.211}$$

  The equilibrium at $x_{j_1}^* = a_{j_1 j_1 s_1} = a_{j_2 j_1 1}$ *is an* (SO,(1:3)II)*-increasing-inflection source infinite-equilibrium.*

- *For* $a_{j_1 j_1 0}(a_{j_1 j_1 s_1} - a_{j_1 j_1 s_2})^2 < 0$ *and* $a_{j_2 j_1 0} > 0$,

$$(a_{j_1 j_1 s_1}, \bar{x}_{j_2}) = \underbrace{(\text{SI}, (1{:}3)\text{DI})}_{(1:3)\text{-decreasing-inflection sink}}. \tag{4.212}$$

  The equilibrium at $x_{j_1}^* = a_{j_1 j_1 s_1} = a_{j_2 j_1 1}$ *is an* (SI,(1:3)DI)*-decreasing-inflection sink infinite-equilibrium.*

- *For* $a_{j_1 j_1 0}(a_{j_1 j_1 s_1} - a_{j_1 j_1 s_2})^2 > 0$ *and* $a_{j_2 j_1 0} < 0$,

$$(a_{j_1 j_1 s_1}, \bar{x}_{j_2}) = \underbrace{(\text{SO}, (1{:}3)\text{DI})}_{(1:3)\text{-decreasing-inflection source}}. \tag{4.213}$$

The equilibrium at $x_{j_1}^* = a_{j_1 j_1 s_1} = a_{j_2 j_1 1}$ *is an* (SO,(1:3)DI)*-decreasing-inflection source infinite-equilibrium.*

- *For* $a_{j_1 j_1 0}(a_{j_1 j_1 s_1} - a_{j_1 j_1 s_2})^2 < 0$ *and* $a_{j_2 j_1 0} < 0$,

$$(a_{j_1 j_1 s_1}, \bar{x}_{j_2}) = \underbrace{(\text{SI}, (1:3)\text{II})}_{(1:3)\text{-increasing-inflection sink}}. \tag{4.214}$$

The equilibrium at $x_{j_1}^* = a_{j_1 j_1 s_1} = a_{j_2 j_1 1}$ *is an* (SI,(1:3)II)*-increasing-inflection sink infinite-equilibrium.*

- *The (1:3)-increasing-inflection and decreasing-inflection source and sink infinite-equilibriums of* $x_{j_1}^* = a_{j_1 j_1 s_1} = a_{j_2 j_1 1}$ *are the appearing bifurcation for the sink and source flows and the up-parabola and down-parabola flows.*
- *The (1:3)-increasing-inflection and decreasing-inflection source and sink infinite-equilibriums of* $x_{j_1}^* = a_{j_2 j_1 l_2} = a_{j_1 j_1 1}$ *are the switching bifurcation for the third-order up-parabola and down-parabola flows and the sink and source flows.*

(x) If

$$\begin{aligned} &\Delta_{j_1 j_1} = 0 \text{ and } \Delta_{j_2 j_1} = 0, \\ &a_{j_1 j_1 s_1} = a_{j_1 j_1 s_2} \equiv a_{j_1 j_1 1}, \\ &a_{j_2 j_1 l_1} = a_{j_2 j_1 l_2} \equiv a_{j_2 j_1 1}, \end{aligned} \tag{4.215}$$

then

$$\begin{aligned} &\dot{x}_{j_1} = a_{j_1 j_1 0}(x_{j_1} - a_{j_1 j_1 1})^3, \\ &\dot{x}_{j_2} = a_{j_2 j_1 0}(x_{j_1} - a_{j_2 j_1 1})^3, \\ &j_1, j_2 \in \{1, 2\}, j_1 \neq j_2. \end{aligned} \tag{4.216}$$

(x$_1$) For $a_{j_1 j_1 1} \neq a_{j_2 j_1 1}$, *the first integral manifold is*

$$\begin{aligned} x_{j_2} - x_{j_2 0} = \frac{a_{j_2 j_1 0}}{a_{j_1 j_1 0}} \Big\{ &(x_{j_1} - x_{j_1 0}) + 3(a_{j_1 j_1 1} - a_{j_2 j_1 1}) \ln \frac{|x_{j_1} - a_{j_1 j_1 1}|}{|x_{j_1 0} - a_{j_1 j_1 1}|} \\ &- 3(a_{j_1 j_1 1} - a_{j_2 j_1 1})^2 \left(\frac{1}{x_{j_1} - a_{j_1 j_1 1}} - \frac{1}{x_{j_1 0} - a_{j_1 j_1 1}}\right) \\ &- \frac{1}{2}(a_{j_1 j_1 1} - a_{j_2 j_1 1})^3 \left[\frac{1}{(x_{j_1} - a_{j_1 j_1 1})^2} - \frac{1}{(x_{j_1 0} - a_{j_1 j_1 1})^2}\right] \Big\}, \\ &j_1, j_2 \in \{1, 2\}, j_1 \neq j_2. \end{aligned} \tag{4.217}$$

(x$_{1a}$) The infinite-equilibrium of $x_{j_1}^* = a_{j_1 j_1 1}$ *has the following properties.*

- *For* $a_{j_1 j_1 0} > 0$ *and* $a_{j_2 j_1 0}(a_{j_1 j_1 1} - a_{j_2 j_1 1})^3 > 0$,

$$(a_{j_1j_11}, \dot{x}_{j_2}) = \underbrace{(3^{\text{rd}}\text{SO}, \text{pF})}_{\text{third-order source flow } (+)}. \tag{4.218}$$

The flow at $x_{j_1}^* = a_{j_1j_11}$ *is a* (3rdSO,pF)*-positive third-order source flow.*

- *For* $a_{j_1j_10} < 0$ *and* $a_{j_2j_10}(a_{j_1j_11} - a_{j_2j_11})^3 > 0$,

$$(a_{j_1j_11}, \dot{x}_{j_2}) = \underbrace{(3^{\text{rd}}\text{SI}, \text{pF})}_{\text{third-order sink flow } (+)}. \tag{4.219}$$

The flow at $x_{j_1}^* = a_{j_1j_11}$ *is a* (3rdSI,pF)*-positive third-order sink flow.*

- *For* $a_{j_1j_10} > 0$ *and* $a_{j_2j_10}(a_{j_1j_11} - a_{j_2j_11})^3 < 0$,

$$(a_{j_1j_11}, \dot{x}_{j_2}) = \underbrace{(3^{\text{rd}}\text{SO}, \text{nF})}_{\text{third-order source flow } (-)}. \tag{4.220}$$

The flow at $x_{j_1}^* = a_{j_1j_11}$ *is a* (3rdSO,nF)*-negative third-order source flow.*

- *For* $a_{j_1j_10} < 0$ *and* $a_{j_2j_10}(a_{j_1j_11} - a_{j_2j_11})^3 < 0$,

$$(a_{j_1j_11}, \dot{x}_{j_2}) = \underbrace{(3^{\text{rd}}\text{SI}, \text{nF})}_{\text{third-order sink flow } (-)}. \tag{4.221}$$

The flow at $x_{j_1}^* = a_{j_1j_11}$ *is a* (3rdSI,nF)*-negative third-order sink flow.*

- *The third-order source and sink flows of* $x_{j_1}^* = a_{j_1j_11}$ *are the appearing bifurcation for the source, sink, and source flows and the sink, source, and sink flows.*
- *The third-order source and sink flows of* $x_{j_1}^* = a_{j_1j_11}$ *are the switching bifurcation for the source, sink, and source flows and the sink, source, and sink flows.*
- *The third-order source and sink flows of* $x_{j_1}^* = a_{j_1j_11}$ *are the switching bifurcation for the upper-saddle and lower-saddle flows and the source and sink flows.*

(x_{1b}*) The infinite-equilibrium of* $x_{j_1}^* = a_{j_2j_11}$ *has the following properties.*

- *For* $a_{j_1j_10}(a_{j_2j_11} - a_{j_1j_11})^3 > 0$ *and* $a_{j_2j_10} > 0$,

$$(\dot{x}_{j_1}, a_{j_2j_11}) = \underbrace{(\text{pF}, 3^{\text{rd}}\text{UP})}_{\text{third-order up-parabola flow } (+)}. \tag{4.222}$$

The flow at $x_{j_1}^* = a_{j_2j_11}$ *is a* (pF,3rdUP)*-positive third-order up-parabola flow.*

- *For* $a_{j_1j_10}(a_{j_2j_11} - a_{j_1j_11})^3 < 0$ *and* $a_{j_2j_10} > 0$,

$$(\dot{x}_{j_1}, a_{j_2 j_1 1}) = \underbrace{(\mathrm{nF}, 3^{\mathrm{rd}}\mathrm{DP})}_{\text{third-order down-parabola flow } (-)}. \tag{4.223}$$

The flow at $x_{j_1}^* = a_{j_2 j_1 1}$ *is an* (nF,3^rd^DP)*-negative third-order down-parabola flow.*

- *For* $a_{j_1 j_1 0}(a_{j_2 j_1 1} - a_{j_1 j_1 1})^3 > 0$ *and* $a_{j_2 j_1 0} < 0$,

$$(\dot{x}_{j_1}, a_{j_2 j_1 1}) = \underbrace{(\mathrm{pF}, 3^{\mathrm{rd}}\mathrm{DP})}_{\text{third-order down-parabola flow } (+)}. \tag{4.224}$$

The flow at $x_{j_1}^* = a_{j_2 j_1 1}$ *is a* $(\mathrm{pF},3^{\mathrm{rd}}\mathrm{UP})$*-positive third-order down-parabola flow.*

- *For* $a_{j_1 j_1 0}(a_{j_2 j_1 1} - a_{j_1 j_1 1})^3 < 0$ *and* $a_{j_2 j_1 0} < 0$,

$$(\dot{x}_{j_1}, a_{j_2 j_1 1}) = \underbrace{(\mathrm{nF}, 3^{\mathrm{rd}}\mathrm{UP})}_{\text{third-order up-parabola flow } (-)}. \tag{4.225}$$

The flow at $x_{j_1}^* = a_{j_2 j_1 1}$ *is an* $(\mathrm{nF},3^{\mathrm{rd}}\mathrm{UP})$*-negative third-order up-parabola flow.*

– *The third-order up-parabola and down-parabola flows of* $x_{j_1}^* = a_{j_2 j_1 1}$ *are the appearing bifurcation for the up-parabola, down-parabola, and up-parabola flows and the down-parabola, up-parabola, and down-parabola flows.*
– *The third-order up-parabola and down-parabola flows of* $x_{j_1}^* = a_{j_2 j_1 1}$ *are the switching bifurcation for the up-parabola, down-parabola, and up-parabola flows and the down-parabola, up-parabola, and down-parabola flows.*
– *The third-order up-parabola and down-parabola flows of* $x_{j_1}^* = a_{j_2 j_1 1}$ *are the switching bifurcation for the increasing-inflection and decreasing-inflection flows and the down-parabola and up-parabola flows.*

(x_3) For $a_{j_1 j_1 1} = a_{j_2 j_1 1}$*, the first integral manifold is*

$$x_{j_2} - x_{j_2 0} = \frac{a_{j_2 j_1 0}}{a_{j_1 j_1 0}}(x_{j_1} - x_{j_1 0}),$$
$$j_1, j_2 \in \{1,2\}, j_1 \neq j_2. \tag{4.226}$$

The infinite-equilibrium of $x_{j_1}^* = a_{j_2 j_1 1} = a_{j_1 j_1 1}$ *with* $\overline{x}_{j_2} \in (-\infty, \infty)$ *has the following properties.*

- *For* $a_{j_1 j_1 0} > 0$ *and* $a_{j_2 j_1 0} > 0$,

$$(a_{j_1 j_1 1}, \overline{x}_{j_2}) = \underbrace{(3^{\mathrm{rd}}\mathrm{SO}, (3:3)\mathrm{II})}_{\text{third-order increasing-inflection source}}. \tag{4.227}$$

The equilibrium at $x_{j_1}^* = a_{j_2 j_1 1} = a_{j_1 j_1 1}$ *is an* (3$^{\rm rd}$SO,(3:3)II)-*third-order increasing-inflection source infinite-equilibrium.*

- *For* $a_{j_1 j_1 0} < 0$ *and* $a_{j_2 j_1 0} > 0$,

$$(a_{j_1 j_1 1}, \bar{x}_{j_2}) = \underbrace{(3^{\rm rd}{\rm SI}, (3:3){\rm DI})}_{\text{third-order decreasing-inflection sink}}. \tag{4.228}$$

The equilibrium at $x_{j_1}^* = a_{j_2 j_1 1} = a_{j_1 j_1 1}$ *is an* (3$^{\rm rd}$SI,(3:3)DI)-*third-order decreasing-inflection sink infinite-equilibrium.*

- *For* $a_{j_1 j_1 0} > 0$ *and* $a_{j_2 j_1 0} < 0$,

$$(a_{j_1 j_1 1}, \bar{x}_{j_2}) = \underbrace{(3^{\rm rd}{\rm SO}, (3:3){\rm DI})}_{\text{third-order decreasing-inflection source}}. \tag{4.229}$$

The equilibrium at $x_{j_1}^* = a_{j_2 j_1 1} = a_{j_1 j_1 1}$ *is an* (3$^{\rm rd}$SO,(3:3)DI)-*third-order decreasing-inflection source infinite-equilibrium.*

- *For* $a_{j_1 j_1 0} < 0$ *and* $a_{j_2 j_1 0} < 0$,

$$(a_{j_1 j_1 1}, \bar{x}_{j_2}) = \underbrace{(3^{\rm rd}{\rm SO}, (3:3){\rm II})}_{\text{third-order increasing-inflection source}}. \tag{4.230}$$

The equilibrium at $x_{j_1}^* = a_{j_2 j_1 1} = a_{j_1 j_1 1}$ *is an* (3$^{\rm rd}$SI,(3:3)II)-*third-order increasing-inflection sink infinite-equilibrium.*

– *The third-order increasing-inflection and decreasing-inflection source and sink infinite-equilibriums of* $x_{j_1}^* = a_{j_2 j_1 1} = a_{j_1 j_1 1}$ *are the appearing bifurcations for the source, sink and source flows and the sink, source and sink flows with the up-parabola, down-parabola, and up-parabola flows and the down-parabola, up-parabola, and down-parabola flows.*
– *The third-order increasing-inflection and decreasing-inflection source and sink infinite-equilibriums of* $x_{j_1}^* = a_{j_2 j_1 1} = a_{j_1 j_1 1}$ *are the switching bifurcations for the source, sink, and source flows and the sink, source, and sink flows with the up-parabola, down-parabola, and up-parabola flows and the down-parabola, up-parabola, and down-parabola flows.*
– *The third-order increasing-inflection and decreasing-inflection source and sink infinite-equilibriums of* $x_{j_1}^* = a_{j_2 j_1 1} = a_{j_1 j_1 1}$ *are the switching bifurcations for the source and sink flows and the upper-saddle and lower-saddle flows with the up-parabola and up-parabola flows and the increasing-inflection and decreasing-inflection flows.*

(xi) If

$$\Delta_{j_1j_1} > 0 \text{ and } \Delta_{j_2j_1} = 0, \tag{4.231}$$

then

$$\begin{aligned} \dot{x}_{j_1} &= a_{j_1j_10}(x_{j_1} - a_{j_1j_11})(x_{j_1} - a_{j_1j_12})(x_{j_1} - a_{j_1j_13}), \\ \dot{x}_{j_2} &= a_{j_2j_10}(x_{j_1} - a_{j_2j_1l_1})(x_{j_1} - a_{j_2j_1l_2})^2, \end{aligned} \tag{4.232}$$

where

$$\begin{aligned} &\{a_{j_1j_11}, a_{j_1j_12}, a_{j_1j_13}\} = \text{sort}\{b_{j_1j_11}, b_{j_1j_12}, b_{j_1j_13}\}, a_{j_1j_11} < a_{j_1j_12} < a_{j_1j_13}; \\ &b_{j_1j_12}, b_{j_1j_13} = -\frac{1}{2}(B_{j_1j_11} \pm \sqrt{\Delta_{j_1j_1}}), \\ &\{a_{j_2j_11}, a_{j_2j_12}\} = \text{sort}\{b_{j_2j_11}, b_{j_2j_12}\}, b_{j_2j_12} = -\frac{1}{2}B_{j_2j_1}; \\ &j_1, j_2 \in \{1, 2\}; j_1 \neq j_2. \end{aligned} \tag{4.233}$$

(xi_1) For $x^*_{j_2} = a_{j_1j_1s_1} \neq a_{j_1j_1l_1}$ (s_1, s_2, $s_3 \in \{1, 2, 3\}$, $s_1 \neq s_2 \neq s_3$), the first integral manifold is

$$\begin{aligned} x_{j_2} - x_{j_20} = \frac{a_{j_2j_10}}{a_{j_1j_10}} \Bigg[&(x_{j_1} - x_{j_10}) + \frac{(a_{j_1j_1s_2} - a_{j_2j_1l_2})^2}{(a_{j_1j_1s_2} - a_{j_1j_1s_3})} \ln \frac{|x_{j_1} - a_{j_1j_1s_2}|}{|x_{j_10} - a_{j_1j_1s_2}|} \\ &+ \frac{(a_{j_1j_1s_3} - a_{j_2j_1l_2})^2}{(a_{j_1j_1s_3} - a_{j_1j_1s_2})} \ln \frac{|x_{j_1} - a_{j_1j_1s_3}|}{|x_{j_10} - a_{j_1j_1s_3}|} \\ &+ (a_{j_1j_1s_1} - a_{j_2j_1l_1}) \sum_{\bar{s}_1=1}^{3} \frac{(a_{j_1j_1\bar{s}_1} - a_{j_2j_1l_2})^2}{\prod_{\bar{s}_2=1, \bar{s}_2 \neq \bar{s}_1}^{3} (a_{j_1j_1\bar{s}_1} - a_{j_1j_1\bar{s}_2})} \ln \frac{|x_{j_1} - a_{j_1j_1\bar{s}_1}|}{|x_{j_10} - a_{j_1j_1\bar{s}_1}|} \Bigg]. \end{aligned} \tag{4.234}$$

(xi_{1a}) The equilibrium of $x^*_{j_1} = a_{j_1j_1s_1} \neq a_{j_2j_11}$ (s_1, $s_2 \in \{1, 2, 3\}$; $s_1 \neq s_2$) in the x_{j_1}-direction possesses the following properties.

- *For $a_{j_1j_10}\prod_{s_2=1, s_2 \neq s_1}^{3}(a_{j_1j_1s_1} - a_{j_1j_1s_2}) > 0$ and $a_{j_2j_10}(a_{j_1j_1s_1} - a_{j_2j_1l_1})(a_{j_1j_1s_1} - a_{j_2j_1l_2})^2 > 0$,*

$$(a_{j_1j_1s_1}, \dot{x}_{j_2}) = \underbrace{(\text{SO}, \text{pF})}_{\text{source flow } (+)}. \tag{4.235}$$

The flow at $x^*_{j_1} = a_{j_1j_1s_1}$ is an (SO,pF)*-positive source-flow.*

- *For* $a_{j_1j_10}\prod_{s_2=1,s_2\neq s_1}^{3}(a_{j_1j_1s_1}-a_{j_1j_1s_2})<0$ *and* $a_{j_2j_10}(a_{j_1j_1s_1}-a_{j_2j_1l_1})(a_{j_1j_1s_1}-a_{j_2j_1l_2})^2>0$,

$$(a_{j_1j_1s_1},\dot{x}_{j_2})=\underbrace{(\mathrm{SI},\mathrm{pF})}_{\text{sink flow }(+)}. \tag{4.236}$$

The flow at $x_{j_1}^*=a_{j_1j_1s_1}$ *is an* (SI,pF)*-positive sink-flow.*

- *For* $a_{j_1j_10}\prod_{s_2=1,s_2\neq s_1}^{3}(a_{j_1j_1s_1}-a_{j_1j_1s_2})>0$ *and* $a_{j_2j_10}(a_{j_1j_1s_1}-a_{j_2j_1l_1})(a_{j_1j_1s_1}-a_{j_2j_1l_2})^2<0$,

$$(a_{j_1j_1s_1},\dot{x}_{j_2})=\underbrace{(\mathrm{SO},\mathrm{nF})}_{\text{source flow }(-)}. \tag{4.237}$$

The flow at $x_{j_1}^*=a_{j_1j_1s_1}$ *is an* (SO,nF)*-negative source-flow.*

- *For* $a_{j_1j_10}\prod_{s_2=1,s_2\neq s_1}^{3}(a_{j_1j_1s_1}-a_{j_1j_1s_2})<0$ *and* $a_{j_2j_10}(a_{j_1j_1s_1}-a_{j_2j_1l_1})(a_{j_1j_1s_1}-a_{j_2j_1l_2})^2<0$,

$$(a_{j_1j_1s_1},\dot{x}_{j_2})=\underbrace{(\mathrm{SI},\mathrm{nF})}_{\text{sink flow }(-)}. \tag{4.238}$$

The flow at $x_{j_1}^*=a_{j_1j_1s_1}$ *is an* (SI,nF)*-negative sink-flow.*

(xi$_{1b}$) The equilibrium of $x_{j_1}^*=a_{j_2j_1l_2}\neq a_{j_1j_1s_1},a_{j_1j_1s_2}$ $(s_1,\ s_2\in\{1,2,3\};\ s_1\neq s_2)$ *in the* x_{j_2}*-direction possesses the following properties:*

- *For* $a_{j_1j_10}\prod_{s_1=1}^{3}(a_{j_2j_1l_2}-a_{j_1j_1s_1})>0$ *and* $a_{j_2j_10}(a_{j_2j_1l_2}-a_{j_2j_1l_1})>0$,

$$(\dot{x}_{j_1},a_{j_2j_1l_2})=\underbrace{(\mathrm{pF},\mathrm{II})}_{\text{increasing-inflection flow }(+)}. \tag{4.239}$$

The flow at $x_{j_1}^*=a_{j_2j_1l_2}$ *is a* (pF,II)*-positive increasing-inflection flow.*

- *For* $a_{j_1j_10}\prod_{s_1=1}^{3}(a_{j_2j_1l_2}-a_{j_1j_1s_1})<0$ *and* $a_{j_2j_10}(a_{j_2j_1l_2}-a_{j_2j_1l_1})>0$,

$$(\dot{x}_{j_1},a_{j_2j_1l_2})=\underbrace{(\mathrm{nF},\mathrm{DI})}_{\text{decreasing-inflection flow }(-)}. \tag{4.240}$$

The flow at $x_{j_1}^*=a_{j_2j_1l_2}$ *is an* (nF, DI)*-negative decreasing-inflection flow.*

- *For* $a_{j_1j_10}\prod_{s_1=1}^{3}(a_{j_2j_1l_2}-a_{j_1j_1s_1})>0$ *and* $a_{j_2j_10}(a_{j_2j_1l_2}-a_{j_2j_1l_1})<0$,

$$(\dot{x}_{j_1},a_{j_2j_1l_2})=\underbrace{(\mathrm{pF},\mathrm{DI})}_{\text{decreasing-inflection flow }(+)}. \tag{4.241}$$

The flow at $x_{j_1}^*=a_{j_2j_1l_2}$ *is a* (pF, DI)*-positive decreasing-inflection flow.*

- *For* $a_{j_1j_10}\prod_{s_1=1}^{3}(a_{j_2j_1l_2}-a_{j_1j_1s_1})<0$ *and* $a_{j_2j_10}(a_{j_2j_1l_2}-a_{j_2j_1l_1})<0$,

$$(\dot{x}_{j_1},a_{j_2j_1l_2})=\underbrace{(\mathrm{nF,II})}_{\text{increasing-inflection flow }(-)}. \tag{4.242}$$

The flow at $x_{j_1}^{*}=a_{j_2j_1l_2}$ *is an* (nF,II)*-negative increasing-inflection flow.*

– *The decreasing-inflection and increasing-inflection flows are the appearing bifurcations of the up-parabola and down-parabola flows.*

(xi$_{1c}$) The equilibrium of $x_{j_1}^{*}=a_{j_2j_1l_1}\neq a_{j_1j_1s_1},a_{j_1j_1s_2}$ $(s_1,\ s_2\in\{1,2,3\};\ s_1\neq s_2)$ *in the* x_{j_2}*-direction possesses the following properties:*

- *For* $a_{j_1j_10}\prod_{s_1=1}^{3}(a_{j_2j_1l_1}-a_{j_1j_1s_1})>0$ *and* $a_{j_2j_10}(a_{j_2j_1l_1}-a_{j_2j_1l_2})^2>0$,

$$(\dot{x}_{j_1},a_{j_2j_1l_1})=\underbrace{(\mathrm{pF,UP})}_{\text{up-parabola flow }(+)}. \tag{4.243}$$

The flow at $x_{j_1}^{*}=a_{j_2j_1l_1}$ *is a* (pF,UP)*-positive up-parabola flow.*

- *For* $a_{j_1j_10}\prod_{s_1=1}^{3}(a_{j_2j_1l_1}-a_{j_1j_1s_1})<0$ *and* $a_{j_2j_10}(a_{j_2j_1l_1}-a_{j_2j_1l_2})^2>0$,

$$(\dot{x}_{j_1},a_{j_2j_1l_1})=\underbrace{(\mathrm{nF,DP})}_{\text{down-parabola flow }(-)}. \tag{4.244}$$

The flow at $x_{j_1}^{*}=a_{j_2j_1l_1}$ *is an* (nF,DP)*-positive down-parabola flow.*

- *For* $a_{j_1j_10}\prod_{s_1=1}^{3}(a_{j_2j_1l_1}-a_{j_1j_1s_1})>0$ *and* $a_{j_2j_10}(a_{j_2j_1l_1}-a_{j_2j_1l_2})^2<0$,

$$(\dot{x}_{j_1},a_{j_2j_1l_1})=\underbrace{(\mathrm{pF,DP})}_{\text{down-parabola flow }(+)}. \tag{4.245}$$

The flow at $x_{j_1}^{*}=a_{j_2j_1l_1}$ *is a* (pF,DP)*-positive down-parabola flow.*

- *For* $a_{j_1j_10}\prod_{s_1=1}^{3}(a_{j_2j_1l_1}-a_{j_1j_1s_1})<0$ *and* $a_{j_2j_10}(a_{j_2j_1l_1}-a_{j_2j_1l_2})^2>0$,

$$(\dot{x}_{j_1},a_{j_2j_1l_1})=\underbrace{(\mathrm{nF,UP})}_{\text{up-parabola flow }(-)}. \tag{4.246}$$

The flow at $x_{j_1}^{*}=a_{j_2j_1l_1}$ *is an* (nF,DP)*-positive up-parabola flow.*

(xi$_2$) For $a_{j_2j_1l_2}=a_{j_1j_1s_1}$ $(s_1\in\{1,2,3\})$*.*

$$\begin{aligned}&\dot{x}_{j_1}=a_{j_1j_10}(x_{j_1}-a_{j_1j_11})(x_{j_1}-a_{j_1j_12})(x_{j_1}-a_{j_1j_13}),\\&\dot{x}_{j_2}=a_{j_2j_10}(x_{j_1}-a_{j_2j_1l_1})(x_{j_1}-a_{j_2j_1l_2})^2,\\&j_1,j_2\in\{1,2\},j_1\neq j_2;\end{aligned} \tag{4.247}$$

and the first integral manifold is

$$\begin{aligned} x_{j_2} - x_{j_2 0} = \frac{a_{j_2 j_1 0}}{a_{j_1 j_1 0}} \Big[& (x_{j_1} - x_{j_1 0}) + (a_{j_1 j_1 s_3} - a_{j_2 j_1 l_2}) \ln \frac{|x_{j_1} - a_{j_1 j_1 s_3}|}{|x_{j_1 0} - a_{j_1 j_1 s_3}|} \\ & + (a_{j_1 j_1 s_2} - a_{j_2 j_1 l_1}) \sum_{\bar{s}_2 = 1, \bar{s}_2 \neq \bar{s}_1}^{3} \frac{(a_{j_1 j_1 \bar{s}_2} - a_{j_2 j_1 l_2})}{\prod_{\bar{s}_3 = 1, \bar{s}_3 \neq s_1, \bar{s}_2}^{3} (a_{j_1 j_1 \bar{s}_2} - a_{j_1 j_1 \bar{s}_3})} \ln \frac{|x_{j_1} - a_{j_1 j_1 \bar{s}_2}|}{|x_{j_1 0} - a_{j_1 j_1 \bar{s}_2}|} \Big]. \end{aligned} \tag{4.248}$$

The infinite-equilibrium of $x^*_{j_1} = a_{j_1 j_1 s_1} = a_{j_2 j_1 l_2}$ $(s_1 \in \{1, 2, 3\})$ with $\bar{x}_{j_2} \in (-\infty, \infty)$ *possesses the following cases.*

- *For* $a_{j_1 j_1 0} \prod_{s_2 = 1, s_2 \neq s_1}^{3} (a_{j_1 j_1 s_1} - a_{j_1 j_1 s_2}) > 0$ *and* $a_{j_2 j_1 0}(a_{j_2 j_1 l_2} - a_{j_2 j_1 l_1}) > 0$,

$$(a_{j_1 j_1 s_1}, \bar{x}_{j_2}) = \underbrace{(\text{SO, UP})}_{\text{up-parabola source}}. \tag{4.249}$$

  The equilibrium at $x^*_{j_1} = a_{j_1 j_1 s_1} = a_{j_2 j_1 l_2}$ *is an* (SO,UP)-*up-parabola source infinite-equilibrium.*

- *For* $a_{j_1 j_1 0} \prod_{s_2 = 1, s_2 \neq s_1}^{3} (a_{j_1 j_1 s_1} - a_{j_1 j_1 s_2}) < 0$ *and* $a_{j_2 j_1 0}(a_{j_2 j_1 l_2} - a_{j_2 j_1 l_1}) > 0$,

$$(a_{j_1 j_1 s_1}, \bar{x}_{j_2}) = \underbrace{(\text{SI, DP})}_{\text{down-parabola sink}}. \tag{4.250}$$

  The equilibrium at $x^*_{j_1} = a_{j_1 j_1 s_1} = a_{j_2 j_1 l_2}$ *is an* (SI,DP)-*down-parabola sink infinite-equilibrium.*

- *For* $a_{j_1 j_1 0} \prod_{s_2 = 1, s_2 \neq s_1}^{3} (a_{j_1 j_1 s_1} - a_{j_1 j_1 s_2}) > 0$ *and* $a_{j_2 j_1 0}(a_{j_2 j_1 l_2} - a_{j_2 j_1 l_1}) < 0$,

$$(a_{j_1 j_1 s_1}, \bar{x}_{j_2}) = \underbrace{(\text{SO, DP})}_{\text{down-parabola source}}. \tag{4.251}$$

  The equilibrium at $x^*_{j_1} = a_{j_1 j_1 s_1} = a_{j_2 j_1 l_2}$ *is an* (SO,DP)-*down-parabola source infinite-equilibrium.*

- *For* $a_{j_1 j_1 0} \prod_{s_2 = 1, s_2 \neq s_1}^{3} (a_{j_1 j_1 s_1} - a_{j_1 j_1 s_2}) < 0$ *and* $a_{j_2 j_1 0}(a_{j_2 j_1 l_2} - a_{j_2 j_1 l_1}) < 0$,

$$(a_{j_1 j_1 s_1}, \bar{x}_{j_2}) = \underbrace{(\text{SI, UP})}_{\text{up-parabola sink}}. \tag{4.252}$$

  The equilibrium at $x^*_{j_1} = a_{j_1 j_1 s_1} = a_{j_2 j_1 l_2}$ *is an* (SI,UP)-*up-parabola sink infinite-equilibrium.*

– *The up-parabola and down-parabola sink, and source infinite-equilibriums are the appearing and switching bifurcations for the up-parabola and down-parabola flows with sink or source flows.*

(xi₃) For $a_{j_2j_1l_1} = a_{j_1j_1s_1}$ $(s_1 \in \{1,2,3\})$.

$$\begin{aligned}
&\dot{x}_{j_1} = a_{j_1j_10}(x_{j_1} - a_{j_1j_11})(x_{j_1} - a_{j_1j_12})(x_{j_1} - a_{j_1j_13}),\\
&\dot{x}_{j_2} = a_{j_2j_10}(x_{j_1} - a_{j_2j_1l_1})(x_{j_1} - a_{j_2j_1l_2})^2,\\
&j_1, j_2 \in \{1,2\}, j_1 \neq j_2;
\end{aligned} \tag{4.253}$$

and the first integral manifold is

$$\begin{aligned}
x_{j_2} - x_{j_20} = \frac{a_{j_2j_10}}{a_{j_1j_10}}\Bigg[&(x_{j_1} - x_{j_10}) + \frac{(a_{j_1j_1s_2} - a_{j_2j_1l_2})^2}{(a_{j_1j_1s_2} - a_{j_1j_1s_3})}\ln\frac{|x_{j_1} - a_{j_1j_1s_3}|}{|x_{j_10} - a_{j_1j_1s_3}|}\\
&+\frac{(a_{j_1j_1s_3} - a_{j_2j_1l_2})^2}{(a_{j_1j_1s_3} - a_{j_1j_1s_2})}\ln\frac{|x_{j_1} - a_{j_1j_1s_2}|}{|x_{j_10} - a_{j_1j_1s_2}|}\Bigg].
\end{aligned} \tag{4.254}$$

The infinite-equilibrium of $x^*_{j_1} = a_{j_1j_1s_1} = a_{j_2j_1l_1}$ $(s_1 \in \{1,2,3\})$ with $\bar{x}_{j_2} \in (-\infty, \infty)$ *possesses the following cases.*

- *For* $a_{j_1j_10}\prod_{s_2=1, s_2\neq s_1}^{3}(a_{j_1j_1s_1} - a_{j_1j_1s_2}) > 0$ *and* $a_{j_2j_10}(a_{j_2j_1l_1} - a_{j_2j_1l_2})^2 > 0$,

$$(a_{j_1j_1s_1}, \bar{x}_{j_2}) = \underbrace{(\text{SO}, \text{II})}_{\text{increasing-inflection source}}. \tag{4.255}$$

  The equilibrium at $x^*_{j_1} = a_{j_1j_1s_1} = a_{j_2j_1l_1}$ *is an* (SO,II)*-increasing-inflection source infinite-equilibrium.*

- *For* $a_{j_1j_10}\prod_{s_2=1, s_2\neq s_1}^{3}(a_{j_1j_1s_1} - a_{j_1j_1s_2}) < 0$ *and* $a_{j_2j_10}(a_{j_2j_1l_1} - a_{j_2j_1l_2})^2 > 0$,

$$(a_{j_1j_1s_1}, \bar{x}_{j_2}) = \underbrace{(\text{SI}, \text{DI})}_{\text{decreasing-inflection sink}}. \tag{4.256}$$

  The equilibrium at $x^*_{j_1} = a_{j_1j_1s_1} = a_{j_2j_1l_1}$ *is an* (SI,II)*-increasing-inflection sink infinite-equilibrium.*

- *For* $a_{j_1j_10}\prod_{s_2=1, s_2\neq s_1}^{3}(a_{j_1j_1s_1} - a_{j_1j_1s_2}) > 0$ *and* $a_{j_2j_10}(a_{j_2j_1l_1} - a_{j_2j_1l_2})^2 < 0$,

$$(a_{j_1j_1s_1}, \bar{x}_{j_2}) = \underbrace{(\text{SO}, \text{DI})}_{\text{decreasing-inflection source}}. \tag{4.257}$$

  The equilibrium at $x^*_{j_1} = a_{j_1j_1s_1} = a_{j_2j_1l_1}$ *is an* (SO,II)*-decreasing-inflection source infinite-equilibrium.*

- *For* $a_{j_1j_10}\prod_{s_2=1,s_2\neq s_1}^{3}(a_{j_1j_1s_1}-a_{j_1j_1s_2})<0$ *and* $a_{j_2j_10}(a_{j_2j_1l_1}-a_{j_2j_1l_2})^2<0$,

$$(a_{j_1j_1s_1},\bar{x}_{j_2})=\underbrace{(\text{SI, II})}_{\text{increasing-inflection source}}. \tag{4.258}$$

The equilibrium at $x_{j_1}^*=a_{j_1j_1s_1}=a_{j_2j_1l_1}$ *is an* (SI,II)*-increasing-inflection sink infinite-equilibrium.*

– *The increasing-inflection and decreasing-inflection sink and source infinite-equilibriums are the appearing and switching bifurcations for the up-parabola and down-parabola flows with sink or source flows.*

(xii) If

$$\begin{aligned}&\Delta_{j_1j_1}>0 \text{ and } \Delta_{j_2j_1}=0,\\&a_{j_2j_1l_1}=a_{j_2j_1l_2}=a_{j_2j_11}\end{aligned} \tag{4.259}$$

then

$$\begin{aligned}&\dot{x}_{j_1}=a_{j_1j_10}(x_{j_1}-a_{j_1j_11})(x_{j_1}-a_{j_1j_12})(x_{j_1}-a_{j_1j_13}),\\&\dot{x}_{j_2}=a_{j_2j_10}(x_{j_1}-a_{j_2j_11})^3,\\&j_1,j_2\in\{1,2\},j_1\neq j_2.\end{aligned} \tag{4.260}$$

(xii₁) For $a_{j_2j_11}\neq a_{j_1j_1s_1}$ $(s_1=1,2,3)$, *the first integral manifold is*

$$\begin{aligned}x_{j_2}-x_{j_20}=\frac{a_{j_2j_10}}{a_{j_1j_10}}\Bigg[&(x_{j_1}-x_{j_10})+(a_{j_1j_1s_3}-a_{j_2j_11})\ln\frac{|x_{j_1}-a_{j_1j_1s_3}|}{|x_{j_10}-a_{j_1j_1s_3}|}\\&+\sum_{s_2=1,\,s_2\neq s_1}^{3}\frac{(a_{j_1j_1s_2}-a_{j_2j_11})^2}{\prod_{s_3=1,s_3\neq s_1,s_2}^{3}(a_{j_1j_1s_2}-a_{j_1j_1s_3})}\ln\frac{|x_{j_1}-a_{j_1j_1s_2}|}{|x_{j_10}-a_{j_1j_1s_2}|}\\&+\sum_{s_1=1}^{3}\frac{(a_{j_1j_1s_1}-a_{j_2j_11})^3}{\prod_{s_2=1,s_2\neq s_1}^{3}(a_{j_1j_1s_1}-a_{j_1j_1s_2})}\ln\frac{|x_{j_1}-a_{j_1j_1s_1}|}{|x_{j_10}-a_{j_1j_1s_1}|}\Bigg].\end{aligned} \tag{4.261}$$

(xii$_{1a}$) The equilibrium of $x_{j_1}^*=a_{j_2j_11}\neq a_{j_1j_1s_1}$ $(s_1\in\{1,2,3\})$ *possesses the following cases.*

- *For* $a_{j_1j_10}\prod_{s_1=1}^{3}(a_{j_2j_11}-a_{j_1j_1s_1})>0$ *and* $a_{j_2j_10}>0$,

$$(\dot{x}_{j_1},a_{j_2j_11})=\underbrace{(\text{pF, 3}^{\text{rd}}\text{UP})}_{\text{third-order up-parabola flow }(+)}. \tag{4.262}$$

The equilibrium at $x_{j_1}^*=a_{j_2j_11}$ *is a* (pF,3rdUP)*-third-order positive up-parabola flow.*

- *For* $a_{j_1j_1 0}\prod_{s_1=1}^{3}(a_{j_2j_1 1}-a_{j_1j_1 s_1})<0$ *and* $a_{j_2j_1 0}>0$,

$$(\dot{x}_{j_1},a_{j_2j_1 1})=\underbrace{(\text{nF},3^{\text{rd}}\text{DP})}_{\text{third-order down-parabola flow }(-)}. \tag{4.263}$$

The equilibrium at $x_{j_1}^{*}=a_{j_2j_1 1}$ *is an* (nF,3^rd^UP)*-third-order negative down-parabola flow.*

- *For* $a_{j_1j_1 0}\prod_{s_1=1}^{3}(a_{j_2j_1 1}-a_{j_1j_1 s_1})>0$ *and* $a_{j_2j_1 0}<0$,

$$(\dot{x}_{j_1},a_{j_2j_1 1})=\underbrace{(\text{pF},3^{\text{rd}}\text{DP})}_{\text{third-order down-parabola flow }(+)}. \tag{4.264}$$

The equilibrium at $x_{j_1}^{*}=a_{j_2j_1 1}$ *is a* ($\text{pF},3^{\text{rd}}\text{DP}$)*-third-order positive down-parabola flow.*

- *For* $a_{j_1j_1 0}\prod_{s_1=1}^{3}(a_{j_2j_1 1}-a_{j_1j_1 s_1})<0$ *and* $a_{j_2j_1 0}<0$,

$$(\dot{x}_{j_1},a_{j_2j_1 1})=\underbrace{(\text{nF},3^{\text{rd}}\text{UP})}_{\text{third-order up-parabola flow }(-)}. \tag{4.265}$$

The equilibrium at $x_{j_1}^{*}=a_{j_2j_1 1}$ *is an* ($\text{nF},3^{\text{rd}}\text{UP}$)*-third-order negative up-parabola flow.*

– *The third-order up-parabola and down-parabola flows are the appearing bifurcations for the up-parabola, down-parabola, and up-parabola flows and the down-parabola, up-parabola, and down-parabola flows.*

(xii$_{1b}$*) The equilibrium of* $x_{j_1}^{*}=a_{j_1j_1 s_1}\neq a_{j_2j_1 1}$ has the following properties.

- *For* $a_{j_1j_1 0}\prod_{s_2=1,s_2\neq s_1}^{3}(a_{j_1j_1 s_1}-a_{j_1j_1 s_2})>0$ *and* $a_{j_2j_1 0}(a_{j_1j_1 s_1}-a_{j_2j_1 1})^3>0$,

$$(a_{j_1j_1 s_1},\dot{x}_{j_2})=\underbrace{(\text{SO},\text{pF})}_{\text{source flow }(+)}. \tag{4.266}$$

The flow at $x_{j_1}^{*}=a_{j_1j_1 s_1}$ *is an* (SO,pF)*-positive source flow.*

- *For* $a_{j_1j_1 0}\prod_{s_2=1,s_2\neq s_1}^{3}(a_{j_1j_1 s_1}-a_{j_1j_1 s_2})<0$ *and* $a_{j_2j_1 0}(a_{j_1j_1 s_1}-a_{j_2j_1 1})^3>0$,

$$(a_{j_1j_1 s_1},\dot{x}_{j_2})=\underbrace{(\text{SI},\text{pF})}_{\text{sink flow }(+)}. \tag{4.267}$$

The flow at $x_{j_1}^{*}=a_{j_1j_1 s_1}$ *is an* (SI,pF)*-positive sink flow.*

- *For* $a_{j_1j_10}\prod_{s_2=1,s_2\neq s_1}^{3}(a_{j_1j_1s_1}-a_{j_1j_1s_2})<0$ *and* $a_{j_2j_10}(a_{j_1j_1s_1}-a_{j_2j_11})^3>0$,

$$(a_{j_1j_1s_1},\dot{x}_{j_2})=\underbrace{(\mathrm{SI},\mathrm{nF})}_{\text{sink flow }(-)}. \tag{4.268}$$

The flow at $x_{j_1}^{*}=a_{j_1j_1s_1}$ *is an* (SI,nF)*-negative sink flow.*

- *For* $a_{j_1j_10}\prod_{s_2=1,s_2\neq s_1}^{3}(a_{j_1j_1s_1}-a_{j_1j_1s_2})<0$ *and* $a_{j_2j_10}(a_{j_1j_1s_1}-a_{j_2j_11})^3<0$,

$$(a_{j_1j_1s_1},\dot{x}_{j_2})=\underbrace{(\mathrm{SO},\mathrm{nF})}_{\text{source flow }(-)}. \tag{4.269}$$

The flow at $x_{j_1}^{*}=a_{j_1j_1s_1}$ *is an* (SO,nF)*-negative source flow.*

(xii$_2$) For $a_{j_2j_11}=a_{j_1j_1s_1}$ $(s_1\in\{1,2,3\})$*, the first integral manifold is*

$$\begin{aligned}x_{j_2}-x_{j_20}=\frac{a_{j_2j_10}}{a_{j_1j_10}}\Bigg[&(x_{j_1}-x_{j_10})\\&+\sum_{s_2=1,\,s_2\neq s_1}^{3}\frac{(a_{j_1j_1s_2}-a_{j_2j_11})^2}{\prod_{s_3=1,s_3\neq s_1,s_2}^{3}(a_{j_1j_1s_2}-a_{j_1j_1s_3})}\ln\frac{|\,x_{j_1}-a_{j_1j_1s_2}\,|}{|\,x_{j_10}-a_{j_1j_1s_2}\,|}\Bigg].\end{aligned} \tag{4.270}$$

The infinite-equilibrium of $x_{j_1}^{*}=a_{j_2j_11}=a_{j_1j_1s_1}$ $(s_1\in\{1,2,3\})$ with $\bar{x}_{j_2}\in(-\infty,\infty)$ *possesses the following cases.*

- *For* $a_{j_1j_10}\prod_{s_2=1,s_2\neq s_1}^{3}(a_{j_1j_1s_1}-a_{j_1j_1s_2})>0$ *and* $a_{j_2j_10}>0$,

$$(a_{j_1j_1s_1},\bar{x}_{j_2})=\underbrace{(\mathrm{SO},(1{:}3)\mathrm{II})}_{(1:3)\text{-increasing-inflection source}}. \tag{4.271}$$

The equilibrium at $x_{j_1}^{*}=a_{j_2j_11}=a_{j_1j_1s_1}$ *is an* (SO,(1:3)II)*-increasing-inflection source infinite-equilibrium.*

- *For* $a_{j_1j_10}\prod_{s_2=1,s_2\neq s_1}^{3}(a_{j_1j_1s_1}-a_{j_1j_1s_2})<0$ *and* $a_{j_2j_10}>0$,

$$(a_{j_1j_1s_1},\bar{x}_{j_2})=\underbrace{(\mathrm{SI},(1{:}3)\mathrm{DI})}_{(1:3)\text{-decreasing-inflection sink}}. \tag{4.272}$$

The equilibrium at $x_{j_1}^{*}=a_{j_2j_11}=a_{j_1j_1s_1}$ *is an* (SI,(1:3)DI)*-decreasing-inflection sink infinite-equilibrium.*

- *For* $a_{j_1j_10}\prod_{s_2=1,s_2\neq s_1}^{3}(a_{j_1j_1s_1}-a_{j_1j_1s_2})>0$ *and* $a_{j_2j_10}<0$,

$$(a_{j_1j_1s_1},\bar{x}_{j_2})=\underbrace{(\mathrm{SO},(1:3)\mathrm{DI})}_{(1:3)\text{-decreasing-inflection source}}.\tag{4.273}$$

The equilibrium at $x_{j_1}^*=a_{j_2j_11}=a_{j_1j_1s_1}$ *is an* (SO,(1:3)DI)*-decreasing-inflection source infinite-equilibrium.*

- *For* $a_{j_1j_10}\prod_{s_2=1,s_2\neq s_1}^{3}(a_{j_1j_1s_1}-a_{j_1j_1s_2})<0$ *and* $a_{j_2j_10}<0$,

$$(a_{j_1j_1s_1},\bar{x}_{j_2})=\underbrace{(\mathrm{SI},(1:3)\mathrm{II})}_{(1:3)\text{-increasing-inflection sink}}.\tag{4.274}$$

The equilibrium at $x_{j_1}^*=a_{j_2j_11}=a_{j_1j_1s_1}$ *is an* (SI,(1:3)II)*-increasing-inflection sink infinite-equilibrium.*

- *The (1:3)-increasing-inflection and decreasing-inflection source and sink infinite-equilibriums are the appearing bifurcations for the up-parabola, down-parabola, and up-parabola flows and the down-parabola, up-parabola, and down-parabola flows with sink and source flows.*
- *The (1:3)-increasing-inflection and decreasing-inflection source and sink infinite-equilibriums are the switching bifurcations for the up-parabola, down-parabola, and up-parabola flows and the down-parabola, up-parabola, and down-parabola flows with sink and source flows.*
- *The (1:3)-increasing-inflection and decreasing-inflection source and sink infinite-equilibriums are the switching bifurcations for the third-order up-parabola and down-parabola flows with sink and source flows.*

(xiii) If

$$\Delta_{j_1j_1}>0 \text{ and } \Delta_{j_2j_1}>0,\tag{4.275}$$

then

$$\begin{aligned}\dot{x}_{j_1}&=a_{j_1j_10}(x_{j_1}-a_{j_1j_11})(x_{j_1}-a_{j_1j_12})(x_{j_1}-a_{j_1j_13}),\\ \dot{x}_{j_2}&=a_{j_2j_10}(x_{j_1}-a_{j_2j_11})(x_{j_1}-a_{j_2j_12})(x_{j_1}-a_{j_2j_13}),\end{aligned}\tag{4.276}$$

where

$$\begin{aligned}&\{a_{j_1j_11},a_{j_1j_12},a_{j_1j_13}\}=\mathrm{sort}\{b_{j_1j_11},b_{j_1j_12},b_{j_1j_13}\},a_{j_1j_11}<a_{j_1j_12}<a_{j_1j_13};\\ &\{a_{j_2j_11},a_{j_2j_12},a_{j_2j_13})=\mathrm{sort}\{b_{j_2j_11},b_{j_2j_12},b_{j_2j_13}\},a_{j_2j_11}<a_{j_2j_12}<a_{j_2j_13};\\ &b_{j_1j_12},b_{j_1j_13}=-\frac{1}{2}(B_{j_1j_11}\pm\sqrt{\Delta_{j_1j_1}}),\\ &b_{j_2j_12},b_{j_2j_13}=-\frac{1}{2}(B_{j_2j_11}\pm\sqrt{\Delta_{j_2j_1}});\\ &j_1,j_2\in\{1,2\};j_1\neq j_2.\end{aligned}\tag{4.277}$$

(xiii$_1$) For $a_{j_1j_1s_1} \neq a_{j_2j_1l_1}$ (s_1, $s_2 \in \{1,2,3\}$; $s_1 \neq s_2$) and (l_1, $l_2 \in \{1,2,3\}$; $l_1 \neq l_2$), the first integral manifold is

$$x_{j_2} - x_{j_20} = \frac{a_{j_2j_10}}{a_{j_1j_10}}\left[(x_{j_1} - x_{j_10}) + \sum_{s_1=1}^{3} \frac{\prod_{l_1=1}^{3}(a_{j_1j_1s_1} - a_{j_2j_1l_1})}{\prod_{s_2=1,s_2\neq s_1}^{3}(a_{j_1j_1s_1} - a_{j_1j_1s_2})} \ln \frac{|x_{j_1} - a_{j_1j_1s_1}|}{|x_{j_10} - a_{j_1j_1s_1}|}\right]. \tag{4.278}$$

or

$$\begin{aligned} x_{j_2} - x_{j_20} = \frac{a_{j_2j_10}}{a_{j_1j_10}}\Bigg[&(x_{j_1} - x_{j_10}) + \sum\nolimits_{s_3=1}^{3}(a_{j_1j_1s_3} - a_{j_2j_1l_3}) \ln \frac{|x_{j_1} - a_{j_1j_1s_3}|}{|x_{j_10} - a_{j_1j_1s_3}|} \\ &+ (a_{j_1j_1s_2} - a_{j_2j_1l_2}) \sum_{\bar{s}_2=1,\,\bar{s}_2\neq s_1}^{3} \frac{\prod_{\bar{l}_3=1,\bar{l}_3\neq l_1,l_2}^{3}(a_{j_1j_1\bar{s}_2} - a_{j_2j_1\bar{l}_3})}{\prod_{\bar{s}_3=1,\bar{s}_3\neq s_1,\bar{s}_2}^{3}(a_{j_1j_1\bar{s}_2} - a_{j_1j_1\bar{s}_3})} \ln \frac{|x_{j_1} - a_{j_1j_1\bar{s}_2}|}{|x_{j_10} - a_{j_1j_1\bar{s}_2}|} \\ &+ (a_{j_1j_1s_1} - a_{j_2j_1l_1}) \sum_{\bar{s}_1=1}^{3} \frac{\prod_{\bar{l}_2=1,\bar{l}_2\neq l_1}^{3}(a_{j_1j_1\bar{s}_1} - a_{j_2j_1\bar{l}_2})}{\prod_{\bar{s}_2=1,\bar{s}_2\neq\bar{s}_1}^{3}(a_{j_1j_1\bar{s}_1} - a_{j_1j_1\bar{s}_2})} \ln \frac{|x_{j_1} - a_{j_1j_1\bar{s}_1}|}{|x_{j_10} - a_{j_1j_1\bar{s}_1}|}\Bigg]. \end{aligned} \tag{4.279}$$

(xiii$_{1a}$) The equilibriums of $x_{j_1}^* = a_{j_1j_1s_1} \neq a_{j_2j_1l_1}, a_{j_2j_1l_2}$ (s_1, $s_2 \in \{1,2,3\}$; $s_1 \neq s_2$) and (l_1, $l_2 \in \{1,2,3\}$; $l_1 \neq l_2$) are as follows.

- *For $a_{j_1j_10}\prod_{s_2=1,s_2\neq s_1}^{3}(a_{j_1j_1s_1} - a_{j_1j_1s_2}) > 0$ and $a_{j_2j_10}\prod_{l_1=1}^{3}(a_{j_1j_1s_1} - a_{j_2j_1l_1}) > 0$,*

$$(a_{j_1j_1s_1}, \dot{x}_{j_2}) = \underbrace{(\text{SO}, \text{pF})}_{\text{source flow }(+)}. \tag{4.280}$$

The flow at $x_{j_1}^* = a_{j_1j_1s_1}$ is an (SO,pF)*-positive source-flow.*

- *For $a_{j_1j_10}\prod_{s_2=1,s_2\neq s_1}^{3}(a_{j_1j_1s_1} - a_{j_1j_1s_2}) < 0$ and $a_{j_2j_10}\prod_{l_1=1}^{3}(a_{j_1j_1s_1} - a_{j_2j_1l_1}) > 0$,*

$$(a_{j_1j_1s_1}, \dot{x}_{j_2}) = \underbrace{(\text{SI}, \text{pF})}_{\text{sink flow }(+)}. \tag{4.281}$$

The flow at $x_{j_1}^* = a_{j_1j_1s_1}$ is an (SI,pF)*-positive sink-flow.*

- *For $a_{j_1j_10}\prod_{s_2=1,s_2\neq s_1}^{3}(a_{j_1j_1s_1} - a_{j_1j_1s_2}) > 0$ and $a_{j_2j_10}\prod_{l_1=1}^{3}(a_{j_1j_1s_1} - a_{j_2j_1l_1}) < 0$,*

$$(a_{j_1j_1s_1}, \dot{x}_{j_2}) = \underbrace{(\text{SO}, \text{nF})}_{\text{source flow }(-)}. \tag{4.282}$$

The flow at $x_{j_1}^* = a_{j_1j_1s_1}$ is an (SO,nF)*-negative source-flow.*

- *For* $a_{j_1j_10}\prod_{s_2=1,s_2\neq s_1}^{3}(a_{j_1j_1s_1}-a_{j_1j_1s_2})<0$ *and* $a_{j_2j_10}\prod_{l_1=1}^{3}(a_{j_1j_1s_1}-a_{j_2j_1l_1})<0$,

$$(a_{j_1j_1s_1},\dot{x}_{j_2})=\underbrace{(\mathrm{SI},\mathrm{nF})}_{\text{sink flow }(-)}. \tag{4.283}$$

The flow at $x_{j_1}^{*}=a_{j_1j_1s_1}$ *is an* (SI,nF)*-negative sink-flow.*

(xiii$_{1b}$) The equilibrium of $x_{j_1}^{*}=a_{j_1j_1l_1}\neq a_{j_2j_1s_1},a_{j_2j_1s_2},a_{j_2j_1s_3}$ $(l_1,l_2\in\{1,2\};l_1\neq l_2)$ *and* $(s_1,s_2,s_3\in\{1,2,3\};s_1\neq s_2\neq s_2)$ *has the following properties.*

- *For* $a_{j_1j_10}\prod_{s_1=1}^{3}(a_{j_2j_1l_1}-a_{j_1j_1s_1})>0$ *and* $a_{j_2j_10}\prod_{l_2=1,l_2\neq l_1}^{3}(a_{j_2j_1l_1}-a_{j_2j_1l_2})>0$,

$$(\dot{x}_{j_1},a_{j_1j_1l_1})=\underbrace{(\mathrm{pF},\mathrm{UP})}_{\text{up-parabola flow }(+)}. \tag{4.284}$$

The flow at $x_{j_1}^{*}=a_{j_1j_1l_1}$ *is a* (pF,UP)*-positive up-parabola flow.*

- *For* $a_{j_1j_10}\prod_{s_1=1}^{3}(a_{j_2j_1l_1}-a_{j_1j_1s_1})<0$ *and* $a_{j_2j_10}\prod_{l_2=1,l_2\neq l_1}^{3}(a_{j_2j_1l_1}-a_{j_2j_1l_2})>0$,

$$(\dot{x}_{j_1},a_{j_1j_1l_1})=\underbrace{(\mathrm{nF},\mathrm{DP})}_{\text{down-parabola flow }(-)}. \tag{4.285}$$

The flow at $x_{j_1}^{*}=a_{j_1j_1l_1}$ *is an* (nF,DP)*-negative down-parabola flow.*

- *For* $a_{j_1j_10}\prod_{s_1=1}^{3}(a_{j_2j_1l_1}-a_{j_1j_1s_1})>0$ *and* $a_{j_2j_10}\prod_{l_2=1,l_2\neq l_1}^{3}(a_{j_2j_1l_1}-a_{j_2j_1l_2})<0$,

$$(\dot{x}_{j_1},a_{j_1j_1l_1})=\underbrace{(\mathrm{pF},\mathrm{DP})}_{\text{down-parabola flow }(+)}. \tag{4.286}$$

The flow at $x_{j_1}^{*}=a_{j_1j_1l_1}$ *is a* (pF,DP)*-positive down-parabola flow.*

- *For* $a_{j_1j_10}\prod_{s_1=1}^{3}(a_{j_2j_1l_1}-a_{j_1j_1s_1})<0$ *and* $a_{j_2j_10}\prod_{l_2=1,l_2\neq l_1}^{3}(a_{j_2j_1l_1}-a_{j_2j_1l_2})<0$,

$$(\dot{x}_{j_1},a_{j_1j_1l_1})=\underbrace{(\mathrm{nF},\mathrm{UP})}_{\text{up-parabola flow }(-)}. \tag{4.287}$$

The flow at $x_{j_1}^{*}=a_{j_1j_1l_1}$ *is an* (nF,UP)*-negative up-parabola flow.*

(xiii$_2$) For $a_{j_1j_1s_1}=a_{j_2j_1l_1}$ $(s_1,s_2,s_3\in\{1,2,3\};s_1\neq s_2\neq s_3)$ and $(l_1,l_2\in\{1,2\};l_1\neq l_2)$,

$$\begin{aligned}\dot{x}_{j_1}&=a_{j_1j_10}(x_{j_1}-a_{j_1j_11})(x_{j_1}-a_{j_1j_12})(x_{j_1}-a_{j_1j_13}),\\ \dot{x}_{j_2}&=a_{j_2j_10}(x_{j_1}-a_{j_2j_11})(x_{j_1}-a_{j_2j_12})(x_{j_1}-a_{j_2j_13}).\end{aligned} \tag{4.288}$$

The first integral manifold is

$$x_{j_2} - x_{j_2 0} = \frac{a_{j_2 j_1 0}}{a_{j_1 j_1 0}} \left[(x_{j_1} - x_{j_1 0}) + (a_{j_1 j_1 s_3} - a_{j_1 j_1 l_3}) \ln \frac{|x_{j_1} - a_{j_1 j_1 s_3}|}{|x_{j_1 0} - a_{j_1 j_1 s_3}|} \right.$$
$$\left. + (a_{j_1 j_1 s_2} - a_{j_1 j_1 l_2}) \sum_{\bar{s}_2 = 1, \bar{s}_2 \neq s_1}^{3} \frac{\prod_{\bar{l}_2 = 1, \bar{l}_2 \neq l_1}^{3} (a_{j_1 j_1 \bar{s}_2} - a_{j_2 j_1 \bar{l}_2})}{\prod_{\bar{s}_3 = 1, \bar{s}_3 \neq s_1, \bar{s}_2}^{3} (a_{j_1 j_1 \bar{s}_2} - a_{j_1 j_1 \bar{s}_3})} \ln \frac{|x_{j_1} - a_{j_1 j_1 \bar{s}_2}|}{|x_{j_1 0} - a_{j_1 j_1 \bar{s}_2}|} \right]. \tag{4.289}$$

The infinite-equilibrium of $x_{j_1}^* = a_{j_1 j_1 s_1} = a_{j_2 j_1 l_1}$ with $\bar{x}_{j_2} \in (-\infty, \infty)$ *has the following properties.*

- *For* $a_{j_1 j_1 0} \prod_{s_2 = 1, s_2 \neq s_1}^{3} (a_{j_1 j_1 s_1} - a_{j_1 j_1 s_2}) > 0$ *and* $a_{j_2 j_1 0} \prod_{l_2 = 1, l_2 \neq l_1}^{3} (a_{j_2 j_1 l_1} - a_{j_2 j_1 l_2}) > 0$,

$$(a_{j_1 j_1 s_1}, \bar{x}_{j_2}) = \underbrace{(\text{SO}, \text{II})}_{\text{increasing-inflection source}}. \tag{4.290}$$

  The equilibrium at $x_{j_1}^* = a_{j_1 j_1 s_1} = a_{j_2 j_1 l_1}$ *is an* (SO,II)*-increasing-inflection source infinite-equilibrium.*

- *For* $a_{j_1 j_1 0} \prod_{s_2 = 1, s_2 \neq s_1}^{3} (a_{j_1 j_1 s_1} - a_{j_1 j_1 s_2}) < 0$ *and* $a_{j_2 j_1 0} \prod_{l_2 = 1, l_2 \neq l_1}^{3} (a_{j_2 j_1 l_1} - a_{j_2 j_1 l_2}) > 0$,

$$(a_{j_1 j_1 s_1}, \bar{x}_{j_2}) = \underbrace{(\text{SI}, \text{DI})}_{\text{decreasing-inflection sink}}. \tag{4.291}$$

  The equilibrium at $x_{j_1}^* = a_{j_1 j_1 s_1} = a_{j_2 j_1 l_1}$ *is an* (SI,DI)*-decreasing-inflection sink infinite-equilibrium.*

- *For* $a_{j_1 j_1 0} \prod_{s_2 = 1, s_2 \neq s_1}^{3} (a_{j_1 j_1 s_1} - a_{j_1 j_1 s_2}) > 0$ *and* $a_{j_2 j_1 0} \prod_{l_2 = 1, l_2 \neq l_1}^{3} (a_{j_2 j_1 l_1} - a_{j_2 j_1 l_2}) < 0$,

$$(a_{j_1 j_1 s_1}, \bar{x}_{j_2}) = \underbrace{(\text{SO}, \text{DI})}_{\text{decreasing-inflection source}}. \tag{4.292}$$

  The equilibrium at $x_{j_1}^* = a_{j_1 j_1 s_1} = a_{j_2 j_1 l_1}$ *is an* (SO,DI)*-increasing-inflection source infinite-equilibrium.*

- *For* $a_{j_1 j_1 0} \prod_{s_2 = 1, s_2 \neq s_1}^{3} (a_{j_1 j_1 s_1} - a_{j_1 j_1 s_2}) < 0$ *and* $a_{j_2 j_1 0} \prod_{l_2 = 1, l_2 \neq l_1}^{3} (a_{j_2 j_1 l_1} - a_{j_2 j_1 l_2}) < 0$,

$$(a_{j_1j_1s_1}, \bar{x}_{j_2}) = \underbrace{(\mathrm{SI}, \mathrm{II})}_{\text{increasing-inflection sink}}. \tag{4.293}$$

The equilibrium at $x^*_{j_1} = a_{j_1j_1s_1} = a_{j_2j_1l_1}$ *is an* (SI,II)*-increasing-inflection sink infinite-equilibrium.*

- *The infinite-equilibriums are the switching bifurcations for the sink and source flows with the sink and source flows.*

(xiv) If

$$\Delta_{j_1j_1} = 0 \text{ and } \Delta_{j_2j_1} > 0, \tag{4.294}$$

then

$$\begin{aligned} \dot{x}_{j_1} &= a_{j_1j_10}(x_{j_1} - a_{j_1j_1s_1})(x_{j_1} - a_{j_1j_1s_2})^2, \\ \dot{x}_{j_2} &= a_{j_2j_10}(x_{j_1} - a_{j_2j_1l_1})(x_{j_1} - a_{j_2j_1l_2})(x_{j_1} - a_{j_2j_1l_3}), \end{aligned} \tag{4.295}$$

where

$$\begin{aligned} &\{a_{j_1j_11}, a_{j_1j_12}\} = \mathrm{sort}\{b_{j_1j_11}, b_{j_1j_12}\}, a_{j_1j_11} < a_{j_1j_12}; \\ &\{a_{j_2j_11}, a_{j_2j_12}, a_{j_2j_13}) = \mathrm{sort}\{b_{j_2j_11}, b_{j_2j_12}, b_{j_2j_13}\}, a_{j_2j_11} < a_{j_2j_12} < a_{j_2j_13}; \\ &b_{j_1j_12} = -\frac{1}{2}B_{j_1j_11}, b_{j_2j_12}, b_{j_2j_13} = -\frac{1}{2}(B_{j_2j_11} \pm \sqrt{\Delta_{j_2j_1}}); \\ &j_1, j_2 \in \{1,2\}; j_1 \neq j_2. \end{aligned} \tag{4.296}$$

(xiv$_1$) For $a_{j_1j_1s_1} \neq a_{j_2j_1l_1}$ $(s_1, s_2 \in \{1,2\};\ s_1 \neq s_2)$ and $(l_1, l_2 \in \{1,2,3\};\ l_1 \neq l_2)$, *the first integral manifold is*

$$\begin{aligned} x_{j_2} - x_{j_20} = \frac{a_{j_2j_10}}{a_{j_1j_10}} \Big\{ &(x_{j_1} - x_{j_10}) + \frac{\prod_{l_1=1}^{3}(a_{j_1j_1s_1} - a_{j_2j_1l_1})}{(a_{j_1j_1s_1} - a_{j_1j_1s_2})^2} \ln \frac{|\, x_{j_1} - a_{j_1j_1s_1} \,|}{|\, x_{j_10} - a_{j_1j_1s_1} \,|} \\ &- \frac{\prod_{l_1=1}^{3}(a_{j_1j_1s_1} - a_{j_2j_1l_3})}{(a_{j_1j_1s_2} - a_{j_1j_1s_1})}\Big(\frac{1}{x_{j_1} - a_{j_1j_1s_2}} - \frac{1}{x_{j_10} - a_{j_1j_1s_2}}\Big) \\ &+ \Big[\frac{\sum_{l_1=1}^{3}\prod_{l_2=1, l_2 \neq l_1}^{3}(a_{j_1j_1s_2} - a_{j_2j_1l_2})}{(a_{j_1j_1s_2} - a_{j_1j_1s_1})} - \frac{\prod_{l_1=1}^{3}(a_{j_1j_1s_1} - a_{j_2j_1l_3})}{(a_{j_1j_1s_2} - a_{j_1j_1s_1})^2}\Big] \\ &\times \ln \frac{|\, x_{j_1} - a_{j_1j_1s_2} \,|}{|\, x_{j_10} - a_{j_1j_1s_2} \,|} \Big\}. \end{aligned} \tag{4.297}$$

or

$$\begin{aligned}
x_{j_2}-x_{j_20}=\frac{a_{j_2j_10}}{a_{j_1j_10}}\Big(\Big[(x_{j_1}-x_{j_10})+(\sum\nolimits_{l_2=1,l_2\neq l_1}^{3}(a_{j_1j_1s_2}-a_{j_2j_1l_2})\ln\frac{|x_{j_1}-a_{j_1j_1s_2}|}{|x_{j_10}-a_{j_1j_1s_2}|}\\
-\prod\nolimits_{l_2=1,l_2\neq l_1}^{3}(a_{j_1j_1s_2}-a_{j_2j_1l_2})(\frac{1}{x_{j_1}-a_{j_1j_1s_2}}-\frac{1}{x_{j_10}-a_{j_1j_1s_2}})\Big]\\
+(a_{j_1j_1s_1}-a_{j_2j_1l_1})\Big\{\frac{\prod_{l_2=1,l_2\neq l_1}^{3}(a_{j_1j_1s_1}-a_{j_2j_1l_2})}{(a_{j_1j_1s_1}-a_{j_1j_1s_2})^2}\ln\frac{|x_{j_1}-a_{j_1j_1s_1}|}{|x_{j_10}-a_{j_1j_1s_1}|}\\
+\frac{\prod_{l_2=1,l_2\neq l_1}^{3}(a_{j_1j_1s_2}-a_{j_2j_1l_2})}{(a_{j_1j_1s_2}-a_{j_1j_1s_1})}(\frac{1}{x_{j_1}-a_{j_1j_1s_2}}-\frac{1}{x_{j_1}-a_{j_1j_1s_2}})\\
+\Big[\frac{\sum_{l_2=1,l_2\neq l_1}^{3}(a_{j_1j_1s_2}-a_{j_2j_1l_2})}{(a_{j_1j_1s2}-a_{j_1j_1s_1})}-\frac{\prod_{l_2=1,l_2\neq l_1}^{3}(a_{j_1j_1s_2}-a_{j_2j_1l_2})}{(a_{j_1j_1s_2}-a_{j_1j_1s_1})^2}\Big]\\
\times\ln\frac{|x_{j_1}-a_{j_1j_1s_2}|}{|x_{j_10}-a_{j_1j_1s_2}|}\Big\}\Big).
\end{aligned}\tag{4.298}$$

(xiv$_{1a}$) The equilibriums of $x_{j_1}^{*}=a_{j_1j_1s_1}\neq a_{j_2j_1l_1},a_{j_2j_1l_2}$ (s_1, $s_2\in\{1,2,3\}$; $s_1\neq s_2$) and (l_1, $l_2\in\{1,2\}$; $l_1\neq l_2$) are as follows.

- *For $a_{j_1j_10}(a_{j_1j_1s_1}-a_{j_1j_1s_2})^2>0$ and $a_{j_2j_10}\prod_{l_2=1}^{3}(a_{j_1j_1s_1}-a_{j_2j_1l_1})>0$,*

$$(a_{j_1j_1s_1},\dot{x}_{j_2})=\underbrace{(\mathrm{SO},\mathrm{pF})}_{\text{source flow }(+)}.\tag{4.299}$$

  The flow at $x_{j_1}^{*}=a_{j_1j_1s_1}$ is an (SO,pF)*-positive source-flow.*

- *For $a_{j_1j_10}(a_{j_1j_1s_1}-a_{j_1j_1s_2})^2<0$ and $a_{j_2j_10}\prod_{l_2=1}^{3}(a_{j_1j_1s_1}-a_{j_2j_1l_1})>0$,*

$$(a_{j_1j_1s_1},\dot{x}_{j_2})=\underbrace{(\mathrm{SI},\mathrm{pF})}_{\text{sink flow }(+)}.\tag{4.300}$$

  The flow at $x_{j_1}^{*}=a_{j_1j_1s_1}$ is an (SI,pF)*-positive sink-flow.*

- *For $a_{j_1j_10}(a_{j_1j_1s_1}-a_{j_1j_1s_2})^2>0$ and $a_{j_2j_10}\prod_{l_2=1}^{3}(a_{j_1j_1s_1}-a_{j_2j_1l_1})<0$,*

$$(a_{j_1j_1s_1},\dot{x}_{j_2})=\underbrace{(\mathrm{SO},\mathrm{nF})}_{\text{source flow }(-)}.\tag{4.301}$$

  The flow at $x_{j_1}^{*}=a_{j_1j_1s_1}$ is an (SO,nF)*-negative source-flow.*

- *For* $a_{j_1j_10}(a_{j_1j_1s_1}-a_{j_1j_1s_2})^2<0$ *and* $a_{j_2j_10}\prod_{l_2=1}^{3}(a_{j_1j_1s_1}-a_{j_2j_1l_1})<0$,

$$(a_{j_1j_1s_1},\dot{x}_{j_2})=\underbrace{(\mathrm{SI},\mathrm{nF})}_{\text{sink flow }(-)}\ . \tag{4.302}$$

The flow at $x_{j_1}^*=a_{j_1j_1s_1}$ *is an* (SI,nF)*-negative sink-flow.*

(xiv$_{1b}$*) The equilibriums of* $x_{j_1}^*=a_{j_1j_1s_2}\neq a_{j_2j_1l_1},a_{j_1j_1s_2}$ $(s_1,\ s_2\in\{1,2\};\ s_1\neq s_2)$ *and* $(l_1=1,\ 2,\ 3)$ *are as follows.*

- *For* $a_{j_1j_10}(a_{j_1j_1s_2}-a_{j_1j_1s_1})>0$ *and* $a_{j_2j_10}\prod_{l_2=1}^{3}(a_{j_1j_1s_2}-a_{j_1j_1l_1})>0$,

$$(a_{j_1j_1s_2},\dot{x}_{j_2})=\underbrace{(\mathrm{US},\mathrm{pF})}_{\text{upper-saddle flow }(+)}\ . \tag{4.303}$$

The flow at $x_{j_1}^*=a_{j_1j_1s_2}$ *is a* (US,pF)*-positive upper-saddle flow.*

- *For* $a_{j_1j_10}(a_{j_1j_1s_2}-a_{j_1j_1s_1})<0$ *and* $a_{j_2j_10}\prod_{l_2=1}^{3}(a_{j_1j_1s_2}-a_{j_1j_1l_1})>0$,

$$(a_{j_1j_1s_2},\dot{x}_{j_2})=\underbrace{(\mathrm{LS},\mathrm{pF})}_{\text{lower-saddle flow }(+)}\ . \tag{4.304}$$

The flow at $x_{j_1}^*=a_{j_1j_1s_2}$ *is an* (LS,pF)*-positive lower-saddle flow.*

- *For* $a_{j_1j_10}(a_{j_1j_1s_2}-a_{j_1j_1s_1})>0$ *and* $a_{j_2j_10}\prod_{l_2=1}^{3}(a_{j_1j_1s_2}-a_{j_1j_1l_1})<0$,

$$(a_{j_1j_1s_2},\dot{x}_{j_2})=\underbrace{(\mathrm{US},\mathrm{nF})}_{\text{upper-saddle flow }(-)}\ . \tag{4.305}$$

The flow at $x_{j_1}^*=a_{j_1j_1s_2}$ *is a* (US,nF)*-negative upper-saddle flow.*

- *For* $a_{j_1j_10}(a_{j_1j_1s_2}-a_{j_1j_1s_1})<0$ *and* $a_{j_2j_10}\prod_{l_2=1}^{3}(a_{j_1j_1s_2}-a_{j_1j_1l_1})<0$,

$$(a_{j_1j_1s_2},\dot{x}_{j_2})=\underbrace{(\mathrm{LS},\mathrm{nF})}_{\text{lower-saddle flow }(-)}\ . \tag{4.306}$$

The flow at $x_{j_1}^*=a_{j_1j_1s_2}$ *is an* (LS,pF)*-positive lower-saddle flow.*

(xiv$_{1c}$*) The equilibrium of* $x_{j_1}^*=a_{j_2j_1l_1}\neq a_{j_1j_1s_1},a_{j_2j_1l_2}$ $(l_1,\ l_2\in\{1,2,3\};\ l_1\neq l_2)$ *and* $(s_1=1,\ 2)$ *has the following properties.*

- *For* $a_{j_1j_10}(a_{j_2j_1l_1}-a_{j_1j_1s_1})(a_{j_2j_1l_1}-a_{j_1j_1s_2})^2>0$ *and* $a_{j_2j_10}\prod_{l_2=1,l_2\neq l_1}^{3}(a_{j_2j_1l_1}-a_{j_2j_1l_2})>0$,

$$(\dot{x}_{j_1}, a_{j_2 j_1 l_1}) = \underbrace{(\text{pF}, \text{UP})}_{\text{up-parabola flow }(+)}. \tag{4.307}$$

The flow at $x_{j_1}^* = a_{j_1 j_1 l_1}$ *is a* (pF,UP)*-positive up-parabola flow.*

- *For* $a_{j_1 j_1 0}(a_{j_2 j_1 l_1} - a_{j_1 j_1 s_1})(a_{j_2 j_1 l_1} - a_{j_1 j_1 s_2})^2 < 0$ *and* $a_{j_2 j_1 0}\prod_{l_2=1, l_2 \neq l_1}^{3}(a_{j_2 j_1 l_1} - a_{j_2 j_1 l_2}) > 0$,

$$(\dot{x}_{j_1}, a_{j_2 j_1 l_1}) = \underbrace{(\text{nF}, \text{DP})}_{\text{down-parabola flow }(-)}. \tag{4.308}$$

The flow at $x_{j_1}^* = a_{j_2 j_1 l_1}$ *is an* (nF,DP)*-negative down-parabola flow.*

- *For* $a_{j_1 j_1 0}(a_{j_2 j_1 l_1} - a_{j_1 j_1 s_1})(a_{j_2 j_1 l_1} - a_{j_1 j_1 s_2})^2 > 0$ *and* $a_{j_2 j_1 0}\prod_{l_2=1, l_2 \neq l_1}^{3}(a_{j_2 j_1 l_1} - a_{j_2 j_1 l_2}) < 0$,

$$(\dot{x}_{j_1}, a_{j_2 j_1 l_1}) = \underbrace{(\text{pF}, \text{DP})}_{\text{down-parabola flow }(+)}. \tag{4.309}$$

The flow at $x_{j_1}^* = a_{j_2 j_1 l_1}$ *is a* (pF,DP)*-positive down-parabola flow.*

- *For* $a_{j_1 j_1 0}(a_{j_2 j_1 l_1} - a_{j_1 j_1 s_1})(a_{j_2 j_1 l_1} - a_{j_1 j_1 s_2})^2 < 0$ *and* $a_{j_2 j_1 0}\prod_{l_2=1, l_2 \neq l_1}^{3}(a_{j_2 j_1 l_1} - a_{j_2 j_1 l_2}) < 0$,

$$(\dot{x}_{j_1}, a_{j_2 j_1 l_1}) = \underbrace{(\text{nF}, \text{UP})}_{\text{up-parabola flow }(-)}. \tag{4.310}$$

The flow at $x_{j_1}^* = a_{j_2 j_1 l_1}$ *is an* (nF,UP)*-negative up-parabola flow.*

(xiv$_{2a}$) For $a_{j_1 j_1 s_1} = a_{j_2 j_1 l_1}$ $(s_1, s_2, s_3 \in \{1,2,3\};\ s_1 \neq s_2 \neq s_3)$ and $(l_1, l_2 \in \{1,2\};\ l_1 \neq l_2)$,

$$\begin{aligned} \dot{x}_{j_1} &= a_{j_1 j_1 0}(x_{j_1} - a_{j_1 j_1 s_1})(x_{j_1} - a_{j_1 j_1 s_2})^2, \\ \dot{x}_{j_2} &= a_{j_2 j_1 0}(x_{j_1} - a_{j_2 j_1 l_1})(x_{j_1} - a_{j_2 j_1 l_2})(x_{j_1} - a_{j_2 j_1 l_3}). \end{aligned} \tag{4.311}$$

The first integral manifold is

$$\begin{aligned} x_{j_2} - x_{j_2 0} = \frac{a_{j_2 j_1 0}}{a_{j_1 j_1 0}}\Big[(x_{j_1} - x_{j_1 0}) &+ \sum\nolimits_{l_2=1, l_2 \neq l_1}^{3}(a_{j_1 j_1 s_2} - a_{j_2 j_1 l_2}) \ln \frac{|x_{j_1} - a_{j_1 j_1 s_2}|}{|x_{j_1 0} - a_{j_1 j_1 s_2}|} \\ &- \prod\nolimits_{l_1=1}^{3}(a_{j_1 j_1 s_2} - a_{j_2 j_1 l_3})\Big(\frac{1}{x_{j_1} - a_{j_1 j_1 s_2}} - \frac{1}{x_{j_1 0} - a_{j_1 j_1 s_2}}\Big)\Big]. \end{aligned} \tag{4.312}$$

The infinite-equilibrium of $x_{j_1}^* = a_{j_1 j_1 s_1} = a_{j_2 j_1 l_1}$ *with* $\overline{x}_{j_2} \in (-\infty, \infty)$ *has the following properties.*

- *For* $a_{j_1j_10}(a_{j_1j_1s_1}-a_{j_1j_1s_2})^2>0$ *and* $a_{j_2j_10}\prod_{l_2=1,l_2\neq l_1}^{3}(a_{j_2j_1l_1}-a_{j_2j_1l_2})>0$,

$$(a_{j_1j_1s_1},\bar{x}_{j_2})=\underbrace{(\mathrm{SO,II})}_{\text{increasing-inflection source}}. \tag{4.313}$$

The equilibrium at $x_{j_1}^*=a_{j_1j_1s_1}=a_{j_2j_1l_1}$ *is an* (SO,II)*-increasing-inflection source infinite-equilibrium.*

- *For* $a_{j_1j_10}(a_{j_1j_1s_1}-a_{j_1j_1s_2})^2<0$ *and* $a_{j_2j_10}\prod_{l_2=1,l_2\neq l_1}^{3}(a_{j_2j_1l_1}-a_{j_2j_1l_2})>0$,

$$(a_{j_1j_1s_1},\bar{x}_{j_2})=\underbrace{(\mathrm{SI,DI})}_{\text{decreasing-inflection sink}}. \tag{4.314}$$

The equilibrium at $x_{j_1}^*=a_{j_1j_1s_1}=a_{j_2j_1l_1}$ *is an* (SI,DI)*-decreasing-inflection sink infinite-equilibrium.*

- *For* $a_{j_1j_10}(a_{j_1j_1s_1}-a_{j_1j_1s_2})^2>0$ *and* $a_{j_2j_10}\prod_{l_2=1,l_2\neq l_1}^{3}(a_{j_2j_1l_1}-a_{j_2j_1l_2})<0$,

$$(a_{j_1j_1s_1},\bar{x}_{j_2})=\underbrace{(\mathrm{SO,DI})}_{\text{decreasing-inflection source}}. \tag{4.315}$$

The equilibrium at $x_{j_1}^*=a_{j_1j_1s_1}=a_{j_2j_1l_1}$ *is an* (SO,DI)*-increasing-inflection source infinite-equilibrium.*

- *For* $a_{j_1j_10}(a_{j_1j_1s_1}-a_{j_1j_1s_2})^2<0$ *and* $a_{j_2j_10}\prod_{l_2=1,l_2\neq l_1}^{3}(a_{j_2j_1l_1}-a_{j_2j_1l_2})<0$,

$$(a_{j_1j_1s_1},\bar{x}_{j_2})=\underbrace{(\mathrm{SI,II})}_{\text{increasing-inflection sink}}. \tag{4.316}$$

The equilibrium at $x_{j_1}^*=a_{j_1j_1s_1}=a_{j_2j_1l_1}$ *is an* (SI,II)*-increasing-inflection sink infinite-equilibrium.*

(xiv$_{2b}$) For $a_{j_1j_1s_2}=a_{j_2j_1l_1}$ ($s_1,\ s_2\in\{1,2\};\ s_1\neq s_2\neq s_3$) and ($l_1,\ l_2,\ l_3\in\{1,2,3\};\ l_1\neq l_2$),

$$\begin{aligned}\dot{x}_{j_1}&=a_{j_1j_10}(x_{j_1}-a_{j_1j_1s_1})(x_{j_1}-a_{j_1j_1s_2})^2,\\ \dot{x}_{j_2}&=a_{j_2j_10}(x_{j_1}-a_{j_2j_1l_1})(x_{j_1}-a_{j_2j_1l_2})(x_{j_1}-a_{j_2j_1l_3}).\end{aligned} \tag{4.317}$$

The first integral manifold is

$$\begin{aligned}x_{j_2}-x_{j_20}=\frac{a_{j_2j_10}}{a_{j_1j_10}}\Bigg[&(x_{j_1}-x_{j_10})+\frac{\prod_{l_2=1,l_2\neq l_1}^{3}(a_{j_1j_1s_1}-a_{j_2j_1l_2})}{(a_{j_1j_1s_1}-a_{j_1j_1s_2})}\ln\frac{|x_{j_1}-a_{j_1j_1s_1}|}{|x_{j_10}-a_{j_1j_1s_1}|}\\ &+\frac{\prod_{l_2=1,l_2\neq l_1}^{3}(a_{j_1j_1s_2}-a_{j_2j_1l_2})}{(a_{j_1j_1s_2}-a_{j_1j_1s_1})}\ln\frac{|x_{j_1}-a_{j_1j_1s_2}|}{|x_{j_10}-a_{j_1j_1s_2}|}\Bigg].\end{aligned} \tag{4.318}$$

The infinite-equilibrium of $x^*_{j_1} = a_{j_1j_1s_2} = a_{j_2j_1l_1}$ *with* $\overline{x}_{j_2} \in (-\infty, \infty)$ *has the following properties.*

- *For* $a_{j_1j_10}(a_{j_1j_1s_2} - a_{j_1j_1s_1}) > 0$ *and* $a_{j_2j_10}\prod_{l_2=1,l_2\neq l_1}^{3}(a_{j_2j_1l_1} - a_{j_2j_1l_2}) > 0$,

$$(a_{j_1j_1s_2}, \overline{x}_{j_2}) = \underbrace{(\text{US}, \text{DU})}_{\text{down-up upper-saddle}}. \tag{4.319}$$

  The equilibrium at $x^*_{j_1} = a_{j_1j_1s_2} = a_{j_2j_1l_1}$ *is a* (US,DU)*-down-up upper-saddle infinite-equilibrium.*

- *For* $a_{j_1j_10}(a_{j_1j_1s_2} - a_{j_1j_1s_1}) < 0$ *and* $a_{j_2j_10}\prod_{l_2=1,l_2\neq l_1}^{3}(a_{j_2j_1l_1} - a_{j_2j_1l_2}) > 0$,

$$(a_{j_1j_1s_2}, \overline{x}_{j_2}) = \underbrace{(\text{LS}, \text{UD})}_{\text{up-down lower-saddle}}. \tag{4.320}$$

  The equilibrium at $x^*_{j_1} = a_{j_1j_1s_2} = a_{j_2j_1l_1}$ *is an* (LS,UD)*-up-down lower-saddle infinite-equilibrium.*

- *For* $a_{j_1j_10}(a_{j_1j_1s_2} - a_{j_1j_1s_1}) > 0$ *and* $a_{j_2j_10}\prod_{l_2=1,l_2\neq l_1}^{3}(a_{j_2j_1l_1} - a_{j_2j_1l_2}) < 0$,

$$(a_{j_1j_1s_2}, \overline{x}_{j_2}) = \underbrace{(\text{US}, \text{UD})}_{\text{up-down upper-saddle}}. \tag{4.321}$$

  The equilibrium at $x^*_{j_1} = a_{j_1j_1s_2} = a_{j_2j_1l_1}$ *is a* (US,UD)*-up-down upper-saddle infinite-equilibrium.*

- *For* $a_{j_1j_10}(a_{j_1j_1s_2} - a_{j_1j_1s_1}) < 0$ *and* $a_{j_2j_10}\prod_{l_2=1,l_2\neq l_1}^{3}(a_{j_2j_1l_1} - a_{j_2j_1l_2}) < 0$,

$$(a_{j_1j_1s_2}, \overline{x}_{j_2}) = \underbrace{(\text{LS}, \text{DU})}_{\text{down-up lower-saddle}}. \tag{4.322}$$

  The equilibrium at $x^*_{j_1} = a_{j_1j_1s_2} = a_{j_2j_1l_1}$ *is an* (LS,DU)*-down-up lower-saddle infinite-equilibrium.*

(xv) If

$$\begin{aligned} &\Delta_{j_1j_1} = 0 \text{ and } \Delta_{j_2j_1} > 0, \\ &a_{j_1j_1s_1} = a_{j_1j_1s_2} = a_{j_1j_11} \end{aligned} \tag{4.323}$$

then

$$\begin{aligned} \dot{x}_{j_1} &= a_{j_1j_10}(x_{j_1} - a_{j_1j_11})^3, \\ \dot{x}_{j_2} &= a_{j_2j_10}(x_{j_1} - a_{j_2j_1l_1})(x_{j_1} - a_{j_2j_1l_2})(x_{j_1} - a_{j_2j_1l_3}), \end{aligned} \tag{4.324}$$

where

$$\begin{array}{l}\{a_{j_1j_1 1},a_{j_1j_1 2}\}=\mathrm{sort}\{b_{j_1j_1 1},b_{j_1j_1 2}\},a_{j_1j_1 1}=a_{j_1j_1 2};\\ \{a_{j_2j_1 1},a_{j_2j_1 2},a_{j_2j_1 3})=\mathrm{sort}\{b_{j_2j_1 1},b_{j_2j_1 2},b_{j_2j_1 3}\},a_{j_2j_1 1}<a_{j_2j_1 2}<a_{j_2j_1 3};\\ b_{j_1j_1 2}=-\frac{1}{2}B_{j_1j_1 1};b_{j_2j_1 2},b_{j_2j_1 3}=-\frac{1}{2}(B_{j_2j_1 1}\pm\sqrt{\Delta_{j_2j_1}});\\ j_1,j_2\in\{1,2\};j_1\neq j_2.\end{array}\tag{4.325}$$

(xv_1) For $a_{j_1j_1 1}\neq a_{j_2j_1 l_1}$ ($l_1=1, 2, 3$), the first integral manifold is

$$\begin{aligned}x_{j_2}-x_{j_2 0}=\frac{a_{j_2j_1 0}}{a_{j_1j_1 0}}\Big\{&(x_{j_1}-x_{j_1 0})\\ &-\frac{1}{2}\prod\nolimits_{l_1=1}^{3}(a_{j_1j_1 1}-a_{j_2j_1 l_1})\Big[\frac{1}{(x_{j_1}-a_{j_1j_1 1})^2}-\frac{1}{(x_{j_1 0}-a_{j_1j_1 1})^2}\Big]\\ &-\sum\nolimits_{l_1=1}^{3}\prod\nolimits_{l_2=1,l_2\neq l_1}^{3}(a_{j_1j_1 1}-a_{j_2j_1 l_2})\Big[\frac{1}{(x_{j_1}-a_{j_1j_1 1})}-\frac{1}{(x_{j_1 0}-a_{j_1j_1 1})}\Big]\\ &+\sum\nolimits_{l_1=1}^{3}\sum\nolimits_{l_2=1,l_2\neq l_1}^{3}\prod\nolimits_{l_1=1,l_3\neq l_1,l_2}^{3}(a_{j_1j_1 1}-a_{j_2j_1 l_3})\ln\frac{|x_{j_1}-a_{j_1j_1 1}|}{|x_{j_1 0}-a_{j_1j_1 1}|}\Big\}.\end{aligned}\tag{4.326}$$

(xv_{1a}) The equilibriums of $x_{j_1}^*=a_{j_1j_1 1}\neq a_{j_2j_1 l_1}$ ($l_1=1, 2, 3$) are as follows.

- *For $a_{j_1j_1 0}>0$ and $a_{j_2j_1 0}\prod_{l_1=1}^{3}(a_{j_1j_1 1}-a_{j_2j_1 l_1})>0$,*

$$(a_{j_1j_1 1},\dot{x}_{j_2})=\underbrace{(3^{\mathrm{rd}}\mathrm{SO,pF})}_{\text{third-order source flow }(+)}.\tag{4.327}$$

The flow at $x_{j_1}^*=a_{j_1j_1 1}$ is a (3^{rd}SO,pF)*-positive third-order source flow.*

- *For $a_{j_1j_1 0}<0$ and $a_{j_2j_1 0}\prod_{l_1=1}^{3}(a_{j_1j_1 1}-a_{j_2j_1 l_1})>0$,*

$$(a_{j_1j_1 1},\dot{x}_{j_2})=\underbrace{(3^{\mathrm{rd}}\mathrm{SI,pF})}_{\text{third-order sink flow }(+)}.\tag{4.328}$$

The flow at $x_{j_1}^*=a_{j_1j_1 1}$ is a (3^{rd}SI,pF)*-positive third-order sink-flow.*

- *For $a_{j_1j_1 0}>0$ and $a_{j_2j_1 0}\prod_{l_1=1}^{3}(a_{j_1j_1 1}-a_{j_2j_1 l_1})<0$,*

$$(a_{j_1j_1 1},\dot{x}_{j_2})=\underbrace{(3^{\mathrm{rd}}\mathrm{SO,nF})}_{\text{third-order source flow }(-)}.\tag{4.329}$$

The flow at $x_{j_1}^*=a_{j_1j_1 1}$ is a (3^{rd}SO,nF)*-negative third-order source flow.*

- *For* $a_{j_1j_1 0} < 0$ *and* $a_{j_2j_1 0}\prod_{l_1=1}^{3}(a_{j_1j_1 1} - a_{j_2j_1 l_1}) < 0$,

$$(a_{j_1j_1 1}, \dot{x}_{j_2}) = \underbrace{(3^{\text{rd}}\text{SI}, \text{nF})}_{\text{third-order sink flow }(-)}. \tag{4.330}$$

The flow at $x_{j_1}^* = a_{j_1j_1 1}$ *is a* (3^{rd}SI,nF)*-negative third-order sink flow.*

(xv$_{1b}$) The equilibrium of $x_{j_1}^* = a_{j_2j_1 l_1} \neq a_{j_1j_1 1}, a_{j_2j_1 l_2}$ (l_1, $l_2 \in \{1,2,3\}$; $l_1 \neq l_2$) *has the following properties.*

- *For* $a_{j_1j_1 0}(a_{j_2j_1 l_1} - a_{j_1j_1 1})^3 > 0$ *and* $a_{j_2j_1 0}\prod_{l_2=1, l_2\neq l_1}^{3}(a_{j_2j_1 l_1} - a_{j_2j_1 l_2}) > 0$,

$$(\dot{x}_{j_1}, a_{j_2j_1 l_1}) = \underbrace{(\text{pF}, \text{UP})}_{\text{up-parabola flow }(+)}. \tag{4.331}$$

The flow at $x_{j_1}^* = a_{j_1j_1 l_1}$ *is a* (pF,UP)*-positive up-parabola flow.*

- *For* $a_{j_1j_1 0}(a_{j_2j_1 l_1} - a_{j_1j_1 1})^3 < 0$ *and* $a_{j_2j_1 0}\prod_{l_2=1, l_2\neq l_1}^{3}(a_{j_2j_1 l_1} - a_{j_2j_1 l_2}) > 0$,

$$(\dot{x}_{j_1}, a_{j_2j_1 l_1}) = \underbrace{(\text{nF}, \text{DP})}_{\text{down-parabola flow }(-)}. \tag{4.332}$$

The flow at $x_{j_1}^* = a_{j_2j_1 l_1}$ *is an* (nF,DP)*-negative down-parabola flow.*

- *For* $a_{j_1j_1 0}(a_{j_2j_1 l_1} - a_{j_1j_1 1})^3 > 0$ *and* $a_{j_2j_1 0}\prod_{l_2=1, l_2\neq l_1}^{3}(a_{j_2j_1 l_1} - a_{j_2j_1 l_2}) < 0$,

$$(\dot{x}_{j_1}, a_{j_2j_1 l_1}) = \underbrace{(\text{pF}, \text{DP})}_{\text{down-parabola flow }(+)}. \tag{4.333}$$

The flow at $x_{j_1}^* = a_{j_2j_1 l_1}$ *is a (pF,DP)-positive down-parabola flow.*

- *For* $a_{j_1j_1 0}(a_{j_2j_1 l_1} - a_{j_1j_1 1})^3 < 0$ *and* $a_{j_2j_1 0}\prod_{l_2=1, l_2\neq l_1}^{3}(a_{j_2j_1 l_1} - a_{j_2j_1 l_2}) < 0$,

$$(\dot{x}_{j_1}, a_{j_2j_1 l_1}) = \underbrace{(\text{nF}, \text{UP})}_{\text{up-parabola flow }(-)}. \tag{4.334}$$

The flow at $x_{j_1}^* = a_{j_2j_1 l_1}$ *is an* (nF,UP)*-negative up-parabola flow.*

(xv$_2$) For $a_{j_1j_1 1} = a_{j_2j_1 l_1}$ ($l_1 = 1, 2, 3$),

$$\begin{aligned} \dot{x}_{j_1} &= a_{j_1j_1 0}(x_{j_1} - a_{j_1j_1 1})^3, \\ \dot{x}_{j_2} &= a_{j_2j_1 0}(x_{j_1} - a_{j_2j_1 l_1})\prod_{l_2=1, l_2\neq l_1}^{3}(x_{j_1} - a_{j_2j_1 l_2}). \end{aligned} \tag{4.335}$$

The first integral manifold is

$$
\begin{aligned}
x_{j_2} - x_{j_2 0} = \frac{a_{j_2 j_1 0}}{a_{j_1 j_1 0}} \Big\{ & (x_{j_1} - x_{j_1 0}) \\
& - \prod\nolimits_{l_2=1, l_2 \neq l_1}^{3} (a_{j_2 j_1 l_1} - a_{j_2 j_1 l_2}) \left[\frac{1}{(x_{j_1} - a_{j_1 j_1 1})} - \frac{1}{(x_{j_1 0} - a_{j_1 j_1 1})} \right] \\
& + \sum\nolimits_{l_2=1, l_2 \neq l_1}^{3} \prod\nolimits_{l_3=1, l_3 \neq l_1, l_2}^{3} (a_{j_2 j_1 l_1} - a_{j_2 j_1 l_3}) \ln \frac{| x_{j_1} - a_{j_1 j_1 1} |}{| x_{j_1 0} - a_{j_1 j_1 1} |} \Big\}.
\end{aligned}
\tag{4.336}
$$

The infinite-equilibrium of $x_{j_1}^* = a_{j_1 j_1 1} = a_{j_2 j_1 l_1}$ with $\bar{x}_{j_2} \in (-\infty, \infty)$ *has the following properties.*

- *For* $a_{j_1 j_1 0} > 0$ *and* $a_{j_2 j_1 0} \prod_{l_2=1, l_2 \neq l_1}^{3} (a_{j_2 j_1 l_1} - a_{j_2 j_1 l_2}) > 0$,

$$
(a_{j_1 j_1 1}, \bar{x}_{j_2}) = \underbrace{(3^{\text{rd}}\text{SO}, (3:1)\text{IC})}_{\text{third-order increasing-concave source}}. \tag{4.337}
$$

  The equilibrium at $x_{j_1}^* = a_{j_1 j_1 1} = a_{j_2 j_1 l_1}$ *is an* (3rdSO,(3:1)IC)*-third-order increasing-concave source infinite-equilibrium.*

- *For* $a_{j_1 j_1 0} < 0$ *and* $a_{j_2 j_1 0} \prod_{l_2=1, l_2 \neq l_1}^{3} (a_{j_2 j_1 l_1} - a_{j_2 j_1 l_2}) > 0$,

$$
(a_{j_1 j_1 1}, \bar{x}_{j_2}) = \underbrace{(3^{\text{rd}}\text{SI}, (3:1)\text{DC})}_{\text{third-order decreasing-concave sink}}. \tag{4.338}
$$

  The equilibrium at $x_{j_1}^* = a_{j_1 j_1 1} = a_{j_2 j_1 l_1}$ *is an* (3rdSI,(3:1)DC)*-third-order decreasing-concave sink infinite-equilibrium.*

- *For* $a_{j_1 j_1 0} > 0$ *and* $a_{j_2 j_1 0} \prod_{l_2=1, l_2 \neq l_1}^{3} (a_{j_2 j_1 l_1} - a_{j_2 j_1 l_2}) < 0$,

$$
(a_{j_1 j_1 1}, \bar{x}_{j_2}) = \underbrace{(3^{\text{rd}}\text{SO}, (3:1)\text{DC})}_{\text{third-order decreasing-asympotic source}}. \tag{4.339}
$$

  The equilibrium at $x_{j_1}^* = a_{j_1 j_1 1} = a_{j_2 j_1 l_1}$ *is an* (3rdSO,(3:1)DC)*-third-order decreasing-concave sink infinite-equilibrium.*

- *For* $a_{j_1 j_1 0} < 0$ *and* $a_{j_2 j_1 0} \prod_{l_2=1, l_2 \neq l_1}^{3} (a_{j_2 j_1 l_1} - a_{j_2 j_1 l_2}) < 0$,

$$
(a_{j_1 j_1 1}, \bar{x}_{j_2}) = \underbrace{(3^{\text{rd}}\text{SI}, (3:1)\text{IC})}_{\text{third-order increasing-concave sink}}. \tag{4.340}
$$

  The equilibrium at $x_{j_1}^* = a_{j_1 j_1 1} = a_{j_2 j_1 l_1}$ *is an* (3rdSI,(3:1)IC)*-third-order increasing-concave sink infinite-equilibrium.*

(xvi) If

$$\Delta_{j_1j_1} < 0 \text{ and } \Delta_{j_2j_1} > 0, \tag{4.341}$$

then

$$\begin{aligned} \dot{x}_{j_1} &= a_{j_1j_10}(x_{j_1} - a_{j_1j_11})\left[(x_{j_1} - a_{j_1j_1})^2 + b_{j_1j_1}\right], \\ \dot{x}_{j_2} &= a_{j_2j_10}(x_{j_1} - a_{j_2j_1l_1})(x_{j_1} - a_{j_2j_1l_2})(x_{j_1} - a_{j_2j_1l_3}), \end{aligned} \tag{4.342}$$

where

$$\begin{aligned} &a_{j_1j_1} = -\frac{1}{2}B_{j_1j_11}, a_{j_1j_1} = -\frac{1}{4}\Delta_{j_1j_1}; \\ &\{a_{j_2j_11}, a_{j_2j_12}, a_{j_2j_13}\big) = \mathrm{sort}\{b_{j_2j_11}, b_{j_2j_12}, b_{j_2j_13}\}, a_{j_2j_11} < a_{j_2j_12} < a_{j_2j_13}; \\ &a_{j_1j_11} = b_{j_1j_11}; b_{j_2j_12}, b_{j_2j_13} = -\frac{1}{2}(B_{j_2j_11} \pm \sqrt{\Delta_{j_2j_1}}); \\ &j_1, j_2 \in \{1, 2\}; j_1 \neq j_2. \end{aligned} \tag{4.343}$$

(xvi$_1$) For $a_{j_1j_11} \neq a_{j_2j_1l_1}$ $(l_1 = 1, 2, 3)$, *the first integral manifold is*

$$\begin{aligned} x_{j_2} - x_{j_20} = \frac{a_{j_2j_10}}{a_{j_1j_10}}\Big\{&(x_{j_1} - x_{j_10}) \\ &+\frac{1}{2}\sum\nolimits_{l_2=1, l_2\neq l_1}^{3}(a_{j_1j_1} - a_{j_2j_1l_2})\ln\frac{|(x_{j_1} - a_{j_1j_1})^2 + b_{j_1j_1}|}{|(x_{j_10} - a_{j_1j_1})^2 + b_{j_1j_1}|} \\ &\frac{1}{\sqrt{b_{j_1j_1}}} + \left[\prod\nolimits_{l_2=1, l_2\neq l_1}^{3}(a_{j_1j_1} - a_{j_2j_1l_2}) - b_{j_1j_1}\right] \\ &\times(\arctan\frac{x_{j_1} - a_{j_1j_1}}{\sqrt{b_{j_1j_1}}} - \arctan\frac{x_{j_10} - a_{j_1j_1}}{\sqrt{b_{j_1j_1}}}) \\ &+\frac{1}{2}(a_{j_1j_11} - a_{j_2j_1l_1})\ln\frac{|(x_{j_1} - a_{j_1j_1})^2 + b_{j_1j_1}|}{|(x_{j_10} - a_{j_1j_1})^2 + b_{j_1j_1}|} \\ &+(a_{j_1j_11} - a_{j_2j_1l_1})\left[(a_{j_1j_1} - a_{j_2j_1l_3}) + (a_{j_1j_11} - a_{j_2j_1l_2})\right] \\ &\times\frac{1}{\sqrt{b_{j_1j_1}}}(\arctan\frac{x_{j_1} - a_{j_1j_1}}{\sqrt{b_{j_1j_1}}} - \arctan\frac{x_{j_10} - a_{j_1j_1}}{\sqrt{b_{j_1j_1}}}) \\ &+\frac{\prod_{l_2=1}^{3}(a_{j_1j_11} - a_{j_2j_1l_2})}{(a_{j_1j_11} - a_{j_1j_1})^2 + b_{j_1j_1}}\Big[\ln\frac{|x_{j_1} - a_{j_1j_11}|}{|x_{j_10} - a_{j_1j_11}|} \\ &-\frac{1}{2}\ln\frac{|(x_{j_1} - a_{j_1j_1})^2 + b_{j_1j_1}|}{|(x_{j_10} - a_{j_1j_1})^2 + b_{j_1j_1}|} \\ &-\frac{(a_{j_1j_11} - a_{j_1j_1})}{\sqrt{b_{j_1j_1}}}\left(\arctan\frac{x_{j_1} - a_{j_1j_1}}{\sqrt{b_{j_1j_1}}} - \arctan\frac{x_{j_10} - a_{j_1j_1}}{\sqrt{b_{j_1j_1}}}\right)\Big]\Big\}. \end{aligned} \tag{4.344}$$

(xvi_{1a}) The equilibriums of $x_{j_1}^* = a_{j_1 j_1 1} \neq a_{j_2 j_1 l_1}$ ($l_1 = 1, 2, 3$) possess the following properties.

- *For $a_{j_1 j_1 0} > 0$ and $a_{j_2 j_1 0} \prod_{l_1 = 1}^{3} (a_{j_1 j_1 1} - a_{j_2 j_1 l_1}) > 0$,*

$$(a_{j_1 j_1 1}, \dot{x}_{j_2}) = \underbrace{(\text{SO}, \text{pF})}_{\text{source flow } (+)}. \tag{4.345}$$

The flow at $x_{j_1}^* = a_{j_1 j_1 1}$ is an (SO,pF)*-positive source flow.*

- *For $a_{j_1 j_1 0} < 0$ and $a_{j_2 j_1 0} \prod_{l_1 = 1}^{3} (a_{j_1 j_1 1} - a_{j_2 j_1 l_1}) > 0$,*

$$(a_{j_1 j_1 1}, \dot{x}_{j_2}) = \underbrace{(\text{SI}, \text{pF})}_{\text{sink flow } (+)}. \tag{4.346}$$

The flow at $x_{j_1}^* = a_{j_1 j_1 1}$ is an (SI,pF)*-positive sink flow.*

- *For $a_{j_1 j_1 0} > 0$ and $a_{j_2 j_1 0} \prod_{l_1 = 1}^{3} (a_{j_1 j_1 1} - a_{j_2 j_1 l_1}) < 0$,*

$$(a_{j_1 j_1 1}, \dot{x}_{j_2}) = \underbrace{(\text{SO}, \text{nF})}_{\text{source flow } (-)}. \tag{4.347}$$

The flow at $x_{j_1}^* = a_{j_1 j_1 1}$ is an (SO,nF)*-negative source flow.*

- *For $a_{j_1 j_1 0} < 0$ and $a_{j_2 j_1 0} \prod_{l_1 = 1}^{3} (a_{j_1 j_1 1} - a_{j_2 j_1 l_1}) < 0$,*

$$(a_{j_1 j_1 1}, \dot{x}_{j_2}) = \underbrace{(\text{SI}, \text{nF})}_{\text{sink flow } (-)}. \tag{4.348}$$

The flow at $x_{j_1}^* = a_{j_1 j_1 1}$ is an (SI,nF)*-negative sink flow.*

(xvi_{1b}) The equilibrium of $x_{j_1}^* = a_{j_2 j_1 l_1} \neq a_{j_1 j_1 1}, a_{j_2 j_1 l_2}$ ($l_1, l_2 \in \{1,2\}$; $l_1 \neq l_2$) and ($s_1 = 1, 2, 3$) has the following properties.

- *For $a_{j_1 j_1 0}(a_{j_2 j_1 l_1} - a_{j_1 j_1 1}) > 0$ and $a_{j_2 j_1 0} \prod_{l_2 = 1, l_2 \neq l_1}^{3} (a_{j_2 j_1 l_1} - a_{j_2 j_1 l_2}) > 0$,*

$$(\dot{x}_{j_1}, a_{j_2 j_1 l_1}) = \underbrace{(\text{pF}, \text{UP})}_{\text{up-parabola flow } (+)}. \tag{4.349}$$

The flow at $x_{j_1}^* = a_{j_1 j_1 l_1}$ is a (pF,UP)*-positive up-parabola flow.*

- *For $a_{j_1 j_1 0}(a_{j_2 j_1 l_1} - a_{j_1 j_1 1}) < 0$ and $a_{j_2 j_1 0} \prod_{l_2 = 1, l_2 \neq l_1}^{3} (a_{j_2 j_1 l_1} - a_{j_2 j_1 l_2}) > 0$,*

$$(\dot{x}_{j_1}, a_{j_2 j_1 l_1}) = \underbrace{(\text{nF}, \text{DP})}_{\text{down-parabola flow } (-)}. \tag{4.350}$$

The flow at $x_{j_1}^* = a_{j_2 j_1 l_1}$ is an (nF,DP)*-negative down-parabola flow.*

- *For* $a_{j_1j_10}(a_{j_2j_1l_1} - a_{j_1j_11}) > 0$ *and* $a_{j_2j_10}\prod_{l_2=1,l_2\neq l_1}^{3}(a_{j_2j_1l_1} - a_{j_2j_1l_2}) < 0$,

$$(\dot{x}_{j_1}, a_{j_2j_1l_1}) = \underbrace{(\text{pF}, \text{DP})}_{\text{down-parabola flow }(+)}. \tag{4.351}$$

The flow at $x_{j_1}^* = a_{j_2j_1l_1}$ *is a* (pF,DP)*-positive down-parabola flow.*

- *For* $a_{j_1j_10}(a_{j_2j_1l_1} - a_{j_1j_11}) < 0$ *and* $a_{j_2j_10}\prod_{l_2=1,l_2\neq l_1}^{3}(a_{j_2j_1l_1} - a_{j_2j_1l_2}) < 0$,

$$(\dot{x}_{j_1}, a_{j_2j_1l_1}) = \underbrace{(\text{nF}, \text{UP})}_{\text{up-parabola flow }(-)}. \tag{4.352}$$

The flow at $x_{j_1}^* = a_{j_2j_1l_1}$ *is an* (nF,UP)*-negative up-parabola flow.*

(xvi$_2$*) For* $a_{j_1j_11} = a_{j_2j_1l_1}$ $(l_1, l_2, l_3 \in \{1,2,3\};\ l_1 \neq l_2 \neq l_3)$,

$$\begin{aligned} \dot{x}_{j_1} &= a_{j_1j_10}(x_{j_1} - a_{j_1j_11})\left[(x_{j_1} - a_{j_1j_1})^2 + b_{j_1j_1}\right], \\ \dot{x}_{j_2} &= a_{j_2j_10}(x_{j_1} - a_{j_2j_1l_1})(x_{j_1} - a_{j_2j_1l_2})(x_{j_1} - a_{j_2j_1l_3}). \end{aligned} \tag{4.353}$$

The first integral manifold is

$$\begin{aligned} x_{j_2} - x_{j_20} = \frac{a_{j_2j_10}}{a_{j_1j_10}}\Big\{&(x_{j_1} - x_{j_10}) \\ &+ \frac{1}{2}\sum\nolimits_{l_2=1,l_2\neq l_1}^{3}(a_{j_1j_1} - a_{j_2j_1l_2})\ln\frac{|(x_{j_1} - a_{j_1j_1})^2 + b_{j_1j_1}|}{|(x_{j_10} - a_{j_1j_1})^2 + b_{j_1j_1}|} \\ &+ \left[\prod\nolimits_{l_2=1,l_2\neq l_1}^{3}(a_{j_1j_1} - a_{j_2j_1l_2}) - b_{j_1j_1}\right] \\ &\times \frac{1}{\sqrt{b_{j_1j_1}}}\Big(\arctan\frac{x_{j_1} - a_{j_1j_1}}{\sqrt{b_{j_1j_1}}} - \arctan\frac{x_{j_10} - a_{j_1j_1}}{\sqrt{b_{j_1j_1}}}\Big)\Big\}. \end{aligned} \tag{4.354}$$

The infinite-equilibrium of $x_{j_1}^* = a_{j_1j_11} = a_{j_2j_1l_1}$ *with* $\bar{x}_{j_2} \in (-\infty, \infty)$ *has the following properties.*

- *For* $a_{j_1j_10} > 0$ *and* $a_{j_2j_10}\prod_{l_2=1,l_2\neq l_1}^{3}(a_{j_2j_1l_1} - a_{j_2j_1l_2}) > 0$,

$$(a_{j_1j_11}, \bar{x}_{j_2}) = \underbrace{(\text{SO}, \text{II})}_{\text{increasing-inflection source}}. \tag{4.355}$$

The equilibrium at $x_{j_1}^* = a_{j_1j_11} = a_{j_2j_1l_1}$ *is an* (SO,II)*-increasing-inflection source infinite-equilibrium.*

- *For* $a_{j_1j_10} < 0$ *and* $a_{j_2j_10}\prod_{l_2=1,l_2\neq l_1}^{3}(a_{j_2j_1l_1} - a_{j_2j_1l_2}) > 0$,

$$(a_{j_1j_1 1}, \bar{x}_{j_2}) = \underbrace{(\text{SI, DI})}_{\text{decreasing-inflection sink}}. \tag{4.356}$$

The equilibrium at $x^*_{j_1} = a_{j_1j_1 1} = a_{j_2j_1l_1}$ *is an* (SI,DI)*-decreasing-inflection sink infinite-equilibrium.*

- *For* $a_{j_1j_10} > 0$ *and* $a_{j_2j_10}\prod_{l_2=1,l_2\neq l_1}^{3}(a_{j_2j_1l_1} - a_{j_2j_1l_2}) < 0$,

$$(a_{j_1j_1 1}, \bar{x}_{j_2}) = \underbrace{(\text{SO, DI})}_{\text{decreasing-inflection source}}. \tag{4.357}$$

The equilibrium at $x^*_{j_1} = a_{j_1j_1 1} = a_{j_2j_1l_1}$ *is an* (SO,DI)*-decreasing-inflection sink infinite-equilibrium.*

- *For* $a_{j_1j_10} < 0$ *and* $a_{j_2j_10}\prod_{l_2=1,l_2\neq l_1}^{3}(a_{j_2j_1l_1} - a_{j_2j_1l_2}) < 0$,

$$(a_{j_1j_1 1}, \bar{x}_{j_2}) = \underbrace{(\text{SI, II})}_{\text{increasing-inflection sink}}. \tag{4.358}$$

The equilibrium at $x^*_{j_1} = a_{j_1j_1 1} = a_{j_2j_1l_1}$ *is an* (SI,II)*-increasing-inflection sink infinite-equilibrium.*

4.2 Proof of Theorem 4.1

Consider

$$\begin{aligned}
\dot{x}_{j_2} &= a_{j_2j_10}(x_{j_1} - b_{j_2j_1 1})(x_{j_1}^2 + B_{j_2j_1 1}x_{j_1} + C_{j_2j_1 1}),\\
\dot{x}_{j_1} &= a_{j_1j_10}(x_{j_1} - b_{j_1j_1 1})(x_{j_1}^2 + B_{j_1j_1 1}x_{j_1} + C_{j_1j_1 1})
\end{aligned}$$

with

$$\Delta_{jj_1} = (B_{jj_1 1})^2 - 4C_{jj_1 1} \ \text{ for } j \in \{j_1, j_2\}, j_1, j_2 \in \{1, 2\} j_1 \neq j_2.$$

(i) If

$$\Delta_{j_1j_1} < 0 \text{ and } \Delta_{j_2j_1} < 0,$$

then

$$\begin{aligned}
&\dot{x}_{j_2} = a_{j_2j_10}(x_{j_1} - b_{j_2j_1 1})\left[(x_{j_1} - a_{j_2j_1})^2 + b_{j_2j_1}\right],\\
&\dot{x}_{j_1} = a_{j_1j_10}(x_{j_1} - b_{j_1j_1 1})\left[(x_{j_1} - a_{j_1j_1})^2 + b_{j_1j_1}\right],\\
&\text{with } a_{sj_1} = -\frac{1}{2}B_{sj_1 1}, b_{sj_1} = \frac{1}{4}(-\Delta_{sj_1}), s = j_1, j_2.
\end{aligned}$$

(i) For $a_{j_1j_11} \neq a_{j_2j_11}$, in phase space, we have

$$
\begin{aligned}
\frac{dx_{j_2}}{dx_{j_1}} &= \frac{a_{j_2j_10}(x_{j_1}-a_{j_2j_11})\left[(x_{j_1}-a_{j_2j_1})^2+b_{j_2j_1}\right]}{a_{j_1j_10}(x_{j_1}-a_{j_1j_11})\left[(x_{j_1}-a_{j_1j_1})^2+b_{j_1j_1}\right]} \\
&= \frac{a_{j_2j_10}}{a_{j_1j_10}}\Big(1+(a_{j_1j_1}-a_{j_2j_1})\frac{2(x_{j_1}-a_{j_1j_1})}{(x_{j_1}-a_{j_1j_1})^2+b_{j_1j_1}} \\
&\quad+\frac{(a_{j_1j_1}-a_{j_2j_1})^2+b_{j_2j_1}-b_{j_1j_1}}{(x_{j_1}-a_{j_1j_1})^2+b_{j_1j_1}}+(a_{j_1j_11}-a_{j_2j_11})\Big\{\frac{(x_{j_1}-a_{j_1j_1})}{(x_{j_1}-a_{j_1j_1})^2+b_{j_1j_1}} \\
&\quad+\frac{a_{j_1j_1}+a_{j_1j_11}-2a_{j_2j_1}}{(x_{j_1}-a_{j_1j_1})^2+b_{j_1j_1}}+\frac{(a_{j_1j_11}-a_{j_2j_1})^2+b_{j_2j_1}}{(a_{j_1j_11}-a_{j_1j_1})^2+b_{j_1j_1}}\Big[\frac{1}{x_{j_1}-a_{j_1j_11}} \\
&\quad-\frac{(x_{j_1}-a_{j_1j_1})}{(x_{j_1}-a_{j_1j_1})^2+b_{j_1j_1}}-\frac{a_{j_1j_11}-a_{j_1j_1}}{(x_{j_1}-a_{j_1j_1})^2+b_{j_1j_1}}\Big]\Big\}\Big).
\end{aligned}
$$

With an initial condition (x_{j_10}, x_{j_20}), the integration of the foregoing equation gives the first integral manifold as

$$
\begin{aligned}
x_{j_2}-x_{j_20} &= \frac{a_{j_2j_10}}{a_{j_1j_10}}\Big(x_{j_1}-x_{j_10}+(a_{j_1j_1}-a_{j_2j_1})\ln\frac{|(x_{j_1}-a_{j_1j_1})^2+b_{j_1j_1}|}{|(x_{j_10}-a_{j_1j_1})^2+b_{j_1j_1}|} \\
&\quad+\frac{(a_{j_1j_1}-a_{j_2j_1})^2+b_{j_2j_1}-b_{j_1j_1}}{\sqrt{b_{j_1j_1}}}(\arctan\frac{x_{j_1}-a_{j_1j_1}}{\sqrt{b_{j_1j_1}}}-\arctan\frac{x_{j_10}-a_{j_1j_1}}{\sqrt{b_{j_1j_1}}}) \\
&\quad+(a_{j_1j_11}-a_{j_2j_11})\Big\{\frac{1}{2}\ln\frac{|(x_{j_1}-a_{j_1j_1})^2+b_{j_1j_1}|}{|(x_{j_10}-a_{j_1j_1})^2+b_{j_1j_1}|} \\
&\quad+\frac{a_{j_1j_11}+a_{j_1j_1}-2a_{j_2j_1}}{\sqrt{b_{j_1j_1}}}(\arctan\frac{x_{j_1}-a_{j_1j_1}}{\sqrt{b_{j_1j_1}}}-\arctan\frac{x_{j_10}-a_{j_1j_1}}{\sqrt{b_{j_1j_1}}}) \\
&\quad+\frac{(a_{j_1j_11}-a_{j_2j_1})^2+b_{j_2j_1}}{(a_{j_1j_11}-a_{j_1j_1})^2+b_{j_1j_1}}\Big[\ln\frac{|x_{j_1}-a_{j_1j_11}|}{|x_{j_10}-a_{j_1j_11}|} \\
&\quad-\frac{1}{2}\ln\frac{|(x_{j_1}-a_{j_1j_1})^2+b_{j_1j_1}|}{|(x_{j_10}-a_{j_1j_1})^2+b_{j_1j_1}|} \\
&\quad-\frac{a_{j_1j_11}-a_{j_1j_1}}{\sqrt{b_{j_1j_1}}}(\arctan\frac{x_{j_1}-a_{j_1j_1}}{\sqrt{b_{j_1j_1}}}-\arctan\frac{x_{j_10}-a_{j_1j_1}}{\sqrt{b_{j_1j_1}}})\Big]\Big\}\Big), \\
&\quad j_1,j_2\in\{1,2\}, j_1\neq j_2.
\end{aligned}
$$

($\mathrm{i}_{1\mathrm{a}}$) For $x_{j_1}^* = a_{j_1 j_1 1} \neq a_{j_2 j_1 1}$, the variational equation is

$$\begin{aligned}\dot{x}_{j_2} &= a_{j_2 j_1 0}(a_{j_1 j_1 1} - a_{j_2 j_1 1})\left[(a_{j_1 j_1 1} - a_{j_2 j_1})^2 + b_{j_2 j_1}\right],\\ \Delta\dot{x}_{j_1} &= a_{j_1 j_1 0}\left[(a_{j_1 j_1 1} - a_{j_1 j_1})^2 + b_{j_1 j_1}\right]\Delta x_{j_1}.\end{aligned}$$

Due to

$$(a_{j_1 j_1 1} - a_{j_2 j_1})^2 + b_{j_2 j_1} > 0 \text{ and } (a_{j_1 j_1 1} - a_{j_1 j_1})^2 + b_{j_1 j_1} > 0,$$

the flows at $x_{j_1}^* = a_{j_1 j_1 1}$ are source and sink flows for $a_{j_1 j_1 0} > 0$ and $a_{j_1 j_1 0} < 0$ in the x_{j_1} direction, and the flows at $x_{j_1}^* = a_{j_1 j_1 1}$ in the x_{j_2}-direction are positive and negative for $a_{j_2 j_1 0}(a_{j_1 j_1 1} - a_{j_2 j_1 1}) > 0$ and $a_{j_2 j_1 0}(a_{j_1 j_1 1} - a_{j_2 j_1 1}) < 0$.

Therefore, the flow at $x_{j_1}^* = a_{j_1 j_1 1}$ for the system has the following properties as in Eqs. (4.6)–(4.9).

- For $a_{j_1 j_1 0} > 0$ and $a_{j_2 j_1 0}(a_{j_1 j_1 1} - a_{j_2 j_1 1}) > 0$,

$$(a_{j_1 j_1 1}, \dot{x}_{j_2}) = \underbrace{(\mathrm{SO}, \mathrm{pF})}_{\text{source flow }(+)}.$$

  The flow of $x_{j_1}^* = a_{j_1 j_1 1}$ is an (SO,pF)-positive source flow.
- For $a_{j_1 j_1 0} < 0$ and $a_{j_2 j_1 0}(a_{j_1 j_1 1} - a_{j_2 j_1 1}) > 0$,

$$(a_{j_1 j_1 1}, \dot{x}_{j_2}) = \underbrace{(\mathrm{SI}, \mathrm{pF})}_{\text{sink flow }(+)}.$$

  The flow of $x_{j_1}^* = a_{j_1 j_1 1}$ is an (SI, pF)-positive sink flow.
- For $a_{j_1 j_1 0} > 0$ and $a_{j_2 j_1 0}(a_{j_1 j_1 1} - a_{j_2 j_1 1}) < 0$,

$$(a_{j_1 j_1 1}, \dot{x}_{j_2}) = \underbrace{(\mathrm{SO}, \mathrm{nF})}_{\text{source flow }(-)}.$$

  The flow of $x_{j_1}^* = a_{j_1 j_1 1}$ is an (SO, nF)-negative source flow.
- For $a_{j_1 j_1 0} < 0$ and $a_{j_2 j_1 0}(a_{j_1 j_1 1} - a_{j_2 j_1 1}) < 0$,

$$(a_{j_1 j_1 1}, \dot{x}_{j_2}) = \underbrace{(\mathrm{SI}, \mathrm{nF})}_{\text{sink flow }(-)}.$$

  The flow of $x_{j_1}^* = a_{j_1 j_1 1}$ is an (SI,nF)-negative sink flow.

(i_{1b}) For $x_{j_1}^* = a_{j_2 j_1 1} \neq a_{j_1 j_1 1}$, we have

$$\frac{dx_{j_2}}{dx_{j_1}}\bigg|_{x_{j_1}^* = a_{j_2 j_1 1}} = \frac{a_{j_2 j_1 0}}{a_{j_1 j_1 0}} \frac{(x_{j_1} - a_{j_2 j_1 1})\left[(x_{j_1} - a_{j_2 j_1})^2 + b_{j_2 j_1}\right]}{(x_{j_1} - a_{j_1 j_1 1})\left[(x_{j_1} - a_{j_1 j_1})^2 + b_{j_1 j_1}\right]}\bigg|_{x_{j_1}^* = a_{j_2 j_1 1}} = 0.$$

If

$$\frac{d^2 x_{j_2}}{dx_{j_1}^2}\bigg|_{x_{j_1}^* = a_{j_2 j_1 1}} = \frac{a_{j_2 j_1 0}}{a_{j_1 j_1 0}} \frac{\left[(a_{j_2 j_1 1} - a_{j_2 j_1})^2 + b_{j_2 j_1}\right]}{(a_{j_2 j_1 1} - a_{j_1 j_1 1})\left[(a_{j_2 j_1 1} - a_{j_1 j_1})^2 + b_{j_1 j_1}\right]} > 0,$$

the flow at $x_{j_1}^* = a_{j_2 j_1 1} \neq a_{j_1 j_1 1}$ is an up-parabola flow, and if

$$\frac{d^2 x_{j_2}}{dx_{j_1}^2}\bigg|_{x_{j_1}^* = a_{j_2 j_1 1}} = \frac{a_{j_2 j_1 0}}{a_{j_1 j_1 0}} \frac{\left[(a_{j_2 j_1 1} - a_{j_2 j_1})^2 + b_{j_2 j_1}\right]}{(a_{j_2 j_1 1} - a_{j_1 j_1 1})\left[(a_{j_2 j_1 1} - a_{j_1 j_1})^2 + b_{j_1 j_1}\right]} < 0,$$

the flow at $x_{j_1}^* = a_{j_2 j_1 1} \neq a_{j_1 j_1 1}$ is a down-parabola flow.

Due to

$$(a_{j_1 j_1 1} - a_{j_2 j_1})^2 + b_{j_2 j_1} > 0 \text{ and } (a_{j_1 j_1 1} - a_{j_1 j_1})^2 + b_{j_1 j_1} > 0,$$

the flows at $x_{j_1}^* = a_{j_2 j_1 1} \neq a_{j_1 j_1 1}$ are positive and negative for $a_{j_1 j_1 0}(a_{j_2 j_1 1} - a_{j_1 j_1 1}) > 0$ and $a_{j_1 j_1 0}(a_{j_2 j_1 1} - a_{j_1 j_1 1}) < 0$ in the x_{j_1}-direction.

Therefore, the flow at $x_{j_1}^* = a_{j_2 j_1 1}$ for the system has the following properties as in Eqs. (4.10)–(4.13).

- For $a_{j_1 j_1 0}(a_{j_2 j_1 1} - a_{j_1 j_1 1}) > 0$ and $a_{j_2 j_1 0} > 0$,

$$(\dot{x}_{j_1}, a_{j_2 j_1 1}) = \underbrace{(\text{pF, UP})}_{\text{up-parabola flow } (+)}.$$

  The flow of $x_{j_1}^* = a_{j_2 j_1 1}$ is a (pF,UP)-positive up-parabola flow.

- For $a_{j_1 j_1 0}(a_{j_2 j_1 1} - a_{j_1 j_1 1}) < 0$ and $a_{j_2 j_1 0} > 0$,

$$(\dot{x}_{j_1}, a_{j_2 j_1 1}) = \underbrace{(\text{nF, DP})}_{\text{down-parabola flow } (-)}.$$

  The flow of $x_{j_1}^* = a_{j_2 j_1 1}$ is an (nF,DP)-negative down-parabola flow.

- For $a_{j_1 j_1 0}(a_{j_2 j_1 1} - a_{j_1 j_1 1}) > 0$ and $a_{j_2 j_1 0} < 0$,

$$(\dot{x}_{j_1}, a_{j_2 j_1 1}) = \underbrace{(\text{pF, DP})}_{\text{down-parabola flow } (+)}.$$

  The flow of $x_{j_1}^* = a_{j_2 j_1 1}$ is a (pF,DP)-positive down-parabola flow.

- For $a_{j_1j_10}(a_{j_2j_11} - a_{j_1j_11}) < 0$ and $a_{j_2j_10} < 0$,

$$(\dot{x}_{j_1}, a_{j_2j_11}) = \underbrace{(\mathrm{nF}, \mathrm{UP})}_{\text{up-parabola flow } (-)}.$$

The flow of $x_{j_1}^* = a_{j_2j_11}$ is an (nF,UP)-negative up-parabola flow.

(i_2) For $a_{j_2j_11} = a_{j_1j_11}$, in phase space,

$$\begin{aligned}\frac{dx_{j_2}}{dx_{j_1}} &= \frac{a_{j_2j_10}\,(x_{j_1} - a_{j_2j_1})^2 + b_{j_2j_1}}{a_{j_1j_10}\,(x_{j_1} - a_{j_1j_1})^2 + b_{j_1j_1}}\\ &= \frac{a_{j_2j_10}}{a_{j_1j_10}}\left[1 + (a_{j_1j_1} - a_{j_2j_1})\frac{2(x_{j_1} - a_{j_1j_1})}{(x_{j_1} - a_{j_1j_1})^2 + b_{j_1j_1}} + \frac{(a_{j_1j_1} - a_{j_2j_1})^2 + b_{j_2j_1} - b_{j_1j_1}}{(x_{j_1} - a_{j_1j_1})^2 + b_{j_1j_1}}\right].\end{aligned}$$

With an initial condition (x_{j_10}, x_{j_20}), the first integral manifold is

$$\begin{aligned}x_{j_2} - x_{j_20} = \frac{a_{j_2j_10}}{a_{j_1j_10}}&\left[x_{j_1} - x_{j_10} + (a_{j_1j_1} - a_{j_2j_1})\ln\frac{|\,(x_{j_1} - a_{j_1j_1})^2 + b_{j_1j_1}\,|}{|\,(x_{j_10} - a_{j_1j_1})^2 + b_{j_1j_1}\,|}\right.\\ &+\left[(a_{j_1j_1} - a_{j_2j_1})^2 + b_{j_2j_1} - b_{j_1j_1}\right]\\ &\left.\times\frac{1}{\sqrt{b_{j_1j_1}}}(\arctan\frac{x_{j_1} - a_{j_1j_1}}{\sqrt{b_{j_1j_1}}} - \arctan\frac{x_{j_10} - a_{j_1j_1}}{\sqrt{b_{j_1j_1}}})\right],\\ &j_1, j_2 \in \{1,2\}, j_1 \neq j_2.\end{aligned}$$

For $x_{j_1}^* = a_{j_2j_11} = a_{j_1j_11}$, if

$$\left.\frac{dx_{j_2}}{dx_{j_1}}\right|_{x_{j_1}^* = a_{j_2j_11} = a_{j_1j_11}} = \frac{a_{j_2j_10}}{a_{j_1j_10}}\,\frac{(a_{j_2j_11} - a_{j_2j_1})^2 + b_{j_2j_1}}{(a_{j_1j_11} - a_{j_1j_1})^2 + b_{j_1j_1}} > 0,$$

the flow at $x_{j_1}^* = a_{j_2j_11} = a_{j_1j_11}$ is an increasing-inflection flow, and if

$$\left.\frac{dx_{j_2}}{dx_{j_1}}\right|_{x_{j_1}^* = a_{j_2j_11} = a_{j_1j_11}} = \frac{a_{j_2j_10}}{a_{j_1j_10}}\,\frac{(a_{j_2j_11} - a_{j_2j_1})^2 + b_{j_2j_1}}{(a_{j_1j_11} - a_{j_1j_1})^2 + b_{j_1j_1}} < 0,$$

the flow at $x_{j_1}^* = a_{j_2j_11} = a_{j_1j_11}$ is a decreasing-inflection flow.

On the other hand, for $x_{j_1}^* = a_{j_2j_11} = a_{j_1j_11}$, the variational equations are

$$\begin{aligned}\Delta\dot{x}_{j_1} &= a_{j_1j_10}\left[(a_{j_1j_11} - a_{j_1j_1})^2 + b_{j_1j_1}\right]\Delta x_{j_1},\\ \Delta\dot{x}_{j_2} &= a_{j_2j_10}\left[(a_{j_2j_11} - a_{j_2j_1})^2 + b_{j_2j_1}\right]\Delta x_{j_1},\end{aligned}$$

and

$$(a_{j_1j_1 1} - a_{j_1 j_1})^2 + b_{j_1 j_1} > 0,$$
$$(a_{j_2j_1 1} - a_{j_2 j_1})^2 + b_{j_2 j_1} > 0.$$

The flows at $x^*_{j_1} = a_{j_2j_1 1} = a_{j_1j_1 1}$ are source and sink for $a_{j_1j_1 0} > 0$ and $a_{j_1j_1 0} < 0$, respectively.

Therefore, the equilibrium of $x^*_{j_1} = a_{j_2j_1 1} = a_{j_1j_1 1}$ with $\bar{x}_{j_2} \in (-\infty, \infty)$ for the system has the following properties as in Eqs. (4.15)–(4.18).

- For $a_{j_1j_1 0} > 0$ and $a_{j_2j_1 0} > 0$,

$$(a_{j_1j_1 1}, \bar{x}_{j_2}) = \underbrace{(\mathrm{SO}, \mathrm{II})}_{\text{increasing-inflection source}}.$$

  The equilibrium of $x^*_{j_1} = a_{j_1j_1 1} = a_{j_2j_1 1}$ is an (SO,II)-increasing-inflection source infinite-equilibrium.
- For $a_{j_1j_1 0} < 0$ and $a_{j_2j_1 0} > 0$,

$$(a_{j_1j_1 1}, \bar{x}_{j_2}) = \underbrace{(\mathrm{SI}, \mathrm{DI})}_{\text{decreasing-inflection sink}}.$$

  The equilibrium of $x^*_{j_1} = a_{j_1j_1 1} = a_{j_2j_1 1}$ is an (SI,DI)-decreasing-inflection sink infinite-equilibrium.
- For $a_{j_1j_1 0} > 0$ and $a_{j_2j_1 0} < 0$,

$$(a_{j_1j_1 1}, \bar{x}_{j_2}) = \underbrace{(\mathrm{SO}, \mathrm{DI})}_{\text{decreasing-inflection source}}.$$

  The equilibrium of $x^*_{j_1} = a_{j_1j_1 1} = a_{j_2j_1 1}$ is an (SO,DI)-decreasing-inflection source infinite-equilibrium.
- For $a_{j_1j_1 0} < 0$ and $a_{j_2j_1 0} < 0$,

$$(a_{j_1j_1 1}, \bar{x}_{j_2}) = \underbrace{(\mathrm{SI}, \mathrm{II})}_{\text{increasing-inflection sink}}.$$

  The equilibrium at $x^*_{j_1} = a_{j_1j_1 1} = a_{j_2j_1 1}$ is an (SI,II)-increasing-inflection sink infinite-equilibrium.

– The inflection infinite-equilibriums are for the switching bifurcation of a parabola flow with a source (sink) flow.

(ii) If

$$\Delta_{j_1j_1} = 0 \text{ and } \Delta_{j_2j_1} < 0,$$

the dynamical system becomes

$$\begin{aligned}
\dot{x}_{j_1} &= a_{j_1j_10}(x_{j_1} - a_{j_1j_1s_1})(x_{j_1} - a_{j_1j_1s_2})^2, \\
\dot{x}_{j_2} &= a_{j_2j_10}(x_{j_1} - a_{j_2j_11})\big[(x_{j_1} - a_{j_2j_1})^2 + b_{j_2j_1}\big],
\end{aligned}$$

with

$$\begin{aligned}
&\{a_{j_1j_11}, a_{j_1j_12}\} = \text{sort}\{b_{j_1j_11}, a_{j_1j_1}\}, a_{j_1j_11} < a_{j_1j_12}; \\
&a_{j_2j_11} = b_{j_2j_11}; \\
&a_{j_1j_1} = -\frac{1}{2}B_{j_1j_11}, b_{j_1j_1} = \frac{1}{4}(-\Delta_{j_1j_1}) = 0; \\
&a_{j_2j_1} = -\frac{1}{2}B_{j_2j_11}, b_{j_2j_1} = \frac{1}{4}(-\Delta_{j_2j_1}) \neq 0; \\
&j_1, j_2 \in \{1,2\}; j_1 \neq j_2.
\end{aligned}$$

In phase space, we have

$$\begin{aligned}
\frac{dx_{j_2}}{dx_{j_1}} &= \frac{a_{j_2j_10}}{a_{j_1j_10}} \frac{(x_{j_1} - a_{j_2j_11})\big[(x_{j_1} - a_{j_2j_1})^2 + b_{j_2j_1}\big]}{(x_{j_1} - a_{j_1j_1s_1})(x_{j_1} - a_{j_1j_1s_2})^2} \\
&= \frac{a_{j_2j_10}}{a_{j_1j_10}} \Big\{1 + \frac{2(a_{j_1j_1s_2} - a_{j_2j_1})}{x_{j_1} - a_{j_1j_1s_2}} + \frac{(a_{j_1j_1s_2} - a_{j_2j_1})^2 + b_{j_2j_1}}{(x_{j_1} - a_{j_1j_1s_2})^2} \\
&\quad + (a_{j_1j_1s_1} - a_{j_2j_11})\Big[\frac{(x_{j_1} - a_{j_1j_1s_2}) + (a_{j_1j_1s_2} - a_{j_1j_1s_1})}{(x_{j_1} - a_{j_1j_1s_2})^2} \\
&\quad + \frac{2(a_{j_1j_1s_1} - a_{j_2j_1})}{(x_{j_1} - a_{j_1j_1s_2})^2} + \frac{(a_{j_1j_1s_1} - a_{j_2j_1})^2 + b_{j_2j_1}}{(x_{j_1} - a_{j_1j_1s_1})(x_{j_1} - a_{j_1j_1s_2})^2}\Big]\Big\} \\
&= \frac{a_{j_2j_10}}{a_{j_1j_10}} \Big(\Big[1 + \frac{2(a_{j_1j_1s_2} - a_{j_2j_1})}{x_{j_1} - a_{j_1j_1s_2}} + \frac{(a_{j_1j_1s_2} - a_{j_2j_1})^2 + b_{j_2j_1}}{(x_{j_1} - a_{j_1j_1s_2})^2}\Big] \\
&\quad + (a_{j_1j_1s_1} - a_{j_2j_11})\Big\{\frac{(x_{j_1} - a_{j_1j_1s_2})}{(x_{j_1} - a_{j_1j_1s_2})^2} + \frac{a_{j_1j_1} + a_{j_1j_1s_1} - 2a_{j_2j_1}}{(x_{j_1} - a_{j_1j_1s_2})^2} \\
&\quad + \frac{(a_{j_1j_1s_1} - a_{j_2j_1})^2 + b_{j_2j_1}}{(a_{j_1j_1s_1} - a_{j_1j_1s_2})^2}\Big[\frac{1}{x_{j_1} - a_{j_1j_1s_1}} - \frac{(x_{j_1} - a_{j_1j_1s_2})}{(x_{j_1} - a_{j_1j_1s_2})^2} + \frac{a_{j_1j_1s_2} - a_{j_1j_1s_1}}{(x_{j_1} - a_{j_1j_1s_2})^2}\Big]\Big\}\Big).
\end{aligned}$$

With an initial condition (x_{j_10}, x_{j_20}), the first integral manifold is

$$
\begin{aligned}
x_{j_2}-x_{j_2 0}=&\frac{a_{j_2 j_1 0}}{a_{j_1 j_1 0}}\Big(\Big\{x_{j_1}-x_{j_1 0}+2(a_{j_1 j_1 s_2}-a_{j_2 j_1})\ln\frac{|x_{j_1}-a_{j_1 j_1 s_2}|}{|x_{j_1 0}-a_{j_1 j_1 s_2}|}\\
&-\big[(a_{j_1 j_1 s_2}-a_{j_2 j_1})^2+b_{j_2 j_1}\big](\frac{1}{x_{j_1}-a_{j_1 j_1 s_2}}-\frac{1}{x_{j_1 0}-a_{j_1 j_1 s_2}})\Big\}\\
&+(a_{j_1 j_1 s_1}-a_{j_2 j_1 1})\Big\{\ln\frac{|x_{j_1}-a_{j_1 j_1 s_2}|}{|x_{j_1 0}-a_{j_1 j_1 s_2}|}\\
&-(a_{j_1 j_1}+a_{j_1 j_1 s_1}-2a_{j_2 j_1})(\frac{1}{x_{j_1}-a_{j_1 j_1 s_2}}-\frac{1}{x_{j_1 0}-a_{j_1 j_1 s_2}})\\
&+\frac{(a_{j_1 j_1 s_1}-a_{j_2 j_1})^2+b_{j_2 j_1}}{(a_{j_1 j_1 s_1}-a_{j_1 j_1 s_2})^2}\Big[\ln\frac{|x_{j_1}-a_{j_1 j_1 s_1}|}{|x_{j_1 0}-a_{j_1 j_1 s_1}|}-\ln\frac{|x_{j_1}-a_{j_1 j_1 s_2}|}{|x_{j_1 0}-a_{j_1 j_1 s_2}|}\\
&-(a_{j_1 j_1 s_2}-a_{j_1 j_1 s_1})(\frac{1}{x_{j_1}-a_{j_1 j_1 s_2}}-\frac{1}{x_{j_1 0}-a_{j_1 j_1 s_2}})\Big]\Big\}\Big).
\end{aligned}
$$

(ii$_{1a}$) For $x_{j_1}^* = a_{j_1 j_1 s_1}$, the variational equation is

$$
\begin{aligned}
\dot{x}_{j_2} &= a_{j_2 j_1 0}(a_{j_1 j_1 s_1}-a_{j_2 j_1 1})\big[(a_{j_1 j_1 s_1}-a_{j_2 j_1})^2+b_{j_2 j_1}\big],\\
\Delta\dot{x}_{j_1} &= a_{j_1 j_1 0}(a_{j_1 j_1 s_1}-a_{j_1 j_1 s_2})^2\Delta x_{j_1}.
\end{aligned}
$$

For $a_{j_1 j_1 s_1} \neq a_{j_1 j_1 s_2}$, due to

$$
(a_{j_1 j_1 s_1}-a_{j_2 j_1})^2+b_{j_2 j_1}>0 \text{ and } (a_{j_1 j_1 s_1}-a_{j_1 j_1 s_2})^2>0,
$$

the flows at $x_{j_1}^* = a_{j_1 j_1 s_1}$ are source and sink flows for $a_{j_1 j_1 0}>0$ and $a_{j_1 j_1 0}<0$ in the x_{j_1}-direction, and the flows at $x_{j_1}^* = a_{j_1 j_1 s_1}$ in the x_{j_2}-direction are positive and negative for $a_{j_2 j_1 0}(a_{j_1 j_1 s_1}-a_{j_2 j_1 1})>0$ and $a_{j_2 j_1 0}(a_{j_1 j_1 s_1}-a_{j_2 j_1 1})<0$.

Therefore, the equilibrium of $x_{j_1}^* = a_{j_1 j_1 s_1}$ for the x_{j_1}-direction has the following properties as in Eqs. (4.22)–(4.25).

- For $a_{j_1 j_1 0}(a_{j_1 j_1 s_1}-a_{j_1 j_1 s_2})^2>0$ and $a_{j_2 j_1 0}(a_{j_1 j_1 s_1}-a_{j_2 j_1 1})>0$,

$$
(a_{j_1 j_1 s_1}, \dot{x}_{j_2}) = \underbrace{(\text{SO}, \text{pF})}_{\text{source flow }(+)}.
$$

  The flow at $x_{j_1}^* = a_{j_1 j_1 s_1}$ is an (SO,pF)-positive source flow.
- For $a_{j_1 j_1 0}(a_{j_1 j_1 s_1}-a_{j_1 j_1 s_2})^2<0$ and $a_{j_2 j_1 0}(a_{j_1 j_1 s_1}-a_{j_2 j_1 1})>0$,

$$
(a_{j_1 j_1 s_1}, \dot{x}_{j_2}) = \underbrace{(\text{SI}, \text{pF})}_{\text{sink flow }(+)}.
$$

  The flow at $x_{j_1}^* = a_{j_1 j_1 s_1}$ is an (SI,pF)-positive sink flow.

- For $a_{j_1j_10}(a_{j_1j_1s_1} - a_{j_1j_1s_2})^2 < 0$ and $a_{j_2j_10}(a_{j_1j_1s_1} - a_{j_2j_11}) < 0$,

$$(a_{j_1j_1s_1}, \dot{x}_{j_2}) = \underbrace{(\mathrm{SO}, \mathrm{nF})}_{\text{source flow }(-)}.$$

  The flow at $x^*_{j_1} = a_{j_1j_1s_1}$ is an (SO,nF)-negative source flow.
- For $a_{j_1j_10}(a_{j_1j_1s_1} - a_{j_1j_1s_2})^2 < 0$ and $a_{j_2j_10}(a_{j_1j_1s_1} - a_{j_2j_11}) < 0$,

$$(a_{j_1j_1s_1}, \dot{x}_{j_2}) = \underbrace{(\mathrm{SI}, \mathrm{nF})}_{\text{sink flow }(-)}.$$

  The flow at $x^*_{j_1} = a_{j_1j_1s_1}$ is an (SI,nF)-negative sink flow.

(ii$_{1b}$) For $x^*_{j_1} = a_{j_1j_1s_2}$, the variational equation is

$$\begin{aligned}\dot{x}_{j_2} &= a_{j_2j_10}(a_{j_1j_1s_2} - a_{j_2j_11})\left[(a_{j_1j_1s_2} - a_{j_2j_1})^2 + b_{j_2j_1}\right],\\ \Delta\dot{x}_{j_1} &= a_{j_1j_10}(a_{j_1j_1s_2} - a_{j_1j_1s_1})(\Delta x_{j_1})^2.\end{aligned}$$

For $a_{j_1j_1s_1} \neq a_{j_1j_1s_2}$, due to

$$(a_{j_1j_1s_2} - a_{j_2j_1})^2 + b_{j_2j_1} > 0,$$

the flows at $x^*_{j_1} = a_{j_1j_1s_2}$ in the x_{j_1}-direction are upper-saddle and lower-saddle flows for $a_{j_1j_10}(a_{j_1j_1s_2} - a_{j_1j_1s_1}) > 0$ and $a_{j_1j_10}(a_{j_1j_1s_2} - a_{j_1j_1s_1}) < 0$, and the flows at $x^*_{j_1} = a_{j_1j_1s_2}$ are positive and negative for $a_{j_2j_10} > 0$ and $a_{j_2j_10} < 0$ in the x_{j_2}-direction.

Therefore, the equilibrium of $x^*_{j_1} = a_{j_1j_1s_2}$ for the x_{j_1}-direction has the following properties as in Eqs. (4.26)–(4.29):

- For $a_{j_1j_10}(a_{j_1j_1s_2} - a_{j_1j_1s_1}) > 0$ and $a_{j_2j_10}(a_{j_1j_1s_2} - a_{j_2j_11}) > 0$,

$$(a_{j_1j_1s_2}, \dot{x}_{j_2}) = \underbrace{(\mathrm{US}, \mathrm{pF})}_{\text{up-parabola flow }(+)}.$$

  The flow at $x^*_{j_1} = a_{j_1j_1s_2}$ is a (US,pF)-positive upper-saddle flow.

- For $a_{j_1j_1 0}(a_{j_1j_1s_2} - a_{j_1j_1s_1}) < 0$ and $a_{j_2j_1 0}(a_{j_1j_1s_2} - a_{j_2j_1 1}) > 0$,

$$(a_{j_1j_1s_2}, \dot{x}_{j_2}) = \underbrace{(\text{LS, pF})}_{\text{lower-saddle flow } (+)}.$$

The flow at $x_{j_1}^* = a_{j_1j_1s_2}$ is an (LS,pF)-positive lower-saddle flow.

- For $a_{j_1j_1 0}(a_{j_1j_1s_2} - a_{j_1j_1s_1}) > 0$ and $a_{j_2j_1 0}(a_{j_1j_1s_2} - a_{j_2j_1 1}) < 0$,

$$(a_{j_1j_1s_2}, \dot{x}_{j_2}) = \underbrace{(\text{US, nF})}_{\text{upper-saddle flow } (-)}.$$

The flow at $x_{j_1}^* = a_{j_1j_1s_2}$ is a (US,nF)-negative upper-saddle flow.

- For $a_{j_1j_1 0}(a_{j_1j_1s_2} - a_{j_1j_1s_1}) < 0$ and $a_{j_2j_1 0}(a_{j_1j_1s_2} - a_{j_2j_1 1}) < 0$,

$$(a_{j_1j_1s_2}, \dot{x}_{j_2}) = \underbrace{(\text{LS, nF})}_{\text{lower-saddle flow } (-)}.$$

The flow at $x_{j_1}^* = a_{j_1j_1s_2}$ is an (LS,nF)-negative lower-saddle flow.

– The upper-saddle and lower-saddle flow bifurcations are for the sink and source flows.

($\text{ii}_{1\text{c}}$) At $x_{j_1}^* = a_{j_2j_1 1} \neq a_{j_1j_1s_1}, a_{j_1j_1s_2}$, we have

$$\frac{dx_{j_2}}{dx_{j_1}}\bigg|_{x_{j_1}^* = a_{j_2j_1 1}} = \frac{a_{j_2j_1 0}}{a_{j_1j_1 0}} \frac{(x_{j_1} - a_{j_2j_1 1})\left[(x_{j_1} - a_{j_2j_1})^2 + b_{j_2j_1}\right]}{(x_{j_1} - a_{j_1j_1s_1})(x_{j_1} - a_{j_1j_1s_2})^2}\bigg|_{x_{j_1}^* = a_{j_2j_1 1}} = 0.$$

If

$$\frac{d^2x_{j_2}}{dx_{j_1}^2}\bigg|_{x_{j_1}^* = a_{j_2j_1 1}} = \frac{a_{j_2j_1 0}}{a_{j_1j_1 0}} \frac{(a_{j_2j_1 1} - a_{j_2j_1})^2 + b_{j_2j_1}}{(a_{j_2j_1 1} - a_{j_1j_1s_1})(a_{j_2j_1 1} - a_{j_1j_1s_2})^2} > 0,$$

the flow at $x_{j_1}^* = a_{j_2j_1 1}$ is an up-parabola flow, and if

$$\frac{d^2x_{j_2}}{dx_{j_1}^2}\bigg|_{x_{j_1}^* = a_{j_2j_1 1}} = \frac{a_{j_2j_1 0}}{a_{j_1j_1 0}} \frac{(a_{j_2j_1 1} - a_{j_2j_1})^2 + b_{j_2j_1}}{(a_{j_2j_1 1} - a_{j_1j_1s_1})(a_{j_2j_1 1} - a_{j_1j_1s_2})^2} < 0,$$

the flow at $x_{j_1}^* = a_{j_2j_1 1}$ is a down-parabola flow.

On the other hand, in vicinity of $x_{j_1}^* = a_{j_2j_1 1}$, the variational equation is

$$\Delta\dot{x}_{j_2} = a_{j_2j_1 0}\left[(a_{j_2j_1 1} - a_{j_2j_1})^2 + b_{j_2j_1}\right]\Delta x_{j_1},$$
$$\dot{x}_{j_1} = a_{j_1j_1 0}(a_{j_2j_1 1} - a_{j_1j_1s_1})(a_{j_2j_1 1} - a_{j_1j_1s_2})^2.$$

The flows at $x^*_{j_1} = a_{j_2 j_1 1}$ are positive and negative in the x_{j_1}-direction for $a_{j_1 j_1 0}(a_{j_2 j_1 1} - a_{j_1 j_1 s_1}) > 0$ and $a_{j_1 j_1 0}(a_{j_2 j_1 1} - a_{j_1 j_1 s_1}) < 0$, respectively.

Therefore, the flow at $x^*_{j_1} = a_{j_2 j_1 1}$ for the system has the following properties as in Eqs. (4.30)–(4.33).

- For $a_{j_1 j_1 0}(a_{j_2 j_1 1} - a_{j_1 j_1 s_1})(a_{j_2 j_1 1} - a_{j_1 j_1 s_2})^2 > 0$ and $a_{j_2 j_1 0} > 0$,

$$(\dot{x}_{j_1}, a_{j_2 j_1 1}) = \underbrace{(\mathrm{pF}, \mathrm{UP})}_{\text{up-parabola flow } (+)}.$$

  The flow of $x^*_{j_1} = a_{j_2 j_1 1}$ is a (pF,UP)-positive up-parabola flow.
- For $a_{j_1 j_1 0}(a_{j_2 j_1 1} - a_{j_1 j_1 s_1})(a_{j_2 j_1 1} - a_{j_1 j_1 s_2})^2 < 0$ and $a_{j_2 j_1 0} > 0$,

$$(\dot{x}_{j_1}, a_{j_2 j_1 1}) = \underbrace{(\mathrm{nF}, \mathrm{DP})}_{\text{down-parabola flow } (-)}.$$

  The flow of $x^*_{j_1} = a_{j_2 j_1 1}$ is an (nF,DP)-negative down-parabola flow.
- For $a_{j_1 j_1 0}(a_{j_2 j_1 1} - a_{j_1 j_1 s_1})(a_{j_2 j_1 1} - a_{j_1 j_1 s_2})^2 > 0$ and $a_{j_2 j_1 0} < 0$,

$$(\dot{x}_{j_1}, a_{j_2 j_1 1}) = \underbrace{(\mathrm{pF}, \mathrm{DP})}_{\text{down-parabola flow } (+)}.$$

  The flow of $x^*_{j_1} = a_{j_2 j_1 1}$ is a (pF,DP)-positive down-parabola flow.
- For $a_{j_1 j_1 0}(a_{j_2 j_1 1} - a_{j_1 j_1 s_1})(a_{j_2 j_1 1} - a_{j_1 j_1 s_2})^2 < 0$ and $a_{j_2 j_1 0} < 0$,

$$(\dot{x}_{j_1}, a_{j_2 j_1 1}) = \underbrace{(\mathrm{nF}, \mathrm{UP})}_{\text{up-parabola flow } (-)}.$$

  The flow of $x^*_{j_1} = a_{j_2 j_1 1}$ is an (nF,UP)-negative up-parabola flow.

(ii$_2$) For $a_{j_1 j_1 s_1} = a_{j_2 j_1 1}$, in phase space, we have

$$\begin{aligned}
\frac{dx_{j_2}}{dx_{j_1}} &= \frac{a_{j_2 j_1 0}}{a_{j_1 j_1 0}} \frac{(x_{j_1} - a_{j_2 j_1})^2 + b_{j_2 j_1}}{(x_{j_1} - a_{j_1 j_1 s_2})^2} \\
&= \frac{a_{j_2 j_1 0}}{a_{j_1 j_1 0}} \left[1 + \frac{2(a_{j_1 j_1 s_2} - a_{j_2 j_1})}{x_{j_1} - a_{j_1 j_1 s_2}} + \frac{(a_{j_1 j_1 s_2} - a_{j_2 j_1})^2 + b_{j_2 j_1}}{(x_{j_1} - a_{j_1 j_1 s_2})^2}\right].
\end{aligned}$$

With an initial condition $(x_{j_1 0}, x_{j_2 0})$, the first integral manifold is

$$x_{j_2} - x_{j_20} = \frac{a_{j_2j_10}}{a_{j_1j_10}} \Big\{ x_{j_1} - x_{j_10} + 2(a_{j_1j_1s_2} - a_{j_2j_1}) \ln \frac{|\, x_{j_1} - a_{j_1j_1s_2} \,|}{|\, x_{j_10} - a_{j_1j_1s_2} \,|}$$
$$- \big[(a_{j_1j_1s_2} - a_{j_2j_1})^2 + b_{j_2j_1}\big]\big(\frac{1}{x_{j_1} - a_{j_1j_1s_2}} - \frac{1}{x_{j_10} - a_{j_1j_1s_2}}\big) \Big\}.$$

For $x_{j_1}^* = a_{j_1j_1s_1} = a_{j_2j_11} \neq a_{j_1j_1s_2}$, in phase space, if

$$\frac{dx_{j_2}}{dx_{j_1}}\Big|_{x_{j_1}^* = a_{j_1j_1s_1}} = \frac{a_{j_2j_10}}{a_{j_1j_10}} \frac{(a_{j_1j_1s_1} - a_{j_2j_1})^2 + b_{j_2j_1}}{(a_{j_1j_1s_1} - a_{j_1j_1s_2})^2} > 0,$$

there is an increasing-inflection flow, and if

$$\frac{dx_{j_2}}{dx_{j_1}}\Big|_{x_{j_1}^* = a_{j_1j_1s_1}} = \frac{a_{j_2j_10}}{a_{j_1j_10}} \frac{(a_{j_1j_1s_1} - a_{j_2j_1})^2 + b_{j_2j_1}}{(a_{j_1j_1s_1} - a_{j_1j_1s_2})^2} < 0,$$

there is a decreasing-inflection flow.

For $x_{j_1}^* = a_{j_1j_1s_1}$, the variational equation is

$$\Delta\dot{x}_{j_2} = a_{j_2j_10}\big[(a_{j_1j_1s_1} - a_{j_2j_1})^2 + b_{j_2j_1}\big]\Delta x_{j_1},$$
$$\Delta\dot{x}_{j_1} = a_{j_1j_10}(a_{j_1j_1s_1} - a_{j_1j_1s_2})^2 \Delta x_{j_1}.$$

For $a_{j_1j_1s_1} \neq a_{j_1j_1s_2}$, due to

$$(a_{j_1j_1s_1} - a_{j_2j_1})^2 + b_{j_2j_1} > 0 \text{ and } (a_{j_1j_1s_1} - a_{j_1j_1s_2})^2 > 0,$$

the inflection flows at $x_{j_1}^* = a_{j_1j_1s_1}$ are source and sink flows for $a_{j_1j_10} > 0$ and $a_{j_1j_10} < 0$ in the x_{j_1}-direction.

Therefore, the infinite-equilibrium of $x_{j_1}^* = a_{j_1j_1s_1} = a_{j_2j_11}$ with $\bar{x}_{j_2} \in (-\infty, \infty)$ for the x_{j_1}-direction has the following properties as in Eqs. (4.35)–(4.38).

- For $a_{j_1j_10}(a_{j_1j_1s_1} - a_{j_1j_1s_2})^2 > 0$ and $a_{j_2j_10} > 0$,

$$(a_{j_1j_1s_1}, \bar{x}_{j_2}) = \underbrace{(\text{SO, II})}_{\text{increasing-infelction source}}.$$

  The infinite-equilibrium of $x_{j_1}^* = a_{j_1j_1s_1} = a_{j_2j_11}$ is an (SO,II)-increasing-inflection source.

- For $a_{j_1j_10}(a_{j_1j_1s_1} - a_{j_1j_1s_2})^2 < 0$ and $a_{j_2j_10} > 0$,

$$(a_{j_1j_1s_1}, \overline{x}_{j_2}) = \underbrace{(\mathrm{SI}, \mathrm{DI})}_{\text{decreasing-infelction sink}}.$$

The infinite-equilibrium of $x^*_{j_1} = a_{j_1j_1s_1} = a_{j_2j_11}$ is an (SI,DI)-decreasing-inflection sink.

- For $a_{j_1j_10}(a_{j_1j_1s_1} - a_{j_1j_1s_2})^2 > 0$ and $a_{j_2j_10} < 0$,

$$(a_{j_1j_1s_1}, \overline{x}_{j_2}) = \underbrace{(\mathrm{SO}, \mathrm{DI})}_{\text{decreasing-infelction source}}.$$

The infinite-equilibrium of $x^*_{j_1} = a_{j_1j_1s_1} = a_{j_2j_11}$ is an (SO,DI)-decreasing-inflection source.

- For $a_{j_1j_10}(a_{j_1j_1s_1} - a_{j_1j_1s_2})^2 < 0$ and $a_{j_2j_10} < 0$,

$$(a_{j_1j_1s_1}, \overline{x}_{j_2}) = \underbrace{(\mathrm{SI}, \mathrm{II})}_{\text{increasing-infelction sink}}.$$

The infinite-equilibrium of $x^*_{j_1} = a_{j_1j_1s_1} = a_{j_2j_11}$ is an (SI,II)-increasing-inflection sink.

– The inflection source and sink are the switching bifurcations of sink and source flows with parabola flows.

(ii_{1b}) For $a_{j_1j_1s_2} = a_{j_2j_11}$, in phase space, we have

$$\begin{aligned}
\frac{dx_{j_2}}{dx_{j_1}} &= \frac{a_{j_2j_10}}{a_{j_1j_10}} \frac{(x_{j_1} - a_{j_2j_1})^2 + b_{j_2j_1}}{(x_{j_1} - a_{j_1j_1s_1})(x_{j_1} - a_{j_1j_1s_2})} \\
&= \frac{a_{j_2j_10}}{a_{j_1j_10}} \left[1 + \frac{(a_{j_1j_1s_1} - a_{j_2j_1})^2 + b_{j_2j_1}}{(a_{j_1j_1s_1} - a_{j_1j_1s_2})(x_{j_1} - a_{j_1j_1s_1})} \right. \\
&\quad \left. + \frac{(a_{j_1j_1s_2} - a_{j_2j_1})^2 + b_{j_2j_1}}{(a_{j_1j_1s_2} - a_{j_1j_1s_1})(x_{j_1} - a_{j_1j_1s_2})} \right].
\end{aligned}$$

With an initial condition (x_{j_10}, x_{j_20}), the first integral manifold is

$$\begin{aligned}
x_{j_2} - x_{j_20} = \frac{a_{j_2j_10}}{a_{j_1j_10}} &\left[x_{j_1} - x_{j_10} + \frac{(a_{j_1j_1s_1} - a_{j_2j_1})^2 + b_{j_2j_1}}{(a_{j_1j_1s_1} - a_{j_1j_1s_2})} \ln \frac{|x_{j_1} - a_{j_1j_1s_1}|}{|x_{j_10} - a_{j_1j_1s_1}|} \right. \\
&\left. + \frac{(a_{j_1j_1s_2} - a_{j_2j_1})^2 + b_{j_2j_1}}{(a_{j_1j_1s_2} - a_{j_1j_1s_1})} \ln \frac{|x_{j_1} - a_{j_1j_1s_2}|}{|x_{j_10} - a_{j_1j_1s_2}|} \right].
\end{aligned}$$

For $a_{j_1j_1s_2}=a_{j_2j_1 1}$, in phase space, $x_{j_1}=a_{j_1j_1s_2}\pm\varepsilon$,

$$\frac{dx_{j_2}}{dx_{j_1}}\bigg|_{x_{j_1}=a_{j_1j_1s_2}\pm\varepsilon}=\pm\frac{a_{j_2j_10}}{a_{j_1j_10}}\frac{(a_{j_1j_1s_2}-a_{j_2j_1})^2+b_{j_2j_1}}{(a_{j_1j_1s_2}-a_{j_1j_1s_1})\varepsilon}$$

If

$$\frac{a_{j_2j_10}}{a_{j_1j_10}}\frac{(a_{j_1j_1s_2}-a_{j_2j_1})^2+b_{j_2j_1}}{(a_{j_1j_1s_2}-a_{j_1j_1s_1})}>0,$$

there is a down-up asymptotic flow, and if

$$\frac{a_{j_2j_10}}{a_{j_1j_10}}\frac{(a_{j_1j_1s_2}-a_{j_2j_1})^2+b_{j_2j_1}}{(a_{j_1j_1s_2}-a_{j_1j_1s_1})}<0,$$

there is an up-down asymptotic flow.

For $x^*_{j_1}=a_{j_1j_1s_2}=a_{j_2j_1 1}$, the variational equation is

$$\begin{aligned}\Delta\dot{x}_{j_2}&=a_{j_2j_10}\left[(a_{j_1j_1s_2}-a_{j_2j_1})^2+b_{j_2j_1}\right]\Delta x_{j_1},\\ \Delta\dot{x}_{j_1}&=a_{j_1j_10}(a_{j_1j_1s_2}-a_{j_1j_1s_1})(\Delta x_{j_1})^2.\end{aligned}$$

Due to

$$(a_{j_1j_1s_2}-a_{j_2j_1})^2+b_{j_2j_1}>0.$$

The up-down and down-up asymptotic flows at $x^*_{j_1}=a_{j_1j_1s_2}=a_{j_2j_1 1}$ are upper-saddle and lower-saddle flows in the x_{j_1}-direction for $a_{j_1j_10}(a_{j_1j_1s_2}-a_{j_1j_1s_1})>0$ and $a_{j_1j_10}(a_{j_1j_1s_2}-a_{j_1j_1s_1})<0$.

Therefore, the finite-equilibrium of $x^*_{j_1}=a_{j_1j_1s_2}=a_{j_2j_1 1}$ with $\overline{x}_{j_2}\in(-\infty,\infty)$ for the x_{j_1}-direction has the following properties as in Eqs. (4.40)–(4.43).

- For $a_{j_1j_10}(a_{j_1j_1s_2}-a_{j_1j_1s_1})>0$ and $a_{j_2j_10}>0$,

$$(a_{j_1j_1s_2},\overline{x}_{j_2})=\underbrace{(\text{US},\text{DU})}_{\text{down-up upper-saddle}}.$$

  The finite-equilibrium of $x^*_{j_1}=a_{j_1j_1s_2}=a_{j_2j_1 1}$ is a (US,DU)-down-up upper-saddle flow.

- For $a_{j_1j_10}(a_{j_1j_1s_2}-a_{j_1j_1s_1})<0$ and $a_{j_2j_10}>0$,

$$(a_{j_1j_1s_2}, \overline{x}_{j_2}) = \underbrace{(\text{LS}, \text{UD})}_{\text{up-down lower-saddle}}.$$

The finite-equilibrium of $x^*_{j_1} = a_{j_1j_1s_2} = a_{j_2j_1 1}$ is an (LS,UD)-up-down lower-saddle flow.

- For $a_{j_1j_1 0}(a_{j_1j_1s_2} - a_{j_1j_1s_1}) > 0$ and $a_{j_2j_1 0} < 0$,

$$(a_{j_1j_1s_2}, \overline{x}_{j_2}) = \underbrace{(\text{US}, \text{UD})}_{\text{up-down upper-saddle}}.$$

The finite-equilibrium of $x^*_{j_1} = a_{j_1j_1s_2} = a_{j_2j_1 1}$ is an (US,UD)-up-down upper-saddle flow.

- For $a_{j_1j_1 0}(a_{j_1j_1s_2} - a_{j_1j_1s_1}) < 0$ and $a_{j_2j_1 0} < 0$,

$$(a_{j_1j_1s_2}, \overline{x}_{j_2}) = \underbrace{(\text{LS}, \text{DU})}_{\text{down-up lower-saddle}}.$$

The finite-equilibrium of $x^*_{j_1} = a_{j_1j_1s_2} = a_{j_2j_1 1}$ is an (LS,DU)-down-up upper-saddle flow.

– The up-down and down-up saddle bifurcations are for switching bifurcations of saddle flow with parabola flows.

(iii) For $a_{j_1j_1s_2} = a_{j_1j_1s_1} \equiv a_{j_1j_1 1}$,

$$\begin{aligned}
\dot{x}_{j_1} &= a_{j_1j_1 0}(x_{j_1} - a_{j_1j_1 1})^3, \\
\dot{x}_{j_2} &= a_{j_2j_1 0}(x_{j_1} - a_{j_2j_1 1})\left[(x_{j_1} - a_{j_2j_1})^2 + b_{j_2j_1}\right].
\end{aligned}$$

In phase space, we have

$$\begin{aligned}
\frac{dx_{j_2}}{dx_{j_1}} &= \frac{a_{j_2j_1 0}}{a_{j_1j_1 0}} \frac{(x_{j_1} - a_{j_2j_1 1})\left[(x_{j_1} - a_{j_2j_1})^2 + b_{j_2j_1}\right]}{(x_{j_1} - a_{j_1j_1 1})^3} \\
&= \frac{a_{j_2j_1 0}}{a_{j_1j_1 0}} \left\{ 1 + \frac{2(a_{j_1j_1 1} - a_{j_2j_1})}{(x_{j_1} - a_{j_1j_1 1})} + \frac{(a_{j_1j_1 1} - a_{j_2j_1})^2 + b_{j_2j_1}}{(x_{j_1} - a_{j_1j_1 1})^2} \right. \\
&\quad \left. + (a_{j_1j_1 1} - a_{j_2j_1 1}) \left[\frac{1}{(x_{j_1} - a_{j_1j_1 1})} + \frac{2(a_{j_1j_1 1} - a_{j_2j_1})}{(x_{j_1} - a_{j_1j_1 1})^2} + \frac{(a_{j_1j_1 1} - a_{j_2j_1})^2 + b_{j_2j_1}}{(x_{j_1} - a_{j_1j_1 1})^3} \right] \right\}.
\end{aligned}$$

With an initial condition $(x_{j_1 0}, x_{j_2 0})$, the first integral manifold is

$$
\begin{aligned}
x_{j_2}-x_{j_2 0}= & \frac{a_{j_2 j_1 0}}{a_{j_1 j_1 0}}\Big(\Big\{x_{j_1}-x_{j_1 0}+2(a_{j_1 j_1 1}-a_{j_2 j_1})\ln\frac{|x_{j_1}-a_{j_1 j_1 1}|}{|x_{j_1 0}-a_{j_1 j_1 1}|} \\
& -\left[(a_{j_1 j_1 1}-a_{j_2 j_1})^2+b_{j_2 j_1}\right](\frac{1}{x_{j_1}-a_{j_1 j_1 1}}-\frac{1}{x_{j_1 0}-a_{j_1 j_1 1}})\Big\} \\
& +(a_{j_1 j_1 1}-a_{j_2 j_1 1})\Big\{\ln\frac{|x_{j_1}-a_{j_1 j_1 1}|}{|x_{j_1 0}-a_{j_1 j_1 1}|} \\
& -2(a_{j_1 j_1 1}-a_{j_2 j_1})(\frac{1}{x_{j_1}-a_{j_1 j_1 1}}-\frac{1}{x_{j_1 0}-a_{j_1 j_1 1}}) \\
& -\frac{(a_{j_1 j_1 1}-a_{j_2 j_1})^2+b_{j_2 j_1}}{2}\left[\frac{1}{(x_{j_1}-a_{j_1 j_1 1})^2}-\frac{1}{(x_{j_1 0}-a_{j_1 j_1 1})^2}\right]\Big\}\Big), \\
& j_1, j_2\in\{1,2\}, j_1\neq j_2.
\end{aligned}
$$

(iii$_{1a}$) At $x_{j_1}^*=a_{j_1 j_1 1}$, the variational equation is

$$
\begin{aligned}
&\Delta\dot{x}_{j_1}=a_{j_1 j_1 0}(\Delta x_{j_1})^3 \\
&\dot{x}_{j_2}=a_{j_2 j_1 0}(a_{j_1 j_1 1}-a_{j_2 j_1 1})\left[(a_{j_1 j_1 1}-a_{j_2 j_1})^2+b_{j_2 j_1}\right].
\end{aligned}
$$

Due to

$$
(a_{j_1 j_1 1}-a_{j_2 j_1})^2+b_{j_2 j_1}>0,
$$

the flows at $x_{j_1}^*=a_{j_1 j_1 1}$ are the third-order source and sink flows for $a_{j_1 j_1 0}>0$ and $a_{j_1 j_1 0}<0$ in the x_{j_1}-direction, and the flows at $x_{j_1}^*=a_{j_1 j_1 1}$ are positive and negative for $a_{j_2 j_1 0}(a_{j_1 j_1 1}-a_{j_2 j_1 1})>0$ and $a_{j_2 j_1 0}(a_{j_1 j_1 1}-a_{j_2 j_1 1})<0$ in the x_{j_2}-direction.

Therefore, the equilibrium of $x_{j_1}^*=a_{j_1 j_1 1}$ for the x_{j_1}-direction has the following properties as in Eqs. (4.46)–(4.49).

- For $a_{j_1 j_1 0}>0$ and $a_{j_2 j_1 0}(a_{j_1 j_1 1}-a_{j_2 j_1 1})>0$,

$$
(a_{j_1 j_1 1},\dot{x}_{j_2})=\underbrace{(3^{\text{rd}}\text{SO, pF})}_{\text{third-order source flow }(+)}.
$$

The flow of $x_{j_1}^*=a_{j_1 j_1 1}$ is a (3^{rd}SO,pF)-positive third-order source flow.

- For $a_{j_1 j_1 0}<0$ and $a_{j_2 j_1 0}(a_{j_1 j_1 1}-a_{j_2 j_1 1})>0$,

$$
(a_{j_1 j_1 1},\dot{x}_{j_2})=\underbrace{(3^{\text{rd}}\text{SI, pF})}_{\text{third-order sink flow }(+)}.
$$

The flow of $x_{j_1}^*=a_{j_1 j_1 1}$ is a (3^{rd}SI,pF)-positive third-order sink flow.

- For $a_{j_1j_10} > 0$ and $a_{j_2j_10}(a_{j_1j_11} - a_{j_2j_11}) < 0$,

$$(a_{j_1j_11}, \dot{x}_{j_2}) = \underbrace{(3^{\mathrm{rd}}\mathrm{SO, nF})}_{\text{third-order source flow } (-)}.$$

The flow of $x_{j_1}^* = a_{j_1j_11}$ is a (3rdSO,nF)-negative third-order source flow.

- For $a_{j_1j_10} < 0$ and $a_{j_2j_10}(a_{j_1j_11} - a_{j_2j_11}) < 0$,

$$(a_{j_1j_11}, \dot{x}_{j_2}) = \underbrace{(3^{\mathrm{rd}}\mathrm{SI, nF})}_{\text{third-order sink flow } (+)}.$$

The flow of $x_{j_1}^* = a_{j_1j_11}$ is a (3rdSI,nF)-negative third-order sink flow.

– The third-order source and sinks flow bifurcations are the appearing and switching bifurcations for the source-sink-source and sink-source-sink flows.

(iii$_{1b}$) As similar to case (ii$_{1b}$), at $x_{j_1}^* = a_{j_2j_11} \neq a_{j_1j_11}$, we have

$$\frac{dx_{j_2}}{dx_{j_1}}\bigg|_{x_{j_1}^* = a_{j_2j_11}} = \frac{a_{j_2j_10}}{a_{j_1j_10}} \frac{(x_{j_1} - a_{j_2j_11})\left[(x_{j_1} - a_{j_2j_1})^2 + b_{j_2j_1}\right]}{(x_{j_1} - a_{j_1j_11})^3}\bigg|_{x_{j_1}^* = a_{j_2j_11}} = 0.$$

If

$$\frac{d^2x_{j_2}}{dx_{j_1}^2}\bigg|_{x_{j_1}^* = a_{j_2j_11}} = \frac{a_{j_2j_10}}{a_{j_1j_10}} \frac{(a_{j_2j_11} - a_{j_2j_1})^2 + b_{j_2j_1}}{(a_{j_2j_11} - a_{j_1j_11})^3} > 0,$$

the flow at $x_{j_1}^* = a_{j_2j_11}$ is an up-parabola flow, and if

$$\frac{d^2x_{j_2}}{dx_{j_1}^2}\bigg|_{x_{j_1}^* = a_{j_2j_11}} = \frac{a_{j_2j_10}}{a_{j_1j_10}} \frac{(a_{j_2j_11} - a_{j_2j_1})^2 + b_{j_2j_1}}{(a_{j_2j_11} - a_{j_1j_11})^3} < 0,$$

the flow at $x_{j_1}^* = a_{j_2j_11}$ is a down-parabola flow.

On the other hand, in vicinity of $x_{j_1}^* = a_{j_2j_11}$, the variational equation is

$$\begin{aligned}
&\Delta\dot{x}_{j_2} = a_{j_2j_10}\left[(a_{j_2j_11} - a_{j_2j_1})^2 + b_{j_2j_1}\right]\Delta x_{j_1}, \\
&\dot{x}_{j_1} = a_{j_1j_10}(a_{j_2j_11} - a_{j_1j_11})^3.
\end{aligned}$$

The flows at $x_{j_1}^* = a_{j_2j_11}$ are positive and negative in the x_{j_1}-direction for $a_{j_1j_10}(a_{j_2j_11} - a_{j_1j_11})^3 > 0$ and $a_{j_1j_10}(a_{j_2j_11} - a_{j_1j_11})^3 < 0$, respectively.

Therefore, the flow at $x_{j_1}^* = a_{j_2j_11}$ for the system has the following properties as in Eqs. (4.50)–(4.53).

- For $a_{j_1j_10}(a_{j_2j_11} - a_{j_1j_11})^3 > 0$ and $a_{j_2j_10} > 0$,

$$(\dot{x}_{j_1}, a_{j_2j_11}) = \underbrace{(\text{pF}, \text{UP})}_{\text{up-parabola flow } (+)}.$$

The flow of $x_{j_1}^* = a_{j_2j_11}$ is a (pF,UP)-positive up-parabola flow.

- For $a_{j_1j_10}(a_{j_2j_11} - a_{j_1j_11})^3 < 0$ and $a_{j_2j_10} > 0$,

$$(\dot{x}_{j_1}, a_{j_2j_11}) = \underbrace{(\text{nF}, \text{DP})}_{\text{down-parabola flow } (-)}.$$

The flow of $x_{j_1}^* = a_{j_2j_11}$ is an (nF,DP)-negative down-parabola flow.

- For $a_{j_1j_10}(a_{j_2j_11} - a_{j_1j_11})^3 > 0$ and $a_{j_2j_10} < 0$,

$$(\dot{x}_{j_1}, a_{j_2j_11}) = \underbrace{(\text{pF}, \text{DP})}_{\text{down-parabola flow } (+)}.$$

The flow of $x_{j_1}^* = a_{j_2j_11}$ is a (pF,DP)-positive down-parabola flow.

- For $a_{j_1j_10}(a_{j_2j_11} - a_{j_1j_11})^3 < 0$ and $a_{j_2j_10} < 0$,

$$(\dot{x}_{j_1}, a_{j_2j_11}) = \underbrace{(\text{nF}, \text{UP})}_{\text{up-parabola flow } (-)}.$$

The flow of $x_{j_1}^* = a_{j_2j_11}$ is an (nF,UP)-negative up-parabola flow.

(iii$_2$) For $a_{j_1j_11} = a_{j_2j_11}$,

$$\begin{aligned}
\dot{x}_{j_1} &= a_{j_1j_10}(x_{j_1} - a_{j_1j_11})^3, \\
\dot{x}_{j_2} &= a_{j_2j_10}(x_{j_1} - a_{j_2j_11})\left[(x_{j_1} - a_{j_2j_1})^2 + b_{j_2j_1}\right],
\end{aligned}$$

In phase space, we have

$$\begin{aligned}
\frac{dx_{j_2}}{dx_{j_1}} &= \frac{a_{j_2j_10}}{a_{j_1j_10}} \frac{(x_{j_1} - a_{j_2j_11})\left[(x_{j_1} - a_{j_2j_1})^2 + b_{j_2j_1}\right]}{(x_{j_1} - a_{j_1j_11})^3} \\
&= \frac{a_{j_2j_10}}{a_{j_1j_10}} \left[1 + \frac{2(a_{j_1j_11} - a_{j_2j_1})}{(x_{j_1} - a_{j_1j_11})} + \frac{(a_{j_1j_11} - a_{j_2j_1})^2 + b_{j_2j_1}}{(x_{j_1} - a_{j_1j_11})^2}\right]
\end{aligned}$$

With an initial condition (x_{j_10}, x_{j_20}), the first integral manifold is

$$x_{j_2} - x_{j_20} = \frac{a_{j_2j_10}}{a_{j_1j_10}}\Big\{x_{j_1} - x_{j_10} + 2(a_{j_1j_11} - a_{j_2j_1})\ln\frac{|x_{j_1} - a_{j_1j_11}|}{|x_{j_10} - a_{j_1j_11}|}$$
$$- \big[(a_{j_1j_11} - a_{j_2j_1})^2 + b_{j_2j_1}\big](\frac{1}{x_{j_1} - a_{j_1j_11}} - \frac{1}{x_{j_10} - a_{j_1j_11}})\Big\}$$
$$j_1, j_2 \in \{1,2\}, j_1 \neq j_2.$$

From the first integral manifold and singularity, the flows at $x_{j_1}^* = a_{j_1j_11} = a_{j_2j_11}$ are (3:1)-increasing concave and decreasing-concave flows for $a_{j_1j_10}a_{j_2j_10} > 0$ and $a_{j_1j_10}a_{j_2j_10} < 0$, respectively. In vicinity of $x_{j_1}^* = a_{j_1j_11} = a_{j_2j_11}$, the variational equations are

$$\Delta\dot{x}_{j_1} = a_{j_1j_10}(\Delta x_{j_1})^3,$$
$$\Delta\dot{x}_{j_2} = a_{j_2j_10}\big[(a_{j_2j_11} - a_{j_2j_1})^2 + b_{j_2j_1}\big]\Delta x_{j_1}.$$

The flows $x_{j_1}^* = a_{j_1j_11} = a_{j_2j_11}$ are source and sink flows in x_1-direction, respectively.

Therefore, the infinite equilibrium of $x_{j_1}^* = a_{j_1j_11} = a_{j_2j_11}$ for the x_{j_1}-direction has the following properties as in Eqs. (4.55)–(4.58).

- For $a_{j_1j_10} > 0$ and $a_{j_2j_10} > 0$,

$$(a_{j_1j_11}, \bar{x}_{j_2}) = \underbrace{(3^{\text{rd}}\text{SO}, (3:1)\text{IC})}_{\text{third-order increasing-concave source}}.$$

  The infinite-equilibrium at $x_{j_1}^* = a_{j_1j_11} = a_{j_2j_11}$ is a (3^{rd}SO,(3:1)IC)-third-order increasing-concave source.
- For $a_{j_1j_10} < 0$ and $a_{j_2j_10} > 0$,

$$(a_{j_1j_11}, \bar{x}_{j_2}) = \underbrace{(3^{\text{rd}}\text{SI}, (3:1)\text{DC})}_{\text{third-order decreasing-inflection sink}}.$$

  The infinite-equilibrium at $x_{j_1}^* = a_{j_1j_11} = a_{j_2j_11}$ is a (3^{rd}SI,(3:1)DC)-third-order decreasing-concave sink.
- For $a_{j_1j_10} > 0$ and $a_{j_2j_10} < 0$,

$$(a_{j_1j_11}, \bar{x}_{j_2}) = \underbrace{(3^{\text{rd}}\text{SO}, (3:1)\text{DC})}_{\text{third-order decreasing-inflection source}}.$$

  The infinite-equilibrium at $x_{j_1}^* = a_{j_1j_11} = a_{j_2j_11}$ is a (3^{rd}SO,(3:1)DC)-third-order decreasing-concave source.

- For $a_{j_1j_10}<0$ and $a_{j_2j_10}<0$,

$$(a_{j_1j_11},\bar{x}_{j_2})=\underbrace{(3^{\rm rd}\mathrm{SI},(3:1)\mathrm{IC})}_{\text{third-order increasing-concave sink}}.$$

The infinite-equilibrium at $x^*_{j_1}=a_{j_1j_11}=a_{j_2j_11}$ is a (3$^{\rm rd}$SI,(3:1)IC)-third-order increasing-concave sink.

- The third-order increasing and decreasing concave source (sink) infinite-equilibriums are the switching bifurcations of third-order source (sink) flows with a parabola flow.
- The third-order increasing and decreasing concave source (sink) infinite-equilibriums are the appearing bifurcations from source (sink) and parabola flows to source (sink), sink(source), and source (sink) flows with a parabola flow.

(iv) If

$$\Delta_{j_1j_1}>0 \text{ and } \Delta_{j_2j_1}<0,$$

then

$$\begin{aligned}&\dot{x}_{j_1}=a_{j_1j_10}(x_{j_1}-a_{j_1j_11})(x_{j_1}-a_{j_1j_12})(x_{j_1}-a_{j_1j_13}),\\&\dot{x}_{j_2}=a_{j_2j_10}(x_{j_1}-a_{j_2j_11})\left[(x_{j_1}-a_{j_2j_1})^2+b_{j_2j_1}\right],\\&j_1,j_2\in\{1,2\};j_1\neq j_2.\end{aligned}$$

where

$$\begin{aligned}&\{a_{j_1j_11},a_{j_1j_12},a_{j_1j_13}\}=\mathrm{sort}\{b_{j_1j_11},b_{j_1j_12},b_{j_1j_13}\},a_{j_1j_11}<a_{j_1j_12}<a_{j_1j_13};\\&a_{211}=b_{211};b_{j_1j_12},b_{j_1j_13}=-\frac{1}{2}(B_{j_1j_11}\pm\sqrt{\Delta_{j_1j_1}});a_{j_2j_1}=-\frac{1}{2}B_{j_2j_11}.\end{aligned}$$

(iv$_1$) In phase space, we have

$$\begin{aligned}\frac{dx_{j_2}}{dx_{j_1}}&=\frac{a_{j_2j_10}}{a_{j_1j_10}}\frac{(x_{j_1}-a_{j_2j_11})\left[(x_{j_1}-a_{j_2j_1})^2+b_{j_2j_1}\right]}{(x_{j_1}-a_{j_1j_11})(x_{j_1}-a_{j_1j_12})(x_{j_1}-a_{j_1j_13})}\\&=\frac{a_{j_2j_10}}{a_{j_1j_10}}\left[1+\frac{(a_{j_1j_1s_1}-a_{j_2j_1})^2+b_{j_2j_1}}{(a_{j_1j_1s_1}-a_{j_1j_1s_2})}\frac{1}{x_{j_1}-a_{j_1j_1s_1}}\right.\\&\quad+\frac{(a_{j_1j_1s_2}-a_{j_2j_1})^2+b_{j_2j_1}}{(a_{j_1j_1s_2}-a_{j_1j_1s_2})}\frac{1}{(x_{j_1}-a_{j_1j_1s_2})}\\&\quad\left.+(a_{j_1j_11}-a_{j_2j_11})\sum_{s_1=1}^{3}\frac{(a_{j_1j_1s_1}-a_{j_2j_1})^2+b_{j_2j_1}}{\prod_{s_2=1,s_2\neq s_1}^{3}(a_{j_1j_1s_1}-a_{j_1j_1s_2})}\frac{1}{(x_{j_1}-a_{j_1j_1s_1})}\right].\end{aligned}$$

With an initial condition $(x_{j_1 0}, x_{j_2 0})$, the first integral manifold is

$$
\begin{aligned}
x_{j_2} - x_{j_2 0} = & \frac{a_{j_2 j_1 0}}{a_{j_1 j_1 0}}\Big(\Big[x_{j_1} - x_{j_1 0} + \frac{(a_{j_1 j_1 s_1} - a_{j_2 j_1})^2 + b_{j_2 j_1}}{(a_{j_1 j_1 s_1} - a_{j_1 j_1 s_2})} \ln \frac{|x_{j_1} - a_{j_1 j_1 s_2}|}{|x_{j_1 0} - a_{j_1 j_1 s_2}|} \\
& + \frac{(a_{j_1 j_1 s_2} - a_{j_2 j_1})^2 + b_{j_2 j_1}}{(a_{j_1 j_1 s_2} - a_{j_1 j_1 s_1})} \ln \frac{|x_{j_1} - a_{j_1 j_1 s_2}|}{|x_{j_1 0} - a_{j_1 j_1 s_2}|}\Big] \\
& + (a_{j_1 j_1 s_3} - a_{j_2 j_1 1})\Big[\sum_{s_1=1}^{3} \frac{(a_{j_1 j_1 s_1} - a_{j_2 j_1 1})^2 + b_{j_2 j_1}}{\prod_{s_2=1, s_2 \neq s_1}^{3} (a_{j_1 j_1 s_1} - a_{j_1 j_1 s_2})} \ln \frac{|x_{j_1} - a_{j_1 j_1 s_1}|}{|x_{j_1 0} - a_{j_1 j_1 s_1}|}\Big]\Big).
\end{aligned}
$$

(iv$_{1a}$) At $x_{j_1}^* = a_{j_1 j_1 s_1}$ (s_1, $s_2 \in \{1,2,3\}$, $s_1 \neq s_2$), the variational equation is

$$
\begin{aligned}
\Delta \dot{x}_{j_1} &= a_{j_1 j_1 0} \prod_{s_2=1, s_2 \neq s_1}^{3} (a_{j_1 j_1 s_1} - a_{j_1 j_1 s_2}) \Delta x_{j_1} \\
\dot{x}_{j_2} &= a_{j_2 j_1 0}(a_{j_1 j_1 s_1} - a_{j_2 j_1 1})\big[(a_{j_1 j_1 s_1} - a_{j_2 j_1})^2 + b_{j_2 j_1}\big].
\end{aligned}
$$

Due to

$$
(a_{j_1 j_1 s_1} - a_{j_2 j_1})^2 + b_{j_2 j_1} > 0,
$$

the flows at $x_{j_1}^* = a_{j_1 j_1 s_1}$ in the x_{j_1}-direction are source and sink flows for $a_{j_1 j_1 0} \prod_{s_2=1, s_2 \neq s_1}^{3} (a_{j_1 j_1 s_1} - a_{j_1 j_1 s_2}) > 0$ and $a_{j_1 j_1 0} \prod_{s_2=1, s_2 \neq s_1}^{3} (a_{j_1 j_1 s_1} - a_{j_1 j_1 s_2}) < 0$, and the flows at $x_{j_1}^* = a_{j_1 j_1 s_1}$ are positive and negative for $a_{j_2 j_1 0}(a_{j_1 j_1 s_1} - a_{j_2 j_1 1}) > 0$ and $a_{j_2 j_1 0}(a_{j_1 j_1 s_1} - a_{j_2 j_1 1}) < 0$ in the x_{j_2}-direction.

Therefore, the equilibrium of $x_{j_1}^* = a_{j_1 j_1 s_1}$ (s_1, $s_2 \in \{1,2,3\}$, $s_1 \neq s_2$) in the x_{j_1}-direction possesses the following properties as in Eqs. (4.63)–(4.66):

- For $a_{j_1 j_1 0} \prod_{s_2=1, s_2 \neq s_1}^{3} (a_{j_1 j_1 s_1} - a_{j_1 j_1 s_2}) > 0$ and $a_{j_2 j_1 0}(a_{j_1 j_1 s_1} - a_{j_2 j_1 1}) > 0$,

$$
(a_{j_1 j_1 s_1}, \dot{x}_{j_2}) = \underbrace{(\text{SO}, \text{pF})}_{\text{source flow } (+)}.
$$

  The flow at $x_{j_1}^* = a_{j_1 j_1 s_1}$ is an (SO,pF)-positive source-flow.

- For $a_{j_1 j_1 0} \prod_{s_2=1, s_2 \neq s_1}^{3} (a_{j_1 j_1 s_1} - a_{j_1 j_1 s_2}) < 0$ and $a_{j_2 j_1 0}(a_{j_1 j_1 s_1} - a_{j_2 j_1 1}) > 0$,

$$
(a_{j_1 j_1 s_1}, \dot{x}_{j_2}) = \underbrace{(\text{SI}, \text{pF})}_{\text{sink flow } (+)}.
$$

  The flow at $x_{j_1}^* = a_{j_1 j_1 s_1}$ is an (SI,pF)-positive sink-flow.

- For $a_{j_1j_10}\prod_{s_2=1,s_2\neq s_1}^{3}(a_{j_1j_1s_1}-a_{j_1j_1s_2})>0$ and $a_{j_2j_10}(a_{j_1j_1s_1}-a_{j_2j_11})<0$,

$$(a_{j_1j_1s_1},\dot{x}_{j_2})=\underbrace{(\mathrm{SO},\mathrm{nF})}_{\text{source flow }(-)}.$$

The flow at $x_{j_1}^*=a_{j_1j_1s_1}$ is an (SO,nF)-negative source-flow.

- For $a_{j_1j_10}\prod_{s_2=1,s_2\neq s_1}^{3}(a_{j_1j_1s_1}-a_{j_1j_1s_2})<0$ and $a_{j_2j_10}(a_{j_1j_1s_1}-a_{j_2j_11})<0$,

$$(a_{j_1j_1s_1},\dot{x}_{j_2})=\underbrace{(\mathrm{SI},\mathrm{nF})}_{\text{sink flow }(-)}.$$

The flow at $x_{j_1}^*=a_{j_1j_1s_1}$ is an (SI,nF)-negative sink-flow.

($\mathrm{iv}_{1\mathrm{b}}$) For $x_{j_1}^*=a_{j_2j_11}\neq a_{j_1j_1s_1}$, in phase space,

$$\frac{dx_{j_2}}{dx_{j_1}}\bigg|_{x_{j_1}^*=a_{j_2j_11}}=\frac{a_{j_2j_10}}{a_{j_1j_10}}\frac{(x_{j_1}-a_{j_2j_11})\left[(x_{j_1}-a_{j_2j_1})^2+b_{j_2j_1}\right]}{\prod_{s_1=1}^{3}(x_{j_1}-a_{j_1j_1s_1})}\bigg|_{x_{j_1}^*=a_{j_2j_11}}=0.$$

If

$$\frac{d^2x_{j_2}}{dx_{j_1}^2}\bigg|_{x_{j_1}^*=a_{j_2j_11}}=\frac{a_{j_2j_10}}{a_{j_1j_10}}\frac{(a_{j_2j_11}-a_{j_2j_1})^2+b_{j_2j_1}}{\prod_{s_1=1}^{3}(a_{j_2j_11}-a_{j_1j_1s_1})}>0,$$

the flow at $x_{j_1}^*=a_{j_2j_11}\neq a_{j_1j_1s_1}$ is an up-parabola flow, and if

$$\frac{d^2x_{j_2}}{dx_{j_1}^2}\bigg|_{x_{j_1}^*=a_{j_2j_11}}=\frac{a_{j_2j_10}}{a_{j_1j_10}}\frac{(a_{j_2j_11}-a_{j_2j_1})^2+b_{j_2j_1}}{\prod_{s_1=1}^{3}(a_{j_2j_11}-a_{j_1j_1s_1})}<0,$$

the flow at $x_{j_1}^*=a_{j_2j_11}\neq a_{j_1j_1s_1}$ is a down-parabola flow.

In vicinity of $x_{j_1}^*=a_{j_2j_11}\neq a_{j_1j_1s_1}$, the variational equation is

$$\Delta\dot{x}_{j_2}=a_{j_2j_10}\left[(a_{j_2j_11}-a_{j_2j_1})^2+b_{j_2j_1}\right]\Delta x_{j_1},$$
$$\dot{x}_{j_1}=a_{j_1j_10}\prod_{s_1=1}^{3}(a_{j_2j_11}-a_{j_1j_1s_1}).$$

The flows at $x_{j_1}^*=a_{j_2j_11}$ are positive and negative in the x_{j_1}-direction for $a_{j_1j_10}\prod_{s_1=1}^{3}(a_{j_2j_11}-a_{j_1j_1s_1})>0$ and $a_{j_1j_10}\prod_{s_1=1}^{3}(a_{j_2j_11}-a_{j_1j_1s_1})<0$, respectively.

Therefore, the equilibrium of $x_{j_1}^*=a_{j_2j_11}\neq a_{j_1j_1s_1}$ ($s_1=1,2,3$) in the x_{j_2}-direction possesses the following properties as in Eqs. (4.67)–(4.70).

- For $a_{j_1j_10}\prod_{s_1=1}^{3}(a_{j_2j_11}-a_{j_1j_1s_1})>0$ and $a_{j_2j_10}>0$,

$$(\dot{x}_{j_1},a_{j_2j_11})=\underbrace{(\text{pF},\text{UP})}_{\text{up-parabola flow }(+)}.$$

The flow of $x^*_{j_1}=a_{j_2j_11}$ is a (pF,UP)-positive up-parabola flow.

- For $a_{j_1j_10}\prod_{s_1=1}^{3}(a_{j_2j_11}-a_{j_1j_1s_1})<0$ and $a_{j_2j_10}>0$,

$$(\dot{x}_{j_1},a_{j_2j_11})=\underbrace{(\text{nF},\text{DP})}_{\text{down-parabola flow }(-)}.$$

The flow of $x^*_{j_1}=a_{j_2j_11}$ is an (nF,DP)-negative down-parabola flow.

- For $a_{j_1j_10}\prod_{s_1=1}^{3}(a_{j_2j_11}-a_{j_1j_1s_1})>0$ and $a_{j_2j_10}<0$,

$$(\dot{x}_{j_1},a_{j_2j_11})=\underbrace{(\text{pF},\text{DP})}_{\text{down-parabola flow }(+)}.$$

The flow of $x^*_{j_1}=a_{j_2j_11}$ is a (pF,DP)-positive down-parabola flow.

- For $a_{j_1j_10}\prod_{s_1=1}^{3}(a_{j_2j_11}-a_{j_1j_1s_1})<0$ and $a_{j_2j_10}<0$,

$$(\dot{x}_{j_1},a_{j_2j_11})=\underbrace{(\text{nF},\text{UP})}_{\text{up-parabola flow }(-)}.$$

The flow of $x^*_{j_1}=a_{j_2j_11}$ is an (nF,UP)-negative up-parabola flow.

(iv$_2$) For $a_{j_2j_11}=a_{11s_1}$ ($s_1=1,2,3$), in phase space,

$$\begin{aligned}\frac{dx_{j_2}}{dx_{j_1}}&=\frac{a_{j_2j_10}}{a_{j_1j_10}}\frac{(x_{j_1}-a_{j_2j_11})\left[(x_{j_1}-a_{j_2j_1})^2+b_{j_2j_1}\right]}{\prod_{s_1=1}^{3}(x_{j_1}-a_{j_1j_1s_2})}\\&=\frac{a_{j_2j_10}}{a_{j_1j_10}}\frac{(x_{j_1}-a_{j_2j_1})^2+b_{j_2j_1}}{(x_{j_1}-a_{j_1j_1s_2})(x_{j_1}-a_{j_1j_1s_3})}\\&=\frac{a_{j_2j_10}}{a_{j_1j_10}}\left[1+\frac{(a_{j_1j_1s_2}-a_{j_2j_1})}{(x_{j_1}-a_{j_1j_1s_2})}+\frac{(a_{j_1j_1s_3}-a_{j_2j_1})}{(x_{j_1}-a_{j_1j_1s_3})}\right.\\&\quad\left.+\frac{(a_{j_1j_1s_2}-a_{j_2j_1})(a_{j_1j_1s_3}-a_{j_2j_1})+b_{j_2j_1}}{(x_{j_1}-a_{j_1j_1s_2})(x_{j_1}-a_{j_1j_1s_3})}\right].\end{aligned}$$

With an initial condition (x_{j_10},x_{j_20}), the first integral manifold is

$$
\begin{aligned}
x_{j_2} - x_{j_2 0} = & \frac{a_{j_1 j_1 0}}{a_{j_1 j_1 0}} \Big\{ x_{j_1} - x_{j_1 0} + (a_{j_1 j_1 s_2} - a_{j_2 j_1}) \ln \frac{| x_{j_1} - a_{j_1 j_1 s_2} |}{| x_{j_1 0} - a_{j_1 j_1 s_2} |} \\
& + (a_{j_1 j_1 s_3} - a_{j_2 j_1}) \ln \frac{| x_{j_1} - a_{j_1 j_1 s_3} |}{| x_{j_1 0} - a_{j_1 j_1 s_3} |} \\
& + \big[(a_{j_1 j_1 s_2} - a_{j_2 j_1})(a_{j_1 j_1 s_3} - a_{j_2 j_1}) + b_{j_2 j_1} \big] \\
& \times \frac{1}{a_{j_1 j_1 s_2} - a_{j_1 j_1 s_3}} \Big(\ln \frac{| x_{j_1} - a_{j_1 j_1 s_2} |}{| x_{j_1 0} - a_{j_1 j_1 s_2} |} - \ln \frac{| x_{j_1} - a_{j_1 j_1 s_3} |}{| x_{j_1 0} - a_{j_1 j_1 s_3} |} \Big) \Big\}.
\end{aligned}
$$

Further,

$$
\begin{aligned}
x_{j_2} - x_{j_2 0} = & \frac{a_{j_1 j_1 0}}{a_{j_1 j_1 0}} \Big[x_{j_1} - x_{j_1 0} + \frac{(a_{j_1 j_1 s_2} - a_{j_2 j_1})^2 + b_{j_2 j_1}}{(a_{j_1 j_1 s_2} - a_{j_1 j_1 s_3})} \ln \frac{| x_{j_1} - a_{j_1 j_1 s_2} |}{| x_{j_1 0} - a_{j_1 j_1 s_2} |} \\
& + \frac{(a_{j_1 j_1 s_3} - a_{j_2 j_1})^2 + b_{j_2 j_1}}{(a_{j_1 j_1 s_3} - a_{j_1 j_1 s_2})} \ln \frac{| x_{j_1} - a_{j_1 j_1 s_3} |}{| x_{j_1 0} - a_{j_1 j_1 s_3} |} \Big]
\end{aligned}
$$

For $x_{j_1}^* = a_{j_2 j_1 1} = a_{11 s_1}$, in phase space, if

$$
\frac{dx_{j_2}}{dx_{j_1}} \Big|_{x_{j_1}^* = a_{j_2 j_1 1} = a_{11 s_1}} = \frac{a_{j_2 j_1 0}}{a_{j_1 j_1 0}} \frac{(a_{j_2 j_1 1} - a_{j_2 j_1})^2 + b_{j_2 j_1}}{\prod_{s_2 = 1, s_2 \neq s_1}^{3} (a_{j_1 j_1 s_1} - a_{j_1 j_1 s_2})} > 0,
$$

there is an increasing-inflection flow, and if

$$
\frac{dx_{j_2}}{dx_{j_1}} \Big|_{x_{j_1}^* = a_{j_2 j_1 1} = a_{11 s_1}} = \frac{a_{j_2 j_1 0}}{a_{j_1 j_1 0}} \frac{(a_{j_2 j_1 1} - a_{j_2 j_1})^2 + b_{j_2 j_1}}{\prod_{s_2 = 1, s_2 \neq s_1}^{3} (a_{j_1 j_1 s_1} - a_{j_1 j_1 s_2})} < 0,
$$

there is a decreasing-inflection flow. In vicinity of $x_{j_1}^* = a_{j_2 j_1 1} = a_{11 s_1}$, the variational equation is

$$
\begin{aligned}
\Delta \dot{x}_{j_2} &= a_{j_2 j_1 0} \big[(a_{j_2 j_1 1} - a_{j_2 j_1})^2 + b_{j_2 j_1} \big] \Delta x_{j_1}, \\
\Delta \dot{x}_{j_1} &= a_{j_1 j_1 0} \prod_{s_2 = 1, s_2 \neq s_1}^{3} (a_{j_1 j_1 s_1} - a_{j_1 j_1 s_2}) \Delta x_{j_1}.
\end{aligned}
$$

The flows at $x_{j_1}^* = a_{j_2 j_1 1} = a_{11 s_1}$ are source and sink in the x_{j_1}-direction for $a_{j_1 j_1 0} \prod_{s_2 = 1, s_2 \neq s_1}^{3} (a_{j_1 j_1 s_1} - a_{j_1 j_1 s_2}) > 0$ and $a_{j_1 j_1 0} \prod_{s_2 = 1, s_2 \neq s_1}^{3} (a_{j_1 j_1 s_1} - a_{j_1 j_1 s_2}) < 0$.

Therefore, the infinite-equilibrium at $x_{j_1}^* = a_{j_2 j_1 1} = a_{11 s_1}$ with $\bar{x}_{j_2} \in (-\infty, \infty)$ has the following properties as Eqs. (4.72)–(4.75).

- For $a_{j_1j_10}\prod_{s_2=1,s_2\neq s_1}^{3}(a_{j_1j_1s_1}-a_{j_1j_1s_2})>0$ and $a_{j_2j_10}>0$,

$$(a_{j_1j_1s_1},\bar{x}_{j_2})=\underbrace{(\mathrm{SO,II})}_{\text{increasing-inflecton source}}.$$

The infinite-equilibrium at $x_{j_1}^{*}=a_{j_1j_1s_1}$ is an (SO,II)-increasing-inflection source.

- For $a_{j_1j_10}\prod_{s_2=1,s_2\neq s_1}^{3}(a_{j_1j_1s_1}-a_{j_1j_1s_2})<0$ and $a_{j_2j_10}>0$,

$$(a_{j_1j_1s_1},\bar{x}_{j_2})=\underbrace{(\mathrm{SI,DI})}_{\text{decreasing-inflecton sink}}.$$

The infinite-equilibrium at $x_{j_1}^{*}=a_{j_1j_1s_1}$ is an (SI,DI)-decreasing-inflection sink.

- For $a_{j_1j_10}\prod_{s_2=1,s_2\neq s_1}^{3}(a_{j_1j_1s_1}-a_{j_1j_1s_2})>0$ and $a_{j_2j_10}<0$,

$$(a_{j_1j_1s_1},\bar{x}_{j_2})=\underbrace{(\mathrm{SO,DI})}_{\text{decreasing-inflecton source}}.$$

The infinite-equilibrium at $x_{j_1}^{*}=a_{j_1j_1s_1}$ is an (SO,DI)-decreasing-inflection source.

- For $a_{j_1j_10}\prod_{s_2=1,s_2\neq s_1}^{3}(a_{j_1j_1s_1}-a_{j_1j_1s_2})<0$ and $a_{j_2j_10}<0$,

$$(a_{j_1j_1s_1},\bar{x}_{j_2})=\underbrace{(\mathrm{SI,II})}_{\text{increasing-inflecton sink}}.$$

The infinite-equilibrium at $x_{j_1}^{*}=a_{j_1j_1s_1}$ is an (SI,II)-increasing-inflection sink.

– The inflection-source (sink) infinite-equilibriums are the switching bifurcations of a source (sink) flow with a parabola flow.

(v) If

$$\Delta_{j_1j_1}<0 \text{ and } \Delta_{j_2j_1}=0,$$

then

$$\begin{aligned}
\dot{x}_{j_2}&=a_{j_2j_10}(x_{j_1}-a_{j_2j_1l_1})(x_{j_1}-a_{j_2j_1l_2})^2,\\
\dot{x}_{j_1}&=a_{j_1j_10}(x_{j_1}-a_{j_1j_11})\left[(x_{j_1}-a_{j_1j_1})^2+b_{j_1j_1}\right],
\end{aligned}$$

with

$$\begin{aligned}
&a_{j_1j_1}=-\frac{1}{2}B_{j_1j_11},b_{j_1j_1}=\frac{1}{4}(-\Delta_{j_1j_1}),a_{j_1j_11}=b_{j_1j_11};\\
&a_{j_2j_1l_1}=b_{j_2j_11};a_{j_2j_1l_2}=-\frac{1}{2}B_{j_2j_11},b_{j_2j_1}=\frac{1}{4}(-\Delta_{j_2j_1})=0.
\end{aligned}$$

(v_1) For $a_{j_1j_1 1} \neq a_{j_2j_1l_1}, a_{j_2j_1l_2}$, we have

$$\begin{aligned}
\frac{dx_{j_2}}{dx_{j_1}} &= \frac{a_{j_2j_1 0}}{a_{j_1j_1 0}} \frac{(x_{j_1}-a_{j_2j_1l_1})(x_{j_1}-a_{j_2j_1l_2})^2}{(x_{j_1}-a_{j_1j_1 1})\left[(x_{j_1}-a_{j_1j_1})^2+b_{j_1j_1}\right]} \\
&= \frac{a_{j_2j_1 0}}{a_{j_1j_1 0}}\Big(1+\frac{2(a_{j_1j_1}-a_{j_2j_1l_2})(x_{j_1}-a_{j_1j_1})}{\left[(x_{j_1}-a_{j_1j_1})^2+b_{j_1j_1}\right]}+\frac{(a_{j_1j_1}-a_{j_2j_1l_2})^2-b_{j_1j_1}}{\left[(x_{j_1}-a_{j_1j_1})^2+b_{j_1j_1}\right]} \\
&+(a_{j_1j_1 1}-a_{j_2j_1l_1})\Big\{\frac{(x_{j_1}-a_{j_1j_1})}{(x_{j_1}-a_{j_1j_1})^2+b_{j_1j_1}}+\frac{a_{j_1j_1}+a_{j_1j_1 1}-2a_{j_2j_1l_2}}{(x_{j_1}-a_{j_1j_1})^2+b_{j_1j_1}} \\
&+\frac{(a_{j_1j_1 1}-a_{j_2j_1l_2})^2}{(a_{j_1j_1 1}-a_{j_1j_1})^2+b_{j_1j_1}}\Big[\frac{1}{x_{j_1}-a_{j_1j_1 1}}-\frac{(x_{j_1}-a_{j_1j_1})}{(x_{j_1}-a_{j_1j_1})^2+b_{j_1j_1}} \\
&-\frac{a_{j_1j_1 1}-a_{j_1j_1}}{(x_{j_1}-a_{j_1j_1})^2+b_{j_1j_1}}\Big]\Big\}\Big).
\end{aligned}$$

With an initial condition $(x_{j_1 0}, x_{j_2 0})$, the first integral manifold is

$$\begin{aligned}
x_{j_2}-x_{j_2 0} &= \frac{a_{j_2j_1 0}}{a_{j_1j_1 0}}\Big(\Big\{x_{j_1}-x_{j_1 0}+(a_{j_1j_1}-a_{j_2j_1l_2})\ln\frac{|(x_{j_1}-a_{j_1j_1})^2+b_{j_1j_1}|}{|(x_{j_1 0}-a_{j_1j_1})^2+b_{j_1j_1}|} \\
&+\frac{\left[(a_{j_1j_1}-a_{j_2j_1l_2})^2-b_{j_1j_1}\right]}{\sqrt{b_{j_1j_1}}}(\arctan\frac{x_{j_1}-a_{j_1j_1}}{\sqrt{b_{j_1j_1}}}-\arctan\frac{x_{j_1 0}-a_{j_1j_1}}{\sqrt{b_{j_1j_1}}})\Big\} \\
&+(a_{j_1j_1 1}-a_{j_2j_1l_1})\Big\{\frac{1}{2}\ln\frac{|(x_{j_1}-a_{j_1j_1})^2+b_{j_1j_1}|}{|(x_{j_1 0}-a_{j_1j_1})^2+b_{j_1j_1}|} \\
&+\left[2(a_{j_1j_1 1}-a_{j_2j_1 1})+a_{j_1j_1}-a_{j_1j_1 1}\right] \\
&\times\frac{1}{\sqrt{b_{j_1j_1}}}(\arctan\frac{x_{j_1}-a_{j_1j_1}}{\sqrt{b_{j_1j_1}}}-\arctan\frac{x_{j_1 0}-a_{j_1j_1}}{\sqrt{b_{j_1j_1}}}) \\
&+\frac{(a_{j_1j_1 1}-a_{j_2j_1 1})^2}{(a_{j_1j_1 1}-a_{j_1j_1})^2+b_{j_1j_1}}\Big[\ln\frac{|x_{j_1}-a_{j_1j_1 1}|}{|x_{j_1 0}-a_{j_1j_1 1}|} \\
&-\frac{1}{2}\ln\frac{|(x_{j_1}-a_{j_1j_1})^2+b_{j_1j_1}|}{|(x_{j_1 0}-a_{j_1j_1})^2+b_{j_1j_1}|} \\
&-\frac{a_{j_1j_1 1}-a_{j_1j_1}}{\sqrt{b_{j_1j_1}}}(\arctan\frac{x_{j_1}-a_{j_1j_1}}{\sqrt{b_{j_1j_1}}}-\arctan\frac{x_{j_1 0}-a_{j_1j_1}}{\sqrt{b_{j_1j_1}}})\Big]\Big\}\Big), \\
&j_1, j_2 \in \{1,2\}, j_1 \neq j_2.
\end{aligned}$$

(v_{1a}) For $x_{j_1}^* = a_{j_1 j_1 1} \neq a_{j_2 j_1 1}$, the variational equation is

$$\begin{aligned} \dot{x}_{j_2} &= a_{j_2 j_1 0}(a_{j_1 j_1 1} - a_{j_2 j_1 l_1})(a_{j_1 j_1 1} - a_{j_2 j_1 l_2})^2, \\ \Delta \dot{x}_{j_1} &= a_{j_1 j_1 0}\left[(a_{j_1 j_1 1} - a_{j_1 j_1})^2 + b_{j_1 j_1}\right]\Delta x_{j_1}. \end{aligned}$$

Due to

$$(a_{j_1 j_1 1} - a_{j_2 j_1})^2 > 0 \text{ and } (a_{j_1 j_1 1} - a_{j_1 j_1})^2 + b_{j_1 j_1} > 0,$$

the flows at $x_{j_1}^* = a_{j_1 j_1 1}$ are source and sink flows for $a_{j_1 j_1 0} > 0$ and $a_{j_1 j_1 0} < 0$ in the x_{j_1}-direction, and the flows at $x_{j_1}^* = a_{j_1 j_1 1}$ in the x_{j_2}-direction are positive and negative for $a_{j_2 j_1 0}(a_{j_1 j_1 1} - a_{j_2 j_1 l_1}) > 0$ and $a_{j_2 j_1 0}(a_{j_1 j_1 1} - a_{j_2 j_1 l_1}) < 0$.

Therefore, the flow at $x_{j_1}^* = a_{j_1 j_1 1}$ for the system has the following properties as in Eqs. (4.79)–(4.82).

- For $a_{j_1 j_1 0} > 0$ and $a_{j_2 j_1 0}(a_{j_1 j_1 1} - a_{j_2 j_1 l_1})(a_{j_1 j_1 1} - a_{j_2 j_1 l_2})^2 > 0$,

$$(a_{j_1 j_1 1}, \dot{x}_{j_2}) = \underbrace{(\text{SO}, \text{pF})}_{\text{source flow } (+)}.$$

The flow of $x_{j_1}^* = a_{j_1 j_1 1}$ is an (SO,pF)-positive source flow.

- For $a_{j_1 j_1 0} < 0$ and $a_{j_2 j_1 0}(a_{j_1 j_1 1} - a_{j_2 j_1 l_1})(a_{j_1 j_1 1} - a_{j_2 j_1 l_2})^2 > 0$,

$$(a_{j_1 j_1 1}, \dot{x}_{j_2}) = \underbrace{(\text{SI}, \text{pF})}_{\text{sink flow } (+)}.$$

The flow of $x_{j_1}^* = a_{j_1 j_1 1}$ is an (SI, pF)-positive sink flow.

- For $a_{j_1 j_1 0} > 0$ and $a_{j_2 j_1 0}(a_{j_1 j_1 1} - a_{j_2 j_1 l_1})(a_{j_1 j_1 1} - a_{j_2 j_1 l_2})^2 < 0$,

$$(a_{j_1 j_1 1}, \dot{x}_{j_2}) = \underbrace{(\text{SO}, \text{nF})}_{\text{source flow } (-)}.$$

The flow of $x_{j_1}^* = a_{j_1 j_1 1}$ is an (SO, nF)-negative source flow.

- For $a_{j_1 j_1 0} < 0$ and $a_{j_2 j_1 0}(a_{j_1 j_1 1} - a_{j_2 j_1 l_1})(a_{j_1 j_1 1} - a_{j_2 j_1 l_2})^2 < 0$,

$$(a_{j_1 j_1 1}, \dot{x}_{j_2}) = \underbrace{(\text{SI}, \text{nF})}_{\text{sink flow } (-)}.$$

The flow of $x_{j_1}^* = a_{j_1 j_1 1}$ is an (SI,nF)- negative sink flow.

(v_{1b}) At $x_{j_1}^* = a_{j_2j_1l_1} \neq a_{j_1j_1 1}, a_{j_2j_1l_2}$, we have

$$\frac{dx_{j_2}}{dx_{j_1}}\bigg|_{x_{j_1}^* = a_{j_2j_1l_1}} = \frac{a_{j_2j_1 0}}{a_{j_1j_1 0}} \frac{(x_{j_1} - a_{j_2j_1l_1})(x_{j_1} - a_{j_2j_1l_2})^2}{(x_{j_1} - a_{j_1j_1 1})\left[(x_{j_1} - a_{j_1j_1})^2 + b_{j_1j_1}\right]}\bigg|_{x_{j_1}^* = a_{j_2j_1l_1}} = 0.$$

If

$$\frac{d^2x_{j_2}}{dx_{j_1}^2}\bigg|_{x_{j_1}^* = a_{j_2j_1l_1}} = \frac{a_{j_2j_1 0}(a_{j_2j_1l_1} - a_{j_2j_1l_2})^2}{a_{j_1j_1 0}(a_{j_2j_1l_1} - a_{j_1j_1 1})\left[(a_{j_2j_1l_1} - a_{j_1j_1})^2 + b_{j_1j_1}\right]} > 0,$$

the flow at $x_{j_1}^* = a_{j_2j_1l_1}$ is an up-parabola flow in the x_{j_2}-direction, and if

$$\frac{d^2x_{j_2}}{dx_{j_1}^2}\bigg|_{x_{j_1}^* = a_{j_2j_1l_1}} = \frac{2a_{j_2j_1 0}}{a_{j_1j_1 0}(a_{j_2j_1l_1} - a_{j_1j_1 1})\left[(a_{j_2j_1l_1} - a_{j_1j_1})^2 + b_{j_1j_1}\right]} < 0,$$

the flow at $x_{j_1}^* = a_{j_2j_1l_1}$ is a down-parabola flow in the x_{j_2}-direction.

At $x_{j_1}^* = a_{j_2j_1l_1}$, the variational equation is

$$\begin{aligned}
\Delta\dot{x}_{j_2} &= a_{j_2j_1 0}(a_{j_2j_1l_1} - a_{j_1j_1l_2})^2 \Delta x_{j_1}, \\
\dot{x}_{j_1} &= a_{j_1j_1 0}(a_{j_2j_1l_1} - a_{j_1j_1 1})\left[(a_{j_2j_1 1} - a_{j_1j_1})^2 + b_{j_1j_1}\right].
\end{aligned}$$

Due to

$$(a_{j_1j_1 1} - a_{j_1j_1})^2 + b_{j_1j_1} > 0 \text{ and } (a_{j_2j_1l_1} - a_{j_1j_1l_2})^2 > 0,$$

the flows at $x_{j_1}^* = a_{j_2j_1l_1}$ are positive and negative for $a_{j_1j_1 0}(a_{j_2j_1l_1} - a_{j_1j_1 1}) > 0$ and $a_{j_1j_1 0}(a_{j_2j_1l_1} - a_{j_1j_1 1}) < 0$ in the x_{j_1}-direction.

Therefore, the flow at $x_{j_1}^* = a_{j_2j_1l_1}$ for the system has the following properties as in Eqs. (4.83)–(4.86).

- For $a_{j_1j_1 0}(a_{j_2j_1l_1} - a_{j_1j_1 1}) > 0$ and $a_{j_2j_1 0}(a_{j_2j_1l_1} - a_{j_2j_1l_2})^2 > 0$,

$$(\dot{x}_{j_1}, a_{j_2j_1l_1}) = \underbrace{(\mathrm{pF}, \mathrm{UP})}_{\text{up-parabola flow }(+)}.$$

  The flow of $x_{j_1}^* = a_{j_2j_1l_1}$ is a (pF,UP)-positive up-parabola flow.

- For $a_{j_1j_1 0}(a_{j_2j_1l_1} - a_{j_1j_1 1}) < 0$ and $a_{j_2j_1 0}(a_{j_2j_1l_1} - a_{j_2j_1l_2})^2 > 0$,

$$(\dot{x}_{j_1}, a_{j_2j_1l_1}) = \underbrace{(\mathrm{nF}, \mathrm{DP})}_{\text{down-parabola flow }(-)}.$$

  The flow of $x_{j_1}^* = a_{j_2j_1l_1}$ is an (nF,DP)-negative down-parabola flow.

- For $a_{j_1j_10}(a_{j_2j_1l_1}-a_{j_1j_11})>0$ and $a_{j_2j_10}(a_{j_2j_1l_1}-a_{j_2j_1l_2})^2<0$,

$$(\dot{x}_{j_1},a_{j_2j_1l_1})=\underbrace{(\text{pF},\text{DP})}_{\text{down-parabola flow }(+)}.$$

The flow of $x_{j_1}^*=a_{j_2j_1l_1}$ is a (pF, DP)-positive down-parabola flow.

- For $a_{j_1j_10}(a_{j_2j_1l_1}-a_{j_1j_11})<0$ and $a_{j_2j_10}(a_{j_2j_1l_1}-a_{j_2j_1l_2})^2<0$,

$$(\dot{x}_{j_1},a_{j_2j_1l_1})=\underbrace{(\text{nF},\text{UP})}_{\text{up-parabola flow }(-)}.$$

The flow of $x_{j_1}^*=a_{j_2j_1l_1}$ is an (nF,UP)-negative up-parabola flow.

($\text{v}_{1\text{c}}$) At $x_{j_1}^*=a_{j_2j_1l_2}\neq a_{j_1j_11},a_{j_2j_1l_1}$, we have

$$\frac{dx_{j_2}}{dx_{j_1}}\Big|_{x_{j_1}^*=a_{j_2j_1l_2}}=\frac{a_{j_2j_10}}{a_{j_1j_10}}\frac{(x_{j_1}-a_{j_2j_1l_1})(x_{j_1}-a_{j_2j_1l_2})^2}{(x_{j_1}-a_{j_1j_11})\big[(x_{j_1}-a_{j_1j_1})^2+b_{j_1j_1}\big]}\Big|_{x_{j_1}^*=a_{j_2j_1l_2}}=0,$$
$$\frac{d^2x_{j_2}}{dx_{j_1}^2}\Big|_{x_{j_1}^*=a_{j_2j_1l_2}}=0.$$

If

$$\frac{d^3x_{j_2}}{dx_{j_1}^3}\Big|_{x_{j_1}^*=a_{j_2j_1l_2}}=\frac{2a_{j_2j_10}(a_{j_2j_1l_2}-a_{j_2j_1l_1})}{a_{j_1j_10}(a_{j_2j_1l_2}-a_{j_1j_11})\big[(a_{j_1j_11}-a_{j_1j_1})^2+b_{j_1j_1}\big]}>0,$$

the flow at $x_{j_1}^*=a_{j_2j_1l_2}$ is an increasing-inflection flow in the x_{j_2}-direction, and if

$$\frac{d^3x_{j_2}}{dx_{j_1}^3}\Big|_{x_{j_1}^*=a_{j_2j_1l_2}}=\frac{2a_{j_2j_10}(a_{j_2j_1l_2}-a_{j_2j_1l_1})}{a_{j_1j_10}(a_{j_2j_1l_2}-a_{j_1j_11})\big[(a_{j_1j_11}-a_{j_1j_1})^2+b_{j_1j_1}\big]}<0,$$

the flow at $x_{j_1}^*=a_{j_2j_1l_2}$ is a deceasing-inflection flow in the x_{j_2}-direction.

At $x_{j_1}^*=a_{j_2j_1l_2}$, the variational equation is

$$\Delta\dot{x}_{j_2}=a_{j_2j_10}(a_{j_2j_1l_2}-a_{j_2j_1l_1})(\Delta x_{j_1})^2,$$
$$\dot{x}_{j_1}=a_{j_1j_10}(a_{j_2j_1l_2}-a_{j_1j_11})\big[(a_{j_2j_11}-a_{j_1j_1})^2+b_{j_1j_1}\big].$$

Due to

$$(a_{j_1j_11}-a_{j_1j_1})^2+b_{j_1j_1}>0,$$

the flows at $x_{j_1}^*=a_{j_2j_1l_2}$ are positive and negative for $a_{j_1j_10}(a_{j_2j_1l_2}-a_{j_1j_11})>0$ and $a_{j_1j_10}(a_{j_2j_1l_2}-a_{j_1j_11})<0$ in the x_{j_1}-direction.

Therefore, the flow at $x_{j_1}^* = a_{j_2j_1l_2}$ for the system has the following properties as in Eqs. (4.87)–(4.90).

- For $a_{j_1j_10}(a_{j_2j_1l_2} - a_{j_1j_11}) > 0$ and $a_{j_2j_10}(a_{j_2j_1l_2} - a_{j_2j_1l_1}) > 0$,

$$(\dot{x}_{j_1}, a_{j_2j_1l_2}) = \underbrace{(\text{pF}, \text{II})}_{\text{increasing-inflection flow } (+)}.$$

The flow of $x_{j_1}^* = a_{j_2j_1l_2}$ is a (pF,II)-positive increasing-inflection flow.

- For $a_{j_1j_10}(a_{j_2j_1l_2} - a_{j_1j_11}) < 0$ and $a_{j_2j_10}(a_{j_2j_1l_2} - a_{j_2j_1l_1}) > 0$,

$$(\dot{x}_{j_1}, a_{j_2j_1l_2}) = \underbrace{(\text{nF}, \text{DI})}_{\text{decreasing-inflection flow } (-)}.$$

The flow of $x_{j_1}^* = a_{j_2j_1l_2}$ is an (nF,DI) negative decreasing-inflection flow.

- For $a_{j_1j_10}(a_{j_2j_1l_1} - a_{j_1j_11}) > 0$ and $a_{j_2j_10}(a_{j_2j_1l_2} - a_{j_2j_1l_1}) < 0$,

$$(\dot{x}_{j_1}, a_{j_2j_1l_2}) = \underbrace{(\text{pF}, \text{DI})}_{\text{decreasing-inflection flow } (+)}.$$

The flow of $x_{j_1}^* = a_{j_2j_1l_2}$ is a (pF, DI)-positive decreasing-inflection flow.

- For $a_{j_1j_10}(a_{j_2j_1l_2} - a_{j_1j_11}) < 0$ and $a_{j_2j_10}(a_{j_2j_1l_2} - a_{j_2j_1l_1}) < 0$,

$$(\dot{x}_{j_1}, a_{j_2j_1l_2}) = \underbrace{(\text{nF}, \text{II})}_{\text{increasing-inflection flow } (-)}.$$

The flow of $x_{j_1}^* = a_{j_2j_1l_2}$ is an (nF,II)-negative increasing-inflection flow.

– The increasing-inflection and decreasing-inflection flows are the appearing (or switching) bifurcation for the up-parabola and down-parabola flows.

(v_2) For $a_{j_1j_11} = a_{j_2j_1l_1}$, we have

$$\begin{aligned}
\frac{dx_{j_2}}{dx_{j_1}} &= \frac{a_{j_2j_10}}{a_{j_1j_10}} \frac{(x_{j_1} - a_{j_2j_1l_1})(x_{j_1} - a_{j_2j_1l_2})^2}{(x_{j_1} - a_{j_1j_11})\left[(x_{j_1} - a_{j_1j_1})^2 + b_{j_1j_1}\right]} \\
&= \frac{a_{j_2j_10}}{a_{j_1j_10}} \left\{1 + \frac{2(a_{j_1j_1} - a_{j_2j_1l_2})(x_{j_1} - a_{j_1j_1})}{\left[(x_{j_1} - a_{j_1j_1})^2 + b_{j_1j_1}\right]} + \frac{(a_{j_1j_1} - a_{j_2j_1l_2})^2 - b_{j_1j_1}}{\left[(x_{j_1} - a_{j_1j_1})^2 + b_{j_1j_1}\right]}\right\}.
\end{aligned}$$

With an initial condition (x_{j_10}, x_{j_20}),the first integral manifold is

$$
\begin{aligned}
x_{j_2}-x_{j_20} &= \frac{a_{j_2j_10}}{a_{j_1j_10}}\Big\{x_{j_1}-x_{j_10}+(a_{j_1j_1}-a_{j_2j_1l_2})\ln\frac{|(x_{j_1}-a_{j_1j_1})^2+b_{j_1j_1}|}{|(x_{j_10}-a_{j_1j_1})^2+b_{j_1j_1}|}\\
&\quad+\left[(a_{j_1j_1}-a_{j_2j_1l_2})^2-b_{j_1j_1}\right]\\
&\quad\times\frac{1}{\sqrt{b_{j_1j_1}}}(\arctan\frac{x_{j_1}-a_{j_1j_1}}{\sqrt{b_{j_1j_1}}}-\arctan\frac{x_{j_10}-a_{j_1j_1}}{\sqrt{b_{j_1j_1}}})\Big\},\\
&\quad j_1,j_2\in\{1,2\},j_1\neq j_2.
\end{aligned}
$$

In phase space, for $x_{j_1}^*=a_{j_1j_11}=a_{j_2j_1l_1}$, if

$$
\left.\frac{dx_{j_2}}{dx_{j_1}}\right|_{x_{j_1}^*=a_{j_1j_11}=a_{j_2j_1l_1}}=\frac{a_{j_2j_10}}{a_{j_1j_10}}\frac{(a_{j_2j_1l_1}-a_{j_2j_1l_2})^2}{\left[(a_{j_1j_11}-a_{j_1j_1})^2+b_{j_1j_1}\right]}>0,
$$

the flow at $x_{j_1}^*=a_{j_1j_11}=a_{j_2j_1l_1}$ is an increasing-inflection flow in the x_{j_2}-direction, and if

$$
\left.\frac{dx_{j_2}}{dx_{j_1}}\right|_{x_{j_1}^*=a_{j_1j_11}=a_{j_2j_1l_1}}=\frac{a_{j_2j_10}}{a_{j_1j_10}}\frac{(a_{j_2j_1l_1}-a_{j_2j_1l_2})^2}{\left[(a_{j_1j_11}-a_{j_1j_1})^2+b_{j_1j_1}\right]}<0,
$$

the flow at $x_{j_1}^*=a_{j_1j_11}=a_{j_2j_1l_1}$ is a decreasing-inflection flow in the x_{j_2}-direction.

At $x_{j_1}^*=a_{j_1j_11}=a_{j_2j_1l_1}$, the variational equations are

$$
\begin{aligned}
\Delta\dot{x}_{j_2}&=a_{j_2j_10}(a_{j_2j_1l_1}-a_{j_2j_1l_2})^2\Delta x_{j_1},\\
\Delta\dot{x}_{j_1}&=a_{j_1j_10}\left[(a_{j_1j_1}-a_{j_1j_1})^2+b_{j_1j_1}\right]\Delta x_{j_1}.
\end{aligned}
$$

Due to

$$
(a_{j_1j_1}-a_{j_1j_1})^2+b_{j_1j_1}>0,
$$

the flows at $x_{j_1}^*=a_{j_1j_11}=a_{j_2j_1l_1}$ are source and sink for $a_{j_1j_10}>0$ and $a_{j_1j_10}<0$ in the x_{j_1}-direction.

Thus, the infinite-equilibrium of $x_{j_1}^*=a_{j_1j_11}=a_{j_2j_1l_1}$ with $\bar{x}_{j_2}\in(-\infty,\infty)$ has the following properties as in Eqs. (4.92)–(4.95):

- For $a_{j_1j_10}>0$ and $a_{j_2j_10}(a_{j_2j_1l_1}-a_{j_2j_1l_2})^2>0$,

$$
(a_{j_1j_11},\bar{x}_{j_2})=\underbrace{(\mathrm{SO},\mathrm{II})}_{\text{increasing-inflection source}}.
$$

  The equilibrium of $x_{j_1}^*=a_{j_1j_11}=a_{j_2j_1l_1}$ is an (SO,II)-increasing-inflection source infinite-equilibrium.
- For $a_{j_1j_10}<0$ and $a_{j_2j_10}(a_{j_2j_1l_1}-a_{j_2j_1l_2})^2>0$,

$$(a_{j_1j_1 1}, \bar{x}_{j_2}) = \underbrace{(\text{SI}, \text{DI})}_{\text{decreasing-inflection sink}}.$$

The equilibrium of $x_{j_1}^* = a_{j_1j_1 1} = a_{j_2j_1l_1}$ is an (SI,DI)-deceasing-inflection sink infinite-equilibrium.

- For $a_{j_1j_1 0} > 0$ and $a_{j_2j_1 0}(a_{j_2j_1l_1} - a_{j_2j_1l_2})^2 < 0$,

$$(a_{j_1j_1 1}, \bar{x}_{j_2}) = \underbrace{(\text{SO}, \text{DI})}_{\text{decreasing-inflection source}}.$$

The equilibrium of $x_{j_1}^* = a_{j_1j_1 1} = a_{j_2j_1l_1}$ is an (SO,DI)-deceasing-inflection source infinite-equilibrium.

- For $a_{j_1j_1 0} < 0$ and $a_{j_2j_1 0}(a_{j_2j_1l_1} - a_{j_2j_1l_2})^2 < 0$,

$$(a_{j_1j_1 1}, \bar{x}_{j_2}) = \underbrace{(\text{SI}, \text{II})}_{\text{increasing-inflection sink}}.$$

The equilibrium of $x_{j_1}^* = a_{j_1j_1 1} = a_{j_2j_1l_1}$ is an (SI,II)-increasing-inflection sink infinite-equilibrium.

– The (1:1)-increasing-inflection and decreasing-inflection source and sink are the switching bifurcations for the sink and source flows with up-parabola and down-parabola flows.

(v_3) For $a_{j_1j_1 1} = a_{j_2j_1l_2}$, we have

$$\begin{aligned}
\frac{dx_{j_2}}{dx_{j_1}} &= \frac{a_{j_2j_1 0}}{a_{j_1j_1 0}} \frac{(x_{j_1} - a_{j_2j_1l_1})(x_{j_1} - a_{j_2j_1l_2})^2}{(x_{j_1} - a_{j_1j_1 1})\left[(x_{j_1} - a_{j_1j_1})^2 + b_{j_1j_1}\right]} \\
&= \frac{a_{j_2j_1 0}}{a_{j_1j_1 0}} \left[1 + (2a_{j_1j_1} - a_{j_2j_1l_1} - a_{j_2j_1l_2}) \frac{(x_{j_1} - a_{j_1j_1})}{(x_{j_1} - a_{j_1j_1})^2 + b_{j_1j_1}} \right. \\
&\quad \left. + \frac{(a_{j_1j_1} - a_{j_2j_1l_1})(a_{j_1j_1} - a_{j_2j_1l_2}) - b_{j_1j_1}}{(x_{j_1} - a_{j_1j_1})^2 + b_{j_1j_1}} \right].
\end{aligned}$$

With an initial condition $(x_{j_1 0}, x_{j_2 0})$, the first integral manifold is

$$\begin{aligned}
x_{j_2} - x_{j_2 0} &= \frac{a_{j_2 j_1 0}}{a_{j_1 j_1 0}} \Big\{ x_{j_1} - x_{j_1 0} + \frac{1}{2}(2a_{j_1 j_1} - a_{j_2 j_1 l_1} - a_{j_2 j_1 l_2}) \\
&\quad \times \ln \frac{|\,(x_{j_1} - a_{j_1 j_1})^2 + b_{j_1 j_1}\,|}{|\,(x_{j_1 0} - a_{j_1 j_1})^2 + b_{j_1 j_1}\,|} \\
&\quad + \frac{1}{\sqrt{b_{j_1 j_1}}} \left[(a_{j_1 j_1} - a_{j_2 j_1 l_1})(a_{j_1 j_1} - a_{j_2 j_1 l_2}) - b_{j_1 j_1} \right] \\
&\quad \times (\arctan \frac{x_{j_1} - a_{j_1 j_1}}{\sqrt{b_{j_1 j_1}}} - \arctan \frac{x_{j_1 0} - a_{j_1 j_1}}{\sqrt{b_{j_1 j_1}}}) \Big\}; \\
&\quad j_1, j_2 \in \{1,2\}, j_1 \neq j_2.
\end{aligned}$$

In phase space, for $x_{j_1}^* = a_{j_1 j_1 1} = a_{j_2 j_1 l_2}$, we have

$$\frac{dx_{j_2}}{dx_{j_1}}\Big|_{x_{j_1}^* = a_{j_1 j_1 1} = a_{j_2 j_1 l_2}} = \frac{a_{j_2 j_1 0}}{a_{j_1 j_1 0}} \frac{(x_{j_1} - a_{j_2 j_1 l_1})(x_{j_1} - a_{j_2 j_1 l_2})}{\left[(x_{j_1} - a_{j_1 j_1})^2 + b_{j_1 j_1}\right]}\Big|_{x_{j_1}^* = a_{j_1 j_1 1} = a_{j_2 j_1 l_2}} = 0.$$

If

$$\frac{d^2 x_{j_2}}{dx_{j_1}^2}\Big|_{x_{j_1}^* = a_{j_1 j_1 1} = a_{j_2 j_1 l_1}} = \frac{a_{j_2 j_1 0}(a_{j_2 j_1 l_2} - a_{j_2 j_1 l_1})}{a_{j_1 j_1 0}\left[(a_{j_1 j_1 1} - a_{j_1 j_1})^2 + b_{j_1 j_1}\right]} > 0,$$

the flow at $x_{j_1}^* = a_{j_1 j_1 1} = a_{j_2 j_1 l_2}$ is an up-parabola flow in the x_{j_2}-direction, and if

$$\frac{d^2 x_{j_2}}{dx_{j_1}^2}\Big|_{x_{j_1}^* = a_{j_1 j_1 1} = a_{j_2 j_1 l_1}} = \frac{a_{j_2 j_1 0}(a_{j_2 j_1 l_2} - a_{j_2 j_1 l_1})}{a_{j_1 j_1 0}\left[(a_{j_1 j_1 1} - a_{j_1 j_1})^2 + b_{j_1 j_1}\right]} < 0,$$

the flow at $x_{j_1}^* = a_{j_1 j_1 1} = a_{j_2 j_1 l_2}$ is a down-parabola flow in the x_{j_2}-direction.

At $x_{j_1}^* = a_{j_1 j_1 1} = a_{j_2 j_1 l_2}$, the variational equations are

$$\begin{aligned}
\Delta\dot{x}_{j_2} &= a_{j_2 j_1 0}(a_{j_2 j_1 l_2} - a_{j_2 j_1 l_1})(\Delta x_{j_1})^2, \\
\Delta\dot{x}_{j_1} &= a_{j_1 j_1 0}\left[(a_{j_2 j_1 1} - a_{j_1 j_1})^2 + b_{j_1 j_1}\right]\Delta x_{j_1}.
\end{aligned}$$

Due to

$$(a_{j_1 j_1 1} - a_{j_1 j_1})^2 + b_{j_1 j_1} > 0,$$

the flows at $x_{j_1}^* = a_{j_1 j_1 1} = a_{j_2 j_1 l_2}$ are source and sink for $a_{j_1 j_1 0} > 0$ and $a_{j_1 j_1 0} < 0$ in the x_{j_1}-direction.

Thus, the infinite-equilibrium of $x_{j_1}^* = a_{j_1 j_1 1} = a_{j_2 j_1 l_2}$ with $\bar{x}_{j_2} \in (-\infty, \infty)$ has the following properties as in Eqs. (4.97)–(4.100).

- For $a_{j_1j_10}>0$ and $a_{j_2j_10}(a_{j_2j_1l_2}-a_{j_2j_1l_1})>0$,

$$(a_{j_1j_11},\bar{x}_{j_2})=\underbrace{(\text{SO},\text{UP})}_{\text{up-parabola source}}.$$

The equilibrium of $x^*_{j_1}=a_{j_1j_11}=a_{j_2j_1l_2}$ is an (SO,UP)-up-parabola source infinite-equilibrium.

- For $a_{j_1j_10}<0$ and $a_{j_2j_10}(a_{j_2j_1l_2}-a_{j_2j_1l_1})>0$,

$$(a_{j_1j_11},\bar{x}_{j_2})=\underbrace{(\text{SI},\text{DP})}_{\text{down-parabola sink}}.$$

The equilibrium of $x^*_{j_1}=a_{j_1j_11}=a_{j_2j_1l_2}$ is an (SI,DP)-down-parabola sink infinite-equilibrium.

- For $a_{j_1j_10}>0$ and $a_{j_2j_10}(a_{j_2j_1l_2}-a_{j_2j_1l_1})<0$,

$$(a_{j_1j_11},\bar{x}_{j_2})=\underbrace{(\text{SO},\text{DP})}_{\text{down-parabola source}}.$$

The equilibrium of $x^*_{j_1}=a_{j_1j_11}=a_{j_2j_1l_2}$ is an (SO,DP)-down-parabola source infinite-equilibrium.

- For $a_{j_1j_10}<0$ and $a_{j_2j_10}(a_{j_2j_1l_2}-a_{j_2j_1l_1})<0$,

$$(a_{j_1j_11},\bar{x}_{j_2})=\underbrace{(\text{SI},\text{UP})}_{\text{up-parabola sink}}.$$

The equilibrium of $x^*_{j_1}=a_{j_1j_11}=a_{j_2j_1l_2}$ is an (SI,UP)-up-parabola sink infinite-equilibrium.

– The up-parabola and down-parabola source and sink are the switching bifurcations for the sink and source flows with increasing-inflection and decreasing-inflection flows.

(vi) If

$$\begin{aligned}&\Delta_{j_1j_1}<0 \text{ and } \Delta_{j_2j_1}=0,\\&a_{j_2j_1l_2}=a_{j_2j_1l_1}=a_{j_2j_11},\end{aligned}$$

then the dynamical system becomes

$$\begin{aligned}\dot{x}_{j_2}&=a_{j_2j_10}(x_{j_1}-a_{j_2j_11})^3,\\\dot{x}_{j_1}&=a_{j_1j_10}(x_{j_1}-a_{j_1j_11})\left[(x_{j_1}-a_{j_1j_1})^2+b_{j_1j_1}\right].\end{aligned}$$

In phase space,

$$
\begin{aligned}
\frac{dx_{j_2}}{dx_{j_1}} &= \frac{a_{j_2j_10}}{a_{j_1j_10}} \frac{(x_{j_1}-a_{j_2j_11})^3}{(x_{j_1}-a_{j_1j_11})\left[(x_{j_1}-a_{j_1j_1})^2+b_{j_1j_1}\right]} \\
&= \frac{a_{j_2j_10}}{a_{j_1j_10}} \left\{1+\frac{2(x_{j_1}-a_{j_1j_1})}{(x_{j_1}-a_{j_1j_1})^2+b_{j_1j_1}}\left[(a_{j_1j_1}-a_{j_1j_11})+\frac{3}{2}(a_{j_1j_11}-a_{j_2j_11})\right]\right. \\
&\quad +\frac{1}{(x_{j_1}-a_{j_1j_1})^2+b_{j_1j_1}}\left[3(a_{j_1j_11}-a_{j_2j_11})(a_{j_1j_1}-a_{j_2j_11})\right. \\
&\quad \left.+(a_{j_1j_1}-a_{j_1j_11})^2-b_{j_1j_1}\right] \\
&\quad +\frac{(a_{j_1j_11}-a_{j_2j_11})^3}{(a_{j_1j_1}-a_{j_1j_11})^2+b_{j_1j_1}}\left[\frac{1}{(x_{j_1}-a_{j_1j_11})}-\frac{(x_{j_1}-a_{j_1j_1})}{(x_{j_1}-a_{j_1j_1})^2+b_{j_1j_1}}\right. \\
&\quad \left.\left.+\frac{(a_{j_1j_1}-a_{j_1j_11})}{(x_{j_1}-a_{j_1j_1})^2+b_{j_1j_1}}\right]\right\}.
\end{aligned}
$$

With an initial condition (x_{j_10}, x_{j_20}), the first integral manifold is

$$
\begin{aligned}
x_{j_2}-x_{j_20} &= \frac{a_{j_2j_10}}{a_{j_1j_10}}\left\{x_{j_1}-x_{j_10}+\left[\frac{3}{2}(a_{j_1j_11}-a_{j_2j_11})+(a_{j_1j_1}-a_{j_1j_11})\right]\right. \\
&\quad \times \ln\frac{|\,(x_{j_1}-a_{j_1j_1})^2+b_{j_1j_1}\,|}{|\,(x_{j_10}-a_{j_1j_1})^2+b_{j_1j_1}\,|}+\left[3(a_{j_1j_11}-a_{j_2j_11})(a_{j_1j_1}-a_{j_2j_11})\right. \\
&\quad \left.+(a_{j_1j_1}-a_{j_1j_11})^2-b_{j_1j_1}\right] \\
&\quad \times\frac{1}{\sqrt{b_{j_1j_1}}}(\arctan\frac{x_{j_1}-a_{j_1j_1}}{\sqrt{b_{j_1j_1}}}-\arctan\frac{x_{j_10}-a_{j_1j_1}}{\sqrt{b_{j_1j_1}}}) \\
&\quad +\frac{(a_{j_1j_11}-a_{j_2j_11})^3}{(a_{j_1j_11}-a_{j_1j_1})^2+b_{j_1j_1}}\left[\ln\frac{|\,x_{j_1}-a_{j_1j_11}\,|}{|\,x_{j_10}-a_{j_1j_11}\,|}\right. \\
&\quad -\frac{1}{2}\ln\frac{|\,(x_{j_1}-a_{j_1j_1})^2+b_{j_1j_1}\,|}{|\,(x_{j_10}-a_{j_1j_1})^2+b_{j_1j_1}\,|} \\
&\quad \left.\left.+\frac{a_{j_1j_1}-a_{j_1j_11}}{\sqrt{b_{j_1j_1}}}(\arctan\frac{x_{j_1}-a_{j_1j_1}}{\sqrt{b_{j_1j_1}}}-\arctan\frac{x_{j_10}-a_{j_1j_1}}{\sqrt{b_{j_1j_1}}})\right]\right\}, \\
&\quad j_1, j_2\in\{1,2\}, j_1\neq j_2.
\end{aligned}
$$

(vi_{1a}) For $x_{j_1}^*=a_{j_1j_11}\neq a_{j_2j_11}$, the variational equation is

$$
\begin{aligned}
\dot{x}_{j_2} &= a_{j_2j_10}(a_{j_1j_11}-a_{j_2j_11})^3, \\
\Delta\dot{x}_{j_1} &= a_{j_1j_10}\left[(x_{j_1}-a_{j_1j_1})^2+b_{j_1j_1}\right]\Delta x_{j_1}
\end{aligned}
$$

The flow at $x_{j_1}^* = a_{j_1j_1 1} \neq a_{j_2j_1 1}$ is source and sink for $a_{j_1j_1 0} > 0$ and $a_{j_1j_1 0} > 0$ in the x_{j_1}-direction. The flow at $x_{j_1}^* = a_{j_1j_1 1} \neq a_{j_2j_1 1}$ is positive and negative for $a_{j_2j_1 0}(a_{j_1j_1 1} - a_{j_2j_1 1})^3 > 0$ and $a_{j_2j_1 0}(a_{j_1j_1 1} - a_{j_2j_1 1})^3 < 0$ in the x_{j_2}-direction.

Therefore, the flow at $x_{j_1}^* = a_{j_1j_1 1} \neq a_{j_2j_1 1}$ for the system has the following properties as in Eqs. (4.104)–4.107).

- For $a_{j_1j_1 0} > 0$ and $a_{j_2j_1 0}(a_{j_1j_1 1} - a_{j_2j_1 1})^3 > 0$,

$$(a_{j_1j_1 1}, \dot{x}_{j_2}) = \underbrace{(\text{SO}, \text{pF})}_{\text{source flow } (+)}.$$

  The flow of $x_{j_1}^* = a_{j_1j_1 1}$ is an (SO,pF)-positive source flow.
- For $a_{j_1j_1 0} < 0$ and $a_{j_2j_1 0}(a_{j_1j_1 1} - a_{j_2j_1 1})^3 > 0$,

$$(a_{j_1j_1 1}, \dot{x}_{j_2}) = \underbrace{(\text{SI}, \text{pF})}_{\text{sink flow } (+)}.$$

  The flow of $x_{j_1}^* = a_{j_1j_1 1}$ is an (SI, pF)-positive sink flow.
- For $a_{j_1j_1 0} > 0$ and $a_{j_2j_1 0}(a_{j_1j_1 1} - a_{j_2j_1 1})^3 < 0$,

$$(a_{j_1j_1 1}, \dot{x}_{j_2}) = \underbrace{(\text{SO}, \text{nF})}_{\text{source flow } (-)}.$$

  The flow of $x_{j_1}^* = a_{j_1j_1 1}$ is an (SO, nF)-negative source flow.
- For $a_{j_1j_1 0} < 0$ and $a_{j_2j_1 0}(a_{j_1j_1 1} - a_{j_2j_1 1})^3 < 0$,

$$(a_{j_1j_1 1}, \dot{x}_{j_2}) = \underbrace{(\text{SI}, \text{nF})}_{\text{sink flow } (-)}.$$

  The flow of $x_{j_1}^* = a_{j_1j_1 1}$ is an (SI,nF)-negative sink flow.

($\text{vi}_{1\text{b}}$) For $x_{j_1}^* = a_{j_2j_1 1} \neq a_{j_1j_1 1}$, in phase space, we have

$$\frac{dx_{j_2}}{dx_{j_1}}\bigg|_{x_{j_1}^* = a_{j_2j_1 1}} = \frac{a_{j_2j_1 0}}{a_{j_1j_1 0}} \frac{(x_{j_1} - a_{j_2j_1 1})^3}{(x_{j_1} - a_{j_1j_1 1})\left[(x_{j_1} - a_{j_1j_1})^2 + b_{j_1j_1}\right]}\bigg|_{x_{j_1}^* = a_{j_2j_1 1}} = 0,$$

$$\frac{d^2 x_{j_2}}{dx_{j_1}^2}\bigg|_{x_{j_1}^* = a_{j_2j_1 1}} = \frac{d^3 x_{j_2}}{dx_{j_1}^3}\bigg|_{x_{j_1}^* = a_{j_2j_1 1}} = 0.$$

If

$$\frac{d^4 x_{j_2}}{dx_{j_1}^4}\Big|_{x_{j_1}^*=a_{j_2j_1 1}} = \frac{a_{j_2j_1 0}}{a_{j_1j_1 0}} \frac{6}{(a_{j_2j_1 1}-a_{j_1j_1 1})\left[(x_{j_1}-a_{j_1j_1})^2+b_{j_1j_1}\right]} > 0,$$

there is a third-order up-parabola flow in the x_{j_2}-direction, and if

$$\frac{d^4 x_{j_2}}{dx_{j_1}^4}\Big|_{x_{j_1}^*=a_{j_2j_1 1}} = \frac{a_{j_2j_1 0}}{a_{j_1j_1 0}} \frac{6}{(a_{j_2j_1 1}-a_{j_1j_1 1})\left[(x_{j_1}-a_{j_1j_1})^2+b_{j_1j_1}\right]} < 0,$$

there is a third-order down-parabola flow in the x_{j_2}-direction.

At $x_{j_1}^* = a_{j_2j_1 1} \neq a_{j_1j_1 1}$, the variational equation is

$$\begin{aligned}\Delta\dot{x}_{j_2} &= a_{j_2j_1 0}(\Delta x_{j_1})^3,\\ \dot{x}_{j_1} &= a_{j_1j_1 0}(a_{j_2j_1 1}-a_{j_1j_1 1})\left[(x_{j_1}-a_{j_1j_1})^2+b_{j_1j_1}\right].\end{aligned}$$

The flows at $x_{j_1}^* = a_{j_2j_1 1} \neq a_{j_1j_1 1}$ are positive and negative for $a_{j_1j_1 0}(a_{j_2j_1 1}-a_{j_1j_1 1}) > 0$ and $a_{j_1j_1 0}(a_{j_2j_1 1}-a_{j_1j_1 1}) < 0$, respectively.

Therefore, the flow at $x_{j_1}^* = a_{j_2j_1 1}$ for the system has the following properties as in Eqs. (4.108)–and (4.111).

- For $a_{j_1j_1 0}(a_{j_2j_1 1}-a_{j_1j_1 1}) > 0$ and $a_{j_2j_1 0} > 0$,

$$(\dot{x}_{j_1}, a_{j_2j_1 1}) = \underbrace{(\text{pF}, 3^{\text{rd}}\text{UP})}_{\text{third-order up-parabola flow } (+)}.$$

  The flow of $x_{j_1}^* = a_{j_2j_1 1}$ is a (pF,3^{rd}UP)-positive third-order up-parabola flow.

- For $a_{j_1j_1 0}(a_{j_2j_1 1}-a_{j_1j_1 1}) < 0$ and $a_{j_2j_1 0} > 0$,

$$(\dot{x}_{j_1}, a_{j_2j_1 1}) = \underbrace{(\text{nF}, 3^{\text{rd}}\text{DP})}_{\text{third-order down-parabola flow } (-)}.$$

  The flow of $x_{j_1}^* = a_{j_2j_1 1}$ is an (nF,3^{rd}DP)-negative third-order down-parabola flow.

- For $a_{j_1j_1 0}(a_{j_2j_1 1}-a_{j_1j_1 1}) > 0$ and $a_{j_2j_1 0} < 0$,

$$(\dot{x}_{j_1}, a_{j_2j_1 1}) = \underbrace{(\text{pF}, 3^{\text{rd}}\text{DP})}_{\text{third-order down-parabola flow } (+)}.$$

  The flow of $x_{j_1}^* = a_{j_2j_1 1}$ is a (pF,3^{rd}DP)-positive third-order down-parabola flow.

- For $a_{j_1j_10}(a_{j_2j_11}-a_{j_1j_11})<0$ and $a_{j_2j_10}<0$,

$$(\dot{x}_{j_1},a_{j_2j_11})=\underbrace{(\mathrm{nF},3^{\mathrm{rd}}\mathrm{UP})}_{\text{third-order up-parabola flow }(-)}.$$

The flow of $x_{j_1}^*=a_{j_2j_11}$ is an (nF,3rdUP)-negative third-order up-parabola flow.

– The third-order up-parabola and down-parabola flows are the appearing (or switching) bifurcations for the up-parabola, down-parabola, and down-parabola flows and down-parabola, up-parabola, and down-parabola flows.

(vi$_2$) For $a_{j_1j_11}=a_{j_2j_11}$, then the dynamical system becomes

$$\begin{aligned}\dot{x}_{j_2}&=a_{j_2j_10}(x_{j_1}-a_{j_2j_11})^3,\\ \dot{x}_{j_1}&=a_{j_1j_10}(x_{j_1}-a_{j_1j_11})\left[(x_{j_1}-a_{j_1j_1})^2+b_{j_1j_1}\right].\end{aligned}$$

In phase space,

$$\begin{aligned}\frac{dx_{j_2}}{dx_{j_1}}&=\frac{a_{j_2j_10}}{a_{j_1j_10}}\frac{(x_{j_1}-a_{j_2j_11})^3}{(x_{j_1}-a_{j_1j_11})\left[(x_{j_1}-a_{j_1j_1})^2+b_{j_1j_1}\right]}\\ &=\frac{a_{j_2j_10}}{a_{j_1j_10}}\left\{1+\frac{2(x_{j_1}-a_{j_1j_1})}{(x_{j_1}-a_{j_1j_1})^2+b_{j_1j_1}}(a_{j_1j_1}-a_{j_2j_11})\right.\\ &\quad\left.+\frac{1}{(x_{j_1}-a_{j_1j_1})^2+b_{j_1j_1}}\left[(a_{j_1j_1}-a_{j_1j_11})^2-b_{j_1j_1}\right]\right\}.\end{aligned}$$

With an initial condition (x_{j_10},x_{j_20}), the first integral manifold is

$$\begin{aligned}x_{j_2}-x_{j_20}&=\frac{a_{j_2j_10}}{a_{j_1j_10}}\left[x_{j_1}-x_{j_10}+(a_{j_1j_1}-a_{j_2j_11})\ln\frac{|(x_{j_1}-a_{j_1j_1})^2+b_{j_1j_1}|}{|(x_{j_10}-a_{j_1j_1})^2+b_{j_1j_1}|}\right.\\ &\quad\left.+\frac{(a_{j_1j_1}-a_{j_1j_11})^2-b_{j_1j_1}}{\sqrt{b_{j_1j_1}}}\left(\arctan\frac{x_{j_1}-a_{j_1j_1}}{\sqrt{b_{j_1j_1}}}-\arctan\frac{x_{j_10}-a_{j_1j_1}}{\sqrt{b_{j_1j_1}}}\right)\right],\\ &\quad j_1,j_2\in\{1,2\},j_1\neq j_2.\end{aligned}$$

For $x_{j_1}^*=a_{j_1j_11}=a_{j_2j_11}$,

$$\begin{aligned}&\left.\frac{dx_{j_2}}{dx_{j_1}}\right|_{x_{j_1}^*=a_{j_1j_11}=a_{j_2j_11}}=\left.\frac{a_{j_2j_10}}{a_{j_1j_10}}\frac{(x_{j_1}-a_{j_2j_11})^2}{(x_{j_1}-a_{j_1j_1})^2+b_{j_1j_1}}\right|_{x_{j_1}^*=a_{j_1j_11}=a_{j_2j_11}}=0,\\ &\left.\frac{d^2x_{j_2}}{dx_{j_1}^2}\right|_{x_{j_1}^*=a_{j_1j_11}=a_{j_2j_11}}=0.\end{aligned}$$

If

$$\frac{d^3x_{j_2}}{dx_{j_1}^3}\bigg|_{x_{j_1}^*=a_{j_1j_11}=a_{j_2j_11}} = \frac{a_{j_2j_10}}{a_{j_1j_10}}\frac{2}{(a_{j_1j_11}-a_{j_1j_1})^2+b_{j_1j_1}} > 0,$$

there is an increasing-inflection flow in the x_{j_2}-direction, and if

$$\frac{d^3x_{j_2}}{dx_{j_1}^3}\bigg|_{x_{j_1}^*=a_{j_1j_11}=a_{j_2j_11}} = \frac{a_{j_2j_10}}{a_{j_1j_10}}\frac{2}{(a_{j_1j_11}-a_{j_1j_1})^2+b_{j_1j_1}} < 0,$$

there is a decreasing-inflection flow in the x_{j_2}-direction.

At $x_{j_1}^*=a_{j_2j_11}=a_{j_1j_11}$, the variational equations are

$$\begin{aligned}\Delta\dot{x}_{j_2} &= a_{j_2j_10}(\Delta x_{j_1})^3,\\ \Delta\dot{x}_{j_1} &= a_{j_1j_10}\left[(x_{j_1}-a_{j_1j_1})^2+b_{j_1j_1}\right]\Delta x_{j_1}.\end{aligned}$$

The flows at $x_{j_1}^*=a_{j_2j_11}=a_{j_1j_11}$ are source and sink for $a_{j_1j_10}>0$ and $a_{j_1j_10}<0$, respectively.

Therefore, the infinite-equilibrium of $x_{j_1}^*=a_{j_1j_11}=a_{j_2j_11}$ with $\bar{x}_{j_2}\in(-\infty,\infty)$ for the system has the following properties as in Eqs. (4.113)–(4.116).

- For $a_{j_1j_10}(a_{j_2j_11}-a_{j_1j_11})>0$ and $a_{j_2j_10}>0$,

$$(a_{j_1j_11},\bar{x}_{j_2}) = \underbrace{(\text{SO},(1:3)\text{II})}_{(1:3)\text{-increasing-inflection source}}.$$

  The (SO,(1:3)II)-infinite-equilibrium at $x_{j_1}^*=a_{j_2j_11}$ is a (1:3)-increasing-inflection source.

- For $a_{j_1j_10}(a_{j_2j_11}-a_{j_1j_11})<0$ and $a_{j_2j_10}>0$,

$$(a_{j_1j_11},\bar{x}_{j_2}) = \underbrace{(\text{SI},(1:3)\text{DI})}_{(1:3)\text{-decreasing-inflection sink}}.$$

  The (SI,(1:3)DI)-infinite-equilibrium at $x_{j_1}^*=a_{j_2j_11}$ is a (1:3)-decreasing-inflection sink.

- For $a_{j_1j_10}(a_{j_2j_11}-a_{j_1j_11})>0$ and $a_{j_2j_10}<0$,

$$(a_{j_1j_11},\bar{x}_{j_2}) = \underbrace{(\text{SO},(1:3)\text{DI})}_{(1:3)\text{-decreasing-inflection source}}.$$

  The (SO,(1:3)DI)-infinite-equilibrium at $x_{j_1}^*=a_{j_2j_11}$ is a (1:3)-decreasing-inflection source.

- For $a_{j_1j_10}(a_{j_2j_11}-a_{j_1j_11})<0$ and $a_{j_2j_10}<0$,

$$(a_{j_1j_11},\bar{x}_{j_2})=\underbrace{(\text{SI},(1{:}3)\text{II})}_{(1:3)\text{-increasing-inflection sink}}.$$

The (SI,(1:3)II)-infinite-equilibrium at $x_{j_1}^*=a_{j_2j_11}$ is a (1:3)-increasing-inflection sink.

– The (1:3)-increasing-inflection and (1:3)-decreasing-inflection source and sink are the appearing (or switching) bifurcations for the third-order increasing-inflection and decreasing-inflection flows with source and sink flows.

(vii) If

$$\Delta_{j_1j_1}=0,\ \Delta_{j_2j_1}=0,$$

then

$$\begin{aligned}
&\dot{x}_{j_1}=a_{j_1j_10}(x_{j_1}-a_{j_1j_1s_1})(x_{j_1}-a_{j_1j_1s_2})^2,\\
&\dot{x}_{j_2}=a_{j_2j_10}(x_{j_1}-a_{j_2j_1l_1})(x_{j_1}-a_{j_2j_1l_2})^2,\\
&j_1,j_2\in\{1,2\},j_1\neq j_2;s_1,s_2\in\{1,2\},s_1\neq s_2,
\end{aligned}$$

where

$$\begin{aligned}
&\{a_{j_1j_11},a_{j_1j_12}\}=\text{sort}\{b_{j_1j_11},b_{j_1j_12}\},a_{j_1j_11}<a_{j_1j_12}\\
&\{a_{j_2j_11},a_{j_2j_12}\}=\text{sort}\{b_{j_2j_11},b_{j_2j_12}\},a_{j_2j_11}<a_{j_2j_12}\\
&b_{j_1j_12}=-\frac{1}{2}B_{j_1j_11},b_{j_2j_12}=-\frac{1}{2}B_{j_2j_11};\\
&j_1,j_2\in\{1,2\},j_1\neq j_2.
\end{aligned}$$

(vii$_1$) For $a_{j_1j_1s_1}\neq a_{j_2j_1l_1}\neq a_{j_1j_1s_2}\neq a_{j_2j_1l_2}$, we have

$$\begin{aligned}
\frac{dx_{j_2}}{dx_{j_1}}&=\frac{a_{j_2j_10}}{a_{j_1j_10}}\frac{(x_{j_1}-a_{j_2j_1l_1})(x_{j_1}-a_{j_2j_1l_2})^2}{(x_{j_1}-a_{j_1j_1s_1})(x_{j_1}-a_{j_1j_1s_2})^2}\\
&=\frac{a_{j_2j_10}}{a_{j_1j_10}}\Big(1+\frac{2(a_{j_1j_1s_2}-a_{j_2j_1l_2})}{(x_{j_1}-a_{j_1j_1s_2})}+\frac{(a_{j_1j_1s_2}-a_{j_2j_1l_2})^2}{(x_{j_1}-a_{j_1j_1s_2})^2}\\
&+(a_{j_1j_1s_1}-a_{j_2j_1l_1})\Big\{\frac{(a_{j_1j_1s_1}-a_{j_2j_11})^2}{(a_{j_1j_1s_1}-a_{j_1j_1s_2})^2}\frac{1}{x_{j_1}-a_{j_1j_1s_1}}\\
&+\frac{(a_{j_1j_1s_2}-a_{j_2j_1l_1})^2}{(a_{j_1j_1s_2}-a_{j_1j_1s_1})}\frac{1}{(x_{j_1}-a_{j_1j_1s_2})^2}\\
&+\Big[\frac{2(a_{j_1j_1s_2}-a_{j_2j_1l_1})}{(a_{j_1j_1s_2}-a_{j_1j_1s_1})}-\frac{(a_{j_1j_1s_2}-a_{j_2j_1l_1})^2}{(a_{j_1j_1s_2}-a_{j_1j_1s_1})^2}\Big]\frac{1}{x_{j_1}-a_{j_1j_1s_2}}\Big\}\Big).
\end{aligned}$$

With an initial condition (x_{j_10}, x_{j_20}), the first integral manifold is

$$
\begin{aligned}
x_{j_2} - x_{j_20} = {} & \frac{a_{j_2j_10}}{a_{j_1j_10}}\Big(\Big[(x_{j_1} - x_{j_10}) + 2(a_{j_1j_1s_2} - a_{j_2j_1l_2})\ln\frac{|x_{j_1} - a_{j_1j_1s_2}|}{|x_{j_10} - a_{j_1j_1s_2}|} \\
& - (a_{j_1j_1s_2} - a_{j_2j_1l_2})^2(\frac{1}{x_{j_1} - a_{j_1j_1s_2}} - \frac{1}{x_{j_10} - a_{j_1j_1s_2}})\Big] \\
& + (a_{j_1j_1s_1} - a_{j_2j_1l_1})\Big\{\frac{(a_{j_1j_1s_1} - a_{j_2j_11})^2}{(a_{j_1j_1s_1} - a_{j_1j_1s_2})^2}\ln\frac{|x_{j_1} - a_{j_1j_1s_1}|}{|x_{j_10} - a_{j_1j_1s_1}|} \\
& - \frac{(a_{j_1j_1s_2} - a_{j_2j_11})^2}{(a_{j_1j_1s_2} - a_{j_1j_1s_1})}(\frac{1}{x_{j_1} - a_{j_1j_1s_2}} - \frac{1}{x_{j_10} - a_{j_1j_1s_2}}) \\
& + \Big[\frac{2(a_{j_1j_1s_2} - a_{j_2j_11})}{(a_{j_1j_1s_2} - a_{j_1j_1s_1})} - \frac{(a_{j_1j_1s_2} - a_{j_2j_11})^2}{(a_{j_1j_1s_2} - a_{j_1j_1s_1})^2}\Big]\ln\frac{|x_{j_1} - a_{j_1j_1s_2}|}{|x_{j_10} - a_{j_1j_1s_2}|}\Big\}\Big), \\
& j_1, j_2 \in \{1,2\}, j_1 \neq j_2; s_1, s_2 \in \{1,2\}, s_1 \neq s_2.
\end{aligned}
$$

(vii$_{1a}$) For $x_{j_1}^* = a_{j_1j_1s_1} \neq a_{j_2j_1l_1}, a_{j_2j_1l_2}, a_{j_1j_1s_2}$, the variational equation is

$$
\begin{aligned}
\dot{x}_{j_2} &= a_{j_2j_10}(a_{j_1j_1s_1} - a_{j_2j_1l_1})(a_{j_1j_1s_1} - a_{j_2j_1l_2})^2, \\
\Delta\dot{x}_{j_1} &= a_{j_1j_10}(a_{j_1j_1s_1} - a_{j_1j_1s_2})^2\Delta x_{j_1}.
\end{aligned}
$$

Due to

$$
(a_{j_1j_1s_1} - a_{j_2j_11})^2 > 0 \text{ and } (a_{j_1j_1s_1} - a_{j_1j_1s_2})^2 > 0,
$$

the flows at $x_{j_1}^* = a_{j_1j_1s_1} \neq a_{j_2j_11}$ are source and sink flows for $a_{j_1j_10} > 0$ and $a_{j_1j_10} < 0$ in the x_{j_1}-direction, and the flows at $x_{j_1}^* = a_{j_1j_1s_1} \neq a_{j_2j_11}$ are positive and negative for $a_{j_2j_10}(a_{j_1j_1s_1} - a_{j_2j_1l_1}) > 0$ and $a_{j_2j_10}(a_{j_1j_1s_1} - a_{j_2j_1l_1}) < 0$ in the x_{j_2}-direction.

Therefore, the equilibrium of $x_{j_1}^* = a_{j_1j_1s_1} \neq a_{j_2j_1l_1}, a_{j_2j_1l_2}, a_{j_1j_1s_2}$ possesses the following properties as in Eqs. (4.121)–(4.124).

- For $a_{j_1j_10}(a_{j_1j_1s_1} - a_{j_1j_1s_2})^2 > 0$ and $a_{j_2j_10}(a_{j_1j_1s_1} - a_{j_2j_1l_1})(a_{j_1j_1s_1} - a_{j_2j_1l_2})^2 > 0$,

$$
(a_{j_1j_1s_1}, \dot{x}_{j_2}) = \underbrace{(\text{SO}, \text{pF})}_{\text{source flow }(+)}.
$$

  The flow of $x_{j_1}^* = a_{j_1j_1s_1}$ is an (SO,pF)-positive source flow.
- For $a_{j_1j_10}(a_{j_1j_1s_1} - a_{j_1j_1s_2})^2 < 0$ and $a_{j_2j_10}(a_{j_1j_1s_1} - a_{j_2j_1l_1})(a_{j_1j_1s_1} - a_{j_2j_1l_2})^2 > 0$,

$$(a_{j_1j_1s_1}, \dot{x}_{j_2}) = \underbrace{(\mathrm{SI}, \mathrm{pF})}_{\text{sink flow }(+)}.$$

The flow of $x_{j_1}^* = a_{j_1j_1s_1}$ is an (SI,pF)-positive sink flow.

- For $a_{j_1j_10}(a_{j_1j_1s_1} - a_{j_1j_1s_2})^2 > 0$ and $a_{j_2j_10}(a_{j_1j_1s_1} - a_{j_2j_1l_1})(a_{j_1j_1s_1} - a_{j_2j_1l_2})^2 < 0$,

$$(a_{j_1j_1s_1}, \dot{x}_{j_2}) = \underbrace{(\mathrm{SO}, \mathrm{nF})}_{\text{source flow }(-)}.$$

The flow of $x_{j_1}^* = a_{j_1j_1s_1}$ is an (SO,nF)-negative source flow.

- For $a_{j_1j_10}(a_{j_1j_1s_1} - a_{j_1j_1s_2})^2 < 0$ and $a_{j_2j_10}(a_{j_1j_1s_1} - a_{j_2j_1l_1})(a_{j_1j_1s_1} - a_{j_2j_1l_2})^2 < 0$,

$$(a_{j_1j_1s_1}, \dot{x}_{j_2}) = \underbrace{(\mathrm{SI}, \mathrm{nF})}_{\text{sink flow }(-)}.$$

The flow of $x_{j_1}^* = a_{j_1j_1s_1}$ is an (SI,pF)-negative sink flow.

(vii$_{1b}$) For $x_{j_1}^* = a_{j_1j_1s_2} \neq a_{j_2j_1l_1}, a_{j_2j_1l_2}, a_{j_1j_1s_1}$, the variational equation is

$$\begin{aligned}\dot{x}_{j_2} &= a_{j_2j_10}(a_{j_1j_1s_2} - a_{j_1j_1l_1})(a_{j_1j_1s_2} - a_{j_2j_1l_2})^2,\\ \Delta\dot{x}_{j_1} &= a_{j_1j_10}(a_{j_1j_1s_2} - a_{j_1j_1s_1})(\Delta x_{j_1})^2.\end{aligned}$$

Due to

$$(a_{j_1j_1s_2} - a_{j_2j_1l_2})^2 > 0,$$

the flows at $x_{j_1}^* = a_{j_1j_1s_2} \neq a_{j_2j_1l_1}, a_{j_2j_1l_2}, a_{j_1j_1s_1}$ are upper-saddle and lower-saddle flows for $a_{j_1j_10}(a_{j_1j_1s_2} - a_{j_1j_1s_1}) > 0$ and $a_{j_1j_10}(a_{j_1j_1s_2} - a_{j_1j_1s_1}) < 0$ in the x_{j_1}-direction, and the flows at $x_{j_1}^* = a_{j_1j_1s_2} \neq a_{j_2j_1}1$ are positive and negative for $a_{j_2j_10}(a_{j_1j_1s_2} - a_{j_1j_1l_1}) > 0$ and $a_{j_2j_10}(a_{j_1j_1s_2} - a_{j_1j_1l_1}) < 0$ in the x_{j_2}-direction.

Thus, the equilibrium of $x_{j_1}^* = a_{j_1j_1s_2} \neq a_{j_1j_1s_1}, a_{j_2j_1l_1}, a_{j_2j_1l_2}$ possesses the following properties as in Eqs. (4.125)–(4.128).

- For $a_{j_1j_10}(a_{j_1j_1s_2} - a_{j_1j_1s_1}) > 0$ and $a_{j_2j_10}(a_{j_1j_1s_2} - a_{j_2j_1l_1})(a_{j_1j_1s_2} - a_{j_2j_1l_2})^2 > 0$,

$$(a_{j_1j_1s_2}, \dot{x}_{j_2}) = \underbrace{(\mathrm{US}, \mathrm{pF})}_{\text{upper-saddle flow }(+)}.$$

Thus, the flow of $x_{j_1}^* = a_{j_1j_1s_2}$ is a (US,pF)-positive upper-saddle flow.

- For $a_{j_1j_10}(a_{j_1j_1s_2}-a_{j_1j_1s_1})<0$ and $a_{j_2j_10}(a_{j_1j_1s_2}-a_{j_2j_1l_1})(a_{j_1j_1s_2}-a_{j_2j_1l_2})^2>0$,

$$(a_{j_1j_1s_2},\dot{x}_{j_2})=\underbrace{(\text{LS, pF})}_{\text{lower-saddle flow }(+)}.$$

  The flow of $x^*_{j_1}=a_{j_1j_1s_2}$ is an (LS,pF)-positive lower-saddle flow.
- For $a_{j_1j_10}(a_{j_1j_1s_2}-a_{j_1j_1s_1})>0$ and $a_{j_2j_10}(a_{j_1j_1s_2}-a_{j_2j_1l_1})(a_{j_1j_1s_2}-a_{j_2j_1l_2})^2<0$,

$$(a_{j_1j_1s_2},\dot{x}_{j_2})=\underbrace{(\text{US, nF})}_{\text{upper-saddle flow }(-)}.$$

  The flow of $x^*_{j_1}=a_{j_1j_1s_2}$ is a (US,nF)-negative upper-saddle flow.
- For $a_{j_1j_10}(a_{j_1j_1s_2}-a_{j_1j_1s_1})<0$ and $a_{j_2j_10}(a_{j_1j_1s_2}-a_{j_2j_1l_1})(a_{j_1j_1s_2}-a_{j_2j_1l_2})^2<0$,

$$(a_{j_1j_1s_2},\dot{x}_{j_2})=\underbrace{(\text{LS, nF})}_{\text{lower-saddle flow }(-)}.$$

  The flow of $x^*_{j_1}=a_{j_1j_1s_2}$ is an (LS,nF)-negative lower-saddle flow.

– The lower-saddle (LS) and upper-saddle (US) flows are the appearing-switching bifurcation for the sink and source flows.

(vii$_{1c}$) In phase space, for $x^*_{j_1}=a_{j_2j_1l_1}\neq a_{j_2j_1l_2},a_{j_1j_1s_1},a_{j_1j_1s_2}$, we have

$$\frac{dx_{j_2}}{dx_{j_1}}\bigg|_{x^*_{j_1}=a_{j_2j_1l_1}}=\frac{a_{j_2j_10}}{a_{j_1j_10}}\frac{(x_{j_1}-a_{j_2j_1l_1})(x_{j_1}-a_{j_2j_1l_2})^2}{(x_{j_1}-a_{j_1j_1s_1})(x_{j_1}-a_{j_1j_1s_2})^2}\bigg|_{x^*_{j_1}=a_{j_2j_1l_1}}=0.$$

If

$$\frac{d^2x_{j_2}}{dx^2_{j_1}}\bigg|_{x^*_{j_1}=a_{j_2j_1l_1}}=\frac{a_{j_2j_10}}{a_{j_1j_10}}\frac{(a_{j_2j_1l_1}-a_{j_2j_1l_2})^2}{(a_{j_2j_1l_1}-a_{j_1j_1s_1})(a_{j_2j_1l_1}-a_{j_1j_1s_2})^2}>0,$$

the flow at $x^*_{j_1}=a_{j_2j_1l_1}\neq a_{j_2j_1l_2},a_{j_1j_1s_1},a_{j_1j_1s_2}$ is an up-parabola flow in the x_{j_2}-direction, and if

$$\frac{d^2x_{j_2}}{dx^2_{j_1}}\bigg|_{x^*_{j_1}=a_{j_2j_1l_1}}=\frac{a_{j_2j_10}}{a_{j_1j_10}}\frac{(a_{j_2j_1l_1}-a_{j_2j_1l_2})^2}{(a_{j_2j_1l_1}-a_{j_1j_1s_1})(a_{j_2j_1l_1}-a_{j_1j_1s_2})^2}<0,$$

the flow at $x^*_{j_1}=a_{j_2j_1l_1}\neq a_{j_2j_1l_2},a_{j_1j_1s_1},a_{j_1j_1s_2}$ is a down-parabola flow in the x_{j_2}-direction.

At $x^*_{j_1} = a_{j_2 j_1 l_1} \neq a_{j_2 j_1 l_2}, a_{j_1 j_1 s_1}, a_{j_1 j_1 s_2}$, the variational equation is

$$\Delta\dot{x}_{j_2} = a_{j_2 j_1 0}(a_{j_2 j_1 l_1} - a_{j_1 j_1 l_2})^2 \Delta x_{j_1},$$
$$\dot{x}_{j_1} = a_{j_1 j_1 0}(a_{j_2 j_1 l_1} - a_{j_1 j_1 s_1})(a_{j_2 j_1 1} - a_{j_1 j_1 s_2})^2.$$

Due to

$$(a_{j_2 j_1 1} - a_{j_1 j_1 s_2})^2 > 0,$$

the flows at $x^*_{j_1} = a_{j_2 j_1 l_1} \neq a_{j_2 j_1 l_2}, a_{j_1 j_1 s_1}, a_{j_1 j_1 s_2}$ are positive and negative for $a_{j_1 j_1 0}(a_{j_2 j_1 l_1} - a_{j_1 j_1 s_1}) > 0$ and $a_{j_1 j_1 0}(a_{j_2 j_1 l_1} - a_{j_1 j_1 s_1}) < 0$ in the x_{j_1}-direction.

Therefore, the equilibrium of $x^*_{j_1} = a_{j_2 j_1 1} \neq a_{j_1 j_1 s_1}, a_{j_1 j_1 s_2}$ possesses the following properties as in Eqs. (4.129)–(4.132).

- For $a_{j_1 j_1 0}(a_{j_2 j_1 l_1} - a_{j_1 j_1 s_1})(a_{j_2 j_1 l_1} - a_{j_1 j_1 s_2})^2 > 0$ and $a_{j_2 j_1 0}(a_{j_2 j_1 l_1} - a_{j_2 j_1 l_2})^2 > 0$,

$$(\dot{x}_{j_1}, a_{j_2 j_1 l_1}) = \underbrace{(\text{pF}, \text{UP})}_{\text{up-parabola flow } (+)}.$$

  The flow of $x^*_{j_1} = a_{j_2 j_1 l_1}$ is a (pF,UP)-positive up-parabola flow.

- For $a_{j_1 j_1 0}(a_{j_2 j_1 l_1} - a_{j_1 j_1 s_1})(a_{j_2 j_1 l_1} - a_{j_1 j_1 s_2})^2 < 0$ and $a_{j_2 j_1 0}(a_{j_2 j_1 l_1} - a_{j_2 j_1 l_2})^2 > 0$,

$$(\dot{x}_{j_1}, a_{j_2 j_1 l_1}) = \underbrace{(\text{nF}, \text{DP})}_{\text{deown-parabola } (-)}.$$

  The flow of $x^*_{j_1} = a_{j_2 j_1 l_1}$ is an (nF,DP)-negative down-parabola flow.

- For $a_{j_1 j_1 0}(a_{j_2 j_1 l_1} - a_{j_1 j_1 s_1})(a_{j_2 j_1 l_1} - a_{j_1 j_1 s_2})^2 > 0$ and $a_{j_2 j_1 0}(a_{j_2 j_1 l_1} - a_{j_2 j_1 l_2})^2 < 0$,

$$(\dot{x}_{j_1}, a_{j_2 j_1 l_1}) = \underbrace{(\text{pF}, \text{DP})}_{\text{down-parabola flow } (+)}.$$

  The flow of $x^*_{j_1} = a_{j_2 j_1 l_1}$ is a (pF,DP)-positive down-parabola flow.

- For $a_{j_1 j_1 0}(a_{j_2 j_1 l_1} - a_{j_1 j_1 s_1})(a_{j_2 j_1 l_1} - a_{j_1 j_1 s_2})^2 < 0$ and $a_{j_2 j_1 0}(a_{j_2 j_1 l_1} - a_{j_2 j_1 l_2})^2 < 0$,

$$(\dot{x}_{j_1}, a_{j_2 j_1 l_1}) = \underbrace{(\text{nF}, \text{UP})}_{\text{iup-parabola flow } (-)}.$$

  The flow of $x^*_{j_1} = a_{j_2 j_1 l_1}$ is an (nF,UP)-positive increasing-inflection flow.

(vii$_{1d}$) In phase space, for $x_{j_1}^* = a_{j_2j_1l_2} \neq a_{j_1j_1s_1}, a_{j_1j_1s_2}, a_{j_2j_1l_1}$, we have

$$
\begin{aligned}
&\frac{dx_{j_2}}{dx_{j_1}}\bigg|_{x_{j_1}^* = a_{j_2j_1l_2}} = \frac{a_{j_2j_1 0}\,(x_{j_1} - a_{j_2j_1l_1})(x_{j_1} - a_{j_2j_1l_2})^2}{a_{j_1j_1 0}\,(x_{j_1} - a_{j_1j_1s_1})(x_{j_1} - a_{j_1j_1s_2})^2}\bigg|_{x_{j_1}^* = a_{j_2j_1l_2}} = 0,\\
&\frac{d^2x_{j_2}}{dx_{j_1}^2}\bigg|_{x_{j_1}^* = a_{j_2j_1l_2}} = 0.
\end{aligned}
$$

If

$$
\frac{d^3x_{j_2}}{dx_{j_1}^3}\bigg|_{x_{j_1}^* = a_{j_2j_1l_2}} = \frac{a_{j_2j_1 0}}{a_{j_1j_1 0}}\,\frac{2(a_{j_2j_1l_2} - a_{j_2j_1l_1})}{(a_{j_2j_1l_2} - a_{j_1j_1s_1})(a_{j_2j_1l_2} - a_{j_1j_1s_2})^2} > 0,
$$

the flow at $x_{j_1}^* = a_{j_2j_1l_2}$ is an increasing-inflection flow in the x_{j_2}-direction, and if

$$
\frac{d^3x_{j_2}}{dx_{j_1}^3}\bigg|_{x_{j_1}^* = a_{j_2j_1l_2}} = \frac{a_{j_2j_1 0}}{a_{j_1j_1 0}}\,\frac{2(a_{j_2j_1l_2} - a_{j_2j_1l_1})}{(a_{j_2j_1l_2} - a_{j_1j_1s_1})(a_{j_2j_1l_2} - a_{j_1j_1s_2})^2} < 0,
$$

the flow at $x_{j_1}^* = a_{j_2j_1l_2}$ is a decreasing-inflection flow in the x_{j_2}-direction.

At $x_{j_1}^* = a_{j_2j_1l_2} \neq a_{j_1j_1s_1}, a_{j_1j_1s_2}, a_{j_2j_1l_1}$, the variational equation is

$$
\begin{aligned}
\Delta\dot{x}_{j_2} &= a_{j_2j_1 0}(a_{j_2j_1l_2} - a_{j_2j_1l_1})(\Delta x_{j_1})^2,\\
\dot{x}_{j_1} &= a_{j_1j_1 0}(a_{j_2j_1l_2} - a_{j_1j_1s_1})(a_{j_2j_1l_2} - a_{j_1j_1s_2})^2.
\end{aligned}
$$

Due to

$$
(a_{j_2j_1l_2} - a_{j_1j_1s_2})^2 > 0,
$$

the flows at $x_{j_1}^* = a_{j_2j_1l_2}$ are positive and negative for $a_{j_1j_1 0}(a_{j_2j_1l_2} - a_{j_1j_1s_1}) > 0$ and $a_{j_1j_1 0}(a_{j_2j_1l_2} - a_{j_1j_1s_1}) < 0$ in the x_{j_1}-direction.

Therefore, the equilibrium of $x_{j_1}^* = a_{j_2j_1l_2} \neq a_{j_1j_1s_1}, a_{j_1j_1s_2}$ possesses the following properties as in Eqs. (4.133)–(4.136).

- For $a_{j_1j_1 0}(a_{j_2j_1l_2} - a_{j_1j_1s_1})(a_{j_2j_1l_2} - a_{j_1j_1s_2})^2 > 0$ and $a_{j_2j_1 0}(a_{j_2j_1l_2} - a_{j_2j_1l_1}) > 0$,

$$
(\dot{x}_{j_1}, a_{j_2j_1l_2}) = \underbrace{(\text{pF, II})}_{\text{increasing-inflection flow } (+)}.
$$

The flow of $x_{j_1}^* = a_{j_2j_1l_2}$ is a (pF,II)-positive increasing-inflection flow.

- For $a_{j_1j_10}(a_{j_2j_1l_2}-a_{j_1j_1s_1})(a_{j_2j_1l_2}-a_{j_1j_1s_2})^2<0$ and $a_{j_2j_10}(a_{j_2j_1l_2}-a_{j_2j_1l_1})>0$,

$$(\dot{x}_{j_1},a_{j_2j_1l_2})=\underbrace{(\text{nF},\text{DI})}_{\text{decreasing-inflection flow }(-)}.$$

The flow of $x_{j_1}^*=a_{j_2j_1l_2}$ is an (nF,DI)-negative decreasing-inflection flow.

- For $a_{j_1j_10}(a_{j_2j_1l_2}-a_{j_1j_1s_1})(a_{j_2j_1l_2}-a_{j_1j_1s_2})^2>0$ and $a_{j_2j_10}(a_{j_2j_1l_2}-a_{j_2j_1l_1})<0$,

$$(\dot{x}_{j_1},a_{j_2j_1l_2})=\underbrace{(\text{pF},\text{DI})}_{\text{decreasing-inflection flow }(+)}.$$

The flow of $x_{j_1}^*=a_{j_2j_1l_2}$ is a (pF,DI)-positive increasing-inflection flow.

- For $a_{j_1j_10}(a_{j_2j_1l_2}-a_{j_1j_1s_1})(a_{j_2j_1l_2}-a_{j_1j_1s_2})^2<0$ and $a_{j_2j_10}(a_{j_2j_1l_2}-a_{j_2j_1l_1})<0$,

$$(\dot{x}_{j_1},a_{j_2j_1l_2})=\underbrace{(\text{nF},\text{II})}_{\text{increasing-inflection flow }(-)}.$$

The flow of $x_{j_1}^*=a_{j_2j_1l_2}$ is an (nF,II)-positive increasing-inflection flow.

– The increasing-inflection (II) and decreasing-inflection (DI) flows are the appearing-switching bifurcations for up-parabola and down-parabola flows.

(vii$_2$) For $a_{j_1j_1s_2}=a_{j_2j_1l_2}$,

$$\begin{aligned}&\dot{x}_{j_1}=a_{j_1j_10}(x_{j_1}-a_{j_1j_1s_1})(x_{j_1}-a_{j_1j_1s_2})^2,\\&\dot{x}_{j_2}=a_{j_2j_10}(x_{j_1}-a_{j_2j_1l_1})(x_{j_1}-a_{j_2j_1l_2})^2,\\&j_1,j_2\in\{1,2\},j_1\neq j_2.\end{aligned}$$

In phase space, we have

$$\begin{aligned}\frac{dx_{j_2}}{dx_{j_1}}&=\frac{a_{j_2j_10}\,(x_{j_1}-a_{j_2j_1l_1})}{a_{j_1j_10}\,(x_{j_1}-a_{j_1j_1s_1})}\\&=\frac{a_{j_2j_10}}{a_{j_1j_10}}\left[1+\frac{(a_{j_1j_1s_1}-a_{j_2j_1l_1})}{(x_{j_1}-a_{j_1j_1s_1})}\right].\end{aligned}$$

With an initial condition (x_{j_10},x_{j_20}), the first integral manifold is

$$\begin{aligned}x_{j_2}-x_{j_20}&=\frac{a_{j_2j_10}}{a_{j_1j_10}}\left[(x_{j_1}-x_{j_10})+(a_{j_1j_1s_1}-a_{j_2j_1l_1})\ln\frac{|x_{j_1}-a_{j_1j_1s_1}|}{|x_{j_10}-a_{j_1j_1s_1}|}\right],\\&j_1,j_2\in\{1,2\},j_1\neq j_2.\end{aligned}$$

In phase space, for $x_{j_1}^* = a_{j_1 j_1 s_2} = a_{j_2 j_1 l_2}$, if

$$\frac{dx_{j_2}}{dx_{j_1}}\bigg|_{x_{j_1}^* = a_{j_2 j_1 l_2}} = \frac{a_{j_2 j_1 0}}{a_{j_1 j_1 0}} \frac{(a_{j_2 j_1 l_2} - a_{j_2 j_1 l_1})}{(a_{j_1 j_1 s_2} - a_{j_1 j_1 s_1})} > 0,$$

the flow at $x_{j_1}^* = a_{j_1 j_1 s_2} = a_{j_2 j_1 l_2}$ is an increasing-inflection flow in the x_{j_2}-direction, and if

$$\frac{dx_{j_2}}{dx_{j_1}}\bigg|_{x_{j_1}^* = a_{j_2 j_1 l_2}} = \frac{a_{j_2 j_1 0}}{a_{j_1 j_1 0}} \frac{(a_{j_2 j_1 l_2} - a_{j_2 j_1 l_1})}{(a_{j_1 j_1 s_2} - a_{j_1 j_1 s_1})} < 0,$$

the flow at $x_{j_1}^* = a_{j_1 j_1 s_2} = a_{j_2 j_1 l_2}$ is a decreasing-inflection flow in the x_{j_2}-direction.

At $x_{j_1}^* = a_{j_1 j_1 s_2} = a_{j_2 j_1 l_2}$, the variational equation is

$$\begin{aligned}\Delta\dot{x}_{j_2} &= a_{j_2 j_1 0}(a_{j_2 j_1 l_2} - a_{j_1 j_1 l_1})(\Delta x_{j_1})^2,\\ \Delta\dot{x}_{j_1} &= a_{j_1 j_1 0}(a_{j_2 j_1 s_2} - a_{j_1 j_1 s_1})(\Delta x_{j_1})^2.\end{aligned}$$

The flows at $x_{j_1}^* = a_{j_1 j_1 s_2} = a_{j_2 j_1 l_2}$ are upper-saddle and lower-saddle flows for $a_{j_2 j_1 0}(a_{j_2 j_1 l_1} - a_{j_1 j_1 s_1}) > 0$ and $a_{j_2 j_1 0}(a_{j_2 j_1 l_1} - a_{j_1 j_1 s_1}) < 0$ in the x_{j_1}-direction. On the both directions, the second-order singularity exist, thus, (2:2)II and (2:2)DI are used for increasing-inflection and decreasing-inflection.

Therefore, the infinite-equilibrium of $x_{j_1}^* = a_{j_1 j_1 s_2} = a_{j_2 j_1 l_2}$ with $\bar{x}_{j_2} \in (-\infty, \infty)$ has the following properties as in Eqs. (4.139)–(4.142).

- For $a_{j_1 j_1 0}(a_{j_1 j_1 s_2} - a_{j_1 j_1 s_1}) > 0$ and $a_{j_2 j_1 0}(a_{j_2 j_1 l_2} - a_{j_2 j_1 l_1}) > 0$,

$$(a_{j_1 j_1 s_2}, \bar{x}_{j_2}) = \underbrace{(\text{US}, (2:2)\text{II})}_{\text{increasing-inflection upper-saddle}}.$$

  The equilibrium at $x_{j_1}^* = a_{j_1 j_1 s_2} = a_{j_2 j_1 l_2}$ is a (US,(2:2)II)-increasing-inflection upper-saddle infinite-equilibrium.
- For $a_{j_1 j_1 0}(a_{j_1 j_1 s_2} - a_{j_1 j_1 s_1}) < 0$ and $a_{j_2 j_1 0}(a_{j_2 j_1 l_2} - a_{j_2 j_1 l_1}) > 0$,

$$(a_{j_1 j_1 s_2}, \bar{x}_{j_2}) = \underbrace{(\text{LS}, (2:2)\text{DI})}_{\text{decreasing-inflection lower-saddle}}.$$

  The equilibrium at $x_{j_1}^* = a_{j_1 j_1 s_2} = a_{j_2 j_1 l_2}$ is an (LS, (2:2)DI)-decreasing-inflection lower-saddle infinite-equilibrium.
- For $a_{j_1 j_1 0}(a_{j_1 j_1 s_2} - a_{j_1 j_1 s_1}) > 0$ and $a_{j_2 j_1 0}(a_{j_2 j_1 l_2} - a_{j_2 j_1 l_1}) < 0$,

$$(a_{j_1 j_1 s_2}, \bar{x}_{j_2}) = \underbrace{(\text{US}, (2:2)\text{DI})}_{\text{decreasing-inflection upper-saddle}}.$$

The equilibrium at $x_{j_1}^* = a_{j_1 j_1 s_2} = a_{j_2 j_1 l_2}$ is a (US,(2:2)DI)-increasing-inflection upper-saddle infinite-equilibrium.

- For $a_{j_1 j_1 0}(a_{j_1 j_1 s_2} - a_{j_1 j_1 s_1}) < 0$ and $a_{j_2 j_1 0}(a_{j_2 j_1 l_2} - a_{j_2 j_1 l_1}) < 0$,

$$(a_{j_1 j_1 s_2}, \bar{x}_{j_2}) = \underbrace{(\text{LS}, (2:2)\text{II})}_{\text{increasing-inflection lower-saddle}}.$$

The equilibrium at $x_{j_1}^* = a_{j_1 j_1 s_2} = a_{j_2 j_1 l_2}$ is an (LS,(2:2)II)-increasing-inflection lower-saddle infinite-equilibrium.

- The infinite-equilibriums are the switching bifurcations for the increasing-inflection and decreasing-inflection flows with the upper-saddle and lower-saddle flows.
- The infinite-equilibriums are the appearing bifurcations for the sink and source flows and the up-parabola and down-parabola flows.

(vii$_3$) For $a_{j_1 j_1 s_1} = a_{j_2 j_1 l_2}$,

$$\begin{aligned}
\dot{x}_{j_1} &= a_{j_1 j_1 0}(x_{j_1} - a_{j_1 j_1 s_1})(x_{j_1} - a_{j_1 j_1 s_2})^2, \\
\dot{x}_{j_2} &= a_{j_2 j_1 0}(x_{j_1} - a_{j_2 j_1 l_1})(x_{j_1} - a_{j_2 j_1 l_2})^2, \\
& j_1, j_2 \in \{1, 2\}, j_1 \neq j_2.
\end{aligned}$$

In phase space, we have

$$\begin{aligned}
\frac{dx_{j_2}}{dx_{j_1}} &= \frac{a_{j_2 j_1 0}}{a_{j_1 j_1 0}} \frac{(x_{j_1} - a_{j_2 j_1 l_1})(x_{j_1} - a_{j_2 j_1 l_2})}{(x_{j_1} - a_{j_1 j_1 s_2})^2} \\
&= \frac{a_{j_2 j_1 0}}{a_{j_1 j_1 0}} \Big\{ 1 + (a_{j_1 j_1 s_2} - a_{j_2 j_1 l_2}) \frac{1}{x_{j_1} - a_{j_1 j_1 s_2}} \\
&\quad + (a_{j_1 j_1 s_2} - a_{j_2 j_1 l_1}) \Big[\frac{1}{x_{j_1} - a_{j_1 j_1 s_2}} + \frac{(a_{j_1 j_1 s_2} - a_{j_2 j_1 l_2})}{(x_{j_1} - a_{j_1 j_1 s_2})^2} \Big] \Big\}.
\end{aligned}$$

With an initial condition $(x_{j_1 0}, x_{j_2 0})$, the first integral manifold is

$$\begin{aligned}
x_{j_2} - x_{j_2 0} = & \frac{a_{j_2 j_1 0}}{a_{j_1 j_1 0}} \Big\{ x_{j_1} - x_{j_1 0} + (a_{j_1 j_1 s_2} - a_{j_2 j_1 l_2}) \ln \frac{|x_{j_1} - a_{j_1 j_1 s_2}|}{|x_{j_1 0} - a_{j_1 j_1 s_2}|} \\
& + (a_{j_1 j_1 s_2} - a_{j_2 j_1 l_1}) \Big[\ln \frac{|x_{j_1} - a_{j_1 j_1 s_2}|}{|x_{j_1 0} - a_{j_1 j_1 s_2}|} \\
& - (a_{j_1 j_1 s_2} - a_{j_2 j_1 l_2}) \Big(\frac{1}{x_{j_1} - a_{j_1 j_1 s_2}} - \frac{1}{x_{j_1 0} - a_{j_1 j_1 s_2}} \Big) \Big] \Big\}, \\
& j_1, j_2 \in \{1, 2\}, j_1 \neq j_2.
\end{aligned}$$

For $x_{j_1}^* = a_{j_1 j_1 s_1} = a_{j_2 j_1 l_2}$, we have

$$\frac{dx_{j_2}}{dx_{j_1}}\Big|_{x_{j_1}^* = a_{j_2 j_1 l_2}} = \frac{a_{j_2 j_1 0}}{a_{j_1 j_1 0}} \frac{(x_{j_1} - a_{j_2 j_1 l_1})(x_{j_1} - a_{j_2 j_1 l_2})}{(x_{j_1} - a_{j_1 j_1 s_2})^2}\Big|_{x_{j_1}^* = a_{j_2 j_1 l_2}} = 0.$$

If

$$\frac{d^2 x_{j_2}}{dx_{j_1}^2}\Big|_{x_{j_1}^* = a_{j_2 j_1 l_2}} = \frac{a_{j_2 j_1 0}}{a_{j_1 j_1 0}} \frac{(a_{j_2 j_1 l_2} - a_{j_2 j_1 l_1})}{(a_{j_1 j_1 s_1} - a_{j_1 j_1 s_2})^2}\Big|_{x_{j_1}^* = a_{j_2 j_1 l_2}} > 0,$$

the flow at $x_{j_1}^* = a_{j_1 j_1 s_1} = a_{j_2 j_1 l_2}$ is an up-parabola flow in the x_{j_2}-direction, and if

$$\frac{d^2 x_{j_2}}{dx_{j_1}^2}\Big|_{x_{j_1}^* = a_{j_2 j_1 l_2}} = \frac{a_{j_2 j_1 0}}{a_{j_1 j_1 0}} \frac{(a_{j_2 j_1 l_2} - a_{j_2 j_1 l_1})}{(a_{j_1 j_1 s_1} - a_{j_1 j_1 s_2})^2}\Big|_{x_{j_1}^* = a_{j_2 j_1 l_2}} < 0,$$

the flow at $x_{j_1}^* = a_{j_1 j_1 s_1} = a_{j_2 j_1 l_2}$ is a down-parabola flow in the x_{j_2}-direction.

At $x_{j_1}^* = a_{j_1 j_1 s_1} = a_{j_2 j_1 l_2}$, the variational equations are

$$\begin{aligned}
\Delta \dot{x}_{j_2} &= a_{j_2 j_1 0}(a_{j_2 j_1 l_2} - a_{j_2 j_1 l_1})(\Delta x_{j_1})^2, \\
\Delta \dot{x}_{j_1} &= a_{j_1 j_1 0}(a_{j_1 j_1 s_1} - a_{j_1 j_1 s_2})^2 \Delta x_{j_1}.
\end{aligned}$$

The flows at $x_{j_1}^* = a_{j_1 j_1 s_1} = a_{j_2 j_1 l_2}$ are source and sink flows for $a_{j_1 j_1 0} > 0$ and $a_{j_1 j_1 0} < 0$ in the x_{j_1}-direction.

Therefore, the infinite-equilibrium of $x_{j_1}^* = a_{j_1 j_1 s_1} = a_{j_2 j_1 l_2}$ with $\bar{x}_{j_2} \in (-\infty, \infty)$ has the following properties as in Eqs. (4.145)–(4.148).

- For $a_{j_1 j_1 0}(a_{j_1 j_1 s_1} - a_{j_1 j_1 s_2})^2 > 0$ and $a_{j_2 j_1 0}(a_{j_2 j_1 l_2} - a_{j_2 j_1 l_1}) > 0$,

$$(a_{j_1 j_1 s_1}, \bar{x}_{j_2}) = \underbrace{(\text{SO}, \text{UP})}_{\text{up-parabola source}}.$$

  The equilibrium at $x_{j_1}^* = a_{j_1 j_1 s_1} = a_{j_2 j_1 l_2}$ is an (SO,UP)-up-parabola source infinite-equilibrium.
- For $a_{j_1 j_1 0}(a_{j_1 j_1 s_1} - a_{j_1 j_1 s_2})^2 < 0$ and $a_{j_2 j_1 0}(a_{j_2 j_1 l_2} - a_{j_2 j_1 l_1}) > 0$,

$$(a_{j_1 j_1 s_1}, \bar{x}_{j_2}) = \underbrace{(\text{SI}, \text{DP})}_{\text{down-parabola sink}}.$$

  The equilibrium at $x_{j_1}^* = a_{j_1 j_1 s_1} = a_{j_2 j_1 l_2}$ is an (SI,DP)-down-parabola sink infinite-equilibrium.

- For $a_{j_1j_10}(a_{j_1j_1s_1}-a_{j_1j_1s_2})^2>0$ and $a_{j_2j_10}(a_{j_2j_1l_2}-a_{j_2j_1l_1})<0$,

$$(a_{j_1j_1s_1},\overline{x}_{j_2})=\underbrace{(\mathrm{SO},\mathrm{DP})}_{\text{down-parabola source}}.$$

The equilibrium at $x^*_{j_1}=a_{j_1j_1s_1}=a_{j_2j_1l_2}$ is an (SO,DP)-down-parabola source infinite-equilibrium.

- For $a_{j_1j_10}(a_{j_1j_1s_1}-a_{j_1j_1s_2})^2<0$ and $a_{j_2j_10}(a_{j_2j_1l_2}-a_{j_2j_1l_1})<0$,

$$(a_{j_1j_1s_1},\overline{x}_{j_2})=\underbrace{(\mathrm{SI},\mathrm{UP})}_{\text{up-parabola sink}}.$$

The equilibrium at $x^*_{j_1}=a_{j_1j_1s_1}=a_{j_2j_1l_2}$ is an (SI,UP)-up-parabola sink infinite-equilibrium.

– The infinite-equilibriums are the switching bifurcation for sink and source flows with increasing-inflection and decreasing-inflection flows.

(vi$_4$) For $a_{j_1j_1s_1}=a_{j_2j_1l_1}$,

$$\begin{aligned}&\dot{x}_{j_1}=a_{j_1j_10}(x_{j_1}-a_{j_1j_1s_1})(x_{j_1}-a_{j_1j_1s_2})^2,\\&\dot{x}_{j_2}=a_{j_2j_10}(x_{j_1}-a_{j_2j_1l_1})(x_{j_1}-a_{j_2j_1l_2})^2,\\&j_1,j_2\in\{1,2\},j_1\neq j_2.\end{aligned}$$

In phase space, we have

$$\begin{aligned}\frac{dx_{j_2}}{dx_{j_1}}&=\frac{a_{j_2j_10}\,(x_{j_1}-a_{j_2j_1l_2})^2}{a_{j_1j_10}\,(x_{j_1}-a_{j_1j_1s_2})^2}\\&=\frac{a_{j_2j_10}}{a_{j_1j_10}}\left[1+\frac{2(a_{j_1j_1s_2}-a_{j_2j_1l_2})}{(x_{j_1}-a_{j_1j_1s_2})}+\frac{(a_{j_1j_1s_2}-a_{j_2j_1l_2})^2}{(x_{j_1}-a_{j_1j_1s_2})^2}\right].\end{aligned}$$

With an initial condition (x_{j_10},x_{j_20}), the first integral manifold is

$$\begin{aligned}x_{j_2}-x_{j_20}=&\frac{a_{j_2j_20}}{a_{j_1j_10}}\Big[x_{j_1}-x_{j_10}+2(a_{j_1j_1s_2}-a_{j_2j_1l_2})\ln\frac{|\,x_{j_1}-a_{j_1j_1s_2}\,|}{|\,x_{j_10}-a_{j_1j_1s_2}\,|}\\&+(a_{j_1j_1s_2}-a_{j_2j_1l_2})^2\Big(\frac{1}{x_{j_1}-a_{j_1j_1s_2}}-\frac{1}{x_{j_10}-a_{j_1j_1s_2}}\Big)\Big],\\&j_1,j_2\in\{1,2\},j_1\neq j_2.\end{aligned}$$

(v$_{2a}$) In phase space, for $x^*_{j_1}=a_{j_1j_1s_1}=a_{j_2j_1l_1}$, if

$$\frac{dx_{j_2}}{dx_{j_1}}\bigg|_{x_{j_1}^*=a_{j_2j_1l_1}} = \frac{a_{j_2j_10}}{a_{j_1j_10}} \frac{(a_{j_2j_1l_1}-a_{j_2j_1l_2})^2}{(a_{j_1j_1s_1}-a_{j_1j_1s_2})^2} > 0,$$

the flow at $x_{j_1}^*=a_{j_1j_1s_1}=a_{j_2j_1l_1}$ is an increasing-inflection flow in the x_{j_2}-direction, and if

$$\frac{dx_{j_2}}{dx_{j_1}}\bigg|_{x_{j_1}^*=a_{j_2j_1l_1}} = \frac{a_{j_2j_10}}{a_{j_1j_10}} \frac{(a_{j_2j_1l_1}-a_{j_2j_1l_2})^2}{(a_{j_1j_1s_1}-a_{j_1j_1s_2})^2} < 0,$$

the flow at $x_{j_1}^*=a_{j_1j_1s_1}=a_{j_2j_1l_1}$ is a decreasing-inflection flow in the x_{j_2}-direction.

At $x_{j_1}^*=a_{j_1j_1s_1}=a_{j_2j_1l_1}$, the variational equations are

$$\Delta\dot{x}_{j_2} = a_{j_2j_10}(a_{j_2j_1l_1}-a_{j_2j_1l_2})^2\Delta x_{j_1},$$
$$\Delta\dot{x}_{j_1} = a_{j_1j_10}(a_{j_1j_1s_1}-a_{j_1j_1s_2})^2\Delta x_{j_1}.$$

The flows at $x_{j_1}^*=a_{j_1j_1s_1}=a_{j_2j_1l_1}$ are upper-saddle and lower-saddle flows for $a_{j_1j_10}>0$ and $a_{j_1j_10}<0$ in the x_{j_1}-direction.

Therefore, the infinite-equilibrium of $x_{j_1}^*=a_{j_1j_1s_1}=a_{j_2j_1l_1}$ with $\bar{x}_{j_2}\in(-\infty,\infty)$ has the following properties as in Eqs. (4.151)–(4.154).

- For $a_{j_1j_10}(a_{j_1j_1s_1}-a_{j_1j_1s_2})^2>0$ and $a_{j_2j_10}(a_{j_2j_1l_1}-a_{j_2j_1l_2})^2>0$,

$$(a_{j_1j_1s_1},\bar{x}_{j_2}) = \underbrace{(\text{SO, II})}_{\text{increasing-inflection source}}.$$

  The equilibrium at $x_{j_1}^*=a_{j_1j_1s_1}=a_{j_2j_1l_1}$ is an (SO,II)-increasing-inflection source infinite-equilibrium.
- For $a_{j_1j_10}(a_{j_1j_1s_1}-a_{j_1j_1s_2})^2<0$ and $a_{j_2j_10}(a_{j_2j_1l_1}-a_{j_2j_1l_2})^2>0$,

$$(a_{j_1j_1s_1},\bar{x}_{j_2}) = \underbrace{(\text{SI, DI})}_{\text{decreasing-inflection sink}}.$$

  The equilibrium at $x_{j_1}^*=a_{j_1j_1s_1}=a_{j_2j_1l_1}$ is an (SI,DI)-decreasing-inflection sink infinite-equilibrium.
- For $a_{j_1j_10}(a_{j_1j_1s_1}-a_{j_1j_1s_2})^2>0$ and $a_{j_2j_10}(a_{j_2j_1l_1}-a_{j_2j_1l_2})^2<0$,

$$(a_{j_1j_1s_1},\bar{x}_{j_2}) = \underbrace{(\text{SO, DI})}_{\text{decreasing-inflection source}}.$$

  The equilibrium at $x_{j_1}^*=a_{j_1j_1s_1}=a_{j_2j_1l_1}$ is an (SO,DI)-increasing-inflection source infinite-equilibrium.

- For $a_{j_1j_10}(a_{j_1j_1s_1}-a_{j_1j_1s_2})^2<0$ and $a_{j_2j_10}(a_{j_2j_1l_1}-a_{j_2j_1l_2})^2<0$,

$$(a_{j_1j_1s_1},\overline{x}_{j_2}) = \underbrace{(\text{SI, II})}_{\text{increasing-inflection sink}}.$$

The equilibrium at $x_{j_1}^*=a_{j_1j_1s_1}=a_{j_2j_1l_1}$ is an (SI,II)-increasing-inflection sink infinite-equilibrium.

– The infinite-equilibriums are the switching bifurcations for the up-parabola and down-parabola flows with the source and sink flows.

(vii$_5$) For $a_{j_1j_1s_2}=a_{j_2j_1l_1}$,

$$\begin{aligned}
\dot{x}_{j_1} &= a_{j_1j_10}(x_{j_1}-a_{j_1j_1s_1})(x_{j_1}-a_{j_1j_1s_2})^2,\\
\dot{x}_{j_2} &= a_{j_2j_10}(x_{j_1}-a_{j_2j_1l_1})(x_{j_1}-a_{j_2j_1l_2})^2,\\
& j_1,j_2\in\{1,2\}, j_1\neq j_2.
\end{aligned}$$

In phase space, we have

$$\begin{aligned}
\frac{dx_{j_2}}{dx_{j_1}} &= \frac{a_{j_2j_10}}{a_{j_1j_10}}\frac{(x_{j_1}-a_{j_2j_1l_2})^2}{(x_{j_1}-a_{j_1j_1s_1})(x_{j_1}-a_{j_1j_1s_2})}\\
&= \frac{a_{j_2j_10}}{a_{j_1j_10}}\Big[1+(a_{j_1j_1s_2}+a_{j_1j_1l_1}-2a_{j_2j_1l_2})\frac{1}{(x_{j_1}-a_{j_1j_1s_2})}\\
&\quad +\frac{(a_{j_1j_1s_1}-a_{j_2j_1l_2})^2}{a_{j_1j_1s_1}-a_{j_1j_1s_2}}(\frac{1}{x_{j_1}-a_{j_1j_1s_1}}-\frac{1}{x_{j_1}-a_{j_1j_1s_2}})\Big].
\end{aligned}$$

With an initial condition (x_{j_10},x_{j_20}), the first integral manifold is

$$\begin{aligned}
x_{j_2}-x_{j_20} &= \frac{a_{j_2j_10}}{a_{j_1j_10}}\Big[(x_{j_1}-x_{j_10})+(a_{j_1j_1s_2}+a_{j_1j_1l_1}+a_{j_2j_1l_2})\ln\frac{|x_{j_1}-a_{j_1j_1s_2}|}{|x_{j_10}-a_{j_1j_1s_2}|}\\
&\quad +\frac{(a_{j_1j_1s_1}-a_{j_2j_1l_2})^2}{a_{j_1j_1s_1}-a_{j_1j_1s_2}}(\ln\frac{|x_{j_1}-a_{j_1j_1s_1}|}{|x_{j_10}-a_{j_1j_1s_1}|}-\ln\frac{|x_{j_1}-a_{j_1j_1s_2}|}{|x_{j_10}-a_{j_1j_1s_2}|})\Big],\\
& j_1,j_2\in\{1,2\}, j_1\neq j_2.
\end{aligned}$$

From the first integral manifold, if

$$\frac{a_{j_2j_10}}{a_{j_1j_10}}\frac{(a_{j_1j_1s_1}-a_{j_2j_1l_2})^2}{a_{j_1j_1s_1}-a_{j_1j_1s_2}}>0,$$

the flow at $x_{j_1}^*=a_{j_1j_1s_1}=a_{j_2j_1}1$ is a down-up asymptotic flows, and if

$$\frac{a_{j_2j_10}}{a_{j_1j_10}}\frac{(a_{j_1j_1s_1}-a_{j_2j_1l_2})^2}{a_{j_1j_1s_1}-a_{j_1j_1s_2}}<0,$$

the flow at $x_{j_1}^*=a_{j_1j_1s_1}=a_{j_2j_1 1}$ is an up-down asymptotic flow.

At $x_{j_1}^*=a_{j_1j_1s_1}=a_{j_2j_1 1}$, the variational equations are

$$\begin{aligned}\Delta\dot{x}_{j_2}&=a_{j_2j_10}(a_{j_2j_1l_1}-a_{j_2j_1l_2})^2\Delta x_{j_1},\\ \Delta\dot{x}_{j_1}&=a_{j_1j_10}(a_{j_1j_1s_2}-a_{j_1j_1s_1})(\Delta x_{j_1})^2.\end{aligned}$$

The flows at $x_{j_1}^*=a_{j_1j_1s_1}=a_{j_2j_1 1}$ are upper-saddle and lower-saddle flows for $a_{j_1j_10}(a_{j_1j_1s_2}-a_{j_1j_1s_1})>0$ and $a_{j_1j_10}(a_{j_1j_1s_2}-a_{j_1j_1s_1})<0$ in the x_{j_1}-direction.

Therefore, the infinite-equilibrium of $x_{j_1}^*=a_{j_1j_1s_2}=a_{j_2j_1l_1}$ with $\overline{x}_{j_2}\in(-\infty,\infty)$ has the following properties as in Eqs. (4.157)–(4.160).

- For $a_{j_1j_10}(a_{j_1j_1s_2}-a_{j_1j_1s_1})>0$ and $a_{j_2j_10}(a_{j_2j_1l_1}-a_{j_2j_1l_2})^2>0$,

$$(a_{j_1j_1s_2},\overline{x}_{j_2})=\underbrace{(\text{US},\text{DU})}_{\text{down-up upper-saddle}}.$$

  The equilibrium at $x_{j_1}^*=a_{j_1j_1s_2}=a_{j_2j_1l_1}$ is a (US,DU)-down-up asymptotic upper-saddle infinite-equilibrium.

- For $a_{j_1j_10}(a_{j_1j_1s_2}-a_{j_1j_1s_1})<0$ and $a_{j_2j_10}(a_{j_2j_1l_1}-a_{j_2j_1l_2})^2>0$,

$$(a_{j_1j_1s_2},\overline{x}_{j_2})=\underbrace{(\text{LS},\text{UD})}_{\text{up-down lower-saddle}}.$$

  The equilibrium at $x_{j_1}^*=a_{j_1j_1s_2}=a_{j_2j_1l_1}$ is an (LS,UD)-up-down asymptotic lower-saddle infinite-equilibrium.

- For $a_{j_1j_10}(a_{j_1j_1s_2}-a_{j_1j_1s_1})>0$ and $a_{j_2j_10}(a_{j_2j_1l_1}-a_{j_2j_1l_2})^2<0$,

$$(a_{j_1j_1s_2},\overline{x}_{j_2})=\underbrace{(\text{US},\text{UD})}_{\text{up-down upper-saddle}}.$$

  The equilibrium at $x_{j_1}^*=a_{j_1j_1s_2}=a_{j_2j_1l_1}$ is a (US,UD)-up-down asymptotic upper-saddle infinite-equilibrium.

- For $a_{j_1j_10}(a_{j_1j_1s_2}-a_{j_1j_1s_1})<0$ and $a_{j_2j_10}(a_{j_2j_1l_1}-a_{j_2j_1l_2})^2<0$,

$$(a_{j_1j_1s_2},a_{j_2j_1l_1})=\underbrace{(\text{LS},\text{DU})}_{\text{down-up lower-saddle}}.$$

  The equilibrium at $x_{j_1}^*=a_{j_1j_1s_2}=a_{j_2j_1l_1}$ is an (LS,DU)-down-up asymptotic lower-saddle infinite-equilibrium.

– The up-down and down-up asymptotic upper-saddle and lower-saddle are for switching bifurcations of sink and source flows with up-parabola or down-parabola flow.

(viii) If

$$\Delta_{j_1 j_1} = 0 \text{ and } \Delta_{j_2 j_1} = 0,$$
$$a_{j_1 j_1 s_1} = a_{j_1 j_1 s_2} \equiv a_{j_1 j_1 1},$$

then

$$\begin{aligned}
\dot{x}_{j_1} &= a_{j_1 j_1 0}(x_{j_1} - a_{j_1 j_1 1})^3, \\
\dot{x}_{j_2} &= a_{j_2 j_1 0}(x_{j_1} - a_{j_2 j_1 l_1})(x_{j_1} - a_{j_2 j_1 l_2})^2, \\
& j_1, j_2 \in \{1, 2\}, j_1 \neq j_2.
\end{aligned}$$

(viii$_1$) For $a_{j_1 j_1 1} \neq a_{j_2 j_1 l_1}, a_{j_2 j_1 l_2}$, we have

$$\begin{aligned}
\frac{dx_{j_2}}{dx_{j_1}} &= \frac{a_{j_2 j_1 0}}{a_{j_1 j_1 0}} \frac{(x_{j_1} - a_{j_2 j_1 l_1})(x_{j_1} - a_{j_2 j_1 l_2})^2}{(x_{j_1} - a_{j_1 j_1 1})^3} \\
&= \frac{a_{j_2 j_1 0}}{a_{j_1 j_1 0}} \left\{ 1 + \frac{2(a_{j_1 j_1 1} - a_{j_2 j_1 l_2})}{(x_{j_1} - a_{j_1 j_1 1})} + \frac{(a_{j_1 j_1 1} - a_{j_2 j_1 l_2})^2}{(x_{j_1} - a_{j_1 j_1 1})^2} \right. \\
&\left. + (a_{j_1 j_1 1} - a_{j_2 j_1 l_1}) \left[\frac{1}{x_{j_1} - a_{j_1 j_1 1}} + \frac{2(a_{j_1 j_1 1} - a_{j_2 j_1 l_2})}{(x_{j_1} - a_{j_1 j_1 s_2})^2} + \frac{(a_{j_1 j_1 1} - a_{j_2 j_1 l_2})^2}{(x_{j_1} - a_{j_1 j_1 1})^3} \right] \right\}.
\end{aligned}$$

With an initial condition $(x_{j_1 0}, x_{j_2 0})$, the first integral manifold is

$$\begin{aligned}
x_{j_2} - x_{j_2 0} = \frac{a_{j_2 j_1 0}}{a_{j_1 j_1 0}} \Bigg(& \Bigg[(x_{j_1} - x_{j_1 0}) + 2(a_{j_1 j_1 1} - a_{j_2 j_1 l_2}) \ln \frac{| x_{j_1} - a_{j_1 j_1 1} |}{| x_{j_1 0} - a_{j_1 j_1 1} |} \\
& - (a_{j_1 j_1 1} - a_{j_2 j_1 l_2})^2 \Big(\frac{1}{x_{j_1} - a_{j_1 j_1 1}} - \frac{1}{x_{j_1 0} - a_{j_1 j_1 1}} \Big) \Bigg] \\
& + (a_{j_1 j_1 1} - a_{j_2 j_1 l_1}) \Bigg\{ \ln \frac{| x_{j_1} - a_{j_1 j_1 1} |}{| x_{j_1 0} - a_{j_1 j_1 1} |} \\
& - 2(a_{j_1 j_1 1} - a_{j_2 j_1 l_2}) \Big(\frac{1}{x_{j_1} - a_{j_1 j_1 1}} - \frac{1}{x_{j_1 0} - a_{j_1 j_1 1}} \Big) \\
& - \frac{1}{2} (a_{j_1 j_1 1} - a_{j_2 j_1 l_2})^2 \left[\frac{1}{(x_{j_1} - a_{j_1 j_1 1})^2} - \frac{1}{(x_{j_1 0} - a_{j_1 j_1 1})^2} \right] \Bigg\} \Bigg), \\
& j_1, j_2 \in \{1, 2\}, j_1 \neq j_2.
\end{aligned}$$

(viii$_{1a}$) For $x^*_{j_1} = a_{j_1 j_1 1} \neq a_{j_2 j_1 l_1}, a_{j_2 j_1 l_2}$, the variational equation is

$$\dot{x}_{j_2} = a_{j_2 j_1 0}(a_{j_1 j_1 1} - a_{j_2 j_1 l_1})(a_{j_1 j_1 1} - a_{j_2 j_1 l_2})^2,$$
$$\Delta\dot{x}_{j_1} = a_{j_1 j_1 0}(\Delta x_{j_1})^3.$$

The flows at $x^*_{j_1} = a_{j_1 j_1 1} \neq a_{j_2 j_1 l_1}, a_{j_2 j_1 l_2}$ are the third-order source and sink flows for $a_{j_1 j_1 0} > 0$ and $a_{j_1 j_1 0} < 0$ in the x_{j_1}-direction. The flows at $x^*_{j_1} = a_{j_1 j_1 1} \neq a_{j_2 j_1 l_1}, a_{j_2 j_1 l_2}$ are positive and negative for $a_{j_2 j_1 0}(a_{j_1 j_1 1} - a_{j_2 j_1 l_1}) > 0$ and $a_{j_2 j_1 0}(a_{j_1 j_1 1} - a_{j_2 j_1 l_1}) < 0$ in the x_{j_2}-direction.

Therefore, the equilibrium of $x^*_{j_1} = a_{j_1 j_1 1} \neq a_{j_2 j_1 l_1}, a_{j_2 j_1 l_2}$ has the following properties as in Eqs. (4.164)–(4.167).

- For $a_{j_1 j_1 0} > 0$ and $a_{j_2 j_1 0}(a_{j_1 j_1 1} - a_{j_2 j_1 l_1})(a_{j_1 j_1 1} - a_{j_2 j_1 l_2})^2 > 0$,

$$(a_{j_1 j_1 1}, \dot{x}_{j_2}) = \underbrace{(3^{\text{rd}}\text{SO}, \text{pF})}_{\text{third-order source flow }(+)}.$$

  The flow at $x^*_{j_1} = a_{j_1 j_1 1}$ is a (3rdSO,pF)-positive third-order source flow.
- For $a_{j_1 j_1 0} < 0$ and $a_{j_2 j_1 0}(a_{j_1 j_1 1} - a_{j_2 j_1 l_1})(a_{j_1 j_1 1} - a_{j_2 j_1 l_2})^2 > 0$,

$$(a_{j_1 j_1 1}, \dot{x}_{j_2}) = \underbrace{(3^{\text{rd}}\text{SI}, \text{pF})}_{\text{third-order sink flow }(+)}.$$

  The flow at $x^*_{j_1} = a_{j_1 j_1 1}$ is a (3rdSI,pF)-positive third-order sink flow.
- For $a_{j_1 j_1 0} > 0$ and $a_{j_2 j_1 0}(a_{j_1 j_1 1} - a_{j_2 j_1 l_1})(a_{j_1 j_1 1} - a_{j_2 j_1 l_2})^2 < 0$,

$$(a_{j_1 j_1 1}, \dot{x}_{j_2}) = \underbrace{(3^{\text{rd}}\text{SO}, \text{nF})}_{\text{third-order source flow }(-)}.$$

  The flow at $x^*_{j_1} = a_{j_1 j_1 1}$ is a (3rdSO,nF)-negative third-order source flow.
- For $a_{j_1 j_1 0} < 0$ and $a_{j_2 j_1 0}(a_{j_1 j_1 1} - a_{j_2 j_1 l_1})(a_{j_1 j_1 1} - a_{j_2 j_1 l_2})^2 < 0$,

$$(a_{j_1 j_1 1}, \dot{x}_{j_2}) = \underbrace{(3^{\text{rd}}\text{SI}, \text{nF})}_{\text{third-order sink flow }(-)}.$$

  The flow at $x^*_{j_1} = a_{j_1 j_1 1}$ is a (3rdSI,pF)-negative third-order sink flow.

– The third-order source and sink flows bifurcations are the appearing bifurcations for source-sink-source and sink-source-sink flows.

(viii$_{1b}$) For $x_{j_1}^* = a_{j_2j_1l_2} \neq a_{j_1j_1 1}$, we have

$$\frac{dx_{j_2}}{dx_{j_1}}\Big|_{x_{j_1}^* = a_{j_2j_1l_2}} = \frac{a_{j_2j_1 0}}{a_{j_1j_1 0}} \frac{(x_{j_1} - a_{j_2j_1l_1})(x_{j_1} - a_{j_2j_1l_2})^2}{(x_{j_1} - a_{j_1j_1 1})^3}\Big|_{x_{j_1}^* = a_{j_2j_1l_2}} = 0,$$
$$\frac{d^2x_{j_2}}{dx_{j_1}^2}\Big|_{x_{j_1}^* = a_{j_2j_1l_2}} = 0.$$

If

$$\frac{d^3x_{j_2}}{dx_{j_1}^3}\Big|_{x_{j_1}^* = a_{j_2j_1l_2}} = \frac{a_{j_2j_1 0}}{a_{j_1j_1 0}} \frac{2(a_{j_2j_1l_2} - a_{j_2j_1l_1})}{(a_{j_2j_1l_2} - a_{j_1j_1 1})^3}\Big|_{x_{j_1}^* = a_{j_2j_1l_2}} > 0,$$

the flow at $x_{j_1}^* = a_{j_2j_1l_2} \neq a_{j_1j_1 1}$ is an increasing-inflection flow in the x_{j_2}-direction, and if

$$\frac{d^3x_{j_2}}{dx_{j_1}^3}\Big|_{x_{j_1}^* = a_{j_2j_1l_2}} = \frac{a_{j_2j_1 0}}{a_{j_1j_1 0}} \frac{2(a_{j_2j_1l_2} - a_{j_2j_1l_1})}{(a_{j_2j_1l_2} - a_{j_1j_1 1})^3}\Big|_{x_{j_1}^* = a_{j_2j_1l_2}} < 0,$$

the flow at $x_{j_1}^* = a_{j_2j_1l_2} \neq a_{j_1j_1 1}$ is a decreasing-inflection flow in the x_{j_2}-direction.

At $x_{j_1}^* = a_{j_2j_1l_2} \neq a_{j_1j_1 1}$, the variational equation is

$$\Delta\dot{x}_{j_2} = a_{j_2j_1 0}(a_{j_2j_1l_2} - a_{j_2j_1l_1})(\Delta x_{j_1})^2,$$
$$\dot{x}_{j_1} = a_{j_1j_1 0}(a_{j_2j_1l_2} - a_{j_1j_1 1})^3.$$

The flows at $x_{j_1}^* = a_{j_2j_1l_2} \neq a_{j_1j_1 1}$ are positive and negative for $\dot{x}_{j_1} > 0$ and $\dot{x}_{j_1} < 0$ in the x_{j_1}-direction.

Therefore, the equilibrium of $x_{j_1}^* = a_{j_2j_1l_2} \neq a_{j_1j_1 1}$ has the following properties as in Eqs. (4.168)–(4.171).

- For $a_{j_1j_1 0}(a_{j_2j_1l_2} - a_{j_1j_1 1})^3 > 0$ and $a_{j_2j_1 0}(a_{j_2j_1l_2} - a_{j_2j_1l_1}) > 0$,

$$(\dot{x}_{j_1}, a_{j_2j_1 1}) = \underbrace{(\text{pF}, \text{II})}_{\text{increasing-inflection flow } (+)}.$$

The flow at $x_{j_1}^* = a_{j_2j_1l_2}$ is a (pF,II)-positive increasing-inflection flow.

- For $a_{j_1j_1 0}(a_{j_2j_1l_2} - a_{j_1j_1 1})^3 < 0$ and $a_{j_2j_1 0}(a_{j_2j_1l_2} - a_{j_2j_1l_1}) > 0$,

$$(\dot{x}_{j_1}, a_{j_2j_1 1}) = \underbrace{(\text{nF}, \text{DI})}_{\text{decreasing-inflection flow } (-)}.$$

The flow at $x_{j_1}^* = a_{j_2j_1l_2}$ is a (pF,DI)-positive decreasing-inflection flow.

- For $a_{j_1j_10}(a_{j_2j_1l_2}-a_{j_1j_11})^3>0$ and $a_{j_2j_10}(a_{j_2j_1l_2}-a_{j_2j_1l_1})<0$,

$$(\dot{x}_{j_1},a_{j_2j_11})=\underbrace{(\text{pF},\text{DI})}_{\text{decreasing-inflection flow }(+)}.$$

The flow at $x_{j_1}^*=a_{j_2j_1l_2}$ is an (nF,DI)-negative decreasing-inflection flow.

- For $a_{j_1j_10}(a_{j_2j_1l_2}-a_{j_1j_11})^3<0$ and $a_{j_2j_10}(a_{j_2j_1l_2}-a_{j_2j_1l_1})<0$,

$$(\dot{x}_{j_1},a_{j_2j_11})=\underbrace{(\text{nF},\text{II})}_{\text{increasing-inflection flow }(-)}.$$

The flow at $x_{j_1}^*=a_{j_2j_1l_2}$ is an (nF,II)-negative increasing-inflection flow.

– The increasing-inflection and decreasing-inflection flow bifurcations are the appearing bifurcations for the up-parabola and down-parabola flows.

(viii$_{1c}$) For $x_{j_1}^*=a_{j_2j_1l_1}\neq a_{j_1j_11}$, we have

$$\left.\frac{dx_{j_2}}{dx_{j_1}}\right|_{x_{j_1}^*=a_{j_2j_1l_1}}=\frac{a_{j_2j_10}}{a_{j_1j_10}}\left.\frac{(x_{j_1}-a_{j_2j_1l_1})(x_{j_1}-a_{j_2j_1l_2})^2}{(x_{j_1}-a_{j_1j_11})^3}\right|_{x_{j_1}^*=a_{j_2j_1l_1}}=0.$$

If

$$\left.\frac{d^2x_{j_2}}{dx_{j_1}^2}\right|_{x_{j_1}^*=a_{j_2j_1l_1}}=\frac{a_{j_2j_10}}{a_{j_1j_10}}\left.\frac{(a_{j_2j_1l_1}-a_{j_2j_1l_2})^2}{(a_{j_2j_1l_1}-a_{j_1j_11})^3}\right|_{x_{j_1}^*=a_{j_2j_1l_1}}>0,$$

the flow at $x_{j_1}^*=a_{j_2j_1l_1}\neq a_{j_1j_11}$ is an up-parabola flow in the x_{j_2}-direction, and if

$$\left.\frac{d^2x_{j_2}}{dx_{j_1}^2}\right|_{x_{j_1}^*=a_{j_2j_1l_1}}=\frac{a_{j_2j_10}}{a_{j_1j_10}}\left.\frac{(a_{j_2j_1l_1}-a_{j_2j_1l_2})^2}{(a_{j_2j_1l_1}-a_{j_1j_11})^3}\right|_{x_{j_1}^*=a_{j_2j_1l_1}}<0,$$

the flow at $x_{j_1}^*=a_{j_2j_1l_1}\neq a_{j_1j_11}$ is a decreasing-inflection flow in the x_{j_2}-direction.

At $x_{j_1}^*=a_{j_2j_1l_1}\neq a_{j_1j_11}$, the variational equation is

$$\begin{aligned}\Delta\dot{x}_{j_2}&=a_{j_2j_10}(a_{j_2j_1l_1}-a_{j_2j_1l_2})^2\Delta x_{j_1},\\ \dot{x}_{j_1}&=a_{j_1j_10}(a_{j_2j_1l_1}-a_{j_1j_11})^3.\end{aligned}$$

The flows at $x_{j_1}^*=a_{j_2j_1l_1}\neq a_{j_1j_11}$ are positive and negative for $\dot{x}_{j_1}>0$ and $\dot{x}_{j_1}<0$ in the x_{j_1}-direction.

Therefore, the equilibrium of $x_{j_1}^*=a_{j_2j_1l_1}\neq a_{j_1j_11}$ has the following properties as in Eqs. (4.172)–(4.175).

- For $a_{j_1j_1 0}(a_{j_2j_1l_1} - a_{j_1j_1 1})^3 > 0$ and $a_{j_2j_1 0}(a_{j_2j_1l_1} - a_{j_2j_1l_2})^2 > 0$

$$(\dot{x}_{j_1}, a_{j_2j_1l_1}) = \underbrace{(\text{pF}, \text{UP})}_{\text{up-parabola flow } (+)}.$$

The flow at $x_{j_1}^* = a_{j_2j_1l_1}$ is a (pF,UP)-positive up-parabola flow.

- For $a_{j_1j_1 0}(a_{j_2j_1l_1} - a_{j_1j_1 1})^3 < 0$ and $a_{j_2j_1 0}(a_{j_2j_1l_1} - a_{j_2j_1l_2})^2 > 0$,

$$(\dot{x}_{j_1}, a_{j_2j_1l_1}) = \underbrace{(\text{nF}, \text{DP})}_{\text{down-parabola flow } (-)}.$$

The flow at $x_{j_1}^* = a_{j_2j_1l_1}$ is an (nF,DP)-negative down-parabola flow.

- For $a_{j_1j_1 0}(a_{j_2j_1l_1} - a_{j_1j_1 1})^3 > 0$ and $a_{j_2j_1 0}(a_{j_2j_1l_1} - a_{j_2j_1l_2})^2 < 0$,

$$(\dot{x}_{j_1}, a_{j_2j_1l_1}) = \underbrace{(\text{pF}, \text{DP})}_{\text{down-parabola flow } (+)}.$$

The flow at $x_{j_1}^* = a_{j_2j_1l_1}$ is a (pF,DP)-positive down-parabola flow.

- For $a_{j_1j_1 0}(a_{j_2j_1l_1} - a_{j_1j_1 1})^3 < 0$ and $a_{j_2j_1 0}(a_{j_2j_1l_2} - a_{j_2j_1l_2})^2 < 0$,

$$(\dot{x}_{j_1}, a_{j_2j_1l_1}) = \underbrace{(\text{nF}, \text{UP})}_{\text{up-parabola flow } (-)}.$$

The flow at $x_{j_1}^* = a_{j_2j_1l_1}$ is an (nF,UP)-negative up-parabola flow.

(viii$_2$) For $a_{j_2j_1l_2} = a_{j_1j_1 1}$,

$$\begin{aligned}
&\dot{x}_{j_1} = a_{j_1j_1 0}(x_{j_1} - a_{j_1j_1 1})^3,\\
&\dot{x}_{j_2} = a_{j_2j_1 0}(x_{j_1} - a_{j_2j_1l_1})(x_{j_1} - a_{j_2j_1l_2})^2,\\
&j_1, j_2 \in \{1,2\}, j_1 \neq j_2.
\end{aligned}$$

We have

$$\frac{dx_{j_2}}{dx_{j_1}} = \frac{a_{j_2j_1 0}}{a_{j_1j_1 0}} \frac{(x_{j_1} - a_{j_2j_1l_1})}{(x_{j_1} - a_{j_1j_1 1})} = \frac{a_{j_2j_1 0}}{a_{j_1j_1 0}} \left[1 + \frac{(a_{j_1j_1 1} - a_{j_2j_1l_1})}{(x_{j_1} - a_{j_1j_1 1})}\right].$$

With an initial condition $(x_{j_1 0}, x_{j_2 0})$, the first integral manifold is

$$\begin{aligned}
x_{j_2} - x_{j_2 0} &= \frac{a_{j_2j_1 0}}{a_{j_1j_1 0}} \left[(x_{j_1} - x_{j_1 0}) + (a_{j_1j_1 1} - a_{j_2j_1l_1}) \ln \frac{| x_{j_1} - a_{j_1j_1 1} |}{| x_{j_1 0} - a_{j_1j_1 1} |}\right],\\
&j_1, j_2 \in \{1,2\}, j_1 \neq j_2.
\end{aligned}$$

From the first integral manifold, if

$$\frac{a_{j_2j_10}}{a_{j_1j_10}}(a_{j_1j_11}-a_{j_2j_1l_1})>0,$$

the flow at $x_{j_1}^*=a_{j_2j_1l_2}=a_{j_1j_11}$ is a down-logarithmic flow, and if

$$\frac{a_{j_2j_10}}{a_{j_1j_10}}(a_{j_1j_11}-a_{j_2j_1l_1})<0,$$

the flow at $x_{j_1}^*=a_{j_2j_1l_2}=a_{j_1j_11}$ is an up-logarithmic flow.

At $x_{j_1}^*=a_{j_2j_1l_2}=a_{j_1j_11}$, the variational equation is

$$\begin{aligned}\Delta\dot{x}_{j_1}&=a_{j_1j_10}(\Delta x_{j_1})^3,\\ \Delta\dot{x}_{j_2}&=a_{j_2j_10}(a_{j_2j_1l_2}-a_{j_2j_1l_1})(\Delta x_{j_1})^2.\end{aligned}$$

From the first integral manifold, the flow at the equilibrium at $x_{j_1}^*=a_{j_2j_1l_2}=a_{j_1j_11}$ is a logarithmic flow. The flow $x_{j_1}^*=a_{j_2j_1l_2}=a_{j_1j_11}$ is the third-order source and sink flows in the x_{j_1} direction. The singularity at the equilibrium is of the second-order and third-order in the x_{j_2}-direction and the x_{j_1}-direction. Thus, (3:2) is used to identify the singularity of the logarithmic flows.

Therefore, the infinite-equilibrium of $x_{j_1}^*=a_{j_2j_1l_2}=a_{j_1j_11}$ with $\overline{x}_{j_2}\in(-\infty,\infty)$ has the following properties as in Eqs. (4.178)–(4.181).

- For $a_{j_1j_10}>0$ and $a_{j_2j_10}(a_{j_2j_1l_2}-a_{j_2j_1l_1})>0$,

$$(a_{j_1j_11},\overline{x}_{j_2})=\underbrace{(3^{\text{rd}}\text{SO},(3:2)\text{DL})}_{\text{third-order down-logarithmic source}}.$$

  The equilibrium at $x_{j_1}^*=a_{j_2j_1l_2}=a_{j_1j_11}$ is a (3^{rd}SO,(3:2)DL)-third-order down-logarithmic source infinite-equilibrium.

- For $a_{j_1j_10}<0$ and $a_{j_2j_10}(a_{j_2j_1l_2}-a_{j_2j_1l_1})>0$,

$$(a_{j_1j_11},\overline{x}_{j_2})=\underbrace{(3^{\text{rd}}\text{SI},(3:2)\text{UL})}_{\text{third-order up-logarithmic sink}}.$$

  The equilibrium at $x_{j_1}^*=a_{j_2j_1l_2}=a_{j_1j_11}$ is a (3^{rd}SI,(3:2)UL)-third-order up-logarithmic sink infinite-equilibrium.

- For $a_{j_1j_10}>0$ and $a_{j_2j_10}(a_{j_2j_1l_2}-a_{j_2j_1l_1})<0$,

$$(a_{j_1j_11},\overline{x}_{j_2})=\underbrace{(3^{\text{rd}}\text{SO},(3:2)\text{UL})}_{\text{third-order up-logarithmic source}}.$$

  The equilibrium at $x_{j_1}^*=a_{j_2j_1l_2}=a_{j_1j_11}$ is a (3^{rd}SO,(3:2)UL)-third-order up-logarithmic source infinite-equilibrium.

- For $a_{j_1j_10} < 0$ and $a_{j_2j_10}(a_{j_2j_1l_2} - a_{j_2j_1l_1}) < 0$,

$$(a_{j_1j_1 1}, \bar{x}_{j_2}) = \underbrace{(3^{\text{rd}}\text{SI}, (3{:}2)\text{DL})}_{\text{third-order down-logarithmic sink}}.$$

The equilibrium at $x^*_{j_1} = a_{j_2j_1l_2} = a_{j_1j_11}$ is a (3rd SI,(3:2)DL)-third-order down-logarithmic sink infinite-equilibrium.

- The third-order up-logarithmic and down-logarithmic source and sink infinite-equilibriums of $x^*_{j_1} = a_{j_2j_1l_2} = a_{j_1j_11}$ are the appearing bifurcation for the sink and source flows and the up-parabola and down-parabola flows.
- The third-order up-logarithmic and down-logarithmic source and sink infinite-equilibriums of $x^*_{j_1} = a_{j_2j_1l_2} = a_{j_1j_11}$ are the switching bifurcation for the third-order sink and source flows and the increasing-inflection and decreasing-inflection flows.

(viii$_3$) For $a_{j_2j_1l_1} = a_{j_1j_11}$,

$$\begin{aligned}
&\dot{x}_{j_1} = a_{j_1j_10}(x_{j_1} - a_{j_1j_11})^3, \\
&\dot{x}_{j_2} = a_{j_2j_10}(x_{j_1} - a_{j_2j_1l_1})(x_{j_1} - a_{j_2j_1l_2})^2, \\
&j_1, j_2 \in \{1,2\}, j_1 \neq j_2.
\end{aligned}$$

In phase space, we have

$$\begin{aligned}
\frac{dx_{j_2}}{dx_{j_1}} &= \frac{a_{j_2j_10}}{a_{j_1j_10}} \frac{(x_{j_1} - a_{j_2j_1l_2})^2}{(x_{j_1} - a_{j_1j_11})^2} \\
&= \frac{a_{j_2j_10}}{a_{j_1j_10}} \left[1 + \frac{2(a_{j_1j_11} - a_{j_2j_1l_2})}{(x_{j_1} - a_{j_1j_11})} + \frac{(a_{j_1j_11} - a_{j_2j_1l_2})^2}{(x_{j_1} - a_{j_1j_11})^2}\right].
\end{aligned}$$

With an initial condition (x_{j_10}, x_{j_20}), the first integral manifold is

$$\begin{aligned}
x_{j_2} - x_{j_20} = \frac{a_{j_2j_10}}{a_{j_1j_10}} &\left[(x_{j_1} - x_{j_10}) + 2(a_{j_1j_11} - a_{j_2j_1l_2}) \ln \frac{|x_{j_1} - a_{j_1j_11}|}{|x_{j_10} - a_{j_1j_11}|}\right. \\
&\left. - (a_{j_1j_11} - a_{j_2j_1l_2})^2 \left(\frac{1}{x_{j_1} - a_{j_1j_11}} - \frac{1}{x_{j_10} - a_{j_1j_11}}\right)\right], \\
&j_1, j_2 \in \{1,2\}, j_1 \neq j_2.
\end{aligned}$$

From the first integral manifold, if

$$\frac{a_{j_2j_10}}{a_{j_1j_10}}(a_{j_1j_11} - a_{j_2j_1l_1})^2 > 0,$$

the flow at $x^*_{j_1} = a_{j_2j_1l_2} = a_{j_1j_11}$ is an increasing-concave flow, and if

$$\frac{a_{j_2j_10}}{a_{j_1j_10}}(a_{j_1j_11}-a_{j_2j_1l_1})^2<0,$$

the flow at $x_{j_1}^* = a_{j_2j_1l_2} = a_{j_1j_11}$ is a decreasing-concave flow.

At $x_{j_1}^* = a_{j_2j_1l_2} = a_{j_1j_11}$, the variational equations are

$$\begin{aligned}\Delta\dot{x}_{j_1} &= a_{j_1j_10}(\Delta x_{j_1})^3,\\ \Delta\dot{x}_{j_2} &= a_{j_2j_10}(a_{j_2j_1l_2}-a_{j_2j_1l_1})^2\Delta x_{j_1}.\end{aligned}$$

The flow $x_{j_1}^* = a_{j_2j_1l_2} = a_{j_1j_11}$ is the third-order source and sink flows in the x_{j_1}-direction. The singularity at the equilibrium is of the first-order and third-order in the x_{j_2}-direction and the x_{j_1}-direction. Thus, (3:1) is used to identify the singularity of the concave flows.

Therefore, the infinite-equilibrium of $x_{j_1}^* = a_{j_2j_1l_1} = a_{j_1j_11}$ with $\overline{x}_{j_2} \in (-\infty,\infty)$ has the following properties as in Eqs. (4.184)–(4.187).

- For $a_{j_1j_10}>0$ and $a_{j_2j_10}(a_{j_2j_1l_1}-a_{j_2j_1l_2})^2>0$,

$$(a_{j_1j_11},\overline{x}_{j_2}) = \underbrace{(3^{\text{rd}}\text{SO},(3:1)\text{IC})}_{\text{third-order increasing-concave source}}.$$

  The equilibrium at $x_{j_1}^* = a_{j_2j_1l_1} = a_{j_1j_11}$ is a (3^{rd}SO,(3:1)IC)-third-order increasing-concave source infinite-equilibrium.
- For $a_{j_1j_10}<0$ and $a_{j_2j_10}(a_{j_2j_1l_1}-a_{j_2j_1l_2})^2>0$,

$$(a_{j_1j_11},\overline{x}_{j_2}) = \underbrace{(3^{\text{rd}}\text{SI},(3:1)\text{DC})}_{\text{third-order decreasing-concave sink}}.$$

  The equilibrium at $x_{j_1}^* = a_{j_2j_1l_1} = a_{j_1j_11}$ is a (3^{rd}SI,(3:1)DC)-third-order decreasing-concave sink infinite-equilibrium.
- For $a_{j_1j_10}>0$ and $a_{j_2j_10}(a_{j_2j_1l_1}-a_{j_2j_1l_2})^2<0$,

$$(a_{j_1j_11},a_{j_2j_1l_2}) = \underbrace{(3^{\text{rd}}\text{SO},(3:1)\text{DC})}_{\text{third-order decreasing-concave source}}.$$

  The equilibrium at $x_{j_1}^* = a_{j_2j_1l_1} = a_{j_1j_11}$ is a (3^{rd}SO,(3:1)DC)-third-order decreasing-concave source infinite-equilibrium.
- For $a_{j_1j_10}<0$ and $a_{j_2j_10}(a_{j_2j_1l_1}-a_{j_2j_1l_2})^2<0$,

$$(a_{j_1j_11},\overline{x}_{j_2}) = \underbrace{(3^{\text{rd}}\text{SI},(3:1)\text{IC})}_{\text{third-order increasing-concave sink}}.$$

The equilibrium at $x_{j_1}^* = a_{j_2j_1l_1} = a_{j_1j_11}$ is a (3^{rd}SI,(3:1)IC)-third-order increasing-concave source infinite-equilibrium.

- The third-order increasing-concave and decreasing-concave source and sink infinite-equilibriums of $x_{j_1}^* = a_{j_2j_1l_1} = a_{j_1j_11}$ are the appearing bifurcation for the sink and source flows and the up-parabola and down-parabola flows.
- The third-order increasing-concave and decreasing-concave source and sink infinite-equilibriums of $x_{j_1}^* = a_{j_2j_1l_1} = a_{j_1j_11}$ are the switching bifurcation for the third-order sink and source flows and the up-parabola and down-parabola flows.

(ix) If

$$\begin{aligned}
&\Delta_{j_1j_1} = 0 \text{ and } \Delta_{j_2j_1} = 0,\\
&a_{j_2j_1l_1} = a_{j_2j_1l_2} \equiv a_{j_2j_11}, a_{j_1j_1s_1} \neq a_{j_1j_1s_2}
\end{aligned}$$

then

$$\begin{aligned}
&\dot{x}_{j_1} = a_{j_1j_10}(x_{j_1} - a_{j_1j_1s_1})(x_{j_1} - a_{j_1j_1s_2})^2,\\
&\dot{x}_{j_2} = a_{j_2j_10}(x_{j_1} - a_{j_2j_11})^3,\\
&j_1, j_2 \in \{1,2\}, j_1 \neq j_2.
\end{aligned}$$

(ix_1) For $a_{j_1j_11} \neq a_{j_2j_1l_1}, a_{j_2j_1l_2}$, in phase space,

$$\begin{aligned}
\frac{dx_{j_2}}{dx_{j_1}} &= \frac{a_{j_2j_10}}{a_{j_1j_10}} \frac{(x_{j_1} - a_{j_2j_11})^3}{(x_{j_1} - a_{j_1j_1s_1})(x_{j_1} - a_{j_1j_1s_2})^2}\\
&= \frac{a_{j_2j_10}}{a_{j_1j_10}} \Big\{ 1 + \frac{(a_{j_1j_1s_1} - a_{j_2j_11})^3}{(a_{j_1j_1s_1} - a_{j_1j_1s_2})^2} \frac{1}{x_{j_1} - a_{j_1j_1s_1}}\\
&\quad + \frac{(a_{j_1j_1s_2} - a_{j_2j_11})^3}{(a_{j_1j_1s_2} - a_{j_1j_1s_1})} \frac{1}{(x_{j_1} - a_{j_1j_1s_2})^2}\\
&\quad + \Big[\frac{3(a_{j_1j_1s_2} - a_{j_2j_11})^2}{a_{j_1j_1s_2} - a_{j_1j_1s_1}} - \frac{(a_{j_1j_1s_2} - a_{j_2j_11})^3}{(a_{j_1j_1s_2} - a_{j_1j_1s_1})^2} \Big] \frac{1}{x_{j_1} - a_{j_1j_1s_2}} \Big\}
\end{aligned}$$

With an initial condition (x_{j_10}, x_{j_20}), the first integral manifold is obtained as

$$\begin{aligned}
x_{j_2} - x_{j_20} &= \frac{a_{j_2j_10}}{a_{j_1j_10}} \Big\{ (x_{j_1} - x_{j_10}) + \frac{(a_{j_1j_1s_1} - a_{j_2j_11})^3}{(a_{j_1j_1s_1} - a_{j_1j_1s_2})^2} \ln \frac{|x_{j_1} - a_{j_1j_1s_1}|}{|x_{j_10} - a_{j_1j_1s_1}|}\\
&\quad - \frac{(a_{j_1j_1s_2} - a_{j_2j_11})^3}{(a_{j_1j_1s_1} - a_{j_1j_1s_2})} \Big(\frac{1}{x_{j_1} - a_{j_1j_1s_2}} - \frac{1}{x_{j_10} - a_{j_1j_1s_2}} \Big)\\
&\quad + \Big[\frac{3(a_{j_1j_1s_2} - a_{j_2j_11})^2}{a_{j_1j_1s_2} - a_{j_1j_1s_1}} - \frac{(a_{j_1j_1s_2} - a_{j_2j_11})^3}{(a_{j_1j_1s_2} - a_{j_1j_1s_1})^2} \Big] \ln \frac{|x_{j_1} - a_{j_1j_1s_2}|}{|x_{j_10} - a_{j_1j_1s_2}|} \Big\},\\
&\quad s_1, s_2 \in \{1,2\}, s_1 \neq s_2.
\end{aligned}$$

(ix$_{1a}$) For $x_{j_1}^* = a_{j_2 j_1 1} \neq a_{j_1 j_1 s_1}, a_{j_2 j_1 s_2}$, in phase space,

$$
\begin{aligned}
&\frac{dx_{j_2}}{dx_{j_1}}\bigg|_{x_{j_1}^* = a_{j_2 j_1 1}} = \frac{a_{j_2 j_1 0}}{a_{j_1 j_1 0}} \frac{(x_{j_1} - a_{j_2 j_1 1})^3}{(x_{j_1} - a_{j_1 j_1 s_1})(x_{j_1} - a_{j_1 j_1 s_2})^2}\bigg|_{x_{j_1}^* = a_{j_2 j_1 1}} = 0, \\
&\frac{d^2 x_{j_2}}{dx_{j_1}^2}\bigg|_{x_{j_1}^* = a_{j_2 j_1 1}} = \frac{d^3 x_{j_2}}{dx_{j_1}^3}\bigg|_{x_{j_1}^* = a_{j_2 j_1 1}} = 0.
\end{aligned}
$$

If

$$
\frac{d^4 x_{j_2}}{dx_{j_1}^4}\bigg|_{x_{j_1}^* = a_{j_2 j_1 1}} = \frac{a_{j_2 j_1 0}}{a_{j_1 j_1 0}} \frac{6}{(a_{j_2 j_1 1} - a_{j_1 j_1 s_1})(a_{j_2 j_1 1} - a_{j_1 j_1 s_2})^2} > 0,
$$

there is the third-order up-parabola flow, and if

$$
\frac{d^4 x_{j_2}}{dx_{j_1}^4}\bigg|_{x_{j_1}^* = a_{j_2 j_1 1}} = \frac{a_{j_2 j_1 0}}{a_{j_1 j_1 0}} \frac{6}{(a_{j_2 j_1 1} - a_{j_1 j_1 s_1})(a_{j_2 j_1 1} - a_{j_1 j_1 s_2})^2} < 0,
$$

there is the third-order down-parabola flow. Let

$$
\dot{x}_{j_1} = a_{j_1 j_1 0}(a_{j_2 j_1 1} - a_{j_1 j_1 s_1})(a_{j_2 j_1 1} - a_{j_1 j_1 s_2})^2.
$$

The third-order parabola flows are positive and negative in the x_{j_1}-direction for $\dot{x}_{j_1} > 0$ and $\dot{x}_{j_1} < 0$, respectively.

Therefore, the equilibrium of $x_{j_1}^* = a_{j_2 j_1 1} \neq a_{j_1 j_1 s_1}, a_{j_2 j_1 s_2}$ has the following properties as in Eqs. (4.191)–(4.194).

- For $a_{j_1 j_1 0}(a_{j_2 j_1 1} - a_{j_1 j_1 s_1})(a_{j_2 j_1 1} - a_{j_1 j_1 s_2})^2 > 0$ and $a_{j_2 j_1 0} > 0$,

$$
(\dot{x}_{j_1}, a_{j_2 j_1 1}) = \underbrace{(\text{pF}, 3^{\text{rd}}\text{UP})}_{\text{third-order up-parabola flow } (+)}.
$$

  The flow at $x_{j_1}^* = a_{j_2 j_1 1}$ is a (pF, 3rdUP)-positive third-order up-parabola flow.
- For $a_{j_1 j_1 0}(a_{j_2 j_1 1} - a_{j_1 j_1 s_1})(a_{j_2 j_1 1} - a_{j_1 j_1 s_2})^2 < 0$ and $a_{j_2 j_1 0} > 0$,

$$
(\dot{x}_{j_1}, a_{j_2 j_1 1}) = \underbrace{(\text{nF}, 3^{\text{rd}}\text{DP})}_{\text{third-order down-parabola flow } (-)}.
$$

  The flow at $x_{j_1}^* = a_{j_2 j_1 1}$ is an (nF, 3rdDP)-negative third-order down-parabola flow.

- For $a_{j_1j_10}(a_{j_2j_11}-a_{j_1j_1s_1})(a_{j_2j_11}-a_{j_1j_1s_2})^2>0$ and $a_{j_2j_10}<0$,

$$(\dot{x}_{j_1},a_{j_2j_11})=\underbrace{(\text{pF},3^{\text{rd}}\text{DP})}_{\text{third-order down-parabola flow }(+)}.$$

The flow at $x_{j_1}^*=a_{j_2j_11}$ is a (pF, 3^{rd}DP)-negative third-order down-parabola flow.

- For $a_{j_1j_10}(a_{j_2j_11}-a_{j_1j_1s_1})(a_{j_2j_11}-a_{j_1j_1s_2})^2<0$ and $a_{j_2j_10}<0$,

$$(\dot{x}_{j_1},a_{j_2j_11})=\underbrace{(\text{nF},3^{\text{rd}}\text{UP})}_{\text{third-order up-parabola flow }(-)}.$$

The flow at $x_{j_1}^*=a_{j_2j_11}$ is an (nF, 3^{rd}UP)-negative third-order up-parabola flow.

– The third-order up-parabola and down-parabola flows bifurcations are the appearing bifurcations for up-parabola, down-parabola, and up-parabola flows and down-parabola, up-parabola and down-parabola flows.

($\text{ix}_{1\text{b}}$) For $x_{j_1}^*=a_{j_1j_1s_1}\neq a_{j_2j_11}$, the variational equation is

$$\begin{aligned}\Delta\dot{x}_{j_1}&=a_{j_1j_10}(a_{j_1j_1s_1}-a_{j_1j_1s_2})^2\Delta x_{j_1},\\ \dot{x}_{j_2}&=a_{j_2j_10}(a_{j_1j_1s_1}-a_{j_2j_11})^3.\end{aligned}$$

The flows at $x_{j_1}^*=a_{j_1j_1s_1}$ are source and sink for $a_{j_1j_10}(a_{j_1j_1s_1}-a_{j_1j_1s_2})^2>0$ and $a_{j_1j_10}(a_{j_1j_1s_1}-a_{j_1j_1s_2})^2<0$ in the x_{j_1}-direction, respectively. Such source and sink flows are positive and negative in the x_{j_2}-direction for $\dot{x}_{j_2}>0$ and $\dot{x}_{j_2}<0$, respectively.

Therefore, the equilibrium of $x_{j_1}^*=a_{j_1j_1s_1}\neq a_{j_2j_11}$ has the following properties as in Eqs. (4.195)–(4.198).

- For $a_{j_1j_10}(a_{j_1j_1s_1}-a_{j_1j_1s_2})^2>0$ and $a_{j_2j_10}(a_{j_1j_1s_1}-a_{j_2j_11})^3>0$,

$$(a_{j_1j_1s_1},\dot{x}_{j_2})=\underbrace{(\text{SO},\text{pF})}_{\text{source flow }(+)}.$$

The flow at $x_{j_1}^*=a_{j_1j_1s_1}$ is an (SO,pF)-positive source flow.

- For $a_{j_1j_10}(a_{j_1j_1s_1}-a_{j_1j_1s_2})^2<0$ and $a_{j_2j_10}(a_{j_1j_1s_1}-a_{j_2j_11})^3>0$,

$$(a_{j_1j_1s_1},\dot{x}_{j_2})=\underbrace{(\text{SI},\text{pF})}_{\text{sink flow }(+)}.$$

The flow at $x_{j_1}^*=a_{j_1j_1s_1}$ is an (SI,pF)-positive sink flow.

- For $a_{j_1j_10}(a_{j_1j_1s_1}-a_{j_1j_1s_2})^2>0$ and $a_{j_2j_10}(a_{j_1j_1s_1}-a_{j_2j_11})^3<0$,

$$(a_{j_1j_1s_1},\dot{x}_{j_2})=\underbrace{(\mathrm{SI},\mathrm{nF})}_{\text{source flow }(-)}.$$

  The flow at $x^*_{j_1}=a_{j_1j_1s_1}$ is an (SI,nF)-negative source flow.
- For $a_{j_1j_10}(a_{j_1j_1s_1}-a_{j_1j_1s_2})^2<0$ and $a_{j_2j_10}(a_{j_1j_1s_1}-a_{j_2j_11})^3<0$,

$$(a_{j_1j_1s_1},\dot{x}_{j_2})=\underbrace{(\mathrm{SO},\mathrm{nF})}_{\text{source flow }(-)}.$$

  The flow at $x^*_{j_1}=a_{j_1j_1s_1}$ is an (SO,nF)-negative sink flow.

(ix_{1c}) For $x^*_{j_1}=a_{j_1j_1s_2}\neq a_{j_2j_11}$, the variational equation is

$$\begin{aligned}\Delta\dot{x}_{j_1}&=a_{j_1j_10}(a_{j_1j_1s_2}-a_{j_1j_1s_1})(\Delta x_{j_1})^2,\\ \dot{x}_{j_2}&=a_{j_2j_10}(a_{j_1j_1s_2}-a_{j_2j_11})^3.\end{aligned}$$

The flows at $x^*_{j_1}=a_{j_1j_1s_2}$ are upper-saddle and lower-saddle flows in the x_{j_1}-direction for $a_{j_1j_10}(a_{j_1j_1s_2}-a_{j_1j_1s_1})>0$ and $a_{j_1j_10}(a_{j_1j_1s_2}-a_{j_1j_1s_1})<0$, respectively. Such upper-saddle and lower-saddle are positive and negative in the x_{j_2}-direction for $\dot{x}_{j_2}>0$ and $\dot{x}_{j_2}<0$, respectively.

Therefore, the equilibrium of $x^*_{j_1}=a_{j_1j_1s_2}\neq a_{j_2j_11}$ has the following properties as in Eqs. (4.199–(4.202).

- For $a_{j_1j_10}(a_{j_1j_1s_2}-a_{j_1j_1s_1})>0$ and $a_{j_2j_10}(a_{j_1j_1s_2}-a_{j_2j_11})^3>0$,

$$(a_{j_1j_1s_2},\dot{x}_{j_2})=\underbrace{(\mathrm{US},\mathrm{pF})}_{\text{upper-saddle flow }(+)}.$$

  The flow at $x^*_{j_1}=a_{j_1j_1s_2}$ is a (US,pF)-positive upper-saddle flow.
- For $a_{j_1j_10}(a_{j_1j_1s_2}-a_{j_1j_1s_1})<0$ and $a_{j_2j_10}(a_{j_1j_1s_2}-a_{j_2j_11})^3>0$,

$$(a_{j_1j_1s_2},\dot{x}_{j_2})=\underbrace{(\mathrm{LS},\mathrm{pF})}_{\text{lower-saddle flow }(+)}.$$

  The flow at $x^*_{j_1}=a_{j_1j_1s_2}$ is an (LS,pF)-positive lower-saddle flow.
- For $a_{j_1j_10}(a_{j_1j_1s_2}-a_{j_1j_1s_1})>0$ and $a_{j_2j_10}(a_{j_1j_1s_2}-a_{j_2j_11})^3<0$,

$$(a_{j_1j_1s_2},\dot{x}_{j_2})=\underbrace{(\mathrm{US},\mathrm{nF})}_{\text{upper-saddle flow }(-)}.$$

  The flow at $x^*_{j_1}=a_{j_1j_1s_2}$ is a (US,nF)-negative upper-saddle flow.

- For $a_{j_1j_10}(a_{j_1j_1s_2}-a_{j_1j_1s_1})<0$ and $a_{j_2j_10}(a_{j_1j_1s_2}-a_{j_2j_11})^3<0$,

$$(a_{j_1j_1s_2},\dot{x}_{j_2})=\underbrace{(\text{LS},\text{nF})}_{\text{lower-saddle flow }(-)}.$$

The flow at $x^*_{j_1}=a_{j_1j_1s_2}$ is an (LS,pF)-negative lower-saddle flow.

(ix$_2$) For $a_{j_2j_11}=a_{j_1j_1s_2}$, we have

$$\begin{aligned}&\dot{x}_{j_1}=a_{j_1j_10}(x_{j_1}-a_{j_1j_1s_1})(x_{j_1}-a_{j_1j_1s_2})^2,\\&\dot{x}_{j_2}=a_{j_2j_10}(x_{j_1}-a_{j_2j_11})^3,\\&j_1,j_2\in\{1,2\},j_1\neq j_2.\end{aligned}$$

In phase space,

$$\begin{aligned}\frac{dx_{j_2}}{dx_{j_1}}&=\frac{a_{j_2j_10}}{a_{j_1j_10}}\frac{x_{j_1}-a_{j_2j_11}}{x_{j_1}-a_{j_1j_1s_1}}\\&=\frac{a_{j_2j_10}}{a_{j_1j_10}}\left[1+(a_{j_1j_1s_1}-a_{j_2j_11})\frac{1}{x_{j_1}-a_{j_1j_1s_1}}\right].\end{aligned}$$

With an initial condition (x_{j_10},x_{j_20}), the first integral manifold is obtained as

$$\begin{aligned}x_{j_2}-x_{j_20}&=\frac{a_{j_2j_10}}{a_{j_1j_10}}\left[(x_{j_1}-x_{j_10})+(a_{j_1j_1s_1}-a_{j_2j_11})\ln\frac{|\,x_{j_1}-a_{j_1j_1s_1}\,|}{|\,x_{j_10}-a_{j_1j_1s_1}\,|}\right],\\&j_1,j_2\in\{1,2\},j_1\neq j_2.\end{aligned}$$

For $x^*_{j_1}=a_{j_1j_1s_2}=a_{j_2j_11}$, in phase space,

$$\frac{dx_{j_2}}{dx_{j_1}}\Big|_{x^*_{j_1}=a_{j_2j_11}}=\frac{a_{j_2j_10}}{a_{j_1j_10}}\frac{x_{j_1}-a_{j_2j_11}}{x_{j_1}-a_{j_1j_1s_1}}\Big|_{x^*_{j_1}=a_{j_2j_11}}=0.$$

If

$$\frac{d^2x_{j_2}}{dx^2_{j_1}}\Big|_{x^*_{j_1}=a_{j_2j_11}}=\frac{a_{j_2j_10}}{a_{j_1j_10}}\frac{1}{(a_{j_2j_11}-a_{j_1j_1s_1})}>0,$$

there is an up-parabola flow, and if

$$\frac{d^2x_{j_2}}{dx^2_{j_1}}\Big|_{x^*_{j_1}=a_{j_2j_11}}=\frac{a_{j_2j_10}}{a_{j_1j_10}}\frac{1}{(a_{j_2j_11}-a_{j_1j_1s_1})}<0,$$

there is a down-parabola flow.

For $x_{j_1}^* = a_{j_1 j_1 s_2} = a_{j_2 j_1 1}$, the variational equations are

$$\begin{aligned}\Delta\dot{x}_{j_1} &= a_{j_1 j_1 0}(a_{j_1 j_1 s_2} - a_{j_1 j_1 s_1})(\Delta x_{j_1})^2,\\ \Delta\dot{x}_{j_2} &= a_{j_2 j_1 0}(\Delta x_{j_1})^3.\end{aligned}$$

The up-parabola and down-parabola flows in the x_{j_1}-direction are upper-saddle and lower-saddle for $a_{j_1 j_1 0}(a_{j_1 j_1 s_2} - a_{j_1 j_1 s_1}) > 0$ and $a_{j_1 j_1 0}(a_{j_1 j_1 s_2} - a_{j_1 j_1 s_1}) < 0$, respectively. In addition, there is a third-order singularity in the x_{j_2}-direction.

Therefore, the infinite-equilibrium of $x_{j_1}^* = a_{j_1 j_1 s_2} = a_{j_2 j_1 1}$ with $\overline{x}_{j_2} \in (-\infty, \infty)$ has the following properties as in Eqs. (4.205)–(4.208).

- For $a_{j_1 j_1 0}(a_{j_1 j_1 s_2} - a_{j_1 j_1 s_1}) > 0$ and $a_{j_2 j_1 0} > 0$,

$$(a_{j_1 j_1 s_2}, \overline{x}_{j_2}) = \underbrace{(\text{US}, (2:3)\text{UP})}_{\text{up-parabola upper-saddle}}.$$

  The equilibrium at $x_{j_1}^* = a_{j_1 j_1 s_2} = a_{j_2 j_1 1}$ is a (US,(2:3)UP)-up-parabola upper-saddle infinite-equilibrium.
- For $a_{j_1 j_1 0}(a_{j_1 j_1 s_2} - a_{j_1 j_1 s_1}) < 0$ and $a_{j_2 j_1 0} > 0$,

$$(a_{j_1 j_1 s_2}, \overline{x}_{j_2}) = \underbrace{(\text{LS}, (2:3)\text{DP})}_{\text{down-parabola lower-saddle}}.$$

  The equilibrium at $x_{j_1}^* = a_{j_1 j_1 s_2} = a_{j_2 j_1 1}$ is an (LS,(2:3)DP)-down-parabola lower-saddle infinite-equilibrium.
- For $a_{j_1 j_1 0}(a_{j_1 j_1 s_2} - a_{j_1 j_1 s_1}) > 0$ and $a_{j_2 j_1 0} < 0$,

$$(a_{j_1 j_1 s_2}, \overline{x}_{j_2}) = \underbrace{(\text{US}, (2:3)\text{DP})}_{\text{down-parabola upper-saddle}}.$$

  The equilibrium at $x_{j_1}^* = a_{j_1 j_1 s_2} = a_{j_2 j_1 1}$ is a (US,(2:3)DP)-up-logarithmic source infinite-equilibrium.
- For $a_{j_1 j_1 0}(a_{j_1 j_1 s_2} - a_{j_1 j_1 s_1}) < 0$ and $a_{j_2 j_1 0} < 0$,

$$(a_{j_1 j_1 s_2}, \overline{x}_{j_2}) = \underbrace{(\text{LS}, (2:3)\text{UP})}_{\text{up-parabola lower-saddle}}.$$

  The equilibrium at $x_{j_1}^* = a_{j_1 j_1 s_2} = a_{j_2 j_1 1}$ is an (LS,(2:3)UP)-up-parabola lower-saddle infinite-equilibrium.

– The (2:3)-up-parabola and down-parabola, upper-saddle, and lower-saddle infinite-equilibriums of $x_{j_1}^* = a_{j_1 j_1 s_2} = a_{j_2 j_1 1}$ are the appearing bifurcation for the sink and source flows and the up-parabola and down-parabola flows.

– The (2:3)-up-parabola and down-parabola, upper-saddle and lower-saddle infinite-equilibriums of $x_{j_1}^* = a_{j_1 j_1 s_2} = a_{j_2 j_1 1}$ are the switching bifurcation for the third-order up-parabola and down-parabola flows and the upper-saddle and lower-saddle flows.

(ix$_3$) For $a_{j_1 j_1 s_1} = a_{j_2 j_1 1}$,

$$\begin{aligned}
&\dot{x}_{j_1} = a_{j_1 j_1 0}(x_{j_1} - a_{j_1 j_1 s_1})(x_{j_1} - a_{j_1 j_1 s_2})^2,\\
&\dot{x}_{j_2} = a_{j_2 j_1 0}(x_{j_1} - a_{j_2 j_1 1})^3,\\
&j_1, j_2 \in \{1,2\}, j_1 \neq j_2.
\end{aligned}$$

In phase space, for $a_{j_1 j_1 s_1} = a_{j_2 j_1 1}$,

$$\begin{aligned}
\frac{dx_{j_2}}{dx_{j_1}} &= \frac{a_{j_2 j_1 0}}{a_{j_1 j_1 0}} \frac{(x_{j_1} - a_{j_2 j_1 1})^2}{(x_{j_1} - a_{j_1 j_1 s_2})^2}\\
&= \frac{a_{j_2 j_1 0}}{a_{j_1 j_1 0}} \left[1 + \frac{2(a_{j_1 j_1 s_2} - a_{j_2 j_1 1})}{x_{j_1} - a_{j_1 j_1 s_2}} + \frac{(a_{j_1 j_1 s_2} - a_{j_2 j_1 1})^2}{(x_{j_1} - a_{j_1 j_1 s_2})^2} \right].
\end{aligned}$$

With an initial condition $(x_{j_1 0}, x_{j_2 0})$, the first integral manifold is

$$\begin{aligned}
x_{j_2} - x_{j_2 0} &= \frac{a_{j_2 j_1 0}}{a_{j_1 j_1 0}} \left[(x_{j_1} - x_{j_1 0}) + 2(a_{j_1 j_1 s_2} - a_{j_2 j_1 1}) \ln \frac{|x_{j_1} - a_{j_1 j_1 s_2}|}{|x_{j_1 0} - a_{j_1 j_1 s_2}|} \right.\\
&\left. - (a_{j_1 j_1 s_2} - a_{j_2 j_1 1})^2 \left(\frac{1}{x_{j_1} - a_{j_1 j_1 s_2}} - \frac{1}{x_{j_1 0} - a_{j_1 j_1 s_2}} \right) \right],\\
&s_1, s_2 \in \{1,2\}, s_1 \neq s_2.
\end{aligned}$$

For $x_{j_1}^* = a_{j_1 j_1 s_2} = a_{j_2 j_1 1}$, in phase space,

$$\begin{aligned}
&\left. \frac{dx_{j_2}}{dx_{j_1}} \right|_{x_{j_1}^* = a_{j_2 j_1 1}} = \left. \frac{a_{j_2 j_1 0}}{a_{j_1 j_1 0}} \frac{(x_{j_1} - a_{j_2 j_1 1})^2}{(x_{j_1} - a_{j_1 j_1 s_2})^2} \right|_{x_{j_1}^* = a_{j_2 j_1 1}} = 0,\\
&\left. \frac{d^2 x_{j_2}}{dx_{j_1}^2} \right|_{x_{j_1}^* = a_{j_2 j_1 1}} = 0.
\end{aligned}$$

If

$$\left. \frac{d^3 x_{j_2}}{dx_{j_1}^3} \right|_{x_{j_1}^* = a_{j_2 j_1 1}} = \frac{a_{j_2 j_1 0}}{a_{j_1 j_1 0}} \frac{2}{(a_{j_2 j_1 1} - a_{j_1 j_1 s_1})^2} > 0,$$

there is an increasing-inflection flow, and if

$$\frac{d^3 x_{j_2}}{dx_{j_1}^3}\bigg|_{x_{j_1}^*=a_{j_2 j_1 1}} = \frac{a_{j_2 j_1 0}}{a_{j_1 j_1 0}} \frac{2}{(a_{j_2 j_1 1} - a_{j_1 j_1 s_1})^2} < 0,$$

there is a decreasing-inflection flow.

For $x_{j_1}^* = a_{j_1 j_1 s_1} = a_{j_2 j_1 1}$, the variational equations are

$$\begin{aligned} \Delta\dot{x}_{j_1} &= a_{j_1 j_1 0}(a_{j_1 j_1 s_1} - a_{j_1 j_1 s_2})^2 \Delta x_{j_1}, \\ \Delta\dot{x}_{j_2} &= a_{j_2 j_1 0}(\Delta x_{j_1})^3. \end{aligned}$$

The inflection flows are source and sink for $a_{j_1 j_1 0}(a_{j_1 j_1 s_1} - a_{j_1 j_1 s_2})^2 > 0$ and $a_{j_1 j_1 0}(a_{j_1 j_1 s_1} - a_{j_1 j_1 s_2})^2 < 0$ in the x_{j_1}-direction, respectively. In addition, there is a third-order singularity in the x_{j_2}-direction.

Therefore, the infinite-equilibrium of $x_{j_1}^* = a_{j_1 j_1 s_1} = a_{j_2 j_1 1}$ with $\overline{x}_{j_2} \in (-\infty, \infty)$ has the following properties as in Eqs. (4.211)–(4.214).

- For $a_{j_1 j_1 0}(a_{j_1 j_1 s_1} - a_{j_1 j_1 s_2})^2 > 0$ and $a_{j_2 j_1 0} > 0$,

$$(a_{j_1 j_1 s_1}, \overline{x}_{j_2}) = \underbrace{(\text{SO}, (1{:}3)\text{II})}_{(1:3)\text{-increasing-inflection source}}.$$

  The equilibrium at $x_{j_1}^* = a_{j_1 j_1 s_1} = a_{j_2 j_1 1}$ is an (SO,(1:3)II)-increasing-inflection source infinite-equilibrium.
- For $a_{j_1 j_1 0}(a_{j_1 j_1 s_1} - a_{j_1 j_1 s_2})^2 < 0$ and $a_{j_2 j_1 0} > 0$,

$$(a_{j_1 j_1 s_1}, \overline{x}_{j_2}) = \underbrace{(\text{SI}, (1{:}3)\text{DI})}_{(1:3)\text{-decreasing-inflection sink}}.$$

  The equilibrium at $x_{j_1}^* = a_{j_1 j_1 s_1} = a_{j_2 j_1 1}$ is an (SI,(1:3)DI)-decreasing-inflection sink infinite-equilibrium.
- For $a_{j_1 j_1 0}(a_{j_1 j_1 s_1} - a_{j_1 j_1 s_2})^2 > 0$ and $a_{j_2 j_1 0} < 0$,

$$(a_{j_1 j_1 s_1}, \overline{x}_{j_2}) = \underbrace{(\text{SO}, (1{:}3)\text{DI})}_{(1:3)\text{-decreasing-inflection source}}.$$

  The equilibrium at $x_{j_1}^* = a_{j_1 j_1 s_1} = a_{j_2 j_1 1}$ is an (SO,(1:3)DI)-decreasing-inflection source infinite-equilibrium.
- For $a_{j_1 j_1 0}(a_{j_1 j_1 s_1} - a_{j_1 j_1 s_2})^2 < 0$ and $a_{j_2 j_1 0} < 0$,

$$(a_{j_1 j_1 s_1}, \overline{x}_{j_2}) = \underbrace{(\text{SI}, (1{:}3)\text{II})}_{(1:3)\text{-increasing-inflection sink}}.$$

  The equilibrium at $x_{j_1}^* = a_{j_1 j_1 s_1} = a_{j_2 j_1 1}$ is an (SI, (1:3)II)-increasing-inflection sink infinite-equilibrium.

- The (1:3)-increasing-inflection and decreasing-inflection source and sink infinite-equilibriums of $x_{j_1}^* = a_{j_1 j_1 s_1} = a_{j_2 j_1 1}$ are the appearing bifurcation for the sink and source flows and the up-parabola and down-parabola flows.
- The (1:3)-increasing-inflection and decreasing-inflection source and sink infinite-equilibriums of $x_{j_1}^* = a_{j_2 j_1 l_2} = a_{j_1 j_1 1}$ are the switching bifurcation for the third-order up-parabola and down-parabola flows and the sink and source flows.

(x) If

$$\begin{aligned}
&\Delta_{j_1 j_1} = 0 \text{ and } \Delta_{j_2 j_1} = 0, \\
&a_{j_1 j_1 s_1} = a_{j_1 j_1 s_2} \equiv a_{j_1 j_1 1}, \\
&a_{j_2 j_1 l_1} = a_{j_2 j_1 l_2} \equiv a_{j_2 j_1 1},
\end{aligned}$$

then

$$\begin{aligned}
&\dot{x}_{j_1} = a_{j_1 j_1 0}(x_{j_1} - a_{j_1 j_1 1})^3, \\
&\dot{x}_{j_2} = a_{j_2 j_1 0}(x_{j_1} - a_{j_2 j_1 1})^3, \\
&j_1, j_2 \in \{1, 2\}, j_1 \neq j_2.
\end{aligned}$$

(x_1) For $a_{j_1 j_1 1} \neq a_{j_2 j_1 1}$, we have

$$\begin{aligned}
\frac{dx_{j_2}}{dx_{j_1}} &= \frac{a_{j_2 j_1 0}\,(x_{j_1} - a_{j_2 j_1 1})^3}{a_{j_1 j_1 0}\,(x_{j_1} - a_{j_1 j_1 1})^3} \\
&= \frac{a_{j_2 j_1 0}}{a_{j_1 j_1 0}}\left[1 + \frac{3(a_{j_1 j_1 1} - a_{j_2 j_1 1})}{(x_{j_1} - a_{j_1 j_1 1})} + \frac{3(a_{j_1 j_1 1} - a_{j_2 j_1 1})^2}{(x_{j_1} - a_{j_1 j_1 1})^2} + \frac{(a_{j_1 j_1 1} - a_{j_2 j_1 1})^3}{(x_{j_1} - a_{j_1 j_1 1})^3}\right].
\end{aligned}$$

With an initial condition $(x_{j_1 0}, x_{j_2 0})$, the first integral manifold is

$$\begin{aligned}
x_{j_2} - x_{j_2 0} = {}& \frac{a_{j_2 j_1 0}}{a_{j_1 j_1 0}}\Big\{(x_{j_1} - x_{j_1 0}) + 3(a_{j_1 j_1 1} - a_{j_2 j_1 1}) \ln \frac{|\, x_{j_1} - a_{j_1 j_1 1} \,|}{|\, x_{j_1 0} - a_{j_1 j_1 1} \,|} \\
& - 3(a_{j_1 j_1 1} - a_{j_2 j_1 1})^2 \Big(\frac{1}{x_{j_1} - a_{j_1 j_1 1}} - \frac{1}{x_{j_1 0} - a_{j_1 j_1 1}}\Big) \\
& - \frac{1}{2}(a_{j_1 j_1 1} - a_{j_2 j_1 1})^3 \Big[\frac{1}{(x_{j_1} - a_{j_1 j_1 1})^2} - \frac{1}{(x_{j_1 0} - a_{j_1 j_1 1})^2}\Big]\Big\}, \\
& j_1, j_2 \in \{1, 2\}, j_1 \neq j_2.
\end{aligned}$$

(x_{1a}) At $x_{j_1}^* = a_{j_1 j_1 1}$, the variational equation is

$$\begin{aligned}
&\dot{x}_{j_2} = a_{j_2 j_1 0}(a_{j_1 j_1 1} - a_{j_2 j_1 1})^3, \\
&\Delta\dot{x}_{j_1} = a_{j_1 j_1 0}(\Delta x_{j_1})^3.
\end{aligned}$$

The flows at $x_{j_1}^* = a_{j_1j_1 1} \neq a_{j_2j_1 1}$ are the third-order source and sink flows for $a_{j_1j_1 0} > 0$ and $a_{j_1j_1 0} < 0$ in the x_{j_1}-direction. The flows at $x_{j_1}^* = a_{j_1j_1 1} \neq a_{j_2j_1 1}$ are positive and negative for $a_{j_2j_1 0}(a_{j_1j_1 1} - a_{j_2j_1 1})^2 > 0$ and $a_{j_2j_1 0}(a_{j_1j_1 1} - a_{j_2j_1 l_1})^2 < 0$ in the x_{j_2}-direction.

Therefore, the equilibrium of $x_{j_1}^* = a_{j_1j_1 1}$ has the following properties as in Eqs. (4.218)–(4.221).

- For $a_{j_1j_1 0} > 0$ and $a_{j_2j_1 0}(a_{j_1j_1 1} - a_{j_2j_1 1})^3 > 0$,

$$(a_{j_1j_1 1}, \dot{x}_{j_2}) = \underbrace{(3^{\rm rd}\mathrm{SO}, \mathrm{pF})}_{\text{third-order source flow } (+)}.$$

The flow at $x_{j_1}^* = a_{j_1j_1 1}$ is a ($3^{\rm rd}$SO,pF)-third-order positive source flow.

- For $a_{j_1j_1 0} < 0$ and $a_{j_2j_1 0}(a_{j_1j_1 1} - a_{j_2j_1 1})^3 > 0$,

$$(a_{j_1j_1 1}, \dot{x}_{j_2}) = \underbrace{(3^{\rm rd}\mathrm{SI}, \mathrm{pF})}_{\text{third-order sink flow } (+)}.$$

The flow at $x_{j_1}^* = a_{j_1j_1 1}$ is a ($3^{\rm rd}$SI,pF)-third-order positive sink flow.

- For $a_{j_1j_1 0} > 0$ and $a_{j_2j_1 0}(a_{j_1j_1 1} - a_{j_2j_1 1})^3 < 0$,

$$(a_{j_1j_1 1}, \dot{x}_{j_2}) = \underbrace{(3^{\rm rd}\mathrm{SO}, \mathrm{nF})}_{\text{third-order source flow } (-)}.$$

The flow at $x_{j_1}^* = a_{j_1j_1 1}$ is a ($3^{\rm rd}$SO,nF)-third-order negative source flow.

- For $a_{j_1j_1 0} < 0$ and $a_{j_2j_1 0}(a_{j_1j_1 1} - a_{j_2j_1 1})^3 < 0$,

$$(a_{j_1j_1 1}, \dot{x}_{j_2}) = \underbrace{(3^{\rm rd}\mathrm{SI}, \mathrm{nF})}_{\text{third-order sink flow } (-)}.$$

The flow at $x_{j_1}^* = a_{j_1j_1 1}$ is a ($3^{\rm rd}$SI,nF)-third-order negative sink flow.

- The third-order source and sink flows of $x_{j_1}^* = a_{j_1j_1 1}$ are the appearing bifurcation for the source, sink, and source flows and the sink, source, and sink flows.
- The third-order source and sink flows of $x_{j_1}^* = a_{j_1j_1 1}$ are the switching bifurcation for the source, sink, and source flows and the sink, source, and sink flows.
- The third-order source and sink flows of $x_{j_1}^* = a_{j_1j_1 1}$ are the switching bifurcation for the upper-saddle and lower-saddle flows and the source and sink flows.

(x_{1b}) At $x_{j_1}^* = a_{j_2j_1 1} \neq a_{j_1j_1 1}$, in phase space,

$$\frac{dx_{j_2}}{dx_{j_1}}\Big|_{x^*_{j_1}=a_{j_2j_11}} = \frac{a_{j_2j_10}\,(x_{j_1}-a_{j_2j_11})^3}{a_{j_1j_10}\,(x_{j_1}-a_{j_1j_11})^3}\Big|_{x^*_{j_1}=a_{j_2j_11}} = 0,$$
$$\frac{d^2x_{j_2}}{dx^2_{j_1}}\Big|_{x^*_{j_1}=a_{j_2j_11}} = \frac{d^3x_{j_2}}{dx^3_{j_1}}\Big|_{x^*_{j_1}=a_{j_2j_11}} = 0.$$

If

$$\frac{d^4x_{j_2}}{dx^4_{j_1}}\Big|_{x^*_{j_1}=a_{j_2j_11}} = \frac{a_{j_2j_10}}{a_{j_1j_10}}\frac{6}{(a_{j_2j_11}-a_{j_1j_11})^3} > 0,$$

there is a third-order up-parabola flow, and if

$$\frac{d^4x_{j_2}}{dx^4_{j_1}}\Big|_{x^*_{j_1}=a_{j_2j_11}} = \frac{a_{j_2j_10}}{a_{j_1j_10}}\frac{6}{(a_{j_2j_11}-a_{j_1j_11})^3} < 0,$$

there is a third-order down-parabola flow.

At $x^*_{j_1}=a_{j_2j_11}$, the variational equation is

$$\dot{x}_{j_1} = a_{j_1j_10}(a_{j_2j_11}-a_{j_1j_11})^3,$$
$$\Delta\dot{x}_{j_2} = a_{j_2j_10}(\Delta x_{j_1})^3.$$

The flows at $x^*_{j_1}=a_{j_2j_11}\neq a_{j_1j_11}$ are the third-order up-parabola and down-parabola flows in the x_{j_2}-direction. The flows at $x^*_{j_1}=a_{j_2j_11}\neq a_{j_1j_11}$ are positive and negative for $a_{j_1j_10}(a_{j_2j_11}-a_{j_1j_11})^3>0$ and $a_{j_1j_10}(a_{j_2j_11}-a_{j_1j_1l_1})^3<0$ in the x_{j_1}-direction.

Therefore, the equilibrium of $x^*_{j_1}=a_{j_2j_11}$ has the following properties as in Eqs. (4.222)–(4.225).

- For $a_{j_1j_10}(a_{j_2j_11}-a_{j_1j_11})^3>0$ and $a_{j_2j_10}>0$,

$$(\dot{x}_{j_1}, a_{j_2j_11}) = \underbrace{(\text{pF}, 3^{\text{rd}}\text{UP})}_{\text{third-order up-parabola flow } (+)}.$$

  The flow at $x^*_{j_1}=a_{j_2j_11}$ is a (pF,3^{rd}UP)-third-order positive up-parabola flow.
- For $a_{j_1j_10}(a_{j_2j_11}-a_{j_1j_11})^3<0$ and $a_{j_2j_10}>0$,

$$(\dot{x}_{j_1}, a_{j_2j_11}) = \underbrace{(\text{nF}, 3^{\text{rd}}\text{DP})}_{\text{third-order down-parabola flow } (-)}.$$

  The flow at $x^*_{j_1}=a_{j_2j_11}$ is an (nF,3^{rd}DP)-third-order negative down-parabola flow.

- For $a_{j_1j_1 0}(a_{j_2j_1 1}-a_{j_1j_1 1})^3>0$ and $a_{j_2j_1 0}<0$,

$$(\dot{x}_{j_1}, a_{j_2j_1 1}) = \underbrace{(\text{pF}, 3^{\text{rd}}\text{DP})}_{\text{third-order down-parabola flow }(+)}.$$

The flow at $x_{j_1}^*=a_{j_2j_1 1}$ is a (pF,3rdUP)-third-order positive down-parabola flow.

- For $a_{j_1j_1 0}(a_{j_2j_1 1}-a_{j_1j_1 1})^3<0$ and $a_{j_2j_1 0}<0$,

$$(\dot{x}_{j_1}, a_{j_2j_1 1}) = \underbrace{(\text{nF}, 3^{\text{rd}}\text{UP})}_{\text{third-order up-parabola flow }(-)}.$$

The flow at $x_{j_1}^*=a_{j_2j_1 1}$ is an (nF,3rdUP)-third-order negative up-parabola flow.

- The third-order up-parabola and down-parabola flows of $x_{j_1}^*=a_{j_2j_1 1}$ are the appearing bifurcation for the up-parabola, down-parabola, and up-parabola flows and the down-parabola, up-parabola, and down-parabola flows.
- The third-order up-parabola and down-parabola flows of $x_{j_1}^*=a_{j_2j_1 1}$ are the switching bifurcation for the up-parabola, down-parabola, and up-parabola flows and the down-parabola, up-parabola, and down-parabola flows.
- The third-order up-parabola and down-parabola flows of $x_{j_1}^*=a_{j_2j_1 1}$ are the switching bifurcation for the increasing-inflection and decreasing-inflection flows and the down-parabola and up-parabola flows.

(x_3) For $a_{j_1j_1 1}=a_{j_2j_1 1}$, in phase space,

$$\frac{dx_{j_2}}{dx_{j_1}}=\frac{a_{j_2j_1 0}}{a_{j_1j_1 0}},$$

With an initial condition $(x_{j_1 0}, x_{j_2 0})$, the first integral manifold is

$$\begin{aligned} x_{j_2}-x_{j_2 0} &= \frac{a_{j_2j_1 0}}{a_{j_1j_1 0}}(x_{j_1}-x_{j_1 0}),\\ & j_1,j_2\in\{1,2\}, j_1\neq j_2. \end{aligned}$$

If

$$\left.\frac{dx_{j_2}}{dx_{j_1}}\right|_{x_{j_1}^*=a_{j_2j_1 1}}=\frac{a_{j_2j_1 0}}{a_{j_1j_1 0}}>0,$$

there is an increasing-inflection flow, and if

$$\left.\frac{dx_{j_2}}{dx_{j_1}}\right|_{x_{j_1}^*=a_{j_2j_1 1}}=\frac{a_{j_2j_1 0}}{a_{j_1j_1 0}}<0,$$

there is a decreasing-inflection flow.

At $x_{j_1}^* = a_{j_2 j_1 1} = a_{j_1 j_1 1}$, the variational equation is

$$\Delta\dot{x}_{j_1} = a_{j_1 j_1 0}(\Delta x_{j_1})^3,$$
$$\Delta\dot{x}_{j_2} = a_{j_2 j_1 0}(\Delta x_{j_1})^3.$$

The flows at $x_{j_1}^* = a_{j_2 j_1 1} = a_{j_1 j_1 1}$ are the third-order source and sink for $a_{j_2 j_1 0} > 0$ and $a_{j_2 j_1 0} < 0$ in the x_{j_1} -direction. Because of the third-order singularity in both direction, (3:3)-increasing-inflection and decreasing-inflection source and sink flows exist.

Therefore, the infinite-equilibrium of $x_{j_1}^* = a_{j_2 j_1 1} = a_{j_1 j_1 1}$ with $\bar{x}_{j_2} \in (-\infty, \infty)$ has the following properties as in Eqs. (4.227)–(4.230).

- For $a_{j_1 j_1 0} > 0$ and $a_{j_2 j_1 0} > 0$,

$$(a_{j_1 j_1 1}, \bar{x}_{j_2}) = \underbrace{(3^{\text{rd}}\text{SO}, (3:3)\text{II})}_{\text{third-order increasing-inflection source}}.$$

  The equilibrium at $x_{j_1}^* = a_{j_2 j_1 1} = a_{j_1 j_1 1}$ is a (3^{rd}SO,(3:3)II)-third-order increasing-inflection source infinite-equilibrium.
- For $a_{j_1 j_1 0} < 0$ and $a_{j_2 j_1 0} > 0$,

$$(a_{j_1 j_1 1}, \bar{x}_{j_2}) = \underbrace{(3^{\text{rd}}\text{SI}, (3:3)\text{DI})}_{\text{third-order decreasing-inflection sink}}.$$

  The equilibrium at $x_{j_1}^* = a_{j_2 j_1 1} = a_{j_1 j_1 1}$ is a (3^{rd}SI,(3:3)DI)-third-order decreasing-inflection sink infinite-equilibrium.
- For $a_{j_1 j_1 0} > 0$ and $a_{j_2 j_1 0} < 0$,

$$(a_{j_1 j_1 1}, \bar{x}_{j_2}) = \underbrace{(3^{\text{rd}}\text{SO}, (3:3)\text{DI})}_{\text{third-order decreasing-inflection source}}.$$

  The equilibrium at $x_{j_1}^* = a_{j_2 j_1 1} = a_{j_1 j_1 1}$ is a (3^{rd}SO,(3:3)DI)-third-order decreasing-inflection source infinite-equilibrium.
- For $a_{j_1 j_1 0} < 0$ and $a_{j_2 j_1 0} < 0$,

$$(a_{j_1 j_1 1}, \bar{x}_{j_2}) = \underbrace{(3^{\text{rd}}\text{SO}, (3:3)\text{II})}_{\text{third-order increasing-inflection source}}.$$

  The equilibrium at $x_{j_1}^* = a_{j_2 j_1 1} = a_{j_1 j_1 1}$ is a (3^{rd}SI,(3:3)II)-third-order increasing-inflection sink infinite-equilibrium.

– The third-order increasing-inflection and decreasing-inflection source and sink infinite-equilibriums of $x_{j_1}^* = a_{j_2 j_1 1} = a_{j_1 j_1 1}$ are the appearing bifurcations for the source, sink, and source flows and the sink, source, and sink flows with the up-parabola, down-parabola, and up-parabola flows and the down-parabola, up-parabola, and down-parabola flows.

- The third-order increasing-inflection and decreasing-inflection source and sink infinite-equilibriums of $x_{j_1}^* = a_{j_2j_11} = a_{j_1j_11}$ are the switching bifurcations for the source, sink, and source flows and the sink, source, and sink flows with the up-parabola, down-parabola, and up-parabola flows and the down-parabola, up-parabola, and down-parabola flows.
- The third-order increasing-inflection and decreasing-inflection source and sink infinite-equilibriums of $x_{j_1}^* = a_{j_2j_11} = a_{j_1j_11}$ are the switching bifurcations for the source and sink flows and the upper-saddle and lower-saddle flows with the up-parabola and up-parabola flows and the increasing-inflection and decreasing-inflection flows.

(xi) If

$$\Delta_{j_1j_1} > 0 \text{ and } \Delta_{j_2j_1} = 0,$$

then

$$\begin{aligned}
\dot{x}_{j_1} &= a_{j_1j_10}(x_{j_1} - a_{j_1j_11})(x_{j_1} - a_{j_1j_12})(x_{j_1} - a_{j_1j_13}),\\
\dot{x}_{j_2} &= a_{j_2j_10}(x_{j_1} - a_{j_2j_1l_1})(x_{j_1} - a_{j_2j_1l_2})^2,
\end{aligned}$$

where

$$\begin{aligned}
&\{a_{j_1j_11}, a_{j_1j_12}, a_{j_1j_13}\} = \mathrm{sort}\{b_{j_1j_11}, b_{j_1j_12}, b_{j_1j_13}\}, a_{j_1j_11} < a_{j_1j_12} < a_{j_1j_13};\\
&b_{j_1j_12}, b_{j_1j_13} = -\frac{1}{2}(B_{j_1j_11} \pm \sqrt{\Delta_{j_1j_1}}),\\
&\{a_{j_1j_11}, a_{j_1j_12}\} = \mathrm{sort}\{b_{j_2j_11}, b_{j_2j_12}\}, b_{j_2j_12} = -\frac{1}{2}B_{j_2j_11};\\
&j_1, j_2 \in \{1, 2\}; j_1 \neq j_2.
\end{aligned}$$

(xi_1) For $x_{j_1}^* = a_{j_1j_1s_1} \neq a_{j_2j_1l_1}, a_{j_2j_1l_2}$ ($s_1 \in \{1, 2, 3\}$), we have

$$\begin{aligned}
\frac{dx_{j_2}}{dx_{j_1}} &= \frac{a_{j_2j_10}}{a_{j_1j_10}} \frac{(x_{j_1} - a_{j_2j_1l_1})(x_{j_1} - a_{j_2j_1l_2})^2}{(x_{j_1} - a_{j_1j_1s_1})(x_{j_1} - a_{j_1j_1s_2})(x_{j_1} - a_{j_1j_1s_3})}\\
&= \frac{a_{j_2j_10}}{a_{j_1j_10}} \Bigg[1 + \frac{(a_{j_1j_1s_2} - a_{j_2j_1l_2})^2}{(x_{j_1} - a_{j_1j_1s_2})(a_{j_1j_1s_2} - a_{j_1j_1s_3})} + \frac{(a_{j_1j_1s_3} - a_{j_2j_1l_2})^2}{(a_{j_1j_1s_3} - a_{j_1j_1s_2})(x_{j_1} - a_{j_1j_1s_3})}\\
&\quad + (a_{j_1j_1s_1} - a_{j_2j_1l_1}) \sum_{\bar{s}_1=1}^{3} \frac{(a_{j_1j_1\bar{s}_1} - a_{j_2j_1l_2})^2}{\prod_{\bar{s}_2=1, \bar{s}_2 \neq \bar{s}_1}^{3} (a_{j_1j_1\bar{s}_1} - a_{j_1j_1\bar{s}_2})} \frac{1}{x_{j_1} - a_{j_1j_1\bar{s}_1}}\Bigg].
\end{aligned}$$

With an initial condition $(x_{j_1 0}, x_{j_2 0})$, the first integral manifold is

$$x_{j_2} - x_{j_2 0} = \frac{a_{j_2 j_1 0}}{a_{j_1 j_1 0}}\Bigg[(x_{j_1} - x_{j_1 0}) + \frac{(a_{j_1 j_1 s_2} - a_{j_2 j_1 l_2})^2}{(a_{j_1 j_1 s_2} - a_{j_1 j_1 s_3})} \ln \frac{|x_{j_1} - a_{j_1 j_1 s_2}|}{|x_{j_1 0} - a_{j_1 j_1 s_2}|}$$
$$+ \frac{(a_{j_1 j_1 s_3} - a_{j_2 j_1 l_2})^2}{(a_{j_1 j_1 s_3} - a_{j_1 j_1 s_2})} \ln \frac{|x_{j_1} - a_{j_1 j_1 s_3}|}{|x_{j_1 0} - a_{j_1 j_1 s_3}|}$$
$$+ (a_{j_1 j_1 s_1} - a_{j_2 j_1 l_1}) \sum_{\bar{s}_1 = 1}^{3} \frac{(a_{j_1 j_1 \bar{s}_1} - a_{j_2 j_1 l_2})^2}{\prod_{\bar{s}_2 = 1, \bar{s}_2 \neq \bar{s}_1}^{3} (a_{j_1 j_1 \bar{s}_1} - a_{j_1 j_1 \bar{s}_2})} \ln \frac{|x_{j_1} - a_{j_1 j_1 \bar{s}_1}|}{|x_{j_1 0} - a_{j_1 j_1 \bar{s}_1}|}\Bigg].$$

(xi$_{1a}$) At $x_{j_1}^* = a_{j_1 j_1 s_1} \neq a_{j_2 j_1 1}$, the variational equation is

$$\Delta \dot{x}_{j_1} = a_{j_1 j_1 0} \prod_{s_2 = 1, s_2 \neq s_1}^{3} (a_{j_1 j_1 s_1} - a_{j_1 j_1 s_2}) \Delta x_{j_1},$$
$$\dot{x}_{j_2} = a_{j_2 j_1 0}(a_{j_1 j_1 s_1} - a_{j_2 j_1 l_1})(a_{j_1 j_1 s_1} - a_{j_2 j_1 l_2})^2.$$

The flows at $x_{j_1}^* = a_{j_1 j_1 s_1} \neq a_{j_2 j_1 l_1}, a_{j_2 j_1 l_2}$ are source and sink flows in the x_{j_1}-direction for $a_{j_1 j_1 0} \prod_{s_2 = 1, s_2 \neq s_1}^{3} (a_{j_1 j_1 s_1} - a_{j_1 j_1 s_2}) > 0$ and $a_{j_1 j_1 0} \prod_{s_2 = 1, s_2 \neq s_1}^{3} (a_{j_1 j_1 s_1} - a_{j_1 j_1 s_2}) < 0$, and the flows at $x_{j_1}^* = a_{j_1 j_1 s_1} \neq a_{j_2 j_1 1}$ are positive and negative in the x_{j_2}-direction for $a_{j_2 j_1 0}(a_{j_1 j_1 s_1} - a_{j_2 j_1 l_1}) > 0$ and $a_{j_2 j_1 0}(a_{j_1 j_1 s_1} - a_{j_2 j_1 l_1}) < 0$.

Therefore, the equilibrium of $x_{j_1}^* = a_{j_1 j_1 s_1} \neq a_{j_2 j_1 1}$ (s_1, $s_2 \in \{1, 2, 3\}$; $s_1 \neq s_2$) in the x_{j_1}-direction possesses the following properties as in Eqs. (4.235)–(4.238).

- For $a_{j_1 j_1 0} \prod_{s_2 = 1, s_2 \neq s_1}^{3} (a_{j_1 j_1 s_1} - a_{j_1 j_1 s_2}) > 0$ and $a_{j_2 j_1 0}(a_{j_1 j_1 s_1} - a_{j_2 j_1 l_1})(a_{j_1 j_1 s_1} - a_{j_2 j_1 l_2})^2 > 0$,

$$(a_{j_1 j_1 s_1}, \dot{x}_{j_2}) = \underbrace{(\text{SO}, \text{pF})}_{\text{source flow } (+)}.$$

The flow at $x_{j_1}^* = a_{j_1 j_1 s_1}$ is an (SO,pF)-positive source-flow.

- For $a_{j_1 j_1 0} \prod_{s_2 = 1, s_2 \neq s_1}^{3} (a_{j_1 j_1 s_1} - a_{j_1 j_1 s_2}) < 0$ and $a_{j_2 j_1 0}(a_{j_1 j_1 s_1} - a_{j_2 j_1 l_1})(a_{j_1 j_1 s_1} - a_{j_2 j_1 l_2})^2 > 0$,

$$(a_{j_1 j_1 s_1}, \dot{x}_{j_2}) = \underbrace{(\text{SI}, \text{pF})}_{\text{sink flow } (+)}.$$

The flow at $x_{j_1}^* = a_{j_1 j_1 s_1}$ is an (SI,pF)-positive sink-flow.

- For $a_{j_1 j_1 0} \prod_{s_2 = 1, s_2 \neq s_1}^{3} (a_{j_1 j_1 s_1} - a_{j_1 j_1 s_2}) > 0$ and $a_{j_2 j_1 0}(a_{j_1 j_1 s_1} - a_{j_2 j_1 l_1})(a_{j_1 j_1 s_1} - a_{j_2 j_1 l_2})^2 < 0$,

$$(a_{j_1j_1s_1}, \dot{x}_{j_2}) = \underbrace{(\text{SO}, \text{nF})}_{\text{source flow }(-)}.$$

The flow at $x_{j_1}^* = a_{j_1j_1s_1}$ is an (SO,nF)-negative source-flow.

- For $a_{j_1j_1 0}\prod_{s_2=1,s_2\neq s_1}^{3}(a_{j_1j_1s_1} - a_{j_1j_1s_2}) < 0$ and $a_{j_2j_1 0}(a_{j_1j_1s_1} - a_{j_2j_1l_1})(a_{j_1j_1s_1} - a_{j_2j_1l_2})^2 < 0$,

$$(a_{j_1j_1s_1}, \dot{x}_{j_2}) = \underbrace{(\text{SI}, \text{nF})}_{\text{sink flow }(-)}.$$

The flow at $x_{j_1}^* = a_{j_1j_1s_1}$ is an (SI,nF)-negative sink-flow.

(xi_{1b}) For $x_{j_1}^* = a_{j_2j_1l_2} \neq a_{j_1j_1s_1}, a_{j_2j_1l_1}$ ($s_1 \in \{1,2,3\}$), we have

$$\frac{dx_{j_2}}{dx_{j_1}}\bigg|_{x_{j_1}^* = a_{j_2j_1l_2}} = \frac{a_{j_2j_1 0}}{a_{j_1j_1 0}}\frac{(x_{j_1} - a_{j_2j_1l_1})(x_{j_1} - a_{j_2j_1l_2})^2}{\prod_{s_1=1}^{3}(x_{j_1} - a_{j_1j_1s_1})}\bigg|_{x_{j_1}^* = a_{j_2j_1l_2}} = 0,$$

$$\frac{d^2x_{j_2}}{dx_{j_1}^2}\bigg|_{x_{j_1}^* = a_{j_2j_1l_2}} = 0.$$

If

$$\frac{d^3x_{j_2}}{dx_{j_1}^3}\bigg|_{x_{j_1}^* = a_{j_2j_1l_2}} = \frac{2a_{j_2j_1 0}(a_{j_2j_1l_2} - a_{j_2j_1l_1})}{a_{j_1j_1 0}\prod_{s_1=1}^{3}(a_{j_2j_1l_2} - a_{j_1j_1s_1})} > 0,$$

the flow at $x_{j_1}^* = a_{j_2j_1l_2}$ is an increasing-inflection flow, and if

$$\frac{d^3x_{j_2}}{dx_{j_1}^3}\bigg|_{x_{j_1}^* = a_{j_2j_1l_2}} = \frac{2a_{j_2j_1 0}(a_{j_2j_1l_2} - a_{j_2j_1l_1})}{a_{j_1j_1 0}\prod_{s_1=1}^{3}(a_{j_2j_1l_2} - a_{j_1j_1s_1})} < 0,$$

the flow at $x_{j_1}^* = a_{j_2j_1l_2}$ is a decreasing-inflection.

At $x_{j_1}^* = a_{j_2j_1l_2}$, the variational equation is

$$\Delta\dot{x}_{j_2} = a_{j_2j_1 0}(a_{j_1j_1l_2} - a_{j_1j_1l_1})(\Delta x_{j_1})^2,$$
$$\dot{x}_{j_1} = a_{j_1j_1 0}\prod_{s_1=1}^{3}(a_{j_1j_1l_2} - a_{j_1j_1s_1}).$$

The flows at $x_{j_1}^* = a_{j_2j_1l_2}$ are positive and negative for $\dot{x}_{j_1} > 0$ and $\dot{x}_{j_1} < 0$ in the x_{j_1}-direction.

Therefore, the equilibrium of $x_{j_1}^* = a_{j_2j_1l_2} \neq a_{j_1j_1s_1}, a_{j_2j_1l_1}$ ($s_1 = 1, 2, 3$) in the x_{j_2}-direction possesses the following properties as in (4.239)–(4.242).

- For $a_{j_1j_10}\prod_{s_1=1}^{3}(a_{j_2j_1l_2}-a_{j_1j_1s_1})>0$ and $a_{j_2j_10}(a_{j_2j_1l_2}-a_{j_2j_1l_1})>0$,

$$(\dot{x}_{j_1},a_{j_2j_1l_2})=\underbrace{(\text{pF, II})}_{\text{increasing-inflection flow }(+)}.$$

The flow at $x_{j_1}^*=a_{j_2j_1l_2}$ is a (pF II)-positive increasing-inflection flow.

- For $a_{j_1j_10}\prod_{s_1=1}^{3}(a_{j_2j_1l_2}-a_{j_1j_1s_1})<0$ and $a_{j_2j_10}(a_{j_2j_1l_2}-a_{j_2j_1l_1})>0$,

$$(\dot{x}_{j_1},a_{j_2j_1l_2})=\underbrace{(\text{nF, DI})}_{\text{decreasing-inflection flow }(-)}.$$

The flow at $x_{j_1}^*=a_{j_2j_1l_2}$ is an (nF, DI)-negative decreasing-inflection flow.

- For $a_{j_1j_10}\prod_{s_1=1}^{3}(a_{j_2j_1l_2}-a_{j_1j_1s_1})>0$ and $a_{j_2j_10}(a_{j_2j_1l_2}-a_{j_2j_1l_1})<0$,

$$(\dot{x}_{j_1},a_{j_2j_1l_2})=\underbrace{(\text{pF, DI})}_{\text{decreasing-inflection flow }(+)}.$$

The flow at $x_{j_1}^*=a_{j_2j_1l_2}$ is a (pF, DI)-positive decreasing-inflection flow.

- For $a_{j_1j_10}\prod_{s_1=1}^{3}(a_{j_2j_1l_2}-a_{j_1j_1s_1})<0$ and $a_{j_2j_10}(a_{j_2j_1l_2}-a_{j_2j_1l_1})<0$,

$$(\dot{x}_{j_1},a_{j_2j_1l_2})=\underbrace{(\text{nF, II})}_{\text{increasing-inflection flow }(-)}.$$

The flow at $x_{j_1}^*=a_{j_2j_1l_2}$ is an (nF,II)-negative increasing-inflection flow.

– The decreasing-inflection and increasing-inflection flows are the appearing bifurcations of the up-parabola and down-parabola flows.

(xi_{1c}) For $x_{j_1}^*=a_{j_2j_1l_1}\neq a_{j_1j_1s_1},a_{j_2j_1l_2}$ ($s_1=1,2,3$), we have

$$\frac{dx_{j_2}}{dx_{j_1}}\bigg|_{x_{j_1}^*=a_{j_2j_1l_1}}=\frac{a_{j_2j_10}}{a_{j_1j_10}}\frac{(x_{j_1}-a_{j_2j_1l_1})(x_{j_1}-a_{j_2j_1l_2})^2}{\prod_{s_1=1}^{3}(x_{j_1}-a_{j_1j_1s_1})}\bigg|_{x_{j_1}^*=a_{j_2j_1l_1}}=0.$$

If

$$\frac{d^2x_{j_2}}{dx_{j_1}^2}\bigg|_{x_{j_1}^*=a_{j_2j_1l_1}}=\frac{a_{j_2j_10}}{a_{j_1j_10}}\frac{(a_{j_2j_1l_1}-a_{j_2j_1l_2})^2}{\prod_{s_1=1}^{3}(a_{j_2j_1l_1}-a_{j_1j_1s_1})}>0,$$

the flow at $x_{j_1}^*=a_{j_2j_1l_1}$ is an up-parabola flow, and if

$$\frac{d^2 x_{j_2}}{dx_{j_1}^2}\bigg|_{x_{j_1}^* = a_{j_2 j_1 l_1}} = \frac{a_{j_2 j_1 0}}{a_{j_1 j_1 0}} \frac{(a_{j_2 j_1 l_1} - a_{j_2 j_1 l_2})^2}{\prod_{s_1=1}^{3}(a_{j_2 j_1 l_1} - a_{j_1 j_1 s_1})} < 0,$$

the flow at $x_{j_1}^* = a_{j_2 j_1 l_1}$ is a down-parabola flow.

At $x_{j_1}^* = a_{j_2 j_1 1} \neq a_{j_1 j_1 s_1}$ ($s_1 = 1, 2, 3$), the variational equation is

$$\Delta \dot{x}_{j_2} = a_{j_2 j_1 0}(a_{j_2 j_1 l_1} - a_{j_2 j_1 l_2})^2 \Delta x_{j_1},$$
$$\dot{x}_{j_1} = a_{j_1 j_1 0} \prod_{s_1=1}^{3}(a_{j_2 j_1 l_1} - a_{j_1 j_1 s_1}).$$

The flows at $x_{j_1}^* = a_{j_2 j_1 l_1}$ are positive and negative for $\dot{x}_{j_1} > 0$ and $\dot{x}_{j_1} < 0$ in the x_{j_1}-direction.

Therefore, the equilibrium of $x_{j_1}^* = a_{j_2 j_1 l_1} \neq a_{j_1 j_1 s_1}, a_{j_2 j_1 l_2}$ ($s_1 = 1, 2, 3$) in the x_{j_2}-direction possesses the following properties as in Eqs. (4.243)–(4.246).

- For $a_{j_1 j_1 0} \prod_{s_1=1}^{3}(a_{j_2 j_1 l_1} - a_{j_1 j_1 s_1}) > 0$ and $a_{j_2 j_1 0}(a_{j_2 j_1 l_1} - a_{j_2 j_1 l_2})^2 > 0$,

$$(\dot{x}_{j_1}, a_{j_2 j_1 l_1}) = \underbrace{(\mathrm{pF}, \mathrm{UP})}_{\text{up-parabola flow }(+)}.$$

  The flow at $x_{j_1}^* = a_{j_2 j_1 l_1}$ is a (pF,UP)-positive up-parabola flow.

- For $a_{j_1 j_1 0} \prod_{s_1=1}^{3}(a_{j_2 j_1 l_1} - a_{j_1 j_1 s_1}) < 0$ and $a_{j_2 j_1 0}(a_{j_2 j_1 l_1} - a_{j_2 j_1 l_2})^2 > 0$,

$$(\dot{x}_{j_1}, a_{j_2 j_1 l_1}) = \underbrace{(\mathrm{nF}, \mathrm{DP})}_{\text{down-parabola flow }(-)}.$$

  The flow at $x_{j_1}^* = a_{j_2 j_1 l_1}$ is an (nF,DP)-positive down-parabola flow.

- For $a_{j_1 j_1 0} \prod_{s_1=1}^{3}(a_{j_2 j_1 l_1} - a_{j_1 j_1 s_1}) > 0$ and $a_{j_2 j_1 0}(a_{j_2 j_1 l_1} - a_{j_2 j_1 l_2})^2 < 0$,

$$(\dot{x}_{j_1}, a_{j_2 j_1 l_1}) = \underbrace{(\mathrm{pF}, \mathrm{DP})}_{\text{down-parabola flow }(+)}.$$

  The flow at $x_{j_1}^* = a_{j_2 j_1 l_1}$ is a (pF,DP)-positive down-parabola flow.

- For $a_{j_1 j_1 0} \prod_{s_1=1}^{3}(a_{j_2 j_1 l_1} - a_{j_1 j_1 s_1}) < 0$ and $a_{j_2 j_1 0}(a_{j_2 j_1 l_1} - a_{j_2 j_1 l_2})^2 > 0$,

$$(\dot{x}_{j_1}, a_{j_2 j_1 l_1}) = \underbrace{(\mathrm{nF}, \mathrm{UP})}_{\text{up-parabola flow }(-)}.$$

  The flow at $x_{j_1}^* = a_{j_2 j_1 l_1}$ is an (nF,DP)-positive up-parabola flow.

(xi$_2$) For $a_{j_2j_1l_2}=a_{j_1j_1s_1}$ $(s_1\in\{1,2,3\})$,

$$\begin{aligned}
&\dot{x}_{j_1}=a_{j_1j_10}(x_{j_1}-a_{j_1j_11})(x_{j_1}-a_{j_1j_12})(x_{j_1}-a_{j_1j_13}),\\
&\dot{x}_{j_2}=a_{j_2j_10}(x_{j_1}-a_{j_2j_1l_1})(x_{j_1}-a_{j_2j_1l_2})^2,\\
&j_1,j_2\in\{1,2\},j_1\neq j_2.
\end{aligned}$$

In phase space,

$$\begin{aligned}
\frac{dx_{j_2}}{dx_{j_1}}&=\frac{a_{j_2j_10}\,(x_{j_1}-a_{j_2j_1l_1})(x_{j_1}-a_{j_2j_1l_2})}{a_{j_1j_10}\,(x_{j_1}-a_{j_1j_1s_2})(x_{j_1}-a_{j_1j_1s_3})}\\
&=\frac{a_{j_2j_10}}{a_{j_1j_10}}\Bigg[1+\frac{(a_{j_1j_1s_3}-a_{j_2j_1l_2})}{(x_{j_1}-a_{j_1j_1s_3})}\\
&\quad+(a_{j_1j_1s_2}-a_{j_2j_1l_1})\sum_{s_2=1,\,s_2\neq s_1}^{3}\frac{(a_{j_1j_1s_2}-a_{j_2j_1l_2})}{\prod_{s_3=1,s_3\neq s_1,s_2}^{3}(a_{j_1j_1s_2}-a_{j_1j_1s_3})}\frac{1}{x_{j_1}-a_{j_1j_1s_2}}\Bigg].
\end{aligned}$$

With an initial condition (x_{j_10},x_{j_20}), the first integral manifold is

$$\begin{aligned}
x_{j_2}-x_{j_20}&=\frac{a_{j_2j_10}}{a_{j_1j_10}}\Bigg[(x_{j_1}-x_{j_10})+(a_{j_1j_1s_3}-a_{j_2j_1l_2})\ln\frac{|x_{j_1}-a_{j_1j_1s_3}|}{|x_{j_10}-a_{j_1j_1s_3}|}\\
&+(a_{j_1j_1s_2}-a_{j_2j_1l_1})\sum_{s_2=1,\,s_2\neq s_1}^{3}\frac{(a_{j_1j_1s_2}-a_{j_2j_1l_2})}{\prod_{s_3=1,s_3\neq s_1,s_2}^{3}(a_{j_1j_1\bar{s}_2}-a_{j_1j_1\bar{s}_3})}\ln\frac{|x_{j_1}-a_{j_1j_1s_2}|}{|x_{j_10}-a_{j_1j_1s_2}|}\Bigg].
\end{aligned}$$

(xi$_{2a}$) For $x_{j_1}^*=a_{j_1j_1s_1}=a_{j_2j_1l_2}$ $(s_1\in\{1,2,3\})$, we have

$$\frac{dx_{j_2}}{dx_{j_1}}\bigg|_{x_{j_1}^*=a_{j_2j_1l_2}}=\frac{a_{j_2j_10}}{a_{j_1j_10}}\frac{(x_{j_1}-a_{j_2j_1l_1})(x_{j_1}-a_{j_2j_1l_2})}{\prod_{s_2=1,s_2\neq s_1}^{3}(x_{j_1}-a_{j_1j_1s_2})}\bigg|_{x_{j_1}^*=a_{j_2j_1l_2}}=0.$$

If

$$\frac{d^2x_{j_2}}{dx_{j_1}^2}\bigg|_{x_{j_1}^*=a_{j_2j_1l_2}}=\frac{a_{j_2j_10}}{a_{j_1j_10}}\frac{(a_{j_2j_1l_2}-a_{j_2j_1l_1})}{\prod_{s_2=1,s_2\neq s_1}^{3}(a_{j_2j_1l_2}-a_{j_1j_1s_2})}>0,$$

the flow at $x_{j_1}^*=a_{j_1j_1s_1}=a_{j_2j_1l_2}$ $(s_1\in\{1,2,3\})$ is an up-parabola flow, and if

$$\frac{d^2x_{j_2}}{dx_{j_1}^2}\bigg|_{x_{j_1}^*=a_{j_2j_1l_2}}=\frac{a_{j_2j_10}}{a_{j_1j_10}}\frac{(a_{j_2j_1l_2}-a_{j_2j_1l_1})}{\prod_{s_2=1,s_2\neq s_1}^{3}(a_{j_2j_1l_2}-a_{j_1j_1s_2})}<0,$$

the flow at $x_{j_1}^*=a_{j_1j_1s_1}=a_{j_2j_1l_2}$ $(s_1\in\{1,2,3\})$ is a down-parabola flow.

At $x_{j_1}^* = a_{j_1j_1s_1} = a_{j_2j_1l_2}$ ($s_1 \in \{1,2,3\}$), the variational equations are

$$\Delta\dot{x}_{j_2} = a_{j_2j_10}(a_{j_2j_1l_2} - a_{j_2j_1l_1})(\Delta x_{j_1})^2,$$
$$\Delta\dot{x}_{j_1} = a_{j_1j_10}\prod\nolimits_{s_2=1,s_2\neq s_1}^{3}(a_{j_1j_1s_1} - a_{j_1j_1s_2})\Delta x_{j_1}.$$

The flows at $x_{j_1}^* = a_{j_2j_1l_2} = a_{j_1j_1s_1}$ in the x_{j_1}-direction are source and sink flows for $a_{j_1j_10}\prod_{s_2=1,s_2\neq s_1}^{3}(a_{j_1j_1s_1} - a_{j_1j_1s_2}) > 0$ and $a_{j_1j_10}\prod_{s_2=1,s_2\neq s_1}^{3}(a_{j_1j_1s_1} - a_{j_1j_1s_2}) < 0$.

Therefore, the infinite equilibrium of $x_{j_1}^* = a_{j_1j_1s_1} = a_{j_2j_1l_2}$ ($s_1 \in \{1,2,3\}$) with $\overline{x}_{j_2} \in (-\infty, \infty)$ possesses the following cases as in Eqs. (4.249)–(4.252).

- For $a_{j_1j_10}\prod_{s_2=1,s_2\neq s_1}^{3}(a_{j_1j_1s_1} - a_{j_1j_1s_2}) > 0$ and $a_{j_2j_10}(a_{j_2j_1l_2} - a_{j_2j_1l_1}) > 0$,

$$(a_{j_1j_1s_1}, \overline{x}_{j_2}) = \underbrace{(\text{SO, UP})}_{\text{up-parabola source}}.$$

  The equilibrium at $x_{j_1}^* = a_{j_1j_1s_1} = a_{j_2j_1l_2}$ is an (SO,UP)-up-parabola source infinite-equilibrium.
- For $a_{j_1j_10}\prod_{s_2=1,s_2\neq s_1}^{3}(a_{j_1j_1s_1} - a_{j_1j_1s_2}) < 0$ and $a_{j_2j_10}(a_{j_2j_1l_2} - a_{j_2j_1l_1}) > 0$,

$$(a_{j_1j_1s_1}, \overline{x}_{j_2}) = \underbrace{(\text{SI, DP})}_{\text{down-parabola sink}}.$$

  The equilibrium at $x_{j_1}^* = a_{j_1j_1s_1} = a_{j_2j_1l_2}$ is an (SI,DP)-down-parabola sink infinite-equilibrium.
- For $a_{j_1j_10}\prod_{s_2=1,s_2\neq s_1}^{3}(a_{j_1j_1s_1} - a_{j_1j_1s_2}) > 0$ and $a_{j_2j_10}(a_{j_2j_1l_2} - a_{j_2j_1l_1}) < 0$,

$$(a_{j_1j_1s_1}, \overline{x}_{j_2}) = \underbrace{(\text{SO, DP})}_{\text{down-parabola source}}.$$

  The equilibrium at $x_{j_1}^* = a_{j_1j_1s_1} = a_{j_2j_1l_2}$ is an (SO,DP)-down-parabola source infinite-equilibrium.
- For $a_{j_1j_10}\prod_{s_2=1,s_2\neq s_1}^{3}(a_{j_1j_1s_1} - a_{j_1j_1s_2}) < 0$ and $a_{j_2j_10}(a_{j_2j_1l_2} - a_{j_2j_1l_1}) < 0$,

$$(a_{j_1j_1s_1}, \overline{x}_{j_2}) = \underbrace{(\text{SI, UP})}_{\text{up-parabola sink}}.$$

  The equilibrium at $x_{j_1}^* = a_{j_1j_1s_1} = a_{j_2j_1l_2}$ is an (SI,UP)-up-parabola sink infinite-equilibrium.

– The up-parabola and down-parabola sink, and source infinite-equilibriums are the appearing and switching bifurcations for the up-parabola and down-parabola flows with sink or source flows.

(xi$_3$) For $a_{j_2j_1l_1} = a_{j_1j_1s_1}$ $(s_1 \in \{1,2,3\})$.

$$\begin{aligned}
&\dot{x}_{j_1} = a_{j_1j_10}(x_{j_1} - a_{j_1j_11})(x_{j_1} - a_{j_1j_12})(x_{j_1} - a_{j_1j_13}),\\
&\dot{x}_{j_2} = a_{j_2j_10}(x_{j_1} - a_{j_2j_1l_1})(x_{j_1} - a_{j_2j_1l_2})^2,\\
&j_1, j_2 \in \{1,2\}, j_1 \neq j_2.
\end{aligned}$$

In phase space,

$$\begin{aligned}
\frac{dx_{j_2}}{dx_{j_1}} &= \frac{a_{j_2j_10}}{a_{j_1j_10}} \frac{(x_{j_1} - a_{j_2j_1l_2})^2}{(x_{j_1} - a_{j_1j_1s_2})(x_{j_1} - a_{j_1j_1s_3})}\\
&= \frac{a_{j_2j_10}}{a_{j_1j_10}} \left[1 + \frac{(a_{j_1j_1s_2} - a_{j_2j_1l_2})^2}{(a_{j_1j_1s_2} - a_{j_1j_1s_3})} \frac{1}{x_{j_1} - a_{j_1j_1s_2}} + \frac{(a_{j_1j_1s_3} - a_{j_2j_1l_2})^2}{(a_{j_1j_1s_3} - a_{j_1j_1s_2})} \frac{1}{x_{j_1} - a_{j_1j_1s_3}} \right].
\end{aligned}$$

With an initial condition (x_{j_10}, x_{j_20}), the first integral manifold is

$$\begin{aligned}
x_{j_2} - x_{j_20} = &\frac{a_{j_2j_10}}{a_{j_1j_10}} \left[(x_{j_1} - x_{j_10}) + \frac{(a_{j_1j_1s_2} - a_{j_2j_1l_2})^2}{(a_{j_1j_1s_2} - a_{j_1j_1s_3})} \ln \frac{|x_{j_1} - a_{j_1j_1s_2}|}{|x_{j_10} - a_{j_1j_1s_2}|} \right.\\
&\left. + \frac{(a_{j_1j_1s_3} - a_{j_2j_1l_2})^2}{(a_{j_1j_1s_3} - a_{j_1j_1s_2})} \ln \frac{|x_{j_1} - a_{j_1j_1s_3}|}{|x_{j_10} - a_{j_1j_1s_3}|} \right].
\end{aligned}$$

For $x_{j_1}^* = a_{j_1j_1s_1} = a_{j_2j_1l_1}$ $(s_1 \in \{1,2,3\})$, if

$$\left.\frac{dx_{j_2}}{dx_{j_1}}\right|_{x_{j_1}^* = a_{j_2j_1l_1}} = \frac{a_{j_2j_10}}{a_{j_1j_10}} \frac{(a_{j_2j_1l_1} - a_{j_2j_1l_2})^2}{\prod_{s_2=1, s_2 \neq s_1}^{3} (a_{j_2j_1l_1} - a_{j_1j_1s_2})} > 0,$$

the flow at $x_{j_1}^* = a_{j_1j_1s_1} = a_{j_2j_1l_1}$ $(s_1 \in \{1,2,3\})$ is an increasing-inflection flow, and if

$$\left.\frac{dx_{j_2}}{dx_{j_1}}\right|_{x_{j_1}^* = a_{j_2j_1l_1}} = \frac{a_{j_2j_10}}{a_{j_1j_10}} \frac{(a_{j_2j_1l_1} - a_{j_2j_1l_2})^2}{\prod_{s_2=1, s_2 \neq s_1}^{3} (a_{j_2j_1l_1} - a_{j_1j_1s_2})} < 0,$$

the flow at $x_{j_1}^* = a_{j_1j_1s_1} = a_{j_2j_1l_1}$ $(s_1 \in \{1,2,3\})$ is a decreasing-inflection flow.

At $x_{j_1}^* = a_{j_1j_1s_1} = a_{j_2j_1l_1}$ $(s_1 \in \{1,2,3\})$, the variational equations are

$$\begin{aligned}
\Delta\dot{x}_{j_2} &= a_{j_2j_10}(a_{j_2j_1l_1} - a_{j_2j_1l_2})^2 \Delta x_{j_1},\\
\Delta\dot{x}_{j_1} &= a_{j_1j_10} \prod\nolimits_{s_2=1, s_2 \neq s_1}^{3} (a_{j_1j_1s_1} - a_{j_1j_1s_2}) \Delta x_{j_1}.
\end{aligned}$$

The flows at $x_{j_1}^* = a_{j_1j_1s_1} = a_{j_2j_1l_1}$ in the x_{j_1}-direction are source and sink flows for $a_{j_1j_10}\prod_{s_2=1, s_2 \neq s_1}^{3} (a_{j_1j_1s_1} - a_{j_1j_1s_2}) > 0$ and $a_{j_1j_10}\prod_{s_2=1, s_2 \neq s_1}^{3} (a_{j_1j_1s_1} - a_{j_1j_1s_2}) < 0$.

Therefore, the infinite-equilibrium of $x_{j_1}^{*}=a_{j_1j_1s_1}=a_{j_2j_1l_1}$ $(s_1\in\{1,2,3\})$ with $\bar{x}_{j_2}\in(-\infty,\infty)$ possesses the following cases as in Eqs. (4.255)–(4.258).

- For $a_{j_1j_1 0}\prod_{s_2=1,s_2\neq s_1}^{3}(a_{j_1j_1s_1}-a_{j_1j_1s_2})>0$ and $a_{j_2j_1 0}(a_{j_2j_1l_1}-a_{j_2j_1l_2})^2>0$,

$$(a_{j_1j_1s_1},\bar{x}_{j_2})=\underbrace{(\text{SO, II})}_{\text{increasing-inflection source}}.$$

  The equilibrium at $x_{j_1}^{*}=a_{j_1j_1s_1}=a_{j_2j_1l_1}$ is an (SO,II)-increasing-inflection source infinite-equilibrium.
- For $a_{j_1j_1 0}\prod_{s_2=1,s_2\neq s_1}^{3}(a_{j_1j_1s_1}-a_{j_1j_1s_2})<0$ and $a_{j_2j_1 0}(a_{j_2j_1l_1}-a_{j_2j_1l_2})^2>0$,

$$(a_{j_1j_1s_1},\bar{x}_{j_2})=\underbrace{(\text{SI, DI})}_{\text{decreasing-inflection sink}}.$$

  The equilibrium at $x_{j_1}^{*}=a_{j_1j_1s_1}=a_{j_2j_1l_1}$ is an (SI,II)-increasing-inflection sink infinite-equilibrium.
- For $a_{j_1j_1 0}\prod_{s_2=1,s_2\neq s_1}^{3}(a_{j_1j_1s_1}-a_{j_1j_1s_2})>0$ and $a_{j_2j_1 0}(a_{j_2j_1l_1}-a_{j_2j_1l_2})^2<0$,

$$(a_{j_1j_1s_1},\bar{x}_{j_2})=\underbrace{(\text{SO, DI})}_{\text{decreasing-inflection source}}.$$

  The equilibrium at $x_{j_1}^{*}=a_{j_1j_1s_1}=a_{j_2j_1l_1}$ is an (SO,II)-decreasing-inflection source infinite-equilibrium.
- For $a_{j_1j_1 0}\prod_{s_2=1,s_2\neq s_1}^{3}(a_{j_1j_1s_1}-a_{j_1j_1s_2})<0$ and $a_{j_2j_1 0}(a_{j_2j_1l_1}-a_{j_2j_1l_2})^2<0$,

$$(a_{j_1j_1s_1},\bar{x}_{j_2})=\underbrace{(\text{SI, II})}_{\text{increasing-inflection source}}.$$

  The equilibrium at $x_{j_1}^{*}=a_{j_1j_1s_1}=a_{j_2j_1l_1}$ is an (SI,II)-increasing-inflection sink infinite-equilibrium.

– The increasing-inflection and decreasing-inflection sink and source infinite-equilibriums are the appearing and switching bifurcations for the up-parabola and down-parabola flows with sink or source flows.

(xii) If

$$\Delta_{j_1j_1}>0 \text{ and } \Delta_{j_2j_1}=0,$$
$$a_{j_2j_1l_1}=a_{j_2j_1l_2}=a_{j_2j_1 1}$$

then

$$\begin{aligned}
&\dot{x}_{j_1} = a_{j_1 j_1 0}(x_{j_1} - a_{j_1 j_1 1})(x_{j_1} - a_{j_1 j_1 2})(x_{j_1} - a_{j_1 j_1 3}),\\
&\dot{x}_{j_2} = a_{j_2 j_1 0}(x_{j_1} - a_{j_2 j_1 1})^3,\\
&j_1, j_2 \in \{1, 2\}, j_1 \neq j_2;
\end{aligned}$$

(xii$_1$) For $a_{j_2 j_1 1} \neq a_{j_1 j_1 s_1}$ ($s_1 = 1, 2, 3$), we have

$$\begin{aligned}
\frac{dx_{j_2}}{dx_{j_1}} &= \frac{a_{j_2 j_1 0}}{a_{j_1 j_1 0}} \frac{(x_{j_1} - a_{j_2 j_1 1})^3}{(x_{j_1} - a_{j_1 j_1 s_1})(x_{j_1} - a_{j_1 j_1 s_2})(x_{j_1} - a_{j_1 j_1 s_3})}\\
&= \frac{a_{j_2 j_1 0}}{a_{j_1 j_1 0}} \Big[1 + \frac{(a_{j_1 j_1 s_2} - a_{j_2 j_1 1})^2}{(x_{j_1} - a_{j_1 j_1 s_2})(a_{j_1 j_1 s_2} - a_{j_1 j_1 s_3})} + \frac{(a_{j_1 j_1 s_3} - a_{j_2 j_1 1})^2}{(a_{j_1 j_1 s_3} - a_{j_1 j_1 s_2})(x_{j_1} - a_{j_1 j_1 s_3})}\\
&\quad + (a_{j_1 j_1 s_1} - a_{j_2 j_1 1}) \sum_{\bar{s}_1 = 1}^{3} \frac{(a_{j_1 j_1 \bar{s}_1} - a_{j_2 j_1 1})^2}{\prod_{\bar{s}_2 = 1, \bar{s}_2 \neq s_1}^{3} (a_{j_1 j_1 \bar{s}_1} - a_{j_1 j_1 \bar{s}_2})} \frac{1}{x_{j_1} - a_{j_1 j_1 \bar{s}_1}} \Big].
\end{aligned}$$

With an initial condition $(x_{j_1 0}, x_{j_2 0})$, the first integral manifold is

$$\begin{aligned}
x_{j_2} - x_{j_2 0} = \frac{a_{j_2 j_1 0}}{a_{j_1 j_1 0}} \Big[&(x_{j_1} - x_{j_1 0})\\
&+ \sum_{\bar{s}_2 = 1, \bar{s}_2 \neq s_1}^{3} \frac{(a_{j_1 j_1 \bar{s}_2} - a_{j_2 j_1 1})^2}{\prod_{\bar{s}_3 = 1, \bar{s}_3 \neq s_1, \bar{s}_2}^{3} (a_{j_1 j_1 \bar{s}_2} - a_{j_1 j_1 \bar{s}_3})} \ln \frac{| x_{j_1} - a_{j_1 j_1 \bar{s}_2} |}{| x_{j_1 0} - a_{j_1 j_1 \bar{s}_2} |}\\
&+ (a_{j_1 j_1 s_1} - a_{j_2 j_1 1}) \sum_{\bar{s}_1 = 1}^{3} \frac{(a_{j_1 j_1 \bar{s}_1} - a_{j_2 j_1 1})^2}{\prod_{\bar{s}_2 = 1, \bar{s}_2 \neq s_1}^{3} (a_{j_1 j_1 \bar{s}_1} - a_{j_1 j_1 \bar{s}_2})} \ln \frac{| x_{j_1} - a_{j_1 j_1 \bar{s}_1} |}{| x_{j_1 0} - a_{j_1 j_1 \bar{s}_1} |} \Big].
\end{aligned}$$

(xii$_{1a}$) At $x^*_{j_1} = a_{j_2 j_1 1} \neq a_{j_1 j_1 s_1}$, in phase space,

$$\begin{aligned}
&\frac{dx_{j_2}}{dx_{j_1}} \Big|_{x^*_{j_1} = a_{j_2 j_1 1}} = \frac{a_{j_2 j_1 0}}{a_{j_1 j_1 0}} \frac{(x_{j_1} - a_{j_2 j_1 1})^3}{\prod_{s_1 = 1}^{3} (x_{j_1} - a_{j_1 j_1 s_2})} \Big|_{x^*_{j_1} = a_{j_2 j_1 1}} = 0\\
&\frac{d^2 x_{j_2}}{dx^2_{j_1}} \Big|_{x^*_{j_1} = a_{j_2 j_1 1}} = \frac{d^3 x_{j_2}}{dx^3_{j_1}} \Big|_{x^*_{j_1} = a_{j_2 j_1 1}} = 0.
\end{aligned}$$

If

$$\frac{d^4 x_{j_2}}{dx^4_{j_1}} \Big|_{x^*_{j_1} = a_{j_2 j_1 1}} = \frac{a_{j_2 j_1 0}}{a_{j_1 j_1 0}} \frac{6}{\prod_{s_1 = 1}^{3} (a_{j_2 j_1 1} - a_{j_1 j_1 s_2})} \Big|_{x^*_{j_1} = a_{j_2 j_1 1}} > 0,$$

there is a third-order up-parabola flow, and if

$$\frac{d^4 x_{j_2}}{dx_{j_1}^4}\Big|_{x_{j_1}^*=a_{j_2j_1 1}} = \frac{a_{j_2j_1 0}}{a_{j_1j_1 0}} \frac{6}{\prod_{s_1=1}^{3}(a_{j_2j_1 1}-a_{j_1j_1 s_2})}\Big|_{x_{j_1}^*=a_{j_2j_1 1}} < 0,$$

there is a third-order down-parabola flow.

At $x_{j_1}^* = a_{j_2j_1 1}$, the variational equation is

$$\dot{x}_{j_1} = a_{j_1j_1 0}\prod_{s_1=1}^{3}(a_{j_2j_1 1}-a_{j_1j_1 s_2}),$$
$$\Delta\dot{x}_{j_2} = a_{j_2j_1 0}(\Delta x_{j_1})^3.$$

The flows at $x_{j_1}^* = a_{j_2j_1 1} \neq a_{j_1j_1 1}$ are the third-order up-parabola and down-parabola flows in the x_{j_2}-direction. The flows at $x_{j_1}^* = a_{j_2j_1 1} \neq a_{j_1j_1 1}$ are positive and negative for $a_{j_2j_1 0}\prod_{s_1=1}^{3}(a_{j_2j_1 1}-a_{j_1j_1 s_2})>0$ and $a_{j_2j_1 0}\prod_{s_1=1}^{3}(a_{j_2j_1 1}-a_{j_1j_1 s_2})<0$ in the x_{j_1}-direction.

Therefore, the infinite-equilibrium of $x_{j_1}^* = a_{j_2j_1 1} \neq a_{j_1j_1 s_1}$ ($s_1 \in \{1,2,3\}$) possesses the following properties as in Eqs. (4.262)–(4.265).

- For $a_{j_1j_1 0}\prod_{s_1=1}^{3}(a_{j_2j_1 1}-a_{j_1j_1 s_1})>0$ and $a_{j_2j_1 0}>0$,

$$(\dot{x}_{j_1}, a_{j_2j_1 1}) = \underbrace{(\text{pF}, 3^{\text{rd}}\text{UP})}_{\text{third-order up-parabola flow }(+)}.$$

  The equilibrium at $x_{j_1}^* = a_{j_2j_1 1}$ is a (pF,3rdUP)-third-order positive up-parabola flow.
- For $a_{j_1j_1 0}\prod_{s_1=1}^{3}(a_{j_2j_1 1}-a_{j_1j_1 s_1})<0$ and $a_{j_2j_1 0}>0$,

$$(\dot{x}_{j_1}, a_{j_2j_1 1}) = \underbrace{(\text{nF}, 3^{\text{rd}}\text{DP})}_{\text{third-order down-parabola flow }(-)}.$$

  The equilibrium at $x_{j_1}^* = a_{j_2j_1 1}$ is an (nF,3rdUP)-third-order negative down-parabola flow.
- For $a_{j_1j_1 0}\prod_{s_1=1}^{3}(a_{j_2j_1 1}-a_{j_1j_1 s_1})<0$ and $a_{j_2j_1 0}<0$,

$$(\dot{x}_{j_1}, a_{j_2j_1 1}) = \underbrace{(\text{pF}, 3^{\text{rd}}\text{DP})}_{\text{third-order down-parabola flow }(+)}.$$

  The equilibrium at $x_{j_1}^* = a_{j_2j_1 1}$ is a (pF,3rdDP)-third-order positive down-parabola flow.
- For $a_{j_1j_1 0}\prod_{s_1=1}^{3}(a_{j_2j_1 1}-a_{j_1j_1 s_1})<0$ and $a_{j_2j_1 0}<0$,

$$(\dot{x}_{j_1}, a_{j_2j_1 1}) = \underbrace{(\text{nF}, 3^{\text{rd}}\text{UP})}_{\text{third-order up-parabola flow }(-)}.$$

The equilibrium at $x_{j_1}^* = a_{j_2 j_1 1}$ is an (nF,3rdUP)-third-order negative up-parabola flow.

- The third-order up-parabola and down-parabola flows are the appearing bifurcations for the up-parabola, down-parabola, and up-parabola flows and the down-parabola, up-parabola, and down-parabola flows.

(xii$_{1b}$) For $x_{j_1}^* = a_{j_1 j_1 x_1} \neq a_{j_2 j_1 1}$, the variational equation is

$$\Delta \dot{x}_{j_1} = a_{j_1 j_1 0} \prod\nolimits_{x_2=1, x_2=x_1}^{3} (a_{j_1 j_1 x_1} - a_{j_1 j_1 x_2}) \Delta x_{j_1},$$
$$\dot{x}_{j_2} = a_{j_2 j_1 0} (a_{j_1 j_1 x_1} - a_{j_2 j_1 1})^3.$$

The flows at $x_{j_1}^* = a_{j_1 j_1 x_1}$ are source and sink for $a_{j_1 j_1 0} \prod_{x_2=1, x_2=x_1}^{3} (a_{j_1 j_1 x_1} - a_{j_1 j_1 x_2}) > 0$ and $a_{j_1 j_1 0} \prod_{x_2=1, x_2=x_1}^{3} (a_{j_1 j_1 x_1} - a_{j_1 j_1 x_2}) < 0$ in the x_{j_1}-direction, respectively. Such source and sink flows are positive and negative in the x_{j_2}-direction for $\dot{x}_{j_2} > 0$ and $\dot{x}_{j_2} < 0$, respectively.

Therefore, the equilibrium of $x_{j_1}^* = a_{j_1 j_1 x_1} \neq a_{j_2 j_1 1}$ has the following properties as in Eqs. (4.266)–(4.269).

- For $a_{j_1 j_1 0} \prod_{x_2=1, x_2=x_1}^{3} (a_{j_1 j_1 x_1} - a_{j_2 j_1 x_2}) > 0$ and $a_{j_2 j_1 0} (a_{j_1 j_1 x_1} - a_{j_2 j_1 1})^3 > 0$,

$$(a_{j_1 j_1 x_1}, \dot{x}_{j_2}) = \underbrace{(\text{SO}, \text{pF})}_{\text{source flow } (+)}.$$

The flow at $x_{j_1}^* = a_{j_1 j_1 x_1}$ is an (SO,pF)-positive source flow.

- For $a_{j_1 j_1 0} \prod_{x_2=1, x_2=x_1}^{3} (a_{j_1 j_1 x_1} - a_{j_1 j_1 x_2}) < 0$ and $a_{j_2 j_1 0} (a_{j_1 j_1 x_1} - a_{j_2 j_1 1})^3 > 0$,

$$(a_{j_1 j_1 x_1}, \dot{x}_{j_2}) = \underbrace{(\text{SI}, \text{pF})}_{\text{sink flow } (+)}.$$

The flow at $x_{j_1}^* = a_{j_1 j_1 x_1}$ is an (SI,pF)-positive sink flow.

- For $a_{j_1 j_1 0} \prod_{x_2=1, x_2=x_1}^{3} (a_{j_1 j_1 x_1} - a_{j_1 j_1 x_2}) > 0$ and $a_{j_2 j_1 0} (a_{j_1 j_1 x_1} - a_{j_2 j_1 1})^3 < 0$,

$$(a_{j_1 j_1 x_1}, \dot{x}_{j_2}) = \underbrace{(\text{SI}, \text{nF})}_{\text{sink flow } (-)}.$$

The flow at $x_{j_1}^* = a_{j_1 j_1 x_1}$ is an (SI,nF)-negative sink flow.

- For $a_{j_1 j_1 0} \prod_{x_2=1, x_2=x_1}^{3} (a_{j_1 j_1 x_1} - a_{j_1 j_1 x_2}) < 0$ and $a_{j_2 j_1 0} (a_{j_1 j_1 x_1} - a_{j_2 j_1 1})^3 < 0$,

$$(a_{j_1 j_1 x_1}, \dot{x}_{j_2}) = \underbrace{(\text{SO}, \text{nF})}_{\text{source flow } (-)}.$$

The flow at $x_{j_1}^* = a_{j_1 j_1 x_1}$ is an (SO,nF)-negative source flow.

(xii$_2$) For $a_{j_2j_11} = a_{j_1j_1s_1}$ $(s_1 \in \{1,2,3\})$, we have

$$\frac{dx_{j_2}}{dx_{j_1}} = \frac{a_{j_2j_10}}{a_{j_1j_10}} \frac{(x_{j_1} - a_{j_2j_11})^2}{(x_{j_1} - a_{j_1j_1s_2})(x_{j_1} - a_{j_1j_1s_3})}$$
$$= \frac{a_{j_2j_10}}{a_{j_1j_10}} \left[1 + \sum_{s_1=1}^{3} \frac{(a_{j_1j_1s_1} - a_{j_2j_11})^2}{\prod_{s_2=1, s_2 \neq s_1}^{3} (a_{j_1j_1s_1} - a_{j_1j_1s_2})} \frac{1}{x_{j_1} - a_{j_1j_1s_1}}\right].$$

With an initial condition (x_{j_10}, x_{j_20}),the first integral manifold is

$$x_{j_2} - x_{j_20} = \frac{a_{j_2j_10}}{a_{j_1j_10}} \left[(x_{j_1} - x_{j_10}) + \sum_{s_2=1, s_2 \neq s_1}^{3} \frac{(a_{j_1j_1s_2} - a_{j_2j_11})^2}{\prod_{s_3=1, s_3 \neq s_1, s_2}^{3} (a_{j_1j_1s_2} - a_{j_1j_1s_3})} \ln \frac{| x_{j_1} - a_{j_1j_1s_2} |}{| x_{j_10} - a_{j_1j_1s_2} |}\right].$$

For $x_{j_1}^* = a_{j_2j_11} = a_{j_1j_1s_1}$, we have

$$\frac{dx_{j_2}}{dx_{j_1}}\bigg|_{x_{j_1}^* = a_{j_2j_11}} = \frac{a_{j_2j_10}}{a_{j_1j_10}} \frac{(x_{j_1} - a_{j_2j_11})^2}{(x_{j_1} - a_{j_1j_1s_2})(x_{j_1} - a_{j_1j_1s_3})}\bigg|_{x_{j_1}^* = a_{j_2j_11}} = 0,$$
$$\frac{d^2x_{j_2}}{dx_{j_1}^2}\bigg|_{x_{j_1}^* = a_{j_2j_11}} = 0.$$

If

$$\frac{d^3x_{j_2}}{dx_{j_1}^3}\bigg|_{x_{j_1}^* = a_{j_2j_11}} = \frac{a_{j_2j_10}}{a_{j_1j_10}} \frac{2}{(a_{j_1j_1s_1} - a_{j_1j_1s_2})(a_{j_1j_1s_1} - a_{j_1j_1s_3})} > 0,$$

there is an increasing-inflection flow, and if

$$\frac{d^3x_{j_2}}{dx_{j_1}^3}\bigg|_{x_{j_1}^* = a_{j_2j_11}} = \frac{a_{j_2j_10}}{a_{j_1j_10}} \frac{2}{(a_{j_1j_1s_1} - a_{j_1j_1s_2})(a_{j_1j_1s_1} - a_{j_1j_1s_3})} < 0,$$

there is a decreasing-inflection flow.

At $x_{j_1}^* = a_{j_2j_11} = a_{j_1j_1s_1}$, the variational equation is

$$\Delta\dot{x}_{j_1} = a_{j_1j_10} \prod\nolimits_{s_2=1, s_2 \neq s_1}^{3} (a_{j_1j_1s_1} - a_{j_1j_1s_2}) \Delta x_{j_1},$$
$$\Delta\dot{x}_{j_2} = a_{j_2j_10} (\Delta x_{j_1})^3.$$

The inflection flows at $x_{j_1}^* = a_{j_2j_11} = a_{j_1j_1s_1}$ for $a_{j_1j_10} \prod_{s_2=1, s_2 \neq s_1}^{3} (a_{j_1j_1s_1} - a_{j_1j_1s_2}) > 0$ and $a_{j_1j_10} \prod_{s_2=1, s_2 \neq s_1}^{3} (a_{j_1j_1s_1} - a_{j_1j_1s_2}) < 0$ are source and sink in the x_{j_1} -direction. The flow at $x_{j_1}^* = a_{j_2j_1 1} = a_{j_1j_1s_1}$ are of the first- and third-order in the two directions. Thus, the increasing-inflection and deceasing-inflection flows are named by (1:3).

Therefore, the infinite equilibrium of $x_{j_1}^* = a_{j_2j_1 1} = a_{j_1j_1s_1}$ $(s_1 \in \{1,2,3\})$ with $\overline{x}_{j_2} \in (-\infty, \infty)$ possesses the following cases as in Eqs. (4.271)–(4.274).

- For $a_{j_1j_1 0}\prod_{s_2=1,s_2\neq s_1}^{3}(a_{j_1j_1s_1} - a_{j_1j_1s_2}) > 0$ and $a_{j_2j_1 0} > 0$,

$$(a_{j_1j_1s_1}, \overline{x}_{j_2}) = \underbrace{(\text{SO}, (1{:}3)\text{II})}_{(1:3)\text{-increasing-inflection source}}.$$

  The equilibrium at $x_{j_1}^* = a_{j_2j_1 1} = a_{j_1j_1s_1}$ is an (SO,(1:3)II)-increasing-inflection source infinite-equilibrium.

- For $a_{j_1j_1 0}\prod_{s_2=1,s_2\neq s_1}^{3}(a_{j_1j_1s_1} - a_{j_1j_1s_2}) < 0$ and $a_{j_2j_1 0} > 0$,

$$(a_{j_1j_1s_1}, \overline{x}_{j_2}) = \underbrace{(\text{SI}, (1{:}3)\text{DI})}_{(1:3)\text{-decreasing-inflection sink}}.$$

  The equilibrium at $x_{j_1}^* = a_{j_2j_1 1} = a_{j_1j_1s_1}$ is an (SI,(1:3)DI)-decreasing-inflection sink infinite-equilibrium.

- For $a_{j_1j_1 0}\prod_{s_2=1,s_2\neq s_1}^{3}(a_{j_1j_1s_1} - a_{j_1j_1s_2}) > 0$ and $a_{j_2j_1 0} < 0$,

$$(a_{j_1j_1s_1}, \overline{x}_{j_2}) = \underbrace{(\text{SO}, (1{:}3)\text{DI})}_{(1:3)\text{-decreasing-inflection source}}.$$

  The equilibrium at $x_{j_1}^* = a_{j_2j_1 1} = a_{j_1j_1s_1}$ is an (SO,(1:3)DI)-decreasing-inflection source infinite-equilibrium.

- For $a_{j_1j_1 0}\prod_{s_2=1,s_2\neq s_1}^{3}(a_{j_1j_1s_1} - a_{j_1j_1s_2}) < 0$ and $a_{j_2j_1 0} < 0$,

$$(a_{j_1j_1s_1}, \overline{x}_{j_2}) = \underbrace{(\text{SI}, (1{:}3)\text{II})}_{(1:3)\text{-increasing-inflection sink}}.$$

  The equilibrium at $x_{j_1}^* = a_{j_2j_1 1} = a_{j_1j_1s_1}$ is an (SI,(1:3)II)-increasing-inflection sink infinite-equilibrium.

- The (1:3)-increasing-inflection and decreasing-inflection source and sink infinite-equilibriums are the appearing bifurcations for the up-parabola, down-parabola, and up-parabola flows and the down-parabola, up-parabola, and down-parabola flows with sink and source flows.
- The (1:3)-increasing-inflection and decreasing-inflection source and sink infinite-equilibriums are the switching bifurcations for the up-parabola, down-parabola, and up-parabola flows and the down-parabola, up-parabola, and down-parabola flows with sink and source flows.

– The (1:3)-increasing-inflection and decreasing-inflection source and sink infinite-equilibriums are the switching bifurcations for the third-order up-parabola and down-parabola flows with sink and source flows.

(xiii) If

$$\Delta_{j_1 j_1} > 0 \text{ and } \Delta_{j_2 j_1} > 0,$$

then

$$\begin{aligned}
\dot{x}_{j_1} &= a_{j_1 j_1 0}(x_{j_1} - a_{j_1 j_1 1})(x_{j_1} - a_{j_1 j_1 2})(x_{j_1} - a_{j_1 j_1 3}),\\
\dot{x}_{j_2} &= a_{j_2 j_1 0}(x_{j_1} - a_{j_2 j_1 1})(x_{j_1} - a_{j_2 j_1 2})(x_{j_1} - a_{j_2 j_1 3}),
\end{aligned}$$

where

$$\begin{aligned}
&\{a_{j_1 j_1 1}, a_{j_1 j_1 2}, a_{j_1 j_1 3}\} = \text{sort}\{b_{j_1 j_1 1}, b_{j_1 j_1 2}, b_{j_1 j_1 3}\}, a_{j_1 j_1 1} < a_{j_1 j_1 2} < a_{j_1 j_1 3};\\
&\{a_{j_2 j_1 1}, a_{j_2 j_1 2}, a_{j_2 j_1 3}\} = \text{sort}\{b_{j_2 j_1 1}, b_{j_2 j_1 2}, b_{j_2 j_1 3}\}, a_{j_2 j_1 1} < a_{j_2 j_1 2} < a_{j_2 j_1 3};\\
&b_{j_2 j_1 2}, b_{j_2 j_1 3} = -\frac{1}{2}(B_{j_1 j_1 1} \pm \sqrt{\Delta_{j_1 j_1}}),\\
&b_{j_2 j_1 2}, b_{j_2 j_1 3} = -\frac{1}{2}(B_{j_2 j_1 1} \pm \sqrt{\Delta_{j_2 j_1}});\\
&j_1, j_2 \in \{1, 2\}; j_1 \neq j_2.
\end{aligned}$$

(xiii$_1$) For $a_{j_1 j_1 s_1} \neq a_{j_2 j_1 l_1}$ ($s_1 \in \{1,2,3\}$) and ($l_1 \in \{1,2,3\}$),

$$\frac{dx_{j_2}}{dx_{j_1}} = \frac{a_{j_2 j_1 0}}{a_{j_1 j_1 0}}\left[1 + \sum_{s_1=1,}^{3} \frac{\prod_{l_1=1}^{3}(a_{j_1 j_1 s_1} - a_{j_2 j_1 l_1})}{\prod_{s_2=1, s_2 \neq s_1}^{3}(a_{j_1 j_1 s_1} - a_{j_1 j_1 s_2})} \frac{1}{x_{j_1} - a_{j_1 j_1 s_1}}\right].$$

or

$$\begin{aligned}
\frac{dx_{j_2}}{dx_{j_1}} = \frac{a_{j_2 j_1 0}}{a_{j_1 j_1 0}}&\left[1 + \frac{(a_{j_1 j_1 s_3} - a_{j_2 j_1 l_3})}{x_{j_1} - a_{j_1 j_1 s_3}} + (a_{j_1 j_1 s_2} - a_{j_2 j_1 l_2})\right.\\
&\sum_{\bar{s}_2=1, \bar{s}_2 \neq s_1}^{3} \frac{\prod_{\bar{l}_3=1, \bar{l}_3 \neq l_1, l_2}^{3}(a_{j_1 j_1 \bar{s}_2} - a_{j_2 j_1 \bar{l}_3})}{\prod_{\bar{s}_3=1, \bar{s}_3 \neq s_1, \bar{s}_2}^{3}(a_{j_1 j_1 \bar{s}_2} - a_{j_1 j_1 \bar{s}_3})} \frac{1}{x_{j_1} - a_{j_1 j_1 \bar{s}_2}}\\
&\left. + (a_{j_1 j_1 s_1} - a_{j_2 j_1 l_1}) \sum_{\bar{s}_1=1}^{3} \frac{\prod_{\bar{l}_2=1, \bar{l}_2 \neq l_1}^{3}(a_{j_1 j_1 \bar{s}_1} - a_{j_2 j_1 \bar{l}_2})}{\prod_{\bar{s}_2=1, \bar{s}_2 \neq \bar{s}_1}^{3}(a_{j_1 j_1 \bar{s}_1} - a_{j_1 j_1 \bar{s}_2})} \frac{1}{x_{j_1} - a_{j_1 j_1 \bar{s}_1}}\right].
\end{aligned}$$

With an initial condition $(x_{j_1 0}, x_{j_2 0})$, the first integral manifold is

$$x_{j_2}-x_{j_20}=\frac{a_{j_2j_10}}{a_{j_1j_10}}\left[(x_{j_1}-x_{j_10})+\sum_{s_1=1,}^{3}\frac{\prod_{l_1=1}^{3}(a_{j_1j_1s_1}-a_{j_2j_1l_1})}{\prod_{s_2=1,s_2\neq s_1}^{3}(a_{j_1j_1s_1}-a_{j_1j_1s_2})}\ln\frac{|x_{j_1}-a_{j_1j_1s_1}|}{|x_{j_10}-a_{j_1j_1s_1}|}\right].$$

or

$$\begin{aligned}x_{j_2}-x_{j_20}=\frac{a_{j_2j_10}}{a_{j_1j_10}}&\left[(x_{j_1}-x_{j_10})+(a_{j_1j_1s_3}-a_{j_2j_1l_3})\ln\frac{|x_{j_1}-a_{j_1j_1s_3}|}{|x_{j_10}-a_{j_1j_1s_3}|}\right.\\&+(a_{j_1j_1s_2}-a_{j_2j_1l_2})\sum_{\bar{s}_2=1,\bar{s}_2\neq s_1}^{3}\frac{\prod_{\bar{l}_3=1,\bar{l}_3\neq l_1,l_2}^{3}(a_{j_1j_1\bar{s}_2}-a_{j_2j_1\bar{l}_3})}{\prod_{\bar{s}_3=1,\bar{s}_3\neq s_1,\bar{s}_2}^{3}(a_{j_1j_1\bar{s}_2}-a_{j_1j_1\bar{s}_3})}\ln\frac{|x_{j_1}-a_{j_1j_1\bar{s}_2}|}{|x_{j_10}-a_{j_1j_1\bar{s}_2}|}\\&\left.+(a_{j_1j_1s_1}-a_{j_2j_1l_1})\sum_{\bar{s}_1=1}^{3}\frac{\prod_{\bar{l}_2=1,\bar{l}_2\neq l_1}^{3}(a_{j_1j_1\bar{s}_1}-a_{j_2j_1\bar{l}_2})}{\prod_{\bar{s}_2=1,\bar{s}_2\neq\bar{s}_1}^{3}(a_{j_1j_1\bar{s}_1}-a_{j_1j_1\bar{s}_2})}\ln\frac{|x_{j_1}-a_{j_1j_1\bar{s}_1}|}{|x_{j_10}-a_{j_1j_1\bar{s}_1}|}\right].\end{aligned}$$

(xiii_{1a}) For the equilibriums of $x_{j_1}^*=a_{j_1j_1s_1}\neq a_{j_2j_1l_1},a_{j_2j_1l_2}$ (s_1, $s_2\in\{1,2,3\}$; $s_1\neq s_2$) and (l_1, $l_2\in\{1,2\}$; $l_1\neq l_2$), the variational equation is

$$\begin{aligned}\Delta\dot{x}_{j_1}&=a_{j_1j_10}\prod_{s_2=1,s_2\neq s_1}^{3}(a_{j_1j_1s_1}-a_{j_1j_1s_2})\Delta x_{j_1},\\\dot{x}_{j_2}&=a_{j_2j_10}\prod_{l_1=1}^{3}(a_{j_1j_1s_1}-a_{j_2j_1l_1}).\end{aligned}$$

The flows at $x_{j_1}^*=a_{j_1j_1s_1}$ are source and sink flows in the x_{j_1}-direction for $a_{j_1j_10}\prod_{s_2=1,s_2\neq s_1}^{3}(a_{j_1j_1s_1}-a_{j_1j_1s_2})>0$ and $a_{j_1j_10}\prod_{s_2=1,s_2\neq s_1}^{3}(a_{j_1j_1s_1}-a_{j_1j_1s_2})<0$. The flow at $x_{j_1}^*=a_{j_1j_1s_1}$ are positive and negative in the x_{j_2}-direction for $\dot{x}_{j_2}>0$ and $\dot{x}_{j_2}<0$.

Therefore, the equilibriums of $x_{j_1}^*=a_{j_1j_1s_1}\neq a_{j_2j_1l_1}$ ($s_1=1,2,3$) and ($l_1=1,2,3$) possess the following properties as in Eqs. (4.280)–(4.283).

- For $a_{j_1j_10}\prod_{s_2=1,s_2\neq s_1}^{3}(a_{j_1j_1s_1}-a_{j_1j_1s_2})>0$ and $a_{j_2j_10}\prod_{l_1=1}^{3}(a_{j_1j_1s_1}-a_{j_2j_1l_1})>0$,

$$(a_{j_1j_1s_1},\dot{x}_{j_2})=\underbrace{(\text{SO},\text{pF})}_{\text{source flow }(+)}.$$

The flow at $x_{j_1}^*=a_{j_1j_1s_1}$ is an (SO,pF)-positive source-flow.

- For $a_{j_1j_10}\prod_{s_2=1,s_2\neq s_1}^{3}(a_{j_1j_1s_1}-a_{j_1j_1s_2})<0$ and $a_{j_2j_10}\prod_{l_1=1}^{3}(a_{j_1j_1s_1}-a_{j_2j_1l_1})>0$,

$$(a_{j_1j_1s_1},\dot{x}_{j_2})=\underbrace{(\text{SI},\text{pF})}_{\text{sink flow }(+)}.$$

The flow at $x_{j_1}^*=a_{j_1j_1s_1}$ is an (SI,pF)-positive sink-flow.

- For $a_{j_1j_10}\prod_{s_2=1,s_2\neq s_1}^{3}(a_{j_1j_1s_1}-a_{j_1j_1s_2})>0$ and $a_{j_2j_10}\prod_{l_1=1}^{3}(a_{j_1j_1s_1}-a_{j_2j_1l_1})<0$,

$$(a_{j_1j_1s_1},\dot{x}_{j_2})=\underbrace{(\mathrm{SO},\mathrm{nF})}_{\text{source flow }(-)}.$$

The flow at $x_{j_1}^*=a_{j_1j_1s_1}$ is an (SO,nF)-negative source-flow.

- For $a_{j_1j_10}\prod_{s_2=1,s_2\neq s_1}^{3}(a_{j_1j_1s_1}-a_{j_1j_1s_2})<0$ and $a_{j_2j_10}\prod_{l_1=1}^{3}(a_{j_1j_1s_1}-a_{j_2j_1l_1})<0$,

$$(a_{j_1j_1s_1},\dot{x}_{j_2})=\underbrace{(\mathrm{SI},\mathrm{nF})}_{\text{sink flow }(-)}.$$

The flow at $x_{j_1}^*=a_{j_1j_1s_1}$ is an (SI,nF)-negative sink-flow.

(xiii_{1b}) For the equilibrium of $x_{j_1}^*=a_{j_2j_1l_1}\neq a_{j_1j_1s_1},a_{j_2j_1l_2}$ (l_1, $l_2\in\{1,2,3\}$; $l_1\neq l_2$) and (s_1 = 1, 2, 3), we have

$$\frac{dx_{j_2}}{dx_{j_1}}\Big|_{x_{j_1}^*=a_{j_1j_1l_1}}=\frac{a_{j_2j_10}}{a_{j_1j_10}}\frac{(x_{j_1}-a_{j_2j_1l_1})\prod_{l_2=1,l_2\neq l_1}^{3}(x_{j_1}-a_{j_2j_1l_2})}{\prod_{s_1=1}^{3}(x_{j_1}-a_{j_1j_1s_1})}\Big|_{x_{j_1}^*=a_{j_1j_1l_1}}=0.$$

If

$$\frac{d^2x_{j_2}}{dx_{j_1}^2}\Big|_{x_{j_1}^*=a_{j_1j_1l_1}}=\frac{a_{j_2j_10}}{a_{j_1j_10}}\frac{\prod_{l_2=1,l_2\neq l_1}^{3}(a_{j_2j_1l_1}-a_{j_2j_1l_2})}{\prod_{s_1=1}^{3}(a_{j_2j_1l_1}-a_{j_1j_1s_1})}>0,$$

the flow at $x_{j_1}^*=a_{j_1j_1l_1}$ is an up-parabola flow, and if

$$\frac{d^2x_{j_2}}{dx_{j_1}^2}\Big|_{x_{j_1}^*=a_{j_1j_1l_1}}=\frac{a_{j_2j_10}}{a_{j_1j_10}}\frac{\prod_{l_2=1,l_2\neq l_1}^{3}(a_{j_2j_1l_1}-a_{j_2j_1l_2})}{\prod_{s_1=1}^{3}(a_{j_2j_1l_1}-a_{j_1j_1s_1})}<0,$$

the flow at $x_{j_1}^*=a_{j_1j_1l_1}$ is a down-parabola flow. Let

$$\dot{x}_{j_1}=a_{j_1j_10}\prod_{s_1=1,}^{3}(a_{j_2j_1l_1}-a_{j_1j_1s_1}).$$

The flows at $x_{j_1}^*=a_{j_1j_1l_1}$ are positive and negative for $\dot{x}_{j_1}>0$ and $\dot{x}_{j_1}<0$.

Therefore, the equilibrium of $x_{j_1}^*=a_{j_2j_1l_1}\neq a_{j_1j_1s_1},a_{j_1j_1l_2}$ (l_1, $l_2\in\{1,2\}$; $l_1\neq l_2$) and (s_1 = 1, 2, 3) has the following properties as in Eqs. (4.284)–(4.287).

- For $a_{j_1j_10}\prod_{s_1=1}^{3}(a_{j_2j_1l_1}-a_{j_1j_1s_1})>0$ and $a_{j_2j_10}\prod_{l_2=1,l_2\neq l_1}^{3}(a_{j_2j_1l_1}-a_{j_2j_1l_2})>0$,

$$(\dot{x}_{j_1}, a_{j_1 j_1 l_1}) = \underbrace{(\text{pF}, \text{UP})}_{\text{up-parabola flow } (+)}.$$

The flow at $x^*_{j_1} = a_{j_1 j_1 l_1}$ is a (pF,UP)-positive up-parabola flow.

- For $a_{j_1 j_1 0}\prod_{s_1=1}^{3}(a_{j_2 j_1 l_1} - a_{j_1 j_1 s_1}) < 0$ and $a_{j_2 j_1 0}\prod_{l_2=1, l_2 \neq l_1}^{3}(a_{j_2 j_1 l_1} - a_{j_2 j_1 l_2}) > 0$,

$$(\dot{x}_{j_1}, a_{j_1 j_1 l_1}) = \underbrace{(\text{nF}, \text{DP})}_{\text{down-parabola flow } (-)}.$$

The flow at $x^*_{j_1} = a_{j_1 j_1 l_1}$ is an (nF,DP)-negative down-parabola flow.

- For $a_{j_1 j_1 0}\prod_{s_1=1}^{3}(a_{j_2 j_1 l_1} - a_{j_1 j_1 s_1}) > 0$ and $a_{j_2 j_1 0}\prod_{l_2=1, l_2 \neq l_1}^{3}(a_{j_2 j_1 l_1} - a_{j_2 j_1 l_2}) < 0$,

$$(\dot{x}_{j_1}, a_{j_1 j_1 l_1}) = \underbrace{(\text{pF}, \text{DP})}_{\text{down-parabola flow } (+)}.$$

The flow at $x^*_{j_1} = a_{j_1 j_1 l_1}$ is a (pF,DP)-positive down-parabola flow.

- For $a_{j_1 j_1 0}\prod_{s_1=1}^{3}(a_{j_2 j_1 l_1} - a_{j_1 j_1 s_1}) < 0$ and $a_{j_2 j_1 0}\prod_{l_2=1, l_2 \neq l_1}^{3}(a_{j_2 j_1 l_1} - a_{j_2 j_1 l_2}) < 0$,

$$(\dot{x}_{j_1}, a_{j_1 j_1 l_1}) = \underbrace{(\text{nF}, \text{UP})}_{\text{up-parabola flow } (-)}.$$

The flow at $x^*_{j_1} = a_{j_1 j_1 l_1}$ is an (nF,UP)-negative up-parabola flow.

(xiii$_2$) For $a_{j_1 j_1 s_1} = a_{j_2 j_1 l_1}$ (s_1, s_2, $s_3 \in \{1,2,3\}$; $s_1 \neq s_2 \neq s_3$) and (l_1, $l_2 \in \{1,2,3\}$; $l_1 \neq l_2$),

$$\begin{aligned}
\dot{x}_{j_1} &= a_{j_1 j_1 0}(x_{j_1} - a_{j_1 j_1 1})(x_{j_1} - a_{j_1 j_1 2})(x_{j_1} - a_{j_1 j_1 3}),\\
\dot{x}_{j_2} &= a_{j_2 j_1 0}(x_{j_1} - a_{j_2 j_1 1})(x_{j_1} - a_{j_2 j_1 2})(x_{j_1} - a_{j_2 j_1 3}).
\end{aligned}$$

In phase space, we have,

$$\begin{aligned}
\frac{dx_{j_2}}{dx_{j_1}} = \frac{a_{j_2 j_1 0}}{a_{j_1 j_1 0}}\Bigg[& 1 + (a_{j_1 j_1 s_3} - a_{j_1 j_1 l_3})\frac{1}{(x_{j_1} - a_{j_1 j_1 s_3})}\\
& +(a_{j_1 j_1 s_2} - a_{j_1 j_1 l_2})\sum_{\bar{s}_2=1, \bar{s}_2 \neq s_1}^{3}\frac{\prod_{\bar{l}_2=1, \bar{l}_2 \neq l_1}^{3}(a_{j_1 j_1 \bar{s}_2} - a_{j_2 j_1 \bar{l}_2})}{\prod_{\bar{s}_3=1, \bar{s}_3 \neq s_1, \bar{s}_2}^{3}(a_{j_1 j_1 \bar{s}_2} - a_{j_1 j_1 \bar{s}_3})}\frac{1}{x_{j_1} - a_{j_1 j_1 \bar{s}_2}}\Bigg].
\end{aligned}$$

With an initial condition $(x_{j_1 0}, x_{j_2 0})$, the first integral manifold is

$$x_{j_2}-x_{j_20}=\frac{a_{j_2j_10}}{a_{j_1j_10}}\Bigg[(x_{j_1}-x_{j_10})+(a_{j_1j_1s_3}-a_{j_1j_1l_3})\ln\frac{|x_{j_1}-a_{j_1j_1s_3}|}{|x_{j_10}-a_{j_1j_1s_3}|}$$
$$+(a_{j_1j_1s_2}-a_{j_1j_1l_2})\sum_{\bar{s}_2=1,\bar{s}_2\neq s_1}^{3}\frac{\prod_{\bar{l}_2=1,\bar{l}_2\neq l_1}^{3}(a_{j_1j_1\bar{s}_2}-a_{j_2j_1\bar{l}_2})}{\prod_{\bar{s}_3=1,\bar{s}_3\neq s_1,\bar{s}_2}^{3}(a_{j_1j_1\bar{s}_2}-a_{j_1j_1\bar{s}_3})}\ln\frac{|x_{j_1}-a_{j_1j_1s_2}|}{|x_{j_10}-a_{j_1j_1s_2}|}\Bigg].$$

At $x_{j_1}^{*}=a_{j_1j_1s_1}=a_{j_2j_1l_1}$, if

$$\frac{dx_{j_2}}{dx_{j_1}}\bigg|_{x_{j_1}^{*}=a_{j_1j_1s_1}=a_{j_2j_1l_1}}=\frac{a_{j_2j_10}}{a_{j_1j_10}}\frac{\prod_{l_2=1,l_2\neq l_1}^{3}(a_{j_2j_1l_1}-a_{j_2j_1l_2})}{\prod_{s_2=1,s_2\neq s_1}^{3}(a_{j_1j_1s_1}-a_{j_1j_1s_2})}>0,$$

the flow is an increasing-inflection flow, and if

$$\frac{dx_{j_2}}{dx_{j_1}}\bigg|_{x_{j_1}^{*}=a_{j_1j_1s_1}=a_{j_2j_1l_1}}=\frac{a_{j_2j_10}}{a_{j_1j_10}}\frac{\prod_{l_2=1,l_2\neq l_1}^{3}(a_{j_2j_1l_1}-a_{j_2j_1l_2})}{\prod_{s_2=1,s_2\neq s_1}^{3}(a_{j_1j_1s_1}-a_{j_1j_1s_2})}<0,$$

the flow is a decreasing-inflection flow.

At $x_{j_1}^{*}=a_{j_1j_1s_1}=a_{j_2j_1l_1}$, the variational equations are

$$\Delta\dot{x}_{j_1}=a_{j_1j_10}\prod_{s_2=1,s_2\neq s_1}^{3}(a_{j_1j_1s_1}-a_{j_1j_1s_2})\Delta x_{j_1},$$
$$\Delta\dot{x}_{j_2}=a_{j_2j_10}\prod_{l_2=1,l_2\neq l_1}^{3}(a_{j_1j_1l_1}-a_{j_2j_1l_2})\Delta x_{j_1}.$$

the flows at $x_{j_1}^{*}=a_{j_1j_1s_1}=a_{j_2j_1l_1}$ are source and sink flows in the x_{j_1}-direction for $a_{j_1j_10}\prod_{s_2=1,s_2\neq s_1}^{3}(a_{j_1j_1s_1}-a_{j_1j_1s_2})>0$ and $a_{j_1j_10}\prod_{s_2=1,s_2\neq s_1}^{3}(a_{j_1j_1s_1}-a_{j_1j_1s_2})<0$.

Therefore, the infinite-equilibrium of $x_{j_1}^{*}=a_{j_1j_1s_1}=a_{j_2j_1l_1}$ with $\bar{x}_{j_2}\in(-\infty,\infty)$ has the following properties as in Eqs. (4.290)–(4.293).

- For $a_{j_1j_10}\prod_{s_2=1,s_2\neq s_1}^{3}(a_{j_1j_1s_1}-a_{j_1j_1s_2})>0$ and $a_{j_2j_10}\prod_{l_2=1,l_2\neq l_1}^{3}(a_{j_2j_1l_1}-a_{j_2j_1l_2})>0$,

$$(a_{j_1j_1s_1},\bar{x}_{j_2})=\underbrace{\text{(SO, II)}}_{\text{increasing-inflection source}}.$$

The equilibrium at $x_{j_1}^{*}=a_{j_1j_1s_1}=a_{j_2j_1l_1}$ is an (SO,II)-increasing-inflection source infinite-equilibrium.

- For $a_{j_1j_10}\prod_{s_2=1,s_2\neq s_1}^{3}(a_{j_1j_1s_1}-a_{j_1j_1s_2})<0$ and $a_{j_2j_10}\prod_{l_2=1,l_2\neq l_1}^{3}(a_{j_2j_1l_1}-a_{j_2j_1l_2})>0$,

$$(a_{j_1j_1s_1},\bar{x}_{j_2})=\underbrace{(\text{SI},\text{DI})}_{\text{decreasing-inflection sink}}.$$

The equilibrium at $x_{j_1}^*=a_{j_1j_1s_1}=a_{j_2j_1l_1}$ is an (SI,DI)-decreasing-inflection sink infinite-equilibrium.

- For $a_{j_1j_10}\prod_{s_2=1,s_2\neq s_1}^{3}(a_{j_1j_1s_1}-a_{j_1j_1s_2})>0$ and $a_{j_2j_10}\prod_{l_2=1,l_2\neq l_1}^{3}(a_{j_2j_1l_1}-a_{j_2j_1l_2})<0$,

$$(a_{j_1j_1s_1},\bar{x}_{j_2})=\underbrace{(\text{SO},\text{DI})}_{\text{decreasing-inflection source}}.$$

The equilibrium at $x_{j_1}^*=a_{j_1j_1s_1}=a_{j_2j_1l_1}$ is an (SO,DI)-increasing-inflection source infinite-equilibrium.

- For $a_{j_1j_10}\prod_{s_2=1,s_2\neq s_1}^{3}(a_{j_1j_1s_1}-a_{j_1j_1s_2})<0$ and $a_{j_2j_10}\prod_{l_2=1,l_2\neq l_1}^{3}(a_{j_2j_1l_1}-a_{j_2j_1l_2})<0$,

$$(a_{j_1j_1s_1},\bar{x}_{j_2})=\underbrace{(\text{SI},\text{II})}_{\text{increasing-inflection sink}}.$$

The equilibrium at $x_{j_1}^*=a_{j_1j_1s_1}=a_{j_2j_1l_1}$ is an (SI,II)-increasing-inflection sink infinite-equilibrium.

– The infinite-equilibriums are the switching bifurcations for the sink and source flows with the sink and source flows.

(xiv) If

$$\Delta_{j_1j_1}=0 \text{ and } \Delta_{j_2j_1}>0,$$

then

$$\begin{aligned}\dot{x}_{j_1}&=a_{j_1j_10}(x_{j_1}-a_{j_1j_1s_1})(x_{j_1}-a_{j_1j_1s_2})^2,\\ \dot{x}_{j_2}&=a_{j_2j_10}(x_{j_1}-a_{j_2j_1l_1})(x_{j_1}-a_{j_2j_1l_2})(x_{j_1}-a_{j_2j_1l_3}),\end{aligned}$$

where

$$\{a_{j_1j_11}, a_{j_1j_12}\} = \text{sort}\{b_{j_1j_11}, b_{j_1j_12}\}, a_{j_1j_11} < a_{j_1j_12};$$
$$\{a_{j_2j_11}, a_{j_2j_12}, a_{j_2j_13}) = \text{sort}\{b_{j_2j_11}, b_{j_2j_12}, b_{j_2j_13}\}, a_{j_2j_11} < a_{j_2j_12} < a_{j_2j_13};$$
$$b_{j_1j_12} = -\frac{1}{2}B_{j_1j_11}; b_{j_2j_12}, b_{j_2j_13} = -\frac{1}{2}(B_{j_2j_11} \pm \sqrt{\Delta_{j_2j_1}});$$
$$j_1, j_2 \in \{1,2\}; j_1 \neq j_2.$$

(xiv$_1$) For $a_{j_1j_1s_1} \neq a_{j_2j_1l_1}$ (s_1, $s_2 \in \{1,2\}$; $s_1 \neq s_2$) and (l_1, $l_2 \in \{1,2,3\}$; $l_1 \neq l_2$),

$$\frac{dx_{j_2}}{dx_{j_1}} = \frac{a_{j_2j_10}\prod_{l_1=1}^{3}(x_{j_1} - a_{j_2j_1l_1})}{a_{j_1j_10}(x_{j_1} - a_{j_1j_1s_1})(x_{j_1} - a_{j_1j_1s_2})^2}$$
$$= \frac{a_{j_2j_10}}{a_{j_1j_10}}\left\{1 + \frac{\prod_{l_1=1}^{3}(a_{j_1j_1s_1} - a_{j_2j_1l_1})}{(a_{j_1j_1s_1} - a_{j_1j_1s_2})^2}\frac{1}{x_{j_1} - a_{j_1j_1s_1}}\right.$$
$$+ \frac{\prod_{l_1=1}^{3}(a_{j_1j_1s_1} - a_{j_2j_1l_3})}{(a_{j_1j_1s_2} - a_{j_1j_1s_1})}\frac{1}{(x_{j_1} - a_{j_1j_1s_2})^2}$$
$$\left.+ \left[\frac{\sum_{l_1=1}^{3}\prod_{l_2=1,l_2\neq l_1}^{3}(a_{j_1j_1s_2} - a_{j_2j_1l_2})}{(a_{j_1j_1s_2} - a_{j_1j_1s_1})} - \frac{\prod_{l_1=1}^{3}(a_{j_1j_1s_1} - a_{j_2j_1l_3})}{(a_{j_1j_1s_2} - a_{j_1j_1s_1})^2}\right]\frac{1}{x_{j_1} - a_{j_1j_1s_2}}\right\}.$$

With an initial condition (x_{j_10}, x_{j_20}), the first integral manifold is

$$x_{j_2} - x_{j_20} = \frac{a_{j_2j_10}}{a_{j_1j_10}}\left\{(x_{j_1} - x_{j_10}) + \frac{\prod_{l_1=1}^{3}(a_{j_1j_1s_1} - a_{j_2j_1l_1})}{(a_{j_1s_1} - a_{j_1j_1s_2})^2}\ln\frac{|x_{j_1} - a_{j_1j_1s_1}|}{|x_{j_10} - a_{j_1j_1s_1}|}\right.$$
$$- \frac{\prod_{l_1=1}^{3}(a_{j_1j_1s_1} - a_{j_2j_1l_3})}{(a_{j_1j_1s_1} - a_{j_1j_1s_2})^2}\left(\frac{1}{x_{j_1} - a_{j_1j_1s_2}} - \frac{1}{x_{j_10} - a_{j_1j_1s_2}}\right)$$
$$+ \left[\frac{\sum_{l_1=1}^{3}\prod_{l_2=1,l_2\neq l_1}^{3}(a_{j_1j_1s_2} - a_{j_2j_1l_2})}{(a_{j_1j_1s_1} - a_{j_1j_1s_2})}\right.$$
$$\left.\left. - \frac{\prod_{l_2=1}^{3}(a_{j_1j_1s_1} - a_{j_2j_1l_3})}{(a_{j_1j_1s_2} - a_{j_1j_1s_1})}\right]\ln\frac{|x_{j_1} - a_{j_1j_1\bar{s}_1}|}{|x_{j_10} - a_{j_1j_1\bar{s}_1}|}\right\}.$$

or

$$
\begin{aligned}
\frac{dx_{j_2}}{dx_{j_1}} &= \frac{a_{j_2j_10}(x_{j_1}-a_{j_2j_1l_1})\prod_{l_2=1,l_2\neq l_1}^{3}(x_{j_1}-a_{j_2j_1l_2})}{a_{j_1j_10}(x_{j_1}-a_{j_1j_1s_1})(x_{j_1}-a_{j_1j_1s_2})^2}\\
&= \frac{a_{j_2j_10}}{a_{j_1j_10}}\Big(\Big\{1+\sum_{l_2=1,l_2\neq l_1}^{3}(a_{j_1j_1s_2}-a_{j_2j_1l_2})\frac{1}{x_{j_1}-a_{j_1j_1s_2}}\\
&\quad+\prod_{l_2=1,l_2\neq l_1}^{3}(a_{j_1j_1s_2}-a_{j_2j_1l_2})\frac{1}{(x_{j_1}-a_{j_1j_1s_2})^2}\\
&\quad+(a_{j_1j_1s_1}-a_{j_2j_1l_1})\Big\{\frac{\prod_{l_2=1,l_2\neq l_1}^{3}(a_{j_1j_1s_1}-a_{j_2j_1l_2})}{(a_{j_1j_1s_1}-a_{j_1j_1s_2})^2}\frac{1}{x_{j_1}-a_{j_1j_1s_1}}\\
&\quad+\frac{\prod_{l_2=1,l_2\neq l_1}^{3}(a_{j_1j_1s_2}-a_{j_2j_1l_2})}{(a_{j_1j_1s_2}-a_{j_1j_1s_1})}\frac{1}{(x_{j_1}-a_{j_1j_1s_2})^2}\\
&\quad+\Big[\frac{\sum_{l_2=1,l_2\neq l_1}^{3}(a_{j_1j_1s_2}-a_{j_2j_1l_2})}{(a_{j_1j_1s_2}-a_{j_1j_1s_1})}-\frac{\prod_{l_2=1,l_2\neq l_1}^{3}(a_{j_1j_1s_2}-a_{j_2j_1l_2})}{(a_{j_1j_1s_2}-a_{j_1j_1s_1})^2}\Big]\\
&\quad\times\frac{1}{x_{j_1}-a_{j_1j_1s_2}}\Big\}\Big).
\end{aligned}
$$

With an initial condition (x_{j_10}, x_{j_20}), the first integral manifold is

$$
\begin{aligned}
x_{j_2}-x_{j_20} &= \frac{a_{j_2j_10}}{a_{j_1j_10}}\Big(\Big\{(x_{j_1}-x_{j_10})+\sum_{l_2=1,l_2\neq l_1}^{3}(a_{j_1j_1s_2}-a_{j_2j_1l_2})\ln\frac{|x_{j_1}-a_{j_1j_1s_2}|}{|x_{j_10}-a_{j_1j_1s_2}|}\\
&\quad-\prod_{l_2=1,l_2\neq l_1}^{3}(a_{j_1j_1s_2}-a_{j_2j_1l_2})\big(\frac{1}{x_{j_1}-a_{j_1j_1s_2}}-\frac{1}{x_{j_10}-a_{j_1j_1s_2}}\big)\Big\}\\
&\quad+(a_{j_1j_1s_1}-a_{j_2j_1l_1})\Big\{\frac{\prod_{l_2=1,l_2\neq l_1}^{3}(a_{j_1j_1s_1}-a_{j_2j_1l_2})}{(a_{j_1j_1s_1}-a_{j_1j_1s_2})^2}\ln\frac{|x_{j_1}-a_{j_1j_1s_1}|}{|x_{j_10}-a_{j_1j_1s_1}|}\\
&\quad+\frac{\prod_{l_2=1,l_2\neq l_1}^{3}(a_{j_1j_1s_2}-a_{j_2j_1l_2})}{(a_{j_1j_1s_2}-a_{j_1j_1s_1})}\big(\frac{1}{x_{j_1}-a_{j_1j_1s_2}}-\frac{1}{x_{j_1}-a_{j_1j_1s_2}}\big)\\
&\quad+\Big[\frac{\sum_{l_2=1,l_2\neq l_1}^{3}(a_{j_1j_1s_2}-a_{j_2j_1l_2})}{(a_{j_1j_1s2}-a_{j_1j_1s_1})}-\frac{\prod_{l_2=1,l_2\neq l_1}^{3}(a_{j_1j_1s_2}-a_{j_2j_1l_2})}{(a_{j_1j_1s_2}-a_{j_1j_1s_1})^2}\Big]\\
&\quad\times\ln\frac{|x_{j_1}-a_{j_1j_1s_2}|}{|x_{j_10}-a_{j_1j_1s_2}|}\Big\}\Big).
\end{aligned}
$$

(xiv$_{1a}$) For $x_{j_1}^{*}=a_{j_1j_1s_1}\neq a_{j_2j_1l_1}, a_{j_1j_1s_2}$ $(s_1, s_2\in\{1,2,3\}; s_1\neq s_2)$ and $(l_1=1, 2, 3)$, the variational equation is

$$\Delta\dot{x}_{j_1} = a_{j_1j_10}(a_{j_1j_1s_1} - a_{j_1j_1s_2})^2\Delta x_{j_1},$$
$$\dot{x}_{j_2} = a_{j_2j_10}\prod\nolimits_{l_1=1}^{3}(a_{j_1j_1s_1} - a_{j_2j_1l_1}).$$

The flows at $x_{j_1}^* = a_{j_1j_1s_1}$ are source and sink flows in the x_{j_1}-direction for $a_{j_1j_10} > 0$ and $a_{j_1j_10} < 0$, respectively. The flows are positive and negative for $\dot{x}_{j_2} > 0$ and $\dot{x}_{j_2} < 0$, respectively.

Therefore, the equilibriums of $x_{j_1}^* = a_{j_1j_1s_1} \neq a_{j_2j_1l_1}, a_{j_1j_1s_2}$ $(s_1, s_2 \in \{1, 2, 3\}; s_1 \neq s_2)$ and $(l_1 = 1, 2, 3)$ possess the following properties as in Eqs. (4.299)–(4.302)

- For $a_{j_1j_10}(a_{j_1j_1s_1} - a_{j_1j_1s_2})^2 > 0$ and $a_{j_2j_10}\prod_{l_2=1}^{3}(a_{j_1j_1s_1} - a_{j_2j_1l_1}) > 0$,

$$(a_{j_1j_1s_1}, \dot{x}_{j_2}) = \underbrace{(\text{SO, pF})}_{\text{source flow }(+)}.$$

The flow at $x_{j_1}^* = a_{j_1j_1s_1}$ is an (SO,pF)-positive source-flow.

- For $a_{j_1j_10}(a_{j_1j_1s_1} - a_{j_1j_1s_2})^2 < 0$ and $a_{j_2j_10}\prod_{l_2=1}^{3}(a_{j_1j_1s_1} - a_{j_2j_1l_1}) > 0$,

$$(a_{j_1j_1s_1}, \dot{x}_{j_2}) = \underbrace{(\text{SI, pF})}_{\text{sink flow }(+)}.$$

The flow at $x_{j_1}^* = a_{j_1j_1s_1}$ is an (SI,pF)-positive sink-flow.

- For $a_{j_1j_10}(a_{j_1j_1s_1} - a_{j_1j_1s_2})^2 > 0$ and $a_{j_2j_10}\prod_{l_2=1}^{3}(a_{j_1j_1s_1} - a_{j_2j_1l_1}) < 0$,

$$(a_{j_1j_1s_1}, \dot{x}_{j_2}) = \underbrace{(\text{SO, nF})}_{\text{source flow }(-)}.$$

The flow at $x_{j_1}^* = a_{j_1j_1s_1}$ is an (SO,nF)-negative source-flow.

- For $a_{j_1j_10}(a_{j_1j_1s_1} - a_{j_1j_1s_2})^2 < 0$ and $a_{j_2j_10}\prod_{l_2=1}^{3}(a_{j_1j_1s_1} - a_{j_2j_1l_1}) < 0$,

$$(a_{j_1j_1s_1}, \dot{x}_{j_2}) = \underbrace{(\text{SI, nF})}_{\text{sink flow }(-)}.$$

The flow at $x_{j_1}^* = a_{j_1j_1s_1}$ is an (SI,nF)-negative sink-flow.

($\text{xiv}_{1\text{b}}$) For $x_{j_1}^* = a_{j_1j_1s_2} \neq a_{j_2j_1l_1}, a_{j_1j_1s_1}$ $(s_1 \in \{1, 2, 3\})$, the variational equation is

$$\Delta\dot{x}_{j_1} = a_{j_1j_10}(a_{j_1j_1s_2} - a_{j_1j_1s_1})(\Delta x_{j_1})^2,$$
$$\dot{x}_{j_2} = a_{j_2j_10}\prod\nolimits_{l_1=1}^{3}(a_{j_1j_1s_1} - a_{j_2j_1l_1}),$$

The flows at $x_{j_1}^* = a_{j_1 j_1 s_1}$ are upper-saddle and lower-saddle flows in the x_{j_1}-direction for $a_{j_1 j_1 0}(a_{j_1 j_1 s_2} - a_{j_1 j_1 s_1}) > 0$ and $a_{j_1 j_1 0}(a_{j_1 j_1 s_2} - a_{j_1 j_1 s_1}) < 0$, respectively. The flows are positive and negative for $\dot{x}_{j_2} > 0$ and $\dot{x}_{j_2} < 0$, respectively.

Therefore, the equilibriums of $x_{j_1}^* = a_{j_1 j_1 s_2} \neq a_{j_2 j_1 l_1}, a_{j_1 j_1 s_2}$ (s_1, $s_2 \in \{1, 2\}$; $s_1 \neq s_2$) and (l_1 = 1, 2, 3) possess the following properties as in Eqs. (4.303)–(4.306).

- For $a_{j_1 j_1 0}(a_{j_1 j_1 s_2} - a_{j_1 j_1 s_1}) > 0$ and $a_{j_2 j_1 0}\prod_{l_2=1}^{3}(a_{j_1 j_1 s_2} - a_{j_1 j_1 l_1}) > 0$,

$$(a_{j_1 j_1 s_2}, \dot{x}_{j_2}) = \underbrace{(\mathrm{US}, \mathrm{pF})}_{\text{upper-saddle flow }(+)} .$$

  The flow at $x_{j_1}^* = a_{j_1 j_1 s_2}$ is a (US,pF)-positive upper-saddle flow.

- For $a_{j_1 j_1 0}(a_{j_1 j_1 s_2} - a_{j_1 j_1 s_1}) < 0$ and $a_{j_2 j_1 0}\prod_{l_2=1}^{3}(a_{j_1 j_1 s_2} - a_{j_1 j_1 l_1}) > 0$,

$$(a_{j_1 j_1 s_2}, \dot{x}_{j_2}) = \underbrace{(\mathrm{LS}, \mathrm{pF})}_{\text{lower-saddle flow }(+)} .$$

  The flow at $x_{j_1}^* = a_{j_1 j_1 s_2}$ is an (LS,pF)-positive lower-saddle flow.

- For $a_{j_1 j_1 0}(a_{j_1 j_1 s_2} - a_{j_1 j_1 s_1}) > 0$ and $a_{j_2 j_1 0}\prod_{l_2=1}^{3}(a_{j_1 j_1 s_2} - a_{j_1 j_1 l_1}) < 0$,

$$(a_{j_1 j_1 s_2}, \dot{x}_{j_2}) = \underbrace{(\mathrm{US}, \mathrm{nF})}_{\text{upper-saddle flow }(-)} .$$

  The flow at $x_{j_1}^* = a_{j_1 j_1 s_2}$ is a (US,nF)-negative upper-saddle flow.

- For $a_{j_1 j_1 0}(a_{j_1 j_1 s_2} - a_{j_1 j_1 s_1}) < 0$ and $a_{j_2 j_1 0}\prod_{l_2=1}^{3}(a_{j_1 j_1 s_2} - a_{j_1 j_1 l_1}) < 0$,

$$(a_{j_1 j_1 s_2}, \dot{x}_{j_2}) = \underbrace{(\mathrm{LS}, \mathrm{nF})}_{\text{lower-saddle flow }(-)} .$$

  The flow at $x_{j_1}^* = a_{j_1 j_1 s_2}$ is an (LS,pF)-positive lower-saddle flow.

(xiv$_{1c}$) For $x_{j_1}^* = a_{j_2 j_1 l_1} \neq a_{j_1 j_1 s_1}, a_{j_2 j_1 l_2}$ (l_1, $l_2 \in \{1, 2, 3\}$; $l_1 \neq l_2$) and (s_1, $s_2 \in \{1, 2\}$, $s_1 \neq s_2$),

$$\left.\frac{dx_{j_2}}{dx_{j_1}}\right|_{x_{j_1}^* = a_{j_2 j_1 l_1}} = \left.\frac{a_{j_2 j_1 0}(x_{j_1} - a_{j_2 j_1 l_1})\prod_{l_2=1, l_2 \neq l_1}^{3}(x_{j_1} - a_{j_2 j_1 l_2})}{a_{j_1 j_1 0}(x_{j_1} - a_{j_1 j_1 s_1})(x_{j_1} - a_{j_1 j_1 s_2})^2}\right|_{x_{j_1}^* = a_{j_2 j_1 l_1}} = 0.$$

If

$$\frac{d^2 x_{j_2}}{dx_{j_1}^2}\bigg|_{x_{j_1}^*=a_{j_2j_1l_1}} = \frac{a_{j_2j_1 0}\prod_{l_2=1,l_2\neq l_1}^{3}(a_{j_2j_1l_1}-a_{j_2j_1l_2})}{a_{j_1j_1 0}(a_{j_2j_1l_1}-a_{j_1j_1s_1})(a_{j_2j_1l_1}-a_{j_1j_1s_2})^2}\bigg|_{x_{j_1}^*=a_{j_2j_1l_1}} > 0,$$

there is an up-parabola flow in the x_{j_2}-direction, if

$$\frac{d^2 x_{j_2}}{dx_{j_1}^2}\bigg|_{x_{j_1}^*=a_{j_2j_1l_1}} = \frac{a_{j_2j_1 0}\prod_{l_2=1,l_2\neq l_1}^{3}(a_{j_2j_1l_1}-a_{j_2j_1l_2})}{a_{j_1j_1 0}(a_{j_2j_1l_1}-a_{j_1j_1s_1})(a_{j_2j_1l_1}-a_{j_1j_1s_2})^2}\bigg|_{x_{j_1}^*=a_{j_2j_1l_1}} < 0,$$

there is a down-parabola flow in the x_{j_2}-direction.

Because of

$$\dot{x}_{j_1} = a_{j_1j_1 0}(a_{j_2j_1l_1}-a_{j_1j_1s_1})(a_{j_2j_1l_1}-a_{j_1j_1s_2})^2,$$

the parabola flows $x_{j_1}^* = a_{j_2j_1l_1}$ are positive and negative for $\dot{x}_{j_1} > 0$ and $\dot{x}_{j_1} < 0$, respectively.

Therefore, the equilibrium of $x_{j_1}^* = a_{j_2j_1l_1} \neq a_{j_1j_1s_1}, a_{j_2j_1l_2}$ $(l_1, l_2 \in \{1,2\};\, l_1 \neq l_2)$ and $(s_1 = 1, 2, 3)$ has the following properties as in Eqs. (4.307)–(4.310).

- For $a_{j_1j_1 0}(a_{j_2j_1l_1}-a_{j_1j_1s_1})(a_{j_2j_1l_1}-a_{j_1j_1s_2})^2 > 0$ and $a_{j_2j_1 0}\prod_{l_2=1,l_2\neq l_1}^{3}(a_{j_2j_1l_1}-a_{j_2j_1l_2})$ >0,

$$(\dot{x}_{j_1}, a_{j_2j_1l_1}) = \underbrace{(\text{pF}, \text{UP})}_{\text{up-parabola flow }(+)}.$$

The flow at $x_{j_1}^* = a_{j_1j_1l_1}$ is a (pF,UP)-positive up-parabola flow.

- For $a_{j_1j_1 0}(a_{j_2j_1l_1}-a_{j_1j_1s_1})(a_{j_2j_1l_1}-a_{j_1j_1s_2})^2 < 0$ and $a_{j_2j_1 0}\prod_{l_2=1,l_2\neq l_1}^{3}(a_{j_2j_1l_1}-a_{j_2j_1l_2})$ > 0,

$$(\dot{x}_{j_1}, a_{j_2j_1l_1}) = \underbrace{(\text{nF}, \text{DP})}_{\text{down-parabola flow }(-)}.$$

The flow at $x_{j_1}^* = a_{j_2j_1l_1}$ is an (nF,DP)-negative down-parabola flow.

- For $a_{j_1j_1 0}(a_{j_2j_1l_1}-a_{j_1j_1s_1})(a_{j_2j_1l_1}-a_{j_1j_1s_2})^2 > 0$ and $a_{j_2j_1 0}\prod_{l_2=1,l_2\neq l_1}^{3}(a_{j_2j_1l_1}-a_{j_2j_1l_2})$ <0,

$$(\dot{x}_{j_1}, a_{j_2j_1l_1}) = \underbrace{(\text{pF}, \text{DP})}_{\text{down-parabola flow }(+)}.$$

The flow at $x_{j_1}^* = a_{j_2j_1l_1}$ is a (pF,DP)-positive down-parabola flow.

- For $a_{j_1j_10}(a_{j_2j_1l_1}-a_{j_1j_1s_1})(a_{j_2j_1l_1}-a_{j_1j_1s_2})^2<0$ and $a_{j_2j_10}\prod_{l_2=1,l_2\neq l_1}^{3}(a_{j_2j_1l_1}-a_{j_2j_1l_2})<0$,

$$(\dot{x}_{j_1},a_{j_2j_1l_1})=\underbrace{(\text{nF},\text{UP})}_{\text{up-parabola flow }(-)}.$$

The flow at $x_{j_1}^*=a_{j_2j_1l_1}$ is an (nF,UP)-negative up-parabola flow.

(xiv$_{2a}$) For $a_{j_1j_1s_1}=a_{j_2j_1l_1}$ (s_1, s_2, $s_3\in\{1,2,3\}$; $s_1\neq s_2\neq s_3$) and (l_1, $l_2\in\{1,2\}$; $l_1\neq l_2$),

$$\begin{aligned}\dot{x}_{j_1}&=a_{j_1j_10}(x_{j_1}-a_{j_1j_1s_1})(x_{j_1}-a_{j_1j_1s_2})^2,\\ \dot{x}_{j_2}&=a_{j_2j_10}(x_{j_1}-a_{j_2j_1l_1})(x_{j_1}-a_{j_2j_1l_2})(x_{j_1}-a_{j_2j_1l_3}).\end{aligned}$$

In phase space,

$$\begin{aligned}\frac{dx_{j_2}}{dx_{j_1}}&=\frac{\prod_{l_2=1,l_2\neq l_1}^{3}(x_{j_1}-a_{j_2j_1l_2})}{a_{j_1j_10}(x_{j_1}-a_{j_1j_1s_2})^2}\\ &=\frac{a_{j_2j_10}}{a_{j_1j_10}}\Big[1+\sum\nolimits_{l_2=1,l_2\neq l_1}^{3}(a_{j_1j_1s_2}-a_{j_2j_1l_2})\frac{1}{x_{j_1}-a_{j_1j_1s_2}}\\ &\quad+\prod\nolimits_{l_2=1,l_2\neq l_1}^{3}(a_{j_1j_1s_2}-a_{j_2j_1l_2})\frac{1}{(x_{j_1}-a_{j_1j_1s_2})^2}\Big].\end{aligned}$$

With an initial condition (x_{j_10},x_{j_20}), the first integral manifold is

$$\begin{aligned}x_{j_2}-x_{j_20}&=\frac{a_{j_2j_10}}{a_{j_1j_10}}\Big[(x_{j_1}-x_{j_10})+\sum\nolimits_{l_2=1,l_2\neq l_1}^{3}(a_{j_1j_1s_2}-a_{j_2j_1l_2})\ln\frac{|x_{j_1}-a_{j_1j_1s_2}|}{|x_{j_10}-a_{j_1j_1s_2}|}\\ &\quad-\prod\nolimits_{l_1=1}^{3}(a_{j_1j_1s_2}-a_{j_2j_1l_3})\Big(\frac{1}{x_{j_1}-a_{j_1j_1s_2}}-\frac{1}{x_{j_10}-a_{j_1j_1s_2}}\Big)\Big].\end{aligned}$$

At $x_{j_1}^*=a_{j_1j_1s_1}=a_{j_2j_1l_1}$, if

$$\frac{dx_{j_2}}{dx_{j_1}}=\frac{\prod_{l_2=1,l_2\neq l_1}^{3}(a_{j_2j_1l_1}-a_{j_2j_1l_2})}{a_{j_1j_10}(a_{j_1j_1s_1}-a_{j_1j_1s_1})^2}>0,$$

there is an increasing-inflection flow, and if

$$\frac{dx_{j_2}}{dx_{j_1}} = \frac{\prod_{l_2=1,l_2\neq l_1}^{3}(a_{j_2j_1l_1} - a_{j_2j_1l_2})}{a_{j_1j_10}(a_{j_1j_1s_1} - a_{j_1j_1s_1})^2} < 0,$$

there is an increasing-inflection flow.

The variational equations at $x_{j_1}^* = a_{j_1j_1s_1} = a_{j_2j_1l_1}$ are

$$\Delta\dot{x}_{j_2} = \prod_{l_2=1,l_2\neq l_1}^{3}(a_{j_2j_1l_1} - a_{j_2j_1l_2})\Delta x_{j_1},$$
$$\Delta\dot{x}_{j_1} = a_{j_1j_10}(a_{j_1j_1s_1} - a_{j_1j_1s_1})^2\Delta x_{j_1}.$$

The flows at $x_{j_1}^* = a_{j_1j_1s_1} = a_{j_2j_1l_1}$ are source and sink for $a_{j_1j_10} > 0$ and $a_{j_1j_10} < 0$.

Therefore, the infinite-equilibrium of $x_{j_1}^* = a_{j_1j_1s_1} = a_{j_2j_1l_1}$ with $\bar{x}_{j_2} \in (-\infty, \infty)$ has the following properties as in Eqs. (4.313)–(4.316).

- For $a_{j_1j_10}(a_{j_1j_1s_1} - a_{j_1j_1s_2})^2 > 0$ and $a_{j_2j_10}\prod_{l_2=1,l_2\neq l_1}^{3}(a_{j_2j_1l_1} - a_{j_2j_1l_2}) > 0$,

$$(a_{j_1j_1s_1}, \bar{x}_{j_2}) = \underbrace{(\text{SO, II})}_{\text{increasing-inflection source}}.$$

  The equilibrium at $x_{j_1}^* = a_{j_1j_1s_1} = a_{j_2j_1l_1}$ is an (SO,II)-increasing-inflection source infinite-equilibrium.

- For $a_{j_1j_10}(a_{j_1j_1s_1} - a_{j_1j_1s_2})^2 < 0$ and $a_{j_2j_10}\prod_{l_2=1,l_2\neq l_1}^{3}(a_{j_2j_1l_1} - a_{j_2j_1l_2}) > 0$,

$$(a_{j_1j_1s_1}, \bar{x}_{j_2}) = \underbrace{(\text{SI, DI})}_{\text{decreasing-inflection sink}}.$$

  The equilibrium at $x_{j_1}^* = a_{j_1j_1s_1} = a_{j_2j_1l_1}$ is an (SI,DI)-decreasing-inflection sink infinite-equilibrium.

- For $a_{j_1j_10}(a_{j_1j_1s_1} - a_{j_1j_1s_2})^2 > 0$ and $a_{j_2j_10}\prod_{l_2=1,l_2\neq l_1}^{3}(a_{j_2j_1l_1} - a_{j_2j_1l_2}) < 0$,

$$(a_{j_1j_1s_1}, \bar{x}_{j_2}) = \underbrace{(\text{SO, DI})}_{\text{decreasing-inflection source}}.$$

  The equilibrium at $x_{j_1}^* = a_{j_1j_1s_1} = a_{j_2j_1l_1}$ is an (SO,DI)-increasing-inflection source infinite-equilibrium.

- For $a_{j_1j_10}(a_{j_1j_1s_1} - a_{j_1j_1s_2})^2 < 0$ and $a_{j_2j_10}\prod_{l_2=1,l_2\neq l_1}^{3}(a_{j_2j_1l_1} - a_{j_2j_1l_2}) < 0$,

$$(a_{j_1j_1s_1}, \bar{x}_{j_2}) = \underbrace{(\text{SI, II})}_{\text{increasing-inflection sink}}.$$

The equilibrium at $x_{j_1}^* = a_{j_1j_1s_1} = a_{j_2j_1l_1}$ is an (SI,II)-increasing-inflection sink infinite-equilibrium.

(xiv$_{2b}$) For $a_{j_1j_1s_2} = a_{j_2j_1l_1}$ (s_1, s_2, $s_3 \in \{1,2,3\}$; $s_1 \neq s_2 \neq s_3$) and (l_1, $l_2 \in \{1,2\}$; $l_1 \neq l_2$),

$$\begin{aligned}
\dot{x}_{j_1} &= a_{j_1j_10}(x_{j_1} - a_{j_1j_1s_1})(x_{j_1} - a_{j_1j_1s_2})^2,\\
\dot{x}_{j_2} &= a_{j_2j_10}(x_{j_1} - a_{j_2j_1l_1})\prod\nolimits_{l_2=1,l_2\neq l_1}^{3}(x_{j_1} - a_{j_2j_1l_2}).
\end{aligned}$$

In phase space,

$$\begin{aligned}
\frac{dx_{j_2}}{dx_{j_1}} &= \frac{\prod_{l_2=1,l_2\neq l_1}^{3}(x_{j_1} - a_{j_2j_1l_2})}{(x_{j_1} - a_{j_1j_1s_1})(x_{j_1} - a_{j_1j_1s_2})}\\
&= \frac{a_{j_2j_10}}{a_{j_1j_10}}\left\{(x_{j_1} - x_{j_10}) + \frac{\prod_{l_2=1,l_2\neq l_1}^{3}(a_{j_1j_1s_1} - a_{j_2j_1l_2})}{(a_{j_1j_1s_1} - a_{j_1j_1s_2})}\frac{1}{(x_{j_1} - a_{j_1j_1s_1})}\right.\\
&\quad\left. + \frac{\prod_{l_2=1,l_2\neq l_1}^{3}(a_{j_1j_1s_2} - a_{j_2j_1l_2})}{(a_{j_1j_1s_2} - a_{j_1j_1s_1})}\frac{1}{(x_{j_1} - a_{j_1j_1s_2})}\right\}.
\end{aligned}$$

With an initial condition (x_{j_10}, x_{j_20}), the first integral manifold is

$$\begin{aligned}
x_{j_2} - x_{j_20} &= \frac{a_{j_2j_10}}{a_{j_1j_10}}\left\{(x_{j_1} - x_{j_10}) + \frac{\prod_{l_2=1,l_2\neq l_1}^{3}(a_{j_1j_1s_1} - a_{j_2j_1l_2})}{(a_{j_1j_1s_1} - a_{j_1j_1s_2})}\ln\frac{|\,x_{j_1} - a_{j_1j_1s_1}\,|}{|\,x_{j_10} - a_{j_1j_1s_1}\,|}\right.\\
&\quad\left. + \frac{\prod_{l_2=1,l_2\neq l_1}^{3}(a_{j_1j_1s_2} - a_{j_2j_1l_2})}{(a_{j_1j_1s_2} - a_{j_1j_1s_1})}\ln\frac{|\,x_{j_1} - a_{j_1j_1s_2}\,|}{|\,x_{j_10} - a_{j_1j_1s_2}\,|}\right\}.
\end{aligned}$$

The variational equations are

$$\begin{aligned}
\Delta\dot{x}_{j_1} &= a_{j_1j_10}(a_{j_1j_1s_2} - a_{j_1j_1s_1})(\Delta x_{j_1})^2,\\
\Delta\dot{x}_{j_2} &= a_{j_2j_10}\prod\nolimits_{l_2=1,l_2\neq l_1}^{3}(a_{j_2j_1l_1} - a_{j_2j_1l_2})\Delta x_{j_1}.
\end{aligned}$$

The flows at $x_{j_1}^* = a_{j_1j_1s_1} = a_{j_2j_1l_1}$ in the x_{j_1}-direction are upper-saddle and lower-saddle for $a_{j_1j_10}(a_{j_1j_1s_2} - a_{j_1j_1s_1}) > 0$ and $a_{j_1j_10}(a_{j_1j_1s_2} - a_{j_1j_1s_1}) < 0$, respectively. From the first integral manifolds, the down-up and up-down asymptotic flows exist for $a_{j_2j_10}\prod_{l_2=1,l_2\neq l_1}^{3}(a_{j_2j_1l_1} - a_{j_2j_1l_2}) > 0$ and $a_{j_2j_10}\prod_{l_2=1,l_2\neq l_1}^{3}(a_{j_2j_1l_1} - a_{j_2j_1l_2}) < 0$.

Therefore, the infinite-equilibrium of $x_{j_1}^* = a_{j_1j_1s_2} = a_{j_2j_1l_1}$ with $\bar{x}_{j_2} \in (-\infty, \infty)$ has the following properties as in Eqs. (4.319)–(4.322).

- For $a_{j_1j_10}(a_{j_1j_1s_2}-a_{j_1j_1s_1})>0$ and $a_{j_2j_10}\prod_{l_2=1,l_2\neq l_1}^{3}(a_{j_2j_1l_1}-a_{j_2j_1l_2})>0$,

$$(a_{j_1j_1s_2},\bar{x}_{j_2})=\underbrace{(\text{US},\text{DU})}_{\text{down-up upper-saddle}}.$$

  The equilibrium at $x_{j_1}^{*}=a_{j_1j_1s_2}=a_{j_2j_1l_1}$ is a (US,DU)-down-up upper-saddle infinite-equilibrium.

- For $a_{j_1j_10}(a_{j_1j_1s_2}-a_{j_1j_1s_1})<0$ and $a_{j_2j_10}\prod_{l_2=1,l_2\neq l_1}^{3}(a_{j_2j_1l_1}-a_{j_2j_1l_2})>0$,

$$(a_{j_1j_1s_2},\bar{x}_{j_2})=\underbrace{(\text{LS},\text{UD})}_{\text{up-down lower-saddle}}.$$

  The equilibrium at $x_{j_1}^{*}=a_{j_1j_1s_2}=a_{j_2j_1l_1}$ is an (LS,UD)-up-down lower-saddle infinite-equilibrium.

- For $a_{j_1j_10}(a_{j_1j_1s_2}-a_{j_1j_1s_1})>0$ and $a_{j_2j_10}\prod_{l_2=1,l_2\neq l_1}^{3}(a_{j_2j_1l_1}-a_{j_2j_1l_2})<0$,

$$(a_{j_1j_1s_2},\bar{x}_{j_2})=\underbrace{(\text{US},\text{UD})}_{\text{up-down upper-saddle}}.$$

  The equilibrium at $x_{j_1}^{*}=a_{j_1j_1s_2}=a_{j_2j_1l_1}$ is a (US,UD)-up-down upper-saddle infinite-equilibrium.

- For $a_{j_1j_10}(a_{j_1j_1s_2}-a_{j_1j_1s_1})<0$ and $a_{j_2j_10}\prod_{l_2=1,l_2\neq l_1}^{3}(a_{j_2j_1l_1}-a_{j_2j_1l_2})<0$,

$$(a_{j_1j_1s_2},\bar{x}_{j_2})=\underbrace{(\text{LS},\text{DU})}_{\text{down-up lower-saddle}}.$$

  The equilibrium at $x_{j_1}^{*}=a_{j_1j_1s_2}=a_{j_2j_1l_1}$ is an (LS,DU)-down-up lower-saddle infinite-equilibrium.

(xv) If

$$\begin{gathered}\Delta_{j_1j_1}=0\text{ and }\Delta_{j_2j_1}>0,\\ a_{j_1j_1s_1}=a_{j_1j_1s_2}=a_{j_1j_11}\end{gathered}$$

then

$$\begin{aligned}\dot{x}_{j_1}&=a_{j_1j_10}(x_{j_1}-a_{j_1j_11})^3,\\ \dot{x}_{j_2}&=a_{j_2j_10}\prod_{l_1=1}^{3}(x_{j_1}-a_{j_2j_1l_1}),\end{aligned}$$

where

$$
\begin{aligned}
&\{a_{j_1j_1 1}, a_{j_1j_1 2}\} = \text{sort}\{b_{j_1j_1 1}, b_{j_1j_1 2}\}, a_{j_1j_1 1} = a_{j_1j_1 2};\\
&\{a_{j_2j_1 1}, a_{j_2j_1 2}, a_{j_2j_1 3}) = \text{sort}\{b_{j_2j_1 1}, b_{j_2j_1 2}, b_{j_2j_1 3}\}, a_{j_2j_1 1} < a_{j_2j_1 2} < a_{j_2j_1 3};\\
&b_{j_1j_1 2} = -\frac{1}{2}B_{j_1j_1 1}; b_{j_2j_1 2}, b_{j_2j_1 3} = -\frac{1}{2}(B_{j_2j_1 1} \pm \sqrt{\Delta_{j_2j_1}});\\
&j_1, j_2 \in \{1,2\}; j_1 \neq j_2.
\end{aligned}
$$

(xv$_1$) For $a_{j_1j_1 1} \neq a_{j_2j_1 l_1}$ ($l_1 = 1, 2, 3$), we have

$$
\begin{aligned}
\frac{dx_{j_2}}{dx_{j_1}} &= \frac{\prod_{l_1=1}^{3}(x_{j_1} - a_{j_2j_1 l_1})}{(x_{j_1} - a_{j_1j_1 1})^3}\\
&= \frac{a_{j_2j_1 0}}{a_{j_1j_1 0}}\Big[1 + \prod\nolimits_{l_1=1}^{3}(a_{j_1j_1 1} - a_{j_2j_1 l_1})\frac{1}{(x_{j_1} - a_{j_1j_1 1})^3}\\
&\quad + \sum\nolimits_{l_1=1}^{3}\prod\nolimits_{l_2=1, l_2\neq l_1}^{3}(a_{j_1j_1 1} - a_{j_2j_1 l_2})\frac{1}{(x_{j_1} - a_{j_1j_1 1})^2}\\
&\quad + \sum\nolimits_{l_1=1}^{3}\sum\nolimits_{l_2=1, l_2\neq l_1}^{3}\prod\nolimits_{l_1=1, l_3\neq l_1, l_2}^{3}(a_{j_1j_1 1} - a_{j_2j_1 l_3})\frac{1}{(x_{j_1} - a_{j_1j_1 1})}\Big].
\end{aligned}
$$

With an initial condition $(x_{j_1 0}, x_{j_2 0})$, the first integral manifold is

$$
\begin{aligned}
x_{j2} - x_{j2}0 &= \frac{a_{j_2j_1 0}}{a_{j_1j_1 0}}\Big\{(x_{j1} - x_{j_1 0})\\
&\quad -\frac{1}{2}\prod\nolimits_{l_1=1}^{3}(a_{j_1j_1 1} - a_{j_2j_1 l_1})\Big[\frac{1}{(x_{j_1} - a_{j_1j_1 1})^2} - \frac{1}{(x_{j_1 0} - a_{j_1j_1 1})^2}\Big]\\
&\quad -\sum\nolimits_{l_1=1}^{3}\prod\nolimits_{l_2=1, l_2\neq l_1}^{3}(a_{j_1j_1 1} - a_{j_2j_1 l_2})\Big[\frac{1}{(x_{j_1} - a_{j_1j_1 1})} - \frac{1}{(x_{j_1 0} - a_{j_1j_1 1})}\Big]\\
&\quad +\sum\nolimits_{l_1=1}^{3}\sum\nolimits_{l_2=1, l_2\neq l_1}^{3}\prod\nolimits_{l_1=1, l_3\neq l_1, l_2}^{3}(a_{j_1j_1 1} - a_{j_2j_1 l_3})\ln\frac{|x_{j_1} - a_{j_1j_1 s_1}|}{|x_{j_1} - a_{j_1j_1 s_1}|}\Big\}.
\end{aligned}
$$

(xv$_{1a}$) For $x_{j_1}^* = a_{j_1j_1 1} \neq a_{j_2j_1 l_1}$ ($l_1 = 1, 2, 3$), the variational equation is

$$
\begin{aligned}
\Delta\dot{x}_{j_1} &= a_{j_1j_1 0}(\Delta x_{j_1})^3,\\
\dot{x}_{j_2} &= a_{j_2j_1 0}\prod\nolimits_{l_1=1}^{3}(a_{j_1j_1 1} - a_{j_2j_1 l_1}).
\end{aligned}
$$

The flows at $x_{j_1}^* = a_{j_1j_1 1} \neq a_{j_2j_1 l_1}$ are the third-order source and sink flows in the x_{j_1}-direction for $a_{j_1j_1 0} > 0$ and $a_{j_1j_1 0} < 0$. The flows are positive and negative for $\dot{x}_{j_2} > 0$ and $\dot{x}_{j_2} < 0$, respectively.

Therefore, the equilibriums of $x_{j_1}^* = a_{j_1j_1 1} \neq a_{j_2j_1 l_1}$ ($l_1 = 1, 2, 3$) are as follows as in Eqs. (4.327)–(4.330).

- For $a_{j_1j_10}>0$ and $a_{j_2j_10}\prod_{l_1=1}^{3}(a_{j_1j_11}-a_{j_2j_1l_1})>0$,

$$(a_{j_1j_11},\dot{x}_{j_2})=\underbrace{(3^{\mathrm{rd}}\mathrm{SO},\mathrm{pF})}_{\text{third-order source flow }(+)}.$$

The flow at $x_{j_1}^{*}=a_{j_1j_11}$ is a (3rdSO,pF)-positive third-order source flow.

- For $a_{j_1j_10}<0$ and $a_{j_2j_10}\prod_{l_1=1}^{3}(a_{j_1j_11}-a_{j_2j_1l_1})>0$,

$$(a_{j_1j_11},\dot{x}_{j_2})=\underbrace{(3^{\mathrm{rd}}\mathrm{SI},\mathrm{pF})}_{\text{third-order sink flow }(+)}.$$

The flow at $x_{j_1}^{*}=a_{j_1j_11}$ is a (3rdSI,pF)-positive third-order sink-flow.

- For $a_{j_1j_10}>0$ and $a_{j_2j_10}\prod_{l_1=1}^{3}(a_{j_1j_11}-a_{j_2j_1l_1})<0$,

$$(a_{j_1j_11},\dot{x}_{j_2})=\underbrace{(3^{\mathrm{rd}}\mathrm{SO},\mathrm{nF})}_{\text{third-order source flow }(-)}.$$

The flow at $x_{j_1}^{*}=a_{j_1j_11}$ is a (3rdSO,nF)-negative third-order source flow.

- For $a_{j_1j_10}<0$ and $a_{j_2j_10}\prod_{l_1=1}^{3}(a_{j_1j_1s_1}-a_{j_2j_1l_1})<0$,

$$(a_{j_1j_11},\dot{x}_{j_2})=\underbrace{(3^{\mathrm{rd}}\mathrm{SI},\mathrm{nF})}_{\text{third-order sink flow }(-)}.$$

The flow at $x_{j_1}^{*}=a_{j_1j_11}$ is a (3rdSI,nF)-negative third-order sink flow.

(xv$_{1b}$) For $x_{j_1}^{*}=a_{j_2j_1l_1}\neq a_{j_1j_11},a_{j_2j_1l_2}$ $(l_1,\ l_2\in\{1,2,3\};\ l_1\neq l_2)$, in phase space, at $x_{j_1}^{*}=a_{j_2j_1l_1}\neq a_{j_1j_11},a_{j_2j_1l_2}$, we have

$$\frac{dx_{j_2}}{dx_{j_1}}\bigg|_{x_{j_1}^{*}=a_{j_2j_1l_1}}=\frac{(x_{j_1}-a_{j_2j_1l_1})\prod_{l_2=1}^{3}(x_{j_1}-a_{j_2j_1l_2})}{(x_{j_1}-a_{j_1j_11})^3}=0.$$

If

$$\frac{d^2x_{j_2}}{dx_{j_1}^2}\bigg|_{x_{j_1}^{*}=a_{j_2j_1l_1}}=\frac{\prod_{l_2=1}^{3}(a_{j_2j_1l_1}-a_{j_2j_1l_2})}{(a_{j_2j_1l_1}-a_{j_1j_11})^3}>0,$$

the flow at $x_{j_1}^{*}=a_{j_2j_1l_1}$ is an up-parabola flow. If

$$\frac{d^2x_{j_2}}{dx_{j_1}^2}\bigg|_{x_{j_1}^{*}=a_{j_2j_1l_1}}=\frac{\prod_{l_2=1}^{3}(a_{j_2j_1l_1}-a_{j_2j_1l_2})}{(a_{j_2j_1l_1}-a_{j_1j_11})^3}<0,$$

the flow at $x_{j_1}^{*}=a_{j_2j_1l_1}$ is a down-parabola flow. Let

$$\dot{x}_{j_2} = (a_{j_2 j_1 l_1} - a_{j_1 j_1 1})^3.$$

The parabola flows are positive and negative for $(a_{j_2 j_1 l_1} - a_{j_1 j_1 1})^3 > 0$ and $(a_{j_2 j_1 l_1} - a_{j_1 j_1 1})^3 < 0$, respectively.

Therefore, the equilibrium of $x^*_{j_1} = a_{j_2 j_1 l_1} \neq a_{j_1 j_1 1}, a_{j_2 j_1 l_2}$ (l_1, $l_2 \in \{1, 2, 3\}$; $l_1 \neq l_2$) has the following properties as in Eqs. (4.331)–(4.334).

- For $a_{j_1 j_1 0}(a_{j_2 j_1 l_1} - a_{j_1 j_1 1})^3 > 0$ and $a_{j_2 j_1 0}\prod_{l_2=1, l_2 \neq l_1}^{3}(a_{j_2 j_1 l_1} - a_{j_2 j_1 l_2}) > 0$,

$$(\dot{x}_{j_1}, a_{j_2 j_1 l_1}) = \underbrace{(\mathrm{pF}, \mathrm{UP})}_{\text{up-parabola flow } (+)}.$$

  The flow at $x^*_{j_1} = a_{j_1 j_1 l_1}$ is a (pF,UP)-positive up-parabola flow.

- For $a_{j_1 j_1 0}(a_{j_2 j_1 l_1} - a_{j_1 j_1 1})^3 < 0$ and $a_{j_2 j_1 0}\prod_{l_2=1, l_2 \neq l_1}^{3}(a_{j_2 j_1 l_1} - a_{j_2 j_1 l_2}) > 0$,

$$(\dot{x}_{j_1}, a_{j_2 j_1 l_1}) = \underbrace{(\mathrm{nF}, \mathrm{DP})}_{\text{down-parabola flow } (-)}.$$

  The flow at $x^*_{j_1} = a_{j_2 j_1 l_1}$ is an (nF,DP)-negative down-parabola flow.

- For $a_{j_1 j_1 0}(a_{j_2 j_1 l_1} - a_{j_1 j_1 1})^3 > 0$ and $a_{j_2 j_1 0}\prod_{l_2=1, l_2 \neq l_1}^{3}(a_{j_2 j_1 l_1} - a_{j_2 j_1 l_2}) < 0$,

$$(\dot{x}_{j_1}, a_{j_2 j_1 l_1}) = \underbrace{(\mathrm{pF}, \mathrm{DP})}_{\text{down-parabola flow } (+)}.$$

  The flow at $x^*_{j_1} = a_{j_2 j_1 l_1}$ is a (pF,DP)-positive down-parabola flow.

- For $a_{j_1 j_1 0}(a_{j_2 j_1 l_1} - a_{j_1 j_1 1})^3 < 0$ and $a_{j_2 j_1 0}\prod_{l_2=1, l_2 \neq l_1}^{3}(a_{j_2 j_1 l_1} - a_{j_2 j_1 l_2}) < 0$,

$$(\dot{x}_{j_1}, a_{j_2 j_1 l_1}) = \underbrace{(\mathrm{nF}, \mathrm{UP})}_{\text{up-parabola flow } (-)}.$$

  The flow at $x^*_{j_1} = a_{j_2 j_1 l_1}$ is an (nF,UP)-negative up-parabola flow.

(xiv$_2$) For $a_{j_1 j_1 1} = a_{j_2 j_1 l_1}$ ($l_1 = 1, 2, 3$),

$$\dot{x}_{j_1} = a_{j_1 j_1 0}(x_{j_1} - a_{j_1 j_1 1})^3,$$
$$\dot{x}_{j_2} = a_{j_2 j_1 0}(x_{j_1} - a_{j_2 j_1 l_1})\prod_{l_2=1, l_2 \neq l_1}^{3}(x_{j_1} - a_{j_2 j_1 l_2}).$$

In phase space,

$$\begin{aligned}\frac{dx_{j_2}}{dx_{j_1}} &= \frac{a_{j_2j_10}\prod_{l_2=1,l_2\neq l_1}^{3}(x_{j_1}-a_{j_2j_1s_2})}{(x_{j_1}-a_{j_1j_11})^2}\\ &= \frac{a_{j_2j_10}}{a_{j_1j_10}}\Big[1+\prod_{l_2=1,l_2\neq l_1}^{3}(a_{j_2j_1l_1}-a_{j_2j_1l_2})\frac{1}{(x_{j_1}-a_{j_1j_11})^2}\\ &\quad+\sum_{l_2=1,l_2\neq l_1}^{3}\prod_{l_1=1,l_3\neq l_1,l_2}^{3}(a_{j_2j_1l_1}-a_{j_2j_1l_3})\frac{1}{(x_{j_1}-a_{j_1j_11})}\Big].\end{aligned}$$

With an initial condition (x_{j_10}, x_{j_20}), the first integral manifold is

$$\begin{aligned}x_{j_2}-x_{j_20} &= \frac{a_{j_2j_10}}{a_{j_1j_10}}\Big\{(x_{j_1}-x_{j_10})\\ &\quad-\prod_{l_2=1,l_2\neq l_1}^{3}(a_{j_2j_1l_1}-a_{j_2j_1l_2})\Big[\frac{1}{(x_{j_1}-a_{j_1j_11})}-\frac{1}{(x_{j_10}-a_{j_1j_11})}\Big]\\ &\quad+\sum_{l_2=1,l_2\neq l_1}^{3}\prod_{l_1=1,l_3\neq l_1,l_2}^{3}(a_{j_2j_1l_1}-a_{j_2j_1l_3})\ln\frac{|x_{j_1}-a_{j_1j_11}|}{|x_{j_10}-a_{j_1j_11}|}\Big\}.\end{aligned}$$

For $x_{j_1}^* = a_{j_1j_11} = a_{j_2j_1l_1}$, the variational equations are

$$\begin{aligned}\Delta\dot{x}_{j_1} &= a_{j_1j_10}(\Delta x_{j_1})^3,\\ \Delta\dot{x}_{j_2} &= a_{j_2j_10}\prod_{l_2=1,l_2\neq l_1}^{3}(a_{j_2j_1l_1}-a_{j_2j_1l_2})\Delta x_{j_1}.\end{aligned}$$

The flows at $x_{j_1}^* = a_{j_1j_11} = a_{j_2j_1l_1}$ are the third-order source and sink for $a_{j_1j_10} > 0$ and $a_{j_1j_10} < 0$, respectively. For the first integral manifold, if

$$\frac{a_{j_2j_10}\prod_{l_2=1,l_2\neq l_1}^{3}(a_{j_2j_1l_1}-a_{j_2j_1l_2})}{a_{j_1j_10}} > 0$$

there is an increasing-concave flow at $x_{j_1}^* = a_{j_1j_11} = a_{j_2j_1l_1}$ and if

$$\frac{a_{j_2j_10}\prod_{l_2=1,l_2\neq l_1}^{3}(a_{j_2j_1l_1}-a_{j_2j_1l_2})}{a_{j_1j_10}} < 0$$

there is a decreasing-concave flow at $x_{j_1}^* = a_{j_1j_11} = a_{j_2j_1l_1}$.

Therefore, the infinite-equilibrium of $x_{j_1}^* = a_{j_1j_11} = a_{j_2j_1l_1}$ with $\bar{x}_{j_2} \in (-\infty, \infty)$ has the following properties as in Eqs. (4.337)–(4.340).

- For $a_{j_1j_10} > 0$ and $a_{j_2j_10}\prod_{l_2=1,l_2\neq l_1}^{3}(a_{j_2j_1l_1} - a_{j_2j_1l_2}) > 0$,

$$(a_{j_1j_11}, \bar{x}_{j_2}) = \underbrace{(3^{\text{rd}}\text{SO}, (3:1)\text{IC})}_{\text{third-order increasing-concave source}}.$$

The equilibrium at $x_{j_1}^* = a_{j_1j_11} = a_{j_2j_1l_1}$ is an (3rdSO,(3:1)IC)-third-order increasing-concave source infinite-equilibrium.

- For $a_{j_1j_10} < 0$ and $a_{j_2j_10}\prod_{l_2=1,l_2\neq l_1}^{3}(a_{j_2j_1l_1} - a_{j_2j_1l_2}) > 0$,

$$(a_{j_1j_11}, \bar{x}_{j_2}) = \underbrace{(3^{\text{rd}}\text{SI}, (3:1)\text{DC})}_{\text{third-order decreasing-concave sink}}.$$

The equilibrium at $x_{j_1}^* = a_{j_1j_11} = a_{j_2j_1l_1}$ is an (3rdSI,(3:1)DC)-third-order decreasing-concave sink infinite-equilibrium.

- For $a_{j_1j_10} > 0$ and $a_{j_2j_10}\prod_{l_2=1,l_2\neq l_1}^{3}(a_{j_2j_1l_1} - a_{j_2j_1l_2}) < 0$,

$$(a_{j_1j_11}, \bar{x}_{j_2}) = \underbrace{(3^{\text{rd}}\text{SO}, (3:1)\text{DC})}_{\text{third-order decreasing-asympotic source}}.$$

The equilibrium at $x_{j_1}^* = a_{j_1j_11} = a_{j_2j_1l_1}$ is an (3rdSO,(3:1)DC)-third-order decreasing-concave sink infinite-equilibrium.

- For $a_{j_1j_10} < 0$ and $a_{j_2j_10}\prod_{l_2=1,l_2\neq l_1}^{3}(a_{j_2j_1l_1} - a_{j_2j_1l_2}) < 0$,

$$(a_{j_1j_11}, \bar{x}_{j_2}) = \underbrace{(3^{\text{rd}}\text{SI}, (3:1)\text{IC})}_{\text{third-order increasing-concave sink}}.$$

The equilibrium at $x_{j_1}^* = a_{j_1j_11} = a_{j_2j_1l_1}$ is an (3rdSI,(3:1)IC)-third-order increasing-concave sink infinite-equilibrium.

(xvi) If

$$\Delta_{j_1j_1} < 0 \text{ and } \Delta_{j_2j_1} > 0,$$

then

$$\begin{aligned}
\dot{x}_{j_1} &= a_{j_1j_10}(x_{j_1} - a_{j_1j_11})\left[(x_{j_1} - a_{j_1j_1})^2 + b_{j_1j_1}\right],\\
\dot{x}_{j_2} &= a_{j_2j_10}(x_{j_1} - a_{j_2j_1l_1})(x_{j_1} - a_{j_2j_1l_2})(x_{j_1} - a_{j_2j_1l_3}),
\end{aligned}$$

where

$$a_{j_1j_11} = b_{j_1j_11}; a_{j_1j_1} = -\frac{1}{2}B_{j_1j_11}, b_{j_1j_1} = -\frac{1}{4}\Delta_{j_1j_1};$$

$$\{a_{j_2j_11}, a_{j_2j_12}, a_{j_2j_13}) = \text{sort}\{b_{j_2j_11}, b_{j_2j_12}, b_{j_2j_13}\}, a_{j_2j_11} < a_{j_2j_12} < a_{j_2j_13};$$

$$b_{j_2j_12}, b_{j_2j_13} = -\frac{1}{2}(B_{j_2j_11} \pm \sqrt{\Delta_{j_2j_1}});$$

$$j_1, j_2 \in \{1, 2\}; j_1 \neq j_2.$$

(xvi$_1$) For $a_{j_1j_11} \neq a_{j_2j_1l_1}$ ($l_1 = 1, 2, 3$), we have

$$\begin{aligned}
\frac{dx_{j_2}}{dx_{j_1}} &= \frac{a_{j_2j_10}(x_{j_1} - a_{j_2j_1l_1})(x_{j_1} - a_{j_2j_1l_2})(x_{j_1} - a_{j_2j_1l_3})}{a_{j_1j_10}(x_{j_1} - a_{j_1j_11})\left[(x_{j_1} - a_{j_1j_1})^2 + b_{j_1j_1}\right]} \\
&= \frac{a_{j_2j_10}}{a_{j_1j_10}}\Bigg(\Bigg\{1 + \sum\nolimits_{l_2=1, l_2 \neq l_1}^{3}(a_{j_1j_1} - a_{j_2j_1l_2})\frac{(x_{j_1} - a_{j_1j_1})}{(x_{j_1} - a_{j_1j_1})^2 + b_{j_1j_1}} \\
&\quad + \left[\prod\nolimits_{l_2=1, l_2 \neq l_1}^{3}(a_{j_1j_1} - a_{j_2j_1l_2}) - b_{j_1j_1}\right]\frac{1}{(x_{j_1} - a_{j_1j_1})^2 + b_{j_1j_1}}\Bigg\} \\
&\quad + (a_{j_1j_11} - a_{j_2j_1l_1})\Bigg\{\frac{(x_{j_1} - a_{j_1j_1})}{(x_{j_1} - a_{j_1j_1})^2 + b_{j_1j_1}} \\
&\quad + \left[(a_{j_1j_1} - a_{j_2j_1l_3}) + (a_{j_1j_11} - a_{j_2j_1l_2})\right]\frac{1}{(x_{j_1} - a_{j_1j_1})^2 + b_{j_1j_1}}\Bigg\} \\
&\quad + \frac{\prod_{l_2=1}^{3}(a_{j_1j_11} - a_{j_2j_1l_2})}{(a_{j_1j_11} - a_{j_1j_1})^2 + b_{j_1j_1}}\Bigg[\frac{1}{x_{j_1} - a_{j_1j_11}} - \frac{(x_{j_1} - a_{j_1j_1})}{(x_{j_1} - a_{j_1j_1})^2 + b_{j_1j_1}} \\
&\quad - \frac{(a_{j_1j_11} - a_{j_1j_1})}{(x_{j_1} - a_{j_1j_1})^2 + b_{j_1j_1}}\Bigg]\Bigg).
\end{aligned}$$

With an initial condition (x_{j_10}, x_{j_20}), the first integral manifold is

$$
\begin{aligned}
x_{j_2}-x_{j_2 0}=\frac{a_{j_2 j_1 0}}{a_{j_1 j_1 0}}\Big\{&(x_{j_1}-x_{j_1 0})\\
&+\frac{1}{2}\sum\nolimits_{l_2=1,l_2\neq l_1}^{3}(a_{j_1 j_1}-a_{j_2 j_1 l_2})\ln\frac{|(x_{j_1}-a_{j_1 j_1})^2+b_{j_1 j_1}|}{|(x_{j_1 0}-a_{j_1 j_1})^2+b_{j_1 j_1}|}\\
&+\frac{1}{\sqrt{b_{j_1 j_1}}}\Big[\prod\nolimits_{l_2=1,l_2\neq l_1}^{3}(a_{j_1 j_1}-a_{j_2 j_1 l_2})-b_{j_1 j_1}\Big](\arctan\frac{x_{j_1}-a_{j_1 j_1}}{\sqrt{b_{j_1 j_1}}}\\
&-\arctan\frac{x_{j_1 0}-a_{j_1 j_1}}{\sqrt{b_{j_1 j_1}}})+\frac{1}{2}(a_{j_1 j_1 1}-a_{j_2 j_1 l_1})\ln\frac{|(x_{j_1}-a_{j_1 j_1})^2+b_{j_1 j_1}|}{|(x_{j_1 0}-a_{j_1 j_1})^2+b_{j_1 j_1}|}\\
&+(a_{j_1 j_1 1}-a_{j_2 j_1 l_1})\big[(a_{j_1 j_1}-a_{j_2 j_1 l_3})+(a_{j_1 j_1 1}-a_{j_2 j_1 l_2})\big]\\
&\times\frac{1}{\sqrt{b_{j_1 j_1}}}(\arctan\frac{x_{j_1}-a_{j_1 j_1}}{\sqrt{b_{j_1 j_1}}}-\arctan\frac{x_{j_1 0}-a_{j_1 j_1}}{\sqrt{b_{j_1 j_1}}})\\
&+\frac{\prod_{l_2=1}^{3}(a_{j_1 j_1 1}-a_{j_2 j_1 l_2})}{(a_{j_1 j_1 1}-a_{j_1 j_1})^2+b_{j_1 j_1}}\Big[\ln\frac{|x_{j_1}-a_{j_1 j_1 1}|}{|x_{j_1 0}-a_{j_1 j_1 1}|}\\
&-\frac{1}{2}\ln\frac{|(x_{j_1}-a_{j_1 j_1})^2+b_{j_1 j_1}|}{|(x_{j_1 0}-a_{j_1 j_1})^2+b_{j_1 j_1}|}\\
&-\frac{(a_{j_1 j_1 1}-a_{j_1 j_1})}{\sqrt{b_{j_1 j_1}}}(\arctan\frac{x_{j_1}-a_{j_1 j_1}}{\sqrt{b_{j_1 j_1}}}-\arctan\frac{x_{j_1 0}-a_{j_1 j_1}}{\sqrt{b_{j_1 j_1}}})\Big]\Big\}.
\end{aligned}
$$

($\text{xvi}_{1\text{a}}$) For $x_{j_1}^{*}=a_{j_1 j_1 1}\neq a_{j_2 j_1 l_1}, a_{j_2 j_1 l_2}$, the variational equation is

$$
\begin{aligned}
\Delta\dot{x}_{j_1}&=a_{j_1 j_1 0}\big[(a_{j_1 j_1 1}-a_{j_1 j_1})^2+b_{j_1 j_1}\big]\Delta x_{j_1},\\
\dot{x}_{j_2}&=a_{j_2 j_1 0}\prod\nolimits_{l_1=1}^{3}(a_{j_1 j_1 1}-a_{j_2 j_1 l_1}),
\end{aligned}
$$

The flows at $x_{j_1}^{*}=a_{j_1 j_1 1}$ are source and sink for $a_{j_1 j_1 0}>0$ and $a_{j_1 j_1 0}<0$, respectively. The sink and source flows are positive and negative for $\dot{x}_{j_2}>0$ and $\dot{x}_{j_2}<0$. Therefore, the equilibriums of $x_{j_1}^{*}=a_{j_1 j_1 1}\neq a_{j_2 j_1 l_1}$ ($l_1=1,2,3$) are as follows as in Eqs. (4.345)–(4.348).

- For $a_{j_1 j_1 0}>0$ and $a_{j_2 j_1 0}\prod_{l_1=1}^{3}(a_{j_1 j_1 1}-a_{j_2 j_1 l_1})>0$,

$$
(a_{j_1 j_1 s_1},\dot{x}_{j_2})=\underbrace{(\text{SI},\text{pF})}_{\text{source flow }(+)}.
$$

  The flow at $x_{j_1}^{*}=a_{j_1 j_1 1}$ is an (SO,pF)-positive source flow.
- For $a_{j_1 j_1 0}<0$ and $a_{j_2 j_1 0}\prod_{l_1=1}^{3}(a_{j_1 j_1 1}-a_{j_2 j_1 l_1})>0$,

$$(a_{j_1j_11}, \dot{x}_{j_2}) = \underbrace{(\mathrm{SI}, \mathrm{pF})}_{\text{sink flow }(+)}.$$

The flow at $x^*_{j_1} = a_{j_1j_11}$ is an (SI,pF)-positive sink flow.

- For $a_{j_1j_10} > 0$ and $a_{j_2j_10}\prod_{l_1=1}^{3}(a_{j_1j_11} - a_{j_2j_1l_1}) < 0$,

$$(a_{j_1j_11}, \dot{x}_{j_2}) = \underbrace{(\mathrm{SO}, \mathrm{nF})}_{\text{source flow }(-)}.$$

The flow at $x^*_{j_1} = a_{j_1j_11}$ is an (SO,nF)-negative source flow.

- For $a_{j_1j_10} < 0$ and $a_{j_2j_10}\prod_{l_1=1}^{3}(a_{j_1j_11} - a_{j_2j_1l_1}) < 0$,

$$(a_{j_1j_11}, \dot{x}_{j_2}) = \underbrace{(\mathrm{SI}, \mathrm{nF})}_{\text{sink flow }(-)}.$$

The flow at $x^*_{j_1} = a_{j_1j_11}$ is an (SI,nF)-negative sink flow.

(xvi$_{1b}$) For $x^*_{j_1} = a_{j_2j_1l_1} \neq a_{j_1j_11}, a_{j_2j_1l_2}$ (l_1, $l_2 \in \{1,2\}$; $l_1 \neq l_2$), we have

$$\frac{dx_{j_2}}{dx_{j_1}}\bigg|_{x^*_{j_1}=a_{j_2j_1l_1}} = \frac{a_{j_2j_10}(x_{j_1} - a_{j_2j_1l_1})\prod_{l_2=1,l_2\neq l_1}^{3}(x_{j_1} - a_{j_2j_1l_2})}{a_{j_1j_10}(x_{j_1} - a_{j_1j_11})\left[(x_{j_1} - a_{j_1j_1})^2 + b_{j_1j_1}\right]}\bigg|_{x^*_{j_1}=a_{j_2j_1l_1}} = 0.$$

If

$$\frac{d^2x_{j_2}}{dx^2_{j_1}}\bigg|_{x^*_{j_1}=a_{j_2j_1l_1}} = \frac{a_{j_2j_10}\prod_{l_2=1,l_2\neq l_1}^{3}(a_{j_2j_1l_1} - a_{j_2j_1l_2})}{a_{j_1j_10}(a_{j_2j_1l_1} - a_{j_1j_11})\left[(a_{j_2j_1l_1} - a_{j_1j_1})^2 + b_{j_1j_1}\right]} > 0,$$

there is an up-parabola flow at $x^*_{j_1} = a_{j_2j_1l_1}$, and if

$$\frac{d^2x_{j_2}}{dx^2_{j_1}}\bigg|_{x^*_{j_1}=a_{j_2j_1l_1}} = \frac{a_{j_2j_10}\prod_{l_2=1,l_2\neq l_1}^{3}(a_{j_2j_1l_1} - a_{j_2j_1l_2})}{a_{j_1j_10}(a_{j_2j_1l_1} - a_{j_1j_11})\left[(a_{j_2j_1l_1} - a_{j_1j_1})^2 + b_{j_1j_1}\right]} < 0,$$

there is a down-parabola flow at $x^*_{j_1} = a_{j_2j_1l_1}$.

Because

$$\dot{x}_{j_2} = a_{j_1j_10}(a_{j_2j_1l_1} - a_{j_1j_11})\left[(a_{j_2j_1l_1} - a_{j_1j_1})^2 + b_{j_1j_1}\right]$$

and

$$(a_{j_2j_1l_1}-a_{j_1j_1})^2+b_{j_1j_1}>0,$$

The parabola flows at $x^*_{j_1}=a_{j_2j_1l_1}$ in the x_{j_1}-direction are positive and negative for $a_{j_1j_10}(a_{j_2j_1l_1}-a_{j_1j_11})>0$ and $a_{j_1j_10}(a_{j_2j_1l_1}-a_{j_1j_11})<0$.

Therefore, the equilibrium of $x^*_{j_1}=a_{j_2j_1l_1}\neq a_{j_1j_11}, a_{j_2j_1l_2}$ ($l_1, l_2\in\{1,2\}; l_1\neq l_2$) and ($s_1=1, 2, 3$) has the following properties as in Eqs. (4.349)–(4.352).

- For $a_{j_1j_10}(a_{j_2j_1l_1}-a_{j_1j_11})>0$ and $a_{j_2j_10}\prod_{l_2=1,l_2\neq l_1}^{3}(a_{j_2j_1l_1}-a_{j_2j_1l_2})>0$,

$$(\dot{x}_{j_1},a_{j_2j_1l_1})=\underbrace{(\text{pF},\text{UP})}_{\text{up-parabola flow }(+)}.$$

The flow at $x^*_{j_1}=a_{j_1j_1l_1}$ is a (pF,UP)-positive up-parabola flow.

- For $a_{j_1j_10}(a_{j_2j_1l_1}-a_{j_1j_11})<0$ and $a_{j_2j_10}\prod_{l_2=1,l_2\neq l_1}^{3}(a_{j_2j_1l_1}-a_{j_2j_1l_2})>0$,

$$(\dot{x}_{j_1},a_{j_2j_1l_1})=\underbrace{(\text{nF},\text{DP})}_{\text{down-parabola flow }(-)}.$$

The flow at $x^*_{j_1}=a_{j_2j_1l_1}$ is an (nF,DP)-negative down-parabola flow.

- For $a_{j_1j_10}(a_{j_2j_1l_1}-a_{j_1j_11})>0$ and $a_{j_2j_10}\prod_{l_2=1,l_2\neq l_1}^{3}(a_{j_2j_1l_1}-a_{j_2j_1l_2})<0$,

$$(\dot{x}_{j_1},a_{j_2j_1l_1})=\underbrace{(\text{pF},\text{DP})}_{\text{down-parabola flow }(+)}.$$

The flow at $x^*_{j_1}=a_{j_2j_1l_1}$ is a (pF,DP)-positive down-parabola flow.

- For $a_{j_1j_10}(a_{j_2j_1l_1}-a_{j_1j_11})<0$ and $a_{j_2j_10}\prod_{l_2=1,l_2\neq l_1}^{3}(a_{j_2j_1l_1}-a_{j_2j_1l_2})<0$,

$$(\dot{x}_{j_1},a_{j_2j_1l_1})=\underbrace{(\text{nF},\text{UP})}_{\text{up-parabola flow }(-)}.$$

The flow at $x^*_{j_1}=a_{j_2j_1l_1}$ is an (nF,UP)-negative up-parabola flow.

(xvi$_2$) For $a_{j_1j_11}=a_{j_2j_1l_1}$ ($l_1, l_2, l_3\in\{1,2,3\}; l_1\neq l_2\neq l_3$),

$$\begin{aligned}\dot{x}_{j_1}&=a_{j_1j_10}(x_{j_1}-a_{j_1j_11})\left[(x_{j_1}-a_{j_1j_1})^2+b_{j_1j_1}\right],\\ \dot{x}_{j_2}&=a_{j_2j_10}(x_{j_1}-a_{j_2j_1l_1})\prod_{l_2=1,l_2\neq l_1}^{3}(x_{j_1}-a_{j_2j_1l_2}).\end{aligned}$$

In phase space,

$$\begin{aligned}\frac{dx_{j_2}}{dx_{j_1}} &= \frac{a_{j_2j_10}\prod_{l_2=1,l_2\neq l_1}^{3}(x_{j_1}-a_{j_2j_1l_2})}{a_{j_1j_10}\left[(x_{j_1}-a_{j_1j_1})^2+b_{j_1j_1}\right]}\\ &= \frac{a_{j_2j_10}}{a_{j_1j_10}}\Big\{1+\sum\nolimits_{l_2=1,l_2\neq l_1}^{3}(a_{j_1j_1}-a_{j_2j_1l_2})\frac{(x_{j_1}-a_{j_1j_1})}{(x_{j_1}-a_{j_1j_1})^2+b_{j_1j_1}}\\ &\quad+\left[\prod\nolimits_{l_2=1,l_2\neq l_1}^{3}(a_{j_1j_1}-a_{j_2j_1l_2})-b_{j_1j_1}\right]\frac{1}{(x_{j_1}-a_{j_1j_1})^2+b_{j_1j_1}}\Big\}.\end{aligned}$$

With an initial condition (x_{j_10}, x_{j_20}), the first integral manifold is

$$\begin{aligned}x_{j_2}-x_{j_20} &= \frac{a_{j_2j_10}}{a_{j_1j_10}}\Big\{(x_{j_1}-x_{j_10})\\ &\quad+\frac{1}{2}\sum\nolimits_{l_2=1,l_2\neq l_1}^{3}(a_{j_1j_1}-a_{j_2j_1l_2})\ln\frac{|\,(x_{j_1}-a_{j_1j_1})^2+b_{j_1j_1}\,|}{|\,(x_{j_10}-a_{j_1j_1})^2+b_{j_1j_1}\,|}\\ &\quad+\frac{1}{\sqrt{b_{j_1j_1}}}\left[\prod\nolimits_{l_2=1,l_2\neq l_1}^{3}(a_{j_1j_1}-a_{j_2j_1l_2})-b_{j_1j_1}\right]\\ &\quad\times\Big(\arctan\frac{x_{j_1}-a_{j_1j_1}}{\sqrt{b_{j_1j_1}}}-\arctan\frac{x_{j_10}-a_{j_1j_1}}{\sqrt{b_{j_1j_1}}}\Big)\Big\}.\end{aligned}$$

At $x_{j_1}^*=a_{j_1j_11}=a_{j_2j_1l_1}$, if

$$\left.\frac{dx_{j_2}}{dx_{j_1}}\right|_{x_{j_1}^*=a_{j_2j_1l_1}}=\frac{a_{j_2j_10}\prod_{l_2=1,l_2\neq l_1}^{3}(a_{j_2j_1l_1}-a_{j_2j_1l_2})}{a_{j_1j_10}\left[(a_{j_1j_1}-a_{j_1j_1})^2+b_{j_1j_1}\right]}>0,$$

there is an increasing-inflection flow, and if

$$\left.\frac{dx_{j_2}}{dx_{j_1}}\right|_{x_{j_1}^*=a_{j_2j_1l_1}}=\frac{a_{j_2j_10}\prod_{l_2=1,l_2\neq l_1}^{3}(a_{j_2j_1l_1}-a_{j_2j_1l_2})}{a_{j_1j_10}\left[(a_{j_1j_1}-a_{j_1j_1})^2+b_{j_1j_1}\right]}<0,$$

there is a decreasing-inflection flow. At $x_{j_1}^*=a_{j_1j_11}=a_{j_2j_1l_1}$, the variational equations are

$$\begin{aligned}\Delta\dot{x}_{j_1}&=a_{j_1j_10}\left[(a_{j_1j_1}-a_{j_1j_1})^2+\Delta_{j_1j_1}\right]\Delta x_{j_1},\\ \Delta\dot{x}_{j_2}&=a_{j_2j_10}\prod\nolimits_{l_2=1,l_2\neq l_1}^{3}(a_{j_2j_1l_1}-a_{j_2j_1l_2})\Delta x_{j_1}.\end{aligned}$$

The infinite-equilibrium of $x_{j_1}^*=a_{j_1j_11}=a_{j_2j_1l_1}$ is source and sink in the x_{j_1}-direction for $a_{j_1j_10}>0$ and $a_{j_1j_10}<0$.

Therefore, the infinite-equilibrium of $x_{j_1}^*=a_{j_1j_11}=a_{j_2j_1l_1}$ with $\overline{x}_{j_2}\in(-\infty,\infty)$ has the following properties as in Eqs. (4.355)–(4.358).

- For $a_{j_1j_10} > 0$ and $a_{j_2j_10}\prod_{l_2=1,l_2\neq l_1}^{3}(a_{j_2j_1l_1} - a_{j_2j_1l_2}) > 0$,

$$(a_{j_1j_11}, \overline{x}_{j_2}) = \underbrace{(\text{SO}, \text{II})}_{\text{increasing-inflection source}}.$$

The equilibrium at is an (SO,II)-increasing-inflection source infinite-equilibrium.

- For $a_{j_1j_10} < 0$ and $a_{j_2j_10}\prod_{l_2=1,l_2\neq l_1}^{3}(a_{j_2j_1l_1} - a_{j_2j_1l_2}) > 0$,

$$(a_{j_1j_11}, \overline{x}_{j_2}) = \underbrace{(\text{SI}, \text{DI})}_{\text{decreasing-inflection sink}}.$$

The equilibrium at $x_{j_1}^* = a_{j_1j_11} = a_{j_2j_1l_1}$ is an (SI,DI)-decreasing-inflection sink infinite-equilibrium.

- For $a_{j_1j_10} > 0$ and $a_{j_2j_10}\prod_{l_2=1,l_2\neq l_1}^{3}(a_{j_2j_1l_1} - a_{j_2j_1l_2}) < 0$,

$$(a_{j_1j_11}, \overline{x}_{j_2}) = \underbrace{(\text{SO}, \text{DI})}_{\text{decreasing-inflection source}}.$$

The equilibrium at $x_{j_1}^* = a_{j_1j_11} = a_{j_2j_1l_1}$ is an (SO,DI)-decreasing-inflection sink infinite-equilibrium.

- For $a_{j_1j_10} < 0$ and $a_{j_2j_10}\prod_{l_2=1,l_2\neq l_1}^{3}(a_{j_2j_1l_1} - a_{j_2j_1l_2}) < 0$,

$$(a_{j_1j_11}, \overline{x}_{j_2}) = \underbrace{(\text{SI}, \text{II})}_{\text{increasing-inflection sink}}.$$

The equilibrium at $x_{j_1}^* = a_{j_1j_11} = a_{j_2j_1l_1}$ is an (SI,II)-increasing-inflection sink infinite-equilibrium.

In the end, from the above local analysis and the first integral of such a dynamical system, Theorem 4.1 is proved. ■

4.3 One-Dimensional Singular Dynamics and Bifurcations

In this section, dynamical systems with two cubic univariate vector fields with a single variable will be discussed. The corresponding stability and bifurcations of equilibriums will be presented.

Consider

$$\begin{aligned} \dot{x}_1 &= f_1(x_1, \mathbf{p}) = a_{110}(x_1 - a_{111})(x_1 - a_{112})(x_1 - a_{113}), \\ \dot{x}_2 &= f_2(x_1, \mathbf{p}) = a_{210}(x_1 - a_{211})^3. \end{aligned} \tag{4.359}$$

The first integral manifold is

$$\begin{aligned} x_2 - x_{20} = \ & \frac{a_{210}}{a_{110}} \left[(x_1 - x_{10}) + \frac{(a_{112} - a_{211})^2}{(a_{112} - a_{113})} \ln \frac{|x_1 - a_{112}|}{|x_{10} - a_{112}|} \right. \\ & + \frac{(a_{113} - a_{211})^2}{(a_{113} - a_{112})} \ln \frac{|x_1 - a_{113}|}{|x_{10} - a_{113}|} + (a_{111} - a_{211}) \\ & \left. \sum_{\bar{s}_1=1}^{3} \frac{(a_{11\bar{s}_1} - a_{211})^2}{\prod_{\bar{s}_2=1, \bar{s}_2 \neq \bar{s}_1}^{3} (a_{11\bar{s}_1} - a_{11\bar{s}_2})} \ln \frac{|x_1 - a_{11\bar{s}_1}|}{|x_{10} - a_{11\bar{s}_1}|} \right]. \end{aligned} \tag{4.360}$$

The corresponding phase portraits for two-dimensional systems on the x_1-direction with $x_1^* = a_{111}, a_{112}$ and on the x_2-direction with $x_1^* = a_{211} > a_{112}$ is presented in Fig. 4.1 for $a_{211} \in (a_{112}, a_{113})$. In Fig. 4.1a, the negative source, negative sink, and positive source flows at $x_1^* = a_{111}, a_{112}, a_{113}$ and a negative third-order down-parabola flow at $x_1^* = a_{211} \in (a_{112}, a_{113})$ for $a_{110} > 0$ and $a_{210} > 0$ are presented from the first integral manifold. At $x_1^* = a_{111}, a_{113}$, the flows are negative and positive source flows because of $df_1/dx_1|_{x_1^* = a_{111}, a_{113}} > 0$ with $\dot{x}_2 < 0$ and $\dot{x}_2 > 0$, respectively. At $x_1^* = a_{112}$, the flow is a negative sink flow due to $df_1/dx_1|_{x_1^* = a_{112}} < 0$ and $\dot{x}_2 < 0$. $x_1^* = a_{211}$ is a negative third-order down-parabola flow due to $d^4x_2/dx_1^4|_{x_1^* = a_{211}} = 6a_{210}/\prod_{s_1=1}^{3}(a_{211} - a_{11s_1}) < 0$ and $\dot{x}_1 < 0$. Similarly, the negative sink, negative source, and positive sink flows and a positive third-order up-parabola flow ($a_{110} < 0$, $a_{210} > 0$) are sketched in Fig. 4.1b for $x_1^* = a_{111}, a_{112}, a_{113}$ on the x_1-direction and $x_1^* = a_{211} \in (a_{112}, a_{113})$ on the x_2-direction. For $a_{110} > 0$ and $a_{210} < 0$, at $x_1^* = a_{111}, a_{112}, a_{113}$ on the x_1-direction and $x_1^* = a_{211} > a_{112}$ on the x_2-direction are positive source, positive sink and negative source flows and a negative third-order up-parabola flow, as shown in Fig. 4.1c. At $x_1^* = a_{111}, a_{112}$ on the x_1-direction and $x_1^* = a_{211} \in (a_{112}, a_{113})$ on the x_2-direction are the positive sink, positive source, and negative sink flows, and a negative third-order down-parabola flow ($a_{110} < 0$, $a_{210} < 0$) is presented in Fig. 4.1d. The flows at equilibriums are listed as follows.

$$\begin{Bmatrix} (a_{111}, \dot{x}_2) \\ (a_{112}, \dot{x}_2) \\ (\dot{x}_1, a_{211}) \\ (a_{113}, \dot{x}_2) \end{Bmatrix} = \begin{Bmatrix} (\text{SO}, \text{nF}) \\ (\text{SI}, \text{nF}) \\ (\text{nF}, 3^{\text{rd}}\text{DP}) \\ (\text{SO}, \text{pF}) \end{Bmatrix} \text{ for } a_{110} > 0, a_{210} > 0; \tag{4.361}$$

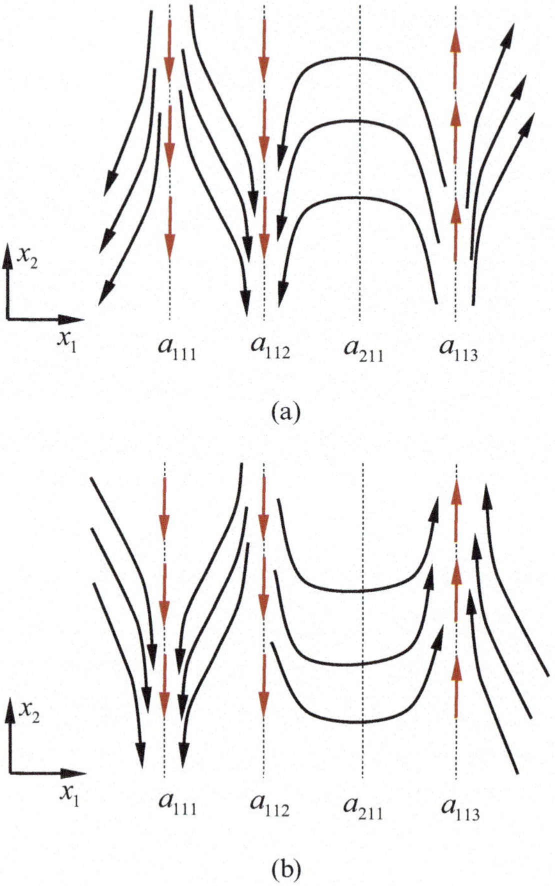

Fig. 4.1 Phase portraits for two-dimensional systems on the x_1-direction with $x_1^* = a_{111}, a_{112}, a_{113}$ and on the x_2-direction with $x_1^* = a_{211} \in (a_{112}, a_{113})$: (**a**) an (SO,nF)-negative source flow, an (SI, nF)-negative sink flow, an (nF,3rdDP)-negative third-order down-parabola flow, and an (SO,pF)-positive source flow ($a_{110} > 0$, $a_{210} > 0$), (**b**) an (SI,nF)- negative sink flow, an (SO,nF)-negative source flow, a (pF, 3rdUP) positive third-order up-parabola flow, and an (SI,pF)-positive sink flow ($a_{110} < 0$, $a_{210} > 0$), (**c**) an (SO,pF)-positive source flow, an (SI,pF)- positive sink flow, an (nF, 3rdUP)-negative third-order up-parabola flow, and an (SO,nF)-negative flow ($a_{110} > 0$, $a_{210} < 0$), (**d**) an (SI,pF)-positive sink flow, an (SO,pF)-positive source flow, an (nF,3rdDP)-negative third-order down-parabola flow, and an (SI,nF)-negative sink flow ($a_{110} < 0$, $a_{210} < 0$)

$$\left\{ \begin{array}{c} (a_{111}, \dot{x}_2) \\ (a_{112}, \dot{x}_2) \\ (\dot{x}_1, a_{211}) \\ (a_{113}, \dot{x}_2) \end{array} \right\} = \left\{ \begin{array}{c} (\text{SI, nF}) \\ (\text{SO, nF}) \\ (\text{pF, 3}^{\text{rd}}\text{UP}) \\ (\text{SI, pF}) \end{array} \right\} \text{ for } a_{110} < 0, a_{210} > 0; \tag{4.362}$$

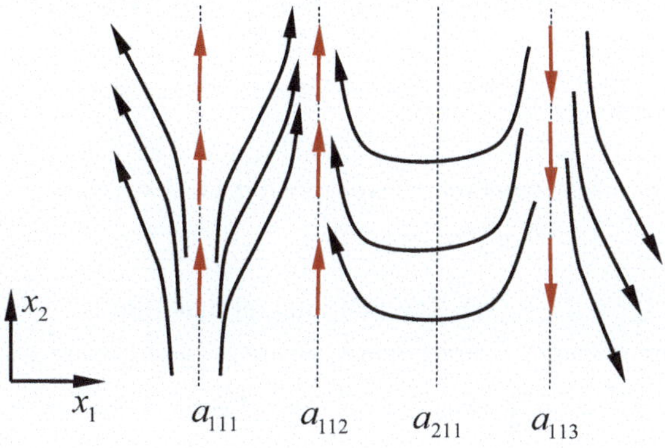

(c)

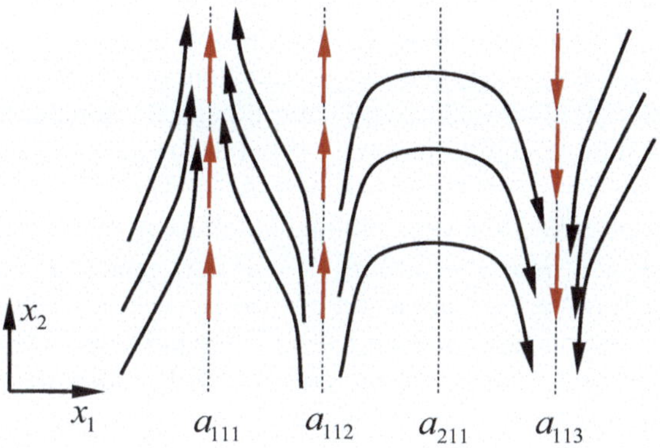

(d)

Fig. 4.1 (continued)

$$\begin{Bmatrix} (a_{111}, \dot{x}_2) \\ (a_{112}, \dot{x}_2) \\ (\dot{x}_1, a_{211}) \\ (a_{113}, \dot{x}_2) \end{Bmatrix} = \begin{Bmatrix} (\text{SO, pF}) \\ (\text{SI, pF}) \\ (\text{nF, 3}^{\text{rd}}\text{UP}) \\ (\text{SO, nF}) \end{Bmatrix} \text{ for } a_{110} > 0, a_{210} < 0; \tag{4.363}$$

$$\begin{Bmatrix} (a_{111}, \dot{x}_2) \\ (a_{112}, \dot{x}_2) \\ (\dot{x}_1, a_{211}) \\ (a_{113}, \dot{x}_2) \end{Bmatrix} = \begin{Bmatrix} (\text{SI, pF}) \\ (\text{SO, pF}) \\ (\text{pF, 3}^{\text{rd}}\text{DP}) \\ (\text{SI, nF}) \end{Bmatrix} \text{ for } a_{110} < 0, a_{210} < 0. \tag{4.364}$$

Similarly, the phase portraits of two-dimensional systems for $x_1^* = a_{111}, a_{112}, a_{113}$ on the x_1-direction with three cases of $x_1^* = a_{211} \in (a_{111}, a_{112})$, $x_1^* = a_{211} < a_{111}$, and $x_1^* = a_{211} > a_{113}$ on the x_2-direction can be presented. The readers can finish drawing the phase portraits as exercise.

Consider a dynamical system as

$$\begin{aligned} \dot{x}_1 &= a_{110}(x_1 - a_{111})(x_1 - a_{112})(x_1 - a_{113}), \\ \dot{x}_2 &= a_{210}(x_1 - a_{211})^3 \end{aligned} \tag{4.365}$$

with $a_{211} = a_{11s_1}$ ($s_1 \in \{1, 2, 3\}$). The first integral manifold is

$$\begin{aligned} x_2 - x_{20} = & \frac{a_{210}}{a_{110}} \left[(x_1 - x_{10}) + \frac{(a_{111} - a_{211})^2}{(a_{111} - a_{113})} \ln \frac{|x_1 - a_{111}|}{|x_{10} - a_{111}|} \right. \\ & \left. + \frac{(a_{113} - a_{211})^2}{(a_{113} - a_{111})} \ln \frac{|x_1 - a_{113}|}{|x_{10} - a_{113}|} \right]. \end{aligned} \tag{4.366}$$

For $a_{211} = a_{112}$, the phase portrait with equilibriums of $x_1^* = a_{111}, a_{112}, a_{113}$ on the x_1-direction and $x_1^* = a_{211} = a_{112}$ on the x_2-direction is presented in Fig. 4.2. In Fig. 4.2a, the flows at $x_1^* = a_{112}, a_{111}, a_{113}$ are a negative source, a (1:3)-decreasing-inflection sink infinite-equilibrium (i.e., (SI,(1:3)DI)-infinite-equilibrium) and a positive source flows for $a_{110} > 0$ and $a_{210} > 0$. At $x_1^* = a_{112} = a_{211}$, the equilibrium is a (1:3)-decreasing-inflection sink infinite-equilibrium due to sink on the x_1-direction with $\Delta\dot{x}_1 = a_{110}(a_{112} - a_{111})(a_{112} - a_{113})\Delta x_1$ and $\Delta\dot{x}_2 = a_{210}(\Delta x_1)^3$. At $x_1^* = a_{111}, a_{113}$, the flows are negative and positive source flows due to $\dot{x}_2 < 0$ and $\dot{x}_2 > 0$. A (1:3)-increasing-inflection source infinite-equilibrium (i.e., (SO, (1:3)II)-infinite-equilibrium) with negative and positive sink flows for $a_{110} < 0$ and $a_{210} > 0$ at $x_1^* = a_{112}, a_{111}, a_{113}$ are presented in Fig. 4.2b. At $x_1^* = a_{211} = a_{112}$ and $x_1^* = a_{111}, a_{113}$ is a (1:3)-increasing-inflection sink infinite-equilibrium (i.e., (SI,(1:3) II)-infinite-equilibrium) with positive and negative source flows for $a_{110} > 0$ and $a_{210} < 0$, as shown in Fig. 4.2c. A (1:3)-decreasing-inflection source infinite-equilibrium (i.e., (SO,(1:3)DI)-infinite-equilibrium) with positive and negative sink flows at $x_1^* = a_{112} = a_{211}$ and $x_1^* = a_{111}, a_{113}$ for $a_{110} < 0$ and $a_{210} < 0$ is presented in Fig. 4.2d. The solid and dashed lines are for sink and source infinite-equilibriums, respectively. The filled and hollow circular symbols are specific equilibriums on the infinite-equilibrium lines, respectively. The flows at equilibriums with $\overline{x}_2 \in (-\infty, \infty)$ are presented as follows.

$$\left\{ \begin{array}{l} (a_{111}, \dot{x}_2) \\ (a_{112}, \overline{x}_2) \\ (a_{113}, \dot{x}_2) \end{array} \right\} = \left\{ \begin{array}{l} (\mathrm{SO}, \mathrm{nF}) \\ (\mathrm{SI}, (1{:}3)\mathrm{DI}) \\ (\mathrm{SO}, \mathrm{pF}) \end{array} \right\} \text{ for } a_{110} > 0, a_{210} > 0; \tag{4.367}$$

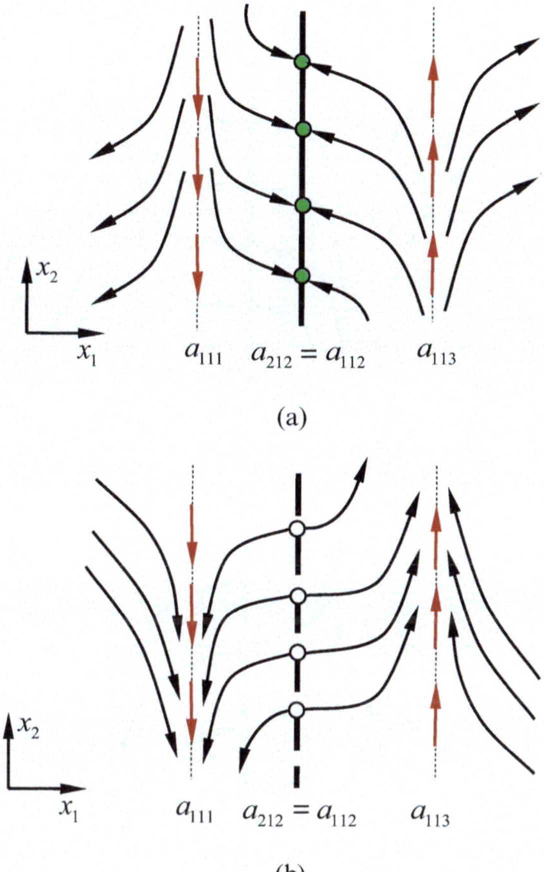

Fig. 4.2 Phase portraits for two-dimensional systems with equilibriums of $x_1^* = a_{111}, a_{112}, a_{113}$ on the x_1-direction and $x_1^* = a_{211} = a_{112}$ on the x_2-direction: (**a**) an (SO,pF)-positive source flow, an (SI,(1:3)DI)-infinite-equilibrium, and an (SO,nF)-positive source flow ($a_{110} > 0$, $a_{210} > 0$), (**b**) an (SI,pF)-positive sink flow, an (SO,(1:3)II)-infinite-equilibrium, and an (SI,pF)-positive sink flow ($a_{110} < 0$, $a_{210} > 0$), (**c**) an (SO,pF)-negative source flow, an (SI,(1:3)II)-infinite-equilibrium and an (SO,nF)-negative source flow ($a_{110} > 0, a_{210} < 0$), (**d**) an (SI,pF)-positive sink flow, an (SO,(1:3)DI)-infinite-equilibrium, and an (SI,pF)-negative sink flow ($a_{110} < 0$, $a_{210} < 0$). The solid and dashed lines are for sink and source infinite-equilibriums, respectively. The filled and hollow circular symbols are specific equilibriums on the infinite-equilibrium lines, respectively

$$\left\{\begin{array}{l}(a_{111},\dot{x}_2)\\(a_{112},\overline{x}_2)\\(a_{113},\dot{x}_2)\end{array}\right\}=\left\{\begin{array}{l}(\text{SI, nF})\\(\text{SO, (1:3)II})\\(\text{SI, pF})\end{array}\right\}\text{ for }a_{110}<0,a_{210}>0; \tag{4.368}$$

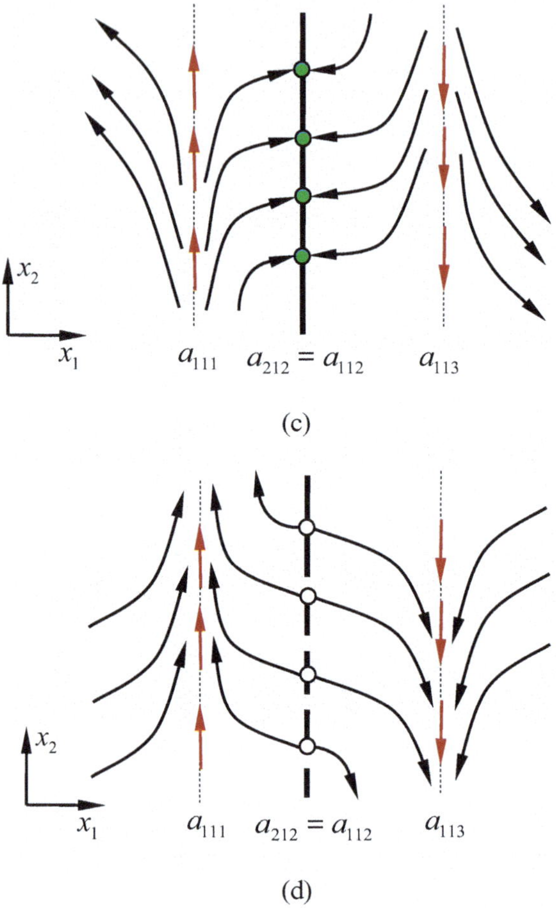

Fig. 4.2 (continued)

$$\left\{\begin{array}{c}(a_{111},\dot{x}_2)\\(a_{112},\bar{x}_2)\\(a_{113},\dot{x}_2)\end{array}\right\}=\left\{\begin{array}{l}(\mathrm{SO},\mathrm{pF})\\(\mathrm{SI},(1{:}3)\mathrm{II})\\(\mathrm{SO},\mathrm{nF})\end{array}\right\}\text{ for } a_{110}>0, a_{210}<0; \tag{4.369}$$

$$\left\{\begin{array}{c}(a_{111},\dot{x}_2)\\(a_{112},\bar{x}_2)\\(a_{113},\dot{x}_2)\end{array}\right\}=\left\{\begin{array}{l}(\mathrm{SO},\mathrm{pF})\\(\mathrm{SI},(1{:}3)\mathrm{DI})\\(\mathrm{SO},\mathrm{nF})\end{array}\right\}\text{ for } a_{110}<0, a_{210}<0. \tag{4.370}$$

Similarly, for cases of $a_{211} = a_{111}$ and $a_{211} = a_{113}$, the phase portraits for two-dimensional systems with $x_1^* = a_{111}, a_{112}$ on the x_1-direction and $x_1^* = a_{211} = a_{112}$ on the x_2-direction can be presented as an exercise for readers to complete.

Consider

$$\begin{aligned}\dot{x}_1 &= a_{110}(x_1 - a_{11s_1})(x_1 - a_{11s_2})^2,\\ \dot{x}_2 &= a_{210}(x_1 - a_{211})^3.\end{aligned} \tag{4.371}$$

The first integral manifold is

$$\begin{aligned}x_2 - x_{20} = \frac{a_{210}}{a_{110}}\Bigg(\Big[&x_1 - x_{10} + 2(a_{11s_2} - a_{211})\ln\frac{|x_1 - a_{11s_2}|}{|x_{10} - a_{11s_2}|}\\ &-(a_{11s_2} - a_{211})^2\Big(\frac{1}{x_1 - a_{11s_2}} - \frac{1}{x_{10} - a_{11s_2}}\Big)\Big]\\ &+(a_{11s_1} - a_{211})\Big\{\frac{(a_{11s_1} - a_{211})^2}{(a_{11s_1} - a_{11s_2})^2}\ln\frac{|x_1 - a_{11s_1}|}{|x_{10} - a_{11s_1}|}\\ &-\frac{(a_{11s_2} - a_{211})^2}{(a_{11s_2} - a_{11s_1})}\Big(\frac{1}{x_1 - a_{11s_2}} - \frac{1}{x_{10} - a_{11s_2}}\Big)\\ &+\Big[\frac{2(a_{11s_2} - a_{211})}{(a_{11s_2} - a_{11s_1})} - \frac{(a_{11s_2} - a_{211})^2}{(a_{11s_2} - a_{11s_1})^2}\Big]\ln\frac{|x_1 - a_{11s_2}|}{|x_{10} - a_{11s_2}|}\Big\}\Bigg).\end{aligned} \tag{4.372}$$

For $a_{211} \in (a_{111}, a_{112})$ with $a_{11s_1} = a_{111}$ and $a_{11s_2} = a_{112}$, the phase portraits for two-dimensional systems on the x_1-direction are presented in Fig. 4.3. For $x_1^* = a_{111}$ and a_{112}, $\dot{x}_1 = 0$ and $\dot{x}_2 = a_{210}(x_1^* - a_{211})^3$. Thus, for $x_1^* = a_{111}$ and a_{112}, the flow is determined by $\dot{x}_2 \neq 0$. For $x_1^* = a_{211}$, $\dot{x}_1 = a_{110}(a_{211} - a_{111})(a_{211} - a_{112})^2$ and $\dot{x}_2 = 0$. Thus, for $x_1^* = a_{211}$, the flow is determined by $\dot{x}_1 \neq 0$. In Fig. 4.3a, the phase portrait for $a_{110} > 0$ and $a_{210} > 0$ is presented from the first integral manifold. At $x_1^* = a_{111}$, the negative source flow is toward the x_2-negative direction due to $\dot{x}_2 = a_{210}(a_{111} - a_{211})^3 < 0$. At $x_1^* = a_{112}$, the upper-saddle flow is toward the x_2-positive direction due to $\dot{x}_2 = a_{210}(a_{112} - a_{211})^3 > 0$. At $x_1^* = a_{211}$, the third-order up-parabola flow is toward the x_1-positive direction because of $\dot{x}_1 = a_{110}(a_{211} - a_{111})(a_{211} - a_{112})^2 > 0$. Similarly, the phase portraits for ($a_{110} < 0$, $a_{210} > 0$), ($a_{110} > 0$, $a_{210} < 0$), and ($a_{110} < 0$, $a_{210} < 0$) are presented in Fig. 4.3b–d, respectively. The flows at the equilibriums are listed as follows.

$$\begin{Bmatrix}(a_{111}, \dot{x}_2)\\ (\dot{x}_1, a_{211})\\ (a_{112}, \dot{x}_2)\end{Bmatrix} = \begin{Bmatrix}(\text{SO, nF})\\ (\text{pF, 3}^{\text{rd}}\text{UP})\\ (\text{US, pF})\end{Bmatrix} \text{ for } a_{110} > 0, a_{210} > 0; \tag{4.373}$$

$$\begin{Bmatrix}(a_{111}, \dot{x}_2)\\ (\dot{x}_1, a_{211})\\ (a_{112}, \dot{x}_2)\end{Bmatrix} = \begin{Bmatrix}(\text{SI, nF})\\ (\text{nF, 3}^{\text{rd}}\text{DP})\\ (\text{LS, pF})\end{Bmatrix} \text{ for } a_{110} < 0, a_{210} > 0; \tag{4.374}$$

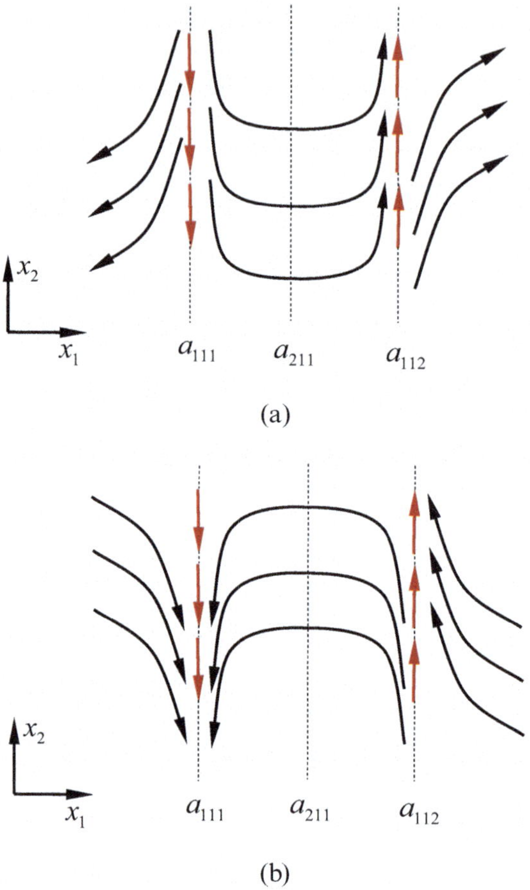

Fig. 4.3 Phase portraits for two-dimensional systems on the x_2-direction with $x_1^* = a_{111}, a_{112}$ and on the x_1-direction with $x_1^* = a_{211} \in (a_{111}, a_{112})$: (**a**) an (SO,nF)-negative source flow, a (pF, 3rdUP)-positive third-order up-parabola flow, and a (US,pF)-positive upper-saddle flow ($a_{110} > 0$, $a_{210} > 0$), (**b**) an (SI,nF)-negative sink flow, an (nF,3rdDP)-neagive third-order down-parabola flow, and an (LS,pF)-positive lower-saddle flow ($a_{110} < 0$, $a_{210} > 0$), (**c**) an (SI,nF)-negative sink flow, a (pF,3rdDPI)-positive decreasing-inflection flow, and a (US,nF)-negative upper-saddle flow ($a_{110} > 0$, $a_{210} < 0$), (**d**) an (SI,pF)-positive sink flow, an (nF,3rdUP)-negative third-order up-parabola flow, and an (LS,nF)-negative lower-saddle flow ($a_{110} < 0$, $a_{210} < 0$)

$$\begin{Bmatrix} (a_{111}, \dot{x}_2) \\ (\dot{x}_1, a_{211}) \\ (a_{112}, \dot{x}_2) \end{Bmatrix} = \begin{Bmatrix} (\text{SO, pF}) \\ (\text{pF, 3}^{\text{rd}}\text{DP}) \\ (\text{US, nF}) \end{Bmatrix} \text{ for } a_{110} > 0, a_{210} < 0; \tag{4.375}$$

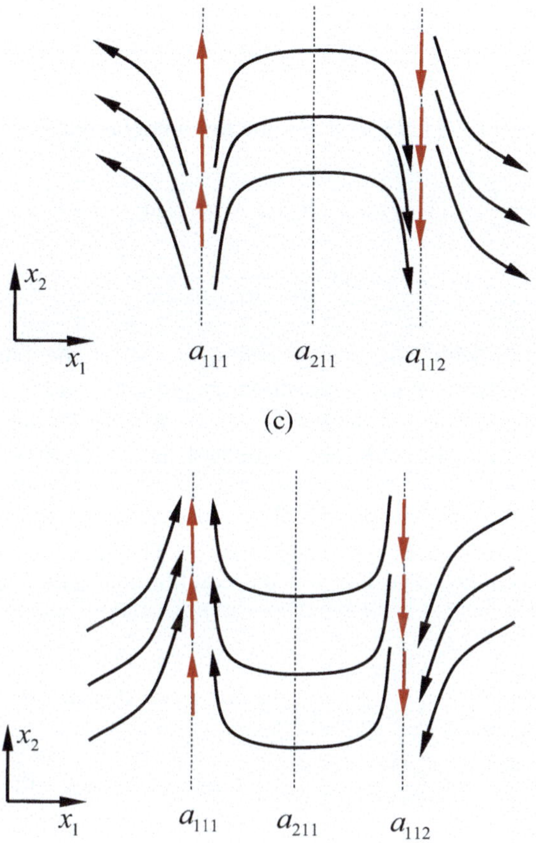

(d)

Fig. 4.3 (continued)

$$\begin{Bmatrix} (a_{111}, \dot{x}_2) \\ (\dot{x}_1, a_{211}) \\ (a_{112}, \dot{x}_2) \end{Bmatrix} = \begin{Bmatrix} (\text{SI}, \text{pF}) \\ (\text{nF}, 3^{\text{rd}}\text{UP}) \\ (\text{LS}, \text{nF}) \end{Bmatrix} \text{ for } a_{110} < 0, a_{210} < 0. \tag{4.376}$$

For $a_{211} < a_{111}$ and $a_{211} > a_{112}$, the phase portraits for two-dimensional systems on the x_1-direction can be sketched for $a_{11s_1} = a_{112}$ and $a_{11s_2} = a_{111}$. The dynamical behaviors for $a_{211} < a_{111}$ and $a_{211} > a_{111}$ are similar. For $a_{11s_1} = a_{111}$ and $a_{11s_2} = a_{112}$, the corresponding phase portraits can be sketched. The upper-saddle and lower-saddle flows are the appearing bifurcations for sink and source flows. The increasing-inflection for decreasing-inflection flows are for the appearing bifurcations for up-parabola and down-parabola flows.

Consider

$$\dot{x}_1 = a_{110}(x_1 - a_{11s_1})(x_1 - a_{11s_2})^2,$$
$$\dot{x}_2 = a_{210}(x_1 - a_{211})^3. \tag{4.377}$$

For $a_{211} = a_{11s_1}$ (s_1, $s_2 \in \{1, 2\}$, $s_1 \neq s_2$), the first integral manifold is

$$\begin{aligned} x_2 - x_{20} = &\frac{a_{210}}{a_{110}}\Big[x_1 - x_{10} + 2(a_{11s_2} - a_{211})\ln\frac{|x_1 - a_{11s_2}|}{|x_{10} - a_{11s_2}|} \\ &- (a_{11s_2} - a_{211})^2\Big(\frac{1}{x_1 - a_{11s_2}} - \frac{1}{x_{10} - a_{11s_2}}\Big)\Big]. \end{aligned} \tag{4.378}$$

For $a_{211} = a_{112}$ with $a_{11s_1} = a_{112}$ and $a_{11s_2} = a_{111}$, the phase portraits for two-dimensional systems on the x_1-direction are presented in Fig. 4.4. In Fig. 4.4a, the phase portrait for $a_{110} > 0$ and $a_{210} > 0$ is presented from the first integral manifold. At $x_1^* = a_{111}$, the sink flow is toward the x_2-negative direction due to $\dot{x}_2 = a_{210}(a_{111} - a_{211})^3 < 0$. At $x_1^* = a_{211} = a_{112}$, the (1:3)-increasing-inflection source infinite-equilibrium exists. Similarly, the phase portraits for ($a_{110} < 0$, $a_{210} > 0$), ($a_{110} > 0$, $a_{210} < 0$), and ($a_{110} < 0$, $a_{210} < 0$) are presented in Fig. 4.4b–d, respectively. The flows at the equilibriums with $\bar{x}_2 \in (-\infty, \infty)$ are listed as follows.

$$\left\{\begin{matrix}(a_{111}, \dot{x}_2)\\(a_{112}, \bar{x}_2)\end{matrix}\right\} = \left\{\begin{matrix}(\text{LS, nF})\\(\text{SO, (1:3)II})\end{matrix}\right\} \text{ for } a_{110} > 0, a_{210} > 0; \tag{4.379}$$

$$\left\{\begin{matrix}(a_{111}, \dot{x}_2)\\(a_{112}, \bar{x}_2)\end{matrix}\right\} = \left\{\begin{matrix}(\text{US, nF})\\(\text{SI, (1:3)DI})\end{matrix}\right\} \text{ for } a_{110} < 0, a_{210} > 0; \tag{4.380}$$

$$\left\{\begin{matrix}(a_{111}, \dot{x}_2)\\(a_{112}, \bar{x}_2)\end{matrix}\right\} = \left\{\begin{matrix}(\text{LS, pF})\\(\text{SO, (1:3)DI})\end{matrix}\right\} \text{ for } a_{110} > 0, a_{210} < 0; \tag{4.381}$$

$$\left\{\begin{matrix}(a_{111}, \dot{x}_2)\\(a_{112}, \bar{x}_2)\end{matrix}\right\} = \left\{\begin{matrix}(\text{US, pF})\\(\text{SI, (1:3)II})\end{matrix}\right\} \text{ for } a_{110} < 0, a_{210} < 0. \tag{4.382}$$

For $a_{211} = a_{111}$, the phase portraits for two-dimensional systems on the x_1-direction can be sketched for $a_{11s_1} = a_{111}$ and $a_{11s_2} = a_{112}$. The up-parabola and down-parabola sink and source are for switching bifurcations for the increasing-inflection and decreasing-inflection flows with sink and source flows.

Consider

$$\dot{x}_1 = a_{110}(x_1 - a_{11s_1})(x_1 - a_{11s_2})^2,$$
$$\dot{x}_2 = a_{210}(x_1 - a_{211})^3. \tag{4.383}$$

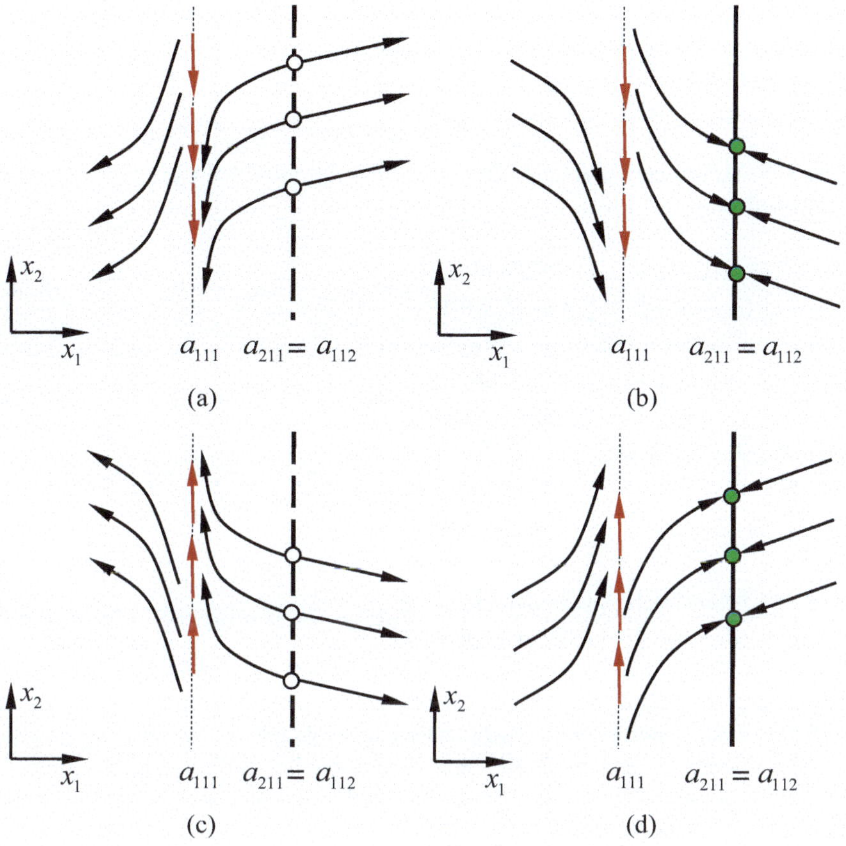

Fig. 4.4 Phase portraits for two-dimensional systems on the x_2-direction with $x_1^* = a_{111}, a_{112}$ and on the x_1-direction with $x_1^* = a_{211} = a_{112}$: (**a**) an (LS,nF)-negative sink flow and an (SO,(1:3)II)-increasing-inflection source, ($a_{110} > 0$, $a_{210} > 0$), (**b**) a (US,nF)-negative source flow and an (SI,(1:3)DI)-decreasing-inflection sink ($a_{110} < 0$, $a_{210} > 0$), (**c**) an (LS,pF)-positive sink flow and an (SO,(1:3)DI)-decreasing-inflection source ($a_{110} > 0$, $a_{210} < 0$), (**d**) an (US,pF)-negative source flow and an (SI,(1;3)II)-increasing-inflection sink ($a_{110} < 0$, $a_{210} < 0$)

For $a_{211} = a_{11s_2}$ ($s_1, s_2 \in \{1, 2\}$, $s_1 \neq s_2$), the first integral manifold is

$$x_2 - x_{20} = \frac{a_{210}}{a_{110}} \left[(x_1 - x_{10}) + (a_{11s_1} - a_{211}) \ln \frac{|\, x_1 - a_{11s_1} \,|}{|\, x_{10} - a_{11s_1} \,|} \right]. \tag{4.384}$$

For $a_{211} = a_{11s_2} = a_{112}$ with $a_{11s_2} = a_{112}$, the phase portraits for two-dimensional systems on the x_1-direction are presented in Fig. 4.5. In Fig. 4.5a, the phase portrait for $a_{110} > 0$ and $a_{210} > 0$ is presented from the first integral manifold. At $x_1^* = a_{111}$, the negative sink flow exists due to $\dot{x}_2 = a_{210}(a_{111} - a_{211})^3 < 0$. At $x_1^* = a_{211} = a_{112}$, the (2:3)-up-parabola source infinite-equilibrium exists because of $\Delta\dot{x}_1 = a_{110}(a_{112} - a_{111})(\Delta x_1)^2$ and $\Delta\dot{x}_2 = a_{210}(\Delta x_1)^3$. Similarly, the phase portraits for ($a_{110} < 0$, $a_{210} > 0$), ($a_{110} > 0$, $a_{210} < 0$), and ($a_{110} < 0$, $a_{210} < 0$) are presented in Fig. 4.5b–d, respectively. The flows at the equilibriums with $\bar{x}_2 \in (-\infty, \infty)$ are listed as follows.

$$\begin{Bmatrix} (a_{111}, \dot{x}_2) \\ (a_{112}, \bar{x}_2) \end{Bmatrix} = \begin{Bmatrix} (\text{SO}, \text{nF}) \\ (\text{US}, (2{:}3)\text{UP}) \end{Bmatrix} \text{ for } a_{110} > 0, a_{210} > 0; \tag{4.385}$$

$$\begin{Bmatrix} (a_{111}, \dot{x}_2) \\ (a_{112}, \bar{x}_2) \end{Bmatrix} = \begin{Bmatrix} (\text{SI}, \text{nF}) \\ (\text{LS}, (2{:}3)\text{DP}) \end{Bmatrix} \text{ for } a_{110} < 0, a_{210} > 0; \tag{4.386}$$

$$\begin{Bmatrix} (a_{111}, \dot{x}_2) \\ (a_{112}, \bar{x}_2) \end{Bmatrix} = \begin{Bmatrix} (\text{SO}, \text{pF}) \\ (\text{US}, (2{:}3)\text{DP}) \end{Bmatrix} \text{ for } a_{110} > 0, a_{210} < 0; \tag{4.387}$$

$$\begin{Bmatrix} (a_{111}, \dot{x}_2) \\ (a_{112}, \bar{x}_2) \end{Bmatrix} = \begin{Bmatrix} (\text{SI}, \text{pF}) \\ (\text{LS}, (2{:}3)\text{UP}) \end{Bmatrix} \text{ for } a_{110} < 0, a_{210} < 0. \tag{4.388}$$

Similarly, for $a_{211} = a_{11s_2} = a_{111}$ with $a_{11s_1} = a_{112}$, the phase portraits for two-dimensional systems on the x_1-direction can be sketched. The (2:3)-up-parabola and down-parabola upper-saddle and lower-saddle infinite-equilibriums are switching bifurcations for the third-order up- and down-parabola flows with lower-saddle and upper-saddle flows.

Consider

$$\begin{aligned} \dot{x}_1 &= a_{110}(x_1 - a_{111})^3, \\ \dot{x}_2 &= a_{210}(x_1 - a_{211})^3 \end{aligned} \tag{4.389}$$

with $a_{111} \neq a_{211}$ and the first integral manifold is

$$\begin{aligned} x_2 - x_{20} = \frac{a_{210}}{a_{110}} \Big\{ & (x_1 - x_{10}) + 3(a_{111} - a_{211}) \ln \frac{|x_1 - a_{111}|}{|x_{10} - a_{111}|} \\ & - 3(a_{111} - a_{211})^2 \Big(\frac{1}{x_1 - a_{111}} - \frac{1}{x_{10} - a_{111}}\Big) \\ & - \frac{1}{2}(a_{111} - a_{211})^3 \Big[\frac{1}{(x_1 - a_{111})^2} - \frac{1}{(x_{10} - a_{111})^2}\Big] \Big\}. \end{aligned} \tag{4.390}$$

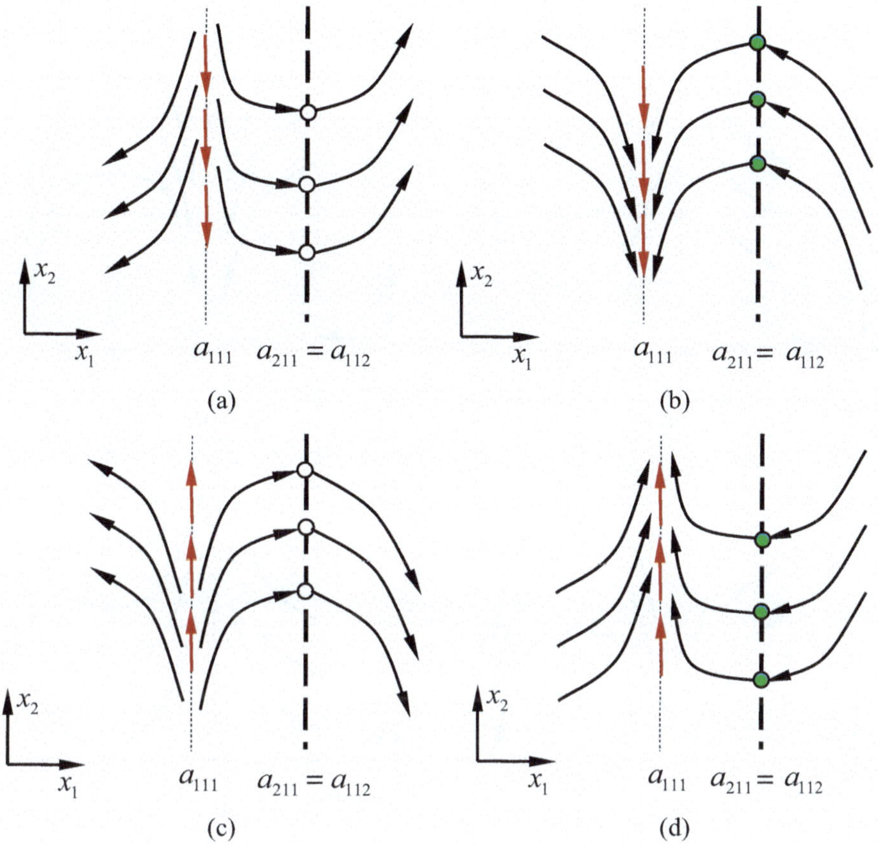

Fig. 4.5 Phase portraits for two-dimensional systems on the x_2-direction with $x_1^* = a_{111}, a_{112}$ and on the x_1-direction with $x_1^* = a_{211} \in (a_{111}, a_{112})$: (**a**) an (LS,pF)-positive upper-saddle flow and an (SO, UP)-up-parabola source, ($a_{110} > 0$, $a_{210} > 0$), (**b**) a (US,pF)-positive lower-saddle flow and an (SI, DP)-down-parabola sink ($a_{110} < 0$, $a_{210} > 0$), (**c**) an (LS,nF)-negative upper-saddle flow and an (SO,DP)-down-parabola source ($a_{110} > 0$, $a_{210} < 0$), (**d**) a (US,nF)-negative upper-saddle flow and an (SI,UP)-up-parabola sink ($a_{110} < 0$, $a_{210} < 0$)

Its phase portraits for two-dimensional systems on the x_1-direction are presented in Fig. 4.6 for $a_{211} > a_{111}$. In Fig. 4.6a, the phase portrait for $a_{110} > 0$ and $a_{210} > 0$ is presented from the first integral manifold. At $x_1^* = a_{111}$, the negative third-order source flow exists because $\dot{x}_2 = a_{210}(a_{111} - a_{211})^3 < 0$ and the positive third-order up-parabola flow exists at $x_1^* = a_{211}$ due to $\dot{x}_1 = a_{110}(a_{211} - a_{111})^3 > 0$. Similarly, in Fig. 4.6b–d, at $x_1^* = a_{111}, a_{211}$, a negative third-order sink flow and a negative third-order down-parabola flow exist for ($a_{110} < 0$, $a_{210} > 0$), a positive third-order source flow and a positive third-order down-parabola flow at $x_1^* = a_{111}, a_{211}$ exist for ($a_{110} > 0$, $a_{210} < 0$), a positive third-order sink flow and a negative third-order

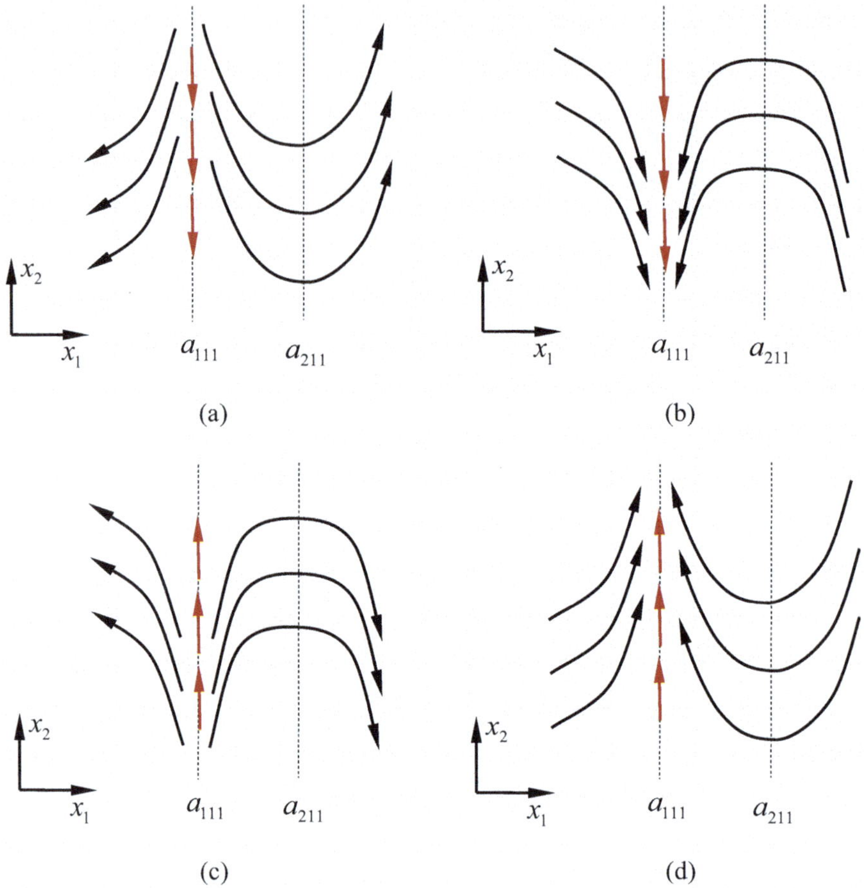

Fig. 4.6 Phase portraits for two-dimensional systems on the x_2-direction with $x_1^* = a_{111}$ and on the x_1-direction with $x_1^* = a_{211} > a_{111}$: (**a**) a (3rd SO,pF)-positive third-order source flow and a (pF,3UP)-positive increasing-inflection flow, ($a_{110} > 0$, $a_{210} > 0$), (**b**) a (3rd SI,pF)-positive third-sink flow and an (nF,3rd DP)-negative decreasing-inflection flow ($a_{110} < 0$, $a_{210} > 0$), (**c**) a (3rd SO, nF)-negative third-order source flow and a (pF,3rd DP)-positive decreasing-inflection flow ($a_{110} > 0$, $a_{210} < 0$), (**d**) a (3rd SI,nF)-negative third-order sink flow and an (nF,3rd UP)-negative increasing-inflection flow ($a_{110} < 0$, $a_{210} < 0$)

up-parabola flow at $x_1^* = a_{111}, a_{211}$ exist for ($a_{110} < 0$, $a_{210} < 0$). The flows at the equilibriums are listed as follows.

$$\begin{Bmatrix} (a_{111}, \dot{x}_2) \\ (\dot{x}_1, a_{211}) \end{Bmatrix} = \begin{Bmatrix} (3^{\text{rd}}\text{SO, nF}) \\ (\text{pF, } 3^{\text{rd}}\text{UP}) \end{Bmatrix} \text{ for } a_{110} > 0, a_{210} > 0; \tag{4.391}$$

$$\begin{Bmatrix} (a_{111}, \dot{x}_2) \\ (\dot{x}_1, a_{211}) \end{Bmatrix} = \begin{Bmatrix} (3^{\text{rd}}\text{SI}, \text{nF}) \\ (\text{nF}, 3^{\text{rd}}\text{DP}) \end{Bmatrix} \text{ for } a_{110} < 0, a_{210} > 0; \tag{4.392}$$

$$\begin{Bmatrix} (a_{111}, \dot{x}_2) \\ (\dot{x}_1, a_{211}) \end{Bmatrix} = \begin{Bmatrix} (3^{\text{rd}}\text{SO}, \text{pF}) \\ (\text{pF}, 3^{\text{rd}}\text{DP}) \end{Bmatrix} \text{ for } a_{110} > 0, a_{210} < 0; \tag{4.393}$$

$$\begin{Bmatrix} (a_{111}, \dot{x}_2) \\ (\dot{x}_1, a_{211}) \end{Bmatrix} = \begin{Bmatrix} (3^{\text{rd}}\text{SI}, \text{pF}) \\ (\text{nF}, 3^{\text{rd}}\text{UP}) \end{Bmatrix} \text{ for } a_{110} < 0, a_{210} < 0. \tag{4.394}$$

Similarly, for $a_{211} < a_{111}$, the phase portraits for two-dimensional systems on the x_1-direction can be sketched, which readers can be as an exercise.

Consider

$$\begin{aligned} \dot{x}_1 &= a_{110}(x_1 - a_{111})^3, \\ \dot{x}_2 &= a_{210}(x_1 - a_{211})^3 \end{aligned} \tag{4.395}$$

with $a_{111} = a_{211}$ and the first integral manifold is

$$x_2 - x_{20} = \frac{a_{210}}{a_{110}}(x_1 - x_{10}). \tag{4.396}$$

Its phase portraits for two-dimensional systems on the x_1-direction are presented in Fig. 4.7 for $a_{211} = a_{111}$. In Fig. 4.7a, the phase portrait for $a_{110} > 0$ and $a_{210} > 0$ is presented from the first integral manifold. At $x_1^* = a_{111} = a_{211}$, the increasing-inflection source exists, which is called a (3rdSO,(3:3)II)-increasing-inflection source infinite-equilibrium. Similarly, a (3rdSI,(3:3)DI)-decreasing-inflection sink infinite-equilibrium ($a_{110} < 0$, $a_{210} > 0$), a (3rdSO,(3:3)DI)-decreasing-inflection source infinite-equilibrium ($a_{110} > 0$, $a_{210} < 0$), a (3rdSI,(3:3)II)-increasing-inflection sink infinite-equilibrium ($a_{110} < 0$, $a_{210} < 0$) are presented in Fig. 4.7b–d. The infinite-equilibrium at $x_1^* = a_{111} = a_{211}$ with $\bar{x}_2 \in (-\infty, \infty)$ is listed as follows.

$$(a_{111}, \bar{x}_2) = \begin{cases} (3^{\text{rd}}\text{SO}, (3:3)\text{II}) \text{ for } a_{110} > 0, a_{210} > 0; \\ (3^{\text{rd}}\text{SI}, (3:3)\text{DI}) \text{ for } a_{110} < 0, a_{210} > 0; \\ (3^{\text{rd}}\text{SO}, (3:3)\text{DI}) \text{ for } a_{110} > 0, a_{210} < 0; \\ (3^{\text{rd}}\text{SI}, (3:3)\text{II}) \text{ for } a_{110} < 0, a_{210} < 0. \end{cases} \tag{4.397}$$

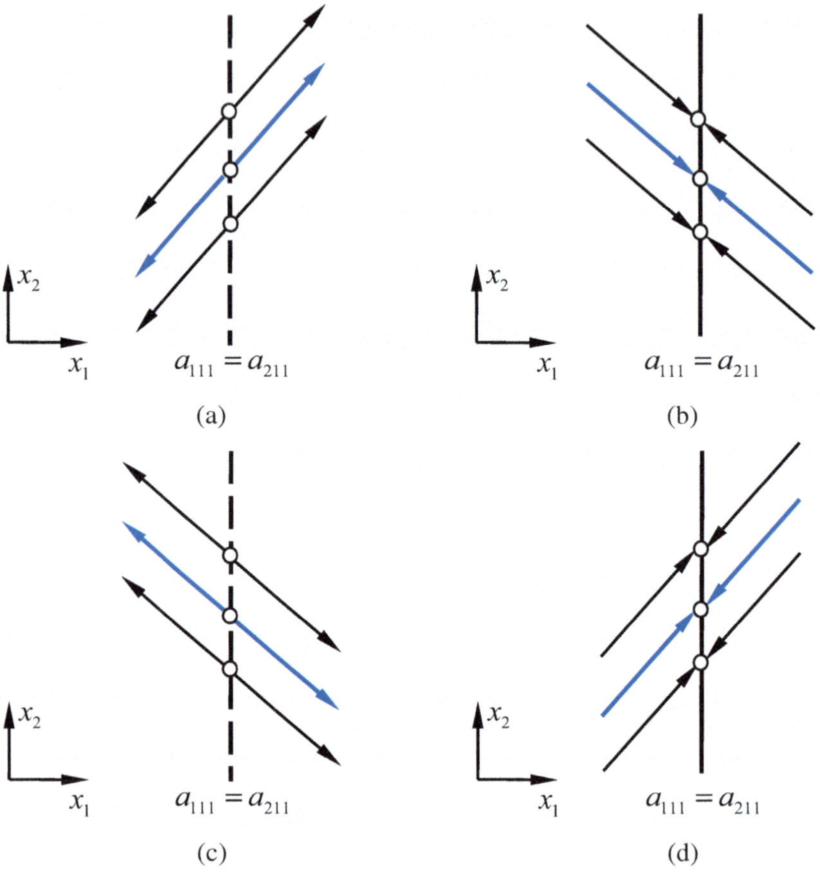

Fig. 4.7 Phase portraits for two-dimensional systems on the x_2-direction with $x_1^* = a_{211}$ and on the x_1-direction with $x_1^* = a_{211} > a_{111}$: (**a**) a ($3^{\text{rd}}$SO,(3:3)II)-increasing-inflection source ($a_{110} > 0$, $a_{210} > 0$), (**b**) a (3^{rd}SI,(3:3)DI)-decreasing-inflection sink ($a_{110} < 0$, $a_{210} > 0$), (**c**) a (3^{rd}SO,(3:3)DI)-decreasing-inflection source ($a_{110} > 0$, $a_{210} < 0$), (**d**) a (3^{rd}SI,(3:3)II)- increasing-inflection sink ($a_{110} < 0$, $a_{210} < 0$)

Index

A. C. J. Luo, *Two-dimensional Single-Variable Cubic Nonlinear Systems, Vol. I*,
https://doi.org/10.1007/978-3-031-48472-8

The manufacturer's authorised representative in the EU is Springer
Nature Customer Service Centre GmbH, Europaplatz 3, 69115 Heidelberg,
Germany. If you have any concerns regarding our products, please
contact ProductSafety@springernature.com

Printed and bound by CPI Group (UK) Ltd, Croydon, CR0 4YY
01/10/2026
02225965-0013